2320

$ 89.00

Methods in Enzymology

Volume 174
BIOMEMBRANES
Part U
Cellular and Subcellular Transport:
Eukaryotic (Nonepithelial) Cells

METHODS IN ENZYMOLOGY

EDITORS-IN-CHIEF

John N. Abelson   Melvin I. Simon

DIVISION OF BIOLOGY
CALIFORNIA INSTITUTE OF TECHNOLOGY
PASADENA, CALIFORNIA

FOUNDING EDITORS

Sidney P. Colowick and Nathan O. Kaplan

Methods in Enzymology

Volume 174

Biomembranes

Part U

Cellular and Subcellular Transport:
Eukaryotic (Nonepithelial) Cells

EDITED BY

Sidney Fleischer

Becca Fleischer

DEPARTMENT OF MOLECULAR BIOLOGY
VANDERBILT UNIVERSITY
NASHVILLE, TENNESSEE

Editorial Advisory Board

ACADEMIC PRESS, INC.

Harcourt Brace Jovanovich, Publishers

San Diego   New York   Berkeley   Boston
London   Sydney   Tokyo   Toronto

ACADEMIC PRESS, INC.
San Diego, California 92101

United Kingdom Edition published by
ACADEMIC PRESS LIMITED
24-28 Oval Road, London NW1 7DX

LIBRARY OF CONGRESS CATALOG CARD NUMBER: 54-9110

ISBN   0-12-182075-0   (alk. paper)

PRINTED IN THE UNITED STATES OF AMERICA
89 90 91 92   9 8 7 6 5 4 3 2 1

Table of Contents

Section I. Transport in Subcellular Organelles of Animals

A. Plasma Membranes and Derived Transporters

B. Intracellular Organelles

Section II. Transport in Plants

A. Higher Plants

B. Lower Plants

C. Organelles

Section III. Transport in Single-Cell Eukaryotes: Fungal Cells

Contributors to Volume 174

Article numbers are in parentheses following the names of contributors.
Affiliations listed are current.

ASHLEY ALLSHIRE (8), *Department of Human Anatomy and Cell Biology, University of Liverpool, Liverpool L69 3BX, England*

WILLIAM J. ARION (7), *Department of Biochemistry, Division of Nutrition Science, Cornell University, Ithaca, New York 14853*

STEPHEN A. BALDWIN (5), *Departments of Biochemistry and Chemistry, and of Protein and Molecular Biology, Royal Free Hospital School of Medicine, London NW3 2PF, England*

M. J. BEILBY (27), *School of Biological Sciences, University of Sydney, Sydney N.S.W. 2006, Australia*

THOMAS BOLLER (31), *Botanisches Institut der Universität Basel, CH-4056 Basel, Switzerland*

GEORGE W. F. H. BORST-PAUWELS (36), *Laboratory of Cell Biology, Katholieke Universiteit, 6525 ED Nijmegen, The Netherlands*

PATRICK J. BOURSIER (19), *School of Biological Sciences, University of Sussex, Brighton, Sussex BN1 9QG, England*

D. J. F. BOWLING (23), *Department of Plant and Soil Science, University of Aberdeen, Aberdeen AB9 2UD, Scotland*

HALVOR N. CHRISTENSEN (13), *Department of Biological Chemistry, University of Michigan Medical School, Ann Arbor, Michigan 48109*

VINCENT P. CIRILLO (37), *Department of Biochemistry, State University of New York at Stony Brook, Stony Brook, New York 11794*

A. H. DE BOER (20), *Department of Plant Physiology, University of Groningen, 9750 AA Haren, The Netherlands*

RICHARD M. DENTON (10), *Department of Biochemistry, University of Bristol Medical School, Bristol BS8 1TD, England*

MATHIAS DÜRR (31), *Botanisches Institut der Universität Basel, CH-4056 Basel, Switzerland*

A. ALAN EDDY (38), *Department of Biochemistry and Applied Molecular Biology, University of Manchester Institute of Science and Technology, Manchester M60 1QD, England*

BECCA FLEISCHER (15), *Department of Molecular Biology, Vanderbilt University, Nashville, Tennessee 37235*

KENNETH GASSER (14), *Departments of Physiology and Biophysics, School of Medicine, Case Western Reserve University, Cleveland, Ohio 44106*

RICHARD GERHARDT (32), *BASF Landwirtschaftliche, 6703 Limburgerhof, Federal Republic of Germany*

D. GRADMANN (30), *Pflanzenphysiologisches Institut der Universität Göttingen, D-3400 Göttingen, Federal Republic of Germany*

ALICE A. GREENE (13), *Department of Pediatrics, University of California, San Diego, La Jolla, California 92093*

RAINER HEDRICH (22), *Pflanzenphysiologisches Institut und Botanischer Garten der Universität Göttingen, D-3400 Göttingen, Federal Republic of Germany*

HANS W. HELDT (32), *Institut für Biochemie der Pflanze, Universität Göttingen, D-3400 Göttingen, Federal Republic of Germany*

MILAN HÖFER (39), *Botanisches Institut der Universität Bonn, 5300 Bonn 1, Federal Republic of Germany*

ULRICH HOPFER (14), *Departments of Physiology and Biophysics, School of Medicine,*

Case Western Reserve University, Cleveland, Ohio 44106

PAMELA HOPKINS (38), Department of Biochemistry and Applied Molecular Biology, University of Manchester Institute of Science and Technology, Manchester M60 1QD, England

PHILLIP E. KISH (2), Mental Health Research Institute, The University of Michigan, Ann Arbor, Michigan 48109

EWALD KOMOR (21, 33), Pflanzenphysiologie, Universität Bayreuth, D-8580 Bayreuth, Federal Republic of Germany

ARNOŠT KOTYK (34, 35), Institute of Physiology, Czechoslovak Academy of Sciences, 142-20 Praha 4, Czechoslovakia

ANDRÉ LÄUCHLI (19), Department of Land, Air and Water Resources, University of California, Davis, Davis, California 95616

OK YOUNG LEE-STADELMANN (17, 18), Department of Horticultural Science and Landscape Architecture, University of Minnesota, St. Paul, Minnesota 55108

DANIEL LEVY (3), Department of Biochemistry, School of Medicine, University of Southern California, Los Angeles, California 90033

GUSTAV E. LIENHARD (5), Department of Biochemistry, Dartmouth Medical School, Hanover, New Hampshire 03756

ROSS McC. LILLEY (32), Department of Biology, University of Wollongong, Wollongong N.S.W. 2500, Australia

WILLIAM J. LUCAS (28), Department of Botany, University of California, Davis, Davis, California 95616

GIAN CARLO LUNAZZI (6), Dipartimento di Biochimica, Biofisica e Chimica delle Macromolecole, Universita di Trieste, Trieste, Italy

JAMES G. McCORMACK (10), Department of Biochemistry, University of Leeds, Leeds LS2 9JT, England

JOHN D. McGIVAN (4), Department of Biochemistry, University of Bristol, Bristol BS8 1TD, England

DAVID G. NICHOLLS (1, 9), Department of Biochemistry, University of Dundee, Dundee DD1 4HN, Scotland

SHOJI OHKUMA (12), Department of Biochemistry, Faculty of Pharmaceutical Sciences, Kanazawa University, Kanazawa, Ishikawa 920, Japan

GABRIELE ORLICH (21), Pflanzenphysiologie, Universität Bayreuth, D-8580 Bayreuth, Federal Republic of Germany

RONALD L. PISONI (13), Department of Biological Chemistry, University of Michigan Medical School, Ann Arbor, Michigan 48109

KLAUS RASCHKE (22), Pflanzenphysiologisches Institut und Botanischer Garten der Universität Göttingen, 3400 Göttingen, Federal Republic of Germany

J. A. RAVEN (25), Department of Biological Sciences, University of Dundee, Dundee DD1 4HN, Scotland

EDUARDO RIAL (9), Centro de Investigaciones Biologicas, C.S.I.C., 28006 Madrid, Spain

DALE SANDERS (28), Department of Biology, University of York, York YO1 5DD, England

NILS-ERIK LEO SARIS (8), Department of Medical Chemistry, University of Helsinki, SF-00170 Helsinki, Finland

N. SAUER (26), Lehrstuhl für Zellbiologie und Pflanzenphysiologie, Universität Regensburg, 8400 Regensburg, Federal Republic of Germany

GENE A. SCARBOROUGH (41), Department of Pharmacology, University of North Carolina at Chapel Hill, Chapel Hill, North Carolina 27599

JERRY A. SCHNEIDER (13), Department of Pediatrics, University of California, San Diego, La Jolla, California 92093

HELMUT SIES (11), Institut für Physiologische Chemie, Universität Düsseldorf, D-4000 Düsseldorf, Federal Republic of Germany

CLIFFORD L. SLAYMAN (40), Department of Cellular and Molecular Physiology, Yale University School of Medicine, New Haven, Connecticut 06510

MARGARET L. SMITH (13), *Department of Pediatrics, University of California, San Diego, La Jolla, California 92093*

SIBYLLE SOBOLL (11), *Institut für Physiologische Chemie, Universität Düsseldorf, D-4000 Düsseldorf, Federal Republic of Germany*

GIAN LUIGI SOTTOCASA (6), *Dipartimento di Biochimica, Biofisica e Chimica delle Macromolecole, Universita di Trieste, Trieste, Italy*

ROGER M. SPANSWICK (29), *Section of Plant Biology, Cornell University, Ithaca, New York 14853*

EDUARD J. STADELMANN (17, 18), *Department of Horticultural Science and Landscape Architecture, University of Minnesota, St. Paul, Minnesota 55108*

ERNST STEUDLE (16), *Lehrstuhl für Pflanzenökologie, Universität Bayreuth, D-8580 Bayreuth, Federal Republic of Germany*

MARK STITT (32), *Institut für Pflanzenphysiologie der Universität Bayreuth, D-8580 Bayreuth, Federal Republic of Germany*

W. TANNER (26), *Lehrstuhl für Zellbiologie und Pflanzenphysiologie, Universität Regensburg, 8400 Regensburg, Federal Republic of Germany*

MARGARET THOM (33), *Hawaiian Sugar Planters' Association, Aiea, Hawaii 96701*

CLAUDIO TIRIBELLI (6), *Istituto di Patologia Medica, Universita di Trieste, Trieste, Italy*

TETSUFUMI UEDA (2), *Department of Pharmacology, Mental Health Research Institute, The University of Michigan, Ann Arbor, Michigan 48109*

PATRICIA VON DIPPE (3), *Department of Biochemistry, School of Medicine, University of Southern California, Los Angeles, California 90033*

ANDRES WIEMKEN (31), *Botanisches Institut der Universität Basel, CH-4056 Basel, Switzerland*

U. ZIMMERMANN (24), *Lehrstuhl für Biotechnologie der Universität Würzburg, 8700 Würzburg, Federal Republic of Germany*

GERALD N. ZUCKIER (40), *Department of Cellular and Molecular Physiology, Yale University School of Medicine, New Haven, Connecticut 06510*

Preface

Biological transport is part of the Biomembranes series of *Methods in Enzymology.* It is a continuation of methodology concerned with membrane function. This is a particularly good time to cover the topic of biological membrane transport because there is now a strong conceptual basis for its understanding. The field of transport has been subdivided into five topics.

1. Transport in Bacteria, Mitochondria, and Chloroplasts
2. ATP-Driven Pumps and Related Transport
3. General Methodology of Cellular and Subcellular Transport
4. Cellular and Subcellular Transport: Eukaryotic (Nonepithelial) Cells
5. Cellular and Subcellular Transport: Epithelial Cells

Topic 1 covered in Volumes 125 and 126 initiated the series. Topic 2 is covered in Volumes 156 and 157, Topic 3 in Volumes 171 and 172, and Topic 4 in Volumes 173 and 174. The remaining topic will be covered in subsequent volumes of the Biomembranes series.

Topic 4 is divided into two parts: this volume (Part U) which covers transport by isolated subcellular organelle fractions, purified components as well as transport in plants, plant organelles, and single cell eukaryotes, and Volume 173 (Part T) which deals mainly with intact cells; major emphasis is on the red cell, derived red cell preparations and the anion transporter.

We are fortunate to have the good counsel of our Advisory Board. Their input insures the quality of these volumes. The same Advisory Board has served for the complete transport series. Valuable input on the outlines of the five topics was also provided by Qais Al-Awqati, Ernesto Carafoli, Halvor Christensen, Isadore Edelman, Joseph Hoffman, Phil Knauf, and Hermann Passow. Additional valuable input for Volumes 173 and 174 was obtained from Ioav Cabanchik, John Exton, Arnošt Kotyk, and Aser Rothstein.

The names of our advisory board members were inadvertently omitted in Volumes 125 and 126. When we noted the omission, it was too late to rectify the problem. For volumes 125 and 126, we are also pleased to acknowledge the advice of Angelo Azzi, Youssef Hatefi, Dieter Oesterhelt, and Peter Pedersen.

The enthusiasm and cooperation of the participants have enriched and made these volumes possible. The friendly cooperation of the staff of Academic Press is gratefully acknowledged.

These volumes are dedicated to Professor Sidney Colowick, a dear friend and colleague, who died in 1985. We shall miss his wise counsel, encouragement, and friendship.

SIDNEY FLEISCHER
BECCA FLEISCHER

METHODS IN ENZYMOLOGY

Section I

Transport in Subcellular Organelles of Animals

A. Plasma Membranes and Derived Transporters
Articles 1 through 6

B. Intracellular Organelles
Articles 7 through 15

[1] Measurement of Transported Calcium in Synaptosomes

By DAVID G. NICHOLLS

The study of Ca^{2+} transport across the plasma membrane of isolated nerve terminals (synaptosomes) and the subsequent distribution of the cation among the various subsynaptosomal compartments is central to the investigation of the Ca^{2+} dependency of neurotransmitter release. Calcium transport in synaptosomes may serve as a model for the integration of cellular Ca^{2+}-transporting pathways in general.[1]

General Methodology

With the possible exception of Na^{+}[2] the accurate quantification of transported Ca^{2+} presents more difficulties than any other common synaptosomal ion or metabolite. There are two main reasons for this. First, the average concentration of total Ca^{2+} within the synaptosome (ignoring compartmentation) is about the same as that in the incubation medium. This is because the plasma membrane actively extrudes Ca^{2+}, and the only significant $^{45}Ca^{2+}$ within the terminal will be that bound, or sequestered by organelles such as mitochondria. Therefore, any technique to separate the synaptosomes from the incubation prior to counting must take accurate account of the contamination from the incubation medium. While it is in theory possible to wash synaptosomes on a filter or in a pellet, there are risks of an artifactual redistribution of the cation across the plasma membrane. Instead we employ a second isotope ([^3H]sucrose), which is excluded from the synaptosome, to give an accurate measure of the contamination from the incubation. In this respect centrifugation through silicone oil is the method of choice, the typical contamination by medium being 2 μl/mg synaptosomal protein.[3]

A second source of error is due to superficially bound Ca^{2+}. Even after correcting for contaminating medium, the amount of $^{45}Ca^{2+}$ bound to the surface of the terminals can easily exceed that transported into the synaptosome itself when incubated in physiological Ca^{2+} concentrations.[3] This

---

[1] D. G. Nicholls and K. E. O. Åkerman, *Philos. Trans. R. Soc. London, Ser. B* **296**, 115 (1981).

[2] K. E. O. Åkerman and D. G. Nicholls, *Eur. J. Biochem.* **117**, 491 (1981).

[3] I. D. Scott, K. E. O. Åkerman, and D. G. Nicholls, *Biochem. J.* **192**, 873 (1980).

can be corrected for by adding EGTA in excess immediately prior to centrifugation, relying on the very rapid off rate constant for Ca^{2+} from these relatively loose binding sites, contrasted with the slow net efflux of Ca^{2+} across membrane itself. From our experience a 10-sec chelation prior to initiating centrifugation is optimal.[3]

A third possible source of error comes from the possibility of Ca^{2+} accumulation by "free" mitochondria contaminating the synaptosomal incubation. It is not possible completely to eliminate free mitochondria from the preparation, so instead it is important to avoid conditions which would favor the development of a membrane potential by these mitochondria, allowing them to accumulate Ca^{2+}. In practice the conditions required for mitochondrial Ca^{2+} uptake differ sufficiently from those suitable for synaptosomes so that little artifact occurs. Brain mitochondria require micromolar Ca^{2+}, exogenous adenine nucleotide, and respiratory substrate to accumulate Ca^{2+}.[4] The absence of a mitochondrial substrate (as opposed to glucose, which can only be utilized by intact synaptosomes), the absence of ADP or ATP to stabilize the mitochondria,[4] and the presence of millimolar Ca^{2+} (which irreversibly inhibits any mitochondria attempting to accumulate the cation[5]) all conspire to prevent this artifact from being of major significance. However, if a mitochondrial substrate is used in a synaptosomal incubation in combination with a lower Ca^{2+} concentration, then the free mitochondrial accumulation can become dominant.[6]

Practical Details

Synaptosomes are prepared[7] by homogenization, within 60 sec of sacrifice, of guinea pig cerebral cortex in 30 ml of isolation medium [310 mM sucrose, 5 mM TES, 1 mM EDTA (Na salt), pH 7.4, 0°]. The homogenate is centrifuged for 5 min at 900 g_{max} and 4°. The supernatant is saved, and the pellet is resuspended in the same medium and recentrifuged. The two supernatants are combined, and centrifuged (10 min at 17,000 g_{max}). The pellet is resuspended in 10 ml isolation medium, and 5 ml is layered onto each of two discontinuous gradients comprising 4 ml of isolation medium containing 12% (w/v) Ficoll (ρ 1.085 g/ml), 1 ml of isolation medium with 9% Ficoll (ρ 1.075 g/ml), and 4 ml of isolation medium with 6% Ficoll (ρ 1.065 g/ml). The gradients are centrifuged in a 6 × 16 ml swing-out rotor for 30 min at 75,000 g_{max}. The myelin, synaptosomal, and mitochondrial fractions form, respectively, above the 6% Ficoll layer, within the 9% layer in a double band, and as a pellet beneath the 12% layer.[7]

---

[4] D. G. Nicholls and I. D. Scott, *Biochem. J.* **186**, 833 (1980).
[5] D. G. Nicholls and K. E. O. Åkerman, *Biochim. Biophys. Acta* **683**, 57 (1982).
[6] G. R. Vickers and M. J. Dowdall, *Exp. Cell Res.* **25**, 429 (1976).
[7] D. G. Nicholls, *Biochem. J.* **170**, 511 (1978).

The synaptosomal layers are carefully aspirated off with a Pasteur pipette and diluted to 12 ml with 250 mM sucrose, 5 mM Na TES, pH 7.4. Protein is determined on this resuspension by the biuret method.[8] Appropriate volumes of the suspension are pipetted into centrifuge tubes to give the synaptosomal protein required for a single experiment. After dilution to 30 ml with 250 mM sucrose, 5 mM Na TES, pH 7.4, the tubes are centrifuged at 27,000 g_{max} for 30 min. The synaptosomal pellets are stored at 0° for up to 3 hr without decanting the supernatant.[9]

Synaptosomes thus prepared are largely depleted of endogenous Ca^{2+}, due to their prolonged exposure to chelator. The guinea pig preparation contains only 0.5 nmol Ca^{2+}/mg protein, less than 10% of the steady-state content when incubated in the presence of physiological Ca^{2+} concentrations.[3] This means that synaptosome studies utilizing $^{45}Ca^{2+}$ are almost free of complexities resulting from incomplete equilibration between isotope and endogenous Ca^{2+}, and it enables $^{45}Ca^{2+}$ and total Ca^{2+} to be equated.

Determination of Transported Ca²⁺

Synaptosomal pellets containing 4.5 mg protein (sufficient for 10 determinations) are suspended by gentle vortexing in 1 ml of incubation medium (pH 7.4, 30°, air equilibrated) containing 122 mM NaCl, 3.1 mM KCl, 1.2 mM MgSO$_4$, 0.4 mM KH$_2$PO$_4$, 5 mM NaHCO$_3$, 20 mM TES (Na salt), 16 μM albumin (fatty acid poor), and 10 mM D-glucose. The suspension is then transferred to a further 2 ml of incubation medium containing additionally 75 μM [³H]sucrose (2.2 μCi/ml). This is to avoid radioactive contamination of the centrifuge tubes storing the pellets. After 5 min preincubation, 1.3 mM ^{45}CaCl$_2$ (0.5 μCi/ml incubation) is added. The initial preincubation period is to allow the synaptosomal plasma membrane to polarize. Otherwise there is a large initial influx of Ca^{2+} through voltage-dependent Ca^{2+} channels.[3]

Eppendorf centrifuge tubes are prepared containing 100 μl of a mixture of Dow-Corning 550 silicone oil fluid and dinonyl phthalate (1 : 1, v/v). This mixture has a density which allows the synaptosomes to pellet reliably, without the risks of inversion of oil and incubation layers. On the surface of the oil, 100 μl of a mixture of 5 mM EGTA (Na salt) and 12 μM ruthenium red is pipetted. Aliquots of the incubation (250 μl) are pipetted into the Eppendorf tubes at precisely defined times, capped, vortexed immediately, and transferred to an Eppendorf Model 5412 microcentrifuge. Exactly 10 sec after the pipetting the centrifuge is started, and centrifugation is continued for 60 sec. The oil layer immediately reforms at the

[8] A. G. Gornall, C. J. Bardawill, and M. M. David, *J. Biol. Chem.* **177**, 751 (1949).
[9] R. Snelling and D. G. Nicholls, *Biochem. J.* **226**, 225 (1985).

bottom of the tube, and a tight pellet of synaptosomes is obtained under the oil. The ruthenium red has no significant effect on the recovered Ca^{2+}; it is present to allow comparison with experiments (described below) in which the synaptosomes are disrupted to determine the Ca^{2+} concentration within intrasynaptosomal mitochondria.

A 50-μl aliquot of the supernatant is pipetted off for determination of the supernatant counts. The remainder of the supernatant is aspirated off, down to the oil, and discarded. One milliliter of water is then added to the surface of the oil and aspirated off to remove any contaminating counts on the walls of the tube. This time the oil is also removed, care being taken to avoid disturbing the pellet, by rotating the tube on its axis such that the pellet is at the top. The oil is allowed to drain while the other tubes are being dealt with, and then any residual oil is removed. This is essential as the oil mixture is a potent quencher of the scintillant. The pellet is solubilized by adding 50 μl of 5% (w/v) sodium dodecyl sulfate (SDS) and heating to 55° for 2 hr. To avoid errors arising from the transfer of the solubilized pellet, 1 ml of scintillant [BDH Scintran T containing 7% (v/v) water] is added directly to the Eppendorf tube. The supernatant sample is added to a similar tube, and 1 ml of scintillant containing additionally 0.25% SDS is added. It is important that the scintillation counter channels can be adjusted to achieve optimal cross-over. Transported Ca^{2+} is calculated after allowing for the contamination of the pellet with extrasynaptosomal medium (estimated from the pellet [^{14}C]sucrose counts).

In view of the small amount of transported Ca^{2+} in relation to the high concentration present in the medium, the utmost precision is required in order to obtain reliable results. The most critical stages are the pipetting of the aliquot, the timing of the EGTA exposure, and the removal of the oil. One caveat with this technique is that the use of EGTA to displace superficial Ca^{2+} is dependent on the maintenance of a relatively low Ca^{2+} permeability across the plasma membrane. This is not so if a Ca^{2+} ionophore such as A23187 is present, when there will be a large artifactual efflux of Ca^{2+} across the plasma membrane in the short time allotted to displacing the surface Ca^{2+}.[10] The chelation should therefore be omitted in experiments involving the ionophore, the magnitude of the superficial binding being estimated from control experiments in the absence of A23187.

A useful variant of this technique is include ^{86}Rb (0.05 μCi/ml) in the incubation medium. As discussed earlier in this series,[11] ^{86}Rb distributes across the synaptosomal membrane in the same way as K^+ and attains a distribution which allows a close approximation to the plasma membrane

[10] K. E. O. Åkerman and D. G. Nicholls, *Eur. J. Biochem.* **115**, 67 (1981).
[11] J. B. Jackson and D. G. Nicholls, this series, Vol. 127, p. 557.

potential to be calculated.[12] A scintillation counter with adjustable gates has no problem in discriminating between ^3H, ^{14}C, and ^{86}Rb in this scintillant. Typically cross-over ratios of 0.1 from ^{86}Rb to ^{14}C, and 0.25 from ^{14}C to ^3H can be attained. In this way plasma membrane potential and transported Ca^{2+} can be determined simultaneously.[13]

Ca^{2+} in Intrasynaptosomal Mitochondria

The mitochondria are major stores of Ca^{2+} within the isolated synaptosomes.[3,14] To determine the proportion of the transported Ca^{2+} which is further translocated into the intrasynaptosomal mitochondria, a rapid disruption of the synaptosomes is carried out using a combination of digitonin and shear under conditions which minimize any artifactual redistribution of Ca^{2+} across the mitochondrial membrane.[14]

Synaptosomes are incubated exactly as described above. Indeed, a nondisrupted control should always be carried out in parallel in order to allow estimation of the nonmitochondrial Ca^{2+} by difference. Aliquots of 500 μl are withdrawn from the incubation and transferred to Eppendorf centrifuge tubes. A 2-ml plastic disposable syringe, bearing a blunted, 23-gauge needle, is preloaded with 1 ml of nonradioactive, Ca^{2+}-free incubation medium containing additionally 7.5 μM ruthenium red, 3 mM EGTA (Na salt), and a specified concentration of digitonin (see below). This loaded syringe is then used to rapidly draw up the 2-ml aliquot of synaptosomal incubation. The EGTA is present this time primarily to lower the free Ca^{2+} in the incubation below 1 μM and prevent artifactual uptake of Ca^{2+} by the mitochondria which are otherwise exposed to the high Ca^{2+} concentration in the incubation medium following the disruption. The ruthenium red is present as a further precaution to inhibit mitochondrial Ca^{2+} uniporter.

The apparatus for the disruption is shown in Fig. 1. An Eppendorf Type 5412 microcentrifuge is modified by removing the lid and replacing it with a Perspex lid containing a guide to allow the accurate positioning of the loaded syringe.[3] Disruption of the synaptosomal plasma membrane occurs by a combination of digitonin action and the shear as the synaptosomes are forced through the 23-gauge needle. To ensure a constant shearing force, the syringe piston is driven in by a 1.8-kg weight (a suitable lead "castle"). To minimize delays arising from loading and starting a conventional centrifuge, the modified Eppendorf is fitted with a purpose-built axial-loading rotor (Fig. 1) which allows the sample to be injected into the axis of

[12] I. D. Scott and D. G. Nicholls, *Biochem. J.* **186,** 21 (1980).
[13] T. S. Sihra, I. G. Scott, and D. G. Nicholls, *J. Neurochem.* **43,** 1624 (1984).
[14] K. E. O. Åkerman and D. G. Nicholls, *Biochim. Biophys. Acta* **645,** 41 (1981).

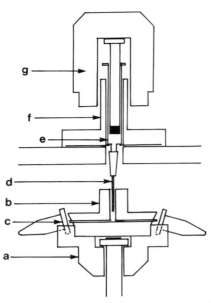

FIG. 1. Cross section of disruptor and axial-loading rotor for the digitonin fractionation of synaptosomes. (a) Extruded aluminum rotor; (b) Teflon insert; (c) swing-out centrifuge tube buckets; (d) 23-gauge needle; (e) 2-ml disposable syringe; (f) Perspex guide; (g) 1.8-kg weight.

the spinning rotor, the sample being equally ejected into two Eppendorf centrifuge tubes containing 100 μl of silicone oil.

The concentration of digitonin must be sufficient to ensure complete disruption of the plasma membrane with minimal disruption of the mitochondria. Trial disruptions are performed with a range of digitonin concentrations, and the percentage release of lactate dehydrogenase is quantified. Alternatively, ^{86}Rb may be included in the incubation as above; since the cation resides in the cytoplasm the extent of its release is a good monitor of the extent of loss of cytoplasmic Ca^{2+}. It appears likely that this technique releases Ca^{2+} from any nonmitochondrial stores; more than 80% of the pellet Ca^{2+} is dependent on the maintenance of a mitochondrial membrane potential and is not seen when this is abolished prior to disruption by the addition of rotenone together with oligomycin.[3]

Ca^{2+} Cycling across the Synaptosomal Plasma Membrane

The steady-state distribution of Ca^{2+} attained after 20–30 min is the result of a balance between uptake and efflux across the plasma membrane, and it is of some importance to quantify the exchange across the membrane. We have shown that synaptosomes retain the ability to exchange Ca^{2+} out of the matrix when the Na^{+} electrochemical potential is collapsed by veratridine plus ouabain, proving that the terminals have some mechanism other than $Na^{+}-Ca^{2+}$ exchange for extruding Ca^{2+} and supporting a major role for the Ca^{2+}-translocating ATPase.[9]

For inward exchange of $^{45}Ca^{2+}$, synaptosomes are incubated for 30 min as described above to achieve steady-state conditions, but substituting nonradioactive Ca^{2+} for the $^{45}Ca^{2+}$ addition at 5 min. At 30 min, $1 \mu M$ $^{45}Ca^{2+}$ is added, and samples are taken for analysis as described above. A parallel run for the determination of total transported $^{45}Ca^{2+}$ is also performed. Exchange kinetics are quantified by a semilogarithmic plot.

Outward exchange of $^{45}Ca^{2+}$ is followed by incubating synaptosomes in the presence of 1.3 mM $^{45}Ca^{2+}$ (2 μCi/ml) but at a protein concentration of 6 mg/ml instead of the normal 1.5 mg/ml. After 30 min to achieve steady-state conditions, the incubation is diluted with 3 vol incubation medium containing 1.3 mM nonradioactive Ca^{2+} to lower the specific activity of the external Ca^{2+} pool 4-fold. Samples are taken and analyzed as above.

[2] Glutamate Accumulation into Synaptic Vesicles

By PHILLIP E. KISH and TETSUFUMI UEDA

Introduction

Despite the fact that glutamate is such a common biochemical substance involved in diverse biological processes,[1,2] a variety of evidence which has accumulated since the 1960s supports the concept that this acidic amino acid functions as a major excitatory neurotransmitter in the central nervous system.[3-11] It is now evident that there exist multiple, but

[1] J. L. Johnson, *Brain Res.* **37**, 1 (1972).
[2] R. P. Shank and G. LeM. Campbell, in "Handbook of Neurochemistry" (A. Lajtha, ed.), 2nd ed., Vol. 3, p. 381. Plenum, New York, 1983.
[3] C. W. Cotman, A. C. Foster, and T. H. Lanthorn, in "Glutamate as a Neurotransmitter" (G. Di Chiara and G. L. Gessa, eds.), p. 1. Raven, New York, 1981.
[4] J. C. Watkins and R. H. Evans, *Annu. Rev. Pharmacol. Toxicol.* **21**, 165 (1981).
[5] F. Fonnum, *J. Neurochem.* **42**, 1 (1984).
[6] V. Crunelli, S. Forda, and J. S. Kelly, *J. Physiol. (London)* **351**, 327 (1984).
[7] G. E. Fagg, *Trends Neurosci.* **8**, 207 (1985).
[8] A. B. MacDermott and N. Dale, *Trends Neurosci.* **10**, 280, (1987).
[9] T. Ueda, in "Excitatory Amino Acids" (P. J. Roberts, J. Storm-Mathisen, and H. F. Bradford, eds.), p. 173. Macmillan, London, 1986.

pharmacologically distinct, specific receptors for dicarboxylic amino acids [7,8] which have different anatomical distributions in the brain.[9,11] It has been demonstrated that glutamate and aspartate are released from nerve terminals in a calcium-dependent manner upon depolarization.[3,12] Moreover, recent evidence suggests that various neurological diseases are attributable to excessive release of excitatory amino acids.[13-16] It is not established, however, whether the calcium-dependent release of acidic amino acids occurs by an exocytotic mechanism involving synaptic vesicles. Isolated synaptic vesicles failed to reveal a significant enrichment of glutamate or aspartate compared to other subcellular fractions.[17,18] Additional evidence also suggested that the cytoplasmic rather than the vesicular compartment is the immediate source of amino acids released.[19,20]

Recently, synaptic vesicles were highly purified from mammalian brain by use of immunoprecipitation in the final step of purification and were shown to be capable of accumulating glutamate in an ATP-dependent manner,[21] supporting the argument for a role of synaptic vesicles in glutamate transmission. The vesicular glutamate uptake is potentiated by physiologically relevant, low millimolar concentrations of chloride and appears to be driven by electrochemical proton gradients generated by a Mg^{2+}-ATPase proton pump,[22-24] a driving force analogous to that responsible for catecholamine uptake into chromaffin granules.[25] The interesting and perhaps important feature of the vesicular glutamate transport system is its remarkable substrate specificity for glutamate.[21,22] This specific vesicular

[10] C. W. Cotman, D. T. Monaghan, O. P. Ottersen, and J. Storm-Mathisen, *Trends Neurosci.* **10**, 273 (1987).

[11] J. T. Greenamyre, A. B. Young, and J. B. Penney, *J. Neurosci.* **4**, 2133 (1984).

[12] A.-S. Abdul-Ghani, J. Coutinho-Netto, and H. F. Bradford, *in* "Glutamate: Transmitter in the Central Nervous System" (P. J. Roberts, J. Storm-Mathisen, and G. A. R. Johnston, eds.), p. 155. Wiley, Chichester, 1981.

[13] B. C. Meldrum, *Clin. Sci.* **68**, 113 (1985).

[14] S. M. Rothman and J. W. Olney, *Trends Neurosci.* **10**, 299 (1987).

[15] J. T. Greenamyre, *Arch. Neurol.* **43**, 1058 (1986).

[16] M. B. Robinson and J. T. Coyle, *FASEB J.* **1**, 446 (1987).

[17] J. L. Mangan and V. P. Whittaker, *Biochem. J.* **98**, 128 (1966).

[18] P. Kontro, K.-M. Marnela, and S. S. Oja, *Brain Res.* **184**, 129 (1980).

[19] D. K. Rassin, *J. Neurochem.* **19**, 139 (1972).

[20] J. S. De Belleroche and H. F. Bradford, *J. Neurochem.* **29**, 335 (1977).

[21] S. Naito and T. Ueda, *J. Biol. Chem.* **258**, 696 (1983).

[22] S. Naito and T. Ueda, *J. Neurochem.* **44**, 99 (1985).

[23] J. Shioi and T. Ueda, *Biochem. J.* **258**, 499 (1989).

[24] P. R. Maycox, T. Deckerwerth, J. W. Hell, and R. Jahn, *J. Biol. Chem.* **263**, 15423 (1988).

[25] R. G. Johnson, S. Carty, and A. Scarpa, *Fed. Proc., Fed. Am. Soc. Exp. Biol.*, **41**, 2746 (1982).

uptake is in accord with immunocytochemical evidence[26] that glutamate is stored in distinct vesicles, and may provide a basis for the selective release of glutamate. These observations argue for the possibility that synaptic vesicles may play a vital role in glutamate synaptic transmission, a notion also supported by more recent studies by Nicholls and colleagues.[27] Further investigation is necessary, however, to firmly establish the involvement of synaptic vesicles in glutamate transmission.

A critical but rate-limiting step in the study of glutamate transport into synaptic vesicles is preparation of synaptic vesicles which are free of contamination yet retain functional capability. The immunochemical method developed by Naito and Ueda[21] enables one to isolate glutamate-transporting synaptic vesicles of high purity. However, this requires rather large quantities of affinity-purified antibodies to a synaptic vesicle protein, synapsin I, and, despite this requirement, the yield of such a vesicle preparation is too low to permit analysis of vesicular glutamate uptake in small tissues. In an effort to circumvent these difficulties, we have recently devised a simplified procedure[28] for synaptic vesicle preparation which results in sufficient yield and purity to exhibit glutamate transport properties identical or highly similar to those of highly purified vesicles.[21,22] In this chapter, we describe the assay and properties of the glutamate transport system in the synaptic vesicle in three preparations: (1) bovine brain synaptic vesicles highly purified by use of antisynapsin I IgG (Ab-BGPV), (2) a bovine brain synaptic vesicle fraction purified by sucrose density gradient centrifugation (BGPV), and (3) a rat brain crude synaptic vesicle fraction (RCV).

Standard Assay for Vesicular Glutamate Uptake*

The uptake activity of synaptic vesicles is measured by incubating the prepared vesicles with [^{3}H]glutamate in the presence or absence of ATP and chloride, filtering and washing the vesicles, and determining the isotope content remaining on the filter.

Reagents for Standard Glutamate Uptake Assay

Reagent A: 0.1 M Tris-maleate, pH 7.4
Reagent B: 0.1 M MgSO$_4$
Reagent C: 1.0 M sucrose

---

* A substitution of potassium HEPES and MgCl$_2$ for Tris-maleate and MgSO$_4$, respectively with omission of KCl (Reagent F), was found to give a slightly better uptake activity (15%).

[26] J. Storm-Mathisen, A. K. Leknes, A. T. Bore, J. L. Vaaland, P. Eminson, F.-M. S. Haug, and O. P. Ottersen, *Nature (London)* **301**, 517 (1983).

[27] D. G. Nicholls, *J. Neurochem.* **52**, 331 (1989).

[28] P. E. Kish, S. Y. Kim, and T. Ueda, *Neurosci. Lett.* **97**, 185 (1989).

Reagent D: 5 mM potassium glutamate, pH 7.4
Reagent E: 20 mM Tris-ATP, pH 7.4 (neutralized with Tris base)
Reagent F: 80 mM KCl
Reagent G: 0.15 M KCl

Procedure

Preincubation Mixture. For 100 assays the following reagents are mixed in a 13 × 75 mm disposable glass test tube:

0.5 ml Reagent A
0.4 ml Reagent B
2.18 ml Reagent C
3.92 ml distilled water
1.0 ml prepared synaptic vesicles (5 mg protein for RCV, 2 mg protein for BGPV)

Glutamate Mixture. For 100 assays the following reagents are mixed in a 12 × 75 mm disposable glass test tube:

0.2 ml (200 μCi) L-[2,3-^{3}H]glutamate (TRK-445 20–40 Ci/mmol, Amersham, Arlington, IL)
0.1 ml Reagent D
1.0 ml Reagent E
0.5 ml Reagent F
0.2 ml distilled H$_2$O

To test for ATP or chloride dependency, another glutamate mixture is also prepared by substituting distilled water for Reagents E or F, respectively.

Aliquots (80 μl) of the preincubation mixture are incubated in 10 × 75 mm disposable glass test tubes in a shaking waterbath at 30° for 5 min. After the preincubation, 20 μl of the glutamate mixture is added using an Eppendorf repeater (4780, Brinkman Instruments, Westbury, NY), vortexed, and incubated for an additional 1.5 min. Glutamate uptake is terminated by the addition of 2 ml of ice-cold Reagent G, followed by immediate vacuum filtration through Millipore HAWP filters (25 mm, 0.45 μm). The incubation tubes are washed with Reagent G 3 times, and the filters are then washed 4 more times with Reagent G. Each filter is dissolved in 7 ml of scintillation cocktail (ACS, Amersham) and radioactivity retained on the filter determined in a scintillation counter. Assays are normally performed in duplicate or triplicate.

Filter controls are determined by filtering and washing the glutamate mixture in the absence of the preincubation mixture. Radioactivity retained on the filter [600–800 disintegrations/min (dpm)] is subtracted from all values obtained in the uptake experiments described. Glutamate uptake into synaptic vesicles is defined as the ATP-dependent glutamate accumulation. This is obtained by subtracting the value obtained in the absence of ATP (perhaps representing the receptor and nonspecific binding and the uptake into resealed plasma membrane ghosts) from the total glutamate accumulated in the presence of ATP.

Synaptic Vesicle Preparation

Reagents for Vesicle Preparation

Solution A: 0.32 M sucrose, 1 mM NaHCO$_3$, 1 mM MgCl$_2$, 0.5 mM CaCl$_2$, pH 7.2

Solution B: 0.32 M sucrose, 1 mM NaHCO$_3$, 1 mM magnesium acetate, 0.5 mM calcium acetate, pH 7.2

Solution C: 0.32 M sucrose, 1 mM NaHCO$_3$, pH 7.2

Solution D: 0.32 M sucrose, 1 mM NaHCO$_3$, 1 mM dithiothreitol, pH 7.2

Solution E: 6 mM Tris-HCl, pH 8.1

Solution F: 6mM Tris-maleate, pH 8.1

Procedure 1

Isolation of Bovine Synaptic Vesicles by Immunoprecipitation with Antisynapsin I IgG. Synapsin I is purified from bovine cerebral cortices essentially by the same procedure as described previously.[29] Antisynapsin I IgG and preimmune IgG are purified from immune and preimmune rabbit sera by affinity chromatography on synapsin I-conjugated and protein A-conjugated agarose columns, respectively, as described previously.[30] A highly purified synaptic vesicle fraction (0.4 M sucrose layer) is prepared from a purified synaptosome fraction (1.2 M sucrose layer) of the fresh bovine cortex by lysis at pH 8.1, followed by sucrose density gradient (0.4, 0.6, and 0.8 M) centrifugation, as described previously.[31] The synaptic vesicle fraction is suspended in Solution C at a concentration of 2 mg/ml and frozen at $-80°$ for one or two nights until subjected to immunoprecipitation. The yield is 4 to 6 mg from 80 g bovine cortex.

Aliquots (500 μl) of the purified synaptic vesicle fraction are treated, on ice for 1 hr, with 100 μl of antisynapsin I IgG solution (10 mg/ml in 5 mM Tris-HCl, pH 7.4), preimmune IgG solution (10 mg/ml), or 5 mM Tris-

[29] T. Ueda and P. Greengard, *J. Biol. Chem.* **252,** 5155 (1977).

[30] S. Naito and T. Ueda, *J. Biol. Chem.* **256,** 10657 (1981).

[31] T. Ueda, P. Greengard, K. Berzins, R. S. Cohen, F. Blomberg, P. J. Grab, and P. Siekevitz, *J. Cell Biol.* **83,** 308 (1979).

HCl, pH 7.4 (control buffer); following dilution with 600 μl of Solution C, the samples are centrifuged at 200,000 g_{max} for 35 min (4°) to remove unbound IgG. The pellets are suspended in 200 μl of Solution C, diluted with 1 ml of the same solution, and centrifuged as above. This washing procedure is repeated 3 times. The final pellets are homogenized in 400 μl of Solution C in a Teflon–glass homogenizer (5 strokes), and centrifuged at 5,900 g for 10 min; the pellets are washed by suspending in 400 μl of Solution C, followed by centrifugation at 5,900 g for 10 min. The washed pellets are suspended in appropriate volumes of Solution C, and, together with the supernatants, frozen at −80° until use.

Approximately 25% (22.7 ± 1.4%, $n = 9$) of protein originally present in the vesicle fraction (0.4 M sucrose layer) is recovered in the antisynapsin I IgG-precipitated vesicles. Preimmune IgG- or control buffer-treated samples yield small amounts of final pellets after the last low-speed centrifugation (4.8% of antisynapsin I IgG-precipitated vesicles or 1.2% of initial untreated vesicles). The recovery of the ATP-dependent glutamate uptake activity is approximately 13%. The lower recovery of the glutamate uptake is in part due to an apparent denaturation of the transport system observed, even in the control buffer-treated sample, during the repeated suspension and centrifugation procedures.

Procedure 2

Bovine Density Gradient-Purified Synaptic Vesicles (BGPV). Bovine brains are removed at a slaughterhouse and transported on ice to the laboratory. On ice, each cerebral cortex is dissected, trimmed free of excess white matter, and meninges removed. Two 60–80 g samples are obtained per brain. Each sample is washed twice with 300 ml of Solution A (collected on a beaker covered with a double layer of cheesecloth), minced with scissors, and 200 ml of Solution A added. Samples are homogenized with a tight-fitting Teflon–glass homogenizer, Potter–Elvehjem type (250 ml, Howe Scientific, Vineland, NJ), for 1 complete stroke at 1,200 rpm, then for 11 strokes at 900 rpm. Both samples are pooled and diluted to 1,200 ml with Solution A, and centrifuged in a Sorvall GSA rotor at 3,000 rpm (1,465 g_{max}) for 10 min (4°). The supernatants are collected by suction, distributed among four GSA centrifuge bottles, and centrifuged at 11,500 rpm (21,500 g_{max}) for 10 min. The supernatants are then discarded.

The pellets are resuspended by homogenization (3 strokes at 900 rpm, Teflon–glass homogenizer; A. H. Thomas, Philadelphia, PA, Type C) in a final volume of 160 ml of Solution B, and 40-ml aliquots are each layered onto 160 ml of 0.8 M sucrose, then centrifuged at 9,400 rpm (14,500 g_{max}) in a GSA rotor for 20 min. After centrifugation, the 0.32–0.8 M sucrose interface (white, myelin band) is removed by suction and discarded. Originally, only the 0.8 M sucrose layer, rich in synaptosomes,[32] was collected. Later, this was modified to include the soft portion of the pellet. This modification increased the yield without a decrement in purity.

Cold distilled water (1.5 vol of the collected fraction) is slowly added while the sample is being vigorously stirred. The sample is distributed among six GSA centrifuge bottles and centrifuged at 11,700 rpm (22,200 g_{max}) for 30 min, and the supernatant is completely removed and discarded. The synaptosomal pellet is gently resuspended using a Teflon rod in Solution E to a final volume of 280 ml. The sample is constantly stirred with a magnetic stirrer (slow speed to avoid aeration) for 45 min at 0°. After lysing, the sample is distributed among eight centrifuge tubes (Sorvall SS-34) and centrifuged at 19,000 rpm (43,500 g_{max}) for 15 min to pellet the synaptic and mitochondrial membranes. The supernatants are removed, distributed among capped polycarbonate tubes (45 Ti rotor, Beckman Instruments, Palo Alto, CA), and centrifuged at 41,000 rpm (195,000 g_{max}) for 70 min. The supernatants are decanted and the pellets (crude vesicles) resuspended in 6 ml of Solution D (5 strokes of a Teflon–glass homogenizer at 900 rpm).

The suspension is layered onto a discontinuous sucrose gradient (15 ml 0.4 M/8 ml 0.6 M/7 ml 0.8 M) and centrifuged at 28,000 rpm (141,000 g_{max}) for 110 min (Beckman SW28 rotor). The top layer (1–2 ml) is removed by pipette and discarded; the 0.32–0.4 M interface and 0.4 M layer are collected. A half-volume of cold distilled water (final 0.25 M sucrose) is added, and the mixture is distributed among capped centrifuge tubes and centrifuged at 47,000 rpm (200,000 g_{max}) for 55 min (Beckman 50 Ti rotor). The supernatants are decanted, and the pellets, containing synaptic vesicles, are resuspended by homogenizing in Solution D (2 mg/ml final protein concentration, 1–2 ml per preparation). The prepared vesicles are stored in conical O-ring screw-cap tubes (Sardstat, Princeton, NJ) at −80° or in liquid nitrogen.

Procedure 3

Preparation of Rat Crude Synaptic Vesicles (RCV). Rat cortex (1 g) is homogenized as above in 10 vol Solution B. The homogenate is then centrifuged at 10,000 rpm in a Sorvall SS-34 rotor (12,000 g_{max}) for 15 min (4°). The supernatant is removed and discarded, and the pellet is gently resuspended (not homogenized) in 20 vol Solution F. The pellet is allowed to lyse for 45 min on ice. The solution is then centrifuged (Sorvall SS-34) at 19,000 rpm (43,500 g_{max}) for 15 min to pellet synaptic and mitochondrial membranes. The supernatant is collected and distributed

---

[32] F. Hajos, *Brain Res.* **93**, 485 (1985).

among capped polycarbonate tubes and centrifuged at 47,000 rpm (200,000 g_{max}) for 55 min (Beckman 50 Ti rotor). The supernatant is decanted and the pellet resuspended in Solution D (0.5 ml/g of original tissue weight) and then homogenized in a Teflon–glass homogenizer (5 strokes). The prepared crude synaptic vesicles (usually at a protein concentration of 5 mg/ml with a yield of 2.5 mg/g tissue) are stored in conical O-ring screw-cap tubes at −80° or in liquid nitrogen.

Vesicle Storage and Preparation from Frozen Tissues

Prepared vesicles are normally frozen and stored in liquid nitrogen. Vesicles stored in this manner retain original activity for at least 1 year. Vesicle preparations, when stored in a freezer at −80°, are stable for at least 2 weeks.

We have also found that freezing brain tissue at −80° for 1 week has no significant effect on the glutamate uptake activity in synaptic vesicles isolated from these tissues. The stability of the synaptic vesicles against freezing permits the analysis of the vesicular uptake of glutamate in a variety of tissue samples. This observation opens the possibility of investigating various aspects of the vesicular glutamate transport system, including its levels in discrete brain regions of autopsied human subjects with malfunctions of the central nervous system.

Characterization of Vesicular Glutamate Uptake

Time Course

Figure 1 compares glutamate uptake in the three vesicle preparations, which show very similar time courses. In all three preparations, the ATP-dependent glutamate uptake reaches a half-maximal level within 5 min and a maximal level between 10 and 20 min. Decreases in ATP-dependent glutamate uptake at times longer than 20 min are likely due to decreased ATP concentration, as a second addition of ATP at 10 min prolonged the plateau phase (data not shown).

It may be pointed out that there is a significant difference in the degree of ATP dependency (in the presence of chloride) between the highly purified preparations and the crude preparation. The presence of 2 mM ATP caused at least a 30-fold increase in the rate of uptake in Ab-BGPV[21] and a 25-fold increase in BGPV, whereas it produced only a 10-fold increase in RCV. The lower degree of ATP dependency in RCV may reflect a larger contamination of nonvesicular membranes; indeed, the specific activity of ATP-dependent uptake in RCV is 2.5 times lower than that in BGPV or

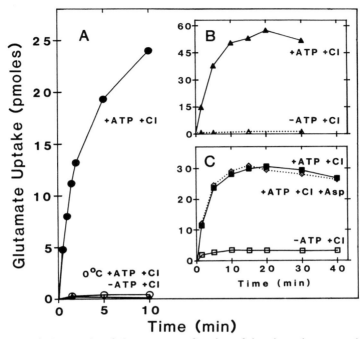

FIG. 1. Vesicular uptake of glutamate as a function of time in various synaptic vesicle preparations. The preparations were incubated for various periods of time as indicated, using the standard assay conditions described in the text. (A) Antisynapsin I IgG-precipitated synaptic vesicles (20 μg protein) were assayed for glutamate uptake at 30° ($\bullet$, $\circ$) and 0° ($\blacktriangle$) in the absence ($\circ$) or presence ($\bullet$, $\blacktriangle$) of 2 mM ATP (from Naito and Ueda[21]). (B) Bovine sucrose density gradient-purified synaptic vesicles (20 μg protein) were assayed for glutamate uptake at 30°, in the presence ($\blacktriangle$) or absence ($\triangle$) of 2 mM ATP. (C) Rat crude synaptic vesicles (50 μg protein) were assayed for glutamate uptake at 30°, in the absence ($\square$) or presence ($\blacksquare$) of 2 mM ATP and in the presence of both ATP and 2 mM aspartate ($\diamond$).

Ab-BGPV. Thus, it is possible that the degree of ATP dependency may serve as an index to the purity of synaptic vesicles.

Based on these and other criteria, the RCV preparation is less pure than the Ab-BGPV or BGPV preparations. Nonetheless, the 10-fold stimulation by ATP is sufficiently large to allow the study of vesicular glutamate uptake in relatively small tissue samples. Moreover, aspartate (2 mM), which was previously shown to cause no inhibition of vesicular glutamate uptake in highly purified synaptic vesicle preparations,[22] had no inhibitory effect on glutamate uptake in the RCV preparation (Fig. 1C); this observation suggests that the RCV preparation has little contamination of the resealed plasma membranes containing the Na^+-dependent glutamate uptake system, as discussed later.

Substrate Specificity

Previous studies showed[21] that the synaptic vesicles purified by immunoprecipitation with antisynapsin I IgG accumulated glutamate in a highly specific manner; thus, none of the other putative amino acid neurotransmitters aspartate, γ-aminobutyric acid (GABA), and glycine was significantly taken up into these vesicles. We have recently analyzed BGPV and RCV preparations for ATP-dependent uptake of these amino acids, and in Table I their uptake activities relative to glutamate uptake are compared with those in the immunoprecipitated synaptic vesicles. Aspartate is hardly accumulated into any of these three preparations. However, GABA is appreciably taken up into RCV but not into BGPV. These results suggest that the RCV preparation may be contaminated with GABA-transporting synaptic vesicles, which could be separated from glutamate-transporting vesicles by sucrose density gradient centrifugation and thus are not observed in BGPV. These observations are all consistent with the idea that the glutamate translocator in the synaptic vesicle is highly specific for glutamate.

TABLE I
ATP-DEPENDENT UPTAKE OF PUTATIVE AMINO ACID
NEUROTRANSMITTERS INTO VARIOUS SYNAPTIC
VESICLE PREPARATIONS[a]

Neurotransmitter	Relative activity (%)		
	Ab-BGPV[b]	BGPV[c]	RCV[c]
Glutamate	100	100	100
Aspartate	0.9 ± 0.2 ($n = 4$)	0	1.3
GABA	1.4 ± 0.5 ($n = 3$)	2.4	11.7

[a] Uptake was determined using standard assay conditions. The concentration of each neurotransmitter was 50 μM. Mean values and standard deviations are expressed as percentages of glutamate uptake.
[b] Data from Naito and Ueda.[21]
[c] Mean values of assays in triplicate, for one or two preparations.

Response to Varying ATP Concentrations

The effects of increasing ATP concentrations on vesicular glutamate uptake in both BGPV and RCV preparations are remarkably similar (Fig. 2). The maximal and half-maximal glutamate uptake activities are observed at 2 and 0.3 mM, respectively; higher concentrations result in a transient decrease in glutamate uptake in both preparations. Previous studies have shown that the ATP hydrolysis carried out by an Mg^{2+}-ATPase is required for the vesicular glutamate uptake.[22] The results shown in Fig. 2

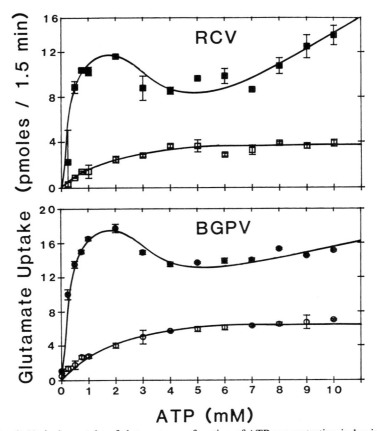

FIG. 2. Vesicular uptake of glutamate as a function of ATP concentration in bovine and rat synaptic vesicle preparations. Rat crude vesicles (RCV; 50 μg protein; ■, □) and bovine sucrose gradient-purified vesicles (BGPV; 20 μg protein; ●, ○) were incubated for 1.5 min in the presence of various concentrations of ATP, in the absence (□, ○) or presence (■, ●) of 4 mM KCl, under the standard assay conditions as described in the text.

suggest that the vesicular Mg^{2+}-ATPase-coupled glutamate uptake in the bovine brain is highly similar to that in the rat brain.

Stimulation by Chloride

An intriguing feature of the vesicular glutamate uptake system is stimulation by physiologically relevant low concentrations of chloride.[19] The effects of chloride on ATP-dependent glutamate uptake are very similar in both BGPV and RCV preparations (Fig. 3). The ATP-dependent uptake in

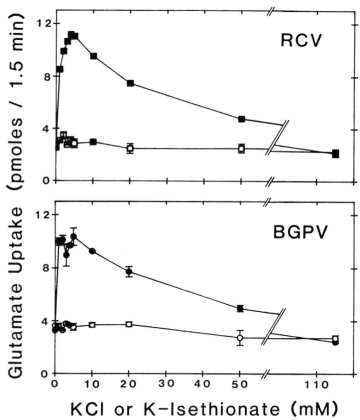

FIG. 3. Effect of various concentrations of chloride and isethionate on ATP-dependent glutamate uptake in rat crude vesicles (RCV) and bovine sucrose density gradient-purified vesicles (BGPV). RCV (50 μg protein; ■, □) and BGPV (20 μg of protein; ●, ○) preparations were incubated for 1.5 min in the presence of various concentrations of KCl (■, ●) or potassium isethionate (□, ○) under the standard assay conditions as described in the text.

TABLE II

EFFECTS OF VARIOUS CONCENTRATIONS OF FCCP
ON GLUTAMATE UPTAKE IN TWO SYNAPTIC
VESICLE PREPARATIONS[a]

FCCP concentration (μM)	Relative activity (%)	
	BGPV	RCV
0 (control)	100	100
0.5	87 ± 3	89 ± 3
1.0	62 ± 2	60 ± 3
3.0	43 ± 0	44 ± 0

[a] FCCP at various concentrations was included in the preincubation mixture. Uptake was measured using standard assay conditions. Mean values and standard deviations are expressed as percentages of control. Assays were performed in triplicate.

both preparations is stimulated approximately 3-fold by 4–5 mM chloride. Isethionate, an impermeant anion, in contrast, causes no such stimulation. Higher concentrations of chloride show less stimulation of glutamate uptake. Evidence suggests that this chloride inhibition may in part be due to competition with glutamate at its binding site,[22] and in part due to a reduction in the membrane potential under these conditions.[24] It was also shown that the stimulatory effect of chloride is a result of an increase in the V_{max} of uptake. However, the mechanism by which chloride causes this increase remains to be elucidated. Whether chloride plays a regulatory role in *in vivo* accumulation of glutamate is also an interesting question to be resolved.

Inhibition by FCCP

Several lines of evidence suggest that glutamate uptake into synaptic vesicles is driven by electrochemical proton gradients generated by a Mg^{2+}-ATPase proton pump. Thus, the protonophore carbonyl cyanide *p*-trifluoromethoxyphenylhydrazone (FCCP) has been shown to inhibit the ATP-dependent glutamate uptake in Ab-BGPV, BGPV, and RCV preparations,[21,22,33] as well as to reduce the membrane potential induced by ATP hydrolysis.[23,24] Shown in Table II is a comparison of the effects of FCCP on

[33] J. K. Disbrow, M. J. Gershten, and J. A. Ruth, *Biochem. Biophys. Res. Commun.* **108**, 1221 (1982).

the ATP-dependent glutamate uptake in BGPV and RCV preparations. The glutamate uptake systems in both preparations are, essentially, equally sensitive to the protonophore.

Response to Glutamate Analogs

The synaptic vesicle glutamate translocator is unique in its remarkably high specificity for L-glutamate.[21,22] This property is observed not only in BGPV but also in RCV preparations (Table III); the response pattern of

TABLE III
EFFECTS OF GLUTAMATE ANALOGS ON
VESICULAR UPTAKE OF GLUTAMATE IN TWO
SYNAPTIC VESICLE PREPARATIONS[a]

	Relative activity (%)	
Test agent (5 mM)	BGPV	RCV
None (control)	100	100
L-Glutamate	27 ± 0	23 ± 2
γ-Methylene DL-glutamate	39 ± 3	45 ± 1
α-Methyl DL-glutamate	51 ± 2	49 ± 2
D-Glutamate	68 ± 1	66 ± 3
L-Glutamine	117 ± 1	108 ± 6
Glutarate	97 ± 4	100 ± 1
α-Ketoglutarate	101 ± 5	99 ± 5
N-Methyl L-glutamate	88 ± 1	—
γ-Methyl L-glutamate	107 ± 2	100 ± 1
γ-Ethyl L-glutamate	115 ± 3	119 ± 4
L-Glutamic acid dimethyl ester	83 ± 1	70 ± 1
L-Glutamic acid diethyl ester	94 ± 9	94 ± 3
2-Amino-4-phosphonobutyrate	95 ± 3	91 ± 2
L-Homocysteate	86 ± 1	89 ± 4
L-Aspartate	104 ± 2	86 ± 5
D-Aspartate	111 ± 1	102 ± 3
N-Methyl D-aspartate	102 ± 1	99 ± 0
α-Amino-D-adipate	81 ± 3	78 ± 3
α-Amino L-adipate	80 ± 3	73 ± 7
α-Amino-DL-pimelate	106 ± 10	88 ± 3
Kainate	97 ± 1	88 ± 4
Ibotenate	102 ± 1	81 ± 5
Quisqualate	99 ± 2	98 ± 4

[a] Test agents (5 mM) were included in the preincubation mixture. Assays were performed using standard
assay conditions. Mean values and standard deviations are expressed as percentages of control. Assays
were performed in triplicate.

RCV to various glutamate analogs is indistinguishable from that of BGPV. The vesicular translocator is insensitive to most of the glutamate analogs tested. Among those tested, only γ-methylene DL-glutamate and α-methyl DL-glutamate showed significant interactions with the glutamate translocator.

Modification of the functional groups of glutamate, namely, the α-amino group or the γ-carboxyl group, abolishes the ability to interact with the glutamate binding site of the translocator; such modifications fail to inhibit glutamate uptake. These modified agents include L-glutamine, glutarate, δ-methyl or ethyl L-glutamate, α-ketoglutarate, N-methyl L-glutamate, L-glutamic acid dimethyl ester, L-glutamic acid diethyl ester, 2-amino-4-phosphonobutyrate, and L-homocysteate. Moreover, slight alteration of the carbon chain length dramatically diminishes interaction with the glutamate translocator. Thus, aspartate, N-methyl D-aspartate, α-aminoadipate, and α-aminopimelate show little or no effect on glutamate uptake. It should also be pointed out that the excitatory glutamate agonists kainate, ibotenate, and quisqualate lack the ability to inhibit L-glutamate uptake into synaptic vesicles in both BGPV and RCV preparations.

These results indicate the exquisite specificity of the vesicular translocator for glutamate. This property renders the vesicular glutamate translocator distinct not only from the sodium-dependent plasma membrane transporter, which does not distinguish between glutamate and aspartate,[34] but also from glutamate receptors on the plasma membrane. This highly restrictive structural requirement for the ligand suggests that the vesicular translocator may play a pivotal role in selecting glutamate for synaptic release.

The differential effects of aspartate on the vesicular and plasma membrane systems for glutamate (namely, no inhibition in the former and a substantial inhibition in the latter) allow one to use the degree of inhibition by aspartate as another index to assess the purity of glutamate-transporting synaptic vesicles; a greater inhibition would reflect a larger plasma membrane contamination. Good vesicle preparations generally exhibit less than 10% inhibition by aspartate. This property of aspartate is also exploited to reduce the background ATP-independent uptake or binding, and thereby to increase the sensitivity of the assay. Aspartate (2 mM) is now routinely included in assay mixtures for vesicular glutamate uptake whenever RCV preparations are used (see Fig. 1C).

[34] W. J. Logan and S. H. Snyder, *Brain Res.* **42**, 413 (1972).

Comparison with Another Crude Vesicle Preparation

Disbrow et al.[33] have prepared crude synaptic vesicles from the rat cerebrum and studied glutamate uptake in the vesicle preparation. However, the stimulation by ATP in their studies is only 2-fold at best, and there was no significant ATP-dependent uptake observed during the initial incubation period, whereas our studies show a substantially greater stimulation by ATP throughout the entire incubation period (see Fig. 1C). Their method of vesicle preparation differs from ours in the manner in which vesicles are released from nerve terminals. Disbrow's technique utilizes homogenization under isosmotic conditions with a glass–glass homogenizer, while ours uses osmotic shock. Another difference is that our method eliminates the low-speed centrifugation step to remove the nuclear and cell debris fraction.

We prepared vesicles from rat brain according to their method as well as ours and compared them with respect to the glutamate uptake measured under our standard assay conditions. It was found that despite the simplicity and rapidness of our method, our RCV preparation exhibited a slightly higher ATP-dependent uptake activity (7.77 ± 0.77 pmol/1.5 min/50 μg protein) than that in their crude vesicle preparation (6.52 ± 0.15 pmol/1.5 min/50 μg protein). Moreover, the degree of ATP dependency (in the presence of aspartate) in our RCV preparation is substantially higher than in theirs. Glutamate uptake observed in the presence of ATP was inhibited by aspartate (2 mM) to a lesser extent in our RCV preparation (21.5%) than in their crude vesicle preparation (37.7%). These results suggest that our RCV preparation contains less contamination than their rat crude vesicle preparation. In addition to the purity, the yield of our RCV preparation (1,870 pmol/1.5 min/g tissue) is better than their crude vesicle preparation (1,180 pmol/1.5 min/g tissue). Hence, it can be concluded that our method of preparation of crude synaptic vesicles is superior to theirs for the study of vesicular glutamate uptake, as judged by simplicity, speed, purity, and yield.

Applications

Beyond chracterization of the properties of glutamate uptake into synaptic vesicles, little research has investigated *in vivo* changes in vesicular uptake. Disbrow and Ruth[35] have shown differences in glutamate uptake into vesicles prepared from ethanol-induced short and long sleep rats. Using the procedure described here for the crude synaptic vesicle preparation, we have studied the ontogeny of the vesicular glutamate uptake

---

[35] J. K. Disbrow and J. A. Ruth, *J. Neurochem.* **45**, 1294 (1985).

system.[28] The ATP-dependent glutamate uptake system was shown to develop in parallel with the time course for synaptogenesis. We have also investigated glutamate levels in certain cerebellar mutant mice, which are deficient in glutamatergic neurons.[36] To support the physiological relevance of the vesicular glutamate uptake system, one could also examine the effect of cerebral ablation on its level in the caudate nucleus, a region known to receive massive innervations from the cerebral cortex. Moreover, the analysis of vesicular glutamate uptake may provide a useful tool for the examination of functional alterations of glutamatergic nerve terminals in certain neurological diseases and in response to the administration of neuroactive drugs.

Acknowledgments

This work was supported by National Institutes of Health Grant NS 15113, National Science Foundation Grants BNS 8207999 and BNS 8509679, and a Biomedical Research Support Grant. P.E.K. was supported by National Institute of Mental Health Training Grant 5T32MHT5794-07. We thank Dr. Carolyn Fischer-Bovenkerk for valuable comments, Soo Kim for expert technical assistance, and Mary Roth for excellent assistance in the preparation of the manuscript.

[36] C. Fischer-Bovenkerk, P. E. Kish, and T. Ueda, *J. Neurochem.* **51**, 1054 (1988).

[3] Identification of Bile Acid Transport Protein in Hepatocyte Sinusoidal Plasma Membranes

By Daniel Levy and Patricia von Dippe

Cell surface receptors and transport proteins have been identified using the technique of photoaffinity labeling.[1,2] This procedure utilizes a light-sensitive biologically active derivative of the natural substrate which is incubated with the membrane or cell system. Irradiation of the membrane protein–photoreactive ligand complex leads to the formation of a highly reactive intermediate which has the potential of covalently labeling membrane components which are in close proximity to the reagent at the

[1] T. H. Ji, *Biochim. Biophys. Acta* **559**, 39 (1979).
[2] V. Chowdry and F. H. Westheimer, *Annu. Rev. Biochem.* **48**, 293 (1979).

time of photoactivation. We have utilized this procedure to label specifically the bile acid transport system in hepatocyte sinusoidal plasma membranes using a photosensitive diazirine derivative of taurocholic acid which has been shown, using kinetic measurements, to be a substrate for the bile acid transport system.[3,4]

Materials

Isolated hepatocytes are prepared from the livers of male Sprague–Dawley rats (200–250 g) using a collagenase perfusion procedure as previously described.[5] Plasma membranes are isolated from intact livers using a sucrose density gradient procedure as previously described.[6] These membrane vesicles are primarily derived from the sinusoidal surface domain as adjudged by the appropriate enzyme markers.[4] Hydroxylamine-O-sulfonic acid is obtained from ICN Pharmaceuticals, Inc., K & K Labs Division (Plainview, NY). [³H]Taurine (12.8 Ci/nmol) is purchased from Amersham Corp. (Arlington Heights, IL).

Synthesis of Photoaffinity Labeling Derivative of Taurocholic Acid: (7,7-Azo-3α,12α-dihydroxy-5β-cholan-24-oyl)-2-amino[1,2,-³H]ethanesulfonic acid (7-ADTC)

Sodium cholate (10 g; 22.6 mmol) is converted to the corresponding 7-keto derivative with N-bromosuccinimide as previously described.[7] The purified product is dried by lyophilization. Conversion to the 7-diaziridine and 7-diazirine derivatives is effected using the method of Church et al.[8] 3α,12α-Dihydroxy-7-ketocholanic acid (1 g; 2.40 mmol) is dissolved in 24 ml of dry methanol saturated with ammonia and stirred at −10° for 30 min in a closed round-bottomed flask fitted with a pressure release valve. Freshly prepared hydroxylamine-O-sulfonic acid (397 mg; 2.95 mmol) in 1 ml of dry methanol is added dropwise and the reaction stirred for 18 hr at 4°. The turbid mixture is filtered, evaporated to dryness, and the resulting residue partitioned between 0.01 N HCl and ethyl acetate. The combined organic layers are dried over magnesium sulfate and evaporated to dryness to afford the 7-diaziridine derivative which is utilized without further purification.

[3] P. von Dippe, P. Drain, and D. Levy, J. Biol. Chem. 258, 8890 (1983).
[4] P. von Dippe and D. Levy, J. Biol. Chem. 258, 8896 (1983).
[5] R. N. Zahlten, F. W. Stratman, and H. A. Lardy, Proc. Natl. Acad. Sci. U.S.A. 70, 3213 (1973).
[6] O'Touster, N. N. Aronson, J. J. Dulaney, and H. Hendrickson, J. Cell Biol. 47, 604 (1970).
[7] L. F. Fieser and S. Rajagopalan, J. Am. Chem. Soc. 71, 3935 (1949).
[8] R. F. R. Church, A. S. Kende, and M. J. Weiss, J. Am. Chem. Soc. 87, 2665 (1965).

$3\alpha,12\alpha$-Dihydroxy-7-diaziridinecholanic acid (1 g; 2.33 mmol) is dissolved in methanol (10 ml). To this solution is added 3 ml of 1 M silver nitrate, and the cloudy mixture is stirred for 20 min at 24°. Sodium hydroxide (2.14 ml, 2.5 N) is then added dropwise over a 10-min period to form a dark brown silver oxide suspension, and the reaction mixture is stirred in the dark for 2 hr at room temperature. The mixture is filtered through Celite, diluted with water (50 ml), acidified with 1 N HCl to pH 2, and extracted with ethyl acetate. The combined organic layers are dried over magnesium sulfate and evaporated to dryness to afford 7,7-azo-$3\alpha,12\alpha$-dihydroxycholanic acid.

The crude product is purified by preparative thin-layer chromatography on a silica gel plate (20 × 20 cm) (0.5 mm) with ultraviolet indicator, developing with benzene/2-propanol/acetic acid (30/10/1). The product zone (R_f 0.48–0.62) is removed from the plate and extracted with 50% aqueous methanol. The resulting product (yield > 40%) afforded one spot on an analytical chromatography plate developed in $CHCl_3$/methanol (6/1). The product is characterized by NMR spectroscopy in D_2O, indicating the presence of one proton at C_{12} (δ 4.1) and one proton at C_3 (δ 3.7). An aqueous solution of this product has a λ_{max} of 350 nm (ϵ 56).

7,7-Azo-$3\alpha,12\alpha$-dihydroxycholanic acid (8.12 mg; 19.0 μmol) is dissolved in 150 μl of redistilled dioxane. Redistilled tri-n-butylamine (21.5 μl; 90 μmol) is added to this solution and stirred at 13° for 20 min. Redistilled ethyl chloroformate (11.1 μl; 117 μmol) is then added and stirred for 15 min at 13°. [1,2-^{3}H]Taurine (5 mCi; 0.687 mg; 5.5 μmol) in 150 μl of 1 N NaOH is added, and the reaction is stirred for 30 min at 13° and then at room temperature for 4 hr. The solvent is removed and the crude product purified by preparative thin-layer chromatography as described above, developing with chloroform/methanol (2/1). The product zone (R_f 0.28-0.34) is extracted from the plate wth 50% aqueous methanol and plate binder removed on a 5 ml silica gel column eluted with chloroform/methanol (2/1). The resulting taurine conjugate (7-ADTC; Fig. 1) (yield > 25%) has an R_f value of 0.52 on an analytical plate, which is developed with chloroform/methanol (2/1). The product is stored at − 10°.

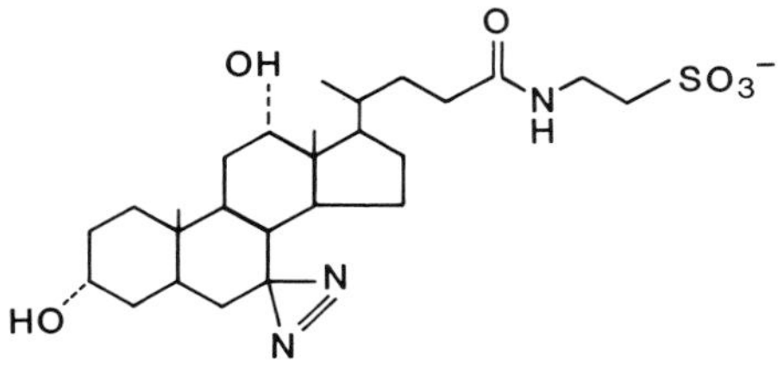

FIG. 1. Structure of 7-ADTC.

Transport Studies

Hepatocytes (2.5×10^6 cells/ml) are incubated with various concentrations of taurocholate or 7-ADTC for 30 sec in Krebs–Ringer phosphate buffer, pH 7.4. Aliquots (400 μl) of the cell suspension are rapidly centrifuged through dibutyl phthalate in a Brinkman 3200 microfuge,[9] and radioactivity associated with the cell pellet is evaluated using a scintillation fluid consisting of 16% v/w Bio-Solv BBS-3 and 0.033% v/w butyl-PBD Fluoralloy in toluene. Transport measurements are carried out in the presence and absence of sodium ion, where NaCl was replaced with choline chloride, in order to evaluate the sodium-dependent and sodium-independent components of the transport process.

Irradiation Procedures

Sinusoidal plasma membranes (1 mg/ml) maintained in suspension with a magnetic stirring bar are preincubated with 100 μM 7-ADTC (100 μCi) for 10 min in Krebs–Ringer phosphate buffer, pH 7.4, in the dark at 24°. Photoaffinity labeling of this membrane system with 7-ADTC is performed in a glass water-jacketed vessel maintained at 24° with a circulating water bath. The system is irradiated for 20 min with a General Electric 400-W medium-pressure mercury arc lamp at a distance of 15 cm. The membranes are subsequently washed in the incubation buffer and prepared for analysis.

SDS–Gel Electrophoretic Analysis

Membrane protein composition and labeling are analyzed on polyacrylamide gels in the presence of sodium dodecyl sulfate (SDS, 10%) as described by O'Farrell.[10] Gels are stained for protein with Coomassie blue and subsequently treated with ENHANCE (New England Nuclear). The gels are dried and exposed to Kodak X-Omat film for various periods of time at −70°. The protein and labeling patterns are quantitated by scanning gels and autoradiograms at 650 nm with a Gilford spectrophotometer.

Analysis of Transport and Membrane Labeling Data

The transport characteristics of taruocholic acid and 7-ADTC were very similar as a function of time, substrate concentration, and sodium dependency. Lineweaver–Burk analysis of the sodium-dependent and so-

---

[9] S. Cheng and D. Levy, *J. Biol. Chem.* **255,** 2637 (1980).
[10] P. H. O'Farrell, *J. Biol. Chem.* **250,** 4007 (1975).

TABLE I
KINETIC PARAMETERS FOR TRANSPORT OF
TAUROCHOLIC ACID AND 7-ADTC BY HEPATOCYTES

Substrate	K_m (μM)	V_{max} (nmol/mg protein/min)
Taurocholic acid		
Na$^+$-dependent	26	0.77
Na$^+$-independent	56	0.15
7-ADTC		
Na$^+$-dependent	25	1.14
Na$^+$-independent	31	0.27

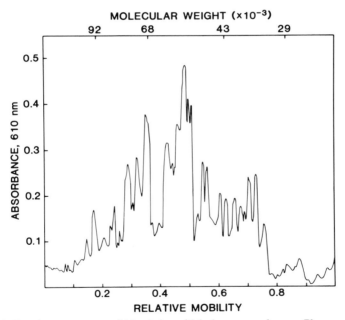

FIG. 2. Protein components of blood sinusoidal plasma membranes. Plasma membrane proteins were analyzed by SDS–gel electrophoresis, stained with Coomassie blue, and quantitated by a densitometer scan at 610 nm. Molecular weight standards used were carbonate dehydratase (29,000), ovalbumin (43,000), bovine serum albumin (68,000), and phosphorylase (92,000).

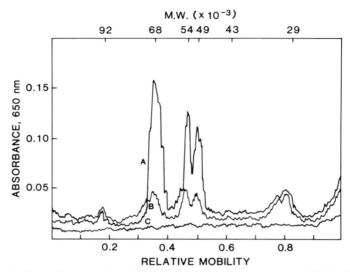

FIG. 3. Photoaffinity labeling of hepatocyte plasma membranes and HTC cells. Hepatocyte plasma membranes were labeled with 7-ADTC in the presence (curve B) and in the absence (curve A) of 100 μM taurocholic acid, and the membrane proteins were analyzed. The resultant 2-week autoradiogram was scanned at 650 nm. The peaks derived from the 54,000 and 49,000 proteins had a total of 1300 cpm. HTC cells (3×10^7) were labeled with 100 μCi of 7-ADTC. Following irradiation, cells were washed and then homogenized in 1 mM NaHCO$_3$, 75 mM NaCl, 2 mM CaCl$_2$, 1 mM EDTA, 10 μg/ml phenylmethylsulfonyl fluoride with 50 strokes in a tight-fitting Dounce homogenizer. The homogenate was centrifuged at 20,000 g for 30 min at 4°, and the resultant pellet was analyzed as above for incorporation of radioactivity. The autoradiogram was developed for 2 months (curve C).

dium-independent transport components afforded linear plots for both bile acids. The kinetic parameters for these two substrates is shown in Table I.

Inhibition studies indicated that these bile acids were competitive inhibitors of each other's sodium-dependent and sodium-independent transport systems. The inhibitory properties of these bile acids in conjunction with the similar transport parameters suggested that both substrates utilized the same membrane carrier system, thus satisfying an important criterion for photoaffinity labeling studies.

Labeled membrane proteins were analyzed by SDS–gel electrophoresis (Fig. 2). Autoradiographic analysis indicated that radioactivity was primarily associated with three proteins (Fig. 3). One of these proteins, with a molecular weight of 68,000, was shown to be membrane-associated serum albumin. Bile acids have previously been shown to bind to serum albumin.[11] The second and third labeled components were shown to be intrin-

[11] W. Kramer, H.-P. Buscher, W. Gerok, and G. Kurz, *Eur. J. Biochem.* **102**, 1 (1979).

sic membrane proteins with molecular weights of 54,000 and 49,000. Similar results were obtained when labeling was performed on intact hepatocytes followed by membrane isolation and SDS–gel analysis. The specificity of the labeling reaction was assessed by carrying out the photolysis of the 7-ADTC–membrane complex in the presence of taurocholic acid. As shown in Fig. 3, the presence of the natural substrate resulted in a large decrease in the labeling of the 68,000, 54,000, and 49,000 proteins. These results suggest a high degree of labeling specificity. Further support for the specificity of this labeling reaction is obtained from studies with hepatoma tissue culture (HTC) cells which have been shown to lack a functional bile acid transport system.[4] When photolysis of this cell system was performed in the presence of 7-ADTC, no significant incorporation of radioactivity could be detected (Fig. 3), further supporting the validity of the labeling results.

These studies indicate that the photoreactive derivative of taurocholic acid (7-ADTC) is a substrate for the hepatocyte bile acid transport system and specifically recognizes and covalently labels two intrinsic membrane proteins in purified sinusoidal plasma membranes as well as in intact hepatocytes. Subsequent studies using monoclonal antibodies[12] have demonstrated that the 49 K protein represents the Na^+-dependent bile acid transport system.

Acknowledgments

This research was supported by National Institutes of Health Grant DK-25836.

[12] M. Ananthanarayanan, P. von Dippe, and D. Levy, *J. Biol. Chem.* **263**, 8338 (1988).

[4] Transport of Alanine across Hepatocyte Plasma Membranes

By JOHN D. McGIVAN

The metabolism of alanine to urea and glucose is a major metabolic function of the liver. The initial step in alanine metabolism is the Na^+-dependent transport of alanine across the cell membrane, and this process has been widely investigated. Alanine transport in liver is inducible in starvation and is subject to both short- and long-term control by hormones. There is evidence that the transport of alanine at physiological concentrations is an important rate-determining step in alanine metabolism. The general properties of the transport of alanine and other amino acids in hepatocytes have been reviewed.[1,2]

The purpose of this chapter is to review the methods which have been used for the study of alanine transport in isolated hepatocytes and in liver plasma membrane vesicles. The principles of the approaches described can be applied to study the transport of other naturally occurring amino acids in suspensions of hepatocytes and in other cell systems. This chapter does not consider the study of the transport of amino acid analogs, the identification of individual transport systems, or the measurement of transport in cultured hepatocytes; these important subjects are dealt with elsewhere in this volume.

Transport of Alanine in Isolated Hepatocytes

The routine isolation of hepatocytes by collagenase digestion of rat liver is described in detail elsewhere.[3] The hepatocytes are washed and suspended in Krebs–Henseleit bicarbonate-buffered medium containing 2% dialyzed bovine serum albumin. The presence of bicarbonate in the medium is essential, since its omission greatly reduces the rate of alanine transport.[4] The cells can be used for transport studies after a short preincubation in the incubation medium, but it is sometimes convenient to store the suspension on ice for up to 2 hr before use. In this case the cells should be incubated at 37° for 20 min before transport studies are initiated in order to allow the reestablishment of ion gradients across the membrane.

For the accurate measurement of alanine transport in hepatocytes, there are three major requirements. First, if radioactive techniques are to be used, the metabolism of alanine must be inhibited so that the radioactivity in the cells accurately represents the internal concentration of alanine. This can be achieved by the addition of 0.5 mM aminooxyacetate, which is a specific inhibitor of transaminase enzymes and does not reduce the ATP content of the cells. It is not necessary to add such an inhibitor if the amino acid used is metabolized only very slowly (as is the case with leucine) or if the incubation period is less than about 1 min (when little metabolism will have occurred). Second, since the cell pellet on centrifugation is contaminated with a considerable quantity of extracellular water, a marker of the extracellular space should be included. Inulin, polyethylene glycol, or sucrose are suitable compounds to use for this purpose. Third, the method of termination of the transport reaction by separation of cells from the me-

[1] M. A. Shotwell, M. S. Kilberg, and D. L. Oxender, *Biochim. Biophys. Acta* **737**, 267 (1983).
[2] M. S. Kilberg, E. F. Barber, and M. E. Handlogten, *Curr. Top. Cell. Regul.* **25**, 133 (1985).
[3] M. S. Kilberg, this series.
[4] J. D. McGivan, *Biochem. J.* **182**, 697 (1979).

dium must also achieve the rapid deproteinization of the cells to prevent any possible metabolism of the substrate, and this is commonly done by centrifuging the cells through a mixture of silicone oil and dinonyl phthalate into perchloric acid.

A suitable protocol for the measurement of alanine transport into hepatocytes is as follows:[5] Tubes for a Beckman Microcentrifuge Model B are prepared by placing 50 μl perchloric acid (15%, v/v) in the bottom of the tube and layering 0.1 ml of a mixture (1 : 1, v/v) of silicone fluid MS550 and dinonyl phthalate above this. The hepatocyte suspension containing 10–12 mg cell protein/ml is incubated with an equal volume of Krebs–Henseleit bicarbonate medium containing [^{3}H]inulin together with the appropriate concentration of alanine, [^{14}C]alanine, and 0.5 mM aminooxyacetate. Transport is terminated by layering a 0.25-ml aliquot of the incubation (containing not more than 3 mg of protein) above the silicone oil layer and centrifuging for a minimum of 10 sec. After withdrawal of a sample of the supernatant layer for counting, the tube is frozen and cut through the silicone oil layer with a sharp knife.

The entire pellet is shaken with scintillation fluid, and the ^{14}C and ^{3}H in the pellet and supernatant samples are determined by dual-label scintillation counting. The intracellular alanine is calculated as the total alanine in the pellet minus that in the extracellular water. If it is required to find the intracellular concentration of alanine, measurements of the internal volume of the cells must be performed in parallel using [carboxy-^{14}C]inulin and ^{3}H$_2$O. The internal volume of hepatocytes determined by this method is 1.8–2.0 μl/mg protein.

Role of Alanine Transport in Regulation of Alanine Metabolism

The transport of alanine across the cell plasma membrane is the first reaction in alanine metabolism, and it is of importance to determine whether there are conditions under which the rate of transport is rate-limiting for the subsequent metabolism of this amino acid. Identification of transport as a rate-limiting step requires the fulfilment of a number of criteria: (1) The rate of transport, measured in the absence of metabolism, must approximate the rate of alanine metabolism at the same alanine concentration. (2) Specific inhibitors of transport must inhibit the rate of metabolism with the same inhibitor concentration dependence. (3) The steady-state concentration ratio of intracellular to extracellular alanine during alanine metabolism should be low and should be greatly increased by the addition of aminooxyacetate to inhibit metabolism. (4) Stimulation of transport should lead to an increase in the intracellular to extracellular alanine concentration ratio associated with an increase in the rate of

[5] S. K. Joseph, N. M. Bradford, and J. D. McGivan, *Biochem. J.* **176,** 827 (1978).

alanine metabolism. If any of the criteria fail to apply under particular conditions, then alanine transport under those conditions is unlikely to exert a significant controlling influence on metabolism.

In order to study this problem in isolated hepatocytes in suspension, alanine metabolism is best measured by the disappearance of substrate rather than by the appearance of products since the endogenous rate of formation of glucose in hepatocytes is considerable. Alanine is conveniently estimated using the NAD-linked bacterial enzyme alanine dehydrogenase.[6] After deproteinization of the cell suspension and removal of the precipitated protein by centrifugation, the supernatant is neutralized by the addition of a small volume of 3 M KOH or 3 M K_2CO_3. After addition of the appropriate buffer and alanine dehydrogenase, NAD^+ reduction is measured by spectrophotometric or fluorometric techniques depending on the sensitivity required.

For the measurement of intracellular alanine concentrations during metabolism, the cell suspension (12 mg protein/ml) is incubated with the appropriate concentration of alanine in the absence of aminooxyacetate together with ^3H_2O and [carboxy-^{14}C]inulin as markers of the total and extracellular spaces, respectively. Aliquots (0.8 ml) of the cell suspension are layered on to 0.5 ml of silicone oil/dinonyl phthalate (1 : 1, v/v) which is itself layered above 0.1 ml of 15% perchloric acid in a tube for the Eppendorf Model 3200 bench centrifuge. The reaction is terminated by the initiation of centrifugation. A sample of the supernatant is immediately acidified and radioactivity and alanine are assayed in the neutralized extract. An aliquot of the acid layer is withdrawn, neutralized with a small volume of K_2CO_3 and similarly assayed for radioactivity and alanine. From these data, the extracellular and intracellular concentrations of alanine can be calculated. In this type of investigation, alanine transport can be inhibited by titration with ouabain or by reducing the extracellular sodium concentration. Transport can be stimulated by the addition of cyclic AMP. Simultaneous measurements of the rate of metabolism and of changes in the alanine distribution can then be made.

Such investigations can also be carried out using the hepatocyte perifusion apparatus of van der Meer and Tager.[7] In a perifusion system a true steady state can be obtained at a constant low external alanine concentration. Centrifugal filtration techniques similar to those described above can be used for the measurement of intracellular alanine concentration. Using such a system, Sips et al.[8] showed that the intracellular concentration of alanine remained low on increasing the extracellular concentration of

[6] D. H. Williamson, O. Lopes Vereira, and B. Walker, *Biochem. J.* **104,** 497 (1967).
[7] R. van der Meer and J. M. Tager, *FEBS Lett.* **67,** 36 (1976).
[8] H. J. Sips, A. K. Groen, and J. M. Tager, *FEBS Lett.* **119,** 271 (1980).

alanine in the physiological range, and the addition of aminooxyacetate greatly increased the intracellular to extracellular concentration ratio.

Using these various approaches, data from several laboratories indicate that alanine transport limits the rate of alanine metabolism at low alanine concentrations (<0.5 mM).[8-10] There is less agreement about the limitation of metabolism by transport at higher concentrations of alanine.

Influence of Cell Plasma Membrane Potential on Alanine Transport

Na$^+$–alanine cotransport is an electrogenic process. Accordingly both the steady-state concentration ratio of alanine in the absence of metabolism as well as the kinetics of alanine transport should be influenced by the cell membrane potential. It is therefore important to be able to observe the effect of varying the cell membrane potential on the initial rate of transport.

The measurement of cell membrane potential is isolated hepatocytes is not straightforward. Direct electrophysiological measurements involving the impalement of cells with microelectrodes has not so far yielded satisfactory results. In a number of cell types, the cell membrane potential has been calculated from the distribution ratio of permeant cations or from the absorption of fluorescence of cationic dyes. This approach is not feasible in hepatocytes, which contain large numbers of mitochondria that accumulate such permeant cations to a high intramitochondrial concentration. In hepatocytes, such measurements tend to reflect the mitochondrial membrane potential rather than that of the cell plasma membrane.

In principle, the membrane potential can be calculated from the distribution of permeant anions, since these are not accumulated by the mitochondria. Thiocyanate has been used for this purpose by Hoek et al.,[11] but these authors found it necessary to correct for the assumed binding of thiocyanate to intracellular constituents. Recently, Edmondson et al.[12] have also used thiocyanate distribution as an indicator of cell membrane potential. In this study, digitonin was used to disrupt the cells and measure the cytosolic space. The thiocyanate technique remains somewhat unsatisfactory because of the various necessary correction factors involved and because of the lack of independent confirmation that changes in thiocyanate distribution in fact accurately measure changes in membrane potential.

An alternative to the use of thiocyanate is to calculate the membrane potential from the distribution of chloride using ^{36}Cl$^-$. It has been shown in

[9] J. D. McGivan, J. C. Ramsell, and J. H. Lacey, *Biochim. Biophys. Acta* **644**, 295 (1981).
[10] P. Fafarnoux, C. Remesy, and C. Demigne, *Biochem. J.* **210**, 645 (1983).
[11] J. B. Hoek, D. G. Nicholls, and J. R. Williamson, *J. Biol. Chem.* **255**, 1458 (1980).
[12] J. W. Edmondson, B. A. Miller, and L. Lumeng, *Am. J. Physiol.* **249**, G427 (1985).

this laboratory that chloride distribution in hepatocytes is not affected by inhibitors which inhibit chloride transport systems in other cell types. The membrane potential calculated from chloride distribution under a number of different conditions is equal to the potential measured directly using microelectrodes in perfused liver under the same conditions.[13] It therefore appears that chloride distributes passively across the cell membrane according to the membrane potential. Using this technique, a good correlation between cell membrane potential and initial rate of alanine transport can be demonstrated, and this relationship is independent of the mechanism by which the potential is varied.

Chloride distribution can be measured by incubating cells in the presence of $^{36}Cl^-$ together with [^3H]inulin to measure the extracellular space. Dual-channel scintillation counting for ^3H plus ^{36}Cl is achieved as for ^3H plus ^{14}C, but quench curves appropriate for ^{36}Cl rather than ^{14}C must be used. While the use of chloride to determine the membrane potential may be preferable to that of thiocyanate, it must be recognized that the validity of both methods requires further independent justification.

Alanine Transport in Liver Plasma Membrane Vesicles

The study of transport in plasma membrane vesicles allows the investigation of electrogenicity of the transport process and its ion dependence and obviates problems arising from possible substrate metabolism. Demonstration of transport in membrane vesicles is also a necessary prerequisite for the eventual identification of the carrier protein. Alanine transport in liver membrane vesicles was first reported by van Amelsvoort et al.,[14] and a modification of their method which is routinely used in this laboratory is detailed below.

A single rat liver is homogenized in a medium containing 0.25 M sucrose, 10 mM K$^+$–HEPES, and 0.2 mM CaCl$_2$ at pH 7.4 and 4° initially by 6 strokes of a Teflon/glass Dounce homogenizer. The homogenate is diluted to 200 ml with the same medium and rehomogenized in 30-ml aliquots using a glass–glass homogenizer with approximately 20 strokes. This second homogenization is critical if satisfactory intact vesicles are to be obtained. After filtration through muslin to remove particles of fat, 1 mM EDTA (final concentration) is added. The homogenate is centrifuged at 1000 g for 10 min, and the pellet is discarded. The supernatant from the centrifugation is recentrifuged at 20,000 g for 30 min, and the pellet is resuspended using the glass–glass homogenizer in 10 ml of homogenization medium to which 1 mM EDTA has been added.

---

[13] N. M. Bradford, M. R. Hayes, and J. D. McGivan, *Biochim. Biophys. Acta* **845,** 10 (1985).
[14] J. M. M. van Amelsvoort, H. J. Sips, and K. van Dam, *Biochem. J.* **174,** 1083 (1978).

Discontinuous sucrose gradients are prepared in tubes for the $3 \times$ 25 ml swing-out head of the M.S.E. 65 ultracentrifuge. Each gradient consists of a lower layer of 10 ml of 46.5% sucrose (w/v) and an upper layer of 10 ml of 21.5% (w/v) sucrose, each solution being made up in 10 mM K^+–HEPES, pH 7.5. Three milliliters of suspension is layered on top of each gradient, and the tubes are centrifuged at 23,000 rpm (50,000 g) for 2.5 h at 4°. The material at the interface between the two sucrose solutions is collected and diluted at least 4 times with the original homogenization medium (containing no EDTA). The resulting suspension is centrifuged at 100,000 g for 40 min, and the pellet is resuspended in approximately 2 ml of homogenization medium using a syringe. The membrane suspension is rapidly frozen in small volumes in liquid nitrogen, and can be kept frozen for some weeks before use in transport experiments. This preparation routinely yields approximately 10 mg of protein, and the specific activity of the plasma membrane marker 5′-nucleotidase is increased 10-fold over the original homogenate.

For measurement of transport, 20 μl of membrane suspension is added to an equal volume of incubation medium to give the following final concentrations: 0.25 M sucrose, 10 mM K^+–HEPES, 0.2 mM $CaCl_2$, 5 mM $MgCl_2$, 100 mM KCNS or 100 mM NaCNS, and 0.1 mM [^{3}H]alanine at pH 7.5. A suitable specific activity for such experiments is 100 Ci/mol, i.e., approximately 200 dpm/pmol. The reaction is terminated by the addition of 1 ml of ice-cold "stopping" solution containing 0.25 M sucrose, 10 mM K^+–HEPES, 0.2 mM $CaCl_2$, plus 0.2 M NaCl at pH 7.5. The diluted suspension is immediately filtered through millipore filters (HAWP 0.45 μm) and washed with 2×1 ml of stopping solution. The filters are dissolved in a suitable scintillator for measurement of radioactivity. In this method, the NaCNS provides the necessary sodium and electrical gradient for Na^+–alanine cotransport. Parallel experiments using KCNS provide a measure of binding of alanine to the membranes, retention of residual radioactivity on the membrane filters, plus any Na^+-independent alanine transport.

As measured by the above method, the maximum uptake of alanine should be 0.25–0.3 nmol/mg protein at an external alanine concentration of 0.1 mM, and this should be attained in 30 sec. Lower values indicate a less than satisfactory membrane preparation. After the vesicles have been thawed the transport activity progressively declines. The transport of alanine in this stem has been extensively investigated.[14-16]

[15] H. J. Sips, J. H. H. van Amelsvoort, and K. van Dam, *Eur. J. Biochem.* **105**, 217 (1980).
[16] H. J. Sips, and K. van Dam, *J. Membr. Biol.* **62**, 231 (1981).

Approaches to Identification of Protein(s) Responsible for Alanine Transport in Liver Cell Membrane

The alanine transport protein or proteins in liver plasma membranes have not yet been identified. The identification of such proteins would be of great interest, in particular because this would allow a detailed investigation of the hormonal induction of synthesis of transport proteins and their insertion into the plasma membrane.

The successful identification of various transport proteins in mammalian cell membranes has relied mainly on the use of specific tight-binding inhibitors which can be used to label the transport protein.[17] No such inhibitor has so far been identified for the alanine carrier in liver. Potential inhibitors of transport are best assessed by a study of their action on alanine transport in plasma membrane vesicles rather than intact cells, since inhibition in cells may be due to nonspecific effects of the inhibitors on cell ATP levels. Using this approach, it has been shown that various sulfhydryl-blocking reagents (e.g., mersalyl, N-ethylmaleimide[18]) inhibit alanine transport in membrane vesicles. The inhibition is due to an effect of the reagents on the carrier molecule rather than to nonspecific membrane damage. However, progress toward carrier identification is dependent on the unambiguous demonstration of protection of the carrier against the inhibitor by high concentrations of substrate, and this has proved difficult to achieve.

A second approach to the identification of the transport protein involves reconstitution of various liver cell membrane fractions into artificial phospholipid membranes. Although such reconstitution has been reported for the alanine-transporting systems of kidney brush border membranes[19] and Ehrlich ascites cells,[20] the methods used do not produce satisfactory results when applied to liver. Modifications of these methods may be appropriate for the liver system, but this has not yet been demonstrated.

In principle, a combination of labeling and reconstitution should lead to the identification of the alanine carrier in liver cell membranes, and other possible approaches to the problem exist. At present, however, the literature contains few reports of work in this area on liver amino acid transport system.

[17] G. Semenza, M. Kessler, M. Hosang, J. Weber, and U. Schmidt, *Biochim. Biophys. Acta* **779**, 343 (1984).
[18] M. R. Hayes and J. D. McGivan, *Biochem. J.* **214**, 489 (1983).
[19] H. Koepsell, K. Korn, D. Ferguson, H. Menuhr, D. Ollig, and W. Haase, *J. Biol. Chem.* **259**, 6548 (1984).
[20] J. I. McCormick, D. Tsang, and R. M. Johnstone, *Arch. Biochem. Biophys.* **231**, 355 (1984).

[5] Purification and Reconstitution of Glucose Transporter from Human Erythrocytes

By STEPHEN A. BALDWIN and GUSTAV E. LIENHARD

Introduction

Most, if not all, animal cells contain a transport system for D-glucose of the facilitated diffusion type. The protein that constitutes this transport system is unusually abundant in the membrane of the human erythrocyte, where it comprises 5% of the protein by weight.[1] In 1977 Kasahara and Hinkle described the purification of this transporter in functionally active form.[2] This chapter describes modified procedures for the purification and reconstitution of the transporter that were developed in our laboratories.[3,4] These procedures give higher amounts of a more active preparation.

The complementary DNA to the messenger RNA for the glucose transporter in the human hepatoma cell line HepG2 has recently been cloned and sequenced.[5] In addition, the amino acid compositions and sequences of a number of peptides from the human erythrocyte transporter have been determined. Comparison of the amino acid sequence given by the complementary DNA with the data from protein chemistry indicates that the transporter polypeptide in the two cell types is probably the same. The molecular weight (M_r) of the polypeptide is 54,116. A speculative model for the folding of the polypeptide chain with respect to the plane of the lipid bilayer, in which there are 12 transmembrane helices, has been proposed on the basis of the hydropathy of segments and results from vectorial labeling. The transporter has been identified as a protein of about M_r 50,000 in several other human cell types and in tissues from a variety of other species.[1,6] The identification has been achieved either through immunoblotting with polyclonal or monoclonal antibodies raised against the purified erythrocyte transporter or through photoaffinity labeling with the ^{3}H-labeled ligand cytochalasin B. To date the transporter has been purified

[1] W. J. Allard and G. E. Lienhard, *J. Biol. Chem.* **260,** 8668 (1985).

[2] M. Kasahara and P. C. Hinkle, *J. Biol. Chem.* **252,** 7384 (1977).

[3] S. A. Baldwin, J. M. Baldwin, and G. E. Lienhard, *Biochemistry* **21,** 3836 (1982).

[4] M. T. Cairns, D. A. Elliot, P. R. Scudder, and S. A. Baldwin, *Biochem. J.* **221,** *179* (1984).

[5] M. Mueckler, C. Caruso, S. A. Baldwin, M. Panico, I. Blench, H. R. Morris, W. J. Allard, G. E. Lienhard, and H. Lodish, *Science* **229,** 941 (1985).

[6] D. W. Schroer, S. C. Frost, R. A. Kohanski, M. D. Lane, and G. E. Lienhard, *Biochim. Biophys. Acta* **885,** 317 (1986), and references therein.

only from human erythrocytes. With the availability of antibodies, it is likely that purification from other sources will be achieved shortly.

Purification and Reconstitution

Purification and reconstitution of the erythrocyte transporter are achieved by the following steps: preparation of erythrocyte membranes by osmotic lysis of erythrocytes; removal of the peripheral (cytoskeletal) proteins from the membranes by treatment with dilute base; partial solubilization of the protein-depleted membranes with the detergent octylglucoside; separation of the transporter and some membrane lipids from this detergent extract by chromatography on DEAE-cellulose; and reconstitution of the transporter into a membrane of these lipids through removal of the detergent by dialysis.[3] All the operations described here are carried out on ice or in a 4° cold room, with solutions prechilled to 0°–4°.

Erythrocyte Membranes.[7] Preparation of erythrocyte membranes starts with a unit (450 ml) of human blood drawn into citrate/phosphate/dextrose. To reduce the risk of acquiring hepatitis or other diseases, we generally use a unit that has been freshly drawn from a known donor. However, recently outdated units from the blood bank (~24 days of storage) yield a transporter preparation of identical properties. The blood is distributed into six 250-ml tubes for the Sorvall GSA rotor and then diluted to the 175-ml mark with 5 mM sodium phosphate, 150 mM NaCl, pH 8.0. The cells are pelleted at 4500 rpm (3300 g_{max}) in a Sorvall RC5 centrifuge for 10 min. The supernatant is carefully aspirated using a Pasteur pipette attached to a water pump until cells begin to be removed. The cells are resuspended to the 175 ml mark in the phosphate buffer by gentle stirring with a glass rod and pelleted again. The supernatant is again aspirated, and this time the buffy coat of white cells is also removed. This washing procedure is repeated 2 more times.

After the last wash, the cells are lysed by the addition of 5 mM sodium phosphate, pH 8.0, to the 175-ml mark, followed by gentle stirring with a glass rod. The membranes are pelleted by centrifugation at 11,500 rpm (21,500 g_{max}) for 25 min. The red supernatant is aspirated, with a portion left behind in order to avoid aspiration of the wispy membranes. The membranes are resuspended in a total volume of 175 ml of 5 mM sodium phosphate, pH 8.0 and pelleted again. This washing procedure is repeated until the membranes are only faintly pink, generally 4 more times. After the first or second wash, the supernatant is sufficiently light red in color that the membranes can be seen clearly. At this point, the hard button of

---

[7] T. L. Steck and J. A. Kant, this series, Vol. 31, p. 172.

intact white cells beneath the lighter pellet of erythrocyte membranes is aspirated by rotating the centrifuge tube so that the membranes slide off the button. Finally the washed membranes are concentrated by transferring them with a little buffer to several 50-ml tubes for a Sorvall SS34 rotor and centrifuging at 20,000 rpm (48,000 g_{max}) for 20 min in a Sorvall RC5 centrifuge. The yield is typically about 150 ml erythrocyte membranes at about 4 mg protein/ml. These are assayed for protein and stored at $-70°$.

Protein-Depleted Erythrocyte Membranes. To a suspension of membranes (40 ml at 3.5–4.0 mg protein/ml) that is being stirred with a magnetic stirrer is added 210 ml of base solution [15 mM NaOH, 2 mM disodium ethylenediaminetetraacetic acid (Na$_2$EDTA), 0.2 mM dithiothreitol]. This solution was previously prepared by dissolving 180 mg of Na$_2$EDTA in 250 ml 15 mM NaOH, chilling the solution, bubbling N$_2$ through it for 5 min to remove O$_2$, and then adding 7.5 mg dithiothreitol just before use. The basic mixture is immediately distributed among eight SS34 tubes and centrifuged at 19,000 rpm (43,500 g_{max}) for 15 min. After careful aspiration of the supernatant, except for a milliliter or so left to avoid removal of the membranes, 25 ml of 50 mM tris(hydroxymethyl)-aminomethane (Tris) chloride, pH 7.4 at 2°, is added to each tube, and the pellet is resuspended with a stirring rod. The protein-depleted membranes are pelleted again by centrifugation at 19,000 rpm for 15 min, and the supernatant is aspirated. At this stage the pellets are tight enough so that virtually all the supernatant can be aspirated. The protein-depleted membranes in each tube are resuspended in 100 μl of 50 mM Tris-Cl, pH 7.4 at 2°, by stirring with a Teflon pestle and transferred to a 20 ml homogenizer vessel (Thomas Scientific). The centrifuge tubes are washed with 3 ml buffer, and this is combined with the membranes in the homogenizer. The protein-depleted membranes are dispersed by several up and down strokes with the Teflon pestle, assayed for protein, and stored at $-70°$. Typically, 160 mg of erythrocyte membranes yield 65 mg of protein-depleted membranes. If two Sorvall centrifuges are available, it is desirable to do a double preparation so as to obtain sufficient material for the typical scale of the final step.

Purified, Reconstituted Transporter. To a stirred suspension of protein-depleted erythrocyte membranes (144 mg protein in 67 ml Tris-Cl, pH 7.4 at 2°) is added to 140 μl of 1 M dithiothreitol, followed by 4.86 ml of 685 mM octylglucoside (octyl-β-D-glucopyranoside, Calbiochem, San Diego, CA). After 20 min of stirring, the mixture is dispensed into 12 polycarbonate tubes for a Beckman 50 Ti rotor and centrifuged at 45,000 rpm (180,000 g_{max}) for 1 hr. The supernatant is then carefully removed with a pipette from the partially loose pellet. Contamination of the supernatant with pellet material leads to a less pure preparation, and thus at this stage the last 0.5 ml or so of the supernatant should be left behind.

The detergent extract ($\sim$ 65 ml) is applied to a column (6.5 $\times$ 2.5 cm, 31 ml total volume) of DEAE-cellulose (Whatman DE-52) that has been equilibrated with 34 mM octylglucoside, 2 mM dithiothreitol, 50 mM Tris-Cl, pH 7.4 at 2° (elution buffer). Because octylglucoside is expensive, the column is first equilibrated with Tris-Cl buffer alone, and then a column volume of the elution buffer is introduced. We use fresh, rather than regenerated, DEAE-cellulose for each preparation. The extract is applied at a flow rate of about 75 ml/hr, and the column is eluted with the elution buffer at the same rate. Fractions (10 ml) are collected, and protein is detected by monitoring the absorbance at 280 nm. The transporter-containing eluent is found between about 30 and 100 ml, measured from the start of application of the extract. These fractions are combined and made 100 mM in NaCl and 1 mM in Na$_2$EDTA by the addition of 25 μl of 4 M NaCl/ml and 5 μl of 200 mM EDTA, pH 7/ml, respectively. The solution is placed in dialysis tubing (Spectra/Por 2, molecular weight cutoff 12,000, Fisher Scientific) and dialyzed against four 2-liter batches of 100 mM NaCl, 1 mM EDTA, 50 mM Tris-Cl, pH 7.4 at 2°, with changes after about 5, 16, and 24 hr. The purified preparation is frozen in liquid nitrogen in 3-ml aliquots and stored at $-70°$.

Recently, one of us introduced a slight modification of this method that increases the yield of protein by about 30% and the specific activity (nanomoles of cytochalasin B sites per milligram protein) by about 40% (see below).[4] In this modification, the supernatant from detergent solubilizaton is made 25 mM in NaCl, and 25 mM NaCl is included in the buffer used to equilibrate and elute the DEAE column. Also the transporter-containing fractions are dialyzed against 100 mM NaCl, 1 mM EDTA, 50 mM sodium phosphate, pH 7.4. This preparation is designated the "modified" preparation hereafter, whereas the other is designated the "original" preparation.

While the transporter is solubilized in octylglucoside, it undergoes a slow denaturation to a form that no longer binds cytochalasin B after reconstitution.[3] Consequently, this final step should be carried out as rapidly as possible; a period of 3–4 hr is required between the solubilization with octylglucoside and the initiation of dialysis. The rate of denaturation is greater at higher temperatures and at higher ratios of octylglucoside to phospholipid. For these reasons, care should be taken to keep the temperature at 0°–4°, and if the scale of the preparation is altered, the concentration of protein-depleted membranes and detergent used for solubilization should not be changed. The volume of DEAE-cellulose should be altered proportionally.

Assays and Properties

Protein. Protein is assayed by a modification of the Lowry procedure used for membrane proteins, with bovine serum albumin as the standard.[8] Because large amounts of Tris interfere with the assay, Tris is removed by dialysis before assay of the original preparation, which has a low protein concentration. The values presented here are the ones from the Lowry assay. The true protein amount for the purified transporter, calculated from amino acid analysis for the stable, fully released amino acids and the amino acid composition, is 1.03 times the value given by the Lowry assay. Some of the values for the original preparation given here differ slightly from those in Baldwin *et al.*[3] because, owing to small errors in the amino acid composition at that time, the protein content by amino acid analysis was taken as 0.88 that by Lowry assay. Typical original and modified preparations have protein concentrations of 150 and 200 μg/ml; the overall yields are 11 and 15 mg, respectively.

It is prudent to examine the preparation by SDS–PAGE. The transporter undergoes partial aggregation in sodium dodecyl sulfate (SDS), to dimers and higher molecular weight aggregates.[3] This aggregation is minimized by preparation of the sample for SDS–PAGE without heating and by the use of a mixture of C_{12}, C_{14}, and C_{16} alkyl sulfates ("lauryl" sulfate from Pierce Chemical) rather than pure C_{12} alkyl sulfate. Under these conditions, in 10% acrylamide gels, the preparations show a predominant broad band between M_r 65,000 and 45,000, with a very weak band of dimer at M_r 120,000 and occasionally one for erythrocyte band 7 at M_r 29,000. About 95% of the Coomassie blue stain is in the main band. The transporter is a glycoprotein consisting of about 15% carbohydrate by weight, and the broadness of this band on SDS–PAGE is due to heterogeneity in the oligosaccharide chains.[9]

Lipid. A substantial portion of the erythrocyte membrane lipid is solubilized with octylglucoside and passes through the DEAE-cellulose in the same fractions as the transporter. The lipid content of the preparations is determined by colorimetric analysis for phospholipid phosphorus and has been calculated assuming an average molecular weight of 750 for a phospholipid molecule.[10] The original and modified preparations have phospholipid to protein ratios by weight that fall in the ranges 2.3–3.5 and 3.5–4.8, respectively. The lipid composition of one original preparation was determined by extraction of the lipids into chloroform–methanol, followed by colorimetric assay for cholesterol and by separation of the

[8] G. L. Peterson, *Anal. Biochem.* **83**, 346 (1977).
[9] G. E. Lienhard, J. H. Crabb, and K. J. Ransome, *Biochim. Biophys. Acta* **769**, 404 (1984).
[10] G. R. Bartlett, *J. Biol. Chem.* **234**, 466 (1959).

phospholipid classes by thin-layer chromatography and subsequent colorimetric assay of each for phosphorus (unpublished results of G. E. Lienhard).[11] It contained about 25% phosphatidylcholine, 28% phosphatidylethanolamine, 17% phosphatidylserine, 9% sphingomyelin, and 14% cholesterol (values as percentage by weight of the total weight of phospholipid and cholesterol). The corresponding values for the human erythrocyte membrane are 20, 20, 9, 18, and 29%.[12] Thus, the transporter preparation is relatively depleted in sphingomyelin and cholesterol. The amount of octylglucoside remaining in the preparation, determined with radiolabeled detergent, is about 1/80 the molar amount of phospholipid.[3]

Physical State. Electron microscopy of the original preparation after negative staining with phosphotungstic acid at pH 7 (unpublished results of Dr. George Rubin and G. E. Lienhard) shows irregularly shaped membranous structures of heterogeneous size (60–300 nm diameter) that, on the basis of the visualization of stain within the vesicles by stereoimaging, are probably unsealed. Biochemical analysis of this preparation with membrane-impermeant enzymes has verified its unsealed nature: 80% of the sialic acid, which is located on the extracellular domain of the transporter, is released by neuraminidase, and 80% of the polypeptide, which is susceptible to trypsin only at the cytoplasmic surface of erythrocytes, is cleaved by trypsin.[13] Because of its membranous nature, the preparation can easily be concentrated by centrifugation. To induce some aggregation, the preparation is frozen in dry ice–acetone and thawed; it is then centrifuged at 45,000 rpm (180,000 g_{max}) in a Beckman 50 Ti rotor for 75 min.

Cytochalasin B Binding. The most convenient assay for functional activity of the purified transporter is measurement of the binding of cytochalasin B. ^{3}H-Labeled cytochalasin B (5–15 Ci/mmol, New England Nuclear, Boston, MA) and unlabeled cytochalasin B (Aldrich Chemical, Milwaukee, WI) are stored as stock solutions in ethanol at −20°. Solutions of the desired concentration and radioactivity in aqueous buffers are prepared by these by placing aliquots in a glass tube, evaporating the ethanol with a nitrogen stream, adding the buffer, and vortexing vigorously. The solubility of cytochalasin B in water at 23° is 46 μM.[14] Cytochalasin B is adsorbed by soft plastics and so should be transferred in glass pipettes or syringes and stored in glass vials.

Binding is measured by equilibrium dialysis in the simple microdialysis

[11] L. L. Rudel and M. D. Morris, *J. Lipid Res.* **14,** 364 (1973); M. Kates, "Techniques of Lipidology," p. 552, solvent system 3. Am. Elsevier, New York, 1972.
[12] L. L. M. Van Deenen and J. De Gier, *in* "The Red Blood Cell" (D. M. Surgenor, ed.), p. 147. Academic Press, New York, 1974.
[13] J. R. Appleman and G. E. Lienhard, *J. Biol. Chem.* **260,** 4575 (1985).
[14] P. G. W. Plagemann, J. C. Graff, and R. M. Wohlhueter, *J. Biol. Chem.* **252,** 4191 (1977).

apparatus described by Uhlenbeck. This apparatus consists of two rectangular Plexiglas pieces, each of which has 6 shallow chambers of 50 μl drilled into it. Each chamber has a small-bore entrance port that fits the needle of a 50-μl Hamilton syringe. The two pieces are held together by wing nuts, with small pieces of dialysis membrane inserted between the opposing chambers. It is not necessary to use grease to obtain a tight seal. Unfortunately, this apparatus is not commercially available. However, a similar apparatus (Model H40317) with larger chambers (1 ml volume) is available from Bel-Art Products (Pequannock, NJ) and presumably would work as well, provided a larger volume of solution is used. Alternatively, an equilibrium microvolume dialyzer (chambers of 50 or 100 μl) of different design that should also work as well is available from Hoefer Scientific. The dialysis membrane employed is Spectra/Por 1 (Fisher Scientific) that has been boiled once in 20 mM Na$_2$CO$_3$, 1 mM Na$_2$EDTA and then 3 times in water. A 40-μl sample of the transporter preparation is introduced into one chamber, and a 40-μl sample containing ^{3}H-labeled cytochalasin B in the same buffer is placed in the opposing chamber. The apparatus is shaken gently at room temperature on an orbital shaker for 14–16 hr, a period that allows the establishment of equilibrium, and then 25-μl samples are removed from each chamber for the measurement of radioactivity.

For routine assay, the binding of cytochalasin B is measured at a single low concentration (initially introduced at 40–80 nM in the chamber without transporter). This concentration is sufficiently below the dissociation constant for cytochalasin B ($\sim$ 150 nM, see below) so that, regardless of the transporter concentration, no more than 20% of the cytochalasin B sites can be occupied.[15] As a consequence, to a good approximation, the ratio of bound cytochalasin B to free cytochalasin B (B/F) is equal to the ratio of the total concentration of cytochalasin B binding sites $[T]_t$ to the dissociation constant for cytochalasin B (K_d).[15] Since K_d is a constant, B/F is proportional to $[T]_t$ and thus is a measure of the concentration of functional transporter. The value of B/F at this low concenration of cytochalasin B is referred to as the "cytochalasin B binding activity." Typically, we measure the B/F value for the undiluted purified preparation. As an example, one original preparation, at 144 μg protein/ml, gave a value of 12.5. Cytochalasin B does partition preferentially into the lipid, and this contribution to the B/F value can be corrected for by measuring the binding to a preparation that has been denatured at 100° for 2 min. However, at the concentration of lipid in the preparation and at the low concentration of cytochalasin B in the assay, this correction is negligibly small; the value of B/F for the denatured preparation is 0.05–0.1.

Specific activity is expressed as the ratio of the B/F value under the above assay conditions to the concentration of protein, in mg/ml. Typical original and modified preparations show specific activities of 75 and 95, respectively. More complete analysis of binding is achieved by determining the B/F value at a series of cytochalasin B concentrations and constructing Scatchard plots, after a small correction of the data for binding to the lipid, as described in Zoccoli et al.[15] These plots reveal that the preparations contain a single class of high-affinity sites.[3,4] The values for the amount of sites and for K_d of typical original and modified preparations are 12.5 and 17.5 nmol/mg protein and 1.5 and 1.9 $\times 10^{-7}$ M, respectively. The original preparation shows no change in its cytochalasin B binding activity on storage at $-70°$ for 1 year.

It is worth noting that detergents at concentrations even below their critical micelle concentrations (cmc) markedly inhibit the binding of cytochalasin B to the transporter, and that consequently binding to detergent-solubilized transporter cannot be measured. The K_i values for octylglucoside and Triton X-100 are 5 and 0.1 mM, respectively.[3,15]

Purity. Because of the broadness of the transporter band on SDS–PAGE, this method does not provide unambiguous information about purity. Treatment of the transporter with endoglycosidase F in Triton X-100 releases the carbohydrate, and the polypeptide then runs as a sharp band of M_r 46,000 on SDS–PAGE.[9] Unfortunately, the incubation in detergent also leads to some aggregation, but this method indicates that a minimum of 65% of the original preparation is transporter. A second estimate of the purity of the original preparation comes from the use of monoclonal antibodies to the transporter. Seventy-five percent of the protein in this preparation is immunoadsorbed by each of three different monoclonal antibodies.[1]

A third method of evaluating purity is the stoichiometry of cytochalasin B binding. The values for the typical preparations described above, calculated from the M_r 54,116 and the protein concentration according amino acid analysis, are 0.66 and 0.92 cytochalasin B site per molecule for the original and modified preparation, respectively. Since the value for pure, fully active transporter is undoubtedly 1.0 and since by SDS–PAGE most impurities have about the same M_r as the transporter, these values are equal to the percentage of functional transporter by weight. It should be emphasized that they are minimum values for the percentage of transporter polypeptide. As noted above, the transporter undergoes slow denaturation while solubilized in octylglucoside to a form that does not bind cytochalasin B on reconstitution, and therefore a portion of it in the purified preparations is probably denatured. On the basis of the values for the rate constants for denaturation of the solubilized transporter under

[15] M. A. Zoccoli, S. A. Baldwin, and G. E. Lienhard, *J. Biol. Chem.* **253**, 6923 (1978).

various conditions (for example, at 8° the half-life in 525 μg/ml phospholipid, 39 mM octylglucoside, 1 mM EDTA, 2 mM dithiothreitol, 100 mM NaCl, 50 mM Tris-Cl, pH 7.4, is 7.5 hr), we estimate about 10% of the original preparation is denatured.[3]

Because of the simplicity of the purification method, it is to be expected that both the original and modified preparations contain a few percentages of several other polypeptides. One of these is almost certainly the nucleoside transporter, a protein also of about M_r 50,000. A procedure for the partial purification of the nucleoside transporter that is very similar to the one for purification of the glucose transporter has been described. Through the use of a binding assay for the nucleoside transporter it was determined that about 3% of the protein in the preparation is nucleoside transporter; most of the remainder is presumably glucose transporter.[16]

Transport. The purified, reconstituted preparations described above are not suitable for measurements of monosaccharide transport because of the unsealed nature of the membrane fragments. The incorporation of the transporter into sealed vesicles requires a higher ratio of lipid to protein, and a variety of procedures involving the introduction of additional membrane lipids have been adopted to achieve this. These various methods yield purified transporter reconstituted into sealed vesicles of various sizes and lipid compositions. Several different methods for the assay of transport have also been employed. The reader interested in pursuing this subject should consult these reports.[17-20]

It is worth noting that most methods of reconstitution probably give vesicles with transporter oriented both right-side out and inside out, with respect to the orientation in the erythrocyte. The fraction of transporter that is oriented inside out can be determined by measuring the fraction of polypeptide that is cleaved by trypsin.[4] Also, where the effects of lipid composition on transporter activity are examined, there is no completely rigorous way to determine whether the effect is on the efficiency of the reconstitution of the transporter into the bilayer in its functional conformation or on the functioning itself of the properly reconstituted transporter. The following is a description of convenient methods for reconstitution into sealed vesicles and assay of transport that we have used.[1]

Stock solutions of buffer (100 mM NaCl, 1 mM EDTA, 50 mM Tris-Cl, pH 7.4 at 2°), the original preparation of purified transporter, erythro-

[16] C.-M. Tse, J. A. Belt, S. M. Jarvis, A. R. P. Patterson, J.-S. Wu, and J. D. Young, *J. Biol. Chem.* **260**, 3506 (1985).
[17] J. M. Baldwin, J. C. Gorga, and G. E. Lienhard, *J. Biol. Chem.* **256**, 3685 (1981).
[18] T. J. Wheeler and P. C. Hinkle, *J. Biol. Chem.* **256**, 8907 (1981).
[19] A. Carruthers and D. L. Melchior, *Biochemistry* **23**, 6901 (1984).
[20] T. J. Connolly, A. Caruthers, and D. L. Melchior, *Biochemistry* **24**, 2865 (1985).

cyte membrane lipids (30 mg phospholipid/ml) dissolved in 170 mM octylglucoside, and 680 mM octylglucoside are mixed in that order to give 6 ml of 60 μg/ml transporter, 3 mg/ml erythrocyte lipids, 62 mM octylglucoside in the buffer. After 10 min, the mixture is centrifuged in a Beckman 50 Ti rotor at 45,000 rpm (180,000 g_{max}) for 1 hr. The supernatant is removed, made 1 mM in dithiothreitol, and dialyzed in Spectra/Por 2 tubing against 2 liters of the above buffer for 15 hr, with a change of buffer after 7 hr. All operations are performed at 0°–3°. The reconstituted vesicles are stored at 5° until use in the transport assay. This method of forming sealed vesicles is based on that described by Mimms et al., who report that the removal of octylglucoside from phosphatidylcholine by dialysis gives vesicles of 230 nm diameter.[21]

The erythrocyte lipids employed for this reconstitution are obtained from human erythrocyte membranes.[22] Membranes are concentrated into a small volume by freezing and thawing 3 times and centrifugation at 20,000 g_{max} for 20 min. These (8 ml at 17 mg protein/ml) are added to 81 ml of methanol, and the mixture is stirred under nitrogen for 30 min. Then chloroform (98 ml) is added; and after 30 min more of stirring, the precipitated protein is removed by centrifugation in glass tubes at 4°. Enough chloroform is added to make the supernatant 2:1 by volume in chloroform–methanol, and this mixture is then extracted with 0.22 vol of 90 mM KCl that has been flushed with nitrogen. The organic layer is centrifuged, any remaining upper layer is aspirated, the solvent is removed by rotary evaporation, and the lipids are stored at −70° under nitrogen in chloroform–methanol (2:1). The yield is about 65 mg of phospholipid.

For measurement of transport,[17] an undiluted aliquot of the reconstituted vesicles is placed in a 10 × 75 mm glass tube at 6°, and uptake is initiated by the addition of 1/20 volume of a stock mixture of D-[^{14}C]- and L-[^{3}H]glucose (New England Nuclear) in 1 mM EDTA, 100 mM NaCl, 50 mM Tris-Cl, pH 7.4, such that the final concentration of each sugar is 0.1 mM, with 150 μCi/ml ^{14}C and 400 μCi/ml ^{3}H. At various times thereafter (0.5, 2 min), 115-μl aliquots are removed and added to 0.5 μl of 4 mM ethanolic cytochalasin B in a 10 × 75 mm glass tube. Within 45 sec thereafter, a 100-μl aliquot of the inhibited sample is applied to a Sephadex column for separation of the vesicles from the medium (see below), and within the subsequent 45 sec centrifugation of the column is begun. To determine the zero time value for uptake, as well as the effect of cytochalasin B on the uptake rate, vesicles are added to a tube containing sufficient

[21] L. T. Mimms, G. Zampighi, Y. Nozaki, C. Tanford, and J. Reynolds, *Biochemistry* **20**, 833 (1981).
[22] M. Kates, "Techniques of Lipidology," Chapter 3. Am. Elsevier, New York, 1972.

solid cytochalasin B to give 20 μM. The tube is vortexed at room temperature and then chilled to 6°. Uptake is initiated and followed as described above, with the exception that the 115-μl samples are added to 0.5 μl ethanol rather than cytochalasin B in ethanol.

The vesicles are separated from the medium on small columns of Sephadex G50 (fine) poured in disposable syringes.[23] The Sephadex is first equilibrated with 0.1 mM D- and L-glucose, 20 μM cytochalasin B, buffer. It is poured to the 1-ml mark in the barrels of 1-ml disposable plastic syringes (5602, Becton Dickinson, Franklin Lakes, NJ) fitted with disks stamped with a machine punch (11/64 inch) from a sheet of porous polyethylene (1.6 mm thick, 70 μm pore size, Bel-Art Products). Excluded medium is removed from the columns by centrifugation at 140 g for 3 min in the horizontal rotor of a clinical centrifuge (setting 3 of a Damon/IEC Model 428 with 221 rotor and 10-ml shields fitted with holed corks to support the syringes). For separation of the vesicles from the medium a 100-μl aliquot is applied to a centrifuged column and allowed to settle into the Sephadex. This application is followed at once by 10 μl of the medium on the Sephadex, and the column is then centrifuged immediately as before, with a 7-ml scintillation vial at the bottom of the bucket to collect the eluate. The ^{3}H and ^{14}C content of the eluate are determined by scintillation spectrometry. All operations are performed in a 4° room.

For time courses, it is necessary to have two centrifuges, since one will be in operation when it is time to begin the centrifugation of another column. Operation is facilitated by the use of an electrical timing device that shuts the centrifuge off after 3 min. For good reproducibility, the centrifuges should always be operated with the same number of columns. This method efficiently separates medium from the vesicles; in a control experiment in which a solution of D-[^{14}C]- and L-[^{3}H]glucose without vesicles was applied to the columns, less than 0.001% of each was found in the eluate.

In a typical experiment the percentage of D-glucose taken up by the reconstituted vesicles rose from 0.01% at zero time to 0.16% after 2 min, whereas that for L-glucose increased from 0.01 to only 0.03%. The uptakes of both in the presence of 20 μM cytochalasin B were about the same as that for L-glucose in the absence of the inhibitor.[1] This experiment illustrates the two conditions that should be established to be certain that the transporter is catalyzing the transport of D-glucose.[17] The uptake of L-glucose, which is a much poorer substrate for the transporter, should be much slower, and cytochalasin B, which potently inhibits transport in erythro-

[23] D. W. Fry, J. C. White, and I. D. Goldman, *Anal. Biochem.* **90**, 809 (1978).

cytes, should inhibit the uptake of D-glucose. In preliminary experiments, in which the reconstitution into vesicles was performed with either dioleoylphosphatidylcholine or with egg yolk phospholipids according to the procedure described above, the rate of D-glucose uptake was much lower and was about the same as that for L-glucose (W. J. Allard and G. E. Lienhard, unpublished results). These reconstituted preparations did exhibit the expected cytochalasin B binding activity, and we have not examined other possible explanations for this result (unsealed vesicles, lower transport activity). In fact, dioleoylphosphatidylcholine does support transport in other methods for reconstitution of the glucose transporter.[17,19] The lack of success with these lipids in the octylglucoside dialysis method led to testing of erythrocyte lipids, which, as described above, allow the assay of the transport function.

Acknowledgments

Research in our laboratories has been supported by grants from the Medical Research Council (to S.A.B.) and the National Institutes of Health (GM22996, to G.E.L.).

[6] Isolation of Bilitranslocase, the Anion Transporter from Liver Plasma Membrane for Bilirubin and Other Organic Anions

By Gian Luigi Sottocasa, Gian Carlo Lunazzi, and Claudio Tiribelli

Bilitranslocase, a carrier protein involved in sulfobromophthalein and other organic anion transport, has been isolated from rat liver plasma membranes.[1,2] Transport activity is directly demonstrated for sulfobromophthalein (BSP), whose movements have been reconstituted *in vitro* using liposomes and the purified protein.[3] The involvement of bilitranslocase in bilirubin uptake has been inferred indirectly by inhibiting *in vivo*

[1] C. Tiribelli, G. C. Lunazzi, M. Luciani, E. Panfili, B. Gazzin, G. Liut, G. Sandri, and G. L. Sottocasa, *Biochim. Biophys. Acta* **532**, 105 (1978).

[2] G. G. Lunazzi, C. Tiribelli, B. Gazzin, and G. L. Sottocasa, *Biochim. Biophys. Acta* **685**, 117 (1982).

[3] G. L. Sottocasa, G. Baldini, G. Sandri, G. C. Lunazzi, and C. Tiribelli, *Biochim. Biophys. Acta* **685**, 123 (1982).

the physiological clearance of the endogenous pigment in the rat.[4] Animals were injected via the portal system with Fab fragments of antibilitranslocase antibodies that had been purified by affinity chromatography. The treatment resulted in a marked unconjugated hyperbilirubinemia. A more direct demonstration of the origin and involvement of bilitranslocase in the transport function has been obtained from immunochemical studies. A monoclonal antibody selected on the basis of its ability to block BSP movements in plasma membrane vesicles binds specifically to bilitranslocase (37kDa band in SDS–PAGE). On the other hand, as expected, purified bilitranslocase binds with high affinity unconjugated bilirubin, as well as a number of other organic anions including dibromosulfophthalein, rifamycin SV, indocyanine green, and nicotinate.

Binding Assay

The only way to trace bilitranslocase during a purification procedure is to follow the BSP binding capacity at the different purification steps. This can be done either by (1) gel filtration or (2) membrane ultrafiltration. In both cases the principle is to separate physically the protein–dye complex and to measure the amount of the dye bound in equilibrium with the free form.

Estimate of BSP Binding to Protein by Gel Filtration on BioGel P-2

Reagents

Solution A: 10 μM sulfobromophthalein (Merck, K$^+$ salt) in 10 mM Potassium phosphate buffer (pH 7.8), containing 10 mM KCl
Solution B: 0.1 M NaOH
BioGel P-2 resin (Bio-Rad, Richmond, CA)
Equipment

Glass columns (1.2 × 30 cm)
Optical unit equipped with a 280-nm filter
Recorder and fraction collector
Visible spectrophotometer

Procedure. A number of identical columns are packed with 30 ml preswollen resin and equilibrated with Solution A overnight. The preequilibration step is important in that the resin binds a certain amount of dye and should be presaturated. Check of the situation can be made by mea-

---

[4] G. L. Sottocasa, C. Tiribelli, M. Luciani, G. C. Lunazzi, and B. Gazzin, *in* "Function and Molecular Aspects of Biomembrane Transport" (E. Quagliariello, F. Palmieri, S. Papa, and M. Klingenberg, eds), p. 451. Elsevier/North-Holland Biomedical Press, Amsterdam, 1979.

suring the absorbance at 580 nm of the solution entering the system and comparing this value with that of the effluent. The measurement should be done at alkaline pH by diluting the sample 1:1 with Solution B. The extinction coefficient of 64 μmol^{-1} cm^2 is used. Samples (0.05–1 ml in volume) are carefully applied onto the column and elution started with Solution A. The effluent is monitored continuously at 280 nm for protein. The peak emerging is collected as a single fraction, the volume is accurately measured, and, after stirring, a sample of 1.5 ml is pipetted into a cuvette containing 1.5 ml of Solution B. A blank sample is also processed in parallel, using the eluent. The difference in absorbance is measured at 580 nm.

Calculation

$$\mu\text{moles bound/mg protein} = \frac{\Delta A_{580} \times 2 \times \text{ml effluent in the peak}}{64 \times \text{mg protein loaded on column}}$$

Monitoring of the effluent at 280 nm is useful to dissect the protein fraction and to avoid the trough which usually follows the binding protein. Fortunately the trough is somewhat delayed due to the dye-absorbing ability of the resin. This, in turn, requires a rather long resaturation of the resin after each run. For this reason the use of more than one column is advisable, especially when dealing with analysis of a large number of samples. An advantage of using gel filtration derives from the properties of the resin which removes potentially inhibitory substances of low molecular weight such as deoxycholate, used in the past for the preparation of the protein.

Estimate of Binding of BSP to Protein by Ultrafiltration on Amicon Centrifree Micropartition System

Reagents

   Solution A: 20 μM BSP in 10 mM potassium phosphate, pH 7.8, containing 10 mM KCl
   Solution B: 0.1 M NaOH
   Solution C: 10 mM potassium phosphate buffer, pH 7.8, containing 10 mM KCl

Equipment

   Amicon Centrifree micropartition System (Amicon, Danvers, MA)
   Fixed-angle (30°–45°) centrifuge with rotor adapters accepting 17 × 100 mm tubes
   Visible spectrophotometer

Procedure. The binding capacity of the filter membrane is marginal. It is advisable, however, to prewash the system with 0.5 ml Solution A diluted 1 : 1 with Solution C and spin the filtering system at 2000 g for 5 min. The rinsing solution is carefully removed from the system, which is then ready to accept the sample. Samples are prepared in test tubes contain the following: 0.5 ml Solution A, x ml protein solution, and $(0.5 - x)$ ml Solution C. After stirring, the samples are transferred into centrifree micropartition systems and spun down for 10 min at 2000 g. Aliquots of the ultrafiltrate after alkalinization (0.3 ml ultrafiltrate + 0.6 ml Solution B) are pipetted into 1-ml microcuvettes and read at 580 nm. Appropriate blanks are run simultaneously.

Calculation

$$\mu \text{moles bound/mg protein} = \frac{\Delta A_{580} \times 3}{64 \times \text{mg protein/sample}}$$

The choice of 0.3-ml aliquots allows the run of samples in duplicate. In the system 10 nmol BSP are available. Good reproducibility and reliability is obtained when 1.5 – 4 nmol are bound per sample.

Comments

The two binding assays share the disadvantage of operating at 10 μM ligand. This value may imply estimation of a certain amount of nonspecific (low-affinity) binding. The disadvantage becomes, however, less and less important as the purification proceeds. On the other hand, the two techniques are relatively simple and may be routinely applied to a large number of fractions.

Purification of Bilitranslocase

Materials

Three rat livers from Wistar albino rats (the animals may be of either sex, 200 – 250 g weight)

Reagents

Rinsing solution: 0.25 M sucrose containing 1 mM MgCl$_2$ and 1 mM NaHCO$_3$

Homogenization medium: 1 mM NaHCO$_3$ – HCl buffer, pH 7.5, containing 0.5 mM CaCl$_2$

Solution A: 20 mM Tris-HCl buffer, pH 8.0, containing 100 mM KCl, 1 mM EDTA, and 0.1% 2-mercaptoethanol

Solution B: 20 mM Tris-HCl buffer, pH 8.0, containing 50 mM KCl and 0.1% 2-mercaptoethanol

Solution C: 20 mM Tris-HCl buffer, pH 8.0, containing 1.3 M KCl and 0.1% 2-mercaptoethanol

Solution D: 20 mM Tris-HCl buffer, pH 8.0, containing 0.1% 2-mer-
captoethanol

Equipment

50 ml Potter homogenizer (A. Thomas, Class C)

Sorvall Preparative Centrifuge RC 5 B equipped with GS3 and SS-34
rotors

450 ml polycarbonate and 450 ml clear nylon bottles (Du Pont,
Newton, MA)

FPLC apparatus, Pharmacia (Stockholm, Sweden)

Fast System, Pharmacia

Step 1: Homogenization and Subfractionation

The technique employed has been described by Ray.[5] Three livers from
fed animals are immersed immediately after sacrifice in approximately
50 ml ice-cold rinsing solution and rapidly cut into small pieces with
stainless steel scissors. The operation is continued for 2–3 min then the
solution is gently decanted and replaced by fresh solution. This treatment
is repeated at least 3 times. At the end the mince consists of tissue fragments, with an average section of 3 × 3 mm, that have lost much of their
blood content. The material is suspended in 10 vol homogenization medium and gently homogenized by means of a motor-driven Teflon–glass
Potter homogenizer. Usually 20 up and down strokes are performed at
about 100 rpm followed by 5 strokes at higher speed (~250 rpm). Small
adjustments of the homogenizing conditions may be necessary depending
on the homogenizer clearance and tissue mechanical properties. A criterion for sufficient homogenization may be the absence of gross tissue
fragments visible between the pestle and the glass tube. Excessive homogenization, however, is to be avoided.

The homogenate is diluted 10 times with homogenization medium in a
large glass beaker, under gentle stirring, on ice. The final volume at this
stage is around 2.4 liters. The stirring is continued for 5 min, and then the
suspension is filtered through 4 layers of cheesecloth. The filtrate is spun at
920 g for 40 min in a Sorvall GS3 fixed-angle rotor in polycarbonate
bottles. The supernatant is gently decanted: the fluffy layers are saved
together with the sediment. The pellets are taken up in a minimal volume
of homogenization medium by the help of a hand-moved Potter homogenizer. The suspension, diluted to approximately 1.2 liters, is then spun
down at 720 g for 15 min. The supernatant is discarded. The fluffy layers
and the tightly packed pellets are washed twice more in the same way with
the same medium. At each step the volume is halved (600 and 300 ml,

[5] T. K. Ray, *Biochim. Biophys. Acta* **196**, 1 (1970).

respectively). The final sediment is saved, but the fluffy layers are now discarded. This fraction corresponds to the third precipitate noted by Ray.[5] The procedure described by Ray can be carried further to obtain the purified plasma membrane preparation. This is not necessary, however, in view of the quality of the protein obtained from the crude material which is identical in all respects to that separated from highly purified fractions.

Step 2: Preparation of the Acetone Powder

The pellet obtained as described above is carefully removed from the centrifuge bottle, by means of a bent glass rod, and suspended in a final volume of 50 ml of the homogenizing medium, after which 50 μl 2-mercaptoethanol is added. Cold acetone (450 ml, $-20°$) is added to the suspension, which is carefully stirred for 3–4 min and then spun down at 4700 g for 5 min at $-10°$ (GS3 rotor, using nylon bottles). The sticky sediment is dispersed in the same volume of anhydrous acetone ($-20°$) by means of a hand-moved Potter homogenizer. After 5 min of stirring the suspension is again spun down at the same speed for the same time. The operation is repeated once or twice more. The powder is ready when it can be easily suspended in acetone by simple shaking of the bottle. Under these conditions the water content is minimal. The sediment in the bottle is placed as soon as possible in a glass desiccator under vacuum in the presence of P_2O_5 and parafin fragments at $0°$–$4°$ overnight. In view of detrimental effects of acetone (especially at room temperature) the last traces of the solvent should be removed as soon as possible. The acetone powder may be stored at $-80°$ for weeks provided it is carefully maintained anhydrous.

Step 3: Solubilization of Protein

Two acetone powder preparations (corresponding to six rat livers, 50–70 g tissue, wet weight) are carefully homogenized (Potter homogenizer) in 20 ml of Solution A. The suspension is spun down in an Sorvall SS-34 rotor for 10 min at 4000 g. The supernatant is decanted and saved. The sediment is washed in the same way and with the same solution 3 times. The combined supernatants may be stored in liquid nitrogen for weeks without appreciable loss of binding activity. The BSP binding capacity of this extract is of the order of 10–16 nmol dye bound/mg protein.

Step 4: Ion-Exchange Chromatography

The extract obtained as described in Step 3, containing up to 400 mg protein, is diluted 1:3 with Solution D. The sample may be loaded onto a Mono S Pharmacia FPLC column, preequilibrated with Solution B. The loading is performed at a flow rate of 1 ml/min, using a 50 ml maxi loop. The bulk of protein is not retained on the column at this pH and is

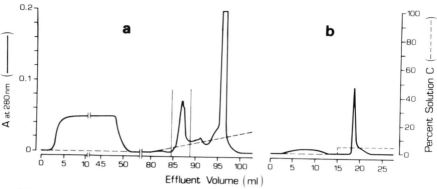

FIG. 1. Ion-exchange chromatography of the acetone powder extract on Mono S column. Experimental conditons were as described in the text. (a) First chromatographic run of the sample, 120 mg extract. (b) Second chromatographic step carried out on the fractions corresponding to the peak delimited by the two vertical lines in (a).

recovered with the effluent. Washing of the column with Solution B is continued until a virtually flat baseline is obtained. Elution of the column proceeds with a linear KCl gradient (0.55% Solution C/min) at 1 ml/min flow rate. As shown in Fig. 1a, two main peaks are eluted. The first one contains bilitranslocase and the second cytochrome c, deriving from mitochondrial contamination of the starting material. Such contamination is minimized starting from the plasma membrane-enriched preparation of Ray.[5] It is worth mentioning in any case that no correlation exists between the yield of bilitranslocase and that of cytochrome c.

The fractions corresponding to the first peak are pooled, diluted 1 : 1 with Solution B, and passed again on the same column, after reconditioning. The elution is now performed stepwise, by raising the concentration of Solution C in the eluent to 4%. Figure 1b shows the elution pattern from the column. The second chromatography results in concentrating the protein sample, and, in addition, it increases the level of purity of the preparation by removing the last traces of acidic protein which may be bound to bilitranslocase electrostatically. The product of this procedure is a homogeneous protein on sodium dodecyl sulfate (SDS)–polyacrylamide slab gradient gel electrophoresis as shown in Fig. 2. By comparing the mobility properties of the protein with that of appropriate standards, a mass of 37 kDa may be derived.

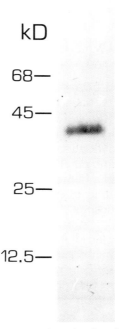

FIG. 2. Polyacrylamide slab gel electrophoresis of purified bilitranslocase. Experimental conditions: PhastSystem apparatus (Pharmacia); PhastGel gradient 10–15% in SDS and 1% 2-mercaptoethanol; staining with Coomassie Brilliant Blue; 80 ng protein in the sample.

Comments

The procedure yields $50-60\ \mu g$ pure protein per liver. Bilitranslocase prepared as described is a basic protein. Its isoelectric point is higher than 9.0 as shown both by its ability to bind to an anionic resin at pH 8.0 and by electrofocusing experiments in which a single protein band is focalized at the highest pH value on ampholine. When measured at $10\ \mu M$ BSP, it binds up to 60 nmol dye/mg protein. The binding properties and the ability to reconstitute transport in liposomes may decay rapidly if the protein is prepared in the absence of 2-mercaptoethanol. In the presence of the thiol the reconstitutive function is maintained at least for 1 week at $0°$. This finding would indicate that at least one SH group is necessary for function and most probably also for maintenance of native structure.

[7] Measurement of Intactness of Rat Liver Endoplasmic Reticulum

By WILLIAM J. ARION

The procedures described below[1] were developed as research tools to aid in defining the role of the membrane in the function of hepatic and renal glucose-6-phosphatase (EC 3.1.3.9). Glucose-6-phosphatase activity is uniformly distributed throughout the granular (rough) and agranular (smooth) elements of the endoplasmic reticulum (ER) and within the nuclear envelope in the hepatic parenchyma.[2-5] The hydrolysis of glucose 6-phosphate by intact[6] hepatic and renal microsomes was shown to involve the coupled functions of three integral components of the ER membrane[7-10]: (1) a glucose 6-phosphate (6-P)-specific translocase (called T_1) that facilitates penetration of the substrate into the ER lumen; (2) a broadly specific phosphohydrolase (referred to as the "enzyme") with its active site facing the lumenal compartment; and (3) a phosphate translocase (called T_2) that facilitates efflux of inorganic phosphate (P_i) and influx of inorganic pyrophosphate and carbamoyl phosphate.[8] D-Glucose, the second product of the hydrolytic reaction, rapidly equilibrates across the membrane by simple diffusion.[8,11] Contrary to earlier reports,[12,13] the general kinetic

---

[1] The original work described in this chapter has been supported by the National Institutes of Health, U.S. Public Health Service Grant AM-19625.

[2] L. W. Tice and R. J. Barnett, *J. Histochem. Cytochem.* **10**, 754 (1962).

[3] A. Leskes, P. Siekevitz, and G. E. Palade, *J. Cell Biol.* **49**, 264 (1971).

[4] H. Glaumann, *Histochemistry* **44**, 169 (1975).

[5] W. J. Arion, L. O. Schulz, and H. E. Walls, *Arch. Biochem. Biophys.* **252**, 467 (1987).

[6] Definitions used: The term *untreated* is used to designate liver homogenates and microsomal vesicles which are prepared and assayed without further treatment. Untreated microsomal preparations are structurally heterogeneous in that glucose-6-phosphatase activity is localized both in intact vesicles, in which the limiting membrane functions as a selective permeability barrier, and in *disrupted* structures, in which selective permeability is artifactually lacking so the phosphohydrolase has free access to substrates and other ions. The glucose-6-phosphatase *system* refers to the phosphohydrolase activity of intact microsomes which is the functional expression of the coupling of T_1, enzyme, and T_2. Activity of the *enzyme* denotes the phosphohydrolase activity in disrupted microsomes where the kinetics of the phosphohydrolase activities are not restricted by substrate transport.

[7] W. J. Arion, B. K. Wallin, A. J. Lange, and L. M. Ballas, *Mol. Cell. Biochem.* **6**, 75 (1975).

[8] W. J. Arion, A. J. Lange, H. E. Walls, and L. M. Ballas, *J. Biol. Chem.* **255**, 10396 (1980).

[9] A. J. Lange, W. J. Arion, and A. L. Beaudet, *J. Biol. Chem.* **255**, 8381 (1980).

[10] R. C. Nordlie, K. A. Sukalski, J. M. Munoz, and J. J. Baldwin, *J. Biol. Chem.* **258**, 9739 (1983).

[11] W. J. Arion, A. J. Lange, and H. E. Walls, *J. Biol. Chem.* **255**, 10387 (1980).

characteristics and substrate specificity of the glucose-6-phosphatase system in isolated microsomes are identical to those determined for the ER *in situ*, i.e., in intact regions of the envelope of isolated nuclei[14] and in the ER of hepatocytes rendered permeable to glucose 6-phosphate by exposure to α-toxin, a pore-forming cytolytic protein secreted by *Staphylococcus aureus*.[15] Finally, the general characteristics of the glucose-6-phosphatase system appear to be quite similar in hepatic and renal microsomes from a variety of species.[7]

Measurement of the Latency of Mannose-6-phosphatase to Assess ER Membrane Intactness

Rationale. The uniform distribution of the glucose-6-phosphatase system throughout the parenchymal ER and the contrast between the broadly specific lumenal enzyme and the glucose-6-P-specific transporter presented the basis for a method to quantify the integrity of the permeability barrier defined by the ER membranes. At 1 to 2 mM, D-mannose-6-P, 2-deoxy-D-glucose-6-P, and 2-deoxy-2-amino-D-glucose-6-P (glucosamine-6-P) are neither bound nor transported by the glucose-6-P transporter, but these hexose phosphates are equivalent to glucose-6-P in their reactivity with the enzyme.[16,17] Several criteria were used to establish that at concentrations below 2 mM the hydrolysis of mannose-6-P by untreated microsomal or nuclear membranes is catalyzed by a subpopulation of enzyme molecules localized in membranes lacking an intact permeability barrier.[9,14,16,17] Thus the ratio of the "low K_m" mannose-6-phosphatase activities determined in untreated[6] and fully disrupted microsomes provides a quantitative measure of the fraction of enzyme localized in membranes lacking selective permeability. Stated in the converse way, the latency[18] of the low K_m mannose-6-phosphatase activity in untreated membranes provides a quantitative measure of the percent intactness of the permeability barrier.[5,17] This approach has been successfully used to evaluate membrane intactness

[12] J. M. Gunderson and R. C. Nordlie, *J. Biol. Chem.* **250**, 3552 (1975).

[13] R. A. Jorgenson and R. C. Nordlie, *J. Biol. Chem.* **255**, 5907 (1980).

[14] W. J. Arion, L. O. Schulz, A. J. Lange, J. N. Telford, and H. E. Walls, *J. Biol. Chem.* **258**, 12661 (1983).

[15] B. F. McEwen and W. J. Arion, *J. Cell Biol.* **100**, 1922 (1985).

[16] W. J. Arion, B. K. Wallin, P. W. Carlson, and A. J. Lange, *J. Biol. Chem.* **247**, 2558 (1972).

[17] W. J. Arion, L. M. Ballas, A. J. Lange, and B. K. Wallin, *J. Biol. Chem.* **251**, 4901 (1976).

[18] Latency is the percentage of activity in fully disrupted microsomes that is not expressed in untreated or intact preparations. Latency is calculated as $100 \times (1 -$ activity in intact or untreated membranes/activity in fully disrupted membranes). Latency measures the extent to which substrate transport restricts the rate of the coupled hydrolytic process.[7-9,11]

in whole liver homogenates,[14] isolated microsomes,[5,17] isolated nuclei,[14] and permeabilized hepatocytes.[15]

Pitfalls and Critical Variables. The principal source of error in determining the latency of mannose-6-phosphatase is the quantitative assay of the phosphohydrolase activity in fully disrupted membranes. The investigator must be aware of three potential pitfalls.

First, the mannose-6-P must be pure. For obvious reasons it cannot be contaminated with glucose-6-P. Glucose-6-P contamination is readily checked using glucose-6-phosphate dehydrogenase.[16,19] Levels of contamination of the mannose-6-P with P_i above 2 mol % will seriously compromise the sensitivity of the assay (see below). Impure substrate should either be purified[20] or returned to the supplier in exchange for a cleaner lot. Tests should be made for the presence of "acid-labile phosphates," [21] like ADP or ATP, since such compounds may interfere with the determination of P_i.

The second potential pitfall is the requirement to assay untreated and fully disrupted preparations under exactly the same conditions with the exception of the amount of each preparation added to the assay medium.[11] For example, if the medium used to suspend the membranes contains an inhibitor of mannose-6-phosphatase activity, then an identical concentration of the inhibitor must be present during the assay of both untreated and fully disrupted membranes (see Ref. 8). There are some restrictions in the choice of buffer. An analysis by H. E. Walls[22,23] in my laboratory showed that the enzyme has essentially identical characteristics when assayed in media containing 25 to 50 mM of any of the following buffers: sodium cacodylate or Tris-cacodylate, HEPES,[24] histidine, imidazole, MES, and MOPS. However, because citrate, succinate, and maleate are relatively good competitive inhibitors of mannose-6-P and glucose-6-P hydrolysis, especially at or below pH 6, buffers containing these anions should be avoided.

Mannose-6-phosphatase can be assayed over the range from pH 5 to 8,[16] but pH 6.5 to 7.0 is preferred. In our experience hydrolysis of mannose-6-P at neutral pH by other cellular phosphatases (i.e., acid and alka-

[19] R. C. Nordlie and W. J. Arion, this series, Vol. 9, p. 619.

[20] M. W. Slein, this series, Vol. 3, p. 154.

[21] L. F. Leloir and C. E. Cardini, this series, Vol. 3, p. 840.

[22] H. E. Walls, Master's Thesis, Cornell University, Ithaca, New York (1978).

[23] H. E. Walls and W. J. Arion, in preparation.

[24] Abbreviations used: CHAPS, 3-[(3-cholamidopropyl)demethylammonio]-1-propane sulfonate; CHAPSO, 3-[(3-cholamidopropyl)dimethylammonio]-2-hydroxy-1-propane sulfonate; ER, endoplasmic reticulum, HEPES, N-2-hydroxyethylpiperazine-N'-2-ethanesulfonic acid; MES, 2-(N-morpholino)ethanesulfonic acid; MOPS, 3-(N-morpholino)propane-sulfonic acid; SH buffer, 0.25 M sucrose, 5 mM HEPES, pH 7.4.

line phosphatases) has not been a problem with the rat liver, even when tissue homogenates were assayed[11,14] (but see below). In media below pH 6 acid phosphatase may be a problem, and its contribution should be assessed by heating the membrane preparation at pH 5 for 10 min at 37°. This treatment completely inactivates glucose-6-phosphatase/mannose-6-phosphatase without affecting acid phosphatase activity assayed at pH 5 with β-glycerol phosphate.[22,25,26] Thus, the contribution of acid phosphatase activity can be assessed by assaying mannose-6-phosphatase before and after heat treatment at pH 5. If quantifying membrane integrity is done to relate intactness to a function of the ER that is assayed between pH 5 and 6, mannose-6-phosphatase must be assayed at the low pH as well, because some disruption of the membranes may occur when microsomes are exposed to media below pH 6.[8,16,26]

The third and potentially most troublesome pitfall arises from the use of detergents to destroy the ER permeability barrier, because detergents simultaneously destabilize the enzyme.[26-32] Thus, enzyme inactivation may occur when detergent-treated preparations are incubated at assay temperature (i.e., 25°–37°). The thermal lability of the enzyme in detergent-treated preparations is greatest when assays are carried out at subsaturating concentrations of phosphate substrates, precisely the conditions required to assay the low K_m mannose-6-phosphatase activity. Therefore, to accomplish membrane permeabilization without enzyme inactivation requires some care in the choice of the detergent and the conditions employed to disrupt membranes with it.

In our experience a variety of nonionic, anionic, and zwitterionic detergents can be used with identical results. These include 0.2% sodium deoxycholate, 0.5% sodium cholate, 0.45% sodium taurocholate, 0.1% Triton X-100, 0.05% Lubrol 12A9, and 0.5% CHAPS.[24] Cetyltrimethylammonium chloride or bromide (cetrimide) and other cationic detergents are unacceptable disrupting agent because they are also potent uncompetitive inhibitors of the phosphohydrolase.[32] Only high-quality (i.e., "enzyme grade") detergents should be used. Satisfactory grades of most are commercially available.

The percentage noted above for each detergent is the concentration

[25] H. G. Hers and C. de Duve, *Bull. Soc. Chim. Biol.* **32**, 20 (1950).
[26] W. J. Arion, A. J. Lange, and L. M. Ballas, *J. Biol. Chem.* **251**, 6784 (1976).
[27] H. Beaufay, H. G. Hers, J. Berthet, and C. de Duve, *Bull. Soc. Chim. Biol.* **36**, 1539 (1954).
[28] M. R. Stetten and F. F. Burnett, *Biochim. Biophys. Acta* **128**, 344 (1966).
[29] M. R. Stetten and F. F. Burnett, *Biochim. Biophys. Acta* **139**, 138 (1967).
[30] R. C. Garland, C. F. Cori, and H. W. Chang, *Proc. Natl. Acad. Sci. U.S.A.* **71**, 3805 (1974).
[31] A. Burchell and B. Burchell, *FEBS Lett.* **118**, 180 (1980).
[32] W. J. Arion, P. W. Carlson, B. K. Wallin, and A. J. Lange, *J. Biol. Chem.* **247**, 2551 (1972).

(w/v) that gives complete and irreversible destruction of the permeability barrier without destabilizing the enzyme, *provided* that treatment of membranes with the detergent is carried out under strictly defined conditions.[9,32-34] These conditions include exposure of microsomes suspended at 1 to 2 mg of protein/ml in a low-salt medium, like SH buffer (0.25 M sucrose in 5 mM HEPES, pH 7.4), at 0° to the indicated final concentration of detergent. For convenience, stock detergent solutions are made at 10 times the desired final concentration, and 1 vol detergent is added to 9 vol membrane suspension. The incubation at 0° is usually for 20 min; however, in the case of sodium cholate, complete disruption takes an hour.[34] The "chaotropic" actions of various salt solutions in combination with a detergent[35] dictate that detergent treatment be carried out in a low ionic strength medium. Since the ratio of detergent to membrane phospholipid (or membrane protein) is a critical factor in the disruption and dissolution of biological membranes,[36] the strategy is to destroy the permeability barrier without causing extensive solubilization and dissociation of integral membrane components which may increase instability of the enzyme.[22,37]

The use of 0.1 M NH$_4$OH, developed by Stetten and Burnett,[28,29] effectively disrupts the membrane and generates a stable form of the enzyme. However, this method is not convenient to use in the assessment of mannose-6-phosphatase latency because the NH$_4$OH-treated preparations must be neutralized before assay,[32-34] with the risk of localized acidification and acid denaturation of the enzyme.[26] Thus, quantitative assays of the enzyme in ammonia-disrupted microsomes are difficult to carry out.

The problems of enzyme destabilization and inhibition by detergents can be overcome either (1) by diluting the detergent-treated preparation 10- to 20-fold with ice-cold SH buffer before assay,[33,34] (2) by adding fatty acid-free bovine serum albumin to the assay media at a final concentration of 10 mg/ml,[16,32] or (3) by a combination of these procedures. The albumin effectively sequesters detergent present in the assay medium. Although dilution is required before assay if initial rates of hydrolysis are to be obtained (see below), there is an added benefit in diluting detergent-treated preparations. We have observed that thermal lability of the enzyme is a function of the concentration of detergent present during heat expo-

[33] O. S. Nilsson, W. J. Arion, J. W. DePierre, G. Dallner, and L. Ernster, *Eur. J. Biochem.* **82**, 627 (1978).

[34] P. W. Carlson, Master's Thesis, Cornell University, Ithaca, New York (1973).

[35] Y. Hatefi and W. G. Hanstein, this series, Vol. 31, p. 770.

[36] D. Lichtenberg, R. J. Robson, and E. A. Dennis, *Biochim. Biophys. Acta* **737**, 285 (1983).

[37] A. Burchell, B. Burchell, M. Monaco, H. E. Walls, and W. J. Arion, *Biochem. J.* **230**, 489 (1985).

sure.[38,39] Enzyme stability comparable to that seen in untreated microsomes[26] or ammonia-disrupted microsomes[28,29] can be restored simply by sedimenting the detergent-treated microsomes and resuspending the membranes in SH buffer.[38,39] Instability returns with the readdition of detergent. For this reason we do not recommend the disruption technique that involves addition of the detergent (usually deoxycholate) to the assay medium to effect an "instantaneous" disruption of the membranes. The concentration of deoxycholate during assay is usually 0.03–0.05%,[40] a level sufficient to cause significant inactivation and/or inhibition.[32,38,41] When microsomes are disrupted by exposure to a nonionic detergent (e.g., Triton X-100 or Lubrol 12A9) followed by dilution, albumin need not be added to the assay media, since at low concentration the nonionic detergents are not inhibitory.[8,32,33] However, the bile salts and especially deoxycholate are highly inhibitory,[32,41] so albumin should be present during assay when these detergents are used.

Some investigators have approached the problem of enzyme instability by assaying mannose-6-phosphatase at lower temperatures, e.g., 20°.[42] While the enzyme is more stable at lower temperatures,[26] the investigator should be aware that we have occasionally observed what appears to be membrane resealing in *untreated* membranes when they are incubated at higher temperatures (i.e., 30° or 37°).[43,44] In such instances, the latency of mannose-6-phosphatase was greater following incubation at the higher temperature than that observed in microsomes kept on ice and assayed at 10° or 20°. The main point here is that if latency of mannose-6-phosphatase is being assessed with reference to an activity in the ER membrane that will be assayed at 30° or 37°, then membrane intactness should also be assessed under exactly the same conditions.

A concern has been raised that dilution combined with the presence of albumin in the assay medium may cause membrane resealing.[42] The basis for this concern is derived from the finding that when microsomes (2 mg protein/ml in SH buffer) were exposed for 1 hr to 4 mM CHAPSO[24] and diluted 10-fold by transfer to an assay medium containing 2 mM mannose-6-P, the observed phosphohydrolase activity was 15% higher when the assay medium also contained 2 mM CHAPSO.[42] Thus it appeared that adding CHAPSO to the assay medium prevented membrane resealing that

[38] A. J. Lange, W. J. Arion, A. Burchell, and B. Burchell, *J. Biol. Chem.* **261**, 101 (1986).
[39] H. E. Walls, A. J. Lange, and W. J. Arion, unpublished observations.
[40] G. Kreibich and D. D. Sabatini, this series, Vol. 31, p. 215.
[41] W. Colilla, W. T. Johnson, and R. C. Nordlie, *Biochim. Biophys. Acta* **364**, 78 (1974).
[42] F. Vanstapel, K. Pua, and N. Blanckaert, *Eur. J. Biochem.* **156**, 73 (1986).
[43] L. M. Ballas and W. J. Arion, unpublished observations.
[44] W. J. Arion and H. E. Walls, *J. Biol. Chem.* **257**, 11217 (1982).

otherwise occurred when CHAPSO-treated microsomes were diluted with the assay medium. We have confirmed these observations using the structurally similar zwitterionic detergent CHAPS.[39] We found, however, that the stimulation of activity by CHAPS in the assay medium resulted from a *decrease* in the K_m for mannose-6-P. V_{max} values for mannose-6-P hydrolysis were identical in the absence and presence of CHAPS, and when CHAPS was not added to the assay media, we did not observe the biphasic kinetics characteristic of microsomes possessing a mixture of intact and disrupted membranes.[17] These results illustrate that destabilization, inhibition, and even activation of enzyme may occur, depending on the nature of the detergent, and they underscore the importance of reducing the detergent concentration in the assay medium, either by dilution prior to assay or by addition of albumin.

We have never observed measurable restoration of the permeability barrier following dilution when the membranes were exposed to the recommended concentrations of detergents. Indeed, we have tried diligently without success to reconstitute intact membranes employing dilution and dialysis techniques.

Assay Procedure. There are many occasions when an investigator should assess the intactness of a microsomal preparation, but the amount of material is limited. The sensitivity of many of the published procedures for determining P_i requires that 0.5 to 2.0 mg of microsomal protein be used to assay mannose-6-phosphatase activity in untreated microsomes. The Burchells[45,46] have modified the phosphate assay method developed by Ames[47] for use as a one-step microvolume assay for glucose-6-phosphatase/mannose-6-phosphatase. The method requires microgram quantities of microsomal protein, is convenient, and generates reliable results (e.g., Refs. 38 and 48). The essence of the modified procedure is the addition of 1% (w/v) sodium dodecyl sulfate to the solution used to the "kill" the reaction, i.e., the "Working Solution." This obviates the necessity of deproteinizing before color development, even when concentrated homogenate preparations are assayed or albumin is added to the assay medium. Details of the assay procedure follow.

The following stock reagents are prepared:

Reagent A: 0.42% ammonium molybdate tetrahydrate in 1 N H_2SO_4 (stable indefinitely at room temperature)

[45] G. F. Bickerstaff and B. Burchell, *Biochem. Soc. Trans.* **8**, 389 (1980).
[46] A. Burchell, R. Hume, and B. Burchell, *Clin. Chim. Acta* **173**, 183 (1988).
[47] B. N. Ames, this series, Vol. 8, p. 115.
[48] W. K. Canfield and W. J. Arion, *J. Biol. Chem.* **263**, 7458 (1988).

Reagent B: 10% (w/v) sodium dodecyl sulfate[49] in water (stable at room temperature)

Reagent C: 10% ascorbic acid in water (keep refrigerated in an amber bottle; make fresh weekly)

Working solution: prepared by combining Reagents A, B, and C in volume proportions of 6:2:1 (stable for 1 day on ice)

The phosphatase reaction is terminated by adding 0.9 ml of Working Solution to 0.1 ml of assay medium. Color is developed by incubating for 20 min at 45° or 1 hr at 37°. The blue reduced phosphomolybdate complex is stable for several hours. Neither glucose-6-P nor mannose-6-P undergo significant hydrolysis under these conditions. However, PP_i and other acid-labile phosphates are substantially hydrolyzed.[47] Sample absorbance is determined at 820 nm; 10 nmol of P_i should yield an OD of about 0.26 cm^{-1}. The method permits determination of 1 to 50 nmol of P_i.

The phosphatase assay medium is prepared so that after addition of 20 μl of enzyme suspension it contains 50 mM buffer (sodium cacodylate or one of the suitable substitute buffers listed above) and 1 or 2 mM mannose-6-P in a final volume of 100 μl. The assay medium is equilibrated at assay temperature, and the phosphatase reaction is initiated by addition of 20 μl of an appropriately diluted enzyme preparation. Typically, 2 to 8 μg of protein from fully disrupted microsomes are added to the assay medium. Assuming 90% intactness, assay tubes for untreated preparations should contain 10 times more protein. Reagent blanks ("zero-time" controls) and standards must be run concurrently.

Membrane disruption can also be carried out on a microscale by adding 5 μl of stock detergent solution to 45 μl of a membrane suspension containing 50 to 100 μg of protein. Enzyme stability is definitively demonstrated by a linear rate of hydrolysis between 2 and 12 min of incubation. As a general rule hydrolysis should consume less than 10% of the mannose-6-P added to the assay tubes. A second useful test both for stability and the presence of inhibitory agents is to compare the rates of hydrolysis of 20 mM glucose-6-P and 1 mM mannose-6-P in fully disrupted microsomes. In our experience with the rat hepatic enzyme, the ratio of the two rates at pH 6.5 is nearly 2:1, if no inhibitors of the enzyme are present and if enzyme inactivation has not occurred, i.e., when initial rates are measured.

If the presence in the assay medium of P_i or other phosphate compounds that may be hydrolyzed by cellular enzymes (e.g., ATP) is unavoid-

---

[49] "Electrophoresis grade" sodium dodecyl sulfate must be used, since most commercial grades are badly contaminated with P_i.

able, mannose-6-phosphatase activity must be assayed by a radiochemical procedure. The latter is based on the measurement of $^{32}P_i$ formed during hydrolysis of mannose-6-^{32}P.[16] Details of a microvolume radiochemical assay are described in Refs. 15 and 48.

Measurements of Latency in Nuclei, Liver Homogenates, Permeabilized Cells, and ER Subfractions of Other Tissues. The latency of mannose-6-phosphatase can be used to estimate ER membrane intactness in whole liver homogenates and isolated hepatic nuclei.[14] However, the investigator should be aware that a lower concentration of detergent is required to effect complete disruption of the nuclear envelope[14] and that the transformation of mannose-6-P to glucose-6-P catalyzed by the sequential actions of phosphomannose isomerase (mannose-6-phosphate isomerase) and phosphoglucose isomerase (glucose-6-phosphate isomerase) in the supernatant may confound the assessment of intactness in whole liver homogenates. Because phosphomannose isomerase activity is very low in rat liver, the latency of mannose-6-phosphatase activity can be accurately assessed in rat liver homogenates. We have, however, observed artifactually high apparent activity of mannose-6-phosphatase in homogenates of human caucasian liver that was traced to a high rate of conversion of mannose-6-P to glucose-6-P, i.e., high phosphomannose isomerase activity.[50] The solution to this problem is to assess membrane intactness using the alternate substrates, glucosamine-6-P or 2-deoxyglucose-6-P.[16,50]

In principle, latency of mannose-6-phosphatase can be used to evaluate intactness of ER membranes from livers of other species and in other tissues with high glucose-6-phosphatase activity in the ER, i.e., renal tubular cells[11] (see Table 13 in Ref. 51), the enterocytes of the small bowel (see Table 14 in Ref. 51), and the β cells of the pancreas.[52] However, with the exception of the rat tissues and human liver,[9,53] little is known about glucose-6-phosphatase systems in tissues from other species.[7] Thus, it will be necessary to determine the optimal conditions for membrane disruption and validate the use of mannose-6-P hydrolysis as a measure of ER membrane intactness, as was done, for example, with human liver.[9]

A Final Caveat. The study of the topography of metabolic systems in the ER has been a major area of application of the latency of mannose-6-

[50] A. J. Lange, Master's Thesis, Cornell University, Ithaca, New York (1983).
[51] R. C. Nordlie, *in* "The Enzymes" (P. D. Boyer, ed.), 3rd ed., Vol. 4, p. 543. Academic Press, New York, 1971.
[52] I. D. Waddell and A. Burchell, *Biochem. J.* **255,** 471 (1988).
[53] A. Burchell, R. T. Jung, C. C. Lang, W. Bennet, and A. N. Shepherd. *Lancet* **1,** 1059 (1987).

phosphatase.[33,54,55] Such studies usually involve treatment of intact microsomes with membrane-impermeant probes, such as chemically reactive agents and proteolytic enzymes. It is conceivable that under the right circumstances a given reagent might modify the structure of the glucose-6-P transporter such that it would bind and transport mannose-6-P, and, as a result, mannose-6-P would be hydrolyzed by the enzyme in intact microsomes. The only unequivocal way to test for this is to conduct an independent assessment of membrane integrity, such as the measurement of EDTA permeability.[14,17,56]

Alternative Methods of Membrane Disruption. Recently two reports have appeared describing the use of specific proteins to destroy the ER permeability barrier. Histone 2A[57] and α-toxin,[58,59] a pore-forming exotoxin secreted by *Staphylococcus aureus,* have been shown to be highly effective in removing the selective permeability barrier. Both proteins eliminate the problem of enzyme destabilization encountered with detergents. Histone 2A can effect complete activation of mannose-6-phosphatase at 0° and above even when the basic protein is added to the assay medium.[57] In contrast, formation of the α-toxin pore requires incubation of the toxin with microsomes at temperatures above 33°.[58,59] In future studies, especially those dealing with previously uncharacterized membrane preparations, histone 2A and α-toxin should be carefully compared with detergents as to their respective efficacies as ER-permeabilizing agents.

[54] T. Hallinan and R. de Brito, *Horm. Cell Regul.* **5,** 73 (1981).
[55] R. A. Coleman and R. M. Bell, *in* "The Enzymes" (P. D. Boyer, ed.), 3rd ed., Vol. 16, p. 605. Academic Press, 1983.
[56] G. Gold and C. C. Widnell, *J. Biol. Chem.* **251,** 1035 (1976).
[57] J. N. R. Blair and A. Burchell, *Biochim. Biophys. Acta* **964,** 161 (1988).
[58] A. J. Lange and W. J. Arion, *Fed. Proc., Fed Am. Soc. Exp. Biol.* **45,** 1833 (1986).
[59] A. J. Lange and W. J. Arion, submitted for publication.

[8] Calcium Ion Transport in Mitochondria

By Nils-Erik Leo Saris and Ashley Allshire

Introduction

Mitochondria are organelles capable of accumulating and releasing Ca^{2+} in the cell. Calcium channels and pumps in the plasma membrane determine the overall cellular Ca^{2+} content. In most cells the endo(sarco)plasmic reticulum acts as the store from which Ca^{2+} can be rapidly released as an intracellular messenger,[1,2] while mitochondria may act as a more sluggish Ca^{2+} buffer and Ca^{2+} sink to damp cytosolic Ca^{2+} activity in the activated cell and prevent excessive concentrations from arising and spreading in the cytosol. In addition the mitochondrial Ca^{2+}-transporting systems will influence the Ca^{2+} activity in the mitochondrial matrix compartment.[3]

It is now well established that mammalian mitochondria accumulate Ca^{2+} mainly by an electrogenic mechanism that drives Ca^{2+} into the matrix at the expense of the membrane potential generated by respiration or hydrolysis of ATP, while efflux of Ca^{2+} occurs mainly by other, electroneutral mechanisms,[1,4-6] Under physiological conditions the distribution of Ca^{2+} between cytosol and mitochondria is thus influenced by the kinetic properties of the influx and efflux mechanisms.[4-6] The Ca^{2+} transport systems of mammalian mitochondria can be summarized as follows:

1. *Ca^{2+} uniporter.*[1,4-6] The Ca^{2+} uniporter is universally present, has low cation specificity (Sr^{2+} also being transported at relatively high rates) and is inhibited by ruthenium red and lanthanide cations. The rate of Ca^{2+} translocation is high above a few micromolar and may exceed the rate of generation of membrane potential; at Ca^{2+} concentrations of 1 μM or below the rate is of the same order of magnitude as that of Ca^{2+} efflux. The

[1] A. K. Campbell, "Intracellular Calcium: Its Universal Role as Regulator." Wiley, New York, 1983.
[2] M. J. Berridge, *Biochem. J.* **220,** 345 (1984).
[3] R. M. Denton and J. G. McCormack, *FEBS Lett.* **119,** 1 (1980).
[4] N.-E. Saris and K. E. O. Åkerman, *Curr. Top. Bioenerg.* **10,** 103 (1980).
[5] D. G. Nicholls and K. E. O. Åkerman, *Biochim. Biophys. Acta* **183,** 57 (1983).
[6] E. Carafoli and G. Sottocasa, *in* "Bioenergetics" (L. Ernster, ed.), p. 269. Elsevier, Amsterdam, 1984.

uptake kinetics often exhibit sigmoidicity indicating cooperativity, and physiological concentrations of Mg^{2+} usually increase the sigmoidicity and inhibit Ca^{2+} influx. Mn^{2+} counteracts the Mg^{2+} effects[7,8]; spermine has a stimulatory effect[9]; efflux via the uniporter occurs when the electrochemical Ca^{2+} gradient is lowered by uncoupling and this efflux is less sensitive to ruthenium red.

2. *$Ca^{2+} - Na^+$ exchanger.* In mitochondria from heart and a number of other excitable tissues the most important mechanism for efflux of Ca^{2+} involves exchange against Na^+.[10] Some activity is present also in liver mitochondria.[11] The exchange is inhibited by lanthanide cations but not by ruthenium red. It is also inhibited by Sr^{2+}.[12] Mobilization of Ca^{2+} from mitochondria could occur by means of this mechanism when the cytosolic concentration of Na^+ is increased.

3. *$Ca^{2+} - 2H^+$ exchange.* While the mechanisms above are relatively well characterized, exchange of Ca^{2+} against H^+ may represent several poorly resolved activities. Coupling between Ca^{2+} and H^+ fluxes may be direct or indirect via anion transport coupled to that of H^+.

a. *Reduction/oxidation-modulated Ca^{2+} efflux pathway.* The reduction/oxidation (red/ox) state of pyridine nucleotides has been shown by Lehninger *et al.*[13,14] to profoundly affect the efflux of Ca^{2+}. However, this effect may be through changes in the red/ox state of membrane sulfhydryl groups (see Siliprandi *et al.*[15]). It is difficult to distinguish between increased Ca^{2+} efflux seen under these conditions and a nonspecific increase in permeability caused by a fall in membrane potential due to Ca^{2+} overload,[16,17] or efflux of Ca^{2+} on the uniporter[18] (see Section 4). However, oxidation of sulfhydryl groups with hydroperoxides induces an efflux of Ca^{2+} that may occur on a separate carrier.[19,20] In the latter case there is no apparent

[7] B. P. Hughes and J. H. Exton, *Biochem. J.* **212**, 773 (1983).

[8] A. Allshire, P. Bernardi, and N.-E. L. Saris, *Biochim. Biophys. Acta* **807**, 202 (1985).

[9] C. V. Nicchitta and J. R. Williamson, *J. Biol. Chem.* **259**, 12978 (1984).

[10] M. Crompton, M. Künzi, and E. Carafoli, *Eur. J. Biochem.* **77**, 549 (1977).

[11] E. J. Harris and J. J. A. Heffron, *Arch. Biochem. Biophys.* **218**, 531 (1982).

[12] N.-E. L. Saris and P. Bernardi, *Biochim. Biophys. Acta* **725**, 19 (1983).

[13] A. L. Lehninger, A. Vercesi, and E. A. Bababunmi, *Proc. Natl. Acad. Sci. U.S.A.* **75**, 1690 (1978).

[14] G. Fiskum and A. L. Lehninger, *J. Biol. Chem.* **254**, 6236 (1979).

[15] D. Siliprandi, N. Siliprandi, and A. Toninello, *Eur. J. Biochem.* **130**, 173 (1983).

[16] D. G. Nicholls and M. D. Brand, *Biochem J.* **188**, 113 (1980).

[17] M. C. Beatrice, J. W. Palmer, and D. R. Pfeiffer, *J. Biol. Chem.* **255**, 8663 (1980).

[18] M. E. Bardsley and M. D. Brand, *Biochem. J.* **202**, 197 (1982).

[19] S. Baumhüter and C. Richter, *FEBS Lett.* **148**, 271 (1982).

[20] G. A. Moore, S. A. Jewell, G. Bellomo, and S. Orrenius, *FEBS Lett.* **153**, 289 (1983).

membrane damage though NAD^+ is hydrolyzed and a membrane protein may be ADP-ribosylated.[21]

b. *Membrane potential-gated efflux pathway.* Bernardi and Azzone[22,23] found that the set point at which liver mitochondria buffer external Ca^{2+} is shifted to lower concentrations when membrane potential is decreased toward a threshold value. The observation was interpreted in terms of inhibition of a potential-gated Ca^{2+} efflux pathway. Whether this putative pathway involves $Ca^{2+}-2H^+$ exchange has not been established.

c. *Ruthenium red-insensitive, lanthanide cation-sensitive efflux pathway.* In liver and kidney mitochondria the ruthenium red-insensitive, lanthanide cation-sensitive efflux pathway has generally been thought to involve $Ca^{2+}-2H^+$ exchange.[4-6] Efflux of Ca^{2+} is induced by an external acid pulse[24] and Ca^{2+} influx when the pH of the matrix compartment is made more acid.[25,26] However, efflux is seen only at relatively high Ca^{2+} loads when a loss of membrane potential in a subpopulation of mitochondria may cause reversal of the uniporter (see Section 4). Influx of Ca^{2+} when the matrix is acidified in nonrespiring mitochondria, by addition of ionophores like nigericin and dianemycin that catalyze an electroneutral exchange of K^+ against H^+, is quite sensitive to ruthenium red.[27,28] Even at relatively high concentrations of ruthenium red uniporter activity still exceeds that of the putative $Ca^{2+}-2H^+$ exchanger.[28] There is thus no evidence of any substantial antiporter activity under these conditions.

4. *Ca^{2+}-induced Ca^{2+} efflux.* In earlier work, attention was seldom paid to the possibility of efflux of Ca^{2+} on the uniporter due to lowering of the membrane potential nor to secondary effects induced by Ca^{2+} in mitochondria such as uncoupling, swelling, or loss of matrix components owing to a general increase in the permeability of the inner membrane.[4,5,29] Mitochondria are more resistant to Ca^{2+} in the presence of ADP or ATP, while P_i increases their sensitivity to Ca^{2+}. This effect may be due to the activation by Ca^{2+} of a mitochondrial phospholipase A_2, leading to an

[21] W. Hofstetter, T. Mühlebach, H.-R. Lötscher, K. Winterhalter, and C. Richter, *Eur. J. Biochem.* **117**, 361 (1981).

[22] P. Bernardi and G. F. Azzone, *FEBS Lett.* **139**, 13 (1982).

[23] P. Bernardi and G. F. Azzone, *Eur. J. Biochem.* **134**, 377 (1983).

[24] K. E. O. Åkerman, *Arch. Biochem. Biophys.* **189**, 256 (1978).

[25] G. Fiskum and R. S. Cockrell, *FEBS Lett.* **92**, 125 (1978).

[26] R. S. Cockrell, *Arch. Biochem. Biophys.* **243**, 70 (1985).

[27] P. Bernardi and G. F. Azzone, *Eur. J. Biochem.* **102**, 555 (1979).

[28] N.-E. L. Saris, *Acta Chem. Scand.* **B41**, 79 (1987).

[29] N.-E. Saris, *Commentat. Phys.-Math., Soc. Sci. Fenn.* **24**(11), 1 (1963).

increase in mitochondrial free fatty acids[30] and lysophospholipids.[31] Free fatty acids, especially unsaturated ones, do increase the Ca^{2+} permeability,[32] in effect acting as $Ca^{2+}-2H^+$ exchangers. Lysophospholipids can also increase the Ca^{2+} permeability of the membrane,[33] though at rather high concentrations at which detergent action may be involved.

Increase in permeability is correlated with transition of an increasing proportion of the mitochondrial population from an aggregated to an orthodox configuration.[34-36] Sr^{2+} does not induce such a transition.[35] It seems likely that the inner membrane is more permeable when the mitochondria are in the orthodox configuration. The transition is inhibited by NADH, ADP, bongkrekate, and Mg^{2+}, while it is stimulated by uncoupling, Ca^{2+}, and atractyloside.[34] It is not clear whether the Ca^{2+} transition or activation of phospholipase A_2 is the primary event in inducing the increased permeability.

Methods for Measuring Calcium Fluxes in Mitochondria

A multitude of approaches have been used for measuring the various Ca^{2+} fluxes. The use of metallochromic indicators and dual-wavelength photometry has been described by Scarpa in earlier chapters of this series.[37,38] Murexide[37] is still useful despite its high K_d (2.7 mM) when relatively high Ca^{2+} (or Sr^{2+}) concentrations (10–200 μM) are used. Antipyrylazo III is suitable for an intermediate range (1–100 μM).[38] Arsenazo III can successfully be used at still lower concentrations, its drawbacks being greater sensitivity to Mg^{2+} and buffering of Ca^{2+}.[38] However, the demonstration that peroxidation of membrane lipids may be stimulated by arsenazo III in the presence of reduced pyridine nucleotides and liver microsomes gives cause for caution, since microsomes are contaminants in mitochondrial preparations.[39]

The use of radiolabeled Ca^{2+} ($^{45}Ca^{2+}$) in transport studies has been described in detail by Reed and Bygrave[40] and by Crompton and Carafoli

[30] D. R. Pfeiffer, P. C. Schmid, M. C. Beatrice, and H. H. O. Schmid, J. Biol. Chem. **254,** 11485 (1979).

[31] D. Siliprandi, M. Rugolo, A. Toninello, and N. Siliprandi, Biochem. Biophys. Res. Commun. **88,** 388 (1979).

[32] I. Roman, P. Gmaj, C. Nowicka, and S. Angielski, Eur. J. Biochem. **102,** 615 (1979).

[33] E. J. Harris and M. B. Cooper, Biochem. Biophys. Res. Commun. **103,** 788 (1981).

[34] D. R. Hunter and R. A. Haworth, Arch Biochem. Biophys. **195,** 453 (1979).

[35] D. R. Hunter and R. A. Haworth, Arch. Biochem. Biophys. **195,** 468 (1979).

[36] M. C. Beatrice, D. L. Stiers, and D. R. Pfeiffer, J. Biol. Chem. **257,** 7161 (1982).

[37] A. Scarpa, this series, Vol. 24, p. 343.

[38] A. Scarpa, this series, Vol. 56, p. 301.

[39] R. Docampo, S. N. J. Moreno, and R. P. Mason, J. Biol. Chem. **258,** 14920 (1983).

in this series.[41] The drawback of this approach is that the mitochondria must be separated from the medium, either by filtration or centrifugation. Thus fast kinetic analyses are difficult. The timing of sampling can be improved by using quench techniques to stop Ca^{2+} flux. Reed and Bygrave[40] advocated using either EGTA alone or with ruthenium red, while Crompton and Carafoli[41] used inhibiting concentrations of Mg^{2+}. Both methods remove surface-bound Ca^{2+} from the mitochondria, but neither stops efflux of Ca^{2+}. Efflux is minimized by cooling but may actually be stimulated by EGTA. $Ca^{2+}-Ca^{2+}$ exchange across the mitochondrial inner membrane can also lead to serious overestimations of slow fluxes measured with $^{45}Ca^{2+}$. Finally, $^{45}Ca^{2+}$ does not report free Ca^{2+}, making it necessary to use calcium buffers.[40,41]

The third common technique for measuring Ca^{2+} transport by mitochondria involves the use of calcium-selective electrodes.[41] The electrode reports Ca^{2+} activity directly. Recently, fluorescent calcium probes, including Fura-1 and Indo-1, have been introduced for measurement of submicromolar concentrations of Ca^{2+}, though their main use is as an indicator of cytosolic Ca^{2+} in intact cells.[42] A potentially useful technique is to introduce Fura-2 into the matrix compartment. This can be done at least with heart mitochondria.[43]

Though use and calibration of calcium-selective electrodes were dealt with briefly earlier in this series,[41] these topics are discussed in more detail below. Use of metallochromic indicators is illustrated with several examples.

Measurements with Calcium-Selective Electrodes

The principles of ion-selective electrode measurements were discussed elsewhere in this series.[44,45] With commercially available electrodes based on neutral exchange resins or ionophores, Ca^{2+} activity can be measured continuously and selectively. Down to $10^{-5} M$ ($= p$Ca 5) or so, electrode response is linearly related to Ca^{2+} activity in accordance with the Nernst equation. At lower activities response becomes increasingly sub-Nernstian but is useful to pCa 8 if calibrated to take into account pH, temperature, ionic strength, and the effects of interfering cations (notably Mg^{2+} and K^+ in this context). In addition, the Philips electrode we routinely use shows a useful sensitivity to Sr^{2+} (about 100-fold less than to Ca^{2+}).

[40] K. C. Reed and F. L. Bygrave, *Eur. J. Biochem.* **55,** 497 (1975).
[41] M. Crompton and E. Carafoli, this series, Vol. 56, p. 338.
[42] G. Grynkiewcz, M. Poenie, and R. Y. Tsien, *J. Biol. Chem.* **260,** 3440 (1985).
[43] G. L. Lukács, A. Kapus, and A. Fonyó, *FEBS Lett.* **229,** 219 (1988).
[44] J. L. Walker, this series, Vol. 56, p. 359.
[45] W. Simon and E. Carafoli, this series, Vol. 56, p. 439.

Calibration. Electrode response can be calibrated either internally with a series of Ca^{2+} additions to the medium[41] or using a set of standards. The former method is suitable for calibration of fluxes while the latter is necessary for establishing the activity of Ca^{2+} in the extramitochondrial space. By diluting solutions of calcium salts in double-distilled water, accurate standards can be prepared down to about 0.1 mM (*p*Ca 4). Beyond this limit adventitious cation in the water and on container surfaces makes buffering of Ca^{2+} necessary.

Typically, buffers comprise a Ca^{2+}–ligand mixture at the pH, temperature, and ionic strength specified by the conditions of the experiment since binding affinity is modified by these factors. Electrode responses to Mg^{2+} and K$^+$ must also be taken into account when preparing appropriate standards. In practice, we make up sets of standards with compositions reflecting the various experimental conditions likely to be used, which are stored frozen in 5-ml portions. For mitochondrial studies nitrilotriacetate (NTA)[40] and *N*-(2-hydroxyethyl)ethylenediamine-*N,N′,N′*-triacetic acid (HEDTA)[41] are suitable, while EGTA may be used below 1 μM Ca^{2+}. The buffering range of HEDTA is 0.4–40 μM and that of NTA 10 μM–1 mM at pH 7.0, 30°, and ionic strength 0.1 M. They also buffer Mg^{2+}.

Absolute stability constants[46,47] describing the affinity of ligand for Ca^{2+} and Mg^{2+} are corrected[48] for effects of pH, temperature, and ionic strength to obtain appropriate apparent constants, K'. Then in a buffer with total ligand concentration [L$_t$], total Ca [Ca$_t$] and Mg [Mg$_t$] required for particular combinations of free Ca^{2+} [Ca$_f$] and Mg^{2+} [Mg$_f$] can be calculated:

$$[L_t] = [L_f] + [LCa] + [LMg] \tag{1}$$
$$[Ca_t] = [Ca_f] + [LCa] \tag{2a}$$
$$[Mg_t] = [Mg_f] + [LMg] \tag{2b}$$

Apparent binding constants, K', may be determined:

$$K'_{Ca} = \frac{[LCa]}{[Ca_f][L_f]} \tag{3a}$$

$$K'_{Mg} = \frac{[LMg]}{[Mg_f][L_f]} \tag{3b}$$

Rearranging Eqs. (3a) and (3b),

$$[LCa] = K'_{Ca} [Ca_f][L_f] \tag{4a}$$
$$[LMg] = K'_{Mg} [Mg_f] [L_f] \tag{4b}$$

[46] L. S. Sillén and A. E. Martell, *Spec. Publ.—Chem. Soc.* **17** (1964).
[47] L. S. Sillén and A. E. Martell, *Spec. Publ.—Chem. Soc.* **25** (1971).
[48] O. Scharff, *Anal. Chim. Acta* **109**, 291 (1979).

and replacing in Eq. (1),

$$[L_t] = [L_f] + K'_{Ca} [Ca_f] [L_f] + K'_{Mg} [Mg_f] [L_f] \tag{5}$$

Thus $[L_f]$ can be obtained and replaced in Eqs. (3a) and (3b) to give [LCa] and [LMg]. Finally, from Eq. (2) $[Ca_t]$ and $[Mg_t]$ are calculated.[49]

It is the *relative* rather than absolute concentrations of Ca and ligand that are important.[50] Errors arising from uncertainty over purity and hydration of ligand as obtained from the manufacturer increase dramatically as the upper limit of the buffering range is reached, i.e., when [Ca] → [L]. Therefore it is advisable to titrate the calcium stock solution against that of the ligand using Ca^{2+} indicator of low affinity such as murexide.

Another way to prepare accurate Ca^{2+} standards involves using a Ca^{2+} electrode to determine the apparent binding constant and ligand concentration empirically,[51] as follows: (1) Using a series of dilutions of a Ca^{2+} stock solution measure electrode output (in millivolts) over the pCa range 2–4 and under appropriate conditions of pH (accurate to 0.01 unit), temperature, ionic strength, and composition. (2) Measure electrode response in a series of Ca-EGTA buffers of corresponding pH, temperature, etc., prepared on the basis of a nominal ligand concentration and an approximate binding constant. (3) From the plot of electrode output versus pCa (step 1 above) electrode response in these buffer solutions is related to free Ca^{2+}, which value is then subtracted from the total calcium, $[Ca_t]$, to obtain $[Ca_{bound}]$. (4) A Scatchard plot of the data, $[Ca_b]/[Ca_f]$ versus $[Ca_b]$, yields a curve with a linear portion over the pCa interval at which electrode response is Nernstian. The slope of this portion gives the apparent binding constant under the conditions used, and the x intercept gives the actual ligand concentration. (5) Using these values, $[Ca_f]$ in the buffers is recalculated and electrode output plotted versus this revised pCa to yield the final calibration curve.

Design of Calcium Transport Measurements in Mitochondria

In this section we describe the design of experiments to study Ca^{2+} uptake by the uniporter using Ca^{2+} or the calcium analog Sr^{2+} with either respiration or a valinomycin-induced K^+ diffusion potential as the driving force. Ca^{2+} fluxes by exchange against Na^+ are treated as well as measurements of mitochondrial set points of isolated organelles or mitochondria in

---

[49] Calculator or computer programs such as those of A. Fabiato and F. Fabiato [*J. Physiol. (Paris)* **75**, 463 (1979)] are useful for routine calculations of apparent binding constants and buffer compositions.

[50] D. Dinjus, R. Klinger, and R. Wetzker, *Biomed. Biochim. Acta* **43**, 1067 (1984).

[51] D. M. Bers, *Am. J. Physiol.* **242**, C404 (1982).

permeabilized cells. We also describe changes in the set point due to modulation of influx and efflux pathways, and we discuss experiments on the Ca^{2+}-induced Ca^{2+} release.

Mitochondrial Ca^{2+} Uniporter

Ca^{2+} Influx through Uniporter. Obviously the choice of method for measuring a Ca^{2+} flux will depend on the experiment, the alternatives being $^{45}Ca^{2+}$ accumulation, use of metallochromic indicators, and use of calcium-selective electrodes (see above). The examples chosen are measurements of Ca^{2+} influx at relatively low concentrations using the electrode technique and at high concentrations using metallochromic indicators.

Measurement with Calcium-Selective Electrodes. Flux measurements with electrodes avoid several of the limitations associated with using indicators or $^{45}Ca^{2+}$. By incubating Ca^{2+}-depleted, deenergized mitochondria in Ca^{2+}-containing medium, passive binding is brought to equilibrium. Then during the first few seconds after reenergization, when efflux is still negligible and P_i in the matrix available to trap Ca^{2+} will not be limiting, net influx measured should be close to the true flux.

Experimental design must ensure that Ca^{2+} flux through the uniporter is the rate-limiting step. In cases where the flux is low, as in heart mitochrondria accumulating Ca^{2+} from a Mg^{2+}-containing medium, respiration is usually fast enough to maintain adequate energization of the membrane. The uptake can then be started by adding a respiratory substrate, such as succinate, to rotenone-inhibited mitochondria. However, fast fluxes, as in liver and brain mitochondria,[52-54] make it necessary to energize the membrane by adding valinomycin to mitochondria suspended in a K^+-free medium in order to generate an outward-directed K^+ diffusion potential.

A sample of rat liver mitochondria from the stock suspension (~40 mg protein/ml) is deenergized by preincubation with 1 nmol rotenone/mg and 0.1 μg antimycin/mg on ice for at least 5 min. Mitochondria are then diluted to a concentration of 1 mg/ml in 1.5 ml medium equilibrated at 30°C and containing 150 mM sucrose, 40 mM choline chloride, 10 mM 3-(N-morpholino)propanesulfonic acid (MOPS) neutralized to pH 7.0 with tris(hydroxymethyl)aminomethane (Tris), 1 mM P_i–Tris, 1 μM rotenone, 1 μg/ml oligomycin (and $MgCl_2$ as required). Addition of 10 μM Ca^{2+} in five equal portions serves as an internal calibration. When electrode output

[52] A. Scarpa and G. F. Azzone, Eur. J. Biochem. 12, 328 (1970).
[53] G. M. Heaton and D. G. Nicholls, Biochem. J. 156, 635 (1976).
[54] D. G. Nicholls, Biochem. J. 170, 511 (1978).

equilibrates, the recorder is adjusted to give a chart advance of 1 cm/sec. To begin Ca^{2+} uptake an excess of valinomycin (4 μM) is rapidly added with a Hamilton syringe.

Clearly, very rapid and smooth stirring of the vessel contents is essential in order to minimize electrode response time, and the electrode tips should be positioned within 5 mm of the upper surface of the follower (Fig. 1). After a lag period of 1–2 sec reflecting residual limits for mixing and electrode response, the rapid phase of Ca^{2+} influx is recorded (Fig 2). By measuring the slope of the tangent to the uptake curve at the same pen deflection on successive traces, comparable measurements of influx are obtained at the Ca^{2+} concentration to which that deflection corresponds. So that slopes can be read accurately the steepest portion of the trace

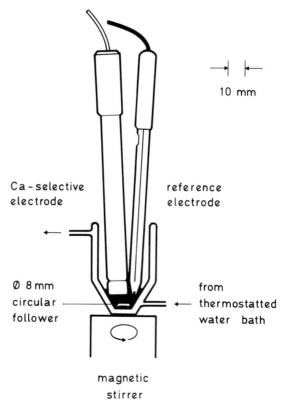

FIG. 1. Potentiometric estimation of Ca^{2+} in mitochondrial suspensions. Electrode output is either amplified in a pH meter and relayed to a strip-chart recorder through a circuit providing additional offset, or transferred directly to a computer.

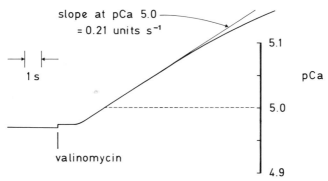

Fig. 2. Ca^{2+} influx measured with a calcium-selective electrode. Rat liver mitochondria preincubated with rotenone and antimycin for 5 min at 4° were added to 1 mg/ml in 1.5 ml of a sucrose–choline chloride–MOPS medium (see text for details). Uptake of Ca^{2+} was begun with a rapid addition of valinomycin (4 μM). Temperature 30°; stirring fast yet smooth. In this particular example, the slope of the uptake trace was 0.21 pCa units/sec at pCa 5.0 (10 μM). This corresponds to a flux of 2.1 nmol/mg/sec.

should be kept within 30–35 degrees of the axis of paper movement through adjustments in mitochondrial concentration used and/or chart speed. The appropriate mitochondrial concentration will also be limited by the response time of the electrode over the pCa range of interest. This can be checked in control experiments under similar stirring conditions with additions of EGTA.

Comments

1. To ensure that the Ca^{2+} content of mitochondria and medium is not too high, Ca^{2+} accumulated by the organelles during the course of their isolation may need to be removed. In control experiments, an estimate of exchangeable Ca^{2+} in the mitochondria can be obtained by measuring net release caused by addition of an uncoupling agent. We use 1 nmol carbonyl cyanide p-trifluoromethoxyphenylhydrazone (FCCP)/mg protein or 5-fold more in media containing bovine serum albumin (BSA) to bind free fatty acids.

Media with reduced endogenous Ca^{2+} are based either on KCl recrystallized from 0.1 mM EDTA or on sucrose or mannitol stock solutions which have been passed 3 times through a protonated Chelex 100 column to remove divalent cations. To deplete mitochondria of exchangeable Ca^{2+}, a sample of the concentrated stock suspension (40–60 mg protein/ml) is diluted to 10 mg/ml in a low-Ca^{2+} medium comprising 210 mM mannitol, 70 mM sucrose, and 10 mM N-2-hydroxyethylpiperazine-N'-ethanesulfonic acid (HEPES)–Tris, pH 7.0 (or some equivalent buffer) with 1 $\mu g/ml$ each of rotenone and oligomycin. After a 10-min incubation at 20°–30° this suspension is further diluted with medium to 5 mg/ml and the mitochondria sedimented by centrifugation. Finally, the pellet is resuspended to the original concentration (~40 mg/ml) in low-Ca^{2+} medium and stored on ice.

In the procedure above there may be a risk of deterioration of mitochondrial integrity during the incubation under deenergized conditions. The following procedure avoids the use of inhibitors and can be used for mitochondria that exhibit Ca^{2+}–Na^{+} exchange activity.[55] The sample of concentrated stock solution is diluted with approximately 75 ml medium comprising 124 mM KCl, 10 mM NaCl, 8 mM HEPES, 100 μM phosphate, 1 mM EGTA, 4 mM potassium succinate, and 100 μM MgCl$_2$ at 21°–23°. After 10–12 min of constant stirring, the suspension is placed in an ice-water bath with stirring for 5 min. The mitochondria are harvested by centrifugation, washed by resuspension in 0.3 M sucrose and 6 mM HEPES, and recentrifuged. After one further washing cycle a new stock suspension is prepared as above. The calcium load of liver mitochondria is reduced by this procedure to a level of 2–3 nmol/mg.

2. The uniporter activity may be stimulated by incubation of mitochondria in the presence of low amounts of Ca^{2+} that may leak from mitochondria on anaerobiosis.[56] Gentle continuous stirring of the mitochondrial stock suspension will keep it aerated.

3. Ca^{2+} uptake can be followed with metallochromic indicators suitable for the concentration range used. These have faster response times, but mixing may still be a rate-limiting factor. Scarpa and Azzone[52] calculated the rate of Ca^{2+} uptake by measuring the corresponding charge-compensating rate of K^{+} efflux with the aid of a K^{+}-sensitive electrode.

4. The valinomycin-induced membrane potential eventually decays because (a) efflux of K^{+} lowers the matrix/extramitochondrial K^{+} gradient and (b) the H^{+} conductance increases at high Ca^{2+} loads. One may use respiration combined with valinomycin to keep the membrane potential as stable as possible. It is preferable to use mitochondrial suspensions as dense as the equipment for mixing and recording allows because of the increased risk of artifacts arising from excessive Ca^{2+} loads. Above room temperature the rate of Ca^{2+} uptake may be so fast that the initial linear phase is missed, and a substantial part of the added Ca^{2+} may already have been accumulated before the recording of uptake has commenced. With high Ca^{2+} loads it may then be necessary to use stopped-flow equipment. The amount of Ca^{2+} taken up before the recording has commenced can be estimated by

[55] D. E. Wingrove and T. E. Gunter, *J. Biol. Chem.* **261**, 15159 (1986).
[56] H. Kröner, *Arch. Biochem. Biophys.* **251**, 525 (1986).

comparison with the trace obtained in the presence of 1 nmol ruthenium red/mg mitochondrial protein to inhibit uptake.

5. In electrolyte-based media without sucrose, choline chloride can be substituted for the 100–130 mM KCl normally used.

Ca²⁺-Induced Ca²⁺ Release

Though mitochondria containing substantial amounts of Ca^{2+} can still buffer external cation to a certain set point efficiently, there is a limit to how much Ca^{2+} can be tolerated. Once this tolerance limit has been exceeded, accumulated Ca^{2+} will be released concomitant with a drop in the membrane potential, and swelling occurs.[4] This aspect of mitochondrial Ca^{2+} function may be studied either by titrating how much Ca^{2+} can be accumulated before Ca^{2+} release takes place,[29] or by subjecting the mitochondria to a critical Ca^{2+} load and measuring the time it takes for Ca^{2+} release to begin. This latter approach is more sensitive and is described below.

First, the critical Ca^{2+} load needed to trigger Ca^{2+} release after a suitable lag time (30–120 sec) is established. Increasing amounts of Ca^{2+} are added to respiring mitochondria suspended in a standard medium containing a respiratory substrate. Ca^{2+} uptake and release are followed with the aid of metallochromic indicators or a Ca^{2+}-selective electrode. When the appropriate load has been found, the effects of varying experimental conditions can be studied. Figure 3 shows experiments relating to the effects of Sr^{2+},

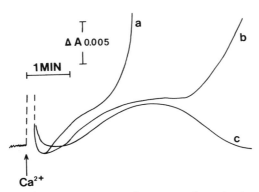

FIG. 3. Sensitivity of mitochondria to a Ca^{2+} load. Rat liver mitochondria (1 mg/ml) were incubated in a sucrose–choline chloride medium as in Fig. 2 in the presence of 4 mM succinate-Tris, 6 μM rotenone, and 40 μM arsenazo III at room temperature (23°). At the point indicated 60 μM Ca^{2+} was then added either alone (trace a), with 30 μM Sr^{2+} (b), or with 60 μM Sr^{2+} (c). Sr^{2+} delayed and inhibited Ca^{2+}-induced release. The wavelength couple 665–685 nm was used.

which is retained well by mitochondria.[35] Arsenazo III (40 μM) reports the uptake and release of added Ca^{2+}. Under the experimental conditions used a load of 60 μM Ca^{2+} (60 nmol/mg) was taken up with a small overshoot (trace a), then the set point drifted toward a slightly higher concentration in the medium. Finally after 2 min an accelerating rate of Ca^{2+} release set in. When 30 μM Sr^{2+} was added with the Ca^{2+} (trace b) the release was delayed for about 4 min. With 60 μM Sr^{2+} (trace c) mitochondria were able to recover the original set point. Not only does Sr^{2+} fail to cause an overload itself, but it protects against the Ca^{2+}-induced release.

Notes

1. The amount of Ca^{2+} per milligram mitochondrial protein is critical. For reproducible results it is essential to pipette the mitochondrial suspension and the Ca^{2+} stock solution with precision.

2. It is important to keep strictly to the experimental protocol in regard to preincubation times, time intervals between additions of various reagents, etc.

3. As the mitochondria age, the amount of Ca^{2+} they can retain progressively decreases. The control experiment should therefore be repeated at intervals.

4. Ruthenium red can be added at various times after the accumulation of Ca^{2+} in order to measure the effluxes. Efflux increases with time elapsed since addition of Ca^{2+}.

5. A drop in membrane potential is a sensitive index of Ca^{2+} overload. It can conveniently be followed by dual-wavelength photometry using safranine[57] or with a TPP^+ (tetraphenylphosphonium) electrode.[58,59]

Use of Sr^{2+} as an Analog of Ca^{2+}

Although Ca^{2+} is the cation of physiological interest it is not necessarily that best suited for studying the uniporter. One complication may be Ca^{2+} overload leading to loss of the membrane potential and the Ca^{2+} taken up. Under some experimental conditions the mitochondria may become more susceptible to Ca^{2+}, and the lowered net accumulation of Ca^{2+} can be misinterpreted as a reduced uniporter activity. Since Sr^{2+} does not induce a state of increased permeability of the inner membrane[35] and is readily transported on the uniporter, it can be a useful Ca^{2+} analog.

Figure 4 shows transport of Ca^{2+} and Sr^{2+} by rat liver mitochondria that have been subjected to hypotonic shock in a procedure designed to extract

[57] K. E. O. Åkerman and M. K. F. Wikström, *FEBS Lett.* **68,** 191 (1976).
[58] N. Kamo, M. Muratsugu, R. Hongoh, and Y. Kobatake, *J. Membr. Biol.* **49,** 105 (1979).
[59] D. G. Nicholls, this volume [1].

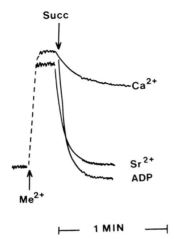

FIG. 4. Use of Sr^{2+} as a substrate for the uniporter. Rat liver mitochondria that had been subjected to a hypotonic shock were tested for their ability to accumulate Ca^{2+} and Sr^{2+} added at the point indicated (Me^{2+}). For experimental details, see text. Sr^{2+} was quickly taken up, but net uptake of Ca^{2+} was low unless the medium contained ADP (100 μM) to inhibit efflux. Succ, Succinate.

the putative Ca^{2+} carrier.[60] The preparation (1.5 mg protein/ml) supplemented with 3 μg/mg cytochrome c was suspended in a medium containing 220 mM mannitol, 10 mM sucrose, 10 mM HEPES–Tris, pH 7.4, 6 mM succinate-Tris, 1 mM P_i–Tris, 5 nM rotenone, 1 μg/mg oligomycin, and 30 μM arsenazo III. Additions of Ca^{2+} and Sr^{2+} were 50 μM. Net Ca^{2+} uptake by the extracted mitochondria was indeed strongly reduced while the rate of Sr^{2+} uptake remained high, indicating that uniporter activity was still present in the inner membrane. In the presence of 20 μM ADP to inhibit Ca^{2+} release, net Ca^{2+} uptake was also stimulated.

Ca^{2+} – 2Na^+ Exchange Activity

Ca^{2+} efflux is conveniently studied by inhibiting the uniporter with ruthenium red and measuring the efflux with any of the available Ca^{2+}-sensing methods.[10] Figure 5 shows an experiment in which brain mitochondria (0.5 mg protein/ml) suspended in a medium containing 130 mM KCl, 10 mM HEPES–Tris, pH 7.0, 30 μM arsenazo III, and 1 nmol/mg rotenone were allowed to take up Ca^{2+} before 1 nmol/mg ruthenium red was added. Inhibition of the uniporter revealed the endogenous efflux.

[60] G. Sandri, G. L. Sottocasa, E. Panfili, and G. F. Liut, *Biochim. Biophys. Acta* **558**, 214 (1979).

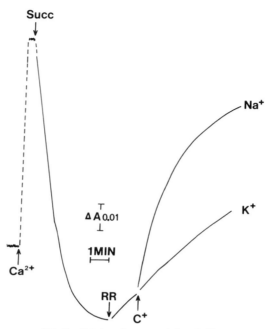

FIG. 5. Measurement of $Ca^{2+}-2Na^+$ exchange activity. Ca^{2+} (15 μM) was added to rotenone-inhibited rat brain mitochondria, followed by succinate (succ) to start uptake and ruthenium red (RR) to stop it. The efflux which ensued was accelerated by 15 mM NaCl (Na^+) but not by addition of 13 mM K^+ to the KCl-based medium. Antipyrylazo III was used as the reporter, and absorbance was recorded at 720–790 nm. For details, see text.

Addition of 15 mM Na^+ markedly stimulated this efflux by activating the exchanger.

$Ca^{2+}-2Na^+$ exchange activity can also be studied by adding Na^+ to mitochondria that have established a steady-state Ca^{2+} distribution (Fig. 6). The set point for external Ca^{2+} will be reestablished at a higher level because of stimulated efflux.

Notes

1. If metallochromic indicators are used one should bear in mind that the sensitivity of the indicator may be decreased by Na^+.

2. Ruthenium red adsorbs avidly to glass and plastic surfaces, and so to obtain reliable final concentrations it should be added directly to the mitochondria in the reaction medium from a freshly prepared stock solution. If disposable cuvettes are not used residual dye can be removed at the end of the experiment by rinsing the vessel with a mitochondrial suspension (3–5 mg/ml).

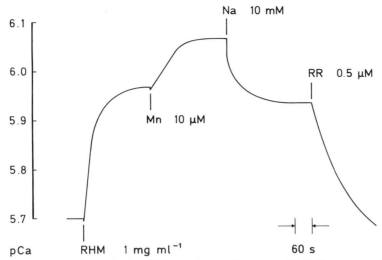

FIG. 6. Response of the mitochondrial set point to adjustment of Ca^{2+} flux, measured with a calcium-selective electrode. Rat heart mitochondria (RHM) were added to a medium containing succinate. For details, see text. Additions of MnCl$_2$, NaCl, and ruthenium red (RR) were made as indicated.

3. In order to approach physiological conditions it may be necessary to lower the Ca^{2+} content of the mitochondria, preferably without deenergization (see Comment 1 under "Measurements with Calcium-Selective Electrodes").

Mitochondrial Set Point

The net distribution of Ca^{2+} across the mitochondrial inner membrane reflects a kinetic equilibrium between simultaneous influx and efflux.[4,5] Measurements of this set point *in vitro* are useful in estimating the likely distribution of Ca^{2+} between the cytosol and mitochondrial matrix in intact cells, provided that the total Ca^{2+} contents of the two systems are similar. Pending direct flux measurements, observations of how the mitochondrial set point responds to particular interventions also provide a sensitive indication of altered Ca^{2+} cycling, i.e., efflux of Ca^{2+} by one or another pathway and reuptake through the uniporter. The electrode technique is particularly suitable for measurements of set point, and electrodes should be calibrated with buffered standards over the pCa range 5.5–7.0.

Isolated Mitochondria. After electrode response has been calibrated, 2 ml of a medium containing 120 mM recrystallized KCl, 10 mM MOPS–Tris (pH 7.0), 1 mM P$_i$–Tris, 1.5 mM MgCl$_2$, 2 mM succinate-Tris, 1 μM rotenone, and 1 μg/ml oligomycin ($\sim$2.5 μM) is equilibrated at

30° in the reaction vessel. Following addition of Ca^{2+}-depleted rat liver mitochondria (1 mg/ml), Ca^{2+} in the medium is reduced to about 1 μM. If 1 μM Ca^{2+} or EGTA is added at this point the mitochondria accumulate or release Ca^{2+} so that their set point is regained. Alternatively, stimulation of Ca^{2+} influx with 10 μM Mn^{2+} [8,61] or 0.1 mM spermine,[9] or inhibition of efflux with 10 μM Sr^{2+},[12] causes a shift of the set point so that Ca^{2+} is maintained at a new, lower level in the medium. Conversely, inhibition of influx with small amounts of ruthenium red (0.04 nmol/mg) or further Mg^{2+}, or stimulation of efflux with 10 mM Na^+ (most apparent in the case of mitochondria from excitable tissues), causes the mitochondria to maintain Ca^{2+} in the medium at higher levels. Examples of such shifts in the set point of heart mitochondria are shown in Fig. 6.

Permeabilized Hepatocytes. For measurement of the mitochondrial set point in cells, advantage is taken of the high cholesterol content of the plasma membrane relative to that of most internal membranes. Addition of small amounts of the detergent digitonin renders the plasma membrane permeable to Ca^{2+},[62] and equilibration between the medium and the cytosol occurs. Thus, external reporters such as a calcium-selective electrode or metallochromic indicator will measure Ca^{2+} maintained by intracellular transport systems. While this model does not provide direct information about cytosolic Ca^{2+} in intact cells or the role played by Ca^{2+} transport across the plasma membrane, it does allow the relative contributions of mitochondria and reticulum to intracellular Ca^{2+} homeostasis to be examined.

Freshly isolated hepatocytes[62] are slowly added to 1.5 ml medium equilibrated at 30°, comprising 120 mM KCl, 10 mM MOPS–Tris (pH 7.0), 1 mM $MgCl_2$, 0.5 mM P_i–Tris, 6 mM succinate-Tris, 1 μM rotenone, and 1 μg/ml oligomycin. Stirring should be gentle ($<$100 rpm) to minimize shearing of the cells. Then 5-μl aliquots of digitonin from a 0.1% (w/v) ethanolic stock solution are added until permeabilization of the plasma membrane is indicated by a drop in Ca^{2+} in the medium to a new level at about 0.5 μM. This latter level probably represents the mitochondrial set point because it responds to manipulation of mitochondrial fluxes. Ca^{2+} in the medium further falls to 0.1–0.2 μM after addition of 0.35 mM ATP-Mg to energize accumulation by nonmitochondrial pools, mainly the endoplasmic reticulum (Fig. 7). This ATP-dependent pool is easily saturated as more Ca^{2+} is added, after which the Ca^{2+} level in the medium reverts to the ATP-independent mitochondrial set point. However, if 0.5 mM sper-

[61] A. Allshire and N.-E. L. Saris, *in* "Manganese in Metabolism and Enzyme Function" (V. L. Schramm and F. C. Wedler, eds.), p. 51. Academic Press, Orlando, Florida, 1986.
[62] E. Murphy, K. Coll, T. L. Rich, and J. R. Williamson, *J. Biol. Chem.* **255**, 6600 (1980).

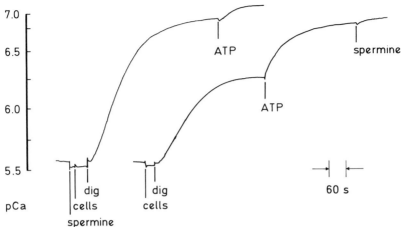

FIG. 7. Measurement of Ca^{2+} fluxes in permeabilized hepatocytes with a calcium-selective electrode. Hepatocytes suspended in a KCl-based medium in the presence of oligomycin at 30°C were made permeable with digitonin. The set points determined in mitochondria with or without spermine are compared with the set point established when ATP-dependent transport systems, mainly in the endoplasmic reticulum, are activated. For details, see text. Additions were as follows: cells, hepatocytes (5 mg protein/ml); dig, 0.002% digitonin; ATP, 0.35 mM ATP-Mg; spermine, 1 mM.

mine is present the mitochondria maintain Ca^{2+} at about 0.2 μM, and little or no further change occurs in response to ATP. Finally, addition of 2 μM, FCCP followed by the Ca^{2+} ionophore A23187 (5 μM) allows exchangeable Ca^{2+} in the mitochondrial and nonmitochondrial pools, respectively, to be estimated.

[9] Measurement of Proton Leakage across Mitochondrial Inner Membranes and Its Relation to Protonmotive Force

By DAVID G. NICHOLLS and EDUARDO RIAL

The efficient transduction of energy from the respiratory chain to the ATP synthase in mitochondria is dependent on the maintenance of a low proton conductivity for the mitochondrial membrane as a whole, thus forcing the bulk of the extruded protons to reenter the mitochondrial matrix through the ATP synthase with the concomitant synthesis of ATP.

In one specific case (thermogenically active brown fat) the physiological functioning of the mitochondria actually requires a raised proton conductance. Thus a specific 32-kDa "uncoupling protein" is induced as an integral inner membrane protein allowing the regulatable dissipation of the protonmotive force (pmf), enabling respiration to proceed uncontrolled by limitations of respiratory control.[1]

However, all mitochondria, even those lacking the uncoupling protein, possess a detectable proton leak, which together with other dissipatory ion cycles is responsible for their State 4 respiration. This chapter deals with two techniques for studying this leak. One, which is mainly qualitative and only applicable at low pmf, follows the rate of swelling of nonrespiring mitochondria driven by diffusion potentials under conditions where swelling is limited by the proton permeability of the membrane. The second technique, which is applicable at high pmf, is based on the influence the membrane proton conductance has on the rate of State 4 respiration.

It is important to distinguish between proton permeability (P_{H^+}), proton current (J_{H^+}), and proton conductance (Cm_{H^+}). A simplistic application of Ohm's law to the mitochondrial membrane allows proton conductance to be defined as the proton current flowing across the membrane per unit of driving force,

$$Cm_{H^+} = J_{H^+}/\text{pmf} \tag{1}$$

where the units are nmol min^{-1} mg^{-1} mV^{-1} for the conductance, nmol min^{-1} mg^{-1} for the current, and mV for the pmf. Where the driving force is ill defined, or varies during the course of the experiment, it is not appropriate to use such quantitative terminology, and the term proton permeability (P_{H^+}) may be used to describe the rate of proton reentry without defining the driving force.

Proton Permeability by Swelling of Nonrespiring Mitochondria

For mitochondria to swell in isosmotic salt media, it is necessary first that both the cation and anion of the major species in the medium are permeable, and second that there should be an overall charge and pH balance during the entry of solute into the matrix. These constraints enable two distinct conditions to be defined where both the cation and anion are permable, but charge and pH balance can only be maintained by H$^+$ diffusion across the membrane. First, electroneutral exchange of K$^+$ for protons, catalyzed by nigericin in combination with a uniport entry of anion (e.g., CNS$^-$), can proceed only if H$^+$ exchanging for K$^+$ can reenter

[1] D. G. Nicholls and R. M. Locke, *Physiol. Rev.* **64**, 1 (1984).

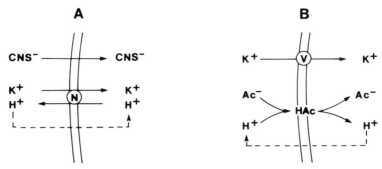

FIG. 1. Ion movements during ionophore-induced mitochondrial swelling. (A) KCNS plus nigericin (N); (B) potassium acetate plus valinomycin (V).

the matrix (Fig. 1A). Alternatively, uniport entry of K^+ with valinomycin, coupled with electroneutral anion entry (e.g., acetate) requires the stoichiometric diffusion of H^+ out of the matrix (Fig. 1B).

Swelling may be followed by the decrease in light scattering of incubations of mitochondria, utilizing either a spectrophotometer whose photodetection system is not immediately adjacent to the cuvette, or a fluorimeter where light scattered through 90° or 135° can be detected. The wavelength chosen should be greater than 500 nm for maximum sensitivity and should avoid major absorption bands, although the changes in directly transmitted light are far larger than those due, for example, to cytochrome absorption. The change in scattering is due to the decrease in refractive index difference as the matrix fills with the external medium. A detailed quantitative treatment is found in Ref. 2. To a first approximation the rate of swelling is linearly related to the rate of light-scattering decrease. In the following, the proton permeability of brown fat mitochondria is determined using an Eppendorf Model 1101M photometer with type 1030 fluorescent attachment. However, the technique may readily be adapted to other systems.

The primary filter (546 nm) and the secondary filter within the fluorimetric attachment (>430 nm) allows any light scattered through 135° to be detected. It is essential that a neutral density filter be incorporated into the secondary filter to cut down the light impinging on the photomultiplier. The output from the photometer is suitably attenuated and backed off for coupling to a chart recorder.

Mitochondria (0.3 mg protein/ml incubation) are suspended in 1 ml of 100 mM potassium acetate, 5 mM K^+–TES, 16 μM albumin, and 1 μM

[2] A. D. Beavis, R. D. Brannan, and K. D. Garlid, *J. Biol. Chem.* **260**, 13424 (1985).

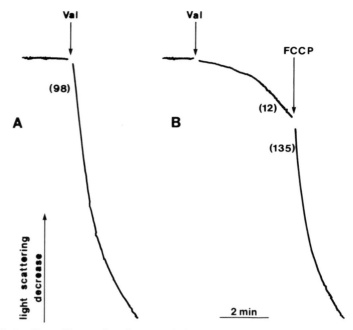

FIG. 2. Swelling of brown fat mitochondria in potassium acetate. Swelling is initiated by the addition of 0.5 μM valinomycin (Val) in the absence (A) or the presence (B) of 0.2 mM GDP. Swelling rates (arbitrary units) are in parentheses.

rotenone, pH 7.0, 30° in a 0.5-cm path-length cuvette. Rotenone is present to inhibit the oxidation of any endogenous substrates which would otherwise extrude protons and interfere with the stoichiometric relationship between proton leakage and swelling. The protein concentration and the dimensions of the cuvette are optimal for reliable light-scattering changes.

Figure 2 shows the results obtained when hamster brown fat mitochondria are incubated in the presence and absence of GDP. The nucleotide is a specific inhibitor of the uncoupling protein.[3] Little swelling occurs at this stage since the entry of K$^+$ is slow in the absence of valinomycin. To initiate proton-limited swelling, 0.5 μM valinomycin is rapidly mixed into the cuvette. Speed is essential here, as the swelling is rapid, and linearity may be maintained for only a few seconds, depending on the magnitude of the proton permeability. The linear portion of the swelling curve is used to define the swelling rate in arbitrary units. Figure 2B shows the effect of 0.2 mM GDP on the swelling rate: GDP inhibits proton efflux, and very

[3] D. G. Nicholls and O. Lindberg, *Eur. J. Biochem.* **31**, 526 (1973).

little swelling occurs; the addition of the protonophore FCCP (carbonyl cyanide-*p*-trifluoromethoxyphenylhydrazone) bypasses the uncoupling protein, and fast swelling takes place.

Proton Conductance from State 4 Respiration

The disadvantage with light-scattering studies of nonrespiring mitochondria is that the driving forces are low, ill defined, and variant (as the ionic inbalance across the membrane decreases during osmotic swelling). It is therefore an essentially qualitative technique which allows permeabilities (P_{H^+}) to be compared, rather than conductance (Cm_{H^+}) to be determined. Also, no information is given as to the behavior of the H^+ leaks at high (physiological) pmf. Finally, it is not possible to compare directly mitochondria from different sources which may differ in size or shape.

The proton conductance Cm_{H^+}, defined as in Eq. (1), requires a simultaneous measurement of J_{H^+} and pmf. The methodologies for pmf determination have been reviewed in previous volumes.[4,5] It is important that the respiratory and pmf measurements are carried out under the same conditions; thus, if valinomycin is present in order to determine $\Delta\psi_m$ (the mitochondrial membrane potential), the ionophore should also be present in the oxygen electrode.

Under steady-state respiring conditions, proton extrusion by the respiratory chain exactly balances the rate at which protons reenter the matrix. After eliminating ATP synthesis, Ca^{2+} cycling (e.g., in the presence of EGTA), and significant proton-linked metabolite transport across the inner membrane, essentially all the proton reentry occurs by electrophoretic leakage across the membrane or, in the case of brown fat mitochondria, through the uncoupling protein. By simply measuring the rate of respiration, one can calculate the inwardly directed proton current flowing across the membrane thus:

$$J_{H^+} = dO/dt \times H^+/O \qquad (2)$$

where dO/dt is the respiratory rate and H^+/O the stoichiometry of proton extrusion by the respiratory chain with the appropriate substrate.

Some typical values for Cm_{H^+} for brown fat mitochondria are shown in Table I. Note that activation of the uncoupling protein by omission of the inhibitory purine nucleotide GDP results in an increase in Cm_{H^+} manifested as a decrease in pmf and an increase in respiration (and hence the proton current).

[4] N. Kamo, T. Racanelli, and L. Packer, this series, Vol. 46, p. 356.
[5] J. B. Jackson and D. G. Nicholls, this series, Vol. 127, p. 557.

TABLE I
TYPICAL VALUES OF BIOENERGETIC PARAMETERS IN BROWN FAT MITOCHONDRIA
RESPIRING WITH α-GLYCEROPHOSPHATE[a]

	pmf (mV)	dO/dt (nmol min^{-1} mg^{-1})	J_{H+} (nmol min^{-1} mg^{-1})	Cm_{H+} (nmol min^{-1} mg^{-1} mV^{-1})
−GDP	55	142	850	15.5
+GDP	218	31	186	0.85

[a] From Nicholls.[6] α-Glycerophosphate (10 mM) was present as substrate.

The rest of this chapter deals with methodologies for quantifying Cm_{H+} as a function of pmf. Two important points must be borne in mind. First, a substrate must be chosen which can generate a sufficient pmf to create a significantly raised conductance. In practice this means that the substrate should donate all, or at least half, of its electrons to the UQ region of the respiratory chain, rather than to NADH dehydrogenase, since Complex I becomes kinetically limited at a lower pmf than the remainder of the respiratory chain.[6] Second, the means of varying the pmf must not by itself alter the membrane conductance. This in practice means that the rate of transfer of electrons into the respiratory chain must be controlled, by rate-limiting infusion of a high-affinity substrate (such as acylcarnitine)[7]; by addition of nonsaturating concentrations of a low-affinity substrate (such as α-glycerophosphate)[6]; by titration with a dehydrogenase inhibitor (e.g., malonate control of succinate respiration)[8]; or by restricting the removal of products (e.g., palmitoyl carnitine oxidation by guinea pig brown fat mitochondria in the presence of limiting malate).[6]

Isotopic Determination of J_{H+} – pmf Plots by Titration with Substrate or Addition of Dehydrogenase Inhibitor

Brown fat mitochondria have an active α-glycerophosphate dehydrogenase, which in the absence of Ca^{2+} has a low affinity for its substrate,[9] permitting the rate of respiration to be varied by adding α-glycerophosphate in the range 0.2–10 mM. If pmf is determined in parallel, using ^{86}Rb, [^{14}C]methylamine, and [^{3}H]acetate as previously described in detail,[8] then the conductance can be calculated as a function of pmf.

Mitochondria from cold-adapted guinea pig brown fat are incubated at

[6] D. G. Nicholls, *Eur. J. Biochem.* **77**, 349 (1977).
[7] E. Rial, A. Poustie, and D. G. Nicholls, *Eur. J. Biochem.* **137**, 197 (1983).
[8] D. G. Nicholls, *Eur. J. Biochem.* **50**, 305 (1974).
[9] L. Buckowiecki and O. Lindberg, *Biochim. Biophys. Acta* **348**, 115 (1974).

0.5 mg protein/ml at 23° in a medium containing 100 mM sucrose, 10 mM TES (Na salt), 0.5 mM KCl, 0.5 mM EDTA (Na salt), 32 μM albumin, 0.5μM valinomycin, 2 μM rotenone, 50 μM [86]Rb (0.05 μCi/ml), 50 μM [14C]methylamine (0.1 μCi/ml), 0.2 mM [3H]acetate (1 μCi/ml), and α-glycerophosphate at final concentrations varying from 0.2 to 10 mM. Respiration and pmf are determined in parallel following a 2-min incubation. Protonmotive force is determined by filtration of aliquots through Sartorius 0.6 μm cellulose nitrate filters, and calculation of the ΔpH and $\Delta\psi_m$ components of the pmf is exactly as described previously.[8] Alternatively, mitochondria can be centrifuged through silicone oil to provide a more complete exclusion of the incubation medium from the pellet.[8]

The experiment shown in Fig. 3 was designed to quantify the ability of exogenous GDP to inhibit the proton conductance through the brown fat uncoupling protein. In the absence of the nucleotide there is a high, ohmic conductance, whereas in the presence of 0.2 mM GDP the conductance is very low until a pmf in excess of 220 mV is attained.

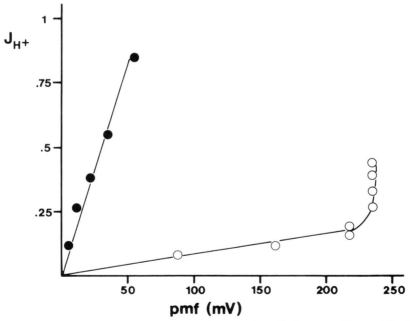

FIG. 3. Effect of GDP on the proton conductance through the uncoupling protein. For experimental details, see text. (●) GDP absent; (○) GDP present at 0.2 mM. Adapted from Nicholls.[6]

Electrode Determination of pmf–J_{H^+} Plots by Restricted Infusion of Substrate

In the presence of high concentrations of P_i, the ΔpH component of the pmf is small, and a reasonable approximation to the total electrochemical potential gradient is given by $\Delta\psi_m$ alone. This allows a tetraphenylphosphonium (TTP)-selective electrode to be employed to provide a continuous readout of $\Delta\psi_m$. Parallel monitoring of respiration allows Cm_{H^+} to be followed continuously.

The construction and use of a TPP electrode to monitor $\Delta\psi_m$ have been described earlier.[5] Again, the following description applies to an investigation of brown fat mitochondria, although the technique is equally applicable to mitochondria from any source.

Membrane potential and respiration are measured in a closed incubation chamber of 1.7 ml capacity fitted with a TPP-selective electrode, reference, and oxygen electrode.[7] A Braun ED2 Perfusor, modified to take a 10 μl Hamilton microsyringe, is connected to the entry port of the incubation chamber with a short length of fine tubing terminating in a stainless steel tube (Fig. 4). The infusion rate may be varied from 0.3 to 16 μl/min. The syringe is filled with 20 mM palmitoyl-L-carnitine dissolved in 1 : 1 (v/v) 250 mM TES (Na salt)–ethanol, pH 8.0. The incubation medium (pH 7.0, 30°) contains 50 mM KCl, 5 mM potassium phosphate, 2 mM EGTA (Na salt), 2 mM malonate (Na salt), 3 mM GDP (to inhibit the uncoupling protein), 16 μM albumin, 12 μg/ml catalase, 5 μM TPP, and 10 mM TES (Na salt).

In the absence of infused substrate, respiration and $\Delta\psi_m$ are both very

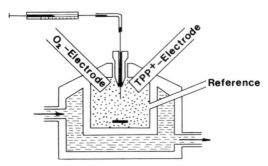

FIG. 4. Electrode assembly for simultaneous monitoring of $\Delta\psi_m$ and respiration. The cannula connected to the syringe allows the infusion of limiting substrate into the incubation chamber.

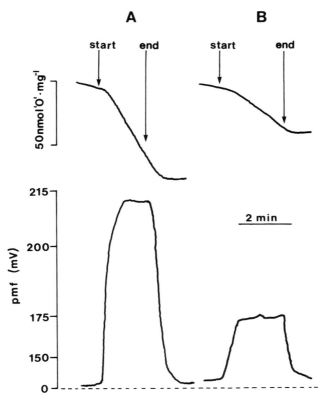

FIG. 5. Respiration and pmf determined during the rate-limiting infusion of palmitoyl-L-carnitine to an incubation of brown fat mitochondria. For experimental details, see text. Infusion rates: (A) 4.8 nmol min^{-1} mg^{-1}; (B) 1.6 nmol min^{-1} mg^{-1}. Adapted from Rial et al.[7]

low (Fig. 5). The ability to use palmitoyl-L-carnitine as a limiting substrate relies on the very high affinity of the mitochondria for this substrate, as seen by the sharp cutoff when a single addition of the substrate is made to the incubation. The experiment is carried out by selecting a rate of substrate supply on the infusion pump. Respiration and $\Delta\psi_m$ reach a steady state within 1 min (Fig. 5). The presence in the medium of catalase enables the incubation to be reoxygenated by the addition of low concentrations of hydrogen peroxide. Figure 5 shows typical results obtained with this method. Palmitoyl-L-carnitine infusion rates of 4.8 and 1.6 nmol min^{-1} mg^{-1} allowed steady values of $\Delta\psi_m$ and respiration to be attained, but it is

[10] V. S. M. Bernson and D. G. Nicholls, Eur. J. Biochem. 47, 517 (1974).

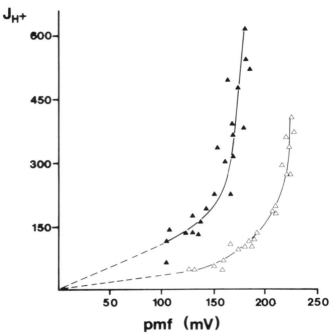

FIG. 6. Effect of palmitate on the J_{H^+}–pmf relationship for brown fat mitochondria. ($\triangle$) No palmitate present; ($\blacktriangle$) 1/3 μM unbound palmitate. For experimental details, see text. Adapted from Rial et al.[7]

notable that a 3-fold decrease in respiration, and hence proton cycling, leads to only a 18% decrease in potential.

On the basis of a H^+/O stoichiometry of 7 for the oxidation of palmitoyl-L-carnitine to acetate,[10] the respiratory rates can be used to calculate the rate of proton cycling across the membrane, in turn enabling J_{H^+} to be plotted as a function of $\Delta\psi_m$ (Fig. 6). The distinct nonohmic nature of the curve is apparent. Of considerable significance for the mechanism by which free fatty acids regulate the uncoupling protein *in situ* is the finding that, in the presence of 1.3 μM unbound palmitate, the nonohmic relationship is still present but is shifted down some 50 mV, to below the level required for respiratory control.

[10] Influence of Calcium Ions on Mammalian Intramitochondrial Dehydrogenases

By James G. McCormack and Richard M. Denton

Introduction

Many of the hormones which activate energy-requiring processes such as contraction or secretion in mammalian cells by causing increases in the cytoplasmic concentration of Ca^{2+} also bring about stimulation of mitochondrial oxidative metabolism.[1] There is now considerable evidence to suggest that the latter complementary response may be due to the activation of three key Ca^{2+}-sensitive intramitochondrial dehydrogenases as the result of accompanying increases in the concentration of Ca^{2+} in the mitochondrial matrix.[1-3] Thus, by extending the messenger role of Ca^{2+} in this way, mammalian cells appear to have evolved an associated mechanism to ensure that ATP synthesis is increased to meet the enhanced demand for ATP without the need to decrease cellular ATP concentrations.[1] Therefore, it is likely that the primary function of the Ca^{2+}-transport system of the mitochondrial inner membrane[4] in mammalian tissues under normal physiological conditions is to relay changes in cytoplasmic Ca^{2+} concentrations into the mitochondrial matrix and thus to control the concentration of Ca^{2+} in this compartment rather than, conversely, to buffer or set extramitochondrial Ca^{2+} concentrations as suggested by others.[5,6]

Ca²⁺-Sensitive Intramitochondrial Enzymes and Their Properties

The Ca^{2+}-activated intramitochondrial enzymes are the pyruvate (PDH), NAD⁺-isocitrate (NAD-ICDH), and 2-oxoglutarate (OGDH) dehydrogenases. Ca^{2+} activates PDH indirectly by causing increases in the amount of active, nonphosphorylated, PDH (PDH$_a$) through activation of PDH-phosphate phosphatase (PDHP-Pase),[7] whereas it activates NAD-

[1] R. M. Denton and J. G. McCormack, *Am. J. Physiol.* **249**, E543 (1985).

[2] J. G. McCormack and R. M. Denton, *Biochem. J.* **218**, 235 (1984).

[3] J. G. McCormack, *Biochem. J.* **231**, 597 (1985).

[4] N.-E. L. Saris and A. Allshire, this volume [8].

[5] G. Fiskum and A. L. Lehninger, *in* "Calcium and Cell Function" (W.Y. Cheung, ed.), Vol. 2, p. 38. Academic Press, New York, 1982.

[6] K. E. O. Åkerman and D. G. Nicholls, *Rev. Physiol. Biochem. Pharmacol.* **95**, 149 (1983).

[7] R. M. Denton, P. J. Randle, and B. R. Martin, *Biochem. J.* **128**, 161 (1972).

ICDH[8] and OGDH[9] more directly by causing decreases in their respective K_m values for *threo*-D_s-isocitrate and 2-oxoglutarate. (It should also be noted that Ca^{2+} may additionally activate PDH through inhibition of PDH_a kinase[10]; however, this enzyme has not been so intensively investigated as the other three in this respect.) In each of these cases the effective Ca^{2+} concentration range is approximately $0.1-10 \ \mu M$. Half-maximal effects ($K_{0.5}$ values) may vary slightly[11,12]; recent estimates in the presence of ADP were PDHP-Pase $0.77 \ \mu M$, NAD-ICDH $5.4 \ \mu M$, and OGDH $0.2 \ \mu M$.[12] Activations of severalfold can be achieved for each enzyme. Sr^{2+} can mimic Ca^{2+} action but at approximately 10-fold higher concentrations[7-9]; this is also true for Ba^{2+} except that it does not affect PDHP-Pase. These Ca^{2+}-sensitive properties have been found in extracts of all vertebrate tissues so far studied but not in nonvertebrates.[13]

The enzymes of the PDH system, NAD-ICDH, and OGDH are all located exclusively within mitochondria in mammalian cells. PDH, NAD-ICDH, and OGDH all catalyze irreversible oxidative decarboxylations to produce NADH, the principal substrate for the respiratory chain and ATP production. All of these steps are generally regarded to be key regulatory sites in mammalian oxidative metabolism[14] and can also be activated by increases in the key metabolite ratios of ADP/ATP and NAD^+/NADH.[9] (In addition both NAD-ICDH[15] and OGDH[9] can also be activated by decreases in pH in the range $7.4-6.6$.) However, this form of "intrinsic" regulation by local metabolite end products appears to be distinct from and largely independent of that exerted by Ca^{2+}, which can thus be viewed as "extrinsic" as it allows a means whereby agents such as hormones could override the intrinsic mechanisms.[16] Therefore, by using Ca^{2+} in this way ATP synthesis could be stimulated without the need for large changes in NAD^+/NADH or ADP/ATP concentration ratios.[16,17] Significantly these intrinsic control mechanisms can also be observed in extracts of nonvertebrate tissues whereas control by Ca^{2+} appears to be restricted to the enzymes from vertebrates.[13]

Details of the methods used to examine the properties (summarized above) of these enzymes in extracts or after purification can be found in the

---

[8] R. M. Denton, D. A. Richards, and J. G. Chin, *Biochem. J.* **176**, 899 (1978).

[9] J. G. McCormack and R. M. Denton, *Biochem. J.* **180**, 533 (1979).

[10] R. H. Cooper, P. J. Randle, and R. M. Denton, *Biochem. J.* **143**, 625 (1974).

[11] R. M. Denton, J. G. McCormack, and N. J. Edgell, *Biochem. J.* **190**, 107 (1980).

[12] G. A. Rutter and R. M. Denton, *Biochem. J.* **252**, 181 (1988).

[13] J. G. McCormack and R. M. Denton, *Biochem. J.* **196**, 619 (1981).

[14] J. R. Williamson and R. H. Cooper, *FEBS Lett.* **117**, K73 (1980).

[15] R. F. Colman, *Adv. Enzyme Regul.* **13**, 413 (1975)

[16] R. M. Denton and J. G. McCormack, *FEBS Lett.* **119**, 1 (1980).

[17] R. G. Hansford, *Rev. Physiol. Biochem. Pharmacol.* **102**, 1 (1985).

appropriate references given above. However, it has been the development of the ability to assay for the Ca^{2+}-sensitive properties of these enzymes within intact mitochondria that has led to the studies from which the main hypotheses given in the Introduction have been formulated, and details of these assays are given below. An essential prerequisite for many of these studies has been the development of methods whereby partially purified mitochondrial fractions can be isolated from rapidly disrupted mammalian tissues under conditions which minimize artifactual Ca^{2+} redistribution.

Mitochondrial Preparation

We have used the same overall strategy in preparing mitochondrial fractions from rat heart,[2] liver,[18] and white adipose tissue,[19] and the process should be applicable to most mammalian tissues. The tissues are rapidly disrupted using a precooled Polytron (Kinematica GmbH, CH-6010 Kriens-Luzern, Switzerland) homogenizer probe (0.5–2 cm diameter) (by employing two bursts of 2–3 sec duration at about Mark 4) into at least 4 vol of ice-cold medium. The medium used comprises 250 mM sucrose, 20 mM Tris (pH 7.4), and 2 mM EGTA (medium A), with the additional presence of 1% (w/v) defatted albumin for heart and liver homogenizations, or 3% defatted albumin and 7.5 mM reduced glutathione for the white adipose tissue homogenizations. The rapidity of homogenization which the Polytron homogenizer allows is especially important when studying the effects on intramitochondrial Ca^{2+} of the prior exposure of tissue preparations to hormones or other agents. Experiments using the addition of trace amounts of ^{45}CaCl$_2$ to the homogenization media have shown that the presence of 2 mM EGTA in at least 4 vol of ice-cold buffer is sufficient to prevent artifactual Ca^{2+} uptake into the mitochondria;[20] the additional presence of ruthenium red (a potent inhibitor of mitochondrial Ca^{2+} uptake[21]) is superfluous unless lower EGTA concentrations are employed.[22] Likewise, experiments employing ^{45}Ca-preloaded mitochondria which then underwent the entire preparation procedure again have established that very little mitochondrial calcium is lost during this process,[2,3] as even in the presence of Na$^+$ (which exchanges for intramitochondrial Ca^{2+} [21]) egress is very slow at 0°–4°; therefore, the additional presence of,

[18] J. G. McCormack, *FEBS Lett.* **180**, 259 (1985).
[19] S. E. Marshall, J. G. McCormack, and R. M. Denton, *Biochem. J.* **218**, 249 (1984).
[20] J. G. McCormack, unpublished observations (1984).
[21] E. Carafoli, *FEBS Lett.* **104**, 1 (1979).
[22] P. H. Reinhart, E. van de Pol, W. M. Taylor, and F. L. Bygrave, *Biochem. J.* **218**, 415 (1984).

e.g., diltiazem (an inhibitor of mitochondrial Na^+-Ca^{2+} exchange[23]) is not necessary unless it is desirable to have higher temperatures during isolation.

The tissue homogenates are diluted 2- to 3-fold with the same medium they were homogenized in, centrifuged at 1000 g for 90 sec, and then the mitochondria sedimented at 10,000 g for 5 min. Each pellet is suspended in 6 ml of medium A, and 1.4 ml of Percoll (Sigma, St. Louis, MO) is added and the mitochondria sedimented as a loose pellet at 10,000 g for 10 min.[24] The mitochondria are washed once by resuspending in medium A and recentrifugation and finally suspended in the same medium to give about 40–80 mg protein/ml. The yield of mitochondria (on the basis of total PDH activity) is approximately 20–40% by this method; however, the main advantage is that the inclusion of the Percoll step allows a purification of the mitochondrial fraction by 1.5- to 6-fold (as assessed by marker enzyme analysis[24-26]). In particular, contamination by endoplasmic reticulum, Golgi apparatus, and plasma membrane fractions is considerably diminished, thus allowing improved detection and confidence in the measurement of mitochondrial parameters. This is especially important in the measurement of the total Ca content after hormone treatment of liver and other tissues following the recent discovery that certain hormones may cause the release of Ca from endoplasmic reticulum fractions through the mediation of *myo*-inositol 1,4,5-trisphosphate.[26,27] Previously, the released Ca (which is responsible for the increases in cytoplasmic $[Ca^{2+}]$ brought about by the hormones) was thought by many workers, on the basis of total Ca measurements on crude fractions, to come from the mitochondria.[28,29]

Assays of Ca^{2+}-Sensitive Properties of Enzymes within Intact Mitochondria

Pyruvate Dehydrogenase System

The effects of intramitochondrial $[Ca^{2+}]$ on PDHP-Pase within intact mitochondria can be assayed indirectly by following increases in steady-state PDH_a content as the extramitochondrial $[Ca^{2+}]$ is increased, by em-

[23] P. L. Vághy, J. D. Johnson, M. A. Matlib, T. Wang, and A. Schwartz, *J. Biol. Chem.* **257**, 6000 (1982).

[24] G. J. Belsham, R. M. Denton, and M. J. Tanner, *Biochem. J.* **192**, 457 (1980).

[25] J. G. McCormack, *Biochem. J.* **231**, 581 (1985).

[26] F. Assimacopoulos-Jeannet, J. G. McCormack, and B. Jeanrenaud, *J. Biol. Chem.* **261**, 8799 (1986).

[27] M. J. Berridge and R. F. Irvine, *Nature (London)* **312**, 315 (1984).

[28] P. H. Reinhart, W. M. Taylor, and F. L. Bygrave, *Biochem. J.* **208**, 619 (1982).

[29] J. H. Exton, *Mol. Cell. Endocrinol.* **23**, 233 (1981).

ploying mitochondrial incubation conditions under which PDH$_a$ kinase activity is likely to be constant, i.e., conditions of high intramitochondrial ATP.[11,30]

Typical incubations of mitochondria are carried out in 1 ml volume at 30° for 5 min (sufficient to achieve steady state) and approximately 0.5 – 1 mg mitochondrial protein/ml in a medium comprising 125 mM KCl, 20 mM Tris (pH 7.3) [or else 20 mM MOPS (4-morpholinepropanesulfonic acid) at pH 7], and 5 mM potassium phosphate (medium B) together with appropriate respiratory substrates (e.g., 5 mM 2-oxoglutarate with 0.2 mM L-malate) to provide intramitochondrial ATP. Such experiments have allowed the demonstration of the Ca^{2+}-dependent activaton of PDH in intact mitochondria isolated from rat heart,[11,31] skeletal muscle,[32,33] white adipose tissue,[19] and brain[34] and also within permeabilized pig lymphocytes.[35]

With intact rat liver mitochondria[25] it was found useful to also add pyruvate, or its analog dichloroacetate (each at about 1 mM), or ADP (about 1 mM) with oligomycin (5 μg/ml), to inhibit PDH$_a$ kinase to some degree,[36] so that the effects of Ca^{2+} on PDHP-Pase could be more easily observed. This is similar to the situation with heart mitochondria from starved or diabetic rats,[36] and presumably it means that, in the absence of these inhibitors of PDH$_a$ kinase, activity of the kinase is greater than that of PDHP-Pase under these conditions; the effects of pyruvate and ADP/oligomycin appear to be additive.[25,36] The effects of Ca^{2+} on PDHP-Pase can also be demonstrated in mitochondria which are uncoupled,[11,25,30] by incubation in medium B (with no respiratory substrates) containing 1 μM FCCP (carbonyl cyanide p-trifluoromethoxyphenylhydrazone) together with ATP-Mg (2 – 5 mM) and oligomycin (5 μg/ml) to allow constant kinase activity, and also, though not always necessary, the ionophores A23187 (2 μg/ml) and valinomycin (1 μM) to allow free Ca^{2+} equilibrium across the inner mitochondrial membrane.

After the incubations the mitochondria are rapidly sedimented (10,000 g for 20 sec) in a microcentrifuge and quickly frozen in liquid N$_2$. Samples can be stored at −70° for several weeks before PDH assay.

The above approaches have some drawbacks, including the following (1) time courses are difficult and tedious to study; (2) both the phosphatase and kinase are usually active simultaneously so strictly only changes in the

[30] J. G. McCormack and R. M. Denton, *Biochem J.* **190,** 95 (1980).
[31] R. G. Hansford, *Biochem. J.* **194,** 721 (1981).
[32] B. Ashour and R. G. Hansford, *Biochem. J.* **214,** 725 (1983).
[33] S. J. Fuller and P. J. Randle, *Biochem J.* **219,** 635 (1984).
[34] R. G. Hansford and F. Castro, *Biochem. J.* **227,** 129 (1985).
[35] E. Baumgarten, M. D. Brand, and T. Pozzan, *Biochem. J.* **216,** 359 (1983).
[36] J. G. McCormack, N. J. Edgell, and R. M. Denton, *Biochem. J.* **202,** 419 (1982).

relative activities of the two interconverting enzymes can be observed; (3) often the exact intramitochondrial concentrations of potential regulators are unknown; and (4) the PDH complex is not catalyzing the conversion of pyruvate to acetyl-CoA as *in vivo*. Recently, toluene permeabilization of mitochondria has been used by Thomas and Denton[37] as a means of surmounting some of these shortcomings. As first described by Matlib *et al.*,[38] mitochondria treated under controlled conditions with toluene become permeable to all small molecules while retaining their structural integrity and full complement of intramitochondrial proteins. With such preparations, the intramitochondrial concentrations of substrates, coenzymes, and potential regulators can be easily manipulated and the kinetic behavior of PDHP-Pase and PDH$_a$ kinase studied individually under conditions very close to those occurring in intact cells. In addition, it is possible to assay the activity of PDH continuously and thus follow accurately the time courses of the effects of changing the activity of either the phosphatase or kinase. Full details of this approach are given in Refs. 37 and 37a.

Extraction of Pyruvate Dehydrogenase Activity from Mitochondria. The frozen mitochondrial pellets are extracted by disruption by passage up and down in a 500-μl microsyringe in about 150–300 μl of ice-cold medium comprising 100 mM potassium phosphate (pH 7.3), 2 mM EDTA, and 0.1% (v/v) Triton X-100 together with 5 mM mercaptoethanol and 50 μl of rat serum/ml added on the day of use, before being dropped into liquid N$_2$; the samples can be stored at $-70°$ or on dry ice before assay for up to 2 days after extraction. The EDTA ensures that PDH$_a$ is not altered as both PDH$_a$ kinase and PDHP-Pase required Mg^{2+} for activity,[39] the Triton disrupts the mitochondrial inner membranes and helps solubilize the enzyme, and the rat serum is a convenient means of preventing proteolysis of the PDH.[40] Prior to assay of PDH, the sample is carefully thawed and then may be centrifuged (10,000 g, 1 min) to precipitate broken mitochondrial membranes.

Basis of Assay of Pyruvate Dehydrogenase. The reaction scheme is given in Eqs. (1) and (2).

[37] A. P. Thomas and R. M. Denton, *Biochem. J.* **238**, 93 (1986).
[37a] P. J. W. Midgely, G. A. Rutter, A. P. Thomas, and R. M. Denton, *Biochem. J.* **241**, 371(1987).
[38] M. A. Matlib, W. A. Shannon, Jr., and P. A. Srere, *Arch. Biochem. Biophys.* **178**, 396 (1977).
[39] R. M. Denton, P. J. Randle, B. J. Bridges, R. H. Cooper, A. L. Kerbey, H. T. Pask, D. L. Severson, D. Stansbie, and S. Whitehouse, *Mol. Cell. Biochem.* **9**, 27 (1975).
[40] A. Lynen, E. Sedlaczek, and O. H. Wieland, *Biochem. J.* **169**, 321 (1978).

$$\text{Pyruvate} + NAD^+ + CoA \xrightarrow{PDH} CO_2 + NADH + \text{acetyl-CoA} \qquad (1)$$

$$\text{Acetyl-CoA} + AABS \xrightarrow{AAT} \text{acetylated AABS} + CoA \qquad (2)$$

AABS is p-(p-aminophenylazo)benzenesulfonic acid which can be obtained from Pfaltz & Bauer (Stamford, CT); its millimolar extinction coefficient is 6.5 at 460 nm wavelength, and it loses color on acetylation. A stock solution of the dye is prepared as the sodium or Tris salt in water at 1 mg/ml; this can be stored indefinitely at room temperature. AAT is arylamine acetyltransferase (EC2.3.1.5) and is prepared from pigeon liver acetone powder after the method of Tabor et al.[41] (see below).

To 1.5 ml of buffer [100 mM Tris, 0.5 mM EDTA, and 1 mM MgSO$_4$ (pH 7.8) containing 20 μl/ml of AABS solution and 0.2 μl/ml of mercaptoethanol added on the day of assay] in a cuvette add the following: (1) 20 μl of substrate mix (36 mg thiamin pyrophosphate, 23 mg NAD$^+$, 9 mg pyruvate, and 7.5 mg CoA dissolved in 1 ml H$_2$O; this can be stored at $-20°$ for up to 1 week); (2) 20 μl of AAT (equivalent to 60–100 mU); and (3) 10–200 μl of mitochondrial sample extract (equivalent to 1–10 mU). The decrease in optical density is then followed at 460 nm and usually 30°.

This method is preferable to simply following NADH production at 340 nm in the presence of rotenone [to inhibit the NADH oxidase (NADH dehydrogenase, ubiquinone; EC 1.6.5.3) activity in the extracts] as both NADH and acetyl-CoA (which will be removed) are potent inhibitors of PDH activity[39] and thus linear rates are achieved for longer with the present system. The assay of $^{14}CO_2$ production from [1-^{14}C]pyruvate also has the problem of acetyl-CoA buildup together with the further problems that many samples have to be taken to check the linearity of the rates, and there can be considerable rates of nonenzymatic decarboxylation of pyruvate (especially in the presence of compounds such as Triton X-100 and mercaptoethanol[42]). Solomon and Stansbie[43] have improved the sensitivity of the present assay up to 10 times by using the AAT to couple acetyl-CoA production to the acetylation of cresyl violet acetate as this causes a shift in this compound's fluorescence spectrum from $\lambda(ex)_{max} = 575$, $\lambda(em)_{max} = 620$ nm to $\lambda(ex)_{max} = 475$, $\lambda(em)_{max} = 575$ nm, and the rate of acetylated dye appearance can thus be followed fluorimetrically. It should also be noted that malate (e.g., carried over from mitochondrial incubations) may transiently interfere with the AAT-based assays because the mitochondrial extracts usually contain high activities of malate dehydrogenase and citrate

[41] H. Tabor, A. H. Mehler, and E. R. Stadtman, J. Biol. Chem. 204, 127, (1953).
[42] G. Constantopoulos and J. A. Barranger, Anal. Biochem. 139, 353 (1984).
[43] M. Solomon and D. Stansbie, Anal. Biochem. 141, 337 (1984).

synthase; there is therefore often a lag in the assay, though steady-state rates should be achieved after 1–5 min.

It is usual to express the amount of PDH existing as PDH$_a$ as a percentage of the total amount of PDH present in the mitochondrial extracts. This is assayed as above after converting all of the PDH-phosphate in the extract to PDH$_a$. This is achieved by incubating some of the mitochondrial extract (up to 50 μl) with 20–40 μl of pig heart PDHP-Pase (free from PDH$_a$ and PDH-phosphate, see below) in the presence of 25 mM MgCl$_2$ and 1 mM CaCl$_2$ (added from a 10 × concentrated stock solution to activate the phosphatase), at 30° until maximum activity is obtained (usually 5–15 min). The mitochondrial fractions prepared as above typically contain between 30 and 120 mU of total PDH/mg protein.

Preparation of Arylamine Acetyltransferase. Briefly, the acetone powder is prepared by homogenizing (using an Atomix or similar) fresh pigeon livers twice with 10 vol of cold (−10°) acetone, filtering with suction through Whatman No. 1 filter paper (in a Büchner funnel) with a dry-ice trap present. Then sufficient diethyl ether to cover the powder (still under vacuum) is added [the ether is pretreated to remove peroxides by passing through a column of aluminum oxide (Neutral, Grade 1)]. The disk of powder is crumbled and dried in air for 1–2 hr and then left in an evacuated desiccator overnight. The powder is stored at −70° and keeps for at least 1 year if well prepared. Note that we have successfully used livers of pigeons killed up to 8 hr previously which were then stored frozen at −70° (for up to 6 months) until sufficient quantities to make a worthwhile powder preparation were achieved. Commercially available pigeon liver acetone powders contain very low amounts of AAT activity and are thus unsuitable.

To prepare AAT the powder (usually 16 g) is homogenized using a Polytron homogenizer (2 cm diameter probe, 1 min, setting 4) in 10 vol of ice-cold H$_2$O containing 10 mg of EDTA per 100 ml before being centrifuged for 15 min at 10,000 g. Acetone (at −10° from a freshly opened bottle, Analytical Grade) is added drop by drop to the (well-stirred) supernatant over 15–30 min to achieve a total of 6.7 ml acetone for each 10 ml of supernatant. After centrifuging as before (but at −10°), a further amount of acetone equal to the volume used in the first stage is added dropwise to the supernatant before repeated centrifugation. The pellet is suspended in a minimum volume (about 4 ml) of 10 mM potassium phosphate (pH 7.5) and dialyzed for at least 2 hr at 0° against 2 liters of 10 mM potassium phosphate (pH 7). If it is cloudy after dialysis, the preparation can be centrifuged (10,000 g, 2 min); activity remains in the supernatant. The enzyme is stored at −70° and is stable for up to 6 months; between 5 and 15 U/ml should be achieved [where a unit (U) is the amount of enzyme converting 1 μmol of substrate to product per minute at 30°]. This represents a yield of about 40%; dilute in phosphate buffer to about 4–5 U/ml. To assay, use the buffer plus dye as for the PDH assay as described above plus 10 μl of 10 mM acetyl-CoA.

Preparation of Partially Purified Pig Heart PDHP-Pase. Preparation of PDHP-Pase is based on the procedure of Severson *et al.*[44] and is described for 8–10 pig hearts, as fresh as possible and kept in ice during transportation from the slaughterhouse. Hearts are trimmed of fat, blood vessels, and smaller chamber walls, cut into 1–2 cm cubes, and homogenized in a Waring blendor or equivalent for 1 min with 3–4 liters of buffer (30 mM potassium phosphate, 1 mM EDTA, pH 7.6, with 5 mM-mercaptoethanol added just before use). Check that the pH has not fallen below 7.0; restore with KOH if necessary. Centrifuge at 5000 g for 20–30 min at 4°. Pour the supernatant through cheesecloth and reextract the pellets with a further 3–4 liters of buffer. Adjust the combined supernatants to pH 5.4 at 4° with 10% (v/v) acetic acid. Centrifuge as before, then resuspend the pellets in 2 liters of ice-cold distilled water and again spin as before. Suspend the washed crude mitochondrial pellets in 0.5 liter of 20 mM potassium phosphate buffer (pH 7) containing 5 mM mercaptoethanol; adjust to pH 7 with KOH.

Transfer the material to four or five round-bottomed Pyrex flasks and break mitochondria by shell-freezing using a liquid N$_2$ bath and thawing 3 times. Then centrifuge the extract at 30,000 g for 2 hr (at 4°); carefully aspirate off the supernatant and incubate it at 30° with MgCl$_2$ added to 10 mM (from a 50 × concentrated stock) to allow conversion of any PDH-phosphate present to PDH$_a$ (monitor by assay as above, usually takes about 15–30 min). The preparation is then cooled (to 4°) and solid (NH$_4$)$_2$SO$_4$ (Analar grade) added slowly over 30 min to give finally 209 g/ liter. Then centrifuge at 10,000 g for 20 min at 4°, dissolve the resulting pellet in a minimum volume (20–30 ml) of 20 mM potassium phosphate, 5 mM mercaptoethanol buffer (pH 7.0, check pH), and dialyze overnight against 2 liters of the same buffer. Adjust the pH (at 0°–4°) to exactly 6.1 with 10% acetic acid and centrifuge at 15,000 g for 20 min (4°). Restore the pH of the supernatant to pH 7.0 with KOH and freeze in samples of 2–5 ml at −20°.

PDHP-Pase is less stable when separated from PDH. When required PDHP-Pase is separated from PDH by centrifuging twice for 1 hr at 150,000 g and 4°; the supernatant is checked to be PDH free and then stored at −20° (for up to 2 months) in small aliquots (100–400 μl) until use. The final preparation should be free of detectable PDH or PDH-phos-

[44] D. L. Severson, R. M. Denton, H. T. Pask, and P. J. Randle, *Biochem. J.* **140**, 225 (1974).

phate and be capable of generating about 5 U of PDH from added pig heart PDH-phosphate/mg protein. Alternative preparations of PDHP-Pase which are more complex and time-consuming but yield a more purified product have been published (e.g., Ref. 45), but these preparations offer little advantage for the assay of total PDH in extracts of tissue or mitochondria.

Oxoglutarate Dehydrogenase

Use of O_2 Electrode. The Ca^{2+}-sensitive properties of OGDH within intact mitochondria from rat heart,[11] kidney,[46] and both brown[30] and white[19] adipose tissue can be readily demonstrated by simply measuring O_2 uptake (with a Clark-type electrode) by the mitochondria when the restrictions on this imposed by respiratory-chain activity are overcome either by using uncoupled mitochondria or by adding an excess of ADP. This is because, in these mitochondria under these conditions, the activity of OGDH appears to be the rate-limiting step for the oxidation of added 2-oxoglutarate[11,19,30,46] and marked effects of Ca^{2+} on the oxidation of nonsaturating concentrations of 2-oxoglutarate can be readily observed; see Fig. 1 for an example of this and the conditions used. It should be noted that this approach is not appropriate for liver mitochondria.[25] With reference to the conditions used in Fig. 1, if rates of O_2 uptake are high in the absence of 2-oxoglutarate then the concentration of malate may be diminished or the malate replaced by malonate. The oxidation of nonsaturating and saturating concentrations of succinate (in the absence of malate) can be employed as a control[11] as its oxidation should not be affected by Ca^{2+}.

Use of Dual-Wavelength Spectrophotometer or Fluorimeter. The Ca^{2+}-sensitive properties of OGDH within intact rat heart[36,47] and liver[25] mitochondria have also been demonstrated by measuring the reduction of $NAD(P)^+$ induced by added 2-oxoglutarate, by using either a fluorimeter [$\lambda(ex)$ 360 nm, $\lambda(em)$ 460 nm] or a dual-wavelength spectrophotometer (wavelength pair 350–370 nm). Each machine is fitted with a thermostatically controlled cuvette unit in which the samples can be continuously stirred. An example of such an experiment using a dual-wavelength spectrophotometer is given in Fig. 2.

Effects of Ca^{2+} on heart mitochondria can be readily demonstrated in the absence or presence of either malonate or malate and with or without ADP or uncoupler, but with liver mitochondria more complex conditions

[45] W. M. Teague, F. H. Pettit, T.-L. Wu, S. R. Silberman, and L. J. Reed, *Biochemistry* **21**, 5585 (1982).

[46] P. Tullson and L. Goldstein, *FEBS Lett.* **150**, 197 (1982).

[47] R. G. Hansford and F. Castro, *Biochem. J.* **198**, 525 (1981).

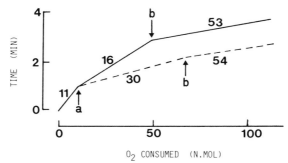

FIG. 1. Effects of Ca^{2+} on O_2 uptake by intact rat heart mitochondria oxidizing 2-oxoglutarate. Mitochondria (equivalent to 0.5 mg protein) were incubated at 30° in 1 ml of medium B (see text) containing 2 mM ADP and 1 mM malate and in the presence of either 5 mM EGTA (solid line; free $[Ca^{2+}]$ <1 nM) or 5 mM EGTA plus 2.5 mM $CaCl_2$ (dashed line; free $[Ca^{2+}]$ ~50 nM). Further additions, at the times indicated were as follows: a, 0.25 mM 2-oxoglutarate; b, 20 mM 2-oxoglutarate. The figures given indicate nanomoles of O_2 taken up per minute. Data are adapted from those in Ref. 11.

are used (see Fig. 2).[25] Respiratory chain or oxidative phosphorylation inhibitors, e.g., oligomycin (up to 5 μg/ml), antimycin (up to 2 μg/ml), or rotenone (up to 0.2 μg/ml), can be used at varying concentrations to enhance NAD(P)H buildup if desired. Conversely, ADP (2 mM), ATP-Mg (0.1–2 mM, as substrate for endogenous ATPase such that "State 3.5" respiration states can be achieved[25]), or NH_4Cl (up to 2 mM) can be used to dissipate the accumulation of NAD(P)H (see Ref. 25) to levels under which changes in NAD(P)⁺ reduction can be more readily observed. It is often advantageous to oxidize all endogenous NAD(P)H first as in Fig. 2. The reduction of NAD(P)⁺ by nonsaturating concentrations of glutamate can also be shown to be Ca^{2+}-sensitive[25] in this way (presumably through effects on OGDH), although the K_m values for glutamate, as expected, were higher than those for 2-oxoglutarate; however, NAD(P)⁺ reduction induced by nonsaturating (or saturating) concentrations of β-hydroxybutyrate or L-malate were unaffected by Ca^{2+} and can be employed as controls.[25]

Use of α-[1-¹⁴C]Oxoglutarate. The use of radiolabeled oxoglutarate was developed to show the effects of Ca^{2+} on OGDH in liver mitochondria incubated in the absence of ADP or uncoupler,[18,25] although it can also be used to demonstrate Ca^{2+} effects with other mitochondria, e.g., heart, in both the absence and presence of ADP or uncoupler.[20] Its main advantages are simplicity, specificity, and applicability to situations where OGDH activity is no longer rate-limiting for O_2 uptake.

Mitochondria are preincubated (up to 5 min) in medium B (above) at

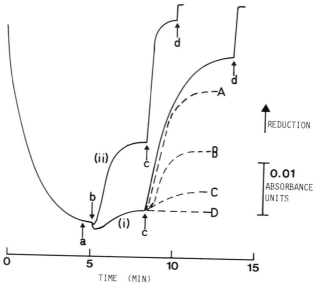

Fig. 2. Effects of Ca^{2+} on the extent of NAD(P)$^+$ reduction induced by 2-oxoglutarate in intact rat liver mitochondria. Mitochondria were incubated for the times shown at 30° and at approximately 1.5 mg protein/ml in 2 ml of medium B (see text) containing initially 0.5 mM EGTA, 0.5 mM malonate, and 0.5 mM NH$_4$Cl. At arrow a either (i) 2 mM EGTA or (ii) 2 mM EGTA plus 2 mM CaCl$_2$ (resultant free [Ca^{2+}] ~200 nM) was added except for the dashed traces, where no addition was made at arrow a. Other additions, at the arrows, were as follows: b, 200 μM 2-oxoglutarate; c, either (solid lines) 2 mM 2-oxoglutarate or (dashed lines) A, 5 mM EGTA plus 5 mM CaCl$_2$ (resultant free [Ca^{2+}] ~470 nM); B, 1.5 mM EGTA plus 1.5 mM CaCl$_2$ (~150 nM); C, 0.5 mM EGTA plus 0.5 mM CaCl$_2$ (~50 nM); or D, 0.5 mM EGTA alone; d, 0.5 μg rotenone. Note that 2-oxoglutarate itself absorbed at the wavelength pair used but that this has been omitted for clarity of presentation. Data are adapted from those in Ref. 25.

0.5–1 mg protein/ml and 30° and then duplicate 100-μl samples of mito-chondrial suspension are added to small soda-glass test tubes (30 × 6 mm) which are resting in 0.5 ml of 2-phenylethylamine in the base of a standard glass scintillation vial. The vial is then sealed with a rubber septum. Nonsaturating (e.g., 50 μM) or saturating (e.g., 2.5 mM) concentrations of α-[1-^{14}C]oxoglutarate are added (by microsyringe) followed 1–5 min later by 50 μl of 20% (v/v) perchloric acid to stop the reaction. The time is chosen so that 5–30% of the oxoglutarate is decarboxylated. The samples are then left gently shaking for at least 1 hr before ^{14}C in the phenylethyl-amine is measured by scintillation counting. Alternatively ^{14}C left as un-converted α-[1-^{14}C]oxoglutarate in the small test tube can be determined.

Isocitrate Dehydrogenase (NAD⁺)

NAD-ICDH is considerably more difficult to study in intact mitochondria than PDH or OGDH. The effects of Ca^{2+} on isocitrate oxidation within intact rat white[19] and brown[30] adipose tissue mitochondria can be demonstrated using an O_2 electrode as described above (Fig. 1) for OGDH except that *threo*-D_s-isocitrate (or citrate) is presented as the substrate; hydroxymalonate can be used to replace malate if required. However, it is not certain that the Ca^{2+} effects observed are only the result of activation of NAD-ICDH as it is possible that the activation of OGDH is also contributing to the observed Ca^{2+} activations of isocitrate oxidation.

It should be noted that NAD-ICDH activity cannot be readily assessed in rat heart mitochondria owing to the very slow transport of tricarboxylic acids into these mitochondria.[48] The $^{14}CO_2$-trapping technique also cannot be used satisfactorily for NAD-ICDH since all mammalian mitochondria have high NADP⁺-linked ICDH activity which may catalyze isotopic exchange.[25] Effects of Ca^{2+} on the reduction of rat liver mitochondrial NAD(P)⁺ induced by nonsaturating concentrations of *threo*-D_s-isocitrate or citrate have been demonstrated,[25] and this approach is probably applicable to other mitochondria; again, however, it is difficult to ascribe the observed effects exclusively and unambiguously to changes in NAD-ICDH activity.

Use of Ca²⁺-Sensitive Properties of Enzymes as Indicators for Intramitochondrial Ca²⁺

Studies on Ca²⁺ Transport across Mitochondrial Inner Membrane

The intramitochondrial [Ca^{2+}] to which the enzymes are exposed is determined by the relative rates of Ca^{2+} uptake and egress across the mitochondrial inner membrane and by the extramitochondrial [Ca^{2+}].[1,4,21] Uptake is by a uniporter mechanism which is driven electrophoretically by the membrane potential set up by proton extrusion by the respiratory chain and is inhibited physiologically by Mg^{2+} or artificially by ruthenium red. The major pathway for Ca^{2+} egress from the matrix of mammalian mitochondria is by electroneutral exchange with Na⁺ (which are then exchanged with H⁺, i.e., this process is again driven by respiration); this can be inhibited physiologically by Ca^{2+} and artificially by diltiazem and other similar drugs.[23] There is also a less well-characterized Na⁺-independent pathway, and the relative activity of this with respect to the Na⁺-de-

---

[48] J. B. Chappell and B. H. Robinson, *Biochem. Soc. Symp.* **27**, 123 (1968).

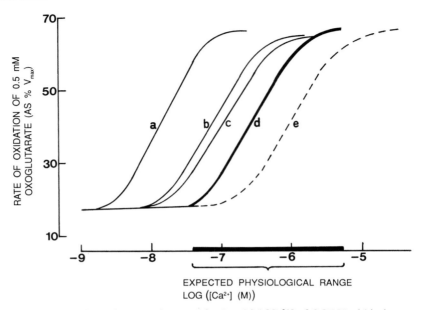

FIG. 3. Sensitivity to increases in extramitochondrial [Ca^{2+}] of OGDH within intact rat heart mitochondria incubated in the absence and presence of Na$^+$, Mg^{2+}, and FCCP. Mitochondria were incubated as for Fig. 1 in the presence of 5 mM EGTA with additions of CaCl$_2$ to give the concentrations of free Ca^{2+} indicated, and with the additional presence of the following: a, no further additions; b, 15 mM NaCl; c, 0.5 mM MgCl$_2$; d, 0.5 mM MgCl$_2$ plus 15 mM NaCl; e, 1 μM FCCP as indicated. Experimental procedure was as for Fig. 1, and rates of oxidation at 0.5 mM 2-oxoglutarate are given as percentages of the rate at a saturating oxoglutarate concentration (20 mM). The rate at 20 mM oxoglutarate was approximately 90 nmol of O$_2$ consumed/min/mg protein. Note especially that in the presence of physiological concentrations of Mg^{2+} and Na$^+$ (heavy line) activation is achieved as extramitochondrial Ca^{2+} is increased over the expected physiological range and that the curve with FCCP (dashed line) is very similar to those obtained with the extracted enzymes.[1] Data are adapted from those in Ref. 11.

pendent pathway appears to vary in mitochondria prepared from different tissues. Spermine is a potent activator of uptake, and perhaps also egress.[49]

The effects of changes in extramitochondrial [Ca^{2+}], and of all of the effectors of mitochondrial Ca^{2+} transport, on intramitochondrial [Ca^{2+}] can be well illustrated by using the Ca^{2+}-sensitive properties of the enzymes, assayed as described above. Examples of this for OGDH and PDH are shown in Figs. 3 and 4. This approach would appear to be the only satisfactory means at present for estimating intramitochondrial free [Ca^{2+}]

[49] C. V. Nicchitta and J. R. Williamson, *J. Biol. Chem.* **254**, 12978 (1984).

in the presence of physiological (and buffered) concentrations of extramitochondrial [Ca^{2+}] and normal physiological loads of mitochondrial Ca^{2+}. K_m values for the effects of Na$^+$ and Mg^{2+} of approximately 1 mM and 60 μM have been found using such techniques with both rat heart and liver mitochondria.[11,25]

It should be noted that in such studies, the extramitochondrial [Ca^{2+}] has to be carefully controlled, usually by using EGTA–Ca^{2+} buffer systems. Free metal ion concentrations are calculated as described previously[8,36] making direct use of published stability constants such as those given by Martell and Smith.[50] However, it is appreciated that this may not yield the absolutely correct concentrations of metal ion, as allowances for the differences between these constants and the chemical activity of both protons and the metal ions are not made; however, there is still considerable controversy over what is the correct allowance.[51,52]

There are two further problems encountered when using EGTA–Ca^{2+} buffers. First, addition of Ca^{2+} to EGTA causes the release of protons, and the binding of Ca^{2+} to EGTA is very sensitive to changes in pH. Second, it is essential that accurate relative concentrations of EGTA and Ca^{2+} are added. These problems can be conveniently minimized by using stock matched solutions of 200 mM EGTA and 200 mM EGTA–Ca^{2+} which are carefully adjusted to the appropriate pH with KOH.[8] The latter solution is prepared by adding CaCl$_2$ (from a 1 M stock solution) to a solution of 400 mM EGTA while continuously adjusting the pH. When there is no perceptible change in pH on adding either more CaCl$_2$ or EGTA then exact equilibration of the EGTA and calcium has been achieved,[8] and the solution is made up to the appropriate volume at 200 mM strength. This can be further checked by using the Ca^{2+} indicator arsenazo III [2-(o-arsonophenylazo)-1,8-dihydroxynaphthalene-3,6-disulfonic acid] or a Ca^{2+}-selective electrode, as described in Ref. 8.

It is also imperative to avoid using Mg^{2+} and Na$^+$ salts of substrates if the effects of these ions are being investigated. Another feature to bear in mind is that when the egress pathway(s) for Ca^{2+} become saturated as matrix Ca^{2+} levels are increased, the mitochondria will then accumulate large amounts of Ca^{2+}[1,25] as the uptake pathway has a much higher maximal velocity than the egress pathway(s). This can lead to progressive damage of mitochondrial function[1,11,25] (Fig. 4). The exact point where such damage begins is again determined by the various effectors of Ca^{2+}

[50] A. E. Martell and R. M. Smith, "Critical Stability Constants," Vol. 1. Plenum, London, 1974.
[51] R. Y. Tsien and T. J. Rink, *Biochim. Biophys. Acta* **599**, 623 (1980).
[52] D. J. Miller and G. L. Smith, *Am. J. Physiol.* **246**, C160 (1984).

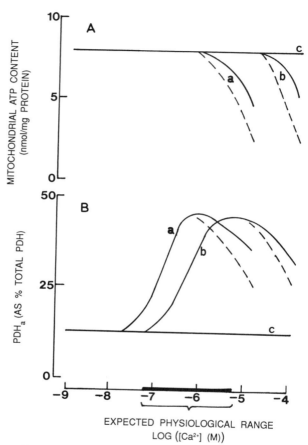

FIG. 4. Effects of increasing extramitochondrial [Ca²⁺] on (A) the ATP content of and (B) the amount of PDH$_a$ in intact rat liver mitochondria incubated in the absence or presence of Na⁺ and Mg²⁺, or ruthenium red. Mitochondria were incubated for either 5 min (solid lines) or 10 min (dashed lines) in medium B (at pH 7) containing 2 mM oxoglutarate, 0.2 mM malate, 1 mM pyruvate, and EGTA–Ca²⁺ buffers to give the concentrations of Ca²⁺ shown. Other additions were as follows: a, none; b, 10 mM NaCl plus 1 mM MgCl₂; c, 1 μM ruthenium red. Where 5- and 10-min samples were similar the latter have been omitted. Data are adapted from those in Ref. 25.

red. Where 5- and 10-min samples were similar the latter have been omitted. Data are adapted from those in Ref. 25.

transport and the amount of Ca^{2+} present. It should be noted that, in contrast, in Fig. 3 (and in Fig. 4, prior to the dashed lines) there is a steady-state relationship between extra- and intramitochondrial Ca^{2+}.

Studies on Effects of Hormones on Intramitochondrial Ca^{2+}

The ability to assay for the Ca^{2+}-sensitive properties of PDH and OGDH within intact rat heart[2] and liver[3] mitochondria has allowed convincing tests to be made of the hypothesis[16] that hormones which are known to increase cytoplasmic [Ca^{2+}] in these tissues also increase intramitochondrial [Ca^{2+}] and so cause the increases in tissue PDH$_a$ which are known to be brought about by these same hormones (see Refs. 2 and 3). This approach involves the rapid preparation of mitochondria from control and treated tissues, under conditions which minimize Ca^{2+} redistribution (as described above)[2,3] followed by their incubation under conditions

TABLE I

EFFECTS OF PRELOADING MITOCHONDRIA WITH Ca^{2+} OR ADRENALINE PRETREATMENT OF TISSUE ON PDH AND OGDH IN SUBSEQUENTLY ISOLATED LIVER MITOCHONDRIA INCUBATED UNDER VARIOUS CONDITIONS[a]

Additions to mitochondrial incubation medium	PDH activity (as % total) after		OGDH activity (as % V_{max}) after	
	No load or hormone	Ca^{2+} loading or adrenaline	No load or hormone	Ca^{2+} loading or adrenaline
(a) Preloading with Ca^{2+}				
None	12	27[b]	4	12[b]
Na$^+$ (10 mM)	10	13[c]	4	5[c]
Na$^+$ plus diltiazem (300 μM)	12	29[b]	4	13[b]
Ca^{2+} (400 nM)	50	49	22	20
(b) Adrenaline pretreatment				
None	12	20[b]	8	13[b]
Na$^+$	13	14[c]	8	8[c]
Na$^+$ plus diltiazem	13	23[b]	9	18[b]
Ca^{2+}	49	51	34	35

[a] Data are taken from Ref. 3; see Ref. 3 for full details. Mitochondria were preloaded and reprepared, or prepared from hormone-treated tissue, as described in the text and then incubated (for 5 min at 30° and in the presence of 0.5 mM EGTA) for the determination of PDH$_a$, PDH total, and OGDH (using α-[1-^{14}C]oxoglutarate decarboxylation in the presence of 0.5 mM malate) as described in the text.

[b] Significant ($p \leq 0.05$) effect of pretreatment.

[c] Significant effect of Na$^+$.

where the Ca^{2+}-sensitive enzymes can be assayed as indicators of hormone-induced changes in intramitochondrial $[Ca^{2+}]$.

To assess the feasibility of this approach, it is first necessary to examine the behavior of these two enzymes within mitochondria which are loaded *in vitro* with Ca^{2+} to achieve similar increases in PDH_a as obtained with hormones acting on tissues (as, e.g., in the experiment shown in Fig. 3) and then "reprepared." With rat heart and liver mitochondria it was found that their behavior in the "Ca^{2+}-preloaded" mitochondria subsequently incubated under a variety of conditions (designed to assess the role of Ca^{2+}) matched that of the enzymes in the mitochondria from hormone-treated tissues (an example of this comparison is shown for rat liver in Table I). Briefly, in both instances [e.g., (a) and (b) in Table I] the enzyme activations are persistent in mitochondria incubated at 30° in the absence of Na^+ but are rapidly lost if Na^+ (to cause Ca^{2+} egress) or enough extramitochondrial Ca^{2+} to saturate the Ca^{2+}-dependent effects on the enzymes is added. Moreover, the effects of Na^+ are blocked by diltiazem. These effects thus establish criteria for determining whether a hormone is affecting intramitochondrial $[Ca^{2+}]$; that is, in incubations of mitochondria prepared from the hormone-treated tissues (as above) the activities of each of the intramitochondrial Ca^{2+}-sensitive enzymes (with respect to those from untreated tissues) should be: (1) enhanced in Na^+-free media (containing EGTA); (2) diminished to control values by incubation with Na^+; (3) enhanced in Na^+-containing media which also contain diltiazem; and (4) maximally stimulated by added Ca^{2+} to the same values as controls can be maximally stimulated.

A more direct approach to monitor the effects of preparation and incubation on mitochondrial Ca content has been to preload the mitochondria with ^{45}Ca and then subject them to these processes.[2,3] Mitochondria are therefore pre-incubated at 30° (at about 4 mg protein/ml) with suitable amounts of ^{45}Ca and unlabeled $CaCl_2$ in 3–5 ml of medium B containing respiratory substrates (e.g., oxoglutarate with malate) and 0.1– 0.5 μCi of ^{45}Ca/ml in the presence of sufficient $CaCl_2$ (usually 50–80 μM) to give near maximal Ca^{2+}-dependent increases in PDH_a content. The appropriate amount of $CaCl_2$ is determined by monitoring the Ca^{2+}-dependent increases by either O_2 uptake[2] or $NAD(P)^+$ reduction[3] in mitochondria which were induced by nonsaturating concentrations of 2-oxoglutarate, since the sensitivity of OGDH to extramitochondrial $[Ca^{2+}]$ is very similar to that of PDH.[11,25] In this manner the specific activity of free $^{45}Ca^{2+}$ can be kept high and variations in the [EGTA] added with the mitochondrial suspension and the endogenous Ca content of solutions allowed for without the use of Ca^{2+}–EGTA buffers. After preincubation for 4 min a large excess (at least 6 vol) of ice-cold medium A is added and the mitochondria resedimented, washed once, and then resuspended as described above.

Table II gives an example of such an experiment using liver mitochondria and shows that after mitochondrial "reisolation" the results obtained on mitochondrial Ca^{2+} handling by monitoring ^{45}Ca egress from ^{45}Ca-loaded mitochondria match very closely those obtained (Table I) by monitoring the Ca^{2+}-sensitive enzymes under the same conditions. It should be noted, first, that very little Ca^{2+} appears to be lost from the mitochondria at 0°, even when Na^+ is present, suggesting that the amounts of Ca^{2+} present in the mitochondria will be maintained during their preparation. It is also clear that in the absence of Na^+, the rate of loss of ^{45}Ca from the preloaded mitochondria is very slow even when they are incubated in medium containing EGTA and respiratory substrates at 30°. The effects of Na^+ in promoting Ca^{2+} egress (and of diltiazem in inhibiting this) are again readily evident at 30°.

TABLE II
EFFECTS OF VARIOUS AGENTS AND INCUBATION CONDITIONS ON EGRESS OF
^{45}Ca FROM PRELOADED RAT LIVER MITOCHONDRIA[a]

Incubation time/conditions	Incubation temperature (°)	^{45}Ca content of "reisolated" mitochondria [as % of initial (zero-time) content]	
		Without Na^+	With 10 mM NaCl
5 min	0	100	100
30 min	0	99	101
5 min	20	101	70[b]
5 min	30	92	52[b]
5 min	37	83	46[b]
2 min	30	99	77[b]
10 min	30	84	28[b]
5 min/300 μM diltiazem	30	96	97

[a] Data are taken from Ref. 3. Mitochondria were preloaded with ^{45}Ca and CaCl₂ ($\sim$ 50 μM) to give approximately 80% of the maximal Ca^{2+}-dependent activation of PDH (see text) and then, after reisolation, incubated ($\sim$1.5 mg protein/ml) in medium B containing 2 mM oxoglutarate, 0.2 mM malate, and 0.5 mM EGTA and other conditions as shown.
[b] Significant effects of Na^+ ($p \leq 0.05$).

Conclusions and Implications for Role of Intramitochondrial Ca²⁺

There is now convincing evidence in heart and liver (Table I) in favor of the view that when the cytoplasmic concentration of Ca^{2+} is increased by hormones or other external stimuli, the intramitochondrial concentration increases in parallel and results in the activation of the Ca^{2+}-sensitive dehydrogenases. It is likely that this role of Ca^{2+} in the regulation of intramitochondrial oxidative metabolism will also be important in other mammalian cell types, and the experimental approaches detailed above should prove useful in establishing this. The stimulation of these intramitochondrial enzymes by Ca^{2+} may greatly contribute to metabolic homeostasis by allowing increased NADH and hence ATP formation to occur within stimulated cells without the need to diminish $NADH/NAD^+$ or ATP/ATP ratios; these key ratios may in fact increase as a result.

The idea that the main role of the Ca^{2+}-transport system in the inner membrane of mammalian mitochondria is to regulate matrix $[Ca^{2+}]$ in the above manner is not compatible with the earlier suggestions that mitochondria act as Ca^{2+} sinks or buffers for extramitochondrial $[Ca^{2+}]$[5,6] (as this behavior requires egress pathway saturation, see Ref. 1) or that they act as a store for calcium which can be mobilized following hormone stimulation.[28,29] Indeed it is likely that the intramitochondrial concentration of Ca^{2+} is only 2–3 times that in the cytoplasm under normal physiological conditions (see Fig. 3) and that $[Ca^{2+}]$ in the two compartments usually changes in parallel. This agrees with the *in situ* measurements of intramitochondrial and cytoplasmic total calcium content using X-ray probe microanalysis,[53] and also with the observations that the total calcium content of rat heart and liver mitochondrial fractions increases as the result of the hormone treatments described in Table I.[26,54]

The techniques described above may also prove useful in studies directed toward assessing any effects that hormones or other agents may have on the activities of components of the Ca^{2+}-transport system of the mitochondrial inner membrane. Some evidence for such changes has already been obtained in studies on both rat heart[55] and liver.[56,57]

[53] A. P. Somlyo, M. Bond, and A. V. Somlyo, *Nature (London)* **314**, 622 (1985).
[54] M. Crompton, P. Kessar, and I. Al-Nassar, *Biochem. J.* **216**, 333 (1983).
[55] P. Kessar, and M. Crompton, *Biochem. J.* **200**, 379 (1981).
[56] T. P. Goldstone, R. J. Duddridge, and M. Crompton, *Biochem. J.* **210**, 463 (1983).
[57] W. M. Taylor, V. Prpic, J. H. Exton, and F. L. Bygrave, *Biochem. J.* **188**, 443 (1980).

Addendum

There has been an important recent development in this subject since the original submission of this chapter. This has been the demonstration in a number of different laboratories that the fluorescent Ca^{2+} indicators Fura-2, Quin 2, and indo-1, and also the fluorescent pH indicator BCECF [biscarboxyethyl-5(6)-carboxyfluorescein], can be successfully entrapped within the matrix of mammalian mitochondria.[58-65] To date this has been most readily accomplished with rat heart mitochondria.[58,59,61,63-65] Liver mitochondria have as yet proved much more difficult to load with the Ca^{2+} indicators[62]; this appears to be the result of partial hydrolysis of the ester forms and as well as reexport from the mitochondria of hydrolyzed and partially hydrolyzed forms.[62,64] There is also a report of Fura-2 being loaded into rat brain mitochondria,[60] but in this study the mitochondria were incubated with millimolar concentrations of Ca^{2+} and were thus likely to be substantially damaged.

With heart mitochondria, all four of the above fluorescent indicators have been successfully loaded, but most studies have employed Fura-2. Appropriate conditions for loading (see below) can be chosen so that the functional bioenergetic capabilities of the mitochondria do not appear to be compromised.[65] This has thus allowed matrix Ca^{2+} (and pH) to be directly and continuously monitored. Significantly the resultant studies have essentially corroborated the conclusions already summarized in this chapter based on the use of the Ca^{2+}-sensitive intramitochondrial enzymes as probes for matrix Ca^{2+}. Fluorescence microscopy of loaded mitochondria attached to coverslips and superfused has also been performed.[64]

The entrapment of Fura-2 into the matrix of rat heart mitochondria involves their simple incubation with the Fura-2 acetoxymethyl ester (Fura 2/AM) (Calbiochem, Cambridge, U.K.), i.e., in much the same way as the now well-developed methodology for loading such compounds into cells.[66] In our laboratory,[65] mitochondria are loaded with Fura-2 by incubation in

[58] G. L. Lukacs and A. Kapus, *Biochem. J.* **248**, 313 (1987).
[59] M. H. Davis, R. A. Altschuld, D. W. Jung, and G. P. Brierley, *Biochem. Biophys. Res. Commun.* **149**, 40 (1987).
[60] H. Komulainen and S. L. Bondy, *Neurochem. Int.* **10**, 55 (1987).
[61] G. L. Lukacs, A. Kapus, and A. Fonyo, *FEBS Lett.* **229**, 219 (1987).
[62] T. E. Gunter, D. Restrepo, and K. K. Gunter, *Am. J. Physiol.* **255**, C304 (1988).
[63] R. Moreno-Sanchez and R. G. Hansford, *Biochem. J.* **256**, 403 (1988).
[64] M. Reers, R. A. Kelly, and T. W. Smith, *Biochem. J.* **257**, 131 (1989).
[65] J. G. McCormack, H. M. Browne, and N. J. Dawes, *Biochim. Biophys. Acta* **973**, 420 (1989).
[66] P. H. Cobbold and T. J. Rink, *Biochem. J.* **248**, 313 (1987).

the final suspension medium (medium A above) at approximately 20–40 mg protein/ml for 5 min at 30° after the addition of 10 μM of Fura 2/AM followed by a brief (1–2 sec) vortex. These conditions allow the generation of suitable amounts of the free acid form of the fluorescent indicator within the matrix, as evident from the appearance of the Ca^{2+}-sensitive fluorescence signal on addition of Triton X-100 (0.1%, v/v) to permeabilize the mitochondria.[58] Under the above conditions, approximately 50–80 pmol of Fura-2/mg protein is loaded (representing 25–40% of the added ester and corresponding to a concentration in the matrix, assuming a volume of 1 μl/mg protein, of around 25–50 μM[65]). This does not result in any substantial additional buffering of the matrix Ca^{2+}. Lower concentrations of Fura 2/AM may not give sufficiently high signal-to-noise ratios, whereas at higher concentrations hydrolysis within the matrix may not be complete within this time. After loading the free acid is retained well by the mitochondria, with less than 10% being lost after 5 hr on ice.[59]

After loading, the mitochondria can be incubated in medium B as described above for the assay of the matrix Ca^{2+}-sensitive dehydrogenases. Indeed, the activities of OGDH and PDH can also be monitored at the same time to get a direct measure of the concentrations of matrix Ca^{2+} to which they respond.[61,65] An example of this is shown in Fig. 5 where the changes in matrix Ca^{2+} are measured using a Perkin-Elmer LS-5 fluorimeter by following the changes in Fura-2 fluorescence at wavelength settings of 340 nm (excitation)/500 nm (emission) or 380 nm (ex)/500 nm (em); OGDH activity can be assessed at subsaturating [2-oxoglutarate] and by using the isosbestic point for Fura-2, i.e., 365 nm (ex)/500 nm (em). The latter signal can then be used to correct the Fura-2 signal in the experiment shown; alternatively, if the experiment is carried out at saturating (10 mM) [2-oxoglutarate], then there is no interfering signal from NAD(P)H changes.[65]

The Fura-2 signals can be calibrated with free Ca^{2+} after permeabilizing the mitochondria either completely [with Triton X-100 (0.1%, v/v)] or selectively [with ionomycin (2 μM) as a nonfluorescent Ca^{2+} ionophore, and the uncoupler FCCP (1 μM) to also equilibrate protons] and then changing the concentration of Ca^{2+} with EGTA–Ca^{2+} buffers[65] (see above). K_D values for Fura-2–Ca^{2+} can thus be established under the required conditions, and thereafter it is only necessary to establish F_{max} and $F_{Mn^{2+}}$ as is shown in Fig. 5[65] to quantify the amount trapped in any particular sample; the K_D is pH-sensitive at physiological pH.[65] This can be exploited as experiments at pH 7 (as opposed to 7.3) will allow measurement of Ca^{2+} up to about 10 μM.[65] The intrinsic fluorescence of the mitochondria has to be corrected for,[58] and ruthenium red (1 μM) can be

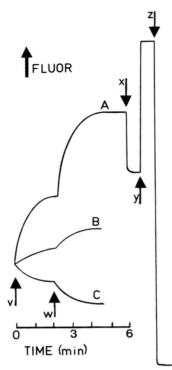

FIG. 5. Effects of increases in extramitochondrial [Ca²⁺] on fluorescence signals from Fura-2-loaded rat heart mitochondria. The loaded mitochondria (∼1 mg protein/ml) were incubated in a fluorimeter with a stirring cuvette unit in medium B (pH 7.3, 30°), containing initially 0.5 mM malonate, 100 μM 2-oxoglutarate, and 1 mM EGTA. Wavelength settings were A, 340 nm (excitation)/500 nm (emission); B, 365/500; or C, 380/500. Following a 2-min stabilization period, additions were made, sequentially at the arrows, as follows: v, 0.5 mM EGTA plus 0.5 mM CaCl₂ (as a buffer solution giving a free final extramitochondrial [Ca²⁺] of ∼26 n*M*); w, 1.5 mM EGTA plus 1.5 mM CaCl₂ (giving 105 nM extramitochondrial [Ca²⁺]); x, 1 μM FCCP plus 2 μM ionomycin; y, 5 mM HEDTA (*N*-hydroxy-ethylethylenediaminetriacetic acid) plus 5 mM CaCl₂ (6.3 μM free Ca²⁺, to give F_{max}); z, 10 mM MnCl₂ (to give $F_{Mn^{2+}}$ or F_0). Data are adapted from those in Ref. 65.

used to correct for any extramitochondrial Fura-2[65]; both of these parameters were found to be less than 2% of the total signal.[65]

Factors influencing the gradient of Ca²⁺ across the mitochondrial inner membrane are readily studied using Fura-2-loaded rat heart mitochondria.[58,59,63–65] In the presence of Na⁺ and Mg²⁺, a large degree of positive cooperativity in the transfer of the Ca²⁺ signal into the matrix is evident[65]; cooperativity is observed with values of matrix Ca^{2+} of around $20-30$ nM (i.e., below the activatory ranges for the enzymes), at 0.1 μM extramitochondrial Ca^{2+}, rising to around $3-5$ μM (and saturation of the enzymes for Ca^{2+}) at about 1.5 μM extramitochondrial Ca^{2+}.[65] These studies suggest an activatory range for PDH and OGDH of around $0.2-2$ μM matrix Ca^{2+} with half-maximal effects around $400-700$ nM.[61,65]

One advantage of the use of Fura-2 in mitochondria when compared with its use in cells is that it is most likely that in the former the compound will report changes in Ca^{2+} in only one compartment, whereas in the latter there is evidence of the Fura-2 locating in different cellular compartments.[66,67] Indeed, this may eventually prove to be of great use in future studies using fluorescence-imaging[66] on the role of mitochondrial Ca^{2+}, as there is already evidence that some Fura-2 fluorescence can be localized to mitochondria within heart[59] and other cell types.[67]

Acknowledgments

Work in the authors' laboratories has been supported by grants from the Medical Research Council, the British Diabetic Association, the British Heart Foundation, and the Percival Waite Salmond Bequest. J. G. McCormack is a Lister Institute Research Fellow.

[67] S. F. Steinberg, J. P. Bilezikian, and Q. Al-Awqati, *Am. J. Physiol.* **253,** C744 (1987).

[11] Effects of Hormones on Mitochondrial Processes

By Sibylle Soboll and Helmut Sies

A short-term effect of hormones such as α-adrenergic agonists, glucagon, vasopressin, and triiodothyronine on mitochondrial metabolism is the stimulation of respiration. The causes for the increase in oxygen uptake are not yet fully elucidated. Respiration may be stimulated by (1) enhancement of biosynthetic processes like gluconeogenesis or ureagenesis, thus increasing the requirement for ATP; (2) direct interaction with the respiratory chain; and (3) increase of substrate supply to the mitochondria. Table I summarizes stimulatory or inhibitory effects of hormones on metabolic processes related to energy metabolism. All the hormones noted in Table I stimulate respiration, gluconeogenesis, and urea synthesis, and they also affect calcium and phosphoinositide metabolism.

α-Adrenergic Agonists and Vasopressin

Enhancement of Flux through Biosynthetic Pathways

α-Adrenergic agonists and vasopressin stimulate not only respiration but also energy-dependent biosynthetic pathways like gluconeogenesis and ureagenesis (hepatocytes,[1,2] perfused liver[3]). Stimulation of gluconeogenesis as well as urea synthesis has a time course similar to that for the stimulation of respiration.[3] The effects are cAMP-independent, sensitive to α_1-antagonists,[1] and dependent on extracellular calcium.[2,4] However, stimulation of respiration seems to be independent of the increase in gluconeogenesis and urea synthesis, since it persists in isolated mitochondria[1] (cf. Ref. 3). An increase in the level of mitochondrial ATP has been discussed as a mechanism for the stimulation of gluconeogenesis and urea synthesis. ATP should activate pyruvate carboxylase and carbamoyl-phosphate synthase.[5] Both pathways can also be stimulated by these hormones in the presence of glutamine, since they activate glutaminase.[6,7] Inhibition of pyruvate kinase may contribute to the stimulation of gluconeogenesis; however, this inhibition is less pronounced with α-adrenergic agonists than with glucagon.[4]

Vasopressin and noradrenaline but not dibutyryl-cAMP enhance the availability of cytosolic ATP,[8] as may be deduced from an increase in gluconeogenesis from dihydroxyacetone and from enhanced cytosolic ATP/ADP in isolated hepatocytes in the presence of carboxyatractyloside. Likewise, vasopressin and α-adrenergic agonists but not dibutyryl-cAMP stimulate glutathione (GSH) export from the liver.[8a] GSH export is likely to stimulate the synthesis of GSH because of its feedback regulation.

Calcium Movements

A direct interaction of α-adrenergic agonists and vasopressin with components of the respiratory chain has not been reported. Rather, it is established that calcium ions mediate the effects of these agonists on cellular

[1] M. A. Titheradge and R. C. Haynes, *Arch. Biochem. Biophys.* **201**, 44 (1980).

[2] A. Binet and M. Claret, *Biochem. J.* **210**, 867 (1983).

[3] W. M. Taylor, E. van de Pol, and F. L. Bygrave, *Eur. J. Biochem.* **155**, 319 (1986).

[4] J. C. Garrison and M. K. Borland, *J. Biol Chem.* **254**, 1129 (1979).

[5] M. A. Titheradge, J. L. Stringer, and R. C. Haynes, *Eur. J. Biochem.* **102**, 117 (1979).

[6] D. Häussinger and H. Sies, *Biochem. J.* **221**, 651 (1984).

[7] A. J. Verhoeven, J. M. Estrela, and A. J. Meijer, *Biochem. J.* **230**, 457 (1985).

[8] M. E. Warnette-Hammond and H. A. Lardy, *J. Biol. Chem.* **260**, 12647 (1985).

[8a] H. Sies and P. Graf, *Biochem. J.* **226**, 1185 (1985).

TABLE I
EFFECTS OF α-ADRENERGIC AGONISTS AND
VASOPRESSIN, GLUCAGON, AND
TRIIODOTHYRONINE (T_3) ON ENERGY
METABOLISM IN RAT LIVER

Increase of flux through metabolic pathways
   Respiration
   Gluconeogenesis
   Urea synthesis
   Citrate cycle
   Malate–aspartate shuttle
Stimulation of mitochondrial enzyme activities
   Pyruvate transport
   Pyruvate carboxylase
   Dehydrogenases (e.g., pyruvate, 2-oxoglutarate,
     isocitrate, and succinate dehydrogenases)
   Transhydrogenase
   Glutaminase
   ATPase
   Adenine nucleotide transport
Increase in mitochondrial parameters
   Proton gradient across inner membrane (ΔpH)
   Membrane potential, protonmotive force
   Adenine nucleotide content[a]
Increase in cellular calcium concentration
   Cytosol
   Mitochondria

[a] For T_3; for references, see text.

bioenergetics.[9,10] In recent years evidence has accumulated that inositol 1,4,5-trisphosphate and diacylglycerol are the signals generated by α-adrenergic agonists and vasopressin. These compounds subsequently induce changes in cellular calcium and activate protein kinase C.[11,12]

Reinhart et al.[10] summarized α-agonist-induced calcium movements, dividing them into three phases. During the first phase (6 sec to 2 min) calcium is mobilized from intracellular stores to increase cytosolic free calcium from 0.2 to 0.6 μM.[12a] The origin of this intracellular calcium was controversial. Whereas according to Ref. 10 at least 50% is of mitochon-

[9] J. R. Williamson, R. H. Cooper, and J. B. Hoek, *Biochim. Biophys. Acta* **639**, 243 (1981).
[10] P. H. Reinhart, W. M. Taylor, and F. L. Bygrave, *Biochem. J.* **223**, 1 (1984).
[11] M. J. Berridge and R. F. Irvine, *Nature (London)* **312**, 315 (1984); L. E. Hokin, *Annu. Rev. Biochem.* **54**, 205 (1985); M. J. Berridge, *ibid.* **56**, 159 (1987).
[12] J. R. Williamson, R. H. Cooper, S. K. Joseph, and A. P. Thomas, *Am. J. Physiol.* **248**, C203 (1985).
[12a] B. Berthon, A. Binet, J. P. Mauger, and M. Claret, *FEBS Lett.* **167**, 19 (1984).

drial origin, others[13-15a] have suggested that endoplasmic reticulum is the primary source of calcium. It has even been reported that mitochondria are not involved in the primary rise in cytosolic calcium,[16,17] especially in the case of vasopressin. The release of calcium from intracellular stores depends on their calcium content, and therefore the conflicting results may be partially due to different experimental conditions.[13] During phase 2 (from 2 min until the end of hormone application) an elevated cytosolic calcium is maintained by increased cycling across the plasma membrane. Therefore, the sustained effect of the α-agonists is dependent on the presence of extracellular calcium. Phase 3 (following the end of hormone application) is characterized by a net uptake of extracellular calcium to refill internal calcium stores.

Recently, calcium movements induced by α-adrenergic agonists were compared to effects of vasopressin and angiotensin.[18-20] Calcium movements were quantitatively and kinetically different, indicating either that the signals generated by these hormones have different effects on calcium transport across cellular membranes or that different calcium channels are involved. The time course of the rise in calcium was dependent on the concentration of the hormone. Interestingly, the free calcium concentration oscillated on hormone addition,[21] and the amplitude and the frequency of oscillation were dependent on the concentration of hormone. Such oscillations were also detectable in the extracellular space.[21a] Oscillations could be a sensitive means of regulating the time course and intensity of effects; the greater the amplitude and frequency of oscillations, the more intense the effect.

Increase in Substrate Supply

Stimulation of pyruvate transport by adrenaline has been reported.[22] Further, an adrenaline-enhanced rate of oxidation of cytosolic NADH in

[13] S. K. Joseph and J. R. Williamson, *J. Biol. Chem.* **258**, 10425 (1983).
[14] A. Binet and M. Claret, *Biochem. J.* **210**, 867 (1983).
[15] S. H. Chueh and D. L. Gill, *J. Biol. Chem.* **261**, 13883 (1986).
[15a] A. Benedetti, P. Graf, R. Fulceri, A. Romani, and H. Sies, *Biochem. Pharmacol.* **38**, 1799 (1989).
[16] S. B. Shears and C. J. Kirk, *Biochem. J.* **220**, 417 (1984).
[17] J. Kleineke and H. D. Soeling, *J. Biol. Chem.* **260**, 1040 (1985).
[18] J. G. Altin and F. L. Bygrave, *Biochem. J.* **232**, 911 (1985).
[19] M. Crompton and T. Goldstone, *FEBS Lett.* **204**, 198 (1986).
[20] M. J. O. Wakelam, G. J. Murphy, V. J. Hruby, and M. D. Houslay, *Nature (London)* **323**, 68 (1986).
[21] N. M. Woods, K. S. R. Cuthbertson, and P. H. Cobbold, *Nature (London)* **319**, 600 (1986).
[21a] P. Graf, S. vom Dahl, and H. Sies, *Biochem. J.* **241**, 933 (1987).
[22] M. A. Titheradge and H. G. Coore, *FEBS Lett.* **71**, 73 (1976).

perifused hepatocytes that was inhibited by aminooxyacetate indicates a higher rate of the malate–aspartate shuttle.[23] These effects and also the stimulation of adenine nucleotide transport discussed above may be secondary to ion movements,[23] especially calcium movements. Thus, tissue perfusion with α-adrenergic agonists induced a significant activation of the mitochondrial calcium uniporter.[24] α-Agonists can produce mitochondrial free calcium concentrations of about 1 μM, stimulating mitochondrial dehydrogenases[25] such as pyruvate dehydrogenase, isocitrate dehydrogenase, and 2-oxoglutarate dehydrogenase[26-30] as well as succinate dehydrogenase.[31] Branched-chain keto acid oxidation is also regulated by α-agonist-mediated calcium fluxes.[32] Consequently, α-agonists can enhance substrate supply to the respiratory chain by stimulating flux through mitochondrial dehydrogenases. After infusion of phenylephrine, $^{14}CO_2$ output from labeled glutamate, 2-oxoglutarate, and fatty acids is stimulated,[6,33-35] and 3-hydroxybutyrate/acetoacetate ratios and NADH levels are elevated.[36] These results show that flux through the tricarboxylic acid cycle[34,35] is stimulated.

Calcium fluxes across the mitochondrial membrane may also be responsible for changes in ΔpH across the mitochondrial inner membrane, as observed in perfused liver.[37] The increase in the ΔpH caused by the hormones (Table II) induces accumulation of dicarboxylic and tricarboxylic acids in the matrix space since they are distributed according to the ΔpH.[38,39] Calcium may also cause a rise in mitochondrial adenine nucleo-

[23] X. M. Leverve, A. J. Verhoeven, A. K. Groen, A. J. Meijer, and J. M. Tager, *Eur. J. Biochem.* **155,** 551 (1986).
[24] T. P. Goldstone, J. Roos, and M. Crompton, *Biochemistry* **26,** 246 (1987).
[25] J. G. McCormack and R. M. Denton, *Trends Biochem. Sci.* **11,** 258 (1986).
[26] J. G. McCormack, *Biochem. J.* **231,** 597 (1985).
[27] R. M. Denton and J. G. McCormack, *Am. J. Physiol.* **249,** E543 (1985).
[28] J. M. Staddon and J. D. McGivan, *Biochem. J.* **225,** 327 (1985).
[29] R. S. Ochs, *J. Biol. Chem.* **259,** 13004 (1985).
[30] F. Assimacopoulos-Jeannet, J. G. McCormack, and B. Jeanrenaud, *J. Biol. Chem.* **261,** 8799 (1986).
[31] M. Yamaguchi and H. Shibano, *Chem. Pharm. Bull.* **35,** 3766 (1987).
[32] D. Buxton, L. L. Barron, and M. Olson, *J. Biol. Chem.* **257,** 14318 (1982).
[33] M. C. Sugden, A. J. Ball, V. Ilic, and D. H. Williamson, *FEBS Lett.* **116,** 37 (1980).
[34] W. M. Taylor, E. van de Pol, and F. L. Bygrave, *Biochem. J.* **233,** 321 (1986).
[35] T. Patel, *Eur. J. Biochem.* **159,** 15 (1986).
[36] P. T. Quinlan and A. P. Halestrap, *Biochem. J.* **236,** 789 (1986).
[37] S. Soboll and R. Scholz, *FEBS Lett.* **205,** 109 (1986).
[38] F. Palmieri, E. Quagliariello, and M. Klingenberg, *Eur. J. Biochem.* **17,** 230 (1970).
[39] S. Soboll, R. Elbers, R. Scholz, and H. W. Heldt, *Hoppe-Seyler's Z. Physiol. Chem.* **361,** 69 (1980).

TABLE II

INFLUENCE OF ADRENALINE, GLUCAGON, AND TRIIODOTHYRONINE ON
MITOCHONDRIAL/CYTOSOLIC PROTON GRADIENT (ΔpH)[a]

Condition	pH gradient (mitochondrial/cytosolic)
Fed, no substrates	0.33 ± 0.02
Fed, adrenaline ($10^{-7} M$)	0.79 ± 0.10
Fed, glucagon ($10^{-8} M$)	0.49 ± 0.04
Fasted, lactate plus pyruvate	0.23 ± 0.05
Fasted, lactate plus pyruvate, glucagon ($10^{-8} M$)	0.49 ± 0.07
Fasted, lactate plus pyruvate, T_3 ($10^{-6} M$)	0.61 ± 0.07

[a] In perfused rat liver. Data ($n = 4-7 \pm$ S.E.M.) are from Ref. 37 except for T_3 (S. Soboll, unpublished results).

tide content by increasing the mitochondrial pyrophosphate content[40] through exchange for extramitochondrial adenine nucleotides by the adenine nucleotide translocator.[41]

Glucagon

Enhancement of Flux through Biosynthetic Pathways

Glucagon stimulates mitochondrial respiration, as demonstrated in isolated mitochondria in state 3[42,43] and in perfused liver[44,45] (for a review, see also Ref. 46). Stimulation of gluconeogenesis by glucagon occurs by cAMP-mediated inhibition of pyruvate kinase.[4,47,48] A stimulation of pyru-

[40] A. Halestrap, P. T. Quinlan, A. E. Armstron, and D. E. Whipps, in "Achievements and Perspectives of Mitochondrial Research" (E. Quagliariello, E. C. Slater, F. Palmieri, C. Saccone, and A. M. Krooy, eds.), Vol. 1, p. 469. Elsevier, Amsterdam, 1985.

[41] J. R. Avrille and G. K. Asimakis, Arch. Biochem. Biophys. 201, 564 (1980).

[42] R. K. Yamazaki, J. Biol. Chem. 250, 7924 (1975).

[43] A. D. Halestrap, Biochem. J. 172, 399 (1978).

[44] R. Kimmig, T. J. Mauch, and R. Scholz, Eur. J. Biochem. 136, 617 (1983).

[45] S. Kimura, T. Suzaki, S. Kobayashi, K. Abe, and E. Ogata, Biochem. Biophys. Res. Commun. 119, 212 (1984).

[46] A. Halestrap, in "Short-Term Regulation of Liver Metabolism" (L. Hue and G. Van de Werve, eds.), p. 389. Elsevier/North-Holland Biomedical Press, Amsterdam, 1981.

[47] J. E. Feliu, L. Hue, and H. G. Hers, Proc. Natl. Acad. Sci. U.S.A. 73, 2762 (1976).

[48] J. C. Garrison, M. K. Borland, U. A. Florio, and O. A. Twible, J. Biol. Chem. 254, 7147 (1979).

vate carboxylase has also been suggested[49] since glucagon-induced stimulation of gluconeogenesis was abolished in biotin-deficient rats. Pyruvate carboxylase effects may result from increases in mitochondrial ATP[37,50,51] and/or acetyl-CoA[51] or from increased pyruvate transport.[52,53] However, no significant effect of glucagon on mitochondrial ATP/ADP has been observed in isolated hepatocytes[54] or perfused liver.[37] In addition, direct stimulation of pyruvate transport does not significantly regulate gluconeogenesis because basal transport activity is high.[52,55] An indirect stimulation of pyruvate transport by increasing the mitochondrial/cytosolic proton gradient[37] could enhance pyruvate supply to the enzyme. Activation of gluconeogenesis by glucagon can also be exerted via mitochondrial glutaminase,[6,56,57] supplying glutamate as precursor. The activation may occur via an interaction with the mitochondrial membrane.[57]

One mechanism by which glucagon increases urea formation may be through stimulation of glutamine breakdown. Thus, ammonia would be channeled directly to carbamoyl-phosphate synthase.[58] In addition, mitochondrial carbamoyl-phosphate synthase may be activated independently of glutamine breakdown by increased N-acetylglutamate.[51,56]

A direct stimulation of both gluconeogenesis and urea synthesis seems to be established and could well be responsible for an enhanced flux through the respiratory chain. However, examination of the time course of mitochondrial events reveals that only gluconeogenesis could cause the early activation of respiration. Ureagenesis increases later[56,59] and could only be responsible for the later phase. An increase of mitochondrial ATP production may be prerequisite to urea synthesis since a direct correlation between mitochondrial ATP and urea synthesis was demonstrated.[1,51]

Studies on the action of glucagon in the fed state also demonstrated a stimulation of respiration. This effect cannot be explained on the basis of an enhanced ATP demand, indicating that the stimulation of respiration is a direct one, possibly related to inhibition of glycolysis.[44]

[49] E. A. Siess, D. G. Brocks, and O. H. Wieland, *Biochem. J.* **172**, 517 (1978).
[50] E. A. Siess, D. G. Brocks, H. K. Lattke, and O. H. Wieland, *Biochem. J.* **166**, 255 (1977).
[51] H. E. S. J. Hensgens, A. J. Verhoeven, and A. J. Meijer, *Eur. J. Biochem.* **107**, 197 (1980).
[52] A. P. Halestrap and A. E. Armston, *Biochem. J.* **323**, 677 (1984).
[53] M. A. Titheradge and H. G. Coore, *FEBS Lett.* **63**, 45 (1976).
[54] E. A. Siess, R. I. Kientsch-Engel, F. Fahimi, and O. H. Wieland, *Eur. J. Biochem.* **141**, 543 (1984).
[55] A. K. Groen, R. C. Verhoeven, R. Van der Meer, and J. M. Tager, *J. Biol. Chem.* **258**, 14346 (1983).
[56] J. M. Staddon, N. M. Bradford, and J. D. McGivan, *Biochem. J.* **217**, 855 (1984).
[57] J. McGivan, M. Vadher, J. Lacey, and N. Bradford, *Eur. J. Biochem.* **148**, 323 (1985).
[58] A. J. Meijer, *FEBS Lett.* **191**, 249 (1985).
[59] E. A. Siess and O. H. Wieland, *Eur. J. Biochem.* **110**, 203 (1980).

Direct Interaction of Glucagon with the Respiratory Chain[46]

Glucagon increases the rate of uncoupler-stimulated and ADP-stimulated oxidation of substrates entering the respiratory chain before cytochrome c_1 and c. The hormone also increases ΔpH, $\Delta\psi$, protonmotive force, and uncoupler-stimulated ATPase. Hence, it seems likely that glucagon causes a higher energization of mitochondria owing to a direct stimulation of respiratory chain components and H^+-ATPase. An increase in mitochondrial volume has also been described,[22,43] an effect that would produce a stimulation of flux through the respiratory chain. However, mitochondrial swelling and some of the effects mentioned above, i.e., those on ΔpH, $\Delta\tilde{\mu}H^+$, and electron flow between cytochrome c_1 and c, were shown to be artifacts arising from the mitochondrial isolation procedure.[60-62] Experiments where such artifacts were avoided produced a stimulation of electron flux within complex III involving the ubiquinone pool.[61] Aging of mitochondria reverses the effect while increasing intramitochondrial volume mimics it. A remarkable effect on ΔpH was observed in perfused liver[37] but not in isolated hepatocytes.[63] Recent studies again indicate that glucagon may affect mitochondrial metabolism by changing matrix volume.[64,65]

Increase in Substrate Supply

Glucagon acts on cellular and mitochondrial metabolism via an increase in cytosolic calcium and, subsequently, mitochondrial calcium concentration by a mechanism that is different from that of α-adrenergic agonists.[19,20,66] Thus, there are differences in the lithium sensitivity of [^{3}H] inositol release into the extracellular space with α-adrenergic agonists, vasopressin, and glucagon.[66a] Glucagon enhances intracellular calcium by stimulating influx into the cell[19,20,66] and the release of intracellular calcium from endoplasmic reticulum, probably via cAMP-stimulated protein kinases.[67] It increases the activities of several dehydrogenases related directly

[60] E. A. Siess, F. M. Fahimi, and O. H. Wieland, *Hoppe-Seyler's Z. Physiol. Chem.* **362**, 1643 (1981).

[61] A. Halestrap, *Biochem. J.* **204**, 37 (1982).

[62] E. A. Siess, *Hoppe-Seyler's Z. Physiol. Chem.* **364**, 279 (1983).

[63] T. Strzelecki, J. A. Thomas, C. D. Koch, and K. F. La Noue, *J. Biol. Chem.* **259**, 4122 (1984).

[64] A. Davidson and A. P. Halestrap, *Biochem. J.* **246**, 715 (1987).

[65] A. P. Halestrap, P. T. Quinlan, D. E. Whipps, and A. Armston, *Biochem. J.* **235**, 779 (1986).

[66] J. G. Altin and F. L. Bygrave, *Biochem. J.* **242**, 43 (1987).

[66a] S. vom Dahl, P. Graf, and H. Sies, *Biochem. J.* **251**, 843 (1988).

[67] J. P. Mauger and M. Claret, *FEBS Lett.* **195**, 106 (1986).

to the respiratory chain, e.g., succinate dehydrogenase[59,68-70] and the energy-linked transhydrogenase,[71,72] or to the citric acid cycle, e.g., pyruvate dehydrogenase (through stimulation of pyruvate dehydrogenase phosphatase,[26,52] cf. Ref. 73), isocitrate dehydrogenase, and 2-oxoglutarate dehydrogenase.[26,59,74]

These changes in activities of dehydrogenases linked to the citric acid cycle are thought to be brought about by an increase in mitochondrial calcium levels (see above). Alternatively, it has been suggested that mitochondrial dehydrogenases are simply activated by an enhanced cytosolic supply of substrate, since glucagon is thought to affect mitochondrial membrane permeability.[46,62,75] This, however, has been ruled out as the sole mechanism of stimulation because the action of the hormone on dehydrogenases was still observed in submitochondrial particles.[76] On the other hand, transport of dicarboxylates and tricarboxylates can indirectly be stimulated by the increased mitochondrial/cytosolic proton gradient as demonstrated with both fed and fasted perfused rat liver[37] (see discussion of adrenaline action). However, glucagon enhances duroquinone oxidation,[77,78] which does not require transport across the mitochondrial membrane, and mitochondrial levels of pyruvate, glutamate, and 2-oxoglutarate are decreased rather than increased with glucagon.[52,59,74] These results argue against stimulated metabolite transport as the only mechanism to increase substrate supply to the respiratory chain.

Stimulation of respiration by enhancing the level of reducing equivalents to the electron transport chain is also suggested by the presence of a more reduced state of pyridine nucleotides in isolated mitochondria,[59,76,78] isolated hepatocytes,[78,79] and intact rat liver in situ.[45] The ability of glucagon to promote retention of mitochondrial calcium[71,80] is also closely connected with a more reduced state of $NADP^+$. A relationship between mitochondrial calcium and the redox potential of $NADP^+$ has been dem-

[68] E. A. Siess and O. H. Wieland, *FEBS Lett.* **93**, 301 (1978).
[69] E. A. Siess and O. H. Wieland, *FEBS Lett.* **101**, 277 (1979).
[70] M. A. Titheradge and R. C. Haynes, *FEBS Lett.* **106**, 330 (1979).
[71] V. Prpic and F. L. Bygrave, *J. Biol. Chem.* **255**, 6193 (1980).
[72] H. C. Hamman and R. C. Haynes, *Biochim. Biophys. Acta* **724**, 241 (1983).
[73] D. Häussinger, W. Gerok, and H. Sies, *Eur. J. Biochem.* **126**, 69 (1982).
[74] F. D. Sistare, R. A. Picking, and R. C. Haynes, *J. Biol. Chem.* **260**, 12744 (1985).
[75] D. E. Wingrove, J. M. Amatruda, and T. E. Gunter, *J. Biol. Chem.* **259**, 9390 (1984).
[76] M. A. Titheradge, S. B. Binder, R. K. Yamazaki, and R. C. Haynes, *J. Biol. Chem.* **253**, 3357 (1978).
[77] M. A. Titheradge and R. C. Haynes, *FEBS Lett.* **106**, 330 (1979).
[78] K. F. La Noue, T. Strzelecki, and F. Finch, *J. Biol. Chem.* **259**, 4116 (1984).
[79] R. S. Balaban and J. J. Blum, *Am. J. Physiol.* **242**, C172 (1982).
[80] V. Prpic, T. L. Spencer, and F. L. Bygrave, *Biochem. J.* **176**, 705 (1978).

onstrated.[81] Moveover, glucagon administration results in an increased rate of calcium uptake in liver mitochondria.[75,76,82,83] These facts are consistent with the role of cellular calcium fluxes in mitochondrial effects of glucagon.

Triiodothyronine (T_3)

Enhancement of Flux through Biosynthetic Pathways

Both long-term (days) and short-term thyroid hormone effects have been reported. These modulate the same metabolic pathways as those affected by glucagon and α-adrenergic agonists (Table I). Thus, respiration, urea synthesis, gluconeogenesis, and also pyruvate oxidation are stimulated within minutes[84-86] of T_3 administration. Little is known about the mechanism of these effects. Since there is a decrease in mitochondrial ATP and acetyl-CoA with T_3,[87] stimulation of pyruvate carboxylase by ATP or acetyl-CoA or stimulation of carbamoyl-phosphate synthase by ATP cannot be responsible. Although initial studies reported an uncoupling effect of the hormone in isolated mitochondria at high doses ($> 10^{-6} M$),[88,89] it is unlikely that the stimulation of respiration observed at lower doses of T_3 ($< 10^{-6} M$) is due to uncoupling, since the rates of urea synthesis and gluconeogenesis are increased.[86] Also, respiration is still sensitive to oligomycin.[86]

Several observations suggest that ATP synthesis is more effective in the presence of T_3: (1) ^{32}P incorporation into ATP is 150% of control levels with $10^{-6} M$ T_3 in isolated mitochondria[84]; (2) the P/O ratios in liver mitochondria from hypothyroid rats are lower than those from euthyroid animals and are restored 15 min after injection of T_3[90,91]; (3) T_3 restores respiratory capacity of liver mitochondria from thyroidectomized rats within 3 hr by about 50% independent of the absolute amount of respira-

[81] C. Richter and B. Frei, *in* "Oxidative Stress" (H. Sies, ed.), p. 221. Academic Press, London, 1985.
[82] R. K. Yamazaki, D. L. Mickey, and M. Story, *Biochim. Biophys. Acta* **592**, 1 (1980).
[83] J. P. Mauger, J. Poggioli, and M. Claret, *J. Biol. Chem.* **260**, 11635 (1985).
[84] K. Sterling, P. O. Milch, M. A. Brenner, and J. H. Lazarus, *Science* **197**, 996 (1977).
[85] T. Kaminski, *Acta Endocrinol. (Copenhagen)* **94**, Suppl. 234, 15 (1980).
[86] M. J. Müller and H. J. Seitz, *Life Sci.* **27**, 827 (1980).
[87] H. J. Seitz, M. J. Müller, and S. Soboll, *Biochem. J.* **227**, 149 (1985).
[88] D. F. Tapley and C. Cooper, *J. Biol. Chem.* **222**, 341 (1956).
[89] J. R. Bronk, *Biochim. Biophys. Acta* **37**, 327 (1960).
[90] R. Palacios-Romero and J. Mowbray, *Biochem. J.* **184**, 527 (1979).
[91] J. Mowbray and A. Crespo-Armas, *Biochem. Soc. Trans.* **13**, 746 (1986).

tory chain components[92]; (4) incubation of mitochondria and submito-chondrial particles with T_3 enhances H^+/O ratios and proton pumping capability, respectively,[90,93] and consequently a higher proton gradient across the mitochondrial membrane is observed.[91] Moreover, T_3 has both long-term as well as short-term effects on adenine nucleotide transport. Hoch[94] reported a decreased rate of uptake of ADP by mitochondria isolated from hypothyroid rats. The lowered transport activity due to hypothyroidism can be restored by T_3 within 15 min.[95] Decreased steady-state mitochondrial ATP/ADP ratios and increased cytosolic ATP/ADP ratios in intact rat livers of hyperthyroid rats are also consistent with a stimulation of adenine nucleotide transport by T_3.[87] Since the efficiency of ATP synthesis and ATP transport is augmented by T_3, the increase in biosynthetic pathways could, at least in part, be mediated by an increased rate of ATP supply, even if the steady-state mitochondrial concentration of ATP is decreased.

Direct Interaction with Mitochondrial Membrane Proteins

The mediators for the rapid action of T_3 on mitochondria are still unknown. A high-affinity receptor for T_3 in mitochondria has been estab-lished.[96,97] The receptor protein has a molecular weight of 28,000, a value fairly close to the molecular weight of a protein of the mitochondrial membrane reported to be phosphorylated by glucagon.[46] Thus, one way of modifying mitochondrial activity could be direct phosphorylation of cer-tain proteins induced by T_3. This may in turn change the ion permeability of the mitochondria. T_3 increased calcium uptake in mitochondria isolated from thyroidectomized rats, and adenine nucleotide translocation was stimulated in the presence of calcium and T_3.[98] Evidence also suggests that ATP synthase and adenine nucleotide translocation are regulated by cal-cium movements.[99] Thermodynamic studies indicate that the control of oxidative phosphorylation exerted by the adenine nucleotide translocator is not changed by the thyroid state of mitochondria[100]; this, however, does

[92] J. R. Bronk, *Science* **153**, 638 (1966).
[93] G. Martino, C. Covello, and R. De Giovanni, *IRCS Med. Sci.* **13**, 1085 (1985).
[94] F. L. Hoch, *Arch. Biochem. Biophys.* **178**, 535 (1977).
[95] J. Mowbray and J. Corrigal, *Eur. J. Biochem.* **139**, 95 (1984).
[96] F. Goglia, J. Torresani, P. Bagli, A. Barletta, and G. Liverini, *Pfluegers Arch.* **390**, 120 (1981).
[97] K. Sterling, G. A. Campbell, and M. A. Brenner, *Acta Endocrinol. (Copenhagen)* **105**, 391 (1984).
[98] P. A. Herd, *Arch. Biochem. Biophys.* **188**, 220 (1978).
[99] R. Moreno-Sanchez, *J. Biol. Chem.* **260**, 12554 (1985).
[100] M. Holness, A. Crespo-Armas, and J. Mowbray, *FEBS Lett.* **177**, 231 (1984).

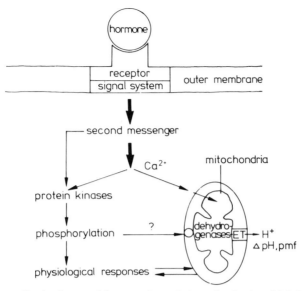

FIG. 1. Generalized scheme of hormonal regulation of mitochondrial functions. ET, Electron transport chain; pmf, protonmotive force.

not exclude stimulation of adenine nucleotide translocation by T_3. However, it is still to be tested whether T_3 can influence respiratory chain components and carriers of the inner mitochondrial membrane directly.

Increase in Substrate Supply to the Respiratory Chain

Stimulation of respiration by T_3 could, in part, be mediated by enhanced substrate supply to the respiratory chain by mechanisms similar to the action of α-adrenergic agonists and glucagon. Thus, T_3 increases cytosolic calcium in isolated hepatocytes and decreases perfusate calcium in perfused rat liver within minutes[101] while also increasing mitochondrial calcium content.[98] Therefore, stimulation of mitochondrial dehydrogenases by enhanced mitochondrial calcium seems conceivable. Also, the mitochondrial/cytosolic proton gradient is increased in perfused liver following infusion of T_3 (Table II), enhancing the driving force for dicarboxylate and tricarboxylate transport into the mitochondria.

[101] H. Hummerich and S. Soboll, *Biochem. J.* **258**, 363 (1989).

Conclusion

As depicted in Fig. 1, the rapid functional responses to hormones seem to arise directly or indirectly from alterations in the degree of phosphorylation of key regulatory proteins brought about by protein kinases and phosphatases sensitive to cAMP (glucagon) or calcium (α-adrenergic agonists, vasopressin,[12,46,102] and T$_3$[98,101]). Whereas for glucagon stimulation of adenylate cyclase and a subsequent increase in cAMP are established, recent work also demonstrated the involvement of calcium in glucagon action,[10,20,66] while α-adrenergic agonists and vasopressin stimulate a membrane-associated phosphodiesterase. Breakdown of phosphatidylinositol 4,5-bisophosphate in the plasma membrane with the production of inositol 1,4,5-trisphosphate (IP$_3$)[11,12] may mediate the non-cAMP effects. IP$_3$ leads to an increase in cytosolic and mitochondrial calcium which stems from the endoplasmic reticulum and from the extracellular space.[10] Calcium then elicits the subsequent changes in mitochondrial metabolism, since there is no indication of a direct mitochondrial effect of IP$_3$.

Phosphorylation of mitochondrial proteins, calcium movements, and/or activation of mitochondrial phospholipases, as well as changes in mitochondrial volume and the mitochondrial/cytosolic proton gradient, may all be mechanisms by which the transport of metabolites into mitochondria is modulated and mitochondrial respiration is activated. In addition, mitochondrial enzymes are directly activated by calcium. T$_3$ may act directly at the mitochondrial membrane through specific receptors[96,97] or as effector of membrane proteins involved in oxidative phosphorylation. Thus, special subunits of respiratory chain components might offer the possibility of allosteric regulation by hormones such as T$_3$.[103]

Acknowledgments

The authors were supported by the Deutsche Forschungsgemeinschaft, Grants No. So 133/5-1 and Si 255/8-1.

[102] P. Cohen, *Eur. J. Biochem.* **151**, 439 (1985).
[103] B. Kadenbach, *J. Bioenerg. Biomembr.* **18**, 39 (1985).

[12] Use of Fluorescein Isothiocyanate-Dextran to Measure Proton Pumping in Lysosomes and Related Organelles

By SHOJI OHKUMA

Introduction

Determination of the ΔpH across a membrane allows calculation of the proton electrochemical potential difference of the protonmotive force. Determination of intraorganellar pH is necessary not only in systems of energy conversion, but also in studies on the control mechanism of organellar and cellular functions. The methods described below should be useful for these purposes in cells, organelles, and membrane vesicles.

Brief Survey of Methods for Determination of pH (ΔpH across Membranes of Cells, Organelles, and Vesicles)

There are several ways of measuring intraorganellar pH, both after isolation of the organelles and in living cells. Techniques for determining ΔpH can be grouped into two classes: (1) ion distribution methods and (2) methods using internal pH indicators. In group (1), the intraorganellar pH is determined from the concentration ratios inside and outside organelles of a weak base (or acid) with a lipophilic neutral form and a hydrophilic ionic form. Methods in group (2) can be subdivided into (a) methods using a microelectrode, (b) optical methods (with colored or fluorescent pH indicator dyes), (c) ^{31}P, ^{19}F, or ^{15}N NMR signals methods, and (d) methods involving measurement of pH-dependent enzyme activities (e.g., lysosomal hydrolases). For further details of these methods, the reader is referred to the literature.[1-3]

Ideally, methods for measuring pH should meet several criteria, among which the most important are (1) specificity for and sensitivity to pH, (2) nondestructiveness (no perturbation), (3) restriction to a specific subcellular compartment, and (4) the ability to obtain continuous, rapid measurements throughout cellular or organellar reactions. Most methods, however,

---

[1] H. Rottenberg, this series, Vol. 55, p. 547.

[2] R. Nuccitelli and D. W. Deamer, eds., "Intracellular pH: Its Measurement, Regulation, and Utilization in Cellular Functions." Liss, New York, 1982.

[3] S. Ohkuma, *in* "Lysosomes; Their Role in Protein Breakdown" (H. Glaumann and F. J. Ballard, eds.), p. 115. Academic Press, Orlando, FL, 1987.

satisfy only some of these criteria. Simple, sensitive, and prompt measurement of the internal pH of organelles can be accomplished with a pH indicator, especially a fluorescent indicator, that is located internally. However, fluorescence is known to be influenced by various factors besides pH. Therefore, many points must be clarified before a fluorescent probe can be used for pH determinations.

Many fluorescent pH indicators are known,[4] among which fluorescein and its derivatives were found to be suitable for the present purpose. This chapter describes details of the fluorescence methods with special emphasis on use of fluorescein isothiocyanate labeled-dextran (FITC-dextran).

Fluorescein as pH Probe

Basic Properties of Fluorescein

Fluorescein has long been used as a fluorescent histochemical agent, but its application for determination of pH was reported only recently.[5-9] Studies on the pH dependence of fluorescein fluorescence were reported in 1971[10] and in more detail in 1975.[11,12]

Fluorescein exists at equilibrium in different molecular forms in aqueous solution. The equilibrium changes depending on ambient pH (Fig. 1). Each form has a different absorption (and, therefore, excitation) spectrum and different quantum yield of fluorescence (Table I). Therefore, the total fluorescence spectrum and intensity represent the sums of those of the different species present. Under physiological conditions, fluorescein exists mainly as a mixture of dianion, monoanion, and neutral molecules. Therefore, its total fluorescence intensity (F) at a given excitation wavelength can theoretically be expressed by Eq. (1), although the fluorescence decreases at extreme alkalinity for some reason.

$$F = \frac{\epsilon_1 q_1 K_1 K_2 + \epsilon_2 q_2 K_1 [H^+] + \epsilon_3 q_3 [H^+]^2}{K_1 K_2 + K_1 [H^+] + [H^+]^2} \cdot [\text{fluorescein}] \cdot V \cdot I \quad (1)$$

[4] R. C. Weast, ed., "CRC Handbook of Chemistry and Physics," 6th ed., p. D149. CRC Press, Cleveland, Ohio, 1976–1977.

[5] S. Ohkuma and B. Poole, *Proc. Natl. Acad. Sci. U.S.A.* **75**, 3327 (1978).

[6] B. Poole and S. Ohkuma, *J. Cell Biol.* **90**, 665 (1981).

[7] M. Eisenbach, H. Garty, E. P. Bakker, G. Klemper, H. Rottenberg, and S. R. Caplan, *Biochemistry* **17**, 4691 (1978).

[8] J. A. Thomas and D. L. Johnson, *Biochem. Biophys. Res. Commun.* **65**, 931 (1975).

[9] J. A. Thomas, R. N. Buchsbaum, A. Zimniak, and E. Racker, *Biochemistry* **18**, 2210 (1979).

[10] H. Leonhardt, L. Gordon, and R. Livingston, *J. Phys. Chem.* **75**, 245 (1971).

[11] M. M. Martin and L. Lindqvist, *J. Lumin.* **10**, 381 (1975).

[12] K. Schauenstein, E. Schauenstein, and G. Wick, *J. Histochem. Cytochem.* **26**, 277 (1978).

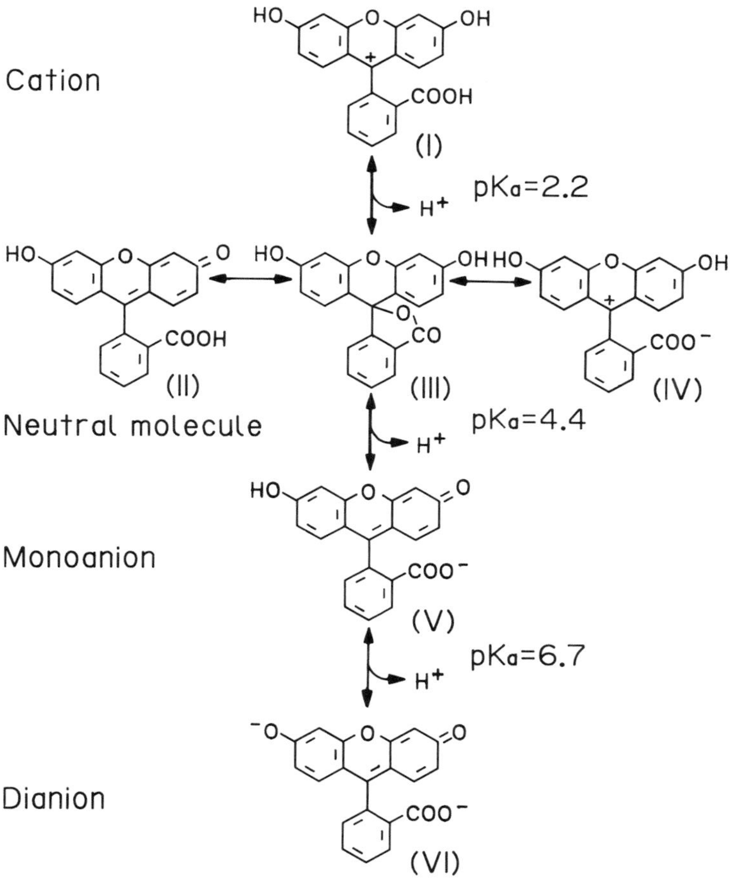

FIG. 1. Structure and equilibrium of fluorescein in aqueous solution.

ϵ_i and q_i represent the molecular absorption coefficient and quantum yield equivalence at a given wavelength of dianion (1), monoanion (2), and neutral (3) species of fluorescein molecules, respectively; K_i represents the proton dissociation constant of neutral (1) and monoanion (2) species of fluorescein molecules, respectively (see Fig. 1 and Table I), and V and I represent the effective sample volume and light intensity, respectively. Equation (1) shows that the environmental pH can be estimated from the fluorescence intensity of fluorescein, provided the concentration of fluorescein is known.

Figure 2A shows the excitation spectra of fluorescence of FITC-dextran (see Fig. 4 for its structure) measured at 519 nm. The spectra exhibit a

TABLE I
PROTON DISSOCIATION CONSTANTS, ABSORPTION MAXIMA, AND QUANTUM YIELDS OF
DIFFERENT FORMS OF FLUORESCEIN[a]

Molecular form[b]	pK_a	Absorption peak (nm)	Molar absorption coefficient (ϵ, M^- cm^{-1})	Quantum yield
Cation (I)	2.2	437	55,000	0.9–1
Neutral molecule (II + III + IV)	4.4	437	16,000	0.20–0.25
Monoanion (V)	6.7	452	30,000	0.25–0.35
		475	31,000	
Dianion (VI)	—	491	88,000	0.93

[a] Modified from Martin and Lindqvist.[11]
[b] See Fig. 1.

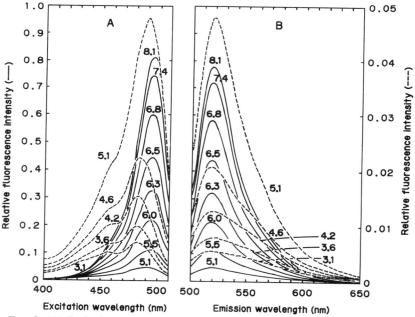

FIG. 2. (A) Excitation and (B) emission fluorescence spectra at 37° of FITC-dextran (FD-70) at different pH. The concentration of FD-70 was 1 μg/ml in 50 mM NaCl and 10 mM phosphate buffer at the pH indicated. Emission was measured at 519 nm (A) and excitation at 495 nm (B) with 5 nm slits on both monochromators. (Modified from Ohkuma and Poole.[5])

strong peak at 495 nm at alkaline pH that changes at lower pH to two peaks of much lower intensity at 480 and 450 nm.

Theoretically, the fluorescence ratio (R) (fluorescence intensity ratio after excitation at two different wavelengths, λ_1 and λ_2; e.g., 495 and 450 nm) can be expressed by Eq. (2), which is independent of the dye concentration.

$$R = \frac{AK_1K_2 + BK_1[H^+] + C[H^+]^2}{EK_1K_2 + FK_1[H^+] + G[H^+]^2} \quad (2)$$

A, B, and C represent ϵ_1q_1, ϵ_2q_2, and ϵ_3q_3 at λ_1, and E, F, and G represent the corresponding values at λ_2. Rearrangement of Eq. (2) gives

$$[H^+] = \frac{-(B - FR)\,K_1 \pm \sqrt{(B - FR)^2K_1^{\,2} - 4(C - GR)(A - ER)K_1K_2}}{2(C - GR)}$$

$$(3)$$

Therefore, the pH can be calculated once the value of R has been obtained by measuring the fluorescence on excitation at λ_1 and λ_2.

Figure 3A shows the pH dependence of the fluorescence intensities of FITC-dextran on excitations at 495 and 450 nm and their ratios. The fluorescence intensity ratio on excitations (ex) at 495 and 450 nm ($F_{ex=495\ nm}/F_{ex=450\ nm}$) varies from 10.0 at pH 8 to about 1.0 at pH 4. At lower pH values, the fluorescence ratio on excitations at 495 and 430 nm is more sensitive to pH. Emission (em) spectra (Fig. 2B) can also be used for this purpose (e.g., ex = 495 nm, em = 519 and 590 nm) as they usually exist in mirror image symmetry to excitation spectra, but their effective pH range is narrow (Fig.3B).

Fluorescein and its derivatives are very suitable for measurements of pH within lysosomes and related acidic organelles, because their absorption and therefore fluorescence excitation spectra change between pH 4 and 8. The modified forms of fluorescein, 4',5'-dimethylfluorescein (pK_a 6.75) (DMF, used as its diacetate or DMF-dextran[13]) and 2',7'-bis(carboxyethyl)-5,6-carboxyfluorescein (BCECF, pK_a 6.97, used as its ester[14]), have different pK values that make them suitable for use in measurement of the pH of cytoplasm.

Fluorescein and its derivatives can be introduced into isolated lysosomes as a (1) lipophilic nonfluorescent diacetate ester [e.g., fluorescein diacetate (FDA), carboxyfluorescein diacetate (CFDA)], which is hydrolyzed inside organelles such as lysosomes to fluorescent and less permeant fluorescein or its derivatives. For suppression of its release from inside

[13] P. Rothenberg, L. Glaser, P. Schlesinger, and D. Cassel, *J. Biol. Chem.* **258**, 4883 (1983).
[14] T. J. Rink, R. Y. Tsien, and T. Pozzan, *J. Cell Biol.* **95**, 189 (1982).

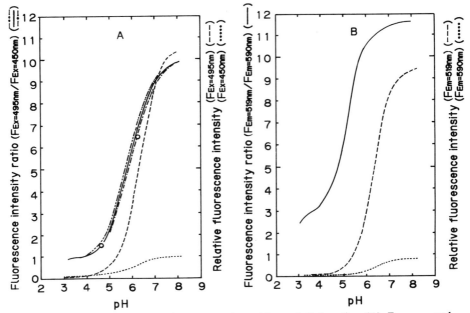

FIG. 3. pH dependence of fluorescence intensities and their ratios. (A) $F_{ex=495nm}$ and $F_{ex=450nm}$ and their ratio ($F_{ex=495nm}/F_{ex=450nm}$) (emission wavelength 519 nm). (B) $F_{em=519nm}$ and $F_{em=590nm}$ and their ratio ($F_{em=519nm}/F_{em=590nm}$) (excitation wavelength 495 nm). The concentration of FD-70 was 1 μg/ml in 10 mM buffer and NaCl ($-\cdot-$, 0 mM; —, ---, $\cdots$, 50 mM; $-\cdot\cdot-$, 100 mM). ○, fluorescence from 30 mg/ml FD-70 in liposomes. (Modified from Ohkuma and Poole.[5])

organelles, the dye may be induced to bind covalently to intraorganellar proteins by use of azido derivatives (e.g., azidofluorescein diacetate,[15] which binds to low molecular weight proteins). (2) Fluorescein can also be coupled to high molecular weight carriers, such as the polysaccharide dextran (FITC-dextran; O-thiocarbamoyl bond)[6] or protein (FITC-ovalbumin; N-thiocarbamoyl bond).[16] These coupled dyes can be introduced into lysosomes and endosomes via the endocytic activity of the cells, into cell cytoplasm by several microinjection methods,[13,16] or into membrane vesicles or isolated organelles by reconstitution or sonication.

[15] A. Rothman and J. Heldman, *FEBS Lett.* **122**, 215 (1980).
[16] J. M. Heiple and D. L. Taylor, *J. Cell Biol.* **86**, 885 (1980).

Application of FITC-Dextran (FD) as pH Indicator

Figure 4 shows the structure of FITC-dextran {O-[(5-fluoresceinyl)thio-carbamoyl]-dextran, fluorescein-dextran, FD}. The pK_a' values for the monoanion of conjugated fluorescein were determined by Geisow[17] from the pH dependence of fluorescence intensity data, assuming a single dissociation constant at physiological pH. Results indicated that the pK_a' values of FD-70 (MW 70,000) and FD-40 (MW 40,000) are 6.40 ± 0.03 and 6.30 ± 0.03, respectively. These values are slightly lower than those of free fluorescein (pK_a' 6.7; Table I) and FITC-ovalbumin (pK_a' 6.80 ± 0.03).

The advantages of FD over other fluorescein derivatives are as follows: (1) It is stable: no evidence of breakage of the carbamoyl–dextran linkage during incubation in plasma at pH values between 4 and 9 could be found at $20° - 30°$ for at least 1 month or at $37°$ for 24 hr; it is also stable for at least 24 hr under physiological conditions *in vivo*.[18–20] (2) It is resistant to most enzymes found *in vivo* including lysosomal enzymes. (3) It is inert and has not yet been found to be toxic. (4) It is hydrophilic and completely impermeant through biological membranes. (5) It has not been shown to bind to proteins *in vivo*.[19]

Various molecular weights of FD are available from Sigma. FD is also easily prepared.[20] For its preparation, a solution of dextran (1 g) in dimethyl sulfoxide (DMSO) (10 ml) is mixed with FITC (100 mg) and dibutyltin dilaurate {dibutylbis(lauroyloxy)tin, $(C_4H_9)_2Sn[OOC(CH_2)_{10}CH_3]$ (20 mg)}, heated at $95°$ for 2 hr, and then precipitated with ethanol several times to separate it from free dye. FD is then filtered off and dried *in vacuo* at $80°$. The yield is 0.9 g, and the degree of substitution is over 0.001 mol FITC/mole glucose residue. Another simple, but less efficient, preparation method is to conjugate dextran (20 mg in 5 ml of 4 mM NaHCO$_3$ – 3.6 mM Na$_2$CO$_3$, pH 8.0) with 1 mg FITC by stirring the mixture overnight at room temperature. The resulting FD shows an $E_{492\ nm}$ of 3.8 at 2 mg/ml in saline.[21] Sometimes the preparations are toxic, probably because of the remaining catalyst, dibutyltin dilaurate, used in the synthesis; this can be removed by extensive dialysis and ethanol precipitation. In a recently reported new method of preparation,[22] cyanogen bromide (CNBr)-activated polysaccharide is coupled with fluoresceinamine to obtain the N-fluoresceinyl imidocarbonate derivative of dextran. The degree of substitu-

[17] M. J. Geisow, *Exp. Cell Res.* **150**, 29 (1984).
[18] U. Schröder, K. E. Arfors, and O. Tangen, *Microvasc. Res.* **11**, 33 (1976).
[19] G. Rutili and K.-E. Arfors, *Microvasc. Res.* **12**, 221 (1976).
[20] A. N. de Belder and K. Granath, *Carbohydr. Res.* **30**, 375 (1973).
[21] V. K. Ghanta, N. M. Hamlin, and R. N. Hiramoto, *Immunochemistry* **10**, 51 (1973).
[22] C. G. Glabe, P. K. Harty, and S. D. Rosen, *Anal. Biochem.* **130**, 287 (1983).

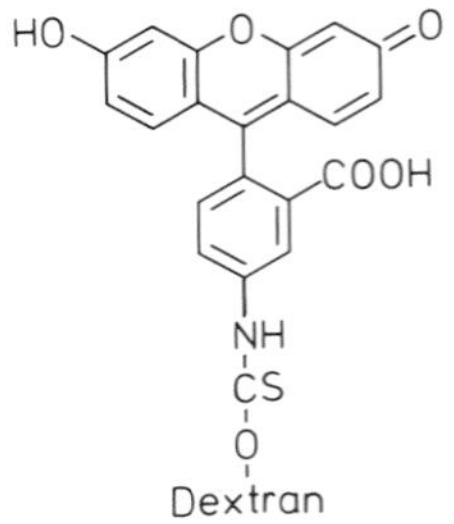

FIG. 4. Structure of FITC-dextran, O-[(5-fluoresceinyl)thiocarbamoyl]-dextran.

tion is 3×10^{-3} to 2.4×10^{-2} mol/mol monosaccharide residue. This product is stable *in vitro,* but its stability *in vivo* has not yet been reported.

Control experiments indicate the pH specificity of the fluorescence intensity ratio of FD (Fig. 3).The ratio is not significantly affected by the type of buffer, nor by the presence of 100 mM NaCl, KCl, NH$_4$Cl, or methylamine. The fluorescence intensity may be affected by interactions of adjacent fluorophores at high concentrations (self-quenching) or by the presence of high concentrations of protein. In fact, Schauenstein *et al.*[23] showed that the fluorescence emissions of fluorescein diacetate, free FITC, and FITC-γ-globulin decrease at dye concentrations above 5 μg/ml (10 μM). However, the fluorescence intensity ratio of FD is independent of the probe concentration from 1 μg/ml to 30 mg/ml (Fig. 3).

The effect of protein on the fluorescence intensity ratio of FITC-dextran has also been examined using 200 mg/ml BSA. The results indicated that protein had only a slight effect on the fluorescence ratio, though it decreased the fluorescence intensity. Moreover, conjugation of the fluorescein moiety to the dextran molecule, a globular polysaccharide with high molecular weight, is expected to prevent self-quenching, which requires excimer formation at intermolecular distances of about 3.5 Å. Another advantage of this dye is that the fluorescein is surrounded by a bulky hydrophilic environment of dextran, which should reduce the undesirable effect of hydrophobic substances, some of which may affect the pK value of fluorescein. FITC-dextran has been reported to interact with the liposome membrane, as judged by fluorescence depolarization.[24] However, the fluorescence intensity ratio was not changed in liposomes (Fig. 3).

A potential problem when using fluorophores with mammalian phagocytes is the active production of H$_2$O$_2$ in the presence of peroxidase, which

[23] K. Schauenstein, E. Schauenstein, and G. Wick, *J. Histochem. Cytochem.* **26,** 277 (1978).
[24] K. Iwamoto and J. Sunamoto, *J. Biochem. (Tokyo)* **91,** 975 (1982).

may cause oxidative loss of fluorescence. However, even with an enzymatic oxidation system *in vitro,* no alteration in the pH indicator properties of fluorescein conjugates was noted. The fluorescence of 2',7'-dichlorofluorescein has also been shown to be unaffected by hydrogen peroxide concentrations up to 3×10^{-4} M.[25] Furthermore, there is little difficulty in pH measurement in phagosomes of phagocytes (see below).

The results obtained so far indicate, therefore, that this method should give a reliable estimate of the environmental pH. Similar extensive studies on the specificity of the fluorescein fluorescence of FITC-ovalbumin for measurement of pH have been performed by Heiple and Taylor.[26]

The fluorescence of a probe within cellular organelles is known, however, to be affected by unpredictable and often undetermined parameters of the cell environment. For example, the fluorescence is affected by solvent hydrogen-bonding effects.[27] Furthermore, FITC-protein indicated that the intracellular excitation spectrum showed a red shift relative to that *in vitro.*[28] Therefore, the curve for fluorescence obtained *in situ* is better for a calibration. The *in situ* calibration curve can be determined by equilibration of the internal pH with the external pH using a high concentration (100 mM) of acetate or ammonium ion, as described by Pollard *et al.,*[29] assuming that the contribution of cellular constituents to the acid–base equilibrium is negligible compared with the large excess of acid or base used to collapse the gradient. Acidic ionophores such as nigericin or monensin have also been applied in *in situ* probe calibration with great success.[17] But final, satisfactory resolution of this calibration problem will obviously depend on corroboration of the values for the standard curve by entirely independent techniques (an electrode method, etc.).

Recently Geisow[17] improved the method by replacing the neutral and monoanion forms of fluorescein, which have low fluorescence, by tetramethylrhodamine (TR), a long-wavelength emitting dye with much stronger fluorescence [molar extinction coefficient at 550 nm = 75,500 M^{-1} cm^{-1}; quantum yield = 0.97 (in ethanol)], and measuring the fluorescence at wavelengths that show maximum fluorescence (the excitation and emission wavelengths are, respectively, about 495 and 530 nm for FITC-dextran, and 550 and 580 nm for TRITC-dextran). TRITC-dextran can be

[25] M. J. Black and R. B. Brandt, *Anal. Biochem.* **58,** 246 (1974).
[26] J. M. Heiple and D. L. Taylor, *in* "Intracellular pH: Its Measurement, Regulation, and Utilization in Cellular Functions" (R. Nuccitelli and D. W. Deamer, eds.), p. 21. Liss, New York, 1982.
[27] M. M. Martin, *Chem. Phys. Lett.* **35,** 105 (1975).
[28] K. G. Romanchuk, *Surv. Ophthalmol.* **26,** 269 (1982).
[29] H. B. Pollard, H. Shindo, C. E. Creutz, C. J. Pazoles, and J. S. Cohen, *J. Biol. Chem.* **254,** 1170 (1979).

prepared as described by Geisow.[17] A mixture of FITC-dextran and TRITC-dextrans (1 mg/ml; $A_{550 nm}^{1 cm} : A_{493 nm}^{1 cm} = 3.0$) is recommended as a probe. Murphy et al.[30] modified the method by using the emission fluorescence to estimate pH. They used double fluorescence flow cytometry to measure the pH of individual cells in a cell suspension as well as the distribution of pH values in the cells.

One advantage of methods for measuring fluorescence change is that cell morphology can be examined microscopically during these measurements. Furthermore, the recent development of silicon image intensifiers and computational image processing techniques allows the estimation of pH of even single vesicles.[31-33] Furthermore, pH changes in these vesicles can be monitored continuously: this approach has been applied in studies on the dynamics of vacuole acidification after endocytosis (see below).

A drawback of this method is that pH is very hard to estimate after drug (basic substance) treatment, as most alkaloids (e.g., chloroquine and atropine) influence the fluorescence of fluorescein, at high concentrations shifting the spectrum and/or pK value. Thus, the calibration curve must be obtained in the presence of different concentrations of these chemicals. But sometimes advantage can be taken of this phenomenon to calculate the pH and concentration of basic substances at the same time.[6]

Measurement of the environmental viscosity (especially in acidic compartments such as lysosomes) from data on flurorescence polarization recovery is not simple with FD, although the use of FD for this purpose in lysosomes has been suggested.[17,34] Difficulty arises because the fluorescence polarization of FD also depends on the pH and is stronger at acidic pH values, suggesting a shorter fluorescence lifetime of monoanion and/or neutral species (S. Ohkuma, unpublished observation).

Determination of pH and Proton Pumping in Cell Cultures

Lysosomes

Intralysosomal pH. Ohkuma and Poole[6] first developed a quantitative method for measuring the internal pH of lysosomes using FD. For measurement of intralysosomal pH, FD can be endocytosed by cells during overnight incubation with them. FITC-dextran is taken up by fluid endocytosis, remains within the cells with little apparent degradation, and can

[30] R. F. Murphy, S. Powers, and C. R. Cantor, *J. Cell Biol.* **98**, 1757 (1984).
[31] B. Tycko and F. M. Maxfield, *Cell (Cambridge, Mass.)* **28**, 643 (1982).
[32] L. Tanasugarn, P. McNeil, G. T. Reynolds, and D. L. Taylor, *J. Cell Biol.* **98**, 717 (1984).
[33] J. M. Heiple and D. L. Taylor, *J. Cell Biol.* **86**, 885 (1980).
[34] M. J. Geisow, P. D'Arcy Hart, and M. R. Young, *J. Cell Biol.* **89**, 645 (1981).

be seen under a fluorescence microscope in the Golgi region as fluorescent dots. Cell fractionation indicates that most of the endocytosed FD is localized predominantly within lysosomes. Incorporation of FD amounts to 2.1–3.4 mM fluorescein or 0.6–0.9 mM (40–64 mg/ml) FD after 24-hr incubation with 1 mg/ml of FD, assuming that lysosomes occupy 2.5% of the cell volume.

The fluorescence of FD incorporated into cells can be measured in (1) cell suspensions, (2) cells on coverslips in a regular fluorescence cuvette, and (3) the same cells under a fluorescence microscope. For case (2), a special holder for the coverslip can be used (Fig. 5). In each case, it is important to reduce the perturbing effect of scattered (stray) light on fluorescence measurement. For this purpose, for case (2), the coverslip is tilted about 30 degrees toward the direction of excitation light. It is very important to fix the angle to reduce the variation of background fluorescence. Furthermore, the difference between the excitation and emission

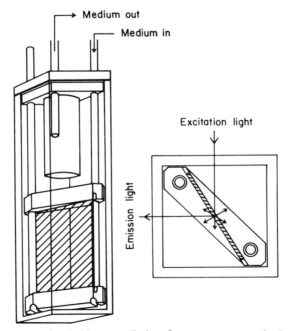

Fig. 5. Apparatus used to hold a coverslip in a fluorescence cuvette in the spectrofluorometer. The coverslip (hatched) was aligned at 30° to the excitation beam to reduce the effect of light reflection. Approximately 30 mm² of the coverslip was illuminated (about 10⁵ cells). (From Ohkuma and Poole.[5])

wavelengths should be adjusted to be sufficiently wide to reduce the scattered light.

The shape of the excitation spectrum of the dye accumulated in the macrophages indicated that the labeled dextran was in an environment with a pH of 4.7–4.8 (cf. Fig. 6). This conclusion was conformed by Geisow et al.[17,34] However, the concentration of FD within lysosomes could possibly reach as high as 50 mM for FD with a molecular weight of 70,000 and a degree of substitution of 0.01 (4 fluorescein residues per molecule) from osmotic pressure requirements. This suggests that the value obtained might show errors arising from concentration quenching, etc., although the pH dependence of the fluorescence spectrum (fluorescence intensity ratio) of FD did not change with up to 30 mg/ml of FD in vitro (Fig. 3). As discussed before, the validity of the fluorescein method should be established by confirmation by independent techniques. It is noteworthy, in this respect, that Hollemans et al.[35] calculated a lysosomal pH of 5.29 for cultured fibroblasts from compartmental analysis using methylamine and chloroquine.

As a modification of this method, McNeil et al.[36] used photon-counting microspectrofluorimetry for measurement of fluorescence from Amoeba proteus with incorporated FITC-RNase. Heterogeneity of lysosomal pH was also recognized in studies with a fluorescence microscope coupled with a silicon image intensifier using the video image-processing technique.[32]

Energy Requirement for Maintenance of the Intralysosomal pH. Direct evidence for the requirement of energy for maintenance of the pH of lysosomes was obtained using FITC-dextran.[6] The lysosomal pH in cultured macrophages increased from 4.8 to 5.0 and to approximately 5.4 within 5 min when the cells were exposed to 50 mM 2-deoxyglucose or to 10 mM NaN$_3$ plus 50 mM 2-deoxyglucose at pH 7.6. A greater inhibitory effect was observed when the cells were incubated at pH 6.6 (the ΔpH increase was 0.82 instead of 0.53–0.57) or when azide was replaced by cyanide (ΔpH was 1.05).[37] A good correlation was found between the cellular ATP level and increases in lysosomal pH: after treatment with 2-deoxyglucose plus NaN$_3$, cellular ATP content was 15% of control values (0.7 mM).[37] These results are consistent with the K_m value (about 0.3 mM) for ATP of lysosomal H$^+$-ATPase (see below). The lysosomal pH returned to normal within 10–20 min after removal of the inhibitors.[6]

[35] M. Hollemans, R. O. Elferink, P. G. de Groot, A. Strijland, and J. M. Tager, Biochim. Biophys. Acta 643, 140 (1981).
[36] P. L. McNeil, L. Tanasugarn, J. B. Meigs, and D. L. Taylor, J. Cell Biol. 97, 692 (1982).
[37] S. Ohkuma and B. Poole, unpublished observation.

Endosomes

Intraendosomal pH. Essentially the same procedure was used to show that the endosomal pH is acidic. There are two ways to restrict the location of FD within endosomes: (1) allowing the cells to endocytose FD at a reduced temperature (18° or lower) to inhibit lysosome–endosome fusion[38] and measuring the fluorescence, and (2) using a fluorescein-labeled ligand against cellular receptors and determining the fluorescence within a short period (e.g., 1–5 min). [Fluorescein-labeled protein can be obtained by incubating proteins with FITC in 0.5 M HCO_3^-, pH 9–10, at 4°–37°. Fluoresceinamine can also be used with the condensing reagent 1-ethyl-3-(dimethylaminopropyl) carbodiimide.[34]] The fluorescence observed indicated that the fluorescein-labeled substrate was in an environment with a pH of 5.4–6.0.[39] This conclusion was confirmed by many studies including those of Geisow and Evans,[40] Murphy *et al.*[30] [with fluorescein-rhodamine-conjugated ligands (insulin, α_2-macroglobulin)], van Renswoude *et al.*[40] (with FITC-transferrin), Tycko and Maxfield[31] (with FITC-α_2-macroglobulin), Heiple and Taylor[42] (with FITC-protein in *Chaos carolinensis*), and McNeil *et al.*[36] (with FITC-protein in *Amoeba proteus*). Figure 6 shows our results. However, in the case of *Entamoeba histolytica*,[43] large vacuoles have been shown to be a nonacidified compartment and to equilibrate rapidly with the external medium.

Using this method, heterogeneity of the endosomal pH was also recognized using a video image-processing technique coupled with a silicon image intensifier.[36] Furthermore, "para Golgi" has been shown to have a pH of 6.5.[44]

Energy Requirement for Maintenance of Intraendosomal pH. Direct evidence for the requirement of energy for maintenance of the intraendosomal pH can also be obtained by use of FD as a pH indicator and metabolic inhibitors. The endosomal pH was shown to increase from 5.4 to approximately 6.4 within 5 min when cells were exposed to cyanide and 2-deoxyglucose or to 10 mM NH_4Cl (cf. our results; Fig. 6) at pH 7.6.[39] The endososomal pH returned to normal within 10–20 min after removal of the inhibitors or amines (cf. Fig. 6).

[38] W. A. Dunn, A. L. Hubbard, and N. N. Aaronson, Jr., *J. Biol. Chem.* **255**, 5971 (1979).

[39] B. Tycko, C. H. Keith, and F. R. Maxfield, *J. Cell Biol.* **97**, 1762 (1982).

[40] M. J. Geisow and W. H. Evans, *Exp. Cell Res.* **150**, 36 (1984).

[41] J. van Renswoude, K. R. Bridges, J. B. Harford, and R. D. Klausner, *Proc. Natl. Acad. Sci. U.S.A.* **79**, 6186 (1982).

[42] J. M. Heiple and D. L. Taylor, *J. Cell Biol.* **94**, 143 (1982).

[43] S. B. Aley, Z. A. Cohn, and W. A. Scott, *J. Exp. Med.* **160**, 724 (1984)

[44] D. J. Yamashiro, B. Tycko, S. R. Fluss, and F. M. Maxfield, *Cell (Cambridge, Mass.)* **37**, 789 (1984).

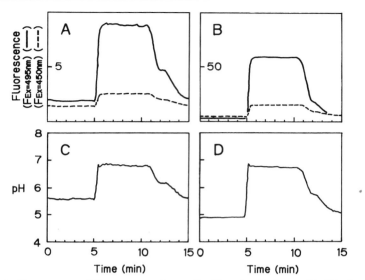

FIG. 6. Time course of fluorescence change (A, B) and pH change (C, D) of cultured macrophages with FITC-dextran in their endosomes (A, C) or lysosomes (B, D). Mouse peritoneal macrophages were fed with 40 mg/ml of FD-70 at 18° for 2 hr (A, C) or at 37° overnight (B, D). Fluorescence from cells in Hanks'–HEPES medium (18°) was measured continuously at 495 and 450 nm by changing the wavelength at fixed (10-sec) intervals in a spectrofluorometer (FS-401, Union Giken Co., Ltd., Osaka, Japan). NH₄Cl (40 mM was added at 5 min and removed at 10 min by perfusion with fresh medium. (Slow recovery of fluorescence after 10 min is due to a slow perfusion rate.) For the estimation of pH, see text. (From S. Ohkuma and T. Takano, in preparation.)

Urinary Bladder

Urinary pH has been shown to be maintained by a kind of exocytosis of acidic plasmalemmal vesicles of epithelial cells of urinary bladder (turtle) or kidney. These vesicles can be labeled with FD during vesicle membrane recycling (i.e., endocytosis). In this way the pH was found to be about 6.0[45] (rabbit kidney) or 5.0 (turtle urinary bladder).[46] The internal pH of these vesicles was found to be increased by addition of ammonium chloride, a combination of metabolic inhibitors (2 mM CN⁻ and 10 mM ICH$_2$CO$_2^-$), or 10 μM CCCP (carbonyl cyanide m-chlorophenylhydrazone), indicating the presence of an H⁺ pump on their membranes.[45,46]

[45] G. J. Schwartz and Q. Al-Awqati, J. Clin. Invest. 75, 1638 (1985).
[46] S. Gluck, C. Cannon, and Q. Al-Awqati, Proc. Natl. Acad. Sci. U.S.A. 79, 4327 (1982).

Phagosomes

Essentially the same procedure can be used for the measurement of intraphagosomal pH. Fluorescein-labeled phagocytosable substrate can be adsorbed onto the membrane of phagocytes via incubation at low temperature (4°). The phagocytic process and the pH change within phagosomes can be synchronized to some extent by increasing the ambient temperature (to 37°). However, for determination of the true intraphagosomal pH, several factors should be considered besides the background fluorescence (without fluorescent particles): (1) the contribution of fluorescence emitted from fluorescent particles outside cells or attached to the invaginations, and (2) the contribution of fluorescein released from fluorescent particles by degradation.[47] These contributory factors should all show fluorescence at the pH of the medium. Measurements can be made in several ways including (1) morphological observations,[47] (2) use of antifluorescein antibody, and (3) rapid change of the external pH by addition of an acidic buffer (see *in vitro* section for details). For calculation of the true fluorescence intensity ratio, the fluorescence at both 495 and 450 nm should be corrected by subtraction of the fluorescence due to contributory factors. Details of the process depend on the method used for estimation of the contribution from the factors mentioned above, and for these details readers are referred to the literature.[47]

Intraphagosomal pH. Decrease in phagosomal pH after phagocytosis of microorganisms was confirmed using fluorescein-labeled yeasts phagocytosed by macrophages.[34] The phagosomal pH also varies depending on the type of materials phagocytosed.[48] However, the interior of phagosomes of *Paramecium* and *Amoeba* was shown not to be acidic and to be in equilibrium with the pH of the medium (see previous section).

Energy Requirement for Intraphagosomal pH Maintenance. Geisow et al.,[34] using mouse peritoneal macrophages containing phagocytosed FD in lysosomes and fluorescein-labeled yeast in phagosomes, showed that during the process of yeast cell phagocytosis (1) the lysosomal pH increases from 4.8 to 5.5 while (2) the phagosomal pH decreases to 5.5 after a transient increase from 6.8 to 7.4. (True pH can be obtained by the correction described above.[47]) This transient increase in intraphagosomal pH was not observed in leukocytes from patients with chronic granulomatous disease[49] and was thought to be related to a metabolic burst (O_2 consumption), namely, activation of NADPH-dependent oxygen con-

[47] P. Cech and R. Lehere, *Blood* **63,** 88 (1984).
[48] C.-F. Bassøe and R. Bjerknes, *J. Med. Microbiol.* **19,** 115 (1985).
[49] A. W. Segal, M. Geisow, R. Garcia, A. Haper, and R. Miller, *Nature (London)* **290,** 406 (1981).

sumption (O_2^- production) in tertiary granules resulting in increase in pH through the production of H_2O_2 from O_2^- ($2\,O_2^- + 2\,H^+ \rightarrow H_2O_2 + O_2$).[50]

Cytoplasm

FD can also be used for measurement of the cytoplasmic pH when it is introduced into the cytoplasm by microinjection.[13] The cytoplasmic pH has been shown to be 6.6 to 7.4 depending on the conditions and the cell type. 4′,5′-Dimethylfluorescein-dextran[13] (A-431 cells), FITC-ovalbumin[33] *(Amoeba),* and FD[32] (3T3 cells) have been used as pH probes, but more frequently used probes are 6-carboxyfluorescein diacetate,[9] BCECF,[14] and quene 1.[51] The cytoplasmic pH has been shown to be maintained by (1) Na^+/H^+ antiport protein and (2) HCO_3^-/OH^- exchanging protein on the plasma membrane.[52] Furthermore, the cytoplasmic pH was shown to be increased by mitogenic stimulation with epidermal growth factor (EGF), hemagglutinin, etc., by activation of the antiport system.

Determination of pH and Proton Pumping in Isolated Organelles

Lysosomes

For measurement of the pH within isolated lysosomes, FD can be introduced into the lysosomes by (1) injection of FD into the peritoneal cavity of animals (rats, mice) and subsequent removal of organs (liver, kidney) for isolation of FD-containing lysosomes,[53] (2) incubation of cells in FD-containing medium overnight and subsequent isolation of lysosomes containing FD,[54,55] or (3) incubation of isolated lysosomes with membrane-permeant fluorescein derivatives.[56]

Intralysosomal pH. After its intraperitoneal injection into animals, FD spreads to the blood plasma via the lymphatics.[57] Cells of organs can then take up FD from the blood plasma by fluid endocytosis. FD is lost from the body by slow destruction and excretion via the kidney in the urine.

[50] A. W. Segal, *J. Clin. Invest.* **19**, 551 (1981).

[51] J. Rogers, T. R. Hesketh, G. A. Smith, and J. C. Metcalfe, *J. Biol. Chem.* **258**, 5994 (1983).

[52] L. Simchowitz and A. Roos, *J. Gen. Physiol.* **85**, 443 (1985).

[53] S. Ohkuma, Y. Moriyama, and T. Takano, *Proc. Natl. Acad. Sci. U.S.A.* **79**, 2758 (1982).

[54] C. J. Galloway, G. E. Dean, M. Marsh, G. Rudnick, and I. Mellman, *Proc. Natl. Acad. Sci. U.S.A.* **80**, 3334 (1983).

[55] M. Merion, P. Schlesinger, R. M. Brooks, J. M. Moehring, T. J. Moehring, and W. S. Sly, *Proc. Natl. Acad. Sci. U.S.A.* **80**, 5315 (1984).

[56] L. Altstiel and D. Branton, *Cell (Cambridge, Mass.)* **32**, 921 (1983).

[57] M. F. Flessner, R. L. Dedrick, and J. S. Schultz, *Am. J. Physiol.* **248**, H15 (1985).

When FD is introduced into the vacuolar system, there is no necessity to take precaution against the contribution of contaminating organelles other than the vacuolar system or to isolate pure membrane vesicles of the type under study, because FD is localized entirely within the organelle of interest. This is another advantage of using FD.

For measurement of the intralysosomal pH using lysosomes containing FD, the pH can be determined from the fluorescence emitted from those FD-containing lysosomes suspended in a suitable buffer (preferably a non-permeant "Good buffer"). However, for measurement of the true intralysosomal pH, corrections should be made for the fluorescence emitted from FD outside the organelles and for the background fluorescence (organelles without fluorescein), as in the case of measurement of the phagosomal pH in cells.

Fluorescence arising from material exposed to the pH of the medium can be measured by (1) use of antifluorescein antibody[41] or (2) rapid change of the external pH by addition of sufficient membrane-impermeant acidic buffer [e.g., 1/20 vol of 1 M MES–tetramethylammonium hydroxide (TMAH), pH 6.0][53,54] to change the pH of the medium (e.g., to pH 7.0). In the antibody technique, the fluorescence intensity at 492 nm at pH 7.0 is decreased about 87% (depending on pH and the lot of antibody) with spectral shift to longer wavelengths.[58] In the pH shift method, the fluorescence intensity derived from FD exposed to the medium (pH 7.0) is promptly decreased about 63% (see Fig. 3), with only gradual, if any, change in intraorganellar pH (see Fig. 7). These values (differences = ΔF) can be used to determine the extraorganellar fluorescence and, therefore, by its subtraction, the true intraorganellar fluorescence. The intraorganellar pH can then be determined from the corrected fluorescence intensity ratio, as shown below. Antifluorescein antibody[58] can be prepared by immunizing rabbits with fluorescein-labeled protein (e.g., γ-globulin) and purifying it with an immunoadsorbent (cellulose attached to fluorescein-labeled bovine serum albumin, etc.)

The true ratio (R) of fluorescence intensities of intraorganellar FD (F^i) can be calculated from the observed fluorescence (F, corrected for background fluorescence) by the following equation:

$$R = \frac{F^i_{495}}{F^i_{450}} = \frac{F_{495} - F^o_{495}}{F_{450} - F^o_{450}}$$

where F^o represents the fluorescence intensity of extraorganellar fluorescein and the subscript number indicates excitation wavelength in nano-

[58] D. E. Lopatin and E. D. Voss, Jr., *Biochemistry* **10**, 208 (1971).

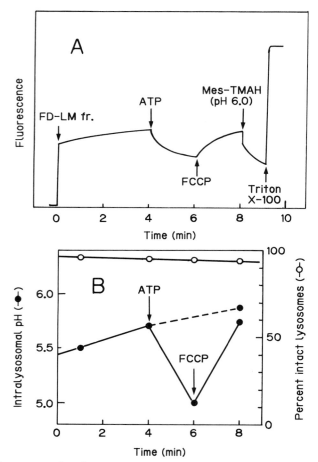

FIG. 7. Time course of (A) fluorescence change and (B) pH and latency changes of isolated lysosomes containing FD-70. FD-LM fractions were incubated with continuous stirring in 2 ml of buffer (0.1 M KCl, 0.2 M sucrose, 10 mM MgCl$_2$, and 20 mM HEPES–TMAH, pH 7.0) at 25°. Fluorescence was recorded at 492 nm excitation and 550 nm emission in a Hitachi 650-10S spectrofluorometer. Where indicated, ATP (1 mM) and FCCP (2.5 μM) were added. MES–TMAH (1M, pH 6.0, 100 μl) and Triton X-100 (2%, w/v, 20 μl) were added to determine the extralysosomal FD fluorescence and to calculate the latency (% intact lysosomes) and intralysosomal pH. For more details, see text. (From S. Ohkuma and T. Takano, in preparation.)

meters, respectively. F^o values can be calculated by the following equations:

$$F^o_{495} = \Delta F_{495} \times \frac{F^s_{495}}{\Delta F^s_{495}}$$

$$F^o_{450} = \Delta F_{450} \times \frac{F^s_{450}}{\Delta F^s_{450}}$$

$$= F^o_{495} \times \frac{F^s_{450}}{F^s_{495}}$$

where F^s represents the fluorescence intensity of standard free FD solution made in medium and ΔF and ΔF^s represent the difference of fluorescence intensities before and after the addition of pH 6.0 buffer or antifluorescein antibody, respectively.

If the correlation between F^i_{495} and F^i_{450} is determined from the *in situ* calibration, the pH change can be determined continuously by recording the fluorescence at 495 nm only. In the pH shift method, however, measurement of the latency of lysosomes and therefore exact pH determination in the presence of weak bases or weak acids or ionophores are difficult, because the pH values inside and outside the organelles are equilibrated instantaneously.

For the results in Fig. 7, rats (200–250 g body weight) were treated with FD (20 mg/100 g body weight) and fasted overnight (12–24 hr). Then the rats were anesthetized and sacrificed by decapitation, and their livers were excised. The lysosome-rich subcellular fraction (light mitochondrial fraction), which included the FD-containing lysosomes (FD-LM fr.), was obtained by differential cell fractionation and 20 μl (50 μg protein) of the fraction was suspended in 2 ml of KCl–sucrose buffer (0.1 M KCl, 0.2 M sucrose, 10 mM MgCl$_2$, 20 mM HEPES–TMAH buffer, pH 7.0) and incubated at 25°. The intralysosomal pH was calculated from the calibration curve obtained *in situ* (Fig. 8). The pH was estimated to be 5.5. This low pH is consistent with previous reports[59-61] obtained by the ion distribution method, and has also been confirmed by the fluorescence method with 6-carboxyfluorescein diacetate (pH 5.2)[56] and FD (pH = 5.5).[55]

There seems to be some reduction of fluorescence intensity within lysosomes, because addition of Triton X-100 increased the fluorescence even after addition of sufficient NH$_4$Cl or acidic ionophores to increase the

[59] D.-J. Reijngoud and J. M. Tager, *Biochim. Biophys. Acta* **297**, 174 (1973).
[60] R. Goldman and H. Rottenberg, *FEBS Lett.* **33**, 233 (1973).
[61] D.-J. Reijngoud and J. M. Tager, *Biochim. Biophys. Acta* **472**, 419 (1977).

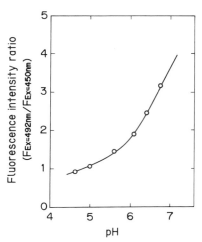

FIG. 8. *In situ* calibration curve for determination of the intralysosomal pH by measurement of the FD-70 fluorescence intensity ratio. Fluorescence intensity ratios of FD-70 fluorescence of FD-LM fractions were determined at different pH values in the presence of 100 mM NH$_4$Cl and/or 100 mM potassium acetate in a Hitachi 650-10S spectrofluorometer at 25°. Corrections were made for extralysosomal FD with the antifluorescein antibody that decreased the FD fluorescence to 13% at pH 7.0 at 492 nm and to 21% at pH 4.5 at 450 nm. (From S. Ohkuma and T. Takano, in preparation.)

pH.[53] This reduction may be caused by self-quenching of FD due to its high concentration.

Lysosomal Proton Pump. Strong evidence for a MgATP-dependent proton pump on rat liver lysosomes was obtained by use of the system described above (Fig. 7).[53] The intralysosomal pH increased in KCl–sucrose buffer, and addition of ATP (in the presence of Mg^{2+}) caused rapid decrease in the lysosomal pH to about 5.0, which was reversed by addition of carbonyl cyanide *p*-trifluoromethoxyphenylhydrazone (FCCP). The K_m value for ATP was 0.2–0.3 mM. AMP, ADP, and nonhydrolyzable analogs of ATP (AMP-CPP, AMP-PCP, and AMP-PNP) had no effect on the lysosomal pH.[53] These results were confirmed by Galloway *et al.*[54] and Merion *et al.*[55] By use of this method, evidence has also been obtained for H$^+$ pump activity in unmodified rat liver lysosomes prepared by the method of Yamada *et al.*,[62] and in autolysosomes produced by injection of leupeptin as described by Furuno *et al.*[63] (unpublished results).

[62] H. Yamada, H. Hayashi, and Y. Natori, *J. Biochem. (Tokyo)* **95**, 1155 (1984).
[63] K. Furuno, T. Ishikawa, and K. Kato, *J. Biochem. (Tokyo)* **91**, 1485 (1982).

Endosomes

Endosomal pH. The endosomal pH can be measured even *in vitro* by the method for measurement of intralysosomal pH described above. In this case, the endosomal fraction containing FD should be separated from lysosomes in some way, such as by Percoll density gradient centrifugation (see Fig. 9). Extraendosomal FD fluorescence can be subtracted as mentioned before. The results obtained show that the endosomal pH is about 6.0. A similar value has been obtained by use of carboxyfluorescein diacetate on uncoated vesicles (about pH 5.4[56]) *in vitro* and on endosomes by Merion *et al.*[55] The pH of urinary bladder plasmalemmal vesicles was also

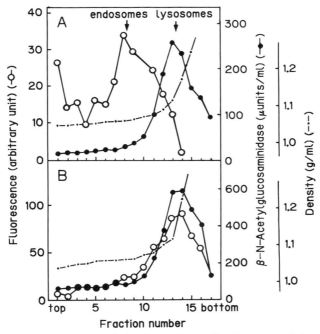

FIG. 9. Separation on a Percoll density gradient of endosomes and lysosomes from macrophages fed with FD-70 in culture. Macrophages were (A) incubated with FD (40 mg/ml) at 18° for 2 hr or (B) fed with FD (1 mg/ml) overnight, in modified Eagle's medium containing 20% fetal bovine serum. Cells were then washed and harvested in 0.25 *M* sucrose, homogenized in Dounce homogenizer, and centrifuged to obtain the postnuclear supernatant (PNS). The PNS was centrifuged on a Percoll density gradient (27% Percoll, 20,000 *g* for 2 hr in a Beckman 50 Ti rotor). (From K. Iinuma, T. Takano, and S. Ohkuma, unpublished results.)

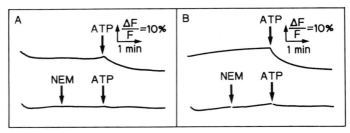

FIG. 10. Proton pump activity of endosomes (A) and lysosomes (B) separated by Percoll density gradient centrifugation as measured by FD fluorescence. See the legend to Fig. 9 for details. (From K. Iinuma, T. Takano, and S. Ohkuma, unpublished results.)

determined with FD to be about pH 6.[64] Percoll should be removed for better fluorescence measurement as it increases light scattering.

Endosomal Proton Pump. ATP-dependent acidification of endosomal vesicles was first demonstrated in studies using FD as a pH indicator.[54] Figure 10 shows results on FD-containing endosomal vesicles isolated from cultured macrophages by Percoll density gradient centrifugation (Fig. 9). The endosomal pH remained at about 6.0 *in vitro.* When ATP was added in the presence of Mg^{2+}, intravesicular acidification was detected by a decrease of fluorescence intensity as in the case of lysosomes.

Proton pumping could also be detected under a fluorescence microscope.[65] For this, cells with an endocytosed fluorescein-labeled ligand were permeabilized by exposure to the detergent digitonin (30–80 μg/ml) for 1–2 min. Then the digitonin was removed by suction, and the cells were rinsed 3 times (2 ml) with medium (150 mM NaCl, 5 mM KCl, 1 mM $CaCl_2$, 30 mM sucrose, and 20 mM HEPES, pH 7.4). Ionophores such as monensin (10 μM) and FCCP (10 μM) were added to dissipate the pH gradient. After complete removal of these ionophores by washing, addition of MgATP caused a decrease in fluorescence intensity, indicating acidification.

A proton pump was also detected on turtle urinary bladder plasmalemmal vesicles after isolation of these vesicles containing FD by Percoll density gradient centrifugation. Furthermore, a similar proton pump was demonstrated on vesicles from kidney cortex of rats treated intravenously with FD (40 mg in 1 ml of Ringer's solution 6–7 min).[64]

[64] I. Sabolic, W. Haase, and G. Burckhardt, *Am. J. Physiol.* **248,** F835 (1985).
[65] D. J. Yamashiro, S. R. Fluss, and F. R. Maxfield, *J. Cell Biol.* **97,** 929 (1983).

Membrane Vesicles Formed in Vitro (Submitochondrial Particles and Reconstituted Vesicles)

The first studies on use of fluorescein fluorescence were made with reconstituted membrane vesicles (*Halobacterium halobium*[7]) and submitochondrial particles.[66] In the former study, a rapid change in extravesicular pH was measured with FD instead of a regular pH electrode. This indicates another advantage of using this pH indicator dye, namely, very rapid changes (microseconds or less) in pH can be measured without the time lag resulting from membrane permeation of permeant dyes or electrode response. This advantage has been used in the study of protonation reactions with Ca^{2+}-ATPase[67]; the pH change in the initial phase of the reaction was determined with FD employing a stopped-flow spectrofluorometer. Measurement of relative changes in fluorescence intensity is sufficient to determine pH changes; the exact pH value can be calculated by addition of a known amount of HCl or NaOH to the medium for calibration.

In the latter study,[66] FD was trapped in submitochondrial particles in medium containing FD (0.166 mg FD/mg protein). After washing the particles several times, a significant fraction of FD was located internally and responded to internal pH changes, while another fraction of FD was apparently bound externally and responded to the external pH. If the external pH is strongly buffered, the observed change of fluorescence is due only to changes in the internal pH. For quantitative evaluation of the change in pH, the internal fraction of the indicator must be determined by inducing a pH change or using antifluorescein antibody. This method has also been used to measure the intravesicular pH in reconstituted ATPase vesicles of the yeast *Saccharomyces cerevisiae*[68] and in reconstituted vesicles containing mitochondrial transhydrogenase [FD (5 mg/ml) was trapped during the reconstitution step].[69] Intravesicularly trapped FD has also been used to demonstrate proton translocation by Na$^+$,K$^+$-ATPASE.[70]

For the study of reconstituted vesicles, fluorescein-labeled phosphatidylethanolamine (FPE) (absorption)[71] can also be used. When the external medium is highly buffered, changes in the pH of only the intravesicular compartment can be measured specifically.

[66] J. Kopecky, E. Glaser, B. Norling, and L. Ernster, *FEBS Lett.* **131**, 208 (1981).
[67] M. Yamaguchi and T. Kanazawa, *J. Biol. Chem.* **260**, 4896 (1985).
[68] G. A. Scarborough, *Biochemistry* **19**, 2925 (1980).
[69] S. R. Earle and R. R. Fisher, *Biochemistry* **19**, 561 (1980).
[70] Y. Hara and M. Nakao, *J. Biol. Chem.* **261**, 12655 (1986).
[71] M. Thelen, P. S. O'Shea, G. Petrone, and A. Azzi, *J. Biol. Chem.* **260**, 3626 (1985).

Scope and Limitations

The advantages of using FD for pH measurements are described above. The limitations of this method are as follows. (1) Some chemicals such as chloroquine affect the fluorescent properties of FD and when present prevent pH determinations from a calibration curve obtained in buffer solution.[6] (2) The pH is difficult to determine in samples showing large light scattering. For such samples, NMR or the pH electrode may be useful. (3) Photobleaching of fluorescein fluorescence is significant. (4) The effective pH range is limited. This last limitation has been partly overcome by use of derivatives of fluorescein with different pK values. However, a more suitable fluorochrome is still required. Of the many fluorescent pH indicators tested, 4-methylesculetin,[2,3,72] pyranine,[2,3] and 5-dimethylaminonaphthalene-1-sulfonyl (dansyl) chromophore[73] all require careful handling. The recently reported compounds quene 1[51] and 1,4-dihydroxyphthalonitrile[74] might be useful for measurement of intracellular and intraorganellar pH values.

[72] D. F. Gerson and A. C. Burton, *J. Cell. Physiol.* **91**, 297 (1977).
[73] W. L. C. Vaz, A. Nicksch, and F. Jähnig, *Eur. J. Biochem.* **83**, 299 (1978).
[74] I. Kurtz and R. S. Balaban, *Biophys. J.* **48**, 499 (1985).

[13] Cystine Exodus from Lysosomes: Cystinosis

By Margaret L. Smith, Alice A. Greene, Jerry A. Schneider, Ronald L. Pisoni, and Halvor N. Christensen

Introduction

The concept of an intracellular transport process for the movement of cystine from the lysosomal space to the cytosol owes its existence to the search for the metabolic defect in the inherited disease cystinosis.[1] In this condition, cystine accumulates within lysosomes. Aside from disulfide exchange reactions, the only subsequent lysosomal pathway for this compound is movement to the cytoplasm. Because the cytoplasmic catabolic pathways for cystine appeared normal in cystinosis, investigators suspected that the defect in this disease involved the escape of cystine from the lysosome to the cytosol. For many years, however, this concept could not

[1] J. A. Schneider and J. D. Schulman, *in* "The Metabolic Basis of Inherited Disease" (J. B. Stanbury, J. B. Wyngaarden, D. S. Fredrickson, J. L. Goldstein, and M. S. Brown, eds.), 5th ed., p. 1844. McGraw-Hill, New York, 1983.

be pursued because it was not possible to load normal lysosomes with cystine.

This problem was solved by Reeves[2] who used the technique of loading lysosomes with methyl esters of amino acids. These methyl esters are rapidly taken up into the lysosome where they are hydrolyzed, resulting in amino acid loading of these organelles. Utilizing this technique, several laboratories have reported that lysosomal cystine transport is defective in cystinotic cells.[3-8] Soon after the report of the cystine transport system, a carrier for cationic amino acids was also demonstrated in the lysosomal membrane.[8] This system is distinct from the corresponding transporter of the plasma membrane.[9] Using the techniques of amino acid loading and exodus as described here, as well as that of amino acid uptake in the presence of analogs, lysosomal transport systems in human fibroblasts have been characterizied for small neutral amino acids,[10] for anionic acids,[11] and for branched and aromatic dipolar amino acids.[12] A carrier, probably related to the latter, serving for tyrosine and other neutral amino acids, as well as one for cystine have been shown to occur in the lysosomes of rat thyroid cells.[13] Systems have also been reported for sialic acid,[14,15] cobalamin,[16] nucleosides,[17] and sugars.[18]

[2] J. P. Reeves, *J. Biol. Chem.* **254**, 8914 (1979).

[3] A. J. Jonas, M. L. Smith, and J. A. Schneider, *J. Biol. Chem.* **257**, 13185 (1982).

[4] A. J. Jonas, M. L. Smith, W. S. Allison, P. K. Laikind, A. A. Greene, and J. A. Schneider, *J. Biol. Chem.* **258**, 11727 (1983).

[5] W. A. Gahl, F. Tietze, N. Bashan, R. Steinherz, and J. D. Schulman, *J. Biol. Chem.* **257**, 9570 (1982).

[6] W. A. Gahl, N. Bashan, F. Tietze, I. Bernardini, and J. D. Schulman, *Science* **217**, 1263 (1982).

[7] W. A. Gahl, F. Tietze, N. Bashan, I. Bernardini, D. Raiford, and J. D. Schulman, *Biochem. J.* **216**, 393 (1983).

[8] R. L. Pisoni, J. G. Thoene, and H. N. Christensen, *J. Biol. Chem.* **260**, 4791 (1985).

[9] R. L. Pisoni, J. G. Thoene, R. M. Lemons, and H. N. Christensen, *J. Biol. Chem.* **262**, 15011 (1987).

[10] R. L. Pisoni, K. S. Flickinger, J. G. Thoene, and H. N. Christensen, *J. Biol. Chem.* **262**, 6010 (1987).

[11] E. J. Collarini, R. L. Pisoni, and H. N. Christensen, *FASEB J.* **2**, A322 (1988).

[12] B. H. Stewart, E. J. Collarini, R. L. Pisoni, and H. N. Christensen, *FASEB J.* **2**, A322 (1988).

[13] J. Bernar, F. Tietze, L. D. Kohn, I. Bernardini, G. S. Harper, E. F. Grollman, and W. A. Gahl, *J. Biol. Chem.* **261**, 17107 (1986).

[14] M. Renlund, F. Tietze, and W. A. Gahl, *Science* **232**, 759 (1986).

[15] A. J. Jonas, *Biochem. Biophys. Res. Commun.* **137**, 175 (1986).

[16] D. S. Rosenblatt, A. Hosack, N. V. Matiaszuk, B. A. Cooper, and R. Laframboise, *Science* **228**, 1319 (1985).

[17] R. L. Pisoni and J. G. Thoene, *J. Biol. Chem.* **264**, 4850 (1989).

[18] G. A. Maguire, K. Docherty, and C. N. Hales, *Biochem. J.* **212**, 211 (1983).

Amino Acid Loading

General Principle. The methyl esters of various amino acids are taken up by lysosomes at $25° - 37°$ and hydrolyzed by endogenous enzymes within the organelle to the ionized form of the acid. This hydrolysis occurs faster than the free acid can exit. If the preparation is then chilled to $4°$, the amino acid will be temporarily retained. On warming, the characteristics of the amino acid exodus may be studied. Radioactive methyl esters may be used or, in the case of cystine, the unlabeled cystine dimethyl ester may also be used because of the availability of a sensitive binding assay for cystine.[19,20] Various combinations of time, temperature, and buffers have been used depending on the particular cell type and amino acid ester. Two representative methods for loading and determination of exodus are given below.

Dimethyl Esters. The unlabeled cystine dimethyl ester is available from Sigma. L-[^{35}S]Cystine dimethyl ester may be prepared as follows. Approximately 3 N methanolic HCl is prepared in the hood by slowly adding 6 ml acetyl chloride to 25 ml cold anhydrous methanol. During addition the mixture is stirred constantly and kept cold in an ice bath. To $10-15$ ml of the above methanolic HCl add approximately 2 mg (1 mCi) of L-[^{35}S]cystine and keep at room temperature for 24 hr or reflux gently for 1 hr. Evaporate as close to dryness as possible with a rotary evaporator, add 1 ml anhydrous methanol, and evaporate under a stream of nitrogen. Repeat 5 times until the compound is absolutely dry. Resuspend in methanol at 1 μmol/ml and store at $-20°$. Purity may be checked by high-voltage electrophoresis on paper. Other radiolabeled amino acids may first be brought to dryness, and the corresponding amount of 3 N methanolic HCl added, the solution allowed to stand overnight at room temperature, and then brought to dryness as above.[3] Amersham is our preferred supplier for L-[^{35}S]cystine because we have found this product to have a higher purity than others we have tested.

Amino Acid Loading in Whole Cell Suspension (Fibroblasts). Fibroblasts are cultured in glass or plastic roller bottles in Coon's modification of Ham's F12 medium with 10% fetal bovine serum.[19] They are removed from their growing surface by a 2 min incubation at $37°$ in 0.25% trypsin with 4% EDTA in phosphate-buffered saline. The action of trypsin is stopped by the addition of an equal volume of medium containing 10% dialyzed fetal bovine serum, and the cells are pelleted by centrifugation at room temperature at 750 g. The fibroblasts are resuspended in prewarmed

[19] R. G. Oshima, R. C. Willis, C. E. Furlong, and J. A. Schneider, *J. Biol. Chem.* **249**, 6033 (1974).

[20] M. Smith, C. E. Furlong, A. A. Greene, and J. A. Schneider, this series, Vol. 143, p. 144.

cystine-free medium containing 10% dialyzed fetal bovine serum (5 ml for every roller bottle of about 2×10^7 cells) and then incubated for 15 min at 37° with gentle agitation in order to recover from trypsinization. The cells are then pelleted as before and resuspended in medium buffered with 10 mM HEPES, pH 7.2, without cystine or serum. The gas phase over the cells is made 10% CO_2 and the closed container incubated for a further 15 min at 37° with gentle agitation. If lysosomes are to be loaded with cystine, cystine dimethyl ester (CDME) is included in this last incubation. Care is taken to adjust the CDME to pH 7.2 prior to addition. Concentrations from 0.1 to 1.0 mM are normally used. Varying the CDME concentration results in a range of intralysosomal cystine concentrations. Cells are repelleted as above, washed twice with ice-cold phosphate-buffered saline, and washed once with ice-cold 0.25 M sucrose. It is important that the preparation be kept cold from this point on if the cystine is to be retained by normal lysosomes. The cells remain viable during the harvest and loading procedure as demonstrated by trypan blue exclusion.

Three or four data points of approximately 150 μg protein each may be obtained from the granular fraction prepared from one roller bottle of about 2×10^7 cells when exodus is followed using the nonradioactive cystine dimethyl ester method described below. In experiments comparing cystine exodus rates of normal and cystinotic lysosomes, the lysosomal cystine content of cystinotic cell strains may be reduced by culturing them in medium without cystine for 24 hr before preparing lysosomes.

Amino Acid Loading in Whole Cell Suspension (Lymphoblasts). The protocol is similar to that for fibroblasts except for the harvest procedure and cell densities. Epstein-Barr virus-transformed lymphoblasts are grown in RPMI 1640 medium with 10% fetal bovine serum and supplemented with 2 mM glutamine. Cells are collected by centrifugation for 10 min at 1000 g, and resuspended at a density of $7-8 \times 10^6$ cells/ml in serum-free Coon's modified Ham's F12 medium without cystine, but containing 10 mM HEPES, pH 7.2. The cells are loaded with cystine dimethyl ester as described for fibroblasts above. About $30-40$ data points of 150 μg of granular fraction protein may be obtained from 10^9 cells.

An apparent maximization of cystine uptake may be reached by the whole cell loading method at $30-50$ mM CDME in the medium. The values may be $1000-1700$ nmol cystine per unit (defined as one micromole of product formed per minute) of β-N-acetylhexosaminidase for fibroblasts and 10-fold greater for lymphoblasts. At this high level, however, the cells exhibit reduced viability, and decreases in the buoyant density of the lysosomes occur. Approximately 90% of the cells remain viable when loaded to levels of 10 and 150 nmol cystine/unit β-N-acetylhexosaminidase for fibroblasts and lymphoblasts, respectively. The CDME concentration required to achieve this level in whole cells varies with cell strain but will be 0.1–1.0 mM CDME in the incubation medium for 15 min at 37° for both fibroblasts and lymphoblasts.

Preparation of Granular Fraction from Loaded Cells. Following the phosphate-buffered saline and sucrose washes described above after loading, the cell pellet is suspended in 5 ml of cold 0.25 M sucrose/ml of cell pellet, and the cells are lysed by repetitive pipetting through a 200 μl pipette tip. A tip with a larger orifice may be used with fibroblasts. Cellular material is kept at 4° throughout the following procedures. Microscopic examination is used to determine when granules have been released from most cells. It will be seen that fibroblasts are much more easily lysed than lymphoblasts. The cell homogenate is diluted to 20 ml and pelleted in a refrigerated centrifuge for 10 min at 750 g. Approximately half of the supernatant is removed and saved without disturbing the loose pellet. The remaining supernatant and pellet are resuspended and the entire lysis procedure repeated, taking only the quantity of supernatant that can be removed without disturbing the pellet. The supernatants are combined and centrifuged for 10 min at 20,000 g. The pellet is resuspended in 10 ml of 0.25 M sucrose, 20 mM HEPES, pH 7.0, and centrifuged for 10 min at 20,000 g. The lysosomal pellet obtained is approximately 3-fold purified and is resuspended in the 20 mM HEPES–0.25 M sucrose buffer.

Amino Acid Loading in the Granular Fraction Using Radiolabeled Methyl Esters. Loading of lysosomes may also be accomplished by incubating the lysosome-enriched granular fraction or a more completely purified lysosomal fraction with methyl esters. The pattern and specificity of loading may be checked using Percoll gradient fractionation. We have observed that, as with whole cells, loading of the lysosomes in the granular fraction changes the buoyant density of lysosomes. In one experiment the ratio of cystine to β-N-acetylhexosaminidase varied in different sections of the gradient. Investigators should be aware of these effects. Although whole cell loading has the advantage of maintaining the organelle under more physiological conditions, the smaller volumes and lower methyl ester concentrations involved with granular fraction loading make this method advantageous when radiolabeled methyl esters are used. An example of radiolabeled methyl ester loading of a granular fraction is given below.

Fibroblasts grown on plates or roller bottles are washed and harvested by scraping cells into ice-cold phosphate-buffered saline. All further steps are performed at 4°. The cell pellet is collected by centrifugation for 10 min at 750 g and washed once with 50 mM MOPS–Tris buffer, pH 7.0, containing 0.25 M sucrose. The cell pellet is resuspended in 3 ml of 10 mM MOPS–Tris buffer, pH 7.6, containing 0.25 M sucrose and 1 mM Na$_2$EDTA, and cells are lysed by repetitive pipetting 12 times through a

5-ml Eppendorf pipette tip. The suspension is centrifuged for 10 min at 750 g and the supernatant transferred to a tube on ice and saved. The lysis is repeated twice more, and the combined 750 g supernatants are pooled and centrifuged for 10 min at 1500 g. The 1500 g supernatant is centrifuged for 12 min at 15,600 g; after the supernatant is decanted, the residual pellet constitutes the crude granular fraction.

Lysosomes present in the crude granular fraction may be loaded at a final concentration of 0.1 mM cystine dimethyl ester. An aliquot of 175 μl of the crude granular fraction suspended in 50 mM MOPS–Tris buffer, pH 7.0, containing 0.25 M sucrose (MST buffer) is incubated with 22.5 nmol of L-[^{35}S]cystine dimethyl ester (400–500 mCi/mmol) suspended in 50 μl of MST buffer. After incubation the mixture is diluted to 1.5 ml with ice-cold buffer and centrifuged for 10 min at 15,600 g at 4° in a Model 5414 Eppendorf microfuge. The wash step is repeated once more, and the final pellet is suspended in an appropriate volume of ice-cold MST buffer for measurement of cystine exodus. The granular fraction from 8×10^7 cells of either normal or cystinotic fibroblasts will provde approximately 15 data points when loaded with L-[^{35}S]cystine dimethyl ester as described. To reduce the amount of radioactivity binding to the lysosomal membrane during loading or during efflux, 2 mM N-ethylmaleimide (NEM) may be added to the buffer. The concentration and specific activity of cystine dimethyl ester may be varied. Concentrations from 0.03 to 0.3 mM have been used. The specific activity of the CDME should be above 100 mCi/mmol.

Exodus from Lysosomes

Exodus is determined from the amount of cystine remaining in the lysosomes at the various time points as shown in Fig. 1. Corresponding cystine appearance may also be measured in the supernatant. In order to exclude cystine loss due to rupture of the lysosomes during incubation, all exodus measurements are reported as nanomoles cystine per unit of β-N-acetylhexosaminidase. Such loss of latency of the lysosomes is normally less than 10% over the course of incubation. Procedures for measuring exodus are given below for both radiolabeled and unlabeled methyl ester-loaded lysosomes.

Measurement of Cystine Exodus (Nonradiolabeled). Aliquots of the loaded granular fraction containing approximately 150 μg of protein are diluted 20-fold into tubes containing 2 ml of various warmed isotonic buffers and incubated at 37°. At various times tubes are removed and rapidly cooled in an ice bath, and lysosomes are pelleted by centrifugation and the supernatant removed. The lysosomes are broken by a 10-sec

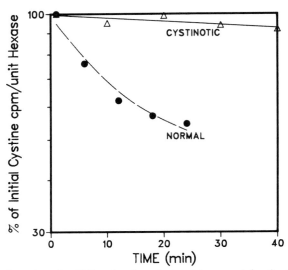

FIG. 1. Basal exodus of L-[^{35}S]cystine from isolated lysosomal fractions of normal and cystinotic human fibroblasts. The crude granular fraction from normal or cystinotic fibroblasts was loaded with 30 μM [^{35}S]cystine as described. The exodus buffer was MST. (From Ref. 8.)

sonication in 20 mM potassium phosphate buffer, pH 7.2, containing 5 mM NEM. An aliquot is removed for the assay of β-N-acetylhexosaminidase. Sulfosalicylic acid is added to the rest of the sample to a final concentration of 4%. After protein is precipitated and pelleted, the acid supernatant is used for the determination of cystine with a binding-protein assay.[19,20] The lysosomal enzyme β-N-acetylhexosaminidase is determined fluorometrically.[21] Protein concentrations are measured spectrophotometrically.[22]

Measurement of Cystine Exodus (Radiolabeled). Aliquots of 100 μl of the cold granular fraction loaded with L-[^{35}S]cystine are added to 720 μl of MST buffer, pH 7.0, prewarmed to 37°, and incubated at 37°. Duplicate aliquots of 80 μl are removed at various times, added to microcentrifuge tubes containing 1.4 ml of ice-cold buffer, and pelleted at 15,600 g for 10 min at 4°. The pellet is washed once with 1.5 ml of cold buffer and resuspended in 80 μl of 10 mM sodium phosphate buffer, pH 7.0, contain-

[21] J. G. Leroy, M. W. Ho, M. C. MacBrinn, K. Zielke, J. Jacob, and J. S. O'Brien, *Pediatr. Res.* **6,** 752 (1972).
[22] O. H. Lowry, N. J. Rosebrough, A. L. Farr, and R. J. Randall, *J. Biol. Chem.* **193,** 265 (1951).

ing 10 mM NEM. The suspension is frozen and thawed in sequence 3 times. A 20-μl aliquot is removed for assay of β-N-acetylhexosaminidase activity, and sulfosalicylic acid is added to the remaining sample to a final concentration of 4%. The tubes are placed at 4° for 30 min and centrifuged at 15,600 g for 7 min at 4°. A 40-μl aliquot from each sulfosalicylic supernatant is spotted on paper along with 30 nmol each of L-cystine and L-cysteine – NEM (prepared by reacting cysteine with a 2-fold molar excess of NEM in 10 mM phosphate buffer, pH 7.0) to serve as internal standards. The paper is subjected to high-voltage electrophoresis in 6% formic acid at 3500 V for 25 min, and the radioactive cystine spots are cut out and counted.

Countertransport and Uptake Studies. The movement of amino acids across the lysosomal membrane can be stimulated or inhibited by the presence of the amino acid on the opposite side of the membrane. Trans-stimulated uptake of cystine, or counterflow, has been demonstrated by Gahl *et al.*, using leukocyte lysosomes.[7] They used this property to determine the effects of various amino acid analogs added to the incubation mixture. They found that lysosomal cystine countertransport was stereospecific for the L-isomer. There was only a very small uptake of cystine into lysosomes not previously loaded with cystine by the methyl ester. Pisoni *et al.*[8] demonstrated trans-stimulation of lysine exodus. Cystine did not stimulate lysine exodus, demonstrating that in human fibroblast lysosomes cystine is not transported by the carrier used by other cationic amino acids.

Effect of MgATP. Cystine exodus from lysosomes occurs at a "basal" rate, at 37° in pH 7 buffer, which is characteristic of each cell strain. A more rapid loss occurs in the presence of 5 mM MgCl$_2$, 2 mM ATP. This we refer to as the "ATP-stimulated" loss. The percent loss is independent of the initial load level, assuming subsaturating loads. Basal and ATP-stimulated cystine loss in 25 min was 40 ± 10 and $78 \pm 6\%$ of initial cystine, respectively, for a set of five experiments with normal fibroblasts, but 7 ± 10 and $14 \pm 7\%$, respectively, for five experiments with cystinotic fibroblasts. The corresponding figures for basal and ATP-stimulated loss in lymphoblast experiments were 33 ± 10 and $72 \pm 6\%$ of initial cystine, respectively, for five normal, but 12 ± 11 and $16 \pm 8\%$, respectively, for five cystinotic experiments.

The ATP-stimulated exodus is eliminated by NEM (shown to inhibit lysosomal membrane proton-translocating ATPase[4]), the basal rate is not. The ATP stimulation of exodus from lymphoblast lysosomes is completely inhibited by 10 μM NEM; exodus from fibroblast lysosomes is inhibited by 1 mM NEM. The ATP-stimulated exodus in lymphoblasts is half-maximal at $0.1 – 0.3$ mM ATP for incubations of 8 min. Additional ATP may be required for longer incubation times because of the rapid loss of ATP in the presence of lymphoblast, but not fibroblast, granular preparations (A. A. Greene, unpublished). We believe the role of ATP is the maintenance of intralysosomal conditions necessary for the transporter activity. Studies in which lysine methyl ester was loaded in the granular fraction have shown that lysine exodus is also stimulated by the presence of ATP either during loading or during the exodus incubation.[8]

Effect of Temperature. Both the ATP-stimulated and the basal exodus of cystine from lysosomes is accelerated by temperature. Energy of activation values for normal lymphoblast exodus measured between 10° and 37° in two different cell lines are as follows: basal, 11.2 and 11.9 kcal/mol; ATP-stimulated, 10.9 and 11.4 kcal/mol. Using a countertransport technique in normal leukocytes, a value of 11.4 kcal/mol was reported.[7] An enhanced rate of cystine-specific egress was observed in lysosomes from cystinotic fibroblasts at 40°–43° by Thoene *et al.*[23]

Cystinosis. The exodus of cystine from the lysosome requires a specific transporter which is aberrant or deficient in cells from individuals with cystinosis. Loss of lysosomal cystine from cultured cystinotic lymphoblasts and fibroblasts is 0–15% of the initial value. The rate is neither stimulated by MgATP nor inhibited by NEM. The cystine present in cystinotic lysosomes can be depleted by incubating the cells with 5 mM cysteamine. This depletion is attributed to the formation of cysteamine–cysteine disulfide which is transported by the cationic amino acid carrier.[8]

[23] J. G. Thoene, R. Lemons, R. Pisoni, and H. Christensen, *Pediatr. Res.* **20,** 273A (1986).

[14] Isolation of Physiologically Responsive Secretory Granules from Exocrine Tissues

By ULRICH HOPFER and KENNETH GASSER

Introduction

Isolated secretory granules have proved to be a useful system for studying many aspects of exocytotic secretion. Information gathered from such systems as chromaffin granules,[1] mast cell granules,[2] or pancreatic[3] and parotid zymogen granules[4] have resulted in a clearer understanding of

[1] H. B. Pollard, C. J. Pazoles, C. E. Creutz, and O. Zinder, *Int. Rev. Cytol.* **58,** 159 (1979).
[2] L. J. Breckenridge and W. Almers, *Nature (London)* **328,** 814 (1987).
[3] K. W. Gasser, J. DiDomenico, and U. Hopfer, *Am J. Physiol.* **254,** G93 (1988).
[4] P. Arvan and J. D. Castle, *J. Cell Biol.* **103,** 1257 (1986).

storage, exocytotic regulation, and primary fluid production. These studies have used a variety of isolation techniques; however, the methods described by Tartakoff and Jamieson,[5] and more recently Meldolesi,[6] represent the style of techniques most commonly used.

Obviously, isolated granules which closely mimic the *in vivo* state will provide the most reliable information on this system. Useful criteria are the behavior of isolated secretory granules suspended in "physiological" saline solutions at the appropriate pH, temperature, and ionic strength. *In vivo*, zymogen granules concentrate, store, and subsequently release the macromolecular product into the lumen, as many histological studies have demonstrated.[7] Therefore, it is reasonable that in the absence of exocytotic stimuli, isolated secretory granules should not lyse or release the secretory product when incubated in artificial solutions with compositions similar to the cytosol. With that in mind, this granule isolation technique is designed to provide granules which readily tolerate physiological salt solutions yet remain osmotically active and physiologically responsive to such intracellular messengers as calcium, pH, and phosphorylation.

Isolation Technique

Secretory granules, from a variety of exocrine tissues and animals, have been isolated by the following technique; most of the experience has been obtained with rat pancreas,[3] rat parotid,[8] and rabbit gastric pepsinogen granules.[9] Although minor variations have been necessary to optimize the isolation from some of these sources, the basic technique remains the same.

Typically, the glands are quickly removed from the animal and placed into ice-cold homogenization buffer which consists of 250 mM sucrose; 50 mM 3-(N-morpholino)propanesulfonic acid (MOPS), pH 7.0; 0.1 mM MgSO$_4$; 0.2 mM ethylene glycol bis(β-aminoethyl ether)-N,N,N',N'-tetraacetic acid (EGTA), free calcium 10^{-7} M; 1.0 mg/ml fatty acid-free bovine serum albumin (BSA); and 0.2 mM phenylmethylsulfonyl fluoride (PMSF) (added immediately before use from a 100 mM stock in anhydrous dimethyl sulfoxide). The exocrine gland is then minced into a coarse paste with a fine scissors or a razor blade, and the tissue is further disrupted by suspending the paste in homogenization buffer (1 : 15 dilution of the wet weight of the starting material) and homogenizing with 5 strokes (500 rpm)

[5] A. Tartakoff and J. Jamieson, this series, Vol. 31, p. 41.

[6] J. Meldolesi, this series, Vol. 98, p. 67.

[7] G. Palade, *Science* **189**, 347 (1975).

[8] K. W. Gasser, J. DiDomenico, and U. Hopfer, *Am. J. Physiol.* **255**, C705 (1988).

[9] B. Sharma, K. W. Gasser, and U. Hopfer, *Gastroenterology* **96**, 1049 (1989).

with a loose-fitting glass–Teflon homogenizer. The crude homogenate contains many intact acinar cells and is therefore subjected to additional homogenization by nitrogen cavitation at 250 psi for most exocrine glands. This procedure disrupts the vast majority of acinar cells but does not damage the secretory granules.

A significant enrichment is achieved by a one-step purification on a Percoll density gradient. The homogenate is supplemented with Percoll buffer to give final concentrations of 250 mM sucrose; 40% Percoll; 50 mM 2-(N-morpholino)ethanesulfonic acid (MES), pH 6.5[10]; 25 mM MOPS (carry over from original homogenate); 2.0 mM EGTA, free calcium 10^{-7} M or lower; 0.2 mM MgSO$_4$; 1 mg/ml BSA; and 0.2 mM PMSF. The Percoll gradient is formed and the granules isolated in one step by centrifuging for 20 min at 20,000 g in a Sorvall SS-90 vertical rotor. Secretory granules are dense (1.13–1.15 g/ml) and typically form a very distinct band toward the bottom of the density gradient.

At this point the zymogen granules (measured as a pelletable form of a specific granule marker enzyme) have been purified approximately 2.5- to 2.8-fold in the pancreas, 4.5- to 5.0-fold in the parotid, and 3.0- to 3.5-fold for the gastric pepsinogen granules. Considering that the zymogen granule-associated protein is very high in most secretory cells ($\sim$20% of the total protein in pancreatic acinar cells), these enrichments are substantial. For example, a theoretical maximum enrichment of 5-fold is possible for the pancreas. The heterogeneity of the gastric mucosa explains the higher pepsinogen granule enrichment which can be attained (9- to 10-fold), since the starting ratio of (chief) cells with granules to other cell types is much lower than in other exocrine glands.

Purification (per milligram protein) can be increased significantly by centrifuging the granules through a second Percoll gradient. This step removes soluble and microsomal protein by dilution and also decreases mitochondrial contamination. The separation can be further enhanced by replacing the sucrose with an equivalent amount of salt (125 mM).[11] Anions which permeate mitochondria but not secretory granules will promote separation owing to differential swelling (Fig. 1).

The second gradient consists of 60% Percoll; 50 mM MES, pH 6.5; 125 mM salt or 250 mM sucrose; 2.0 mM EGTA, free calcium 10^{-7} M or lower; 0.2 mM MgSO$_4$; 0.2 mM PMSF; and 1 mg/ml BSA. The higher Percoll concentration is necessary when salt is included in the gradient, as the shape of the gradient shifts slightly to lower densities. The gradient is again formed by centrifuging in a vertical rotor for 20 min at 20,000 g

---

[10] MES can be replaced with MOPS at pH 7.0 with no significant change in the yield or osmotic stability of the zymogen granules.

[11] K. W. Gasser, J. DiDomenico, and U. Hopfer, *Anal. Biochem.* **171**, 41 (1988).

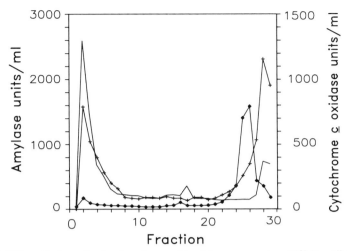

FIG. 1. Effect of salt on the fractionation profile of the pancreas in a 40% Percoll gradient. The tissue is disrupted with 4 strokes of a glass–Teflon homogenizer and nitrogen cavitation, and the gradient is as described in the text. The profiles represent amylase (zymogen granules) fractionated in Percoll plus 250 mM sucrose (—); cytochrome-c oxidase (mitochondria) fractionated in Percoll plus 250 mM sucrose (+); and cytochrome-c oxidase fractionated in Percoll plus 50 mM sucrose and 100 mM sodium succinate (♦). Fraction 0 is the bottom of the gradient.

(4°). As described below, granule preparations can display different membrane permeability characteristics to the point that suspension in a high salt Percoll gradient can result in the isolation of a specific subpopulation with low ion permeabilities. Therefore, the choice of a salt or sucrose in the gradient depends in part on the characteristics of the granules being purified.

Modification for Rat Parotid Granules. A minor change was found to be necessary in the homogenization protocol arising from the difference in texture between the rat parotid and pancreas. The parotid is a physically tougher tissue, and therefore higher nitrogen cavitation pressures were necessary to achieve a comparable degree of cell disruption (500 versus 250 psi).

Modification for Rabbit Pepsinogen Granules. In the case of gastric mucosa, a variety of cell types as well as a rich supply of mucins are present. To optimize gradient centrifugation under these conditions, the homogenization and Percoll buffers also contained 50 mM dithioerythritol for pepsinogen granule isolation. This addition succeeds in reducing large-scale cross-linking of mucins which otherwise cause excessive clumping as well as hinder granule mobility in the gradient.

Colorimetric Assays in the Presence of Percoll. Percoll has been shown to interfere with a number of enzyme and protein determinations, and other assays have been developed which try to deal specifically with the interference problem.[12] However, the majority of the Percoll can easily be removed by diluting the granules ($\sim 2-3$ ml after fractionation of the gradient) with 25 ml of a solution consisting of 320 mM sucrose, 0.1 mM MgSO$_4$, 1.0 mM EGTA, 2.0 mM MES, pH 6.0. This mixture is centrifuged at 1500 g for 20 min (4°). The zymogen granules form a loose pellet, while the Percoll remains in suspension and is decanted. However, it should be noted that pelleting reduces the structural stability of secretory granules and therefore multiple or high-speed pelleting should be avoided if possible.

Critical Elements of the Isolation Procedure

Calcium. The calcium concentration of solutions used in the isolation are maintained at approximately 10^{-7} M using EGTA–Ca^{2+} buffers. Calcium concentrations in the homogenization solutions above 10^{-5} M result in significantly lower yields of zymogen granules in the pancreas; however, concentrations of calcium lower than 10^{-9} M will alter the fractionation profile and also decrease the yield of zymogen granules (Table I).

High calcium decreases yield by causing an increase in granule lysis as determined by the high soluble amylase levels throughout the gradient. The lysis could be initiated by granule fusion, which calcium is thought to influence, granule damage by calcium-activated lipolytic enzymes, or altered membrane ion permeabilities causing granule swelling. We have recently collected evidence for a calcium (10^{-4} M)-activated component of the chloride channel in isolated granules from both the pancreas and parotid.[13] It is reasonable to believe that lower calcium levels (10^{-6} M) would activate similar pathways *in vivo* in the presence of calcium-binding proteins. Low calcium concentrations reduce yields because they cause changes in the gradient fractionation profile. A higher proportion of the granules are trapped in the low-density, low-purity band at the top of the gradient, while the yield is based on the high-purity, high-density fractions.

Surfactants. To ensure long-term osmotic stability at physiological conditions (pH, temperature, ionic strength), extreme care must be taken to avoid exposure to surfactants. It is noteworthy that detergents (at low levels) do not adversely affect zymogen granule yield or gross morphology while in nonionic solutions at a pH below 7.0. However, when these granules are suspended in ionic solution, lysis occurs quickly. Table II[14]

---

[12] M. N. Kahn, R. J. Kahn, and B. I. Posner, *Anal. Biochem.* **117**, 108 (1981).

[13] K. W. Gasser and U. Hopfer, unpublished (1989).

[14] K. W. Gasser and U. Hopfer, *in* "Gastrointestinal and Hepatic Secretion: Mechanisms and Control" (J. S. Davison and E. A. Shaffer, eds.), p. 197. Univ. of Calgary Press, Calgary, Alberta, 1988.

TABLE I
EFFECT OF CALCIUM CONCENTRATION ON
PANCREATIC ZYMOGEN GRANULE YIELD

Calcium concentration (M)	Yield[a] (%)
10^{-5}	21.6 ± 12.4
10^{-7}	39.2 ± 8.3
10^{-9}	24.5 ± 10.1

[a] Yields are based on five determinations of amylase recovery in the high-density band compared to total amylase in the homogenate.

shows the effect of acute exposure of osmotically stable pancreatic zymogen granules to low levels of detergent, with half-lives being dramatically reduced from the basal level of 3–4 hr. Fatty acids produce a qualitatively similar result on isolated zymogen granules.

TABLE II
EFFECT OF LOW-LEVEL DETERGENTS ON
PANCREATIC ZYMOGEN GRANULE STABILITY

Treatment[a]	Concentration[b] (%)	Half-life (hr)
Control	—	3.8
Ethanol	10	0.5
Cholic acid	0.5	0.1
Deoxycholic acid	0.01	0.2
Digitonin	0.001	0.2
Triton X-100	0.001	0.08
Sodium dodecyl sulfate	0.001	0.05

[a] The treatment was an acute exposure of the granules in KCl saline (pH 7.0, 37°) to the detergent, and the change in the rate of lysis was followed. Data are in part from Gasser and Hopfer.[14]

[b] Concentrations are v/v% for ethanol and Triton X-100 and w/v% for all others.

To guard against surfactant exposure, all glass and plasticware is rinsed with 95% ethanol prior to a final rinse with deionized water. As an added precaution against endogenous as well as exogenous surfactants, 1 mg/ml of fatty acid-free bovine serum albumin is added to all solutions as a buffer or detergent scavenger. BSA serves to stabilize the zymogen granules, decreasing the frequency of osmotically unstable preparations. However, the BSA does not detract from the endogenous ion permeability, as shown below.

Homogenization. Three modes of cell disruption have been tested: glass–Teflon homogenization, disruption with high-speed rotors inside a shaft (such as the Tekmar Tissumizer), or low-pressure nitrogen cavitation. As the zymogen granules are very fragile and can be easily damaged, it was necessary to find a technique which would maximize cell disruption but leave organelles unaltered. Low-pressure nitrogen cavitation has proved to be the most reliable method. Owing to the high surface-to-volume ratio of most organelles compared to the intact cell, the nitrogen pressures employed (200–500 psi) will cause cellular lysis without significant damage to organelles, as judged by phase-contrast microscopy, gradient fractionation, and the yield of intact zymogen granules. Glass–Teflon homogenization alone (10 strokes at 500 rpm) or in combination with a Tissumizer (45 sec at half-speed) leave a significant percentage of the acinar cells intact (35–50%). Increasing the homogenization time or revolutions per minute decreased the number of intact cells but also caused an increase (~ 20%) in the soluble granule marker, implying that a higher percentage of the granules had been broken.

Secretagogue Treatment. It has recently been demonstrated that secretagogue and antagonist treatments significantly alter the membrane permeability of zymogen granules.[3] Untreated animals produce granules which display the greatest degree of osmotic variability, while secretagogue exposure drives the membrane ion permeability to the highest level and antagonist exposure to the lowest level. Adrenergic and cholinergic agents are widely effective for promoting secretion in exocrine tissue, and it is therefore necessary to control their levels. Labetalol, a broad-acting adrenergic antagonist, is administered in the drinking water so that the animal receives a dosage of 1–5 mg/kg over a 12-hr period before sacrifice. Atropine is effective when administered intraperitoneally (50 μg/kg) approximately 30 minutes before sacrifice. Secretagogues are given by either intracardiac or intravenous injection 4 min before sacrifice in rats anesthetized with pentobarbitol (75 mg/kg). Figure 2 illustrates the manipulation of the ion permeabilities in pancreatic zymogen granules using secretagogue and antagonist treatments. The parotid, however, is not as easily manipulated by exogenous secretagogues as it is prone to complete degranulation following stimulation.

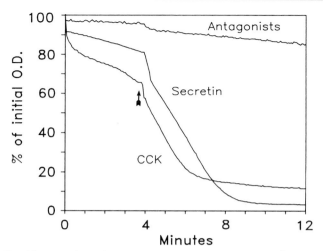

FIG. 2. Effect of antagonist and secretagogue treatments on pancreatic zymogen granule chloride conductance. Antagonists: atropine at 50 μg/kg and labetalol at 4 mg/kg; Secretin: antagonist treatment plus secretin at 140 μg/kg; CCK: antagonist treatment plus cholecystokinin at 50 μg/kg. At the arrow, the cation conductance ionophore nonactin (15 μg/ml) was added to the granule suspension, and the change in light scatter was monitored. (From Gasser et al.,[3] with permission.)

Granule Characterization

Electron micrographs of purified secretory granules from different sources show organelles with a dense core of secretory protein, a total organelle diameter of approximately 1 μm, and insignificant contamination by other particulate structures.[9,15] Biochemical assessments of granule yields and enrichments are presented based on marker enzyme analysis (Table III).[16-18] Mitochondria are the most common contaminant in the gradient isolation owing to a similarity in density; however, the results clearly show that mitochondrial contamination can be dramatically reduced. Granule yields are similar from tissue to tissue, averaging approximately 40% at enrichments (2.5- to 4.0-fold) suitable for many physiological experiments on the intact granule and 20% for the more highly enriched preparations needed for sensitive biochemical analysis. The increase in enrichment of the highest purity fraction is derived mainly from dilution of soluble and microsomal protein.

[15] R. DeLisle, I. Schulz, T. Tryakowski, W. Haase, and U. Hopfer, *Am. J. Physiol.* **246**, G411 (1984).

[16] A. Scarpa and P. Graziotti, *J. Gen. Physiol.* **62**, 756 (1973).

[17] P. Bernfeld, this series, Vol. 1, p. 149.

[18] M. L. Anson and E. A. Mirsky, *J. Gen. Physiol.* **16**, 59 (1932).

TABLE III

PURIFICATION OF EXOCRINE SECRETORY GRANULES
BASED ON RECOVERY OF MARKER ENZYMES AND LOSS OF
MITOCHONDRIAL CONTAMINATION[a]

Source	Cytochrome-c oxidase[b] yield (%)	Marker enzyme[c]	
		Yield (%)	Enrichment (-fold)
Homogenate	100	100	1
Pancreas			
First gradient	20	39	2.6
Second gradient	1	20	4.6
Parotid			
First gradient	7	26	4.8
Pepsinogen			
First gradient	12	39	3.3
Second gradient	0.8	19	8.1

[a] Data are from Gasser et al.[3,11] and Sharma et al.[9]
[b] Cytochrome-c oxidase, a measure of mitochondrial contamination, was quantitated by the method of Scarpa and Graziotti.[16]
[c] The marker enzyme was α-amylase[17] for the pancreas and parotid, and pepsin[18] for pepsinogen granules.

TABLE IV

OSMOTIC STABILITY AND ION TRANSPORT RATE CONSTANTS FROM
EXOCRINE SECRETORY GRANULES

Source	Inverse half-life[a] (hr^{-1}), control	Rate constants[b] (hr^{-1})		
		Cl$^-$ exchange	Cl$^-$ channel	K$^+$ channel
Pancreas	<0.3	35.8	11.2	13.4
Parotid	<0.3	48.7	15.4	30.8
Pepsinogen	<0.3	4.6	4.1	3.7

[a] Amount of time required for 50% of the zymogen granules to lyse in 37°, 150 mM KCl, pH 7.0. The granules were monitored for 1 hr, and the half-life was estimated by extrapolation of the rate of lysis.
[b] Rate constants are the inverse of the half-life (hr) of granule lysis after addition of optimal concentrations of the ionophore probes nigericin (5 μg/ml), valinomycin or nonactin (10 μg/ml), and tripropyltin (10 μg/ml) plus carbonyl cyanide m-chlorophenylhydrazone (3.5 μg/ml), respectively. Data are from Gasser et al.[3,8] and Sharma et al.[9]

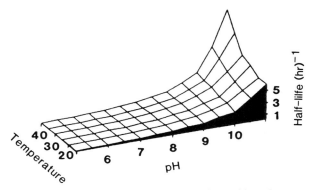

FIG. 3. Effect of temperature and pH on the osmotic stability of pancreatic zymogen granules. The granules were suspended in an isotonic NaCl suspension and the rate of granule lysis followed for 1 hr. The rate was extrapolated to the point where the optical density fell to one-half the original and is reported as the inverse of the half-life. (From Gasser and Hopfer,[14] with permission.)

Light scatter techniques have also proved useful in characterizing exocrine secretory granules, including measures of the membrane ion permeability. The bulk of these studies were carried out on granules harvested after one gradient. The zymogen granules were responsible for 90–95% of the light scatter and therefore constitute a large signal compared to noise.[3] As shown in Table IV, these granules typically display long-term osmotic stability at physiological conditions. Figure 3 illustrates that pancreatic zymogen granules retain their morphological integrity over a broad range of pH and temperature.[14] The osmotic stability in the face of significant gradients attests to the structural integrity of the granule membrane and the regulation of ion transport. Without this transport control the granules would quickly lyse in physiological salt solutions. As illustrated in Fig. 4 the isolated granules retain specific ion permeabilities which can be demonstrated using ionophore probes to override the intrinsic control mechanisms.[3,19] These transport pathways include a K^+ channel, Cl^- channel, and a Cl^-–anion exchanger. Qualitatively, the granules from different exocrine glands are similar although significant quantitative differences exist in the rates of transport (Table IV).

The intragranular environment is thought to be slightly acidic, as has been demonstrated for *in vivo* pancreatic granules using acridine orange[20]

[19] R. DeLisle and U. Hopfer, *Am. J. Physiol.* **250,** G489 (1986).
[20] C. Niederau, R. W. VanDyke, B. F. Scharschmidt, and J. H. Grendell, *Gastroenterology* **91,** 1433 (1986).

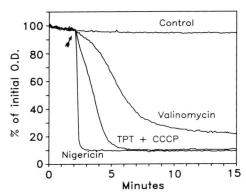

FIG. 4. Electrolyte transport pathways of pancreatic zymogen granules identified using ionophore probes and light scatter techniques. The granules were isolated as described in the text and suspended in an isotonic KCl solution at 37°, pH 7.0. At the arrow, the ionophore probe valinomycin (10 μg/ml) was added to demonstrate a chloride conductance, nigericin (5 μg/ml) for chloride–anion exchange, and tripropyltin (TPT) (10 μg/ml) plus carbonyl cyanide m-chlorophenylhydrazone (CCCP) (3.5 μg/ml) for potassium conductance.

and for isolated parotid granules using the distribution of [³H]acetate and [¹⁴C]methylamine.[21] Recently, using the pH-sensitive probe BCECF [2′,7′-bis(carboxyethyl)-5,6-carboxyfluorescein], we have also recorded an acidic intragranular environment (pH 6.3–6.7) for isolated pancreatic granules.[8] Aside from the recognized importance of pH in the storage and accumulation of product in some granular systems, pH also appears to have a significant role in the control of granule membrane permeability. Driving the intragranular pH toward 7.0 increases both the chloride and potassium channel conductance in parotid and pancreatic zymogen granules.[8,13]

Acknowledgments

These studies were supported in part by U.S. Public Health Service Grants AM-25170, HL-07415, and DK-27651 and by the Cystic Fibrosis Foundation (Grant Z0298).

[21] P. Arvan, G. Rudnick, and J. D. Castle, *J. Biol. Chem.* **259**, 13567 (1984).

[15] Transport of Nucleotides in the Golgi Complex

By BECCA FLEISCHER

Introduction

The Golgi apparatus is a complex, intracellular, smooth membranous organelle present in most eukaryotic cells. Its major function is to transport and to direct proteins and lipids from the endoplasmic reticulum, where they are synthesized, to their destination, both intracellular and extracellular.[1] A wide variety of cellular products including proteins, glycoproteins, glycosaminoglycans, and lipids are segregated within the lumen of the Golgi until they can be directed to their final extracellular, plasma membrane, or lysosomal destination. During this process of secretion, a wide variety of enzymatic modifications are carried out on the secretory products by enzymes localized in the Golgi apparatus membrane. These include the terminal glycosylation of N-asparagine-linked glycoproteins[2] and the formation of O-glycosidic linkages to threonine or serine moieties of proteins.[3] In addition, the Golgi is the site of most of the glycosylations involved in the formation of glycosphingolipids and gangliosides.[4] Other modifications carried out uniquely in the Golgi include sulfation of cerebroside to form sulfatide,[5] sulfation of the sugar moieties of glycosaminoglycans[6] and glycoproteins, and sulfation of tyrosine residues of proteins.[7] All of these modifications involve specific membrane-bound transferases which use nucleotide sugars or 3'-phosphoadenosine 5'-phosphosulfate (PAPS) as donor molecules.

Although large, highly organized, and membranous, the Golgi apparatus can be isolated from a number of mammalian tissues in a fairly intact and recognizable form provided gentle homogenization is employed and osmotic shock is minimal.[8] Even when vesiculation occurs, Golgi vesicles remain largely oriented with their cytoplasmic face toward the medium.[9] Many of the transferases involved in the modification of secretory products

---

[1] M. G. Farquhar, *Annu. Rev. Cell Biol.* **1,** 447 (1985).
[2] R. Kornfeld and S. Kornfeld, *Annu. Rev. Biochem.* **54,** 631 (1985).
[3] C. Abeijon and C. B. Hirschberg, *J. Biol. Chem.* **262,** 4153 (1987).
[4] H. K. M. Yusuf, G. Pohlentz, and K. Sandhoff, *J. Neurosci. Res.* **12,** 161 (1984).
[5] B. Fleischer and F. Zambrano, *J. Biol. Chem.* **249,** 5995 (1974).
[6] E. Brandan and C. B. Hirschberg, *J. Biol. Chem.* **263,** 2417 (1988).
[7] R. N. H. Lee and W. B. Huttner, *Proc. Natl. Acad. Sci. U.S.A.* **82,** 6143 (1985).
[8] B. Fleischer, this series, Vol. 98, p. 60.
[9] B. Fleischer, *J. Cell Biol.* **89,** 246 (1981).

in the Golgi have been shown to be lumenally oriented, while the cellular site of synthesis of their nucleotide cofactors is the cytoplasm, or in the case of CMPNeuAc, the nucleus. Thus, it is clear that mechanisms must exist to allow the passage of nucleotide sugars and PAPS across the Golgi membrane into the lumen as well as to allow the exit of nucleotide products. mainly monophosphatides[10] from the Golgi lumen into the cytoplasm. In this chapter, a method developed in our laboratory for measuring the uptake of substances such as nucleotides or nucleotide sugars by vesicles derived from the Golgi apparatus is described.[11]

Assay Methods

Principle

In this assay, summarized in Fig. 1, concentrated vesicles ($\sim$ 3 mg/ml) are incubated with radioactive nucleotides in a small volume (100 μl) and the uptake stopped by diluting the samples 20-fold with cold isosmotic medium. The vesicles are recovered by rapid Millipore filtration. Binding or incorporation of substrates can be estimated independently of uptake into the lumen of the vesicles by permeabilizing the vesicles by pretreatment with filipin[9] or by including low levels of Triton X-100 in the cold stop solution. Treatment with Triton X-100 is preferable because its action is not restricted to membranes containing significantly high levels of cholesterol as is the case of filipin. The level of detergent used is first determined by measuring the level necessary to release the bulk of the radioactive nucleotide without solubilizing proteins from the vesicles. For each concentration of nucleotide used, binding to filters alone is determined by running samples without Golgi vesicles added. A detailed protocol is given below for measuring uptake of UDPGal, but the method is generaly applicable for other nucleotides or nucleotide sugars as well.

Equipment

Amicon vacuum filtration manifold, Model VFM1 with 10 place cups (or equivalent)
Gas vacuum pump capable of providing vacuum of 500–600 mm Hg with above manifold
0.45 μm HA type Millipore filters (for use with manifold)
12 $\times$ 75 mm disposable glass test tubes

---

[10] E. Brandan and B. Fleischer, *Biochemistry* **21**, 4640 (1982).
[11] B. Fleischer, *J. Histochem. Cytochem.* **31**, 1033 (1983).

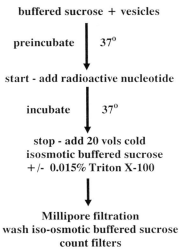

buffered sucrose + vesicles

preincubate | 37°

start - add radioactive nucleotide

incubate | 37°

stop - add 20 vols cold
isosmotic buffered sucrose
+/- 0.015% Triton X-100

Millipore filtration
wash iso-osmotic buffered sucrose
count filters

FIG. 1. Flow diagram of assay for uptake of nucleotides by Golgi vesicles using Millipore filtration.

Pipetman with disposable tips; adjustable volumes: 20 μl, 100 μl; fixed
    volumes: 2 ml, 5 ml
vortex mixer

Solutions

0.25 M sucrose
0.50 M sucrose
1.0 M imidazole-HCl, pH 6.9 at room temperature
Golgi vesicles, 5–10 mg protein/ml in 0.25 M sucrose
25 mM UDP[^{14}C]Gal ($\sim 10^4$ dpm/nmol) in 0.25 M sucrose
Ethylene glycol methyl ether
Aqueous scintillation counting cocktail

Procedure

Golgi vesicles, prepared as described previously[8] and suspended in
0.25 M sucrose at a concentration of 5–10 mg protein/ml, are added to
0.25 M sucrose containing imidazole-HCl, pH 6.9, in the small glass test
tubes to give a final buffer concentration of 0.1 M and 0.2–0.3 mg protein
per tube. For determining the filter bank, 0.25 M sucrose is substituted for
the Golgi preparation. The tubes are preincubated at 37° for 3 min, and
uptake is initiated by adding 5–10 μl UDP[^{14}C]Gal in 0.25 M sucrose with
vortexing to give a final concentration of 2 mM UDPGal and a final assay
volume of 100 μl. The samples are incubated at 37° for the desired time

and the uptake stopped by adding 2.0 ml ice-cold wash solution (0.25 M sucrose, 0.1 M imidazole-HCl, pH 6.9) to each tube. The tubes are rapidly and thoroughly vortexed and immediately filtered using 1.0-ml aliquots on each of two Millipore filters. Before filtering the samples, the filters are each prewashed with 5 ml of ice-cold wash solution. After filtration of the samples each filter is again washed with 5 ml of cold wash solution.

For determination of incorporated radioactivity, the incubated Golgi vesicles are diluted with 2.0 ml cold wash solution containing 0.015% (w/v) Triton X-100, vortexed, and allowed to stand at 0° for several minutes before filtration. After filtration they are washed with 5 ml cold wash solution containing 0.015% Triton X-100. The filters are removed from the apparatus, air dried, and placed in 15-ml glass counting vials, and 2 ml ethylene glycol methyl ether is added. The vials are allowed to stand at room temperature with occasional shaking until the filters are dissolved (about 30 min). Ten milliliters of scintillation fluid (Aqueous Counting Solution, Amersham, or equivalent) are added and the samples counted in a scintillation counter. The uptake rate is expressed as nanomoles UDPGal taken up per (minute × milligram protein) and is calculated after subtracting values for filter blanks and for incorporated [14C]galactose obtained from the Triton-treated samples.

Comments

Golgi vesicles prepared using D_2O–sucrose gradients[8] generally have a better uptake rate for nucleotides than those prepared using H_2O–sucrose gradients, presumably because of their greater intactness.[12] The vesicles can be frozen and stored in liquid nitrogen for later assay provided they are quickly frozen at a high protein concentrations (5–10 mg/ml) and used within 10 days.

The use of Triton X-100 in the stop solution is a convenient method of estimating the amount of endogenous incorporation of sugars from radioactive nucleotide sugar donors labeled in the sugar moiety. The level of Triton X-100 used is obtained by determining the amount which will release nucleotide sugar from the vesicles without releasing proteins into the filtrate. As shown in Fig. 2, when the level of Triton X-100 in the assay is 1 mg/mg protein, CMP[14C]NeuAc is released from the vesicles and appears in the filtrate. Higher levels of Triton X-100 result in release of protein from the vesicles together with additional radioactivity, probably arising from protein-bound [14C]NeuAc. Similar results are obtained when [14C]UMP is used, except that all the nucleotide is released before any

12 B. Fleischer, *Arch. Biochem. Biophys.* **212,** 602 (1981).

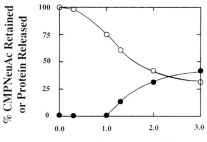

FIG. 2. Effect of Triton X-100 added to the stop and wash solutions on release of CMP[^{14}C]NeuAc (O) or protein (●) from Golgi vesicles after Millipore filtration.

protein is released, since there is no incorporation of the nucleotide into protein. When uptake of UDPGal is assayed using liver cell fractions other than Golgi vesicles, a considerable amount of binding to some fractions such as rough endoplasmic reticulum is seen, but little or no accumulation within the vesicles can be measured (Table I).[13]

Uptake of nucleotide sugars such as UDPGal by Golgi vesicles is not very pH dependent between pH 6 and 7.5 but does fall off above pH 7.5. It is not affected by divalent cations although the rate of endogenous incorporation is stimulated by the addition of Mn^{2+}. The rate of uptake does not appear to be affected by creation of a ΔpH or $\Delta\psi$ across the Golgi vesicle membrane.[14] The rate and extent of uptake of UDPGal by Golgi vesicles can be greatly enhanced by preloading the vesicles with N-acetylglucosamine, an acceptor for N-acetyllactosamine synthase located on the lumenal side of the vesicle membrane. Since the K_m for this galactosyltransferase is very low (about 15 μM), accumulation of UDPGal inside the vesicles is prevented by the presence of high levels of acceptor. The Golgi membranes are relatively impermeable to disaccharides, so that the product of the galactosyltransferase reaction, N-acetyllactosamine, accumulates within the vesicles and can be estimated using the Millipore filtration assay (Fig. 3).

The method has been used to demonstrate uptake of a number of nucleotides and nucleotide sugars into Golgi vesicles.[14] Uptake is saturable with respect to nucleotide concentration and is time and temperature dependent. Uptake is a linear function of protein concentration in the

[13] M. Kervina and S. Fleischer, this series, Vol. 31, p. 6.
[14] B. Fleischer, in "Protein Transfer and Organelle Biogenesis" (R. C. Das and P. W. Robbins, eds.), p. 289. Academic Press, San Diego, California, 1988.

TABLE I
UPTAKE OF UDPGal BY SUBCELLULAR FRACTIONS
OF RAT LIVER[a]

Cell fraction	Transport activity (nmol $\times$ mg protein^{-1} $\times$ min^{-1})
Golgi	0.52
Rough endoplasmic reticulum	0.02
Mitochondria	0.04
Plasma membrane	0.06

[a] Golgi[8] and other cell fractions[13] prepared as described previously were assayed by the procedure described in the text. Total uptake activities were corrected for endogenous incorporation and/or binding of nucleotides using Triton X-100-treated samples as described to obtain the transport activities shown.

range of 1–4 mg protein/ml of assay medium. Low concentrations of protein give poor accuracy in the assay and are not recommended. Similarly, concentrations higher than 4 mg/ml cause slow filtration and poor washing due to clogging of the filters and should be avoided. The best range is 2–3 mg protein/ml. A summary of the apparent kinetic constants obtained for the transport of a number of sugar nucleotides by Golgi vesicles is shown in Table II.

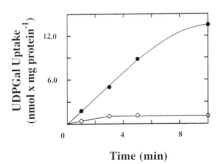

Time (min)

FIG. 3. Effect of galactosyltransferase substrate on UDPGal uptake by Golgi vesicles. Vesicles were preincubated for 1 hr at 0° with (●) or without (○) 20 mM GlcNAc, and the uptake was measured as described in the text. In both cases, endogenous incorporation into protein has been subtracted. When GlcNAc is present, the counts measured represent internal N-acetyl[^{14}C]lactosamine formed. (Data from Ref. 14.)

TABLE II
KINETIC CONSTANTS FOR UPTAKE OF NUCLEOTIDE
SUGARS BY GOLGI VESICLES[a]

Nucleotide sugar	K_m (μM)	V_{max} (nmol $\times$ mg protein^{-1} $\times$ min^{-1})
UDPGal	420	0.60
UDPGlc	37	0.30
UDPGlcNAc	400	1.20
CMPNeuAc	200	0.50

[a] Initial rates of uptake were measured at 37° as described in the text. In all cases, endogenous incorporation and/or binding has been corrected for by subtracting values obtained by releasing intravesicular nucleotide with low levels of Triton X-100. From Ref. 14.

Uptake of nucleotides and nucleotide sugars in Golgi vesicles has also been demonstrated by measuring the distribution of the radioactive solute between the vesicles and the medium after recovery of the vesicles by centrifugation.[15] The assay differs from that described in this chapter in that low (micromolar) concentrations of sugar nucleotides are used so that most of the transported substrates are incorporated rather than accumulating inside the vesicles. Thus, the low (micromolar) apparent K_m observed largely reflects the K_m of the sugar transferase involved rather than the K_m of the transport system itself.

Acknowledgments

The author wishes to thank Kathryn Dewey for technical assistance in this work. The work was supported in part by National Science Foundation Grant DCB 8402370 and by Biomedical Research Support Grant S07 RR07201-9 from Vanderbilt University.

[15] M. Perez and C. B. Hirschberg, this series, Vol. 138, p. 709.

Section II

Transport in Plants

A.   Higher Plants
Articles 16 through 24

B.   Lower Plants
Articles 25 through 30

C.   Organelles
Articles 31 through 33

[16] Water Flow in Plants and Its Coupling to Other Processes: An Overview

By ERNST STEUDLE

Introduction

The water relations of plants have a sound, physical basis in that the water status of plants can be adequately and quantitatively described by the water potential, which is a measure of the free energy of water. Although there is some debate as to whether this concept is going to change or be extended,[1,2] the main principles will last. To a large extent, water potential controls the movement of water in the soil–plant–atmosphere continuum as well as at the levels of cells, tissues, and organs. However, despite the fact that the theory of hydraulic and osmotic properties of plants is rather well developed, there are large gaps in our understanding of the linkage between water relations and plant metabolism. These gaps are most obvious in the area of growth and development of plants where physical (water uptake, cell wall extension) and metabolic (cell wall and protein synthesis, solute uptake) processes cooperate and may both become rate-limiting. Other examples of coupling between water relations and metabolic processes are the adaptations to water deficits, salinity, and cold temperature. Hormones such as abscisic acid (ABA) are thought to play an important role in triggering some of these phenomena and providing the linkage, but the precise mechanisms of the transformation of signals in either direction are not known.

Concerning osmoregulation and turgor maintenance, some evidence has been presented that a hydraulic signal such as turgor may be directly transformed into changes in active transport across cell membranes by means of a compressible turgor sensor in the membrane,[3,4] while in other systems the absolute value of osmotic pressure seems to be sensed. On the other hand, it has been demonstrated that in higher plants, root/shoot communication could well operate without any hydraulic signaling.[5] When the xylem water potential was kept at a constant level, the shoot nevertheless reacted to changes in soil water content.

[1] P. J. Kramer, *Plant, Cell Environ.* **11**, 565 (1988).

[2] E.-D. Schulze, E. Steudle, T. Gollan, and U. Schurr, *Plant, Cell Environ.* **11**, 573 (1988).

[3] H. G. L. Coster, E. Steudle, and U. Zimmermann, *Plant Physiol.* **58**, 636 (1977).

[4] U. Zimmermann, *Annu. Rev. Plant Physiol.* **29**, 121 (1978).

[5] T. Gollan, J. B. Passioura, and R. Munns, *Aust. J. Plant Physiol.* **13**, 459 (1986).

METHODS IN ENZYMOLOGY, VOL. 174

In this chapter I do not intend to review completely the vast field of possible couplings between water transport and other processes as they occur, for example, during gas exchange,[6] during water and nutrient uptake into the root,[7-14] or during the coupled flow of water and assimilates in the phloem.[15-17] Rather, some basic phenomena of the coupling between water and solute transport at the cell and tissue level and coupling phenomena between water transport and cell wall mechanics are considered as they occur during water stress,[18,19] osmoregulation,[4,20] or growth.[21] Emphasis has been given to water–solute interactions in roots, where they are most obvious. In most cases, recent results obtained with pressure probe techniques are reported and compared with others. (A detailed description of the cell pressure probe is given elsewhere in this volume [24].)

Theory

Transport Equations

Transport of matter across membranes is properly dealt with by the theory of irreversible thermodynamics which may also be applied to complex barriers. The appropriate theory has been adequately treated in a number of reviews and textbooks.[22-27] Considering the production of en-

[6] E.-D. Schulze, *Annu. Rev. Plant Physiol.* **37**, 247 (1986).

[7] W. P. Anderson, *Encycl. Plant Physiol. New Ser.* **2**, Part B, 129 (1976).

[8] J. Dainty, *Acta Hortic.* **171**, 21 (1985).

[9] E. L. Fiscus, *Plant Physiol.* **55**, 917 (1975).

[10] P. J. Kramer, "Water Relations of Plants." Academic Press, New York, 1983.

[11] J. B. Passioura, *Annu. Rev. Plant. Physiol.* **39**, 245 (1988).

[12] M. G. Pitman, *Q. Rev. Biophys.* **15**, 481 (1982).

[13] P. E. Weatherley, *Adv. Bot. Res.* **3**, 171 (1970).

[14] P. E. Weatherley, *Encycl. Plant Physiol., New Ser.* **12**, Part B, 79 (1982).

[15] J. A. Milburn, *Encycl. Plant Physiol., New Ser.* **1**, 328 (1975).

[16] D. C. Spanner, *Encycl. Plant Physiol., New Ser.* **1**, 301 (1975).

[17] H. Ziegler, *in* "Biophysik" (W. Hoppe, W. Lohmann, H. Markl, and H. Ziegler, eds.), p. 652. Springer-Verlag, Berlin and New York, 1982.

[18] K. J. Bradford and T. C. Hsiao, *Encycl. Plant Physiol., New Ser.* **12**, Part B, 263 (1982).

[19] M. T. Tyee and P. G. Jarvis, *Encycl. Plant Physiol., New Ser.* **12**, Part B, 35 (1982).

[20] H. Kauss, *Prog. Phythochem.* **5**, 1 (1978).

[21] D. J. Cosgrove, *Annu. Rev. Plant Physiol.* **37**, 377 (1986).

[22] J. Dainty, *Adv. Bot. Res.* **1**, 279 (1963).

[23] J. Dainty, *Encycl. Plant Physiol., New Ser.* **2**, Part A, 12 (1976).

[24] C. R. House, "Water Transport in Cells and Tissues." Arnold, London, 1974.

[25] A. Katchalsky and P. F. Curran, "Nonequilibrium Thermodynamics in Biophysics." Harvard Univ. Press, Cambridge, Massachusetts, 1965.

[26] D. Woermann, *Encycl. Plant Physiol., New Ser.* **3**, 419 (1976).

[27] U. Zimmermann and E. Steudle, *Adv. Bot. Res.* **6**, 45 (1978).

tropy during an irreversible process (passage of water and solutes across a membrane; chemical reaction), theory predicts that *all* flows in a system are functions of *all* driving forces. Provided that forces and flows are sufficiently small, they will be linearly related to each other. The relationships are expressed by certain transport coefficients which allow quantitative expression of the coupling between flows, e.g., between the flow of water and that of a solute. Active transport can be incorporated into the theory.

For a plant cell membrane separating two compartments, a cell interior, i, and a medium, o, the water (volume) flow per unit area, J_V, is given by

$$J_V = -\frac{1}{A}\frac{dV}{dt} = Lp[P - RT(C^i - C^o) - \sigma_s RT(C_s^i - C_s^o)] \qquad (1)$$

V is cell volume; A, cell surface area; Lp, hydraulic conductivity; P, cell turgor; C, concentration of nonpermeating solute; C_s, concentration of permeating solute; σ_s, reflection coefficient. Equation (1) denotes the volume flow in the presence of permeating and nonpermeating solutes. J_V has a hydraulic (driving force: hydrostatic pressure gradient, P) and an osmotic component (driving force: osmotic pressure gradient). In the case of permeating solutes, $\sigma_s < 1$ indicates a contribution of the flow of solute s to the overall volume flow and some frictional interaction (coupling) between water and solutes as they move across the membrane or barrier.

If there are only nonpermeating solutes present, Eq. (1) reduces to

$$J_V = Lp(P - \Delta\pi) \qquad (2)$$

or

$$J_V = Lp\,\Delta\psi \qquad (3)$$

where $\Delta\psi$ and $\Delta\pi$ are the differences in water potential and osmotic pressure, respectively. Equation (3) is an Ohm's law analog and states that the water flow is proportional to $\Delta\psi$. This equation is often used to describe short as well as long distance transport of water in plants (e.g., water flow between root and shoot). It is evident that it can be used only if semipermeable barriers are considered. If matric potentials (τ) also have to be taken into account, Eq. (2) has to be extended:

$$J_V = Lp(P - \Delta\pi - \Delta\tau) \qquad (4)$$

Equation (1) is also used for complex barriers, i.e., for barriers which are composed of different membranelike elements arranged in series and in parallel (e.g., in the root and in the leaf). By analogy to electricity, Kirchhoff's rules are applied to evaluate overall resistances. It is important to note that although this may be a fairly good approach in many cases, deviations may occur if the reflection coefficients of the elements of the composite barrier differ.[28,29]

The flow of a solute s in a two-compartment system [as in Eq. (1)] will be given by

$$J_s = -\frac{1}{A}\frac{dn_s}{dt} = P_s(C_s^i - C_s^o) + (1 - \sigma_s)\overline{C}_s J_V + J_s^* \qquad (5)$$

where P_s is the permeability coefficient; n_s, content of solute in the cell; $\overline{C}_s$, mean concentration of s in the membrane; J_s^* active transport component of s. According to Eq. (5) the solute flow consists of a diffusional $[P_s(C_s^i - C_s^o)]$, a solvent drag $[(1 - \sigma_s)\overline{C}_s J_V]$, and an active (J_s^*) component which is related to metabolic energy.

Active Water Transport

Equations (1)–(5) are incomplete because only nonelectrolytes are considered and direct coupling between metabolic reactions and water flow has been omitted ("active water flow"). To account for the presence of permeating ions, terms for flow of electric current and electrokinetic phenomena such as electroosmosis and streaming potentials can also be incorporated.[24,30] They have been left out here for the sake of simplicity and because the effects of electrokinetic phenomena on water transport across plant cell membranes should be small, although such mechanisms have been proposed for phloem and also for cells.[16,31] Furthermore, there is no evidence for active water transport in plants. The reason active transport mechanisms (if existing) would be rather ineffective is that the water permeability of plant membranes is usually rather high, i.e., the permeability of water will be two to three orders of magnitude larger than that of low molecular weight solutes. Thus, any electroosmotic or active pumping or coupling will be short-circuited by the low hydraulic resistance of the membrane. Hence, the coupling between water and (active or passive) solute flows via changes in $\Delta\pi_s$ or $\Delta\pi$ should be the most relevant in plants.

Dynamic Water Relations of Plant Cells

Equations (1) and (5) are used to evaluate the important transport coefficients Lp, σ_s, and P_s and also to determine changes in cell volume and turgidity. To calculate the coefficients, measurement of turgor pressure

[28] O. Kedem and A. Katchalsky, *Trans. Faraday Soc.* **59**, 1931 (1963).
[29] O. Kedem and A. Katchalsky, *Trans. Faraday Soc.* **59**, 1941 (1963).
[30] N. A. Walker, *Encycl. Plant Physiol., New Ser.* **2**, Part A, 36 (1976).
[31] P. H. Barry, *J. Membr. Biol.* **3**, 335 (1970).

changes is much more convenient than that of cell volume changes because the latter are usually very small. Both volume and turgor are related to each other by the elastic coefficient (elastic modulus) of the cell wall, ϵ, i.e.,

$$\epsilon \equiv V \frac{dP}{dV} \simeq V \frac{\Delta P}{\Delta V} \tag{6}$$

By combining Eqs. (1) and (6) it is easily verified that the time constant, t_c, of shrinking or swelling of a plant cell in response to a change in the water potential of the surroundings will be given by

$$t_c = \frac{T^w_{1/2}}{\ln 2} = \frac{1}{LpA} \frac{V}{\epsilon + \pi^i} \tag{7}$$

where $T^w_{1/2}$ is the half-time; π^i, osmotic pressure of the cell. Hence, by measuring $T^w_{1/2}$ and ϵ and also determining V, A, and π^i, the hydraulic conductivity can be evaluated. In terms of an electrical analog, the water capacitance of the cell $[C_c = V/(\epsilon + \pi^i)]$ is charged via a hydraulic resistance $(R = 1/LpA)$. Equation (7) can be also written as

$$t_c = RC_c \tag{8}$$

Since the response time of a tissue or plant organ is related to that of cells, Eq. (8) is important for physiological and ecological reasons. It is also noteworthy that (besides geometrical factors) hydraulic properties of the membrane *and* mechanical properties of the wall will determine the response.

Dynamics of Solute Relations

Equations (1) and (5) have to be integrated simultaneously in order to work out the influence of solute transport on water flow and on shrinking and swelling of a cell. In a typical experiment, the cell would be in a stationary state prior to the addition of permeating solutes to the medium, and the response in turgor or cell volume would be measured following the change. Under these conditions, an explicit solution for $P(t)$ can be obtained if the solvent drag is neglected and J_s^* is assumed to be constant. Both assumptions will hold for plant cells to a good approximation.[32] The solution is

$$\frac{P - P_o}{\epsilon} = \frac{V - V_o}{V_o} = \frac{\sigma_s \Delta \pi_s^o Lp}{(\epsilon + \pi^i)Lp - P_s}[\exp(-k_w t) - \exp(-k_s t)] \tag{9}$$

where P_o and V_o are original turgor and cell volume, respectively; $\Delta \pi_s^o$,

[32] E. Steudle and S. D. Tyerman, *J. Membr. Biol.* **75**, 85 (1983).

change in osmotic pressure of medium. k_w is the rate constant of water exchange ($= 1/t_c$) and k_s the rate constant of solute exchange ($= P_s A/V_o$). Equation (9) describes a biphasic pressure response: after a decrease in turgor (volume) owing to a rapid water efflux (mainly determined by the first term within the brackets) water is again taken up, because the permeating solutes equilibrate across the membrane (solute phase). The determination of k_s yields P_s directly. The reflection coefficient (σ_s) is evaluated according to its thermodynamic definition at $J_V = 0$, i.e., from the maximum or minimum changes of P. It can be verified from Eq. (9) that at $dP/dt,\ dV/dt = 0$, it should be valid that[32]

$$\sigma_s = \frac{P_o - P_{min}}{\Delta \pi_s^o} \frac{\epsilon + \pi^i}{\epsilon} \exp(k_s t_{min}) \tag{10}$$

where t_{min} is the time to reach the minimum pressure P_{min}. Thus, by following the changes in turgor of individual cells, the coefficients Lp, σ_s, and P_s which characterize the transport properties of cells and the coupling between solute and water flow can be calculated. This approach has been also applied to more complex systems (see section on water and solute relations of roots).

Effects of Unstirred Layers

Two types of unstirred layers may tend to reduce the absolute values of transport coefficients as measured by pressure probe, tracer flow, or other types of experiments as compared with the true values. They are also important in many transport phenomena in plants such as the uptake of nutrients from the soil or during transpiration. In the first type, water flow would concentrate solutes at one side of a membrane or barrier and would enhance the actual concentration, whereas at the other side solutes would be diluted ("sweep-away effect"[22]). The relation of the actual concentration at the membrane (C_s^m) and that of the bulk solution (C_s^b) will be given by

$$C_s^m/C_s^b = \exp(-J_V \delta/D_s) \tag{11}$$

where D_s is the diffusion coefficient of solute s and δ the thickness of unstirred layer.

In the second type of unstirred layers, the permeation of solutes across an osmotic barrier or membrane could also reduce the actual concentration at the more concentrated side and could enhance the concentration on the other. The relation between permeation resistance of the membrane ($1/P_s$) and that of the two unstirred layers on both sides of the membrane (a and b) is important. The measured permeability would be lower because of the additive resistances of the unstirred layers, i.e.,

$$\frac{1}{P_s^{meas}} = \frac{1}{P_s} + \frac{\delta^a}{D_s^a} + \frac{\delta^b}{D_s^b} \tag{12}$$

This type of unstirred layer effect may be called the "gradient-dissipation effect."[33] It is obvious that it could become very important for rapidly permeating solutes (high P_s) such as in tracer experiments with tritiated water or with lipophilic or low molecular weight solutes.

Sometimes the thickness of the unstirred layers is not easy to estimate. Stirring reduces δ but cannot completely eliminate it. Unstirred layers could be of importance in the wall space of tissues, where solutes may diffuse over long distances. The time constants required for these processes (t_D) would depend on the geometry of the tissue and its dimension (x). A rough estimation of t_D would be

$$t_D \approx x^2/D \tag{13}$$

according to the Einstein–Smoluchowski equation (for more rigorous estimates, see Ref. 33). It should be noted that the equations given for unstirred layer effects have been derived for steady-state conditions, but they are also used to estimate effects during dynamic changes because rigorous treatments of the effects under nonstationary conditions are still missing.

Water Transport in Tissues

Philip[34-36] was the first to formulate water transport quantitatively in tissues in terms of a mathematical model. He neglected the apoplasmic water flow (around cells) and assumed that water would travel from cell to cell either across plasmodesmata (symplasmic transport) or by crossing cell membranes (transcellular transport). He did not distinguish between the latter two possibilities and assumed that the flow between cells would be proportional to the difference in their water potentials [Eq. (3)]. This resulted in a diffusion type of kinetics for a change in water potential, turgor, or cell volume. In 1974 Molz and Ikenberry[37] extended this model to allow for an apoplasmic transport, i.e., two parallel pathways were

[33] P. H. Barry and J. M. Diamond, *Physiol. Rev.* **64**, 763 (1984).
[34] J. R. Philip, *Plant Physiol.* **33**, 264 (1958).
[35] J. R. Philip, *Plant Physiol.* **33**, 271 (1958).
[36] J. R. Philip, *Plant Physiol.* **33**, 275 (1958).
[37] F. J. Molz and E. Ikenberry, *Soil Sci. Soc. Am. Proc.* **38**, 699 (1974).

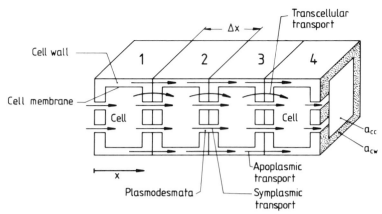

FIG. 1. One-dimensional model of the different water transport pathways in a plant tissue. The apoplasmic path is exterior to the plasmalemma while the symplasmic path is restricted to the flow across plasmodesmata. The transcellular flow denotes the flow across cell membranes which also crosses the walls between adjacent cells. Transcellular and symplasmic flows may be summarized as the cell-to-cell transport of water. a_{cc} and a_{cw} are the mean cross-sectional areas for the cell-to-cell (transcellular plus symplasmic) and apoplasmic paths, respectively; Δx is the thickness of a cell in direction x. [Redrawn from F. J. Molz and J. M. Ferrier, *Plant, Cell Environ.* **5**, 191 (1982).]

considered (Fig. 1). They assumed local water flow equilibrium between each cell and its immediate apoplasmic surroundings and arrived at a diffusion type of process for the dynamics of tissue water relations, the "diffusivity" of tissue, D_t, being

$$D_t = \frac{\Delta x(Lp_{cw}a_{cw} + a_{cc}Lp\Delta x/2)}{C_c + C_{cw}} \tag{14}$$

where Lp_{cw} and Lp are the hydraulic conductivities of the wall path and the cell membrane, respectively, and a_{cw} and a_{cc} denote the cross-sectional areas. Δx is the thickness of cells in the direction of the propagation of the change. C_{cw} and C_c are the storage capacities of the pathways per cell (see below). It can be seen that the conductances of the pathways are additive and increase D_t, whereas increasing capacities damp the propagation in the tissue. It has to be emphasized that D_t is not a measure of a "diffusional mass flow" of water in the tissue, but rather describes the rate at which changes in water potential (free energy), cell volume, or turgor propagate. Thus, the analogy of the equations used by Philip and Molz and Ikenberry with Fick's laws is only formal. It can be seen from Eq. (14) that neglecting the apoplasmic component yields Philip's diffusivity for the cell-to-cell transport (D_c)

$$D_c = \frac{a_{cc}\Delta x^2 Lp}{2C_c} = \frac{a_{cc}\Delta x^2 Lp(\epsilon + \pi^i)}{2V} \approx \frac{1}{T^w_{1/2}} \qquad (15)$$

which is inversely proportional to the time constant of water exchange of individual cells [Eq. (7)].

Equations (14) and (15) govern dynamic water relations properties of tissues. Under stationary conditions (e.g., at steady-state water flows across a root or leaf), the parallel hydraulic conductance $Lp_{cw}a_{cw}$ and Lpa_{cc} as well as the cell dimensions play the important role. Under these conditions, the classical catenary concept of van den Honert[38] is used when describing water transport in the entire plant. This can be extended for nonstationary conditions by adding water capacities to the network.[39,40]

As compared with the cell level, the description of water transport at the tissue and organ level is still incomplete. This is mainly due to the fact that important parameters such as the conductivity of the wall path (Lp_{cw}) or its storage capacitance (C_{cw}) are still not known with sufficient accuracy, owing to technical difficulties. Moreover, quantitative concepts for describing water transport at the level of organs and entire plants are largely missing. They would also incorporate interactions with the movement of solutes.

Water Relations of Plant Cells and Tissues

The hydraulic properties of plant cells (Lp, $T^w_{1/2}$, D_c) are important physiological and ecological parameters. Besides geometric factors and the hydraulic properties of the apoplast, the hydraulic conductivity of tissues will depend on the hydraulic conductivity of the cell membranes. D_c ($T^w_{1/2}$) will be important for the dynamic properties of tissues, i.e., a short half-time at the cell level will also mean a short half-time of water potential equilibration at the tissue level.

Before the pressure probe for higher plant cells was introduced in 1978,[41] data about water relations of individual cells were rather rare. They were mostly determined from tracer flow and by plasmolysis techniques.[23,27,42,43] Since then the pressure probe technique has been applied to quite a number of plant species and tissues. Some of the values for Lp, $T^w_{1/2}$, and D_c obtained so far are listed in Tables I and II.

[38] T. H. van den Honert, *Dicuss. Faraday Soc.* **3**, 146 (1948).

[39] I. R. Cowan, *Planta* **106**, 185 (1972).

[40] P. S. Nobel and P. W. Jordan, *J. Exp. Bot.* **34**, 1379 (1983).

[41] D. Hüsken, E. Steudle, and U. Zimmermann, *Plant Physiol.* **61**, 158 (1978).

[42] E. J. Stadelmann and O. Y. Lee-Stadelmann, this volume [18].

[43] E. Steudle and U. Zimmermann, *in* "Membrane Transport in Plants" (W. J. Cram, K. Janacek, R. Rybova, and K. Sigler, eds.), p. 73. Akademia-Verlag, Prague, 1984.

TABLE I

HYDRAULIC CONDUCTIVITY *(Lp)* AND HALF-TIME OF WATER EXCHANGE $(T^w_{1/2})$ OF
CELLS OF HIGHER PLANTS DETERMINED BY METHODS DIFFERENT FROM THE
CELL PRESSURE PROBE TECHNIQUE[a]

Species	Tissue/ cell type	Technique	Half-time, $T^w_{1/2}$ (sec)	Hydraulic conductivity, Lp (m sec^{-1} MPa^{-1})	Ref.
Allium	Bulb	Plasmolysis	—	$(2-3) \times 10^{-8}$	b
cepa	epidermis	Tracer flow	—	$(1-2) \times 10^{-8}$ (P = 0 MPa)	c
		Tracer flow	—	1.3×10^{-8} $(P \simeq 1.2$ MPa)	c
		External force	9–55	$(2-5) \times 10^{-7}$	d
Helianthus tuberosus	Storage tissue	External force	50	$(6-36) \times 10^{-8}$	e
Beta vulgaris	Storage tissue	External force	55–690	$(2-6) \times 10^{-8}$	f
Elodea nutallii	Leaf cells	NMR	—	$>2 \times 10^{-6}$	g
Hedera helix	Bark cells	NMR	—	2×10^{-6}	h
Zea mays	Root cells	NMR	—	5×10^{-7}	i
		Tracer flow	—	$(3-4) \times 10^{-7}$ (cell-to-cell model)	j
		Tracer flow	—	$\sim 1 \times 10^{-9}$ (isolated cell model)	j
		Tracer flow	—	4×10^{-10} (cortex) 1.4×10^{-9} (stele) (isolated cell model)	k, l, m
Daucus carota	Xylem paren- chyma	Tracer flow	—	6×10^{-8} (cell-to-cell model)	n
		Tracer flow	—	$\simeq 1 \times 10^{-10}$ (isolated cell model)	n

[a] Only a few typical examples are given for the plasmolytic, tracer flow, external force, and NMR techniques. Tracer flow data on the permeability coefficient of water (P_d) have been transformed into hydraulic conductivities *(Lp)*. For these data, the Lp values strongly depend on the model used for the calculation of $Lp(P_d)$, i.e., on whether the tissues would behave like isolated cells or whether water moves from cell to cell within the tissues.

[b] W. G. Url, *Protoplasma* **72**, 427 (1971).

[c] J. P. Palta and E. J. Stadelmann, *J. Membr. Biol.* **33**, 231 (1977).

[d] J. M. Ferrier and J. Dainty, *Can. J. Bot.* **55**, 858 (1977).

[e] J. M. Ferrier and J. Dainty, *Can. J. Bot.* **56**, 22 (1978).

[f] W. N. Green, J. M. Ferrier, and J. Dainty, *Can. J. Bot.* **57**, 981 (1979)

[g] D. G. Stout, R. M. Cotts, and P. L. Steponkus, *Can. J. Bot.* **57**, 1623 (1977).

[h] D. G. Stout, P. L. Steponkus, and R. M. Cotts, *Plant Physiol.* **62**, 636 (1978).

[i] G. Bacic and S. Ratkovic, *Biophys. J.* **45**, 767 (1984).

[j] J. T. Woolley, *Plant Physiol.* **40**, 711 (1965).

[k] C. R. House and P. Jarvis, *J. Exp. Bot.* **19**, 31 (1968).

[l] P. Jarvis and C. R. House, *J. Exp. Bot.* **18**, 695 (1967).

[m] P. Jarvis and C. R. House, *J Exp. Bot.* **20**, 507 (1969).

[n] Z. Glinka and L. Reinhold, *Plant Physiol.* **49**, 602 (1972).

Absolute Value of Lp, $T^w_{1/2}$, and D_c and Their Physiological Meaning

The absolute values of Lp given in Tables I and II range over four orders of magnitude with the lowest values 10^{-10} (tracer flow) and the highest 10^{-6} m sec^{-1} MPa^{-1} (NMR). The large variation seems to be related to the techniques applied. The tracer data will be strongly influenced by unstirred layers and by the use of the proper model for evaluating $P_d(Lp)$ (see Discussion in Ref. 23). Nevertheless, these techniques are still used.

The NMR data, on the other hand, are surprisingly high, although it has been claimed that they still represent underestimates because of unstirred layer effects. The technique[44] suffers from the fact that high external concentrations of a paramagnetic ion (Mn^{2+}) in the wall space have to be used in order to determine the transverse magnetic relaxation of protons. Furthermore, it is assumed that the diffusion of this ion into the cell is negligible.

The data from the plasmolytic experiments are of the order of some 10^{-8} to 10^{-7} m sec^{-1} MPa^{-1}, similar to values found with the pressure probe. However, it has been also shown with the latter technique that Lp could change at low turgor (for discussion, see Ref. 27). Another objection is that unstirred layers could affect measurements in the presence of the high osmotic concentrations necessary for plasmolysis and the fairly large water flows.

Water relations parameters of about 20 species have been measured using the cell pressure probe. Since the amounts of water moved across the cell membranes during measurements are very small, there are no problems with unstirred layers with this technique which measures intact, turgid cells. Effects of puncturing the cells with the microcapillary are negligible. Lp values from pressure probe measurements range between 10^{-8} and 10^{-6} m sec^{-1} MPa^{-1}, i.e., they are in the upper part of the range given above. As a consequence of this high hydraulic conductivity of higher plant cells, the half-times of water exchange of cells ($T^w_{1/2}$) are rather short (~ 10 sec) with the exception of very big cells (see, e.g., data for the giant bladder cells of *Mesembryanthemum crystallinum* in Table II) which exhibit a very large volume to surface area ratio. This also means that D_c usually should be quite high and the rate of water potential equilibration fast in higher plant tissues.

Although the hydraulic conductivity of a higher plant cell is usually of the order of 10^{-7} m sec^{-1} MPa^{-1}, variations of Lp (and D_c) occur. Lp and D_c seem to be larger for growing than for mature tissue,[45–47] which would

[44] T. Conlon and R. Outhred, *Biochim. Biophys. Acta* **288**, 354 (1972).
[45] D. J. Cosgrove and E. Steudle, *Planta* **153**, 343 (1981).

TABLE II
HYDRAULIC CONDUCTIVITY (Lp), HALF-TIME OF WATER EXCHANGE ($T^w_{1/2}$), AND TISSUE DIFFUSIVITY FOR WATER (D_c) AS DETERMINED FROM PRESSURE PROBE EXPERIMENTS[a]

Species	Tissue/cell type	Half-time, $T^w_{1/2}$ (sec)	Diffusivity, D_c (m² sec⁻¹)	Hydraulic conductivity, Lp (m sec⁻¹ MPa⁻¹)	Ref.
Capsicum annuum	Mesophyll of fruit tissue	65–250	$(3–6) \times 10^{-11}$	$(4–6) \times 10^{-8}$	b
	Subepidermal bladder cells of inner pericarp of fruit	1–12	—	$(2–17) \times 10^{-6}$	c
	Tissue cells of inner pericarp	18–54	—	$(1.2–3.4) \times 10^{-7}$	c
Tradescantia virginiana	Leaf epidermis	1–35	$(0.2–6) \times 10^{-10}$	$(0.2–11) \times 10^{-7}$	d, e, f
	Subsidiary cells	3–34	$10^{-11}–10^{-10}$	$(2–35) \times 10^{-8}$	
	Mesophyll cells	55–95	1×10^{-12}	$(4–6) \times 10^{-8}$	
	Isolated epidermis	9–54	$(0.5–3) \times 10^{-11}$	6×10^{-8}	
Kalanchoë daigremontiana	CAM tissue of the leaf	2–9	6×10^{-10}	$(0.2–1.6) \times 10^{-6}$	g
Pisum sativum	Growing epicotyl	1–27 (epidermis); 0.3–1 (cortex)	—	$(0.2–2) \times 10^{-7}$; $(0.4–9) \times 10^{-6}$	h, i
Glycine max	Growing hypocotyl	0.3–5.2 (epidermis); 0.4–15.1 (cortex)	$(1–9) \times 10^{-11}$; $(1–55) \times 10^{-11}$	$(0.7–17) \times 10^{-6}$; $(0.2–10) \times 10^{-6}$	j
Zea mays	Midrib tissue of leaf	1–8	$(0.4–6.1) \times 10^{-10}$	$(0.3–2.5) \times 10^{-6}$	k
Oxalis carnosa	Epidermal bladder cells	22–213 (adaxial); 7–38 (abaxial)	—	4×10^{-7}; 2×10^{-6}	l
Mesembryanthemum crystallinum	Epidermal bladder cells	200–2000	—	2×10^{-7}	m
Salix exigua	Sieve elements of isolated bark strips	110–480	—	5×10^{-9} (lateral hydraulic conductivity)	n

194

				o
				p, q
				r
				q
				s

Hordeum distichon — Root cortex and rhizodermis — $1-21$ — $(0.5-9.5) \times 10^{-11}$ (cortex) $(1-7) \times 10^{-12}$ (rhizodermis) — 1.2×10^{-7} — o

Triticum aestivum — Root hairs, rhizodermis, cortex — $8-12$ — — — 1.2×10^{-7} — p, q

Zea mays — Root cortex — $1-28$ — $(2-53) \times 10^{-12}$ — $(0.5-9) \times 10^{-7}$ — r

— Root cortex, rhizodermis — — — — 1.2×10^{-7} — q

Phaseolus cocineus — Root cortex — $0.4-2.3$ — $(0.3-1.7) \times 10^{-10}$ — 2×10^{-6} — s

[a] D_c values have been determined from $T^{w}_{1/2}$ and cell dimensions and refer to the cell-to-cell path only. For some tissues, rather large ranges are found for the parameters which indicate some variation within a tissue. For further explanation, see text.

[b] D. Hüsken, E. Steudle, and U. Zimmermann, *Plant Physiol.* **61**, 158 (1978).

[c] J. Rygol and U. Lüttge, *Plant, Cell Environ.* **6**, 545 (1983).

[d] A. D. Tomos, E. Steudle, U. Zimmermann, and E.-D. Schulze, *Plant Physiol.* **68**, 1135 (1981).

[e] S. D. Tyerman and E. Steudle, *Aust. J. Plant Physiol.* **9**, 461 (1982).

[f] U. Zimmermann, D. Hüsken, and E.-D. Schulze, *Planta* **149**, 445 (1980).

[g] E. Steudle, J. A. C. Smith, and U. Lüttge, *Plant Physiol.* **66**, 1155 (1980).

[h] D. J. Cosgrove and R. E. Cleland, *Plant Physiol.* **72**, 332 (1983).

[i] D. J. Cosgrove and E. Steudle, *Planta* **153**, 343 (1981).

[j] E. Steudle and J. S. Boyer, *Planta* **164**, 189 (1985).

[k] M. E. Westgate and E. Steudle, *Plant Physiol.* **78**, 183 (1985).

[l] E. Steudle, H. Ziegler, and U. Zimmermann, *Planta* **159**, 38 (1983).

[m] E. Steudle, U. Lüttge, and U. Zimmermann, *Planta* **126**, 229 (1975).

[n] J. P. Wright and D. B. Fisher, *Plant Physiol.* **73**, 1042 (1983).

[o] E. Steudle and W. D. Jeschke, *Planta* **158**, 237 (1983).

[p] H. Jones, A. D. Tomos, R. A. Leigh, and R. G. Wyn Jones, *Planta* **158**, 230 (1983).

[q] H. Jones, R. A. Leigh, R. G. Wyn Jones, and A. D. Tomos, *Planta* **174**, 1 (1988).

[r] E. Steudle, R. Oren, and E.-D. Schulze, *Plant Physiol.* **84**, 1220 (1987).

[s] E. Steudle and E. Brinckmann, *Bot. Acta* **102**, 85 (1989).

be important for a possible limitation of extension growth by water supply (see below). Results from leaves of *Tradescantia virginiana*[48,49] indicate that Lp may vary for different tissues of a plant. This has also been confirmed with fruits of pepper *(Capsicum annuum)*[50] and leaf cells of *Oxalis carnosa*.[51] Also, variations within a tissue are possible. In the midrib tissue of 2- to 3-week-old leaves of maize, Westgate and Steudle[52] observed an increase of Lp from the adaxial toward the abaxial region whereas ϵ decreased (not shown in Table II). This resulted in a decrease of $T^w_{1/2}$ (increase of D_c) in the same direction, which means that cells closer to the peripheral vessels conducted water and potential changes more rapidly than those in the adaxial parts. The reason for this nonhomogeneity in the water relations parameters is not known, but it may reflect the necessity of rapidly supplying water to the tissue which is some distance away from the vessels.

The above examples demonstrate that overall measurements of water relations on tissues or entire organs (e.g., by pressure chamber or by psychrometry) have to be interpreted with caution. An averaging of parameters could lead to conclusions which may misrepresent the real situation. This will be especially important if mechanisms of water movement are considered from these measurements.[53]

The hydraulic conductivity of plant cell membranes of 10^{-6} to 10^{-8} m sec^{-1} MPa^{-1} is larger by 5 to 6 orders of magnitude than that of the cuticles, which may be calculated from tracer data for the permeability coefficient.[54] Thus, for the aboveground parts of higher plants, the cuticle with its stomatal pores will limit water exchange between plant and atmosphere. In the root, a cuticle is missing, and, therefore, changes in soil water potential will be immediately reflected in changes of the root ψ because the tissue dimensions are usually of an order of 1 mm or less for absorbing roots. Effective mechanisms for osmotic regulation are, therefore, required for roots to prevent osmotic damage owing to larger changes of the soil ψ.

The diffusivities (D_c) calculated from $T^w_{1/2}$ and cell dimensions range between 10^{-12} and 10^{-10} m^2 sec^{-1}, which means that the rate at which water potential changes propagate is somewhat smaller or similar to that of

[46] E. Steudle and J. S. Boyer, *Planta* **164**, 189 (1985).
[47] E. Steudle and J. Wieneke, *J. Am. Soc. Hortic. Sci.* **110**, 824 (1985).
[48] A. D. Tomos, E. Steudle, U. Zimmermann, and E.-D. Schulze, *Plant Physiol.* **68**, 1135 (1981).
[49] U. Zimmermann, D. Hüsken, and E.-D. Schulze, *Planta* **149**, 445 (1980).
[50] J. Rygol and U. Lüttge, *Plant, Cell Environ.* **6**, 545 (1983).
[51] E. Steudle, H. Ziegler, and U. Zimmermann, *Planta* **159**, 38 (1983).
[52] M. E. Westgate and E. Steudle, *Plant Physiol.* **78**, 183 (1985).
[53] M. T. Tyree and Y. N. S. Cheung, *Can. J. Bot.* **55**, 2591 (1977).
[54] J. Schönherr, *Encycl. Plant Physiol., New Ser.* **12**, Part B, 153 (1982).

the diffusional mass flow in aqueous solutions (diffusion coefficients 10^{-9} to 10^{-10} m^2 sec^{-1}). The actual diffusivity will strongly depend on cell size and on the contribution of the wall path [Eq. (14)] which is largely unknown. As for $T^w_{1/2}$, there should be also some nonhomogenity in D_c (and D_t) in tissues (see Table II and Ref. 52).

A comparison between D_c and measured values of D_t has been used to work out the contribution of the wall path (apoplast) and, hence, also the contribution of the different pathways to the overall water flow in tissues.[45,46,52] In the presence of hydrostatic pressure gradients it was found that $D_t \gg D_c$, whereas in its absence $D_t \approx D_c$. This suggests that the apoplast has a fairly large potential conductance for water (Lp_{cw}). However, in the presence of matrix or osmotic potential components the driving force within the apoplast would largely be given by $\sigma_s \Delta \pi_s$, which should be rather small because of the extremely small σ_s of this structure. Hence, the effective driving force and the resulting apoplasmic flow should be small.[46,52] Furthermore, when hydrostatic gradients were applied across the tissue, the apoplast was infiltrated with water which may have also increased the measured Lp_{cw} (see Table III).

Pressure Dependence of Hydraulic Conductivity

A strong dependence of the cell Lp on cell turgor has been found for certain algae (Characean species, *Valonia;* for references, see Ref. 27). The pressure dependence of Lp (and of other water relations parameters) have been rigorously studied for the submerged plant *Elodea densa*,[55] for the halophyte *Mesembryanthemum crystallinum*,[56] and for the mesophytic plant *Tradescantia virginiana*.[57] Except for *Elodea*, which exhibited an increase in Lp toward the plasmolytic point (as for the algae), no pressure dependence of Lp was found. Tracer flow measurements on isolated epidermis of onion bulb using tritiated water also showed no pressure dependence of $Lp(P_d)$.[58]

A pressure dependence of Lp (if generally existing in higher plants) would be very important for both the stationary and dynamic properties of cells and tissues and, thus, is of physiological and ecological significance. For example, the variable hydraulic resistance of the root has been discussed in terms of a pressure-dependent Lp of the endodermis.[59] The reason for the pressure dependence is unclear. A coupling between water

[55] E. Steudle, U. Zimmermann, and J. Zillikens, *Planta* **154,** 371 (1982).

[56] E. Steudle, U. Lüttge, and U. Zimmermann, *Planta* **126,** 229 (1975).

[57] E. Brinckmann, S. D. Tyerman, E. Steudle, and E.-D. Schulze, *Oecologia* **62,** 110 (1984).

[58] J. P. Palta and E. J. Stadelmann, *J. Membr. Biol.* **33,** 231 (1977).

[59] D. B. B. Powell, *Plant, Cell Environ.* **1,** 69 (1978).

TABLE III

ESTIMATES OF HYDRAULIC CONDUCTIVITY OF APOPLAST
(CELL WALL) OF ISOLATED INTERNODES OF *Nitella*, OF
WOOD, AND OF PERFUSED INTERNODES OF SOYBEAN
HYPOCOTYL[a]

Species, wall material	Hydraulic conductivity, Lp_{cw} (m^2 sec^{-1} MPa^{-1})	Ref.
Nitella flexilis,	5×10^{-11}	b
wall of internode (wall	1.4×10^{-10}	c
thickness: $7-10\ \mu$m)	$(7-10) \times 10^{-11}$	d
Pinus sp.,	5.6×10^{-10}	e
lignified wall		
Glycine max,	8×10^{-8}	f
cortex of growing		
hypocotyl		

[a] Values for the soybean system are fairly high, because of an infiltration of intercellulars during pressure perfusion. Note that Lp_{cw} refers to unit area *and* length of wall material and thus has dimensions different from that of Lp (Tables I and II).

[b] N. Kamiya, M. Tazawa, and T. Takata, *Plant Cell Physiol.* **3,** 285 (1962).

[c] M. T. Tyree, *J. Exp. Bot.* **20,** 341 (1969).

[d] U. Zimmermann and E. Steudle, *Aust. J. Plant Physiol.* **2,** 1 (1975).

[e] G. E. Briggs, "The Movement of Water in Plants." Blackwell, Oxford, 1967.

[f] E. Steudle and J. S. Boyer, *Planta* **164,** 189 (1985).

and solute flow has been proposed (as for the alga *Valonia*) as well as effects of membrane compression and membrane folding which may affect active and passive transport properties. For a discussion, see Ref. 27.

Hydraulic Properties of the Apoplast

In contrast to the hydraulic conductivity of a membrane *(Lp)* which is given per unit area, the hydraulic conductivity of the cell wall *(Lp_{cw})* is usually given per unit length *and* cross-sectional area of the apoplast. There are only a few estimates of Lp_{cw} in the literature (Table III). Experimentally it is difficult to separate the hydraulic conductivity of the entire cell from that of the wall which forms a layer of only a fraction of a micron around the protoplast. It is difficult to determine the effective, mean cross-sectional area and the path length for apoplasmic transport in tissues. The data summarized in Table III mainly refer to values which were obtained from isolated walls of *Nitella* (which may not be representative for higher plants) and from hydrostatic perfusion experiments on hypocotyls of soybean seedlings (which may be overestimated because of infiltration, see above). At present, a value of about 10^{-10} m^2 sec^{-1} MPa^{-1} may be an appropriate estimate for Lp_{cw} which would result in an "Lp" of the wall of 10^{-4} m sec^{-1} MPa^{-1}, if the wall thickness is 1 μm. This value is much larger than the Lp of the membrane, and, hence, it would be expected that the wall Lp would not influence the cell Lp measured with the aid of the pressure probe or by other techniques.

On the other hand, the contribution of the wall path during the steady flow of water across a tissue would be significant. Using again an Lp_{cw} value of 10^{-10} m^2 sec^{-1} MPa^{-1}, an Lp of 10^{-7} m sec^{-1} MPa^{-1}, and a mean fractional area of the wall path of 2.5%, the total conductance per cell layer (Lp_{total}) would be

$$Lp_{total} = 0.975 \frac{Lp}{2} + 0.025 \frac{Lp_{cw}}{5 \times 10^{-5}}$$

$$Lp_{total} = 0.5 \times 10^{-7} + 0.5 \times 10^{-7} = 1 \times 10^{-7} \qquad \text{m sec}^{-1} \text{ MPa}^{-1}$$

if the cell diameter is 50 μm. Thus, both pathways should contribute similarly to the overall conductance. It has been estimated that during dynamic changes, the cell-to-cell and the apoplasmic path contribute similarly to the overall water movement.[37,60]

Direct Coupling between Water and Solute Flow in Plant Cells: Reflection Coefficients

As already indicated in the theoretical section, the coupling between solute and water flow across plant cell membranes is adequately described by the formalism of irreversible thermodynamics, which may also incorporate active transport. The coupling may be indirect in that water relations are influenced by changes in osmotic pressure due to solute movement or metabolic activity. However, there is also a direct coupling between water and solutes as they move across the cell membrane which results from frictional forces exerted between flows. The reflection coefficient is the quantitative measure describing these interactions.

There are only a few comprehensive determinations of reflection coefficients for turgid plant cells. These data refer to isolated cells of giant algae such as the internodes of Characean species where reflection coefficients

[60] F. J. Molz and J. M. Ferrier, *Plant, Cell Environ.* **5,** 191 (1982).

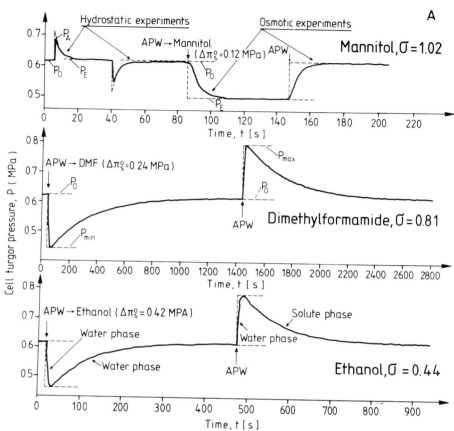

FIG. 2. Pressure relaxation experiments on plant cells using the cell pressure probe. In (A) measurements are shown for an isolated plant cell (giant internode of *Chara corallina*). Hydrostatic experiments and experiments with a nonpermeating solute (mannitol) gave monophasic responses from which the hydraulic conductivity *(Lp)* could be evaluated. Permeating osmotica (dimethylformamide, ethanol) produced "biphasic relaxations" from which the permeability (P_s) and reflection (σ_s) coefficients could be evaluated as well. The second phase represented the permeation of solute into the cell. In (B) similar experiments are shown for a higher plant cell (isolated epidermis of leaves of *Tradescantia virginiana*). Note that for this species negative reflection coefficients could be obtained for certain solutes (e.g., for *n*-propanol), i.e., the cells swelled upon the addition of the osmoticum ("anomalous osmosis"). P_o, Original cell turgor; P_A, initial cell turgor; P_E, final cell turgor; APW, "artificial pond water."

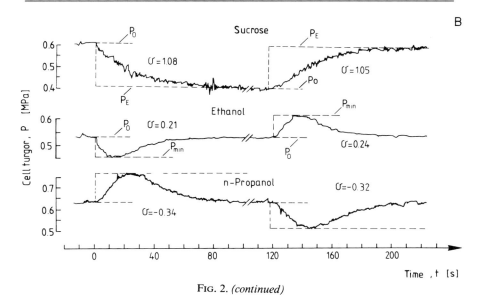

FIG. 2. *(continued)*

have been mainly determined using the technique of transcellular osmosis[61] or the cell pressure probe.[32,62-65] The latter technique has also been used for the isolated epidermis of *Tradescantia virginiana* leaves. In transcellular osmosis experiments, σ_s values have been determined from the steady-state, transcellular water flow at a given osmotic gradient, $\sigma_s \Delta \pi_s$, whereas with the cell pressure probe the data were obtained from biphasic turgor pressure relaxations which also yield the permeability coefficient [see Eq. (9)]. The basis for these measurements has already been given in the theoretical section.

Typical experiments of biphasic relaxations are shown in Fig. 2A,B for *Chara corallina* and *Tradescantia*. Table IV provides most of the data which are currently available for plant cells, and Table V gives some permeability coefficients for the same solutes and cells. It can be seen from Table IV that for polar, low molecular weight nonelectrolytes, such as

[61] J. Dainty and B. Z. Ginzburg, *Biochim. Biophys. Acta* **79**, 129 (1964).
[62] E. Steudle, S. D. Tyerman, and S. Wendler, *in* "Effects of Stress on Photosynthesis" (R. Marcelle, H. Clijsters, and M. van Poucke, eds.), p. 95. Martinus Nijhoff/Dr. W. Junk Publishers, The Hague, 1985.
[63] S. D. Tyerman and E. Steudle, *Aust. J. Plant Physiol.* **9**, 461 (1982).
[64] S. D. Tyerman and E. Steudle, *Plant Physiol.* **74**, 464 (1984).
[65] U. Zimmermann and E. Steudle, *Z. Naturforsch.* **25B**, 500 (1970).

TABLE IV
REFLECTION COEFFICIENTS (σ_s) OF PLANT CELL MEMBRANES FOR SOME NONELECTROLYTES[a]

Solute	Chara corallina[b] σ_s	Chara corallina[b] $1 - \dfrac{P_s \overline{V}_s}{Lp\,RT}$	Nitella translucens[c] σ_s	Nitella translucens[c] σ_{sc}	Nitella flexilis[d]	Valonia uticularis[e]	Valonia ventricosa[f]	Tradescantia virginiana, isolated epidermis[g]
Sucrose	0.95	—	—	—	0.97	1	—	1.04
Mannitol	1.02	—	—	—	—	—	—	1.06
Urea	—	1[c]	1	—	0.91	0.76	≈1	1.06
Acetamide	—	—	—	—	0.91	0.79	≈1	1.02
Formamide	0.99, 1[c]	—	1	—	0.79	—	≈1	0.99
Dimethylformamide	0.76	0.98	—	—	—	—	—	—
Glycerol	—	—	—	—	0.80	0.81	—	0.93
Ethylene glycol	—	1[c]	1	—	0.94	—	≈1	0.99
n-Butanol	0.14	0.91	—	—	—	—	—	—
Isobutanol (2-methyl-1-propanol)	0.21	0.91	—	—	—	—	—	—
n-Propanol	0.24, 0.22[c]	0.93	0.16	—	0.17	—	—	−0.58
2-Propanol	0.45	0.95	0.27	0.40	0.35	—	—	0.26
Ethanol	0.40, 0.27[c]	0.95	0.29	0.44	0.34	—	Negative	0.25
Methanol	0.38, 0.30[c]	0.95	0.25	0.50	0.31	—	−1.32	0.15
Acetone	0.17	0.92	—	—	—	—	—	—

[a] Values obtained by transcellular osmosis and by the cell pressure probe are given for giant algal cells (Nitella, Chara, Valonia) and for a higher plant cell (isolated epidermis of Tradescantia virginiana). For N. translucens, transcellular osmosis data were corrected for unstirred layers (σ_{sc}). For Valonia and Tradescantia negative values of σ_s have also been reported ("anomalous osmosis"). For further explanation, see text.

[b] E. Steudle and S. D. Tyerman, J. Membr. Biol. **75**, 85 (1983).

[c] J. Dainty and B. Z. Ginzburg, Biochim. Biophys. Acta **79**, 129 (1964).

[d] E. Steudle and U. Zimmermann, Biochim. Biophys. Acta **332**, 399 (1974).

[e] U. Zimmermann and E. Steudle, Z. Naturforsch. **25B**, 500 (1970).

[f] J. Gutknecht, Biochim. Biophys. Acta **163**, 20 (1968).

[g] S. D. Tyerman and E. Steudle, Aust. J. Plant Physiol. **9**, 461 (1982).

TABLE V

PERMEABILITY COEFFICIENTS (P_s) OF PLANT CELL
MEMBRANES FOR SOME SOLUTES[a]

Solute	Permeability coefficients, $P_s \times 10^6$ (m sec^{-1})	
	Chara corallina[b,c]	*Nitella translucens*[d]
Formamide	0.1 —	—
Dimethylformamide	0.8 —	—
Methanol	3.3, 4.0[d]	4.8
Ethanol	2.4, 2.8[d]	4.3
n-Propanol	2.6 —	—
2-Propanol	1.9, 2.0[d]	2.1
n-Butanol	2.5 —	—
Isobutanol (2-methyl-1-propanol)	2.3 —	—
Acetone	3.4 —	—

[a] Data in the first column refer to cell pressure probe measurements, whereas the others have been obtained from tracer flow using ^{14}C-labeled alcohols.

[b] E. Steudle and S. D. Tyerman, *J. Membr. Biol.* **75,** 85 (1983).

[c] S. D. Tyerman and E. Steudle, *Plant Physiol.* **74,** 464 (1984).

[d] J. Dainty and B. Z. Ginzburg, *Biochim. Biophys. Acta* **79,** 122 (1964).

sugars, amides, or polyalcohols, the reflection coefficients are close to unity, which also indicates a fairly low permeability, and, thus, the second (solute) phase was practically missing in the pressure probe experiments. However, low molecular weight monoalcohols and organic solvents such as acetone exhibit rather low σ_s values, which suggests a rapid diffusional flow either within the lipid phase of the membrane or through water-filled pores. For an extensive discussion, see Refs. 32, 62, and 63. For the pore model, theory predicts that σ_s would be given by[22,25,27]

$$\sigma_s = 1 - \frac{P_s \overline{V}_s}{Lp\,RT} - \frac{f_{sw} K_s^c}{f_{sw} + f_{sm}} \tag{16}$$

where $\overline{V}_s$ is the partial molar volume of solute s; K_s^c, partition coefficient between water and membrane phase; f_{sw}, frictional coefficient (per mole of solute) between water and solute; f_{sm}, coefficient describing the friction between solute and membrane.

Clearly, for a nonpermeating solute $P_s = 0$ and $f_{sw} = 0$ and, hence, $\sigma_s = 1$, i.e., the reflection coefficient would have its maximum value. For a membrane without pores in which water and solutes move by independent diffusion, $f_{sw} = 0$ and σ_s may be estimated from P_s, $\overline{V}_s$, and Lp, whereby the second term on the right-hand side of Eq. (16) would represent the contribution of the solute flow to the overall volume flow. On the other hand, if there is some interaction between water and solutes, for example, if they move in a common pore, σ_s will be further reduced by the frictional term. In fact, $\sigma_s < 1 - P_s\overline{V}_s/LpRT$ has been interpreted as an indication for the existence of pores. If the third (frictional) term in Eq. (16) is sufficiently large because of high f_{sw}, σ_s could become even negative. This "anomalous osmosis" means that upon addition of the solute, the cell will be swelling instead of shrinking. Anomalous osmosis has been found in pressure probe experiments with certain solutes (see Fig. 2B; Table IV) and strongly suggests some friction between water and solutes as they pass the membrane.

For the internodes of *Chara corallina*, Eq. (16) could be checked by varying the hydraulic conductivity and measuring the changes of σ_s, because Lp depends on the osmotic pressure of the medium. These experiments showed that to a good approximation σ_s was linearly related to $1/Lp$.[32] Furthermore, P_s determined from the concentration dependence of σ_s could be shown to be of the same order as the values obtained directly from the solute phase. The frictional term was significantly different from zero for all solutes used. Both the frictional term and P_s were independent of the external concentration. Thus, at least for *Chara* cells, this result supports the view of a strong interaction between solutes and water which could take place in water-filled pores. Recently, Kiyosawa and Ogata[66] extended this view. They also varied the Lp of *Chara corallina* by varying the external pressure and concomitantly measured the electrical conductance of the membrane which did not change. However, they concluded that water and ions pass across the membrane in different pores which respond differently to changes in osmotic pressure.

Coupling between Water Transport and Mechanical Properties

Plant cells and tissues exhibit mechanical properties due to the rigid wall envelopes surrounding their protoplasts. Usually, elastic (reversible) properties of the wall are considered to be important for mature tissue,

[66] K. Kiyosawa and K. Ogata, *Plant Cell Physiol.* **28,** 1013 (1987).

whereas plastic (irreversible) deformation takes place during elongation growth. This distinction is essentially correct but does not hold strictly, since mature cells will also exhibit some plastic (viscous) properties[67,68] and there will also be an elastic component in the wall of growing cells. The elastic properties of plant cells are expressed by the elastic modulus (ϵ) which is defined by Eq. (6). The elastic modulus is a rather complex parameter and, as already pointed out in the theory section, is mainly related to the dynamic water relations of cells. This section summarizes some recent aspects of this relation rather than giving a detailed summary of the methods used to determine ϵ or a comprehensive overview of its absolute value. This information can be found in other reviews.[19,23,27,43]

Elastic Coefficients of Cells and Tissues

Usually, elastic properties are determined by measuring the water potential as a function of relative water content (psychrometry), by determining pressure–volume curves with the pressure chamber (Scholander bomb), or directly by using the cell pressure probe. The former methods average over quite different tissues and cells in a sample according to their volume ("volume-averaged elastic modulus"), whereas the latter gives direct measures of individual cells. The disadvantage in using the pressure probe is that a rather large number of cells has to be measured in a tissue in order to get a proper statistical mean. The different types of measurements have revealed a pressure dependence of ϵ (ϵ increases with increasing P) which reflects an increase of Young's moduli of the wall material with tension. However, the measurements on individual cells also showed that ϵ was a function of cell volume (size), which would be expected theoretically for the same reason because tension increases with both turgor pressure and cell diameter.

Absolute values of ϵ range between a few and some 10 MPa depending on the species and the absolute value of cell turgor. The physiological meaning of ϵ is twofold. First, ϵ (besides other factors) determines dynamic properties ($T^w_{1/2}$, D_c, D_t, etc.) as discussed above, and, second, ϵ is the important parameter for water storage in plants. From Eq. (7) the storage capacity of an individual cell, i.e., the amount of water taken up or released per unit change of water potential (C_c), will be

$$C_c \equiv \frac{dV}{d\psi} = \frac{V}{\epsilon + \pi^i} \tag{17}$$

[67] N. Kamiya, M. Tazawa, and T. Takata, *Protoplasma* **57**, 501 (1963).
[68] U. Zimmermann and D. Hüsken, *J. Membr. Biol.* **56**, 55 (1980).

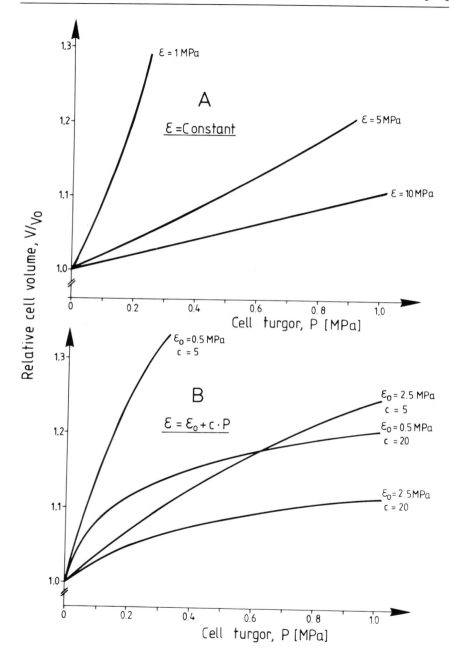

Since usually $\epsilon \gg \pi^i$, C_c will be inversely proportional to ϵ. Furthermore, since

$$\psi = P - \pi^i \tag{18}$$

in the absence of matrix forces, it can be also verified from Eqs. (6) and (18) that the changes in turgidity and osmotic pressure of a tissue subjected to a change in water potential (water stress) will be given by

$$\frac{dP}{d\psi} = \frac{\epsilon}{\epsilon + \pi^i} \tag{19}$$

and

$$\frac{d\pi^i}{d\psi} = -\frac{\pi^i}{\epsilon + \pi^i} \tag{20}$$

This means that for $\epsilon \gg \pi^i$ (low elastic extensibility of the cell) a change in ψ would be nearly completely transformed into a change in turgor rather than into a change of osmotic pressure. On the other hand, for $\epsilon \approx \pi^i$ or even lower, changes in π^i will be substantial.

In Fig. 3A the effects of different constant values of ϵ are shown schematically. Figure 3B demonstrates how a pressure dependence of ϵ would affect the relative volume changes of the cell. Thus, during the osmotic adaptation of plants, two reactions may be distinguished, an *elastic response* where the changes are largely in turgor and an *osmotic response* where the changes are largely in π^i. The relative significance of the effects will depend on the relative magnitude of ϵ as compared with π^i. It should be noted, however, that Eqs. (19) and (20) refer to cells in the absence of a metabolic (active) reaction to water stress (osmoregulation) which would affect π^i as well.[4,18,20]

Coupling of Water Storage with an Active Process during CAM

During osmoregulation under hypertonic conditions, the water status of a tissue may be improved either by import of solutes across the plasma membrane or by synthesis of solutes in the cell, e.g., from stored carbohy-

---

FIG. 3. Pressure (turgor)–volume curves for plant cells calculated for different constant (A) or pressure-dependent (B) values of the elastic modulus of the cell (ϵ). A linear pressure dependence was assumed in (B). In (A) the curves are convex to the pressure axis, whereas in (B) they are concave owing to the increase of ϵ with increasing pressure. For high absolute values of ϵ, the changes in turgor are much larger at a given change in volume than for low ϵ values. In other words, with decreasing ϵ there is a tendency for the cell to maintain turgor, whereas in the opposite case there is a tendency to maintain cell volume and osmotic pressure.

drates (see references given above). Similar effects have been discussed for crassulacean acid metabolism (CAM) plants where malic acid accumulates in large amounts during the night, which increases π^i. The carbon dioxide fixed in this way is used during the day (when the plants have closed stomata) in normal C_3 photosynthesis. It is thought that the increase of π^i and the decrease in ψ would be profitable for the water balance of CAM plants, because the storage of malic acid in the vacuoles would also bind water osmotically.

A quantitative estimate of the water storage capacity of CAM plants has been given by Lüttge,[69] who estimated volume changes of a CAM tissue during the dark period from literature values of water uptake and transpiration. These data (2–11% per dark period) were compared with volume changes measured directly (up to 2.7%)[70] and with those calculated from measurements of ϵ (pressure probe[71]) which gave about 5% volume change during the dark period. Thus, similar results were obtained by different approaches.

A more detailed calculation of the changes in turgor (P), water potential (ψ), and cell volume (V) during CAM yields

$$P - P_o = \frac{\epsilon}{\epsilon + \pi_o^i} \frac{1}{C_c + C_{cw}} \left(RT \frac{\Delta n_s}{V} C_{cw} + \Delta V_{total} \right) \tag{21}$$

$$\psi_{cw} - \psi_{cwo} = \frac{1}{C_c + C_{cw}} \left(\Delta V_{total} - RT \frac{\Delta n_s}{V} C_c \right) \tag{22}$$

$$V - V_o = \frac{C_c}{C_c + C_{cw}} \left(RT \frac{\Delta n_s}{V} C_{cw} + \Delta V_{total} \right) \tag{23}$$

where P_o, ψ_o, and V_o refer to a reference state during CAM. Δn_s is the change in malic acid content of a cell over a given period of time ($\Delta n_s > 0$ during the night and $\Delta n_s < 0$ during the day), and ΔV_{total} is the net gain (or loss) of volume per cell of the tissue (uptake minus transpiration). It can be seen that for the storage of water in the cells [Eq. (23)], changes in both π^i [$RT\Delta n_s/V$] and C_{cw} will be important, because the water available will be shared between the apoplasmic and protoplasmic compartments according to their storage capacities. A small C_{cw}/C_c ratio would strongly reduce the capability of the cells to store water owing to the active process.

Equations (21)–(23) have been derived assuming water flow equilibration within the tissue, an assumption which is usually true (see Tables I and II). The relations should also hold for osmoregulating tissues provided that

[69] U. Lüttge, *Planta* **168**, 287 (1986).
[70] S.-S. Chen and C.C. Black, *Plant Physiol.* **71**, 373 (1983).
[71] E. Steudle, J. A. C. Smith, and U. Lüttge, *Plant Physiol.* **66**, 1155 (1980).

within the tissue water flow equilibration is maintained. For CAM as well as for osmoregulating plants, combined measurements of the cellular water storage capacity (C_c), water uptake, and transpiration are necessary to work out the importance of water storage. A critical figure will be the water storage of the apoplast which is not easily assessed experimentally.

Storage Capacity of the Apoplast

Analogous to the storage capacity of cells [Eq. (17)], the water storage capacity of the apoplast (C_{cw}) will be defined by

$$C_{cw} = \frac{dV_{cw}}{d\psi_{cw}} \tag{24}$$

Changes in the water content of the apoplast (V_{cw}) are not directly measurable but may be derived as follows. The water potential of the apoplast will be identical to that of the tissue. Since there is no hydrostatic pressure component of ψ in the apoplast, the water potential of the apoplast will be

$$\psi_{cw} = -\pi_{cw} - \tau_{cw} \tag{25}$$

Equations (24) and (25), on differentiation of Eq. (25) with respect to V_{cw}, yield

$$\frac{1}{C_{cw}} = -\frac{d\pi_{cw}}{dV_{cw}} - \frac{d\tau_{cw}}{d\psi_{cw}} \tag{26}$$

If the amount of solutes in the wall space is constant, i.e., if $\pi_{cw}V_{cw} = \pi_{cwo}V_{cwo} = $ constant, Eq. (26) may be rewritten as

$$C_{cw} = \frac{V_{cwo}/\pi_{cwo}}{1 - \left(\dfrac{d\tau_{cw}}{dV_{cw}}\right)\left(\dfrac{V_{cwo}}{\pi_{cwo}}\right)} \tag{27}$$

The matrix component should be zero at the reference state of complete water saturation when $\pi_{cw} = \pi_{cwo}$ and $V_{cw} = V_{cwo}$, and $d\tau_{cw}/dV_{cw} < 0$. Therefore, the storage capacity should have its maximum value at water saturation, i.e.,

$$C_{cw}^{max} = \frac{V_{cwo}}{\pi_{cwo}} \tag{28}$$

V_{cwo} may be of the order of 1% of the tissue volume, and π_{cwo} could be as large as few tenths of a megapascal.[46,72] For a typical plant cell with

[72] D. J. Cosgrove and R. E. Cleland, *Plant Physiol.* **72**, 326 (1983).

$(\epsilon + \pi^i) = 7.5$ MPa and $\pi_{cwo} = 0.2$ MPa, the ratio between C_{cw}^{max} and C_c would, thus, be

$$\frac{C_{cw}^{max}}{C_c} = \frac{V_{cwo}}{V} \frac{\epsilon + \pi_o^i}{\pi_{cwo}} = 0.38 \qquad (29)$$

In other words, the water capacity of the apoplast would be only about one-third that of the protoplasts or about one-fourth that of the entire tissue. Molz and Ferrier[60] estimate that C_{cw} would, in fact, be only one-tenth that of C_c. It can be seen that C_{cw} will become smaller with increasing solute concentration in the wall (e.g., in halophytes). Thus, except for tissues with a large apoplasmic water volume, low π_{cw}, and very rigid cells, the water potential in response to changes in water content will be mainly governed by C_c rather than by C_{cw}. This throws some doubt on the thesis that high apoplasmic solute concentrations in growing tissue could by themselves create low water potentials,[72] because the water relations of the apoplast and the symplast are strictly coupled to each other and (owing to its larger capacitance) the symplast will usually damp changes in π_{cw}.

Ecological Significance of ϵ

It has often been proposed that the elastic modulus may vary according to different conditions (e.g., water shortage) and in this way may provide plants with some adaptation to adverse conditions (for discussion and review, see Ref. 19). However, the results obtained are quite ambiguous and difficult to explain. For example, in response to water stress in some cases, increases of ϵ have been reported, whereas in others decreases were indicated. The reason for these disparate results is, perhaps, that the physiological and ecological meanings of average values of ϵ, which are usually measured, are unclear (see above).

A more detailed determination of elastic coefficients would be desirable. Up to now, there has been only one comprehensive study of elastic properties in this context using the pressure probe.[57] For epidermal leaf cells of *Tradescantia virginiana* grown under different water and temperature regimes it was shown that ϵ did vary with the growing conditions. However, the "adaptation" of ϵ to different treatments was due to the pressure dependence of this parameter rather than to changes of the cell wall structure or composition. More detailed measurements of ϵ under a variety of conditions are necessary to work out the effects. These measurements should be made at the cell level and should also include the $\epsilon(P)$ characteristics of different tissues.

Water Transport Coupled with Growth

During expansion growth, the volume changes of plant cells require a large uptake of water in order to maintain turgor which drives the plastic (viscous) deformation of the walls. For water uptake (ΔV_w) and mechanical extension (ΔV_m), two different differential equations are valid which describe the volume changes, i.e.,

$$\frac{dV_w}{dt} = -LpA[P(t) - \Delta\pi(t)] \tag{30}$$

$$\frac{dV_m}{dt} = Vm[P(t) - P_c] \tag{31}$$

where dV_w/dt and dV_m/dt denote the different rates of cell volume changes. Equation (30) is identical to Eq. (2). m is the extensibility of the wall material, which has units of fluidity (inverse viscosity), and P_c is a threshold of turgor which has to be overcome to induce viscous (irreversible) stretching of the walls. In fact, in Eq. (31), $P(t)$ and P_c are measures of the tensions within the cell wall which are usually not known. The terms in brackets on the right-hand sides of both equations denote "driving forces" for the processes which are related to the flows by different coefficients. Equations (30) and (31) represent coupled differential equations which may be solved provided that $P(t)$ and $\Delta\pi(t)$ as well as the coefficients are known.

It is important that dV_w/dt decreases (becomes more negative) with increasing turgor, whereas the mechanical extension increases. Thus, a steady-state "growth rate" will be obtained where $dV_w/dt = dV_m/dt = dV/dt$ and P, $\Delta\pi$ = constant.[73,74] This allows the elimination of P, and, combining Eqs. (30) and (31), yields

$$\left(\frac{1}{Lp\,A} + \frac{1}{Vm}\right)\frac{dV}{dt} = \Delta\pi - P_c \tag{32}$$

Equation (32) is again an Ohm's law analog where $1/LpA$ and $1/Vm$ are the hydraulic and mechanical resistances in series and $\Delta\pi - P_c$ the driving force. Provided that water flow is rapid, $\Delta\pi \simeq P$ will be valid and $1/LpA \ll 1/Vm$. Therefore, Eq. (32) becomes identical to Eq. (31). Otherwise, the hydraulic resistance will limit ($1/Vm \ll 1/LpA$). Thus, for steady growth, Eq. (32) describes the influence of water and mechanical properties

[73] J. A. Lockhart, *J. Theor. Biol.* **8**, 264 (1965).
[74] P. M. Ray, P. B. Green, and R. E. Cleland, *Nature (London)* **239**, 163 (1972).

on the rate of expansion provided that there are no other limiting factors (solute transport, protein and wall synthesis; see below).

Except for the basic studies on isolated internodes of *Nitella*,[75] Eqs. (30)–(32) have usually been applied to higher plants. The extensibility *(m)* and yield threshold (P_c) have turned out not to be real constants, but to depend on environmental conditions and to shift with time.[18,76] Growth regulators act on the growth rate mainly by increasing the extensibility of the tissue. According to the "acid–growth hypothesis,"[77] growth regulators stimulate H^+ extrusion from growing cells into the wall space, which lowers the wall pH and induces wall loosening by a direct cleavage of bonds, some induction of enzymatic cleavage, or other mechanisms.[78]

For tissues, Eq. (32) has to be modified by introducing tissue instead of cell properties (e.g., the parallel cell-to-cell and apoplasmic pathways for water[79]) without changing its basic sense. In tissues and organs such as a coleoptile or internode, serious objections against the use of the simple equations arise from the fact that the physical meaning of the parameters (*m* and P_c) is not well defined in terms of properties of cells because of mechanical interactions between cells and tissues during elongation. Furthermore, the mechanical extension of certain tissues may limit the overall rates while others are compressed.[80,81] For example, the outer epidermal walls of growing plant organs could be under tension and should mainly limit extension, whereas the inner tissue is under compression. It is not known how continuous indole-3-acetic acid (IAA)-dependent growth is maintained under these conditions, which requires some adjustment of the expansive forces created by the inner tissue to the tensile strength of the epidermis in order to coordinate growth.

Hence, the physical meaning of the growth parameters (*m* and P_c) is much less clear than may be expected from Eq. (32), although linear relationships have often been reported. Furthermore, turgor has been determined indirectly in most of the cases. More direct measurements of the

[75] P. B. Green, R. O. Erickson, and J. Buggy, *Plant Physiol.* **47**, 423 (1971).

[76] T. C. Hsiao, W. K. Silk, and J. Jing, *in* "Control of Leaf Growth" (N. R. Baker, W. J. Davies, and C. K. Ong, eds.) p. 239. Cambridge Univ. Press, London and New York, 1985.

[77] A. Hager, K. Menzel, and A. Krauss, *Planta* **100**, 47 (1971).

[78] R. E. Cleland, *in* "Plant Growth Substances" (F. Skoog, ed.), p. 71. Springer-Verlag, New York, 1980.

[79] E. Steudle, *in* "Control of Leaf Growth" (N. R. Baker, W. J. Davies, and C. K. Ong, eds.), p. 35. Cambridge Univ. Press, London and New York, 1985.

[80] U. Kutschera, *in* "Physiology of Cell Expansion during Plant Growth" (D. J. Cosgrove and D. P. Knievel, eds.), p. 215. ASPP Symposium Publication, Rockville, Maryland, 1987.

[81] U. Kutschera, R. Bergfeld, and P. Schopfer, *Planta* **170**, 168 (1987).

parameters at the cell level are undoubtedly necessary. They are, in principle, available using the cell pressure probe.[45,46,82,83]

There has been some debate about the relative importance of hydraulic and mechanical resistances during expansion. Boyer and co-workers proposed that the hydraulic resistances for the water supply of growing tissue *via* the xylem in the stem of, say, a dicot seedling could be significant.[84,85] Using isopiestic psychrometry, Boyer *et al.*[86] have shown in excision experiments with soybean hypocotyls that wall relaxation was fast (i.e., m was high) and was completed within 5 min after cutting off the water supply. The coefficients mV and LpA contributed similarly to the overall rate of enlargement.

Cosgrove and co-workers, on the other hand, observed much smaller rates of wall relaxation in pea seedlings, which required a few hours for completion.[87,88] The reason for this large discrepancy is not known but may be related to whether mature tissue, still attached during the experiments, could have supplied the growing tissue with water, thus damping the relaxation effect.[83] However, there are other results from direct turgor measurements in Cosgrove's experiments which indicate smaller gradients of water potential than those found by Boyer. Because Cosgrove's results still indicate a gradient in water potential, i.e., 0.05 MPa over a distance of only a few hundred microns, there seems to be a considerable gradient also in his experiments.[45]

A more complete modeling of tissue water relations extending the basic theory of Molz and Boyer[85] is undoubtedly necessary. This should take into account the diffusivities of water transport as well as directly measured extensibilities of tissue cells. Modeling will require a more detailed knowledge of the hydraulic resistances in the tissue for the different pathways which can be obtained by combining the cell pressure probe with other techniques.[45,46,52]

Limitations of Growth by Solute Relations

Another difficulty related to the use of Eq. (32) is that, due to the osmotic pressure difference between a cell and its surroundings ($\Delta\pi$), the growth rate could be either coupled to solute import or to the generation of

[82] D. J. Cosgrove and R. E. Cleland, *Plant Physiol.* **72**, 332 (1983).
[83] R. Matyssek, S. Maruyama, and J. S. Boyer, *Plant Physiol.* **86**, 1163 (1988).
[84] J. S. Boyer, *Annu. Rev. Plant Physiol.* **36**, 473 (1985).
[85] F. J. Molz and J. S. Boyer, *Plant Physiol.* **62**, 423 (1978).
[86] J. S. Boyer, A. J. Cavalieri, and E.-D. Schulze, *Planta* **163**, 527 (1985).
[87] D. J. Cosgrove, *Plant Physiol.* **78**, 347 (1985).
[88] D. J. Cosgrove, *Planta* **171**, 266 (1987).

osmotically active substances in the cell, i.e., to metabolic reactions. In fact, some osmoregulation should take place during growth in order to adjust $\Delta\pi$ (π^i) in spite of dilution effects owing to cell expansion.[18,89] The effects are implicitly incorporated in Eq. (32) but are often neglected. For example, for steady growth it is also required that $\Delta\pi$ = constant which means that

$$C^i = \frac{J'_s}{J'_v} = \text{constant} \tag{33}$$

should hold for the concentration in the cell at a constant concentration of the medium.[79] J'_s is the solute influx in mol sec^{-1}; $J'_v = dV/dt$ is the water influx in m^3 sec^{-1}. Incorporating this expression into Eq. (32), on derivation of the external water potential ($-\pi^o$) with respect to J'_v,[79] yields

$$\frac{d\psi^o}{dJ'_v} = \frac{1}{LpA} + \frac{1}{mV} + RT\frac{J'_s}{(J'_v)^2} \tag{34}$$

$d\psi^o/dJ'_v$ would be the apparent resistance to expansion growth, and the third term on the righthand side of Eq. (34) denotes the contribution of dilution to this resistance. It is plausible that under steady-state conditions, the dilution effect will increase with increasing C^i and that the effect will decrease with increasing growth rate (J'_v). It has to be stressed that Eq. (34) incorporates dilution effects at a given constant J'_s only and not effects of changes in J'_s. Usually, it is thought that effects due to dilution and changes in J'_s would only be important on a long-term scale (hours) and not in the short term (minutes). This may be not true in all cases, however, and will depend on the actual difference between $\Delta\pi$ and P_c and on the absolute level of the internal concentration (C^i). For example, during growth under conditions of water stress, C^i may be rather large and changes in J'_v may have a big effect. It is not known whether growing cells have mechanisms to regulate osmotic pressure and turgor via pressure-dependent solute transport as has been shown for algae.[27]

Water and Solute Transport in Roots

Introduction and Background

Besides the uptake of nutrients, the most important function of the root is to take up water which compensates for transpirational water losses in

[89] T. C. Hsiao and J. Jing, in "Physiology of Cell Expansion during Plant Growth" (D. J. Cosgrove and D. P. Knievel, eds.), p. 180. ASPP Symposim Publication, Rockville, Maryland, 1987.

the shoot. In the soil – plant – atmosphere continuum which acts as a series of hydraulic resistances, the uptake of water from the soil is only part of the entire process, and the hydraulic resistance of the root will usually be smaller than that of the stomata. However, the root resistance could be equally important and, at a given rate of transpiration, may control the water status of the shoot. In addition, there will be some control by the hydraulic conductance within the soil depending on its relative magnitude as compared with that of the root. Like the stomata, both the soil and the root represent variable resistances.

The difference in water potential between the soil (or root medium) and the root xylem is the driving force for the uptake of water into the root. For the same reasons discussed for cells (see section on theory), it is very unlikely that active water transport also occurs in the root. In the transpiring plant, transpiration will cause a tension (negative pressure) within the root xylem, and, therefore, the driving force will be mainly a hydrostatic pressure gradient under these conditions. However, at low rates of transpiration the osmotic component of water potential [Eq. (3)] will also be important. This component results from the active uptake and accumulation of solutes in the xylem. At sufficiently low or zero rates of transpiration, this water uptake will result in a positive "root pressure" which will also drive xylem solution up into the shoot. Hence, in the root, water and solute (nutrient) transport are coupled to each other, and the relative importance of the coupling for the water supply of the plant will depend on the transpirational demand. The coupling is most obvious and well-known in phenomena like guttation and the exudation of excised roots.

Recently, this model of the root has been further developed by the "apoplast canal model" of Katou and co-workers[90,91] in extending the standing-gradient osmotic flow hypothesis of Diamond and Bossert[92] for the flow of solution in the narrow channels of animal epithelia. In the canal model, which has been also applied to growing tissue[93] and other types of respiration-dependent water transport in higher plants,[94] the Japanese workers assumed an uptake of solutes from the apoplast into the symplast at the root periphery which is followed by an osmotic water flow. In the stele, the situation is reversed according to the classical standing-gradient hypothesis. The flow equations were solved numerically, and it was claimed that variable water transport as well as apparently low reflection

[90] K. Katou, T. Taura, and M. Furumoto, *Protoplasma* **140**, 123 (1987).

[91] T. Taura, Y. Iwaikawa, M. Furumoto, and K. Katou, *Protoplasma* **144**, 170 (1988).

[92] J. M. Diamond and W. H. Bossert, *J. Gen. Physiol.* **50**, 2061 (1967).

[93] K. Katou and M. Furumoto, *Protoplasma* **133**, 174 (1986).

[94] K. Katou and M. Furumoto, *Protoplasma* **144**, 62 (1988).

coefficients of root cell membranes may be explained by the model in terms of a tight metabolic control of water transport in the root.

Usually, for water and solute uptake, roots are treated as a two-compartment system with the xylem and soil solution separated by an osmotic barrier which has selective, membranelike properties and is commonly identified as the endodermis. This results in a modeling of transport processes in roots analogous to those of cells (see theory section), i.e., roots are treated as osmometers. The linear equations of irreversible thermodynamics used for cells may be applied to roots[7,9,95-97] using the radial water and solute flows per unit surface area of root (J_{vr} and J_{sr}) instead of J_v and J_s in Eqs. (1) and (5) and the concentration in the xylem and the root pressure (P_r) instead of the concentration in the cell and the turgor, respectively.

This approach may be basically correct, but there is a problem in using it because J_{vr} will not be linearly related to the driving force over the entire range of water flows, rather the hydraulic conductivity of the root (Lp_r) will usually increase with increasing J_{vr}. The reasons for this nonlinearity have been discussed at length in the literature.[9-11,98-101] The phenomenon is not completely understood but should be related to solute–water interactions and to the complex structure of the root. In fact, Fiscus[9] and Dalton et al.[95] were the first to show that at least some of the nonlinearities observed were due to water–solute interactions and a dilution effect in the root caused by the uptake of water. Other phenomena such as the dependence of Lp_r on metabolic inhibitors such as KCN and carbonyl cyanide m-chlorophenylhydrazone (CCCP)[102] or on hormones such as ABA[12,103] also indicated that J_{vr} (Lp_r) is likely to be influenced by a variation of solute flow (J_{sr}) which changes osmotic gradients in the root, although direct effects (e.g., on the membrane Lp of root cells) could not be totally excluded.[12]

Besides the variability in the water and solute flows, another important feature of roots is that the transport properties (Lp_r, P_{sr}, σ_{sr}) should vary during root development, for example, during the formation of the endodermis and the Casparian strip which is a prerequisite of the proper function of the root. The development of Casparian bands in the hypodermis

[95] F. N. Dalton, P.A.C. Raats, and W. R. Gardner, *Agron. J.* **67**, 334 (1975).
[96] C. R. House and N. Findley, *J. Exp. Bot.* **17**, 344 (1966).
[97] C. R. House and N. Findley, *J. Exp. Bot.* **17**, 627 (1966).
[98] R. Brouwer, *Proc. K. Ned. Akad. Wet., Ser. C* **57**, 68 (1954).
[99] J. J. Landsberg and N. D. Fowkes, *Ann. Bot. (London) [N.S.]* **42**, 493 (1978).
[100] G. C. Mees and P. E. Weatherley, *Proc. R. Soc. London, Ser. B* **147**, 367 (1957).
[101] G. C. Mees and P. E. Weatherley, *Proc. R. Soc. London, Ser. B* **147**, 381 (1957).
[102] M. G. Pitman, D. Wellfare, and C. Carter, *Plant Physiol.* **67**, 802 (1981).
[103] J. C. Collins and A. P. Kerrigan, *New Phytol.* **73**, 309 (1974).

(exodermis) should further complicate the simple situation.[104-106] The osmotic properties and the coupling between water and solute transport should, therefore, vary along the root axis. There should be also some variation in transport because of a changing contribution of the longitudinal resistance to transport as the root xylem develops. Thus, although the linear relations of irreversible thermodynamics could be applied in analogy to cells, there are remarkable differences between the cell and the root system which have to be kept in mind.

Experimental Techniques

In intact plants, the hydraulic resistance of root systems (including the soil and the xylem) may be estimated by measuring the difference in water potential between soil and leaves (e.g., by tensiometry and pressure bomb) and the rate of transpiration (e.g., from sap flow or weighing). With excised roots, the simple exudation from a segment or the entire root system provides a direct measure of the water flow which can be related to the osmotic pressure difference between medium and exudate.[96,97] These techniques can be also used to work out differences in Lp_r along the root, if they are combined with micropotometers. The method has been extended in pressurized root exudation in which the root is sitting either in a nutrient medium[9] or in soil[11] and is pressurized in steps to yield steady-state exudation rates following the changes in the water potential difference across the root cylinder. In a variation of the exudation technique, stop-flow methods have also been used in which, after the establishment of a steady water flow, either the osmotic pressure of the medium or the hydrostatic pressure in the root xylem was increased to terminate the flow. Thus, the driving force could be calculated.[107,108]

The above techniques have provided some data on Lp_r as well as on P_{sr} and σ_{sr} of roots or parts thereof. However, precise transport models for the root which would explain the pathways of water and solute movements within the root cylinder and the interaction between flows are largely missing. This is because the contributions of the different pathways to the overall transport have not yet been determined. Also, the roles of the different transport barriers such as the exodermis, central cortex, and endodermis have not been sufficiently characterized for technical reasons.

[104] D. T. Clarkson, A. W. Robards, J. E. Stephens, and M. Stark, *Plant, Cell Environ.* **10**, 83 (1987).
[105] C. J. Perumalla and C. A. Peterson, *Can. J. Bot.* **64**, 1873 (1986).
[106] C. A. Peterson and C. J. Perumalla, *J. Exp. Bot.* **35**, 51 (1984).
[107] D. M. Miller, *Can. J. Bot.* **58**, 351 (1980).
[108] M. G. Pitman and D. Wellfare, *J. Exp. Bot.* **29**, 1125 (1978).

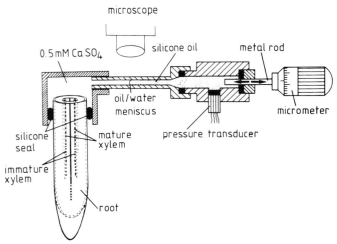

F<small>IG</small>. 4. Root pressure probe for measuring coupled flows of water and solutes in plant roots (schematic). The excised root is tightly connected with the probe by a silicone seal so that root pressure can be built up in the rather incompressible system filled with 10^{-4} M CaSO$_4$ solution and silicone oil. Root pressure is recorded by a pressure transducer and can be changed by a movable rod in order to produce water flows across the root ("hydrostatic experiments"). On the other hand, changes of the osmotic pressure of the medium cause "osmotic flows." From the responses of the root (see Fig. 5), the hydraulic conductivity (Lp_r) as well as permeability (P_{sr}), and reflection (σ_{sr}) coefficients are determined. For further explanation, see text.

Recently, a new approach has been made to measure some of the root parameters mentioned above by developing the cell pressure probe (see this volume [24]) for use with excised roots.[109,110] As for cells, the technique has been shown to provide both water and solute parameters during osmotic processes in roots. Root segments as well as entire root systems can be measured. Furthermore, by combination with the original cell pressure probe, measurements at the cell and tissue (organ) level can be performed to obtain the data necessary for modeling root transport. In the following, this technique is briefly reviewed and some of the results obtained so far are compared with those obtained by other techniques.

Root Pressure Probe Technique

An excised root (segment or intact root system) is attached to the root pressure probe via silicone seals which are pressure-tight but, on the other

---

[109] E. Steudle and W. D. Jeschke, *Planta* **158,** 237 (1983).
[110] E. Steudle, R. Oren, and E.-D. Schulze, *Plant Physiol.* **84,** 1220 (1987).

hand, do not constrain the water flow along the root xylem (Fig. 4). Since the probe contains water or a 10^{-4} M CaSO$_4$ solution on top of the excised root, then silicone oil, root pressure will develop within the fairly incompressible system. This pressure is recorded by a pressure transducer which transforms pressure into a proportional voltage. The meniscus between the aqueous solution and the silicone oil is held within a measuring capillary, and its position serves as a reference point during the measurements. By moving the meniscus with the aid of a movable rod, volume changes can be produced within the system which can be also measured quantitatively

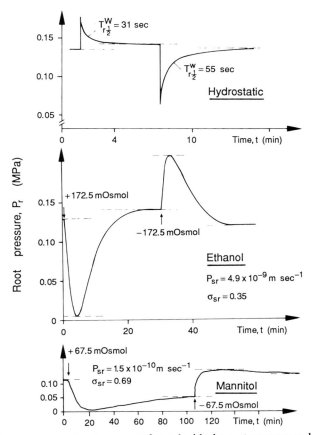

FIG. 5. Root pressure relaxations as performed with the root pressure probe (Fig. 4) on bean *(Phaseolus coccineus)* roots. Analogously to cells (Fig. 2) mono- and biphasic curves are obtained from which transport coefficients (Lp_r, P_{sr}, and σ_{sr}) are evaluated. A rapidly permeating solute (ethanol) exhibited a shorter half-time of solute exchange than a slowly permeating one (mannitol) which also had a higher σ_{sr}.

TABLE VI

COMPARISON BETWEEN ENTIRE ROOT (Lp_r) AND ROOT CELL (Lp) HYDRAULIC
CONDUCTIVITIES OF PLANTS AS DETERMINED BY DIFFERENT TECHNIQUES[a]

Root	Technique	Root $Lp_r \times 10^8$ (m sec^{-1} MPa^{-1})	Cell $Lp \times 10^8$ (m sec^{-1} MPa^{-1})	Ref.
Zea mays	Osmotically induced root exudation	5.7 1–12[b] 2.2 4	— — — —	d e, f g h
	Hydrostatic stop flow	10–21	—	i, j
	Osmotic stop flow	0.9–4.8	12	k
	Root pressure probe:			
	Osmotic exp.	0.1–5	24	l, m
	Hydrostatic exp.	1–46		
Triticum aestivum	Osmotic stop flow	1.6–5.5 0.5	— 12	k n
Hordeum vulgare	Osmotic stop flow	0.9–2.1	—	o, p
Hordeum distichon	Root pressure probe:			
	Osmotic exp.	0.3–4.3	12	q
	Hydrostatic exp.	0.3–4.0		
Phaseolus vulgaris	Exudation of pressurized roots	1–5[c] 30	— —	r s
	Osmotically induced water flow	0.56	—	g
Phaseolus coccineus	Root pressure probe:			
	Osmotic exp.	2–8	30–470	t
	Hydrostatic exp.	3–7		

[a] Lp of root cell membranes was determined with the aid of the cell pressure probe. For maize, the data of hydrostatic Lp_r are of the same order as the cell Lp which indicates some apoplasmic bypass of water in the root cylinder. For barley and bean roots, on the other hand, the comparison points to a dominating cell-to-cell transport. For these species, hydrostatic and osmotic Lp_r are also similar.

[b] Lp_r varies along the root.

[c] Lp_{vr} increases with increasing J_{vr}.

[d] C. R. House and N. Findley, *J. Exp. Bot.* **17**, 344 (1966).

[e] W. P. Anderson, D. P. Aikman, and D. P. Meiri, *Proc. R. Soc. London, Ser. B* **174**, 445 (1970).

[f] M. G. Pitman, D. Wellfare, and C. Carter, *Plant Physiol.* **67**, 802 (1981).

[g] E. I. Newman, *New Phytol.* **72**, 547 (1973).

from the diameter of the capillary and the shift of the position of the meniscus.

Experiments can be performed in two ways, i.e., by changing the root pressure after the establishment of a steady-state value (P_{ro}) and following the change of P_r with time in so-called pressure relaxation experiments analogous to cells (Fig. 5). "Exosmotic" water flows are induced when increasing P_r so that water flows from the xylem into the medium, and "endosmotic" flows result from a reduction of P_r. These types of experiments may be called "hydrostatic" because the hydrostatic component of water potential was changed to cause flows. In the other type of experiment, the osmotic pressure of the bathing medium is changed and the response in root pressure is followed. In these experiments, different osmotica can be used. As with cells (Fig. 2), the relaxations will be biphasic and will reveal the permeability (P_{sr}) and reflection (σ_{sr}) coefficients of the root for the given solute in addition to the hydraulic conductivity (Lp_r). The theory of the processes is similar (though not identical) to that given above for cells. A detailed description is given in Refs. 109 and 110. Typical pressure–time curves are shown in Fig. 5 for bean roots.

Table VI presents some data obtained for the hydraulic conductivity of roots (Lp_r per square meter of root surface area) by the root pressure probe and other techniques. It can be seen that most of the data are of the same order. For corn differences occur when using the pressure probe depending on whether the experiments are hydrostatic or osmotic. For this species, the hydrostatic Lp_r was larger than the osmotic by about an order of magnitude and was similar to the cell Lp in the cortex which had also been determined. This was interpreted in terms of an apoplasmic bypass in parts of the roots, i.e., by a water flow around the cells in the presence of a

[h] J. C. Collins and A. P. Kerrigan, *New Phytol.* **73**, 309 (1974).

[i] D. M. Miller, *Plant Physiol.* **77**, 162 (1985).

[j] D. M. Miller, *Plant Physiol.* **77**, 168 (1985).

[k] H. Jones, R. A. Leigh, R. G. Wyn Jones, and A. D. Tomos, *Planta* **174**, 1 (1988).

[l] E. Steudle and J. Frensch, *Planta* **177**, 281 (1989).

[m] E. Steudle, R. Oren, and E.-D. Schulze, *Plant Physiol.* **84**, 1220 (1987).

[n] H. Jones, A. D. Tomos, R. A. Leigh, and R. G. Wyn Jones, *Planta* **158**, 230 (1983).

[o] M. G. Pitman and D. Wellfare, *J. Exp. Bot.* **29**, 1125 (1978).

[p] M. G. Pitman, D. Wellfare, and C. Carter, *Plant Physiol.* **67**, 802 (1981).

[q] E. Steudle and W. D. Jeschke, *Planta* **158**, 237 (1983).

[r] M. Salim and M. G. Pitman, *J. Exp. Bot.* **35**, 869 (1984).

[s] E. L. Fiscus, *Plant Physiol.* **80**, 752 (1986).

[t] E. Steudle and E. Brinckmann, *Bot. Acta* **102**, 85 (1989).

hydrostatic gradient. In the presence of osmotic gradients, this bypass was not effective, since the reflection coefficient of the apoplast should have been very small and, hence, the driving force [$\sigma_{sr}\Delta\pi_s$; Eq. (1)] should have also been small. These apoplasmic bypasses may be due to secondary root initials as indicated from experiments using apoplasmic dyes.[111] For other species such as barley or bean roots, hydrostatic and osmotic Lp_r values were similar, which suggests differences in root structure between species and may point to differences in the "tightness" of the Casparian strip.

The comparison between cell Lp and root Lp_r supports the view of an apoplasmic bypass in maize and a cell-to-cell transport of water in barley and bean roots, as do the experiments of Jones et al.[112,113] in which Lp_r was determined by an osmotic stop-flow technique, although some variability was found which indicates difficulties with the technique and aging effects. Thus, from the comparison of root Lp_r and cell Lp, it can be concluded that the mechanisms of water transport across the root cylinder may differ depending on the species and on the nature of the driving force. One may speculate on whether this phenomenon could also be related to the variability of Lp_r found in the literature and also to the dependence of Lp_r on J_{Vr} in that at low flow rates the contribution of the osmotic component would be larger than the hydrostatic (see above).

It should be noted that the values given in Table VI refer to root segments or entire root systems but that Lp_r should vary along the root, i.e., the different zones may contribute differently to the overall value. This has, in fact, been found in stop-flow experiments[113] and with the root pressure probe.[114] With the latter technique a detailed analysis of Lp_r along the root was possible which also incorporated the contribution of the hydraulic resistance of the xylem. In young seminal maize roots the longitudinal resistance has been shown to dominate in the apical 20 mm, whereas in the basal parts its contribution was fairly small compared with the radial resistance.[115]

Solute relations (permeability and reflection coefficients) as measured with root pressure probe and other techniques are summarized in Table VII. Permeability coefficients (P_{sr}) obtained from biphasic root pressure relaxations were similar to those obtained from pressurized root exudation. The P_{sr} values of electrolytes, sugars, and polyols were sufficiently small to allow proper functioning of the root. On the other hand, σ_{sr} values were rather low in the pressure probe experiments even for solutes for which cell membranes exhibit a σ_s of approximately 1 (electrolytes, sugars). It can be

[111] C. A. Peterson, M. E. Emanuel, and G. B. Humphreys, Can. J. Bot. 59, 618 (1981).
[112] H. Jones, A. D. Tomos, R. A. Leigh, and R. G. Wyn Jones, Planta 158, 230 (1983).
[113] H. Jones, R. A. Leigh, R. G. Wyn Jones, and A. D. Tomos, Planta 174, 1 (1988).
[114] E. Steudle and J. Frensch, Planta 177, 281 (1989).
[115] J. Frensch and E. Steudle, Plant Physiol. (in press).

TABLE VII

Reflection (σ_{sr}) and Permeability (P_{sr}) Coefficients of Roots Determined by the Root Pressure Probe and Other Techniques[a]

Root	Solute	Reflection coefficient, σ_{sr}	Permeability coefficient, $P_{sr} \times 10^{10}$ (m sec^{-1})	Ref.
Glycine max	Nutrients	0.9	—	b
Lycopersicon esculentum	Nutrients	0.76	—	c
Zea mays,	Sucrose	0.9–1.0	—	d, e
cortical	Urea	0.85	—	d, e
sleeves		1	—	d, e
		0.05	—	f
Zea mays,	Nutrients	0.85	—	g
excised	Ethanol	0.27	60–190	h, i
roots	Mannitol	0.74	—	h, i
	Sucrose	0.54	30	h, i
	PEG 1000	0.82	—	h, i
	NaCl	0.5–0.6	60–140	h, i
	KNO$_3$	0.5–0.7	10–80	h, i
Hordeum distichon	Mannitol	≈0.5	—	j
Phaseolus vulgaris	Nutrients	0.98	2.2	k
Phaseolus coccineus	Methanol	0.16–0.34	27–62	l
	Ethanol	0.15–0.47	44–73	
	Urea	0.41–0.51	11	
	Mannitol	0.68	1.5	
	KCl	0.43–0.54	7–9	
	NaCl	0.59	2	
	NaNO$_3$	0.59	4	

[a] For maize, data are also given for a root preparation (cortical sleeves). σ_{sr} appears to be significantly lower than unity even for solutes for which cell membranes exhibit a σ_s of approximately 1 (electrolytes, sugars, etc.), although P_{sr} values are rather low.

[b] E. L. Fiscus, *Plant Physiol.* **59,** 1013 (1977).
[c] G. C. Mees and P. E. Weatherley, *Proc. R. Soc. London, Ser. B* **147,** 381 (1957).
[d] H. Ginsburg and B. Z. Ginzburg, *J. Exp. Bot.* **21,** 580 (1970).
[e] H. Ginsburg and B. Z. Ginzburg, *J. Exp. Bot.* **21,** 593 (1970).
[f] J. C. Collins, *in* "Membrane Transport in Plants" (U. Zimmermann and J. Dainty, eds.), p. 441. Springer-Verlag, Berlin and New York, 1974.
[g] D. M. Miller, *Plant Physiol.* **77,** 162 (1985).
[h] E. Steudle and J. Frensch, *Planta* **177,** 281 (1989).
[i] E. Steudle, R. Oren, and E.-D. Schulze, *Plant Physiol.* **84,** 1220 (1987).
[j] E. Steudle and W. D. Jeschke, *Planta* **158,** 237 (1983).
[k] E. L. Fiscus, *Plant Physiol.* **80,** 752 (1986).
[l] E. Steudle and E. Brinckmann, *Bot. Acta* **102,** 85 (1989).

seen from Table VII that there are also indications of a σ_{sr} below 1 from other measurements, although in some cases the differences are remarkable (see results for bean roots). These findings seem to be due to technical difficulties.

Unstirred layers have to be considered in the determination of transport coefficients of roots. They would tend to reduce the true values of the coefficients and may, in principle, also result in too low reflection coefficients. However, a rigorous treatment has shown that these effects should be rather small when using the root pressure probe technique.[114] They cannot completely explain the finding of low σ_{sr} values. Low reflection coefficients could be important because they would mean that solutes (nutrients) could be also taken up by solvent drag [cf. Eq. (5)], and polarization effects at the osmotic barrier of the root could be avoided at high transpiration rates. Nevertheless, roots may function properly in spite of a low σ_{sr}, because P_{sr} rather than the reflection coefficient determines the leak rate for nutrients across the root cylinder.

Low σ_{sr} values have been explained by the composite structure of the root which exhibits different parallel and series resistances (barriers) for water and solutes (see above). Parallel pathways are the apoplasmic, transcellular, and symplasmic paths[109] (Fig. 1) which should be characterized by different coefficients and may vary during root development. A serial arrangement should result from the placement of the exodermis, cortex, endodermis, and stele in series. These components (and perhaps also others) may act as "membranes" which would subdivide the entire barrier into a multicompartment system. From the basic theory of composite membranes it is known that the parallel as well as the serial arrangement of membrane elements could result in rather unexpected overall properties such as polarity and low σ_{sr} as they are observed in the experiments.[24,27-29]

Thus, the results indicate that the simple osmometer model of the root should, perhaps, be extended to allow for the complications mentioned. However, in order to model the root completely more data are necessary to quantitatively characterize the variation of transport properties along the root and also of entire root systems. Furthermore, it is also necessary to combine root pressure probe work with the measurement of solute flows in the excised system and, as already mentioned, with measurements of Lp at the cell level. Another important aspect is the measurement of the contribution of the root xylem to the overall hydraulic resistance which can be also performed with the technique using segments from different parts of the root which are also cut at the apical end.[115]

In principle, with the root pressure probe one should also be able to measure active solute movements [J_{sr}^{*}; cf. Eq. (5)] and their coupling with water flow. Changes in J_{sr}^{*} during the biphasic pressure relaxations have to be considered.[116] They may contribute to the measured coefficients, but the effects on Lp_r and σ_{sr} should be small, since the water flow would usually be much more rapid than the solute flow (Fig. 5).

In conclusion, the analysis of water and solute relations of roots indicates a strong coupling between water and solutes which could be, perhaps, described best in terms of an extended osmometer model of the root. The model should take into account the structural complexity of the root as well as active pumping. It is obvious that a better understanding of the basic osmotic processes in roots would also be important for other processes in plants in which water transport interacts with solute transport such as during growth and osmoregulation or in the regulation of stomatal movement. Compared with the root, however, in these areas much less information that could be used for quantitative modeling seems to be available.

Acknowledgments

I thank Dr. Carol A. Peterson, Department of Biology, University of Waterloo, Ontario, Canada, for reading and discussing the manuscript. This work was supported by a grant from the Deutsche Forschungsgemeinschaft, Sonderforschungsbereich 137.

[116] E. Steudle and E. Brinckmann, *Bot. Acta* **102,** 85 (1989).

[17] Plasmolysis and Deplasmolysis

By OkYoung Lee-Stadelmann and Eduard J. Stadelmann

Introduction

In most mature higher plant cells the living protoplasm surrounds a large aqueous central vacuole and forms a thin layer, generally less than 10% of cell volume,[1,2] between the cell wall and vacuole. The protoplasm layer includes two membranes: the plasmalemma (or cell membrane) and the tonoplast, which delimit the protoplasm from the cell wall and from the vacuole, respectively.

The vacuole and protoplasm layer (collectively called the protoplast)

[1] J. P. Palta and O. Y. Lee-Stadelmann, *Plant, Cell Environ.* **6,** 601 (1983).
[2] E. J. Stadelmann, *in* "Handbuch der Pflanzenphysiologie" (W. Ruhland, ed.), Vol. 2, p. 71. Springer-Verlag, Berlin and New York, 1956.

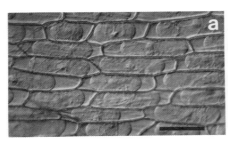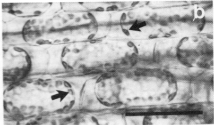

FIG. 1. Final stages of normal plasmolysis (healthy cells with convex plasmolysis forms). (a) *Allium cepa* adaxial epidermal cells of the bulb scale, plasmolyzed stepwise with intermediary steps at 0.2 *M*, 0.5 *M* (for 25 min each), and finally 0.7 *M* (0.77 os*M*) mannitol (after 25 min); differential interference contrast. Bar, 200 μm. (b) *Pisum sativum* (24 days old) subepidermal stem base cells (first internode) plasmolyzed stepwise with intermediary steps at 0.20 *M*, 0.225 *M* (for 40 min each), and finally 0.25 *M* (0.640 os*M*) Ca Cl$_2$ for 40 min, bright field. Chloroplasts (arrows) are equally distributed in the protoplasmic layer. Bar, 100 μm.

form a nearly ideal osmotic system because the membranes possess a high degree of differential permeability. When a hypertonic nonpermeating solute (e.g., sugars or mannitol) is applied in the external solution, the vacuole (and to a limited extent the protoplasm) loses water until the cell water potential equals the water potential of the outside solution. The water loss leads to a decrease of the protoplast volume occupied mainly by the vacuole, resulting in a loss of cell turgor followed by separation of the protoplast from the cell wall. This separation of the living protoplast from the cell wall resulting from water-withdrawing solutions (plasmolytica) is called plasmolysis (Figs. 1 and 2).

For embryonic and meristematic cells and cells rich in protoplasm and without large vacuoles, withdrawal of water causes shrinkage or deswelling of the protoplasm with or without separation of the protoplasm from the cell wall. Shrinkage of isolated or free protoplasts (without walls) in hypertonic solutions is called plasmorrhysis.

Plasmolysis is a unique technique in experimental plant cell physiology to study the physicochemical properties of individual plant cells and their alterations in the living state. Protoplasmic differences between cell types as well as changes caused by developmental stages and environment within the same cell type have been well demonstrated by plasmolysis.[2]

The cell parameters most commonly measured by plasmolytic techniques[2,3] are (1) passive permeability of the protoplasm layer, (2) proto-

---

[3] E. J. Stadelmann, *Methods Cell Physiol.* **2**, 143 (1966).

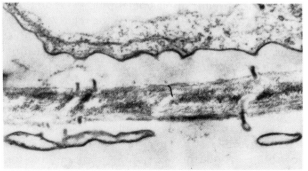

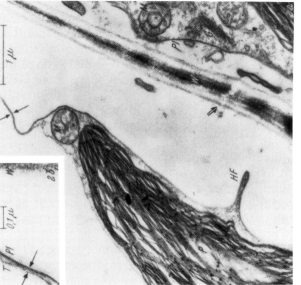

FIG. 2. (Left) Portion of an *Elodea callitrichoides* leaf cell, plasmolyzed in 0.5 *M* sucrose. The thin protoplasm layer contains a chloroplast (P, partially seen), a mitochondrion (M), and Hecht's thread (HF, withdrawing). Outside these inclusions the protoplasm layer is extremly thin and essentially contains the plasmalemma and tonoplast only (see inset). The double arrow indicates plasmodesma. W, Cell wall; Pl, plasmalemma; G, Golgi body; T, tonoplast. Note that the plasmalemma under plasmolysis is smooth and no folding is visible. The inset shows how closely plasmalemma and tonoplast may come together and how thin the protoplasm layer may become. (Right) Portion of two leaf cells of *Elodea callitrichoides*. In the cell on the right the protoplast is retracting from the cell wall. In the cell on the left the protoplast is not visible since it is pulled farther away from the cell wall, but protoplasm remnants connected to the cell wall below or above the picture plane can be seen. (From Ref. 10.)

plasmic viscosity, (3) cell wall attachment of plasmalemma, (4) solute potential of a single cell with large central vacuole, and (5) the relative size of the nonsolvent space in the vacuole. Furthermore, plasmolysis is the most frequently used test for cell viability; it is essential in the procedures for protoplast isolation, probably because it protects the protoplasts from damage by cell wall-digesting enzymes,[4] and may even enhance morphogenesis.[5] The plasmolyticum also prevents bursting of the freed protoplasts.

There exists a vast diversity in plasmolysis behavior of different cell types. Some cell types are inherently sensitive to plasmolysis, whereas others can tolerate or even perform normally in the plasmolyzed state. Plasmolyzed cells may continue cytoplasmic streaming, remain alive for many days, undergo cell division (after deplasmolysis), and regain turgor even after a high degree of plasmolytic contraction.

When plasmolysis is not properly performed, it can disturb normal functioning of cells. Many changes can occur in cell metabolism (see Ref. 2, p. 103),[6] growth, ultrastructure,[7] and locomotion of unicellular organisms following plasmolysis. Depending on the cell type and state of the cell, these alterations may be reversible or result in cell damage and eventual cell death (see Ref. 2, p. 102). For these reasons it is of greatest importance to establish for each cell type a specific protocol for plasmolysis which is most suitable. There exists no general plasmolysis procedure which fits all cell types.

In special situations, the living protoplast separates from the cell wall without contact with an external hypertonic solution. Such conditions include extreme desiccation, intracellular ice formation between the cell wall and protoplast, specific stimuli [e.g., wounding, mechanical alterations, strong light, and ultrasound (see Ref. 2, p. 83)], and cytogenesis in a few cell types (see Ref. 8, p. 29).

This chapter deals with general techniques of plasmolysis, modifying factors, experimental conditions that could counteract inherent problems, and applications of plasmolysis techniques for measuring some cell qualities. Details of techniques for the application of plasmolysis are described elsewhere.[3,8,9] Permeability measurements by the plasmolytic method are discussed elsewhere in this volume [18].

[4] E. C. Cooking, *Annu. Rev. Plant Physiol.* **23**, 29 (1972); D. W. Burger and W. Hackett, *Physiol. Plant.* **56**, 324 (1982).
[5] D. F. Wetherell, *Plant Cell Tissue Organ Cult.* **3**, 221 (1984).
[6] H. Falk, U. Lüttge, and J. Weigl, *Z. Pflanzenphysiol.* **54**, 446 (1966).
[7] C. Pargney, *Z. Pflanzenphysiol.* **107**, 237 (1982).
[8] E. Küster, "Die Pflanzenzelle," 3rd ed. Fischer, Jena, 1956.
[9] E. Küster, "Pathologie der Pflanzenzelle. I. Pathologie des Protoplasmas," Protoplasma-Monogr. Vol. 3. Borntraeger, Berlin, 1929.

Factors Affecting Plasmolysis Form and Time

Internal and external factors greatly affect plasmolysis form and plasmolysis time (i.e., the time from the first contact of the plasmolyticum with the tissue until osmotic equilibrium is reached). The most commonly observed plasmolysis forms are convex, concave, and angular (Fig. 3; see Ref. 2, p. 76). When the wall attachment is not too great and protoplasm and vacuole are sufficiently liquid, concave plasmolysis forms generally become convex. Plasmolysis time is greatly affected by differences in viscosity and wall attachment. Plasmolysis form and time have been extensively used as parameters for characterizing the physiological conditon of plant cells.

Cell Factors

The most important cell factors affecting plasmolysis are cell wall attachment, protoplasmic viscosity, and for some cell species also cell wall pore size. These factors vary greatly with cell type, plant age, and the stage of development.

The contact between cell wall and protoplast may also involve an intimate interweaving between cell wall and protoplasm, mutually penetrating each other,[10] and resulting in an increase of the actual cell membrane area ($\sim 10\%$ in *Valonia* cells[11]). Strong wall attachment of the plasmalemma, high protoplasmic viscosity, and stiffness of chromatophore or large chloroplast number may hinder the withdrawal of the protoplast from the cell wall in hypertonic solutions, and such cells will not act as an ideal osmometer.

External Factors

Plant growth and experimental conditions, such as temperature, light, pH, and the kind of medium (ion composition), all influence plasmolysis considerably (see Ref. 2, p. 81).

Alterations of the Protoplast by Plasmolysis

When plasmolytic conditions are not adequate, pathological alterations can be observed (some are shown in Fig. 3). Leakage of ions and organic

[10] P. Sitte, *Protoplasma* 57, 325 (1963); H. Falk and P. Sitte, *ibid.* p. 290.
[11] U. Zimmermann, R. Benz, and H. Koch, *Planta* 152, 352 (1981); A. D. Thomas, E. Steudle, U. Zimmermann, and E.-D. Schulze, *Plant Physiol.* 68, 1135 (1981).

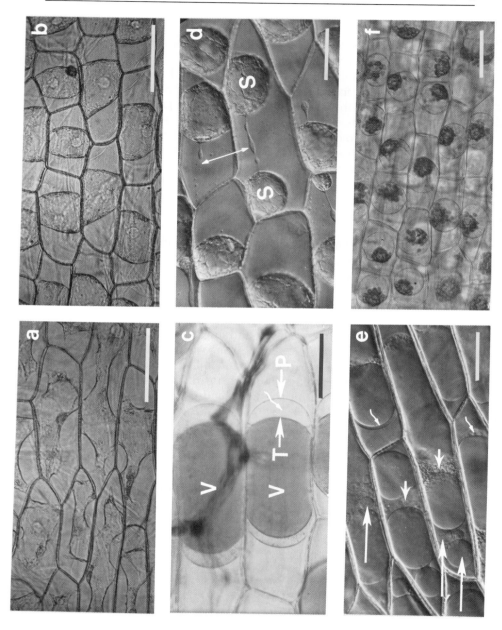

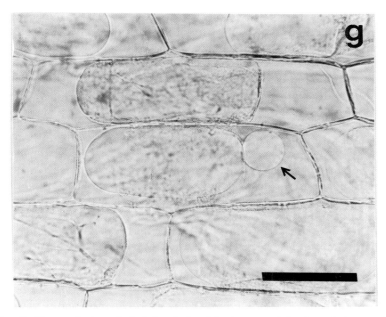

Fig. 3. Abnormal plasmolysis (often induced by inappropriate plasmolysis procedure). (a) Concave plasmolysis caused by direct transfer of the tissue into the plasmolyticum (here 0.8 M mannitol, 0.92 os$M$, after 25 min). Bar, 200 μm. (b) Angular (slightly concave) plasmolysis caused by either high protoplasmic viscosity or a strong cell wall attachment. Cells were plasmolyzed with intermediary steps in 0.2 M, 0.5 M (40 min each), and finally 0.8 M mannitol. Concave or angular plasmolysis becomes convex with time or by treatment with octylguanidine (see text). Bar, 200 μm. (c) Cap plasmolysis caused by extended plasmolysis (42 hr) in 0.8 M KNO$_3$. Note the swollen cytoplasm and its clearly recognizable membranes. P, Plasmalemma; T, tonoplast; wavy arrow, swollen protoplasm (caps); V, vacuole. Bar, 50 μm. (d) Formation of partial protoplasts (subprotoplasts) by osmotic shock and an excessively high concentration of plasmolyticum (observed 12 hr after direct transfer of the tissue into 1.5 M glucose). Partial protoplasts (S) can form immediately after transfer or develop later from severe concave plasmolysis. The arrow shows that droplets of protoplasm of different sizes are often connected by protoplasmic threads between the protoplast and the cell wall or between partial protoplasts. Bar, 100 μm. (e) Degeneration of cytoplasm after extended plasmolysis (12 hr) following osmotic shock (direct transfer into 0.8 M mannitol). Note swollen caps (wavy arrows), granular cytoplasmic degeneration (short arrows), and small vacuoles (tonoplasts, long arrows) released from the degenerated protoplast; v, vacuoles and partial vacuoles. Bar, 200 μm. (f) Systrophy, after extended plasmolysis (12 hr) following stepwise plasmolysis, in 0.25 M CaCl$_2$. Note aggregation of chloroplasts and cytoplasm. This type of systrophy is generally reversible. Bar, 100 μm. (g) Protoplast protuberance (arrow) developed in a few cells 6–7 hr after direct transfer of the tissue into 0.8 M sucrose. Bar, 100 μm. Materials: a, b, d, e, *Allium cepa*, yellow variety, adaxial (inner) epidermis of bulb scale (third fleshy scale from outside); c, *Allium cepa*, yellow variety, abaxial (outer) epidermis of bulb scale (third fleshy scale from outside); f, *Pisum sativum* (24 days old), subepidermal cells of the first internode (stem base); g, *Allium cepa*, red variety, adaxial (inner) epidermis of bulb scale (third fleshy scale from outside). Micrographs a, b, and e were obtained under differential interference contrast; all others, bright field.

compounds into the external solution (see Ref. 2, p. 103)[12] may occur as a result of membrane damage by osmotic shock (excessive concentration difference between the vacuole and external solution into which tissue was transferred), or even by transfer of the tissue into water (especially distilled water!). The degree of leakage varies greatly with cell type and external conditions.[6,13] Osmotic shock often causes mild to severe concave plasmolysis forms which later may become convex.

Prolonged plasmolysis may induce aggregation of plastids and cytoplasm around the nucleus (systrophy) brought about by protoplasmic streaming. Systrophy is generally harmless and reversible.

Plasmolysis times of several hours to several days, especially in sugar solutions, may lead to a transformation of the protoplast surface so that the protoplast can not expand when in contact with a permeating solution (surface stiffening). The protoplast surface then breaks at a weak location on the free protoplast ends, and protuberances (Fig. 3g) are formed which are delimited by a fine new membrane (perhaps an extension of the tonoplast membrane[14]). Protuberances may occur after plasmolysis in mannitol, sucrose, or $CaCl_2$.

Formation of protuberances is a cell type-specific reaction of the protoplast to a specific plasmolyticum. The same plasmolyticum with other protoplasts may not provoke formation of protuberances and vice versa. Some protoplasts are slightly permeable to a specific plasmolyticum (e.g., mannitol) which also causes surface stiffening. Consequently, when the excess osmotic pressure that builds up in the vacuole owing to the slow permeation of the plasmolyticum can no longer be contained by the stiffened protoplast surface, the latter breaks open and a protuberance appears. This mechanism is also responsible for the frequently observed formation of protuberances ("budding") on isolated protoplasts in culture media.

In some cells callose begins to develop after 2–4 h of plasmolysis in various sugars, mainly at the free protoplast ends which are not in contact with the cell wall.[15] After several days, the protoplast may form a new cell wall from which it can be separated by a second plasmolysis, using a stronger plasmolyticum (see Ref. 2, p. 96).

---

[12] L. Amar and L. Reinhold, *Plant Physiol.* **51**, 620 (1973); B. Rubinstein, *ibid.* **69**, 945 (1982).

[13] J. F. Sutcliffe, *J. Exp. Bot.* **5**, 215 (1953); L. Adamec, *Biol. Plant.* **26**, 128 (1984).

[14] R. Eichberger, *Protoplasma* **20**, 606 (1934).

[15] W. Escherich, *Planta* **48**, 578 (1957).

In elongated cells with sufficiently low cytoplasmic viscosity, concave protoplasts generally constrict to form two or more eventually convex partial protoplasts (subprotoplasts) because of protoplast surface tension and/or alteration by osmotic shock. Swelling of the cytoplasm (cap plasmolysis; Fig. 3c) may occur in hypertonic alkali salt solutions such as KCl or KNO_3. Cap plasmolysis develops because of differences in the permeability for K^+ between plasmalemma and tonoplast. Cap plasmolysis may be reversed and can be prevented by adding solutions containing alkali earth ions, such as $CaCl_2$, to the medium. It is seldom observed in sugar solutions.

Deplasmolysis

Deplasmolysis is the opposite process of plasmolysis: when the concentration of the solution external to a plasmolyzed cell is decreased or when solutes permeate from the external solution into the vacuole, water will reenter the vacuole, and the increase in protoplast volume leads to restoration of full turgidity. In some plasmolyzed cells deplasmolysis may also be induced by osmoregulation, i.e., by an increase in solute concentration of the cell sap arising from metabolic activities of the cell. The deplasmolysis rate is the basis of osmotic permeability measurements which are dealt with elsewhere in this volume [18].

Materials and Experimental Procedures

Selection and Preparation of Optimal Osmotica

For selecting the optimal plasmolyticum for a given cell type several plasmolytica with the same osmolar concentration have to be tested in preliminary experiments. The plasmolyticum is chosen which gives the easiest and smoothest separation of protoplasts, with convex plasmolysis, and without cytomorphological alterations (see above).

The plasmolytica used in such test must meet the following general criteria:

1. The osmotically active substance must be able to pass through the cell wall pores[16] while the cell membrane must be impermeable to it.

---

[16] B. Frenzel, *Planta* **8,** 642 (1929); N. Carpita, D. Sabularse, D. Montezinos, and D. P. Delmer, *Science* **205,** 1144 (1979); M. Tepfler, *ibid.* **213,** 761 (1978).

Generally, inorganic ions, sugars, and other small molecules readily pass through cell walls of most mesophytic plants. Large molecules such as proteins and polyethylene glycol (PEG) (molecular weight greater than 6000) pass through the cell wall with difficulty or not at all. Cell walls of some mosses are not permeable to sucrose.[17] Partial or complete impermeability of the cell wall results in little or no separation of the protoplast and generally leads to (sometimes transient) cytorrhysis. When the plasmolyticum is able to pass through the cell wall one should be able to recognize the separation of the protoplast from the cell wall in 1–5 min, depending on the number of cell layers. If no incipient plasmolysis occurs withing 5–10 min in a hypertonic solution, the plasmolyticum is either not able to permeate the cell wall or the cell wall attachment of the protoplast is too high. Prolonged contact of the cell with the plasmolyticum may cause cell damage under these conditions.

2. The plasmolyticum must be chemically inert and nontoxic to the cells. The appearance of the cell and the cytomorphology during and after plasmolysis must be the same as before the plasmolysis. When plasmolyzed protoplasts shrink with time, ion leakage should be suspected due to cell damage.

3. The plasmolyticum must be a nonmetabolite, at least for the duration of the plasmolysis experiment and in addition must exert a protective effect on the cell membrane. This criterion seems to be particularly important for long-term plasmolysis (several days).[1]

Sugars (sucrose, glucose) and sugar alcohols (mannitol, sorbitol) are the most frequently used osmotica and seem to meet these criteria at least for plasmolysis for up to a few hours and for most materials. Sucrose seems to have both osmotic and protective action for many materials. Mannitol does not seem to have a protective effect and may even be damaging to certain cell types after longer exposure or when used without Ca^{2+} in the solution. It may also permeate into some cell types.[18] When the plasmolysis is performed for several hours, or if the cells are specifically sensitive most sugars and sugar derivatives cause alteration of membranes and result in protuberances, when the protoplast is deplasmolyzing (Fig. 3g).[19]

[17] B. Brilliant, *C. R. Acad. Sci. URSS* **1927**, 155 (1927); G. Kressin, Ph.D. Dissertation, Univ. Greifswald, Germany (1935).

[18] V. Kozinka and S. Klenovska, *Biol. Plant.* **7**, 285 (1965); W. J. Cram, *in* "Membrane Transport in Plants" (W. J. Cram, K. Janáček, R. Rybová, and K. Sigler, eds.), p. 483. Akademia-Verlag, Prague, 1984; O. Y. Lee-Stadelmann, *Plant Physiol.* **80**(4), 97 (1986).

[19] I. Chung, O. Y. Lee-Stadelmann, and E. Stadelmann, *Plant Physiol.* **80**(4), 90 (1986).

A small amount of Ca^{2+} is necessary to protect the membranes of many cell types. Salt mixtures, for instance, 3.7 mM $CaCl_2$ plus 25 mM KCl, spring water, or similar ion-balanced media are recommended as solvents for the plasmolyticum. Pure distilled water without balanced ions is detrimental to the cell and should never be used in preparation or for storage of the tissue material.[20]

A mixture of KCl and $CaCl_2$[21] (preferably a 9 : 1 ratio), Ca^{2+} salts alone, concentrated seawater, Brenner solution (see Ref. 2, p. 102), and α-methylglucose[22] have been used successfully as plasmolytica. In some lower organisms single salt solutions (NaCl, KNO_3) have occasionally been used,[23] but such solutions may damage cells of higher plants. Salt mixtures as plasmolytica are necessary for cell types with small cell wall pores or a strong wall attachment to the protoplasm. No deleterious effects have been reported after long exposure to balanced salt mixtures, although they have been noted with sugar solutions.

Selection of osmotica also depends on the purpose of the experiment. In general, sugars and sugar derivatives are recommended for short-term experiments (e.g., for preplasmolysis in permeability experiments). Ca^{2+} salts and ion mixtures are good for long-term plasmolysis and solute potential determination but are not recommended for permeability experiments as ions may alter the cell membrane permeability. Uptake of the plasmolyticum can be easily recognized by the increase in volume of the plasmolyzed protoplast with time. Membrane damage will cause shrinkage of fast expansion of the protoplast owing to leakage of ions from the vacuole or pathological permeability for the plasmolyticum.

Selection of Optimal Plasmolytic Concentrations

Most mesophytic cells will plasmolyze in 0.4–0.7 os$M$ solutions because the cell sap concentration generally ranges between 0.2 and 0.5 os$M$. For most cell physiological studies, the protoplast should be contracted from one-half to two-thirds of its original volume (see Fig. 1). Some of the most frequently used plasmolytica and their osmolarities are shown in Table I.

[20] G. W. Scarth, *Proc. Trans. R. Soc. Can [3]* **18**, Sect. 5, 97 (1924); A. Kaczmarek, *Protoplasma* **6**, 209 (1929); I. de Haan, *Recl. Trav. Bot. Neerl.* **30**, 234 (1933).
[21] W. Url, *Protoplasma* **72**, 427 (1971).
[22] O. Y. Stadelmann, W. R. Bushnell, and E. J. Stadelmann, *Can. J. Bot.* **62**, 1714 (1984).
[23] E. J. Stadelmann, *in* "Environmental Biology" (P. L. Altman and D. S. Dietmer, eds.), p. 541. Fed. Am. Soc. Exp. Biol., Bethesda, Maryland, 1966.

TABLE I
OSMOLARITY TO MOLARITY CONVERSION

OsM (osmol/kg)	Osmotic potential (MPa, at 25°)	Molarity (mol/liter)					
		Mannitol[a]	Sucrose[a]	Glucose[a]	Methyl-glucose[b]	CaCl$_2$[a]	Salt mixture[c]
.050	−.124			.050	.050		
.075	−.186			.074	.075		
.100	−.248	.098	.096	.098	.092		
.125	−.310	.122	.120	.121	.115		
.150	−.372	.146	.143	.145	.135		
.175	−.433	.170	.164	.169	.155	.067	
.200	−.495	.194	.187	.195	.176	.077	1.000
.225	−.557	.217	.210	.218	.196	.086	
.250	−.619	.241	.231	.241	.215	.096	1.294
.275	−.681	.265	.253	.264	.235	.106	
.300	−.743	.288	.273	.287	.254	.116	1.588
.325	−.805	.310	.294	.309	.272	.125	
.350	−.867	.333	.314	.332	.290	.135	1.882
.375	−.929	.355	.335	.354	.310	.145	
.400	−.991	.378	.355	.376	.327	.155	2.158
.425	−1.053	.400	.375	.398	.346	.164	
.450	−1.115	.422	.393	.421	.364	.174	2.421
.475	−1.177	.445	.412	.443	.380	.184	
.500	−1.238	.466	.431	.465	.396	.194	2.684
.525	−1.300	.488	.450	.488	.412	.203	
.550	−1.362	.509	.468	.509	.429	.213	2.947
.575	−1.424	.531	.486	.531	.440	.223	
.600	−1.486	.553	.504	.553	.461	.233	3.235
.625	−1.548	.574	.522	.574	.477	.242	
.650	−1.610	.595	.539	.595	.493	.252	3.529
.675	−1.672	.615	.556	.615	.510	.262	
.700	−1.734	.635	.573	.636	.526	.271	3.824
.725	−1.796	.656	.591	.656	.543	.280	
.750	−1.858	.678	.607	.676	.557	.290	4.105
.775	−1.920	.698	.624	.696	.575	.299	
.800	−1.982	.718	.639	.716	.591	.308	4.368
.825	−2.043	.739	.655	.736	.608	.317	
.850	−2.105	.758	.671	.756	.624	.327	4.632
.875	−2.167	.778	.687	.775	.640	.336	
.900	−2.229	.798	.702	.795	.655	.345	4.895
.925	−2.291	.817	.718	.814	.670	.353	
.950	−2.353	.837	.733	.834	.685	.363	5.167
.975	−2.415	.857	.748	.853	.700	.372	
1.000	−2.477		.763	.872	.715	.380	5.444
1.025	−2.539		.778	.892	.730	.389	
1.050	−2.601		.791	.910	.745	.398	5.722
1.075	−2.663		.805	.930	.760	.406	
1.100	−2.725		.819	.949	.774	.415	6.000
1.125	−2.787		.834	.968	.789	.423	
1.150	−2.848		.847	.986	.804	.432	6.278
1.175	−2.910		.861	1.005	.819	.440	

TABLE I   *(continued)*

OsM (osmol/kg)	Osmotic potential (MPa, at 25°)	Molarity (mol/liter)					
		Mannitol[a]	Sucrose[a]	Glucose[a]	Methyl-glucose[b]	$CaCl_2$[a]	Salt mixture[c]
1.200	−2.972		.874	1.023			6.556
1.225	−3.034		.887	1.041	.835	.449	
1.250	−3.096		.901	1.059	.848	.457	6.833
1.275	−3.158		.914	1.077	.861	.465	
1.300	−3.220		.927	1.095	.875	.473	7.118
1.325	−3.282		.939	1.113	.890	.481	
1.350	−3.344		.952	1.130	.902	.489	7.412
1.375	−3.406		.965	1.148	.915	.498	
1.400	−3.468		.977	1.165	.928	.506	7.706
1.425	−3.530		.990	1.182	.941	.514	
1.450	−3.592		1.002	1.200	.953	.521	8.000
1.475	−3.653		1.013	1.216	.966	.529	
1.500	−3.715		1.024	1.232	.979	.537	8.278
1.525	−3.777		1.036	1.248	.990	.544	
1.550	−3.839		1.048	1.264		.552	8.556
1.575	−3.901		1.060	1.280		.559	
1.600	−3.963		1.071	1.296		.567	8.833
1.625	−4.025		1.082	1.312		.574	
1.650	−4.087		1.093	1.328		.582	9.111
1.675	−4.149		1.103	1.343		.589	
1.700	−4.211		1.114	1.358		.597	
1.725	−4.273		1.125	1.373		.604	
1.750	−4.335		1.136	1.389		.611	
1.775	−4.396		1.144	1.404		.619	
1.800	−4.458		1.154	1.419		.626	
1.825	−4.520		1.164	1.434		.633	
1.850	−4.582		1.175	1.449		.640	
1.875	−4.644		1.185	1.464		.647	
1.900	−4.706		1.196	1.479		.654	
1.925	−4.768		1.206	1.494		.661	
1.950	−4.830		1.216	1.509		.668	
1.975	−4.892		1.226	1.524		.675	
2.000	−4.954		1.235	1.539		.682	
2.025	−5.016		1.244	1.554		.688	
2.050	−5.078		1.254	1.569			
2.075	−5.140		1.263	1.584			
2.100	−5.201		1.273	1.599			
2.125	−5.263		1.282	1.613			
2.150	−5.325		1.292	1.628			
2.175	−5.387		1.300	1.642			
2.200	−5.449		1.309	1.657			
2.225	−5.511		1.317	1.672			
2.250	−5.573		1.326	1.686			
2.275	−5.635		1.334	1.701			
2.300	−5.697		1.343	1.715			
2.325	−5.759		1.351	1.730			
2.350	−5.821		1.360	1.744			
2.375	−5.883		1.369	1.758			

(continued)

TABLE I *(continued)*

OsM (osmol/kg)	Osmotic potential (MPa, at 25°)	Molarity (mol/liter)					
		Mannitol[a]	Sucrose[a]	Glucose[a]	Methyl-glucose[b]	CaCl$_2$[a]	Salt mixture[c]
2.400	−5.945		1.377	1.773			
2.425	−6.006		1.385	1.787			
2.450	−6.068		1.393	1.801			
2.475	−6.130		1.400	1.815			
2.500	−6.192		1.408	1.829			
2.525	−6.254			1.843			
2.550	−6.316			1.858			
2.575	−6.378			1.872			

[a] Values interpolated from data given in R. C. Weast, ed., "CRC Handbook of Chemistry and Physics," 69th ed. CRC Press, Boca Raton, Florida (1988–1989).

[b] From measurements by a vapor pressure osmometer (Hewlett-Packard).

[c] Numbers indicate milliliters of a salt mixture stock solution (900 ml of 1 M KCl and 100 ml of 1 M CaCl$_2$) to which distilled water should be added to make a total volume of 10 ml final solution.[21]

Some cell materials require unusually high concentrations of plasmolytica because of the high solute concentration of the cell sap. Ray parenchyma cells in some woody stems[24] or leaf cells of *Selaginella lepidophylla* require about 1 M CaCl$_2$ (3.26 osM; D. Nelson, O. Y. Lee-Stadelmann, and E. J. Stadelmann, unpublished results). These cells cannot be plasmolyzed in sugar solutions.

Stepwise plasmolysis is applied to avoid large concentration changes which cause osmotic shock. Concentration changes should not be greater than 0.1 M for osmotically sensitive species like barley coleoptile epidermis. Some types of woody and poikilohydric plant cells are less sensitive and can be transferred either using much greater increments or directly into a high concentration of the plasmolyticum. Osmotic shock can often be microscopically recognized by concave plasmolysis or subdivision of the protoplast during plasmolysis.

Cells with inherently strong wall attachment (e.g., mesophyll cells of *Glycine max*) can be plasmolyzed by brief pretreatment with octylguanidine (OG). A 250 μM solution of OG is applied either in isotonic or hypertonic solution, preferably with mannitol as solute, for no longer than

[24] S. G. Hong, E. Sucoff, and O. Y. Lee-Stadelmann, *Bot. Gaz. (Chicago)* **141**, 464 (1980).

1 to 2 min. The material is then transferred into a mannitol solution of the same concentration without OG.[25,26] Some examples of optimal plasmolytica and concentrations are shown in Table II. Solute potential values have so far been compiled only for lower plants.[23]

Preparation of Tissue and Cell Material

Three groups of plant materials can be distinguished on the basis of the different preparation procedures required. Preparations should always be transferred consecutively into two samples of the plasmolyticum, the first one serving as washing solution.

Unicellular Plants and Isolated Cells. Single cells can be concentrated from suspensions by centrifugation. The pellet which accumulates in the centrifugation tube can be transferred with a pipette or similar device into a droplet of the plasmolyticum on a slide.

Filamentous Plants (e.g., Spirogyra). Filamentous material is best transferred into square watch glasses (Carolina Biological Supply) or similar dishes containing solution. The square watch glasses should have ground glass tops and bottoms so they can be stacked to avoid evaporation. Glass hooks should be used for transfer of the algae.

Tissues of Metaphytes (Multicellular Plants). Microscopic observation requires that the tissue be thin ($\sim 20-80 \mu$m). Some organs or parts of metaphytes can be used for plasmolysis experiments wihtout further preparation (e.g., root hairs, *Elodea* leaves, leaves of many Bryophyta). Usually, however, dissection of the material either free hand or with a vibrating microtome (Vibratome, American Scientific Products) is necessary to achieve the proper thickness. Layers one to three cells thick near the surface of an organ may be stripped off by making a shallow incision and removing them with a fine forceps (e.g., epidermis of onion bulb scale). Excessive bending of the tissue during the stripping process and other manipulations should be avoided. For best results, the transfer of the tissue from one solution to the next should be done using a fine camel's hair brush or a mesh.

Vacuum infiltration in the ion-balanced medium (see above) of leaf or stem pieces after incision is necessary to remove air, which can interfere with microscopic observation from intercellular spaces. It also facilitates the stripping off procedure without tissue damage. The tissues are im-

[25] B. E. Gómez-Lepe, O. Y. Lee-Stadelamnn, J. P. Palta, and E. J. Stadelmann, *Plant Physiol.* **64,** 131 (1979).
[26] O. Y. Lee-Stadelmann, I. Chung, and E. J. Stadelmann, *Plant Sci.* **38,** 1 (1985).

TABLE II

PLASMOLYTICA, CONCENTRATIONS, AND CONCENTRATION STEPS FOR PRODUCING ADEQUATE
CONDITIONS FOR PLASMOLYSIS OF SELECTED CELL TYPES

Species and cell type	Cell sap concentration (osM)	Recommended plasmolyticum (osM)	Number of concentration steps	Ref.
Allium cepa, bulb scale epidermis	0.3–0.4	Glucose (1.2), $CaCl_2$ (0.58)	2–5	a
Glycine max, first and second trifoliate leaf mesophyll	0.4–0.6	Mannitol (0.9), with octylguanidine[b]	3–4	c
Hordeum vulgare, coleoptile inner epidermis	0.44	α-Methylglucose (0.4)	5	d
Hordeum vulgare, root cortex parenchyma	0.27–0.36	Sorbitol (0.5)	1–5	e
Lycopersicon esculentum, petiole subepidermis	0.41	$CaCl_2$ (0.78)	4	f
Lycopersicon esculentum, root cortex parenchyma	0.2–0.3	α-Methylglucose (0.64)	3	g
Pisum sativum, stem base subepidermis	0.35–0.48	Mannitol (1.1), $CaCl_2$ (0.5, 0.8)	2–5	h
Quercus rubra, root cortex parenchyma	~0.5	Sucrose (0.59)	4–5	i
Selaginella lepidophylla, leaf mesophyll	~1.2	Mannitol–glucose (1:3 M; 1.65)	4	j
Triticum vulgare, first leaf, leaf sheath subepidermis	0.4–0.5	Sucrose (1.2)	4–5	k
Zea mays, root cortex parenchyma	0.41–0.68	Mannitol (0.67), $CaCl_2$ (0.58)	2	l

[a] J. P. Palta and O. Y. Lee-Stadelmann, *Plant, Cell Environ.* **6,** 601 (1983); E. J. Stadelmann, *Protoplasma* **59,** 14 (1964).

[b] Application of 0.250 mM octylguanidine in the plasmolyticum for 1–2 min.

[c] O. Y. Lee-Stadelmann, I. Chung, and E. J. Stadelmann, *Plant Sci.* **38,** 1 (1985).

[d] O. Y. Lee-Stadelmann, W. R. Bushnell, and E. J. Stadelmann, *Can. J. Bot.* **62,** 1714 (1984).

[e] I. Chung, personal communication.

[f] H. El Attir, Ph.D. Thesis, Univ. of Minnesota, St. Paul, 1986.

[g] C. Redouanne, personal communication.

[h] J. P. Palta and O. Y. Lee-Stadelmann, *Plant, Cell Environ.* **6,** 601 (1983); O. Y. Lee, Ph.D. Thesis, Univ. of Minnesota, St. Paul, 1975.

[i] X.-J. Zhao, E. Sucoff, and E. J. Stadelmann, *Plant Physiol.* **83,** 159 (1987).

[j] D. Nelson, personal communication.

[k] M. M. F. Mansour, O. Y. Lee-Stadelmann, and E. J. Stadelmann, *Physiol. Plant.* (submitted for publication)

[l] M. Carceller, personal communication.

mersed in the medium in a vacuum flask, and infiltrated by application of reduced pressure until all the intercellular air is replaced, usually 2–5 min (see Ref. 27, p. 8).

Cells with cylindrical shape are most suitable for the plasmometric method (see below). Thus, many epidermal cells, subepidermal cells, or parenchyma cells of elongated plant parts such as stems, roots, petioles, or leaf midveins are the most frequently used objects.

Plasmolysis Procedure

Once cell type, optimal osmoticum, and concentration have been decided, plasmolysis is conducted either on a slide or in a perfusion chamber. The progress of plasmolysis and the changes the protoplast undergoes during plasmolytic contraction can be followed by observing while the external medium is exchanged.

Slide Method. The simplest method is to seal two opposite edges of a cover glass placed on a slide with petroleum jelly and to apply a droplet of the new solution on one open side of the cover glass, collecting the previous solution with a filter paper strip on the opposite side.

Perfusion Chamber Method. For better control of the concentration changes a perfusion chamber[28] (see [18] in this volume) connected to a peristaltic pump or similar device is used. Cuttings or strips are usually sufficiently immobilized by the slight pressure of the cover glass. For continuous observation of filamentous plants they must be fixed in their position on the cover glass.[29,30] For unicellular organisms (e.g., desmids or Chrysophyta) and isolated cells of metaphytes, a special perfusion chamber with a wedge-shaped receptacle can be used.[30]

Application of Plasmolysis

Plasmolysis form and time have been used to interpret protoplasmic viscosity and cell wall attachment, and plasmolysis has been used intensively to investigate the protoplasm after alteration of the environment.[2,3] In this section the quantitative measurement of cell solute potential of

[27] S. Strugger, "Praktikum der Zell- und Gewebephysiologie der Pflanze", 2nd ed. Springer-Verlag, Berlin, 1949.
[28] W. Werth, *Protoplasma* **53,** 457 (1961).
[29] W. W. Lepeschkin, *Ber. Dtsch. Bot. Ges.* **26a,** 198 (1908).
[30] W. Gerdenitsch, *Protoplasma* **99,** 79 (1979).

individual cells is described. Furthermore, a procedure to estimate nonsolvent space and protoplasm volume is introduced and the use of plasmolysis as a test for viability briefly mentioned. The measurement of cell permeability using the plasmometric method is explained in the next chapter of this volume [18].

Measurement of Cell Sap Solute Potential

Plasmolysis is the only method available for determining the solute potential of individual cells. Unlike most other methods, plasmolysis measures the true solute potential of the cell sap, excluding interference by solutes present in the cell wall or the intercellular spaces.

Two plasmolytic methods are available for determination of the concentration of osmotically active material in the cell sap in the state of cell relaxation, i.e., when turgor pressure is zero (osmotic ground value O_g). From this value the osmotic pressure P and the solute potential ψ_s are calculated from van't Hoff's law:

$$-\psi_s = P = O_g RT \tag{1}$$

where ψ_s is the solute potential in Pa; O_g, osmotic ground value, in osM m^{-3}; T, absolute temperature in $°K$; R, gas constant in m^3 Pa osM^{-1} $°K^{-1}$; P, osmotic pressure of the cell sap at the temperature T, in Pa.

Method of Incipient Plasmolysis (Plasmolysis Frequency Method). The incipient plasmolysis method can be used for cells of any shape and size which are plasmolyzable. Subsamples of the tissue are exposed to a series of different concentrations of the plasmolyticum. The concentration causing incipient plasmolysis in 50% of the cells when the osmotic equilibrium is reached is considered to be equal to the cell sap concentration.[3] This method cannot be used for cells with an appreciably high wall attachment of the protoplast.

Plasmometric Method. The plasmometric method can only be used for cylindrical cells (parallel longitudinal walls and circular cross section). This method is less sensitive to wall attachment of the protoplasts. The geometry of the cell allows the calculation of cell volume (cylindrical) and volume of the protoplast (cylindrical with two hemishperic ends). Parallel transverse cell walls are desirable but not necessary.

Convex plasmolysis forms are required for this method since protoplast with two hemispheric ends allows measurement of protoplast volume in cylindrical cells. The volume of a cell with a circular cross section is (see Fig. 4)

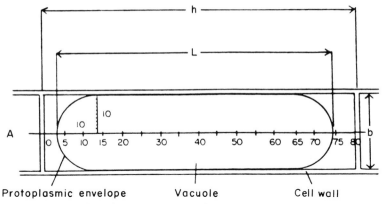

FIG. 4. Schematic view of a cylindrical cell. b, Inner cell width; L, protoplast length; h, inner cell length. (From Ref. 3.)

$$V_z = (\pi b^2/4)h \qquad (2)$$

and the volume of the protoplast, which has two hemispheric ends (see Fig. 4), is

$$V_p = (\pi b^2/4)(L - b/3) \qquad (3)$$

With these values the osmotic ground value becomes

$$O_g = C\frac{V_p}{V_z} = CG = C\frac{L - b/3}{h} \qquad (4)$$

where b is the inner width of the cell; C, concentration of the plasmolyticum; G, degree of plasmolysis (relative protoplast volume); h, inner length of the cell; L, length of the protoplast; O_g, osmotic ground value of the cell; V_p, volume of the protoplast; V_z, volume of the cell.

It is also possible to measure cells containing protoplasts with flat ends or two or more protoplasts. This type of protoplast occurs frequently in elongated cells with a high degree of wall attachment. The formulas for the calculation of the osmotic ground value for these cells are given in Table III.

The solute potential of the cell sap in the relaxed cell state can be calculated from Eq. (1) and the first equation shown in Table III. This method can be accurate to ±0.001–0.002 osM (~0.022–0.044 bars, i.e., 2.2–4.4 Pa, see Ref. 3, p. 161).

TABLE III
FORMULAS FOR THE OSMOTIC GROUND VALUE O_g (CELL SAP CONCENTRATION IN THE
RELAXED CELL) BY THE PLASMOMETRIC METHOD FOR DIFFERENT SHAPES AND
PARTIAL PROTOPLASTS

Form of protoplast[a]	Formula[b]
a	$O_g = C \dfrac{L - \dfrac{b}{3}}{h}$ [c]
b	$O_g = C \dfrac{\sum\limits_{i=1}^{i=n}\left(L_i - \dfrac{b}{3}\right)}{h}$
c	$O_g = C \dfrac{L - \dfrac{b}{6}}{h}$
d	$O_g = C \dfrac{L_1 + L_2 - \dfrac{b}{3}}{h}$
e	$O_g = C \dfrac{L_1 + L_2 - \dfrac{b}{2}}{h}$

[a] a, Single protoplast with two hemispheric ends; b, two partial protoplasts with hemispheric ends: c, protoplast lining the transverse cell wall on one side and with one hemispheric end; d, two partial protoplasts with one hemispheric end each; e, two partial protoplasts, one with one hemispheric end.

[b] C, Concentration of the plasmolyticum in os$M$; L, protoplast length(s); b, inner cell width; h, inner cell length; L, b, and h can be measured in relative units (eyepiece micrometer units); i, number of partial protoplasts.

[c] K. Höfler, Denkschr. Akad. Wiss. Wien, Math. Naturwiss. Kl. **95,** 99 (1918).

Nonsolvent Space and Protoplasm Volume[1]

By plotting protoplast volume versus the inverse osmolarity of plasmolyticum the nonsolvent space (the space which is not occupied by solvent and does not respond to osmotic volume changes) can be determined from the intercept (Fig. 5). Accurate measurements of length and width of the

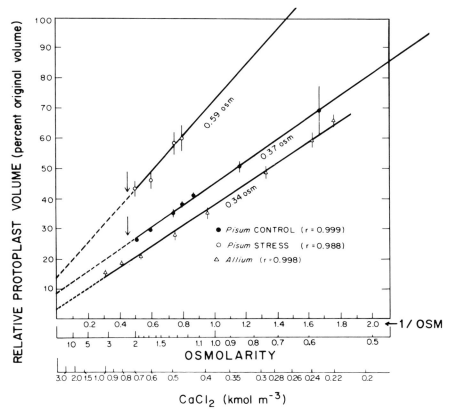

FIG. 5. Boyle–Mariotte–van't Hoff plot of G (relative protoplast volume) versus the inverse osmolar concentration of the plasmolyticum. The linear relationship shows that the protoplast acts as an ideal osmometer. The slope indicates the solute potential, and the intercept the nonsolvent space. Material: adaxial epidermal cells of the bulb scale of *Allium cepa* (mean of eight cells) and subepidermal stem base cells of *Pisum sativum* (mean of five cells). Control, Plants well watered; stress, plants not watered after 1 week of planting. Arrows indicate the concentration at which control cells die and at which protoplasts of cells from stress-adapted plants do not contract further. (From Ref. 1.)

completely rounded protoplast in cylindrical cells of subsamples of the same material are used to calculate G [Eq. (4)] (as shown in Fig. 5).

The nonsolvent space includes the volume occupied by the protoplasm and the nonsolvent space of the vacuole. The protoplasm volume may be directly determined by using centrifugation to accumulate most of the protoplasm as a body of geometrically simple shape with a calculable volume. In some cell types the protoplasm may form a prolapse of approximately spherical shape, containing almost all of the protoplasm, during stronger plasmolysis. Assuming that no change in protoplasm hydration occurs during such experiments, the nonsolvent space of the vacuole can be determined by subtracting the protoplasm volume from the volume of the nonsolvent space of the protoplast.

Cell Viability[31]

Plasmolysis is frequently used to test cell viability. Viable and damaged living cells can be plasmolyzed, but damaged cells will not withstand further osmotic alterations. Therefore, a subsequent deplasmolysis with appropriate decreasing concentration steps is an additional, more reliable proof of viability. Absence of plasmolysis does not necessarily indicate a dead cell. Further confirmation of viability can be obtained through vital staining[32] and demonstration of cytoplasmic streaming. False plasmolysis in dying cells as a result of coagulated protoplasm should not be confused with plasmolysis (see Ref. 9, p. 127).

Acknowledgments

This is Paper No. 2039 of the Scientific Journal Series, Project 80, Agricultural Experiment Station, University of Minnesota, St. Paul, Minnesota 55108. Thanks are due to Insun Chung for technical assistance in preparing the manuscript.

[31] J. P. Palta, J. Levitt, and E. J. Stadelmann, *Cryobiology* 15, 249 (1978).
[32] E. J. Stadelmann and H. Kinzel, *Methods Cell Physiol.* 5, 325 (1972).

[18] Passive Permeability

By EDUARD J. STADELMANN and OKYOUNG LEE-STADELMANN

Introduction

Permeability is a quality of membranes that refers to the ability to permit transport through them. In living cells permeability concerns the passage of water and solutes through cytomembranes. Some solutes, mainly nonelectrolytes, permeate through the membrane passively, i.e., do not require expenditure of metabolic energy within the membrane. Passive permeability of cell membranes for specific nonelectrolytes and water has become a valuable tool for studying the structure of membranes and their function as barriers to transport,[1] but it cannot be measured directly and can only be derived from measuring the rate of the transport process.

Passive permeability largely depends on the cell species. It has been proven to serve as a sensitive indicator for detecting changes in the physiology of the cell, cell alterations by environmental factors (chemical or physical), and cell history, and for determining differences between cell species and varieties.[2,3] Passive permeability is also known to differ for the same cell type in different plant species or varieties.

Only a few methods are available to measure passive permeability of cell membranes. It is generally assumed that the passive permeability of the protoplasm layer surrounding the large central vacuole of a protoplast, is determined mainly by the cell membrane.[4-11] Protoplasts are able to undergo changes in volume in response to osmotic variations in the environment or to permeation of a solute. The volume changes are evaluated to determine the amount of permeator. Osmotic methods have been widely used for water and solute permeability. The recently introduced pressure probe technique[12,13] measures mainly water transport, requires special equipment, and is limited in use to selected plant species and cell types.

This chapter deals with permeability measurements of the protoplasm layer (cell membrane) based on plasmolysis, details of procedures, theoretical treatment, and derivation of formulas can be found elsewhere.[9,10,14-16]

[1] R. Collander, in "Plant Physiology" (F. C. Steward, ed.), Vol. 2, p. 3. Academic Press, New York, 1959.

[2] W. Url, Sitzungsber. Oesterr. Akad. Wiss., Mathem.-Naturwiss. Kl., Abt. 1 171, 259 (1962).

[3] W. Url, Abh. Dtsch. Akad. Wiss. Berlin, Kl. Med. 4a, 18 (1968).

[4] J. Dainty and B. Z. Ginzburg, Biochim. Biophys. Acta 79, 112 (1964).

[5] J. Palta and E. J. Stadelmann, J. Membr. Biol. 33, 231 (1977).

[6] U. Zimmermann and E. Steudle, Biochim. Biophys. Acta 332, 399 (1974); D. E. Hüsken, E. Steudle, and U. Zimmermann, Plant Physiol. 61, 158 (1978).

[7] For earlier reviews of methods for permeability measurements, see V. Grafe, in "Handbuch der biologiche Arbeitsmethoden" (E. Abderhalden, ed.), Vol. 11, Part 2, p. 23. Urban & Schwarzenberg, Berlin, 1924; R. Collander, in "Handbuch der Pflanzenphysiologie" (W. Ruhland, ed.), Vol. 2, p. 196. Springer-Verlag, Berlin and New York, 1956.

[8] K. T. Wieringa, Protoplasma 8, 522 (1930).

[9] E. J. Stadelmann, in "Handbuch der Pflanzenphysiologie" (W. Ruhland, ed.), Vol. 2, p. 139. Springer-Verlag, Berlin and New York, 1956.

[10] E. J. Stadelmann, Protoplasma 46, 692 (1956).

[11] G. W. Scarth, Plant Physiol. 14, 129 (1939).

[12] S. D. Tyerman and E. Steudle, Plant Physiol. 74, 464 (1984).

[13] E. Steudle, this volume [16]; U. Zimmermann, this volume [24].

[14] E. J. Stadelmann, in "Regulation of Cell Membrane Activity in Plants" (E. Marré and O. Ciferri, eds.), p. 3. Elsevier/North-Holland Biomedical Press, Amsterdam, 1977.

[15] E. J. Stadelmann, Methods Cell Physiol. 2, 143 (1966).

[16] E. J. Stadelmann, Eur. Biophys. Congr., Proc., 1st 1971 Vol. 3, p. 147 (1972).

Plasmolytic methods so far have been used to measure endosmotic permeability (permeability of water or solute from outside to the inside of the cell) for a cylindrical cell. In this chapter for the first time, Eqs. (19), (20), and (22) for the exosmotic (from inside to outside) and endosmotic solute and water permeabilities for spherical protoplasts are presented.

Principles of Permeability Measurement

The amount *(m)* of the permeator passing through the membrane in unit time *(t)* is proportional to the driving force *(Z)*, and the membrane area *(A)*. The proportionality characterizes the magnitude of membrane permeability and is called the permeability constant or permeability coefficient (K)[17]:

$$\frac{dm}{dt} = KAZ \quad \text{or} \quad \frac{dm}{dt}\frac{1}{A} = F = KZ \tag{1}$$

where dm is the amount of the permeator migrating through the membrane during the time differential dt and F the flux (dimension product: mol L^{-2} T^{-1}). The dimensions of K are L T^{-1} (length $\times$ time^{-1}), when dm is given in moles, Z in moles L^{-3}, A in L^2, and t in T.

The permeability constant *(K)* for a given membrane and substance is therefore defined as the flux *(F)* under the unit driving force or the amount of permeator migrating through the unit membrane area in unit time under the action of the unit driving force. The basic Eq. (1) can be modified according to the type of driving force, the permeator, and the geometry of the membrane.

Irreversible thermodynamics of membrane transport does not advance insight into membrane function, structure, or composition. It rather considers the membrane as some kind of a black box.[18]

Harmless nonelectrolytes dissolved in water are used as permeators; these solutes permeate almost exclusively through the phospholipid bilayer of the membranes, and the permeability depends on their solubility in the bilayer. Diffusion across the membrane, brought about by the concentration gradient of the permeator, is the only driving force. This gradient is the difference between the concentration of the bulk outside permeator solu-

---

[17] J. Runstroem, *Ark. Zool.* **7**, 1 (1911).
[18] J. G. Kirkwood. *in* "Ion Trnsport across Membranes" (H. T. Clarke and D. Nachmansohn, eds.), p. 119. Academic Press, New York, 1954; G. Eisenman, J. P. Sandblom, and J. L. Walker, Jr., *Science* **155**, 956 (1967); T. Teorell, *Protoplasma* **63**, 336 (1967).

tion, and the (experimentally derived) concentration of the permeator in the vacuole. Its value is the driving force *(Z)* since modifications of the concentration difference, by unstirred layers see (Ref. 1, p. 37; Nernst diffusion layer[19,20]) are negligible for cellular dimensions and the ranges of the water and solute permeability of most plant cells (Ref. 20, p. 779; Ref. 21, p. 305; Ref. 22, p. 295; Ref. 23, p. 143).

The driving force Z has the dimensions moles L^{-3}. More accurately, the dimensions for the driving force are osmoles L^{-3}. Since most permeators act as ideal solutes in the concentration range used in these experiments, there is no difference in their osmotic activity and moles L^{-3} = osmoles L^{-3}. Osmolalities of the solutions used, however, have to be considered for the plasmolyzing solutions (e.g., mannitol and sugars) and also when these solutions are used for producing the driving force in water permeability experiments.

Antropoff (1911) and later Staverman[24] introduced the reflection coefficient as a relative measure for solute permeability. Since there is no interference (coupling) between water and solute flux inside the membrane (because of absence of pores) and the experimental design of a plasmolysis experiment excludes interference outside the membrane (see below), the reflection coefficient becomes essentially a function of the water and solute permeability constants. For most of the permeators used in permeability experiments with the plasmolysis method, the reflection coefficient is too insensitive to be an indicator of solute permeability.[25]

Driving forces for water permeation are gradients of osmotic potential or hydrostatic pressure across the membrane, both of which result in a mass flow of water. Water passes easily through cytomembranes; to a large extent passage is through the phospholipid bilayer or perhaps through both proteins and lipids.[26] In plasmolytic methods the permeated amount of solute or water is inferred from changes in the volume of the protoplast (vacuole with protoplasmic envelope) with time.

[19] W. Nernst, *Z. Phys. Chem.* **47**, 52 (1904).

[20] P. H. Barry and J. M. Diamond, *Physiol. Rev.* **64**, 763 (1984).

[21] J. Dainty, *Adv. Bot. Res.* **1**, 279 (1963).

[22] T. J. Pedley, *J. Fluid Mech.* **107**, 281 (1981).

[23] T. J. Pedley, *Q. Rev. Biophys.* **16**, 115 (1983).

[24] A. V. Antropoff, *Z. Physikal. Chem.* **76**, 721 (1911); A. J. Staverman, *Rec. Trav. Chem. Pays-Bas* **70**, 344 (1951).

[25] J. M. Diamond and E. M. Wright, *Annu. Rev. Physiol.* **31**, 581 (1969).

[26] A. Carruthers and D. L. Melchior, *Biochemistry* **22**, 5797 (1983).

Principles of Plasmolytic Methods

General

When a fully differentiated plant cell is exposed to an external solution of a nonpermeating solute (plasmolyticum) with a solute potential (Ψ_s) lower than the Ψ_s of the cell sap of the relaxed cell, water will leave the vacuole until the vacuole reaches osmotic equilibrium with the external solution. In nonpermeating solute a plant cell, when intact, acts as an almost perfect osmometer.[27] As the protoplast volume decreases, the living protoplast will separate from the cell wall (plasmolysis), and the pressure potential (Ψ_p) becomes zero.

When the plasmolyticum, after reaching osmotic equilibrium, is exchanged with an isosmolal concentration of a permeator (solute), the permeator will enter the vacuole. The permeator will take along water to maintain the previously established equilibrium of the water potential. This water uptake will lead to a volume increase of the vacuole, and this increase indicates the amount of the permeator entering the vacuole. Effects of unstirred layers can be neglected (see above), and the volume contribution of protoplasm is also generally negligible.[27]

Water movement through the membrane does not interfere with solute movement across the membrane, because (1) water and solute use different pathways inside the membrane (molecular diffusion owing to absence of aqueous pores[14]); narrow channels, if present, do not allow interaction of water and solute during passage through the membrane. (2) Such kinds of permeators must be selected so that the water permeability of the membrane is at least 100 times higher than the permeability for the permeator. Thus, the water movement through the membrane is always fast enough to maintain osmotic equilibrium, and the driving force for solute permeability is generated solely by the concentration gradient between vacuole and external solution.

When cells or protoplasts have a simple geometric shape as in many types of plant cells (e.g., cylindrical cells with a plasmolyzed protoplast with two hemispheric ends, plasmometric method; see [17] in this volume) the volume changes of the protoplast can easily be followed. Free protoplasts assume a spherical shape, and the diameter of the sphere and its changes can also be measured with accuracy.

---

[27] J. Palta and O. Y. Lee-Stadelmann, *Plant, Cell Environ.* **6**, 601 (1983); see P. Kornmann, *Protoplasma* **21**, 340 (1934); *ibid.* **23**, 34 (1935).

Assumptions and Limitations of Plasmolytic Methods

The limitations and source of errors have frequently been investigated, and a number of assumptions and conditions have been postulated for assuring reliable permeability values (see Ref. 9, p. 141; Ref. 15, p. 181). The following conditions are occasionally viewed as sources of error: (1) unstirred layers (Ref. 28, p. 482) may be appreciable for thicker tissue pieces, but they are negligible for tissues one or a few cell layers thick of the sort used in plasmolysis experiments (see above); (2) plasmolysis as such was suggested to cause alterations of the permeability values (see Ref. 29, p. 433); evidence indicates, however, that no significant differences in permeability can be found between plasmolyzed and nonplasmolyzed cells.[30] Further support for the absence of permeability changes induced by the plasmolysis process comes from the analysis of protoplast expansion during deplasmolysis in a permeability experiment.

The formula for the permeability constant for endosmotic solute permeability is derived from the assumption that the permeability over the whole protoplast surface is equal, whether or not the protoplast is attached to the cell wall. From this formula it follows that the protoplast expansion must be, for cylindrical cells, proportional with time; the experimental data confirm this time proportionality (Fig. 4b; endosmotic phase, 0–30 min).[31] Water permeability may depend on cell turgor as suggested for *Nitella* and *Valonia*.[32] However, no such turgor dependence was found by other methods or in other species investigated.[5,33]

Experimental Aspects[7]

Selection of Permeators (Nonelectrolytes)

Suitable nonelectrolytes must be selected based on the experimental material and the purpose of the experiment. Most commonly used nonelectrolytes are glycerol, malonamide, the urea series (urea, methylurea, ethylurea, propylurea), and other homologous series.[34] The nonelectrolytes

---

[28] J. S. Boyer, *Annu. Rev. Plant Physiol.* **36**, 473 (1985).

[29] J. Dainty, *in* "The Physiology of Plant Growth and Development" (M. B. Wilkins, ed.), p. 421. McGraw-Hill, New York, 1969.

[30] H. Schmidt, *Jahrb. Wiss. Bot.* **83**, 470 (1936).

[31] E. J. Stadelmann, *Sitzungsber. Oesterr. Akad. Wiss. Math.-Naturwiss. Kl., Abt. 1* **160**, 761 (1951).

[32] U. Zimmermann and E. Steudle, *J. Membr. Biol.* **16**, 331 (1974).

[33] J. Dainty and B. Z. Ginzburg, *Biochim. Biophys. Acta* **79**, 102 (1964); M. Tazawa and N. Kamiya, *Aust. J. Biol. Sci.* **19**, 339 (1966); K. Kiyosawa and M. Tazawa, *Protoplasma* **74**, 257 (1972); E. Steudle, U. Lüttge, and U. Zimmermann, *Planta* **126**, 229 (1975).

[34] E. M. Wright and J. M. Diamond, *Proc. R. Soc. London, Ser. B* **172**, 227 (1969).

must be nontoxic to the cells at hypertonic concentrations and must be nonmetabolites. Some of these substances are available comercially in analytical grade. It is important to use the chemicals without impurities; therefore, for some permeators purification is necessary.[35]

Selection of Tissue and Cell Material

Since permeability measurements by plasmolytic methods require microscopic observation, the cells must be clearly visible. Monolayers (*Allium cepa* adaxial bulb scale epidermis) and tissue strips with two to three cell layers are suitable. The most suitable and frequently used tissues in higher plants are epidermis, subepidermis, parenchyma cells from stem, petiole, midrib tissue of leaf, hypercotyl, coleoptile, and roots. Recently, parenchyma cells from *Triticum vulgare* leaf sheaths were found to be very suitable. Filamentous algae such as *Spirogyra* have often been used for permeability measurements. Leaf mesophyll cells, generally difficult to plasmolyze, will do so after pretreatment with octylguanidine.[36,37] Unicellular ogranisms and isolated protoplasts require immobilization in a special chamber.[38]

Plasmolysis Procedure

Plasmolysis is the basis for permeability measurements; therefore, any cell type selected must be tested for its ability to plasmolyze and for absence of damage. Internal and external conditions affecting the plasmolysis and general techniques are described elsewhere in this volume [17].

Permeability Measurements

Solute Permeability. Solute permeability is most commonly measured with endosmotic permeation. The cells are first plasmolyzed in a nonpermeating solute (preplasmolysis; e.g., $CaCl_2$,[39] sugars, mannitol, equilibrated salt solutions[40]). The tissue section is subsequently transferred to an isosmolal concentration of the permeator (preplasmolysis method), and the rate of deplasmolysis in the permeating solution is recorded. In cylin-

[35] R. Collander, *Physiol. Plant.* **2**, 300 (1949); R. Collander and H. Bärlund, *Acta Bot. Fenn.* **11**, 1 (1933); W. Url, personal communication.

[36] B. E. Gómez-Lepe, O. Y. Lee-Stadelmann, J. P. Palta, and E. J. Stadelmann, *Plant Physiol.* **64**, 131 (1979).

[37] O. Y. Lee-Stadelmann, I. Chung, and E. J. Stadelmann, *Plant Sci.* **38**, 1 (1985).

[38] W. Gerdenitsch, *Protoplasma* **99**, 79 (1979).

[39] J. Levitt and G. W. Scarth, *Can. J. Res.*, **14**, (Sect. C), 267 (1936); F. S. Thatcher, Ph.D. Thesis, McGill University, Montreal (1943); *Am. J. Bot.* **26**, 449 (1939).

[40] W. Url, *Protoplasma* **72**, 427 (1971).

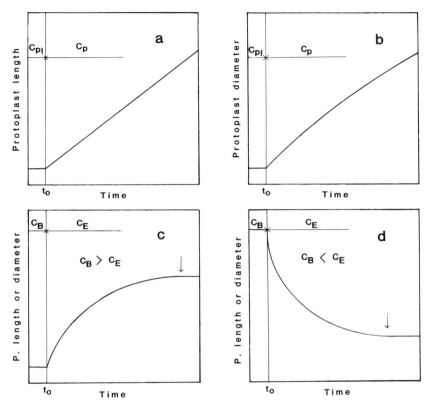

FIG. 1. Schematic diagrams for the change of protoplast length (cylindrical cells) or protoplast diameter (spherical protoplasts) for different kinds of permeability experiments. (a) Time proportionality (protoplast in cylindrical cells); (b) logarithmic without final equilibrium (spherical cells); (c and d) asymptotically reaching final equilibrium. pl, Plasmolyticum or water ($C_p = 0$, in endosmotic water permeability experiments); p, permeator; C_P, permeator solution of the same osmolarity as C_{Pl}; C_B, first concentration of plasmolyticum or concentration of permeator (in exosmotic permeability experiments); C_E second concentration of plasmolyticum. Arrows indicate where final osmotic equilibrium is approached.

drical cells the protoplast length during the expansion phase will increase proportionally with time (Fig. 1a); in spherical protoplasts the expansion will be exponential without attaining a final value (Fig. 1b). Alternatively, the tissue can be directly transferred to a sufficiently hypertonic solution of the permeator (direct transfer method). In this case, the protoplast will first shrink (plasmolyze) according to the osmotic action of the permeator solution until osmotic equilibrium is reached, and then expand (deplasmolyze) with time as the solute enters the vacuole (Fig. 2).[31] Only the expansion phase can be used to calculate the permeability of the permeator

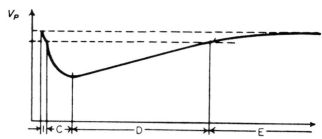

FIG. 2. Schematics of the volume change of the vacuole during an experiment for endosmotic permeation with the plasmolytic method. V_p, Protoplast volume (length); I, initial phase; C, contraction phase; D, expansion phase; E, end phase. Abscissa: time; ordinate: vacuolar volume.[15,40]

because osmotic equilibrium is established between the external permeator solution and the vacuole. The direct transfer method is used for slowly permeating substances such as urea and malonamide; however, it causes unnecessarily long exposure to the permeator. The preplasmolysis method is required for fast permeators (e.g., ethylurea, propylurea) because more measurements can be made during the expansion phase.

For the measurement of exosmotic permeability, it is first necessary to carry out the endosmotic permeability experiment; cells are allowed to deplasmolyze as usual in the endosmotic permeability experiment, in the permeator solution, but not completely: when the protoplast ends in a cylindrical cell almost reach the transverse cell walls, the permeator solution is replaced with an isosmolal solution of a nonpermeating substance (plasmolyticum) to begin the exosmotic permeability experiment. The protoplast will contract asymptotically with time as the permeator diffuses out from the vacuole and the protoplast volume approaches a final value at equilibrium with the concentration of the plasmolyzing solution (Figs. 1d and 4b, exosmotic phase).

For slow permeators (e.g., urea for most cell types) a very simple setup is possible. A well is formed by petroleum jelly rims on a microscope slide containing a droplet of the permeator solution. The epidermis or section, kept in water or preplasmolyzed, is washed on another slide in one or two droplets of the isosmolal permeator solution, and placed in the droplet inside the vaseline well, which is then sealed airtight with a cover glass.

For fast permeators such as methyl- or ethylurea, a perfusion chamber[41,42] is required (Fig. 3). This will allow (1) measurement of the

[41] E. J. Stadelmann, Z. Wiss. Mikrosk. Mikrosk. Tech. 64, 286 (1959).
[42] W. Werth, Protoplasma 53, 457 (1961).

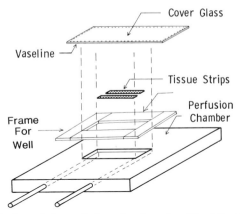

FIG. 3. Exploded view of the perfusion chamber after Werth[42] with two tissue strips. Tissues are held in place by slight pressure of the cover glass.

protoplast length immediately before and after the introduction of the permeator solution and (2) maintenance of a constant external concentration. The flow of solution through the perfusion chamber must be adjustable. Changes of the solution are made by simply changing the flask with the plasmolyticum or permeator solution without stopping the perfusion. In doing so a small air bubble will be introduced into the inlet tube to mark the time when the permeator solution reaches the tissue in the chamber (t_o, Fig. 1).

In measuring the permeability for rapidly permeating solutes, it is important to ensure that the rate of solution flow is sufficiently high (about 4 drops per second at the outflow tube into the waste container) during the first few minutes after exchange of the solution. For endosmotic permeability measurements, the flow rate later can be slow. For exosmotic permeability measurements, a sufficient flow rate of the isosmolar nonpermeating substance must be maintained throughout the entire experiment to keep the outside concentration of the permeator at zero.

Water Permeability. It is helpful first to determine the cell sap concentration of the cell type to be measured. The driving force should be approximately $0.1-0.3$ os$M$ depending on the cell type and the methods selected (plasmometric method or approximation method, see below). The cell or tissue is first brought into osmotic equilibrium with a hypertonic concentration (C_B) of a nonpermeating plasmolyticum, then C_B is rapidly replaced by another concentration (C_E) of the same plasmolyticum. The osmotic gradient will cause a water flow, the rate of which is controlled by the cell dimensions and the water permeability of the protoplasm envelope. For $C_B < C_E$ exosmotic water permeability and for $C_B > C_E$ endosmotic water permeability is determined. The protoplasts asymptotically approach the final volume in concentration C_E (Fig. 1c, d) except in endosmotic experiments with $C_E = 0$ where, for protoplasts in cylindrical cells, the expansion will be proportionate to time (Fig. 1a). Water as deplasmolyticum, however, should be avoided in such experiments because of the great concentration gradient between cell sap and the water as external medium. For all water permeability experiments a perfusion chamber must be used with a sufficiently high solution flow rate.

Partial Concentration Method for Solute Permeability

Rapidly permeating nonelectrolytes (e.g., formamide) often cause cell damage owing to too rapid deplasmolysis in the permeator solution (e.g., 2–3 min). Deplasmolysis time can be considerably extended by mixing the nonelectrolyte solution with a nonpermeating solution (e.g., glucose or mannitol) in such a way that the latter solution is hypertonic (partial concentration method[43]; Ref. 16, p. 149; Ref. 44, p. 191). For example, a cell has a cell sap concentration at incipient plasmolysis corresponding to 0.35 os$M$. This cell is plasmolyzed in 0.8 os$M$ glucose as nonpermeating solute (using concentration steps; see this volume [17]). After reaching osmotic equilibrium, the 0.8 os$M$ glucose solution is replaced by a mixture of 0.5 os$M$ glucose and 0.3 os$M$ formamide. The protoplast will expand until it reaches a volume corresponding to the 0.5 os$M$ outside solution of the nonpermeating solute (glucose). The protoplast length–time diagram becomes asymptotic instead of linear (Fig. 1c).

When the partial concentration method is not sufficient to reduce cell damage and/or recording is still difficult because of rapid changes of the protoplast length, a cold stage may be used since permeability decreases with temperature. Such permeability values, of course, can only be compared with those obtained at the same temperature.

Recording of protoplast length (or protoplast diameter) data points should be sufficiently frequent to obtain good time resolution of the change in protoplast length. For asymptotically progressing changes in protoplast length or diameter (Fig. 1c, d), the intervals can be wider as the final equilibrium is approached (Fig. 4a). Protoplast length or diameter, inner cell width, and cell length should be measured to 3 digits of accuracy. For doing so, the outermost contours of the protoplast ends must be in focus. Several cells can often be measured from a single experiment.

[43] L. Hofmeister, *Bibl. Bot.* **113**, 1 (1935).
[44] O. Y. Lee, Ph.D. Thesis, University of Minnesota, St. Paul (1975).

For slowly permeating substances, where total deplasmolysis time is longer than 20 min, visual recording is satisfactory, if necessary in connection with a tape recorder. For rapidly permeating substances, for simultaneous measurement of a sufficiently large number of cells, and for permanent records, time-lapse microphotography is recommended.[45]

Evaluation of Experimental Data and Calculation of the Permeability Constant

From the protoplast length (in cylindrical cells) or diameter (in spherical protoplasts) and time values a protoplast length–time diagram (L–t diagram) or protoplast diameter–time diagram is plotted as the basis for calculation of the permeability constant. For solute endosmotic permeability with cylindrical cells, the L–t diagram is a straight line unless the cell is damaged or unstable. The line can be drawn either visually or by regression analysis using a suitable portion of the data points. Sample L–t diagrams are shown in Fig. 4. Sometimes the L–t diagram is a straight line only at the beginning of the expansion and becomes irregular later. In these cells the straight portion of the L–t diagram may still be evaluated. Cells with an irregular L–t diagram cannot be used.

For water permeability, the L–t diagram is asymptotic (Fig. 4a).[46] Osmotic equilibrium between protoplast and external solution is usually reached over a time period of 5–20 min, yielding a constant protoplast length. When the protoplast length changes continuously with time (either increasing or decreasing) and does not reach equilibrium, either the cell is damaged or the solute (plasmolyticum) is entering the cell. Normally a curve can be interpolated for the asymptotic graph (Fig. 4a). The values for L_1, t_1, and L_2, t_2 to be used in the formula to calculate the permeability constant are taken from this interpolated curve, instead of actual points, and are generally 25–33% and 67–80% of the values for full protoplast expansion.[47]

Formulas for the Permeability Constant. The geometry of the protoplast of cylindrical cells or spherical protoplasts, free or encased in the cell wall, lead to simple formulas for surface area and volume.

I. Cylindrical protoplasts with circular cross section and two hemispheric ends (see this volume [17]):

$$A = \pi bL \qquad \text{[see Ref. 48]} \qquad (2)$$

[45] E. J. Stadelmann, *Protoplasma* **59**, 14 (1964).
[46] E. J. Stadelmann, *in* "Food, Fiber and the Arid Lands" (W. G. McGinnies, B. J. Godman, and P. Paylore, eds.), p. 337. Univ. of Arizona Press, Tucson, 1971.
[47] V. Pedeliski and E. J. Stadelmann, *Protoplasma* **82**, 379 (1974).

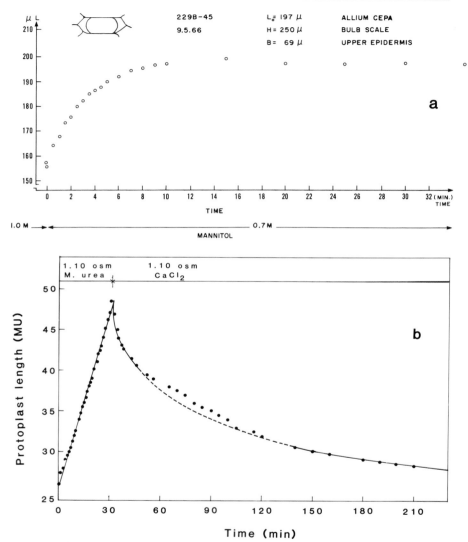

FIG. 4. Sample $L-t$ diagrams for protoplasts of cylindrical cells. (a) Endosmotic water permeability experiment[46]; (b) solute permeability [combined endosmotic and exosmotic experiment with methylurea (M. urea); subepidermal cells from the sheath of the first leaf of *Triticum aestivum;* unpublished data from L.-K. Guo and R.-G. Zhou.

$$V = \frac{\pi b^2}{4} (L - b/3) \tag{3}$$

II. Spherical protoplasts

$$A = \pi d^2 \tag{4}$$

$$V = \pi d^3 \tag{5}$$

With these formulas the dependent variables dm (moles), A (cm^2), and Z (moles/cm^3) from Eq. (1) can be calculated and expressed by a single variable (L or d); Thus, Eq. (1) can be integrated for the specific permeation process, and the permeability constant can be derived.

I. Cylindrical protoplasts with circular cross section and two hemispheric ends (see this volume [17])

A. Endosmotic solute permeation. Use Eq. (2) and

$$dm = C \frac{\pi b^2}{4} dL \qquad \text{[see Ref. 40]} \tag{6}$$

$$Z = C \left(\frac{L_o - b/3}{L - b/3} \right) \qquad \text{[see Ref. 40]} \tag{7}$$

$$K_s = \frac{b}{4} \frac{L_2 - L_1 - (b/3) \ln (L_2/L_1)}{(L_o - b/3)(t_2 - t_1)} \qquad \text{[see Ref. 40]} \tag{8}$$

B. Endosmotic solute permeation, partial concentration method. Use

$$dm = (C + E) \frac{\pi b^2}{4} dL \tag{9}$$

$$Z = E \left(\frac{L_E - b/3}{L - b/3} - 1 \right) \tag{10}$$

$$K_s = \frac{b}{4} \frac{C + E}{EL_E} \frac{1}{t_2 - t_1} \left[(L_E - b/3) \ln \frac{L_E - L_1}{L_E - L_2} - (b/3) \ln(L_2/L_1) \right]$$
$$\text{[see Ref. 49]} \tag{11}$$

C. Exosmotic solute permeation. Use Eqs. (2), (6), and

$$Z = k = C \left(1 - \frac{L_o - b/3}{L - b/3} \right) \tag{12}$$

[48] W. W. Lepeschkin, *Ber. Dtsch. Bot. Ges.* **26a**, 198 (1908); K. Höfler, *Jahrb. Wiss. Bot.* **73**, 300 (1930); H. Schmidt, *ibid.* **83**, 470 (1936).

[49] From Ref. 16, L_E substituted for R; corrected and rearranged.

$$K_s = \frac{b}{4} \frac{1}{L_o(t_2 - t_1)} \left[(L_o - b/3) \ln \frac{L_2 - L_o}{L_1 - L_o} + (b/3) \ln(L_2/L_1) \right] \quad (13)$$

D. Endosmotic and exosmotic water permeation
1. For $C_E > 0$ and $C_B > C_E$ or $C_B < C_E$, use

$$dm = \frac{\pi b^2}{4} \frac{dL}{18} \quad \text{[see Ref. 50]} \quad (14)$$

$$Z = C_E \left(1 - \frac{L_E - b/3}{L - b/3} \right) \quad \text{[see Ref. 50]} \quad (15)$$

$$K_w = \frac{1}{72} \frac{b}{C_E L_E(t_2 - t_1)} \left[(L_E - b/3) \ln \frac{L_1 - L_E}{L_2 - L_E} + (b/3) \ln(L_1/L_2) \right]$$
$$\text{[see Ref. 50]} \quad (16)$$

2. For $C_E = 0$, use

$$Z = C_o \left(\frac{L_o - b/3}{L - b/3} \right) \quad (17)$$

and with Eqs. (2) and (14),

$$K_w = \frac{1}{72} \frac{b}{C_o} \left[\frac{L_2 - L_1 - (b/3) \ln(L_2/L_1)}{(L_o - b/3)(t_2 - t_1)} \right] \quad \text{[see Ref. 50]} \quad (18)$$

II. Spherical protoplasts
A. Endosmotic solute permeation. Use Eqs. (4), (5), and

$$Z = C \frac{d_0^3}{d^3}$$

$$K_s = \frac{1}{8} \frac{d_2^4 - d_1^4}{d_0^3(t_2 - t_1)} \quad \text{[see Ref. 11]} \quad (19)$$

B. Exosmotic solute permeation. Use Eqs. (4), (5), and

$$Z = C_E \left(1 - \frac{d_E^3}{d^3} \right)$$

$$K_s = \frac{1}{2(t_2 - t_1)} \left[d_2 - d_1 + \frac{d_E}{3} \left(\frac{B}{2} + A \right) \right] \quad (20)$$

For A and B, see Eqs. (23) and (24).
C. Endosmotic and exosmotic water permeation. Use Eq. (4) and

$$Z_E = C_E \left(1 - \frac{d_E^3}{d^3} \right) \quad (21)$$

[50] E. J. Stadelmann, *Protoplasma* 57, 660 (1963).

$$K_w = \frac{1}{36} \frac{1}{C_E(t_2 - t_1)} \left[d_2 - d_1 + \frac{d_E}{3} \left(\frac{B}{2} + A \right) \right] \qquad \text{[see Ref. 51]} \qquad (22)$$

$$A = \sqrt{3} \left(tg^{-1} \frac{2d_2 + d_E}{d_E\sqrt{3}} - tg^{-1} \frac{2d_1 + d_E}{d_E\sqrt{3}} \right) \qquad (23)$$

$$B = \ln \left[\frac{(d_2 - d_E)^2}{(d_1 - d_E)^2} \frac{d_E^2 + d_E d_1 + d_1^2}{d_E^2 + d_E d_2 + d_2^2} \right] \qquad (24)$$

In the above equations dm is the amount of permeator (solute or water), in moles; A, surface area of the protoplast, in cm³; b, inner cell width in cylindrical cells, in cm; C, external concentration of the permeating solute, in moles/cm³; C_E, external concentration of the plasmolyticum at which the protoplast expansion (or contraction) will be observed and a final osmotic equilibrium will be reached, for water permeability and exosmotic solute permeability experiments, in moles/cm³; d, diameter of spherical protoplast, in cm; d_E, final diameter of a spherical protoplast in the concentration C_E, in cm; d_o, diameter of a spherical protoplast in a concentration of plasmolyticum which is isosmolar to the concentration of the permeator used, in cm; d_1 and d_2, diameter of spherical protoplasts at t_1 and t_2, in cm; E, concentration of the nonpermeating component (plasmolyticum) of a solution mixture containing a permeator and a plasmolyticum, in moles/cm³; k, concentration of the original osmotically active material in the cell sap vacuole, in moles/cm³; K_s, solute permeability constant, in cm/sec, indicating the amount of the permeator (in moles) penetrating through the unit membrane area (1 cm²) in unit time (1 sec) when the driving force is unity (1 mole/cm³); K_w, water permeability constant, in cm/sec, indicating the amount of water (in moles) penetrating through the unit membrane area (1 cm²) in unit time (1 sec) when the driving force is unity (1 mole/cm³, equivalent to 1315 atm; see Ref. 20, p. 788); L, protoplast length in cm; ln, natural logarithm; L_E, protoplast length reached in the final equilibrium in a water permeability experiment or an endosmotic solute permeability experiment, in cm; L_o, protoplast length in a cylindrical cell in a concentration of a plasmolyticum which is isosmolal to the concentration of the permeator used, in cm; L_1 and L_2, protoplast lengths in a cylindrical cell, measured at the times t_1 and t_2 during protoplast expansion (in exosmotic permeability experiments, contraction), in cm; t_1 and t_2, times of the measurement, in sec; V, protoplast volume, in cm³; Z, driving force, in moles/cm³; ¹/₁₈, ¹/₃₆, ¹/₇₂, numerical factors with units moles/cm³.

⁵¹ B. Lucké, H. K. Hartline, and M. McCutcheon, *J. Gen. Physiol.* **14**, 405 (1931).

Approximation Methods

Endosmotic Solute Permeability. For homogeneous cell material, same driving force, and measuring times $t_1 = t_o$ (time of first contact of the cell with the permeator solution) and $t_2 = T$ (deplasmolysis time), the magnitudes of $L_2 - L_1 = L_T - L_o$, L_o, and b will be the same in all experiments, and therefore solute permeability calculated by Eq. (8) will be inversely proportional to the deplasmolysis time (Fig. 5). The latter can be used as a

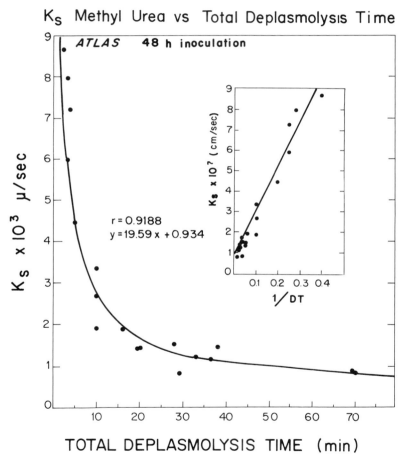

FIG. 5. Correlation between solute permeability constant and deplasmolysis time (inner epidermal cells of the coleoptile of *Hordeum vulgare* 'Atlas'; permeator: methylurea; O. Y. Lee-Stadelmann, unpublished, 1985). The inset shows the proportionality between the inverse of the deplasmolysis time and the permeability constant. From the regression line an empirical formula can be derived to calculate the permeability constant from the deplasmolysis time observed experimentally in the same material.

relative measurement of permeability. Introducing t_o, T, L_o, $L_T = h + \frac{b}{3}$ [see Eq. (25) and Fig. 6], K_s can be calculated with Eq. (8).

Water Permeability. For approximate values of K_w the total deplasmolysis time method can be used; this method does not require recording the time course of deplasmolysis (Fig. 6). In Eq. (16) the deplasmolysis time T is substituted for $t_2 - t_1$ (Fig. 6). Correspondingly, $L_1 = L_B$, the protoplast length in the plasmolyzing solution at $t_1 = 0$, and

$$L_2 = h + \frac{b}{3} \tag{25}$$

the equivalent protoplast length for the cell volume, have to be introduced. The protoplast length in the final concentration C_E, L_E, can be calculated from the protoplast length in the plasmolyticum C_B:

$$L_E = (L_B - b/3)\frac{C_B}{C_E} + b/3 \tag{26}$$

Equation (16) therefore becomes modified for the deplasmolysis time method

$$K_w = \frac{1}{72}\frac{b}{C_E L_E T}\left[(L_E - b/3)\ln\frac{L_B - L_E}{L_2 - L_E} + (b/3)\ln(L_B/L_2)\right]$$

$$\text{[see Ref. 44]} \quad (27)$$

where L_2 and L_E are substituted from Eqs. (25) and (26).

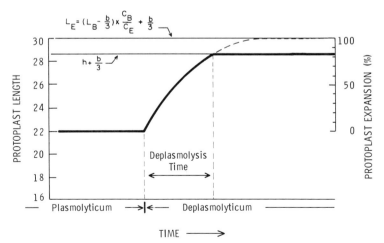

FIG. 6. Schematic protoplast length–time diagram for the total deplasmolysis time method.[44]

With uniform cell sizes the average deplasmolysis time, average cell length *(h)*, average protoplast width *(b)*, and average protoplast length (L_B and L_E) can be determined, and an average water permeability constant can be calculated for the whole tissue. Otherwise the deplasmolysis time T, h, L_B, and L_E must be determined individually for each cell.

Sample Experiment and Calculation for Determination of Endosomotic Permeability[15]

Material. Outer epidermis of *Hordeum vulgare* coleoptile, 7 days old, with the inner epidermis stripped off, is used. The tissue is kept in 10 mM Ca(NO$_3$)$_2$ for 24 hr and stepwise plasmolyzed in methylglucose (0.1, 0.2, 0.3, 0.4 M for 20 min in each concentration), finally in a 0.5 M (0.56 os$M$) solution for 30 min. Methylglucose is replaced by methylurea (0.6 os$M$) at $t = 0$. The slight discrepancy in the concentrations between the final plasmolyzing solution (0.56 os$M$) and the permeator solution (0.6 os$M$) is unimportant since it will take only a few minutes to reestablish the osmotic equilibrium and L_o is extrapolated from the regression line. The first measurement begins at $t = 3.42$ min and is recorded by time-lapse microphotography.[45] The data for cell #10 are tabulated below.

Frame no.	Time (min)	Length (cm)*
1	3.42	3.24
2	6.50	3.29
3	9.33	3.40
4	12.5	3.50
5	15.0	3.60
6	20.0	3.70
7	23.25	3.77
8	27.75	3.89
9	34.17	4.08

* Length is measured on the enlargements prepared from the frames. A length of 1 cm on the enlargement corresponds to 103 μm.

These values, plotted as an $L-t$ diagram to check the regularity of expansion (e.g., see Fig. 4b, left portion, for 1.10 os$M$ urea) yield from the regression line values to use in Eq. (8):

t_1	L_1	t_2	L_2	L_0	b^*
13.67	3.52	23.92	3.80	3.15	0.42

so that for the permeability constant K_s the result is $K_s = 1.57 \times 10^{-7}$ cm sec^{-1} = 1.57×10^{-9} m sec^{-1}.

TABLE I

PERMEABILITY CONSTANTS (K_s) FOR SOME COMMONLY USED
NONELECTROLYTES AND WATER (K_w)[a]

Permeator	MW	P^{52}	$K (\times 10^9)$ in m sec^{-1}			
			Allium cepa[53] bulb scale epidermis	Pisum sativum[44] stem base subepidermis	Hordeum vulgare[54] coleoptile epidermis	Quercus rubra[55] root cortex
Urea	60.1	0.00776	0.64	0.486	0.227	0.229
Methyl urea	74.0	0.0615	10.2	1.93	1.09	2.60
Ethyl urea	88.1	0.182	—	11.9	4.3	10.95
Propyl urea	102.1	0.741	—	—	36.2	—
Water	18.0	—	5,790	25,600	29,500	35,400

[a] Measured by the plasmometric method (endosmotic permeability); MW, molecular weight; P, partition coefficient in octanol–water.

Permeability Constants

Table I[52–55] gives sample values for permeability constants for frequently used permeators.

Analysis of Membrane Structure Based on Permeability Measurements

The permeability constants (K_s and K_w) are sensitive probes of the permeability of the protoplasm layer. Preliminary results suggest that the main barrier to passive permeation is the plasmalemma. When measured for the same permeator the permeability constants often reveal impressive differences between cells of different kinds, from different plant species,[3] and between cells under different internal and external conditions. Specifically, permeability is a sensitive indicator of harmful environmental factors, since cell damage generally first results in membrane damage which leads to a drastic change in permeability. In plasmolyzed cylindrical cells irregularities in the time course of the protoplast dilatation (i.e., lack of proportionality with time in an experiment for endosmotic solute permeability) clearly indicate membrane damage.

[52] C. Hansch and A. Leo, "Substituent Constants for Correlation Analysis in Chemistry and Biology." Wiley, New York, 1979; C. Hansch, personal communication.
[53] J. P. Palta and E. J. Stadelmann, Physiol. Plant. **50,** 83 (1980).
[54] O. Y. Lee-Stadelmann, W. R. Bushnell, C. M. Curran, and E. J. Stadelmann, unpublished.
[55] X.-J. Zhao, E. Sucoff, and E. J. Stadelmann, Plant Physiol. **83,** 159 (1987).

Thermodynamic parameters (activation energy,[56] incremental free energy, entropy, and enthalpy), and partiality of the membrane[55,57,58] can be derived from permeability experiments using a permeator series which may provide additional information on membrane composition. Furthermore, attempts have been made to derive the molecular mechanisms of the permeation process, the kind of membrane lipids involved, and their fluidity[59] from the permeability for a homologous series of selected permeators, such as the solutes in Table I, varying in their degree of polarity or lipophily. Graphing the inverse of the permeability constant versus the inverse distribution coefficient lipid–water (e.g., octanol–water) may allow one to discriminate between the contribution of the lipid–water interphase and the lipid core portion of the membrane toward the total permeation resistance.[55,57]

The existence of two membranes, plasmalemma and tonoplast in series, no longer presents the absolute impediment to attributing an observed permeability alteration to the plasmalemma or the tonoplast[4,60]; methods are now available for the selective destruction of one of the membranes (plasmalemma) *in situ*. Furthermore, testing the effect on permeability for substances which are known to bind specifically with membrane surface proteins without permeating through the membrane (e.g., PCMBS, DIDIT) may allow analysis of the interaction between membrane proteins and lipids and between permeator and membrane molecules in more detail, and enhance understanding of molecular mechanisms of passive membrane permeability in plant cells *in vivo*.[61] Eventually it should become possible to obtain information on membrane lipid composition from permeability experiments with a few selected test substances as permeators.

Acknowledgments

This is Paper No. 14944 of the Scientific Journal Series, Project 80, Agricultural Experiment Station, University of Minnesota, St. Paul, Minnesota 55108. The authors wish to thank Insun Chung for technical and editorial assistance in preparing the manuscript and L.-K. Guo and R.-G. Zhou, Hebei Academy of Agriculture, Shijiazhuang, Hebei, China, for the data for Fig. 4b.

[56] V. Wartiovaara and R. Collander, *Protoplasmatologia* **2,** Part C8d (1960); S. P. Leibo, *J. Membr. Biol.* **53,** 179 (1980); N. Binslev and E. M. Wright, *J. Membr. Biol.* **29,** 265 (1976).
[57] A. Fennel and P. H. Li, *Plant Physiol.* **80,** 470 (1986).
[58] B. J. Zwolinski, H. Eyring, and C. E. Reese, *J. Phys. Colloid Chem.* **53,** 1426 (1949).
[59] J. M. Diamond and E. M. Wright, *Proc. R. Soc. London, Ser. B* **172,** 273 (1969); J. A. Dix, D. Kivelson, and J. M. Diamond, *J. Membr. Biol.* **40,** 315 (1978); E. Orbach and A. Finkelstein, *J. Gen. Physiol.* **75,** 427 (1980).
[60] S. Wendler and U. Zimmermann, *J. Membr. Biol.* **85,** 121 (1985).
[61] I. Chung, O. Y. Lee-Stadelmann, and E. J. Stadelmann, *Plant Physiol.* **80**(4), 90 (1986).

[19] Compartmentation in Root Cells and Tissues: X-Ray Microanalysis of Specific Ions

By André Läuchli and Patrick J. Boursier

Introduction

Electron probe X-ray microanalysis is a method of elemental analysis at the microscopic level. Since the early 1960s it has been used to determine the location and compartmentation of major inorganic ions and to estimate their concentrations in cells, and more recently in subcellular compartments, of both plant and animal systems. For convenience this method is usually referred to as X-ray microanalysis. The principles of the method and instrumentation are treated in the monograph by Heinrich.[1] X-Ray microanalysis is used to determine the elements with atomic number 6 (C) to 92 (U). Using X-ray microanalysis in conjunction with a scanning electron microscope equipped with an energy-dispersive detector system, the range of detectable elements in biological specimens extends essentially from Na to Fe and metals of similar atomic number, and encompasses particularly the biologically important elements Na, K, Mg, Ca, P, S, and Cl.

The absolute sensitivity of the method is of the order of 10^{-19} g. For frozen hydrated plant specimens the sensitivity limits expressed in units of concentration have been estimated to be about 1 mol m^{-3} for K$^+$ and Cl$^-$ and about 5 mol m^{-3} for Na$^+$.[2] The range of spatial resolution depends on both the electron beam diameter at the specimen surface and depth of electron penetration and lateral electron spread in the specimen, the latter being a function of accelerating voltage of the electron gun and specimen density. At an accelerating voltage of 10 kV (frequently used for bulk, frozen hydrated specimens) the spatial resolution was estimated to be about 2 to 3 μm for the biologically important elements.[2] This estimate of resolution has essentially been confirmed in other studies.[3,4] Thus, the resolution of X-ray microanalysis provides the opportunity for analyses in

[1] K. F. J. Heinrich, "Electron Beam X-Ray Microanalysis." Van Nostrand-Reinhold, New York, 1981.
[2] M. G. Pitman, A. Läuchli, and R. Stelzer, *Plant Physiol.* **68**, 673 (1981).
[3] A. T. Marshall and R. J. Condron, *J. Microsc. (Oxford)* **140**, 109 (1985).
[4] A. W. Robards and K. Oates, *J. Exp. Bot.* **37**, 940 (1986).

plant cell compartments, i.e., cytoplasm, vacuole, and chloroplast, but measurements at the level of the cell wall are at the limits of resolution.

The technique of specimen preparation used is crucial to the application of X-ray microanalysis to biological specimens. Ideally specimens should be prepared to minimize loss and redistribution of diffusible elements and to maintain cellular structures in the *in vivo* condition throughout the preparation and analysis. In addition, a relatively flat specimen topography must be obtained. Topographical features on the fracture face of frozen specimens severely limit quantitative analysis and give rise to spurious X-ray signals.[5,6] Plant specimens pose a particular problem because of their high water content and heterogeneity of cell structures (cell walls, vacuoles).

There are basically three preparation techniques in use for X-ray microanalysis of biological specimens, i.e., (1) preparation of frozen hydrated bulk specimens,[7] (2) preparation of frozen hydrated sections,[8] and (3) freeze substitution.[9] The frozen hydrated bulk method has been used extensively and successfully on plant specimens (see, for example, Table I). It is by far the simplest and fastest preparation method and maintains the specimens in the hydrated state.[10] The main disadvantage of the frozen hydrated bulk method arises from limitations due to uneven surface topography created during the preparation of the sample. Although the use of frozen hydrated thin sections would provide improved spatial resolution in comparison with frozen hydrated bulk specimens, frozen thin sectioning in general has not been applied successfully to plant samples. Freeze substitution permits the preparation of thin, dehydrated sections for high-resolution X-ray microanalysis, but the possibility of redistribution of diffusible elements during freeze substitution in such solvents as acetone or ether cannot be ruled out entirely.

We have developed a cryofracturing technique for the preparation of frozen hydrated bulk plant specimens for X-ray microanalysis. This new technique has been found to preserve subcellular structures, reduce significantly the occurrence of topographical inhomogeneities, and has been applied extensively to both root and shoot samples.

[5] A. Boekestein, A. L. H. Strols, and A. M. Stadhouders, *Scanning Electron Microsc.* **2**, 321 (1980).
[6] P. Echlin and S. E. Taylor, *J. Microsc. (Oxford)* **141**, 329 (1986).
[7] A. T. Marshall, *in* "X-Ray Microanalysis in Biology" (M. A. Hayat, ed.), p. 167. University Park Press, Baltimore, Maryland, 1980.
[8] A. J. Sauberman, P. Echlin, P. D. Peters, and R. Beeuwkes III, *J. Cell Biol.* **88**, 257 (1981).
[9] D. M. R. Harvey, *J. Microsc. (Oxford)* **127**, 209 (1982).
[10] P. Echlin, C. Lai, and T. L. Hayes, *Scanning Electron Microsc.* **2**, 489 (1981).

TABLE I
Ion Compartmentation in Root Cells and Tissues by Means of X-Ray Microanalysis

Species	Elements determined	Preparation technique	Compartments	Data presentation	Ref.
Hordeum vulgare	Na, Mg, P, S, Cl, K	Frozen hydrated bulk	Vacuole, cytoplasm	$p - b$,[a] element ratios	b
Lemna minor	Na, Mg, P, S, Cl, K, Ca	Frozen hydrated bulk	Cell	p/b, element ratios	c
Hordeum vulgare	Na, Mg, P, S, Cl, K, Ca	Frozen hydrated bulk	Cell	$p - b$, % of total counts	d
Lupinus luteus	Na, P, S, Cl, K	Frozen hydrated bulk	Vacuole	$p - b$, element ratios	e
Atriplex spongiosa	Na, P, S, Cl, K	Frozen hydrated bulk	Vacuole, cytoplasm	$p - b$, element ratios	f
Zea mays	Na, Cl, K	Freeze substitution, dry sections (~ 100 nm)	Vacuole, cytoplasm, cell wall	$p - b$, organic standards, ion concentrations (mol m^{-3})	g
Plantago coronopus	Na, Mg, P, S, Cl, K, Ca	Frozen hydrated bulk	Vacuole	$p - b$, % of total counts	h
Citrus genotypes	Na, Cl, K	Frozen hydrated bulk	Vacuole	"Plant tissue standards," ion concentrations (mol m^{-3})	i

[a] p, Peak; b, background.
[b] M. G. Pitman, A. Läuchli, and R. Stelzer, *Plant Physiol.* **68**, 673 (1981).
[c] P. Echlin, C. E. Lai, and T. L. Hayes, *J. Microsc. (Oxford)* **126**, 285 (1982).
[d] A. H. Markhart III and A. Läuchli, *Plant Sci. Lett.* 25, 29 (1982).
[e] R. F. M. Van Steveninck, M. E. Van Steveninck, R. Stelzer, and A. Läuchli, *Physiol. Plant.* **56**, 465 (1982).
[f] R. Storey, M. G. Pitman, and R. Stelzer, *J. Exp. Bot.* **34**, 1196 (1983).
[g] D. M. R. Harvey, *Planta* **165**, 242 (1985).
[h] D. M. R. Harvey, R. Stelzer, R. Brandtner, and D. Kramer, *Physiol. Plant.* **66**, 328 (1986).
[i] R. Storey and R. R. Walker, *J. Exp. Bot.* **38**, 1769 (1987).

Cryofracturing Technique for Preparation of Frozen Hydrated Bulk Plant Specimens for X-Ray Microanalysis

Specimen Preparation. Small pieces of balsa wood are cut with a clean razor blade, immersed in Tissue-Tek (Miles Scientific), an embedding medium for frozen tissue which does not penetrate the specimens, and placed under moderate vacuum. After 24 hr the wood is saturated with the embedding medium. To prepare plant samples a piece of balsa wood is placed into a groove machined into the top of the copper specimen holder and then secured into place by applying a small drop of glue (Krazy-Glue). A longitudinal slit is cut into the wood with a clean surgical blade (No. 11) and subsequently filled with the embedding medium using a 30-gauge hypodermic needle. Plant tissue is excised and cut into small segments ($\sim 1.5 \times 25.0$ mm), and the segments are quickly placed into the embedding medium-filled groove in the balsa wood. A bead of Tissue-Tek is applied around the plant specimens on top of the wood. Two clean surgical blades are set onto the flat portion of the copper holder and gently slid up against the plant tissue. The blades are quickly secured into place by applying a single drop of glue.

The bead of embedding medium situated between the tips of the blades and the wood supports the plant tissue below the tip of the blades and directs the movement of the blades during subsequent fracturing of the tissue. The glue holds the surgical blades securely to the copper holder at room temperature, but only a minimal force is required to release the surgical blades from the holder once all the components of the tissue holder are at liquid N_2 temperature. The blades are positioned on each side of the plant specimen at the same height thus forcing the fracture line during subsequent fracturing to occur precisely between the two tips of the blades. In addition, this configuration allows the fracture to proceed through the tissue at any desired location of the specimen.

A schematic diagram of the specimen holder in cross section (Fig. 1) shows the positioning of the balsa wood, plant specimen, and surgical blades. The primary function of the piece of wood is to support the plant specimen in an upright position and a manner which would not compress cells but facilitate fracturing the tissue in a horizontal plane. In addition, individual wood pieces may be altered to hold specimens of different shapes and sizes by varying slit width and depth. Saturating the wood with Tissue-Tek increases the electrical conductivity of the wood, keeps the specimen in good thermal contact with the copper block, and maintains it in a rigid position after freezing. Balsa wood treated in this manner is stable in the presence of the primary electron beam. Tissue-Tek alone did not work sufficiently well as an embedding medium, and droplets of it tended

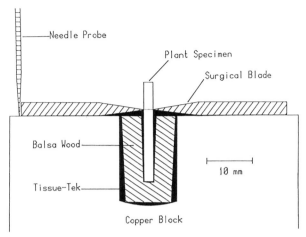

FIG. 1. Schematic diagram of the assembly for cryofracturing frozen hydrated bulk plant specimens.

to contract to a random shape in liquid N_2, providing little or no control over the eventual fracture plane through the tissue. Plant specimens not properly supported may break below the surface, causing insufficient electrical and thermal conductivity and thus charging of the specimen.

After the plant specimens are satisfactorily positioned in the specimen holder, the copper block is quickly submersed into a liquid N_2 slurry ($-196°$) and transferred under vacuum to the working chamber of the equipment for specimen preparation (Emscope SP 2000). This equipment is fitted with devices for freezing, fracturing, sublimating, coating, and transferring specimens under vacuum to the scanning electron microscope for eventual X-ray microanalysis.

Once the specimen holder is inside the working chamber maintained at liquid N_2 temperature, a stainless steel needle probe is placed adjacent to one of the surgical blades (see Fig. 1). The handle of the needle probe is given a sharp push that causes the end of the probe inside the chamber to cleave the surgical blade from its secured position, thus forcing a line of fracture through the plant specimen. The second surgical blade is cleaved in a similar manner. The fracture face of the plant samples is then lightly etched using a substage heater ($-95°$) to remove superficial ice layers. Finally, the samples are coated with a thin film of evaporated chromium to prevent surface charging.[6]

Coating frozen biological specimens is critical. In addition to chromium, carbon coating is used frequently. The latter method is preferable to chromium coating if elements are to be determined whose X-ray energy maxima are close to that of chromium. For light element X-ray microanalysis at low temperature, beryllium is ideal for coating, and it has higher electrical and thermal conductivity and lower absorption of X-rays than carbon, and chromium in particular.[11] Alternatively, disks of beryllium instead of copper holders can be used, because the contribution of beryllium to the continuum radiation (= background radiation) is low, and excellent thermal and electrical conductivity of the specimen is achieved.[12] However, if disks of beryllium are used, coating of the specimens, for example, with beryllium or carbon, is still necessary. In addition, adequate precautions have to be taken in working with beryllium because of its toxicity.

Analytical Conditions. Following coating the specimens are transferred under vacuum to the cold stage of a Hitachi S-800 scanning electron microscope equipped with a field emission gun and kept at liquid N_2 temperature throughout analysis. The scanning electron microscope is also equipped with a Kevex 8000 energy-dispersive X-ray detector system.

The samples are viewed and analyzed at 12.0 kV accelerating voltage and a magnification of 3000× using the reduced raster area (4.0 μm). Counting rates are generally between 150 and 250 counts/sec (cps), and counts are collected for a total counting time of 200 sec. Working distance is 15.0 mm, and the emission current is approximately 10.0μA. The detector angle is at 10° above horizontal, and the specimen is tilted 45° to the primary electron beam. X-Ray energies between 0.000 and 10.110 KeV are recorded and stored using a minidisk memory system (Kevex 8000, PDP-11 microcomputer).

Structural Preservation and Analytical Resolution. Blade tissue from leaves and root specimens of maize were prepared for X-ray microanalysis using the cryofracturing technique described above. A consistently improved structural preservation of the specimens was obtained in comparison with earlier published reports. A typical example of a scanning electron micrograph from the fracture face of a maize leaf specimen maintained at liquid N_2 temperature is shown in Fig. 2. In this leaf cell are clearly delineated the cell wall, the vacuole with the tonoplast surrounding it, and the cytoplasm with several organelles (chloroplasts, nucleus, mitochondria). The chloroplast (Fig. 2, upper right) was fractured longitudinally and exhibits suborganelle structures, probably regions of densely stacked thylakoids. To the best of our knowledge, this is the first documentation of a cryofractured plant specimen prepared for X-ray microanalysis with struc-

---

[11] A. T. Marshall and D. Carde, *J. Microsc. (Oxford)* **134**, 113 (1984).
[12] A. Läuchli, A. R. Spurr, and R. W. Wittkopp, *Planta* **95**, 341 (1970).

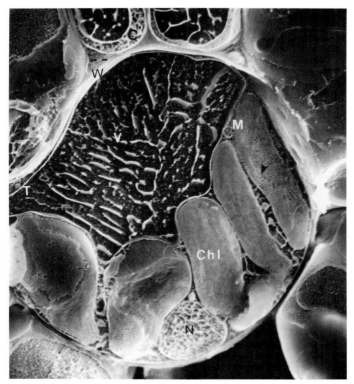

Fig. 2. Secondary electron image of a cell from a cryofractured, lightly etched, frozen hydrated bulk maize leaf specimen. W, Cell wall; C, cytoplasm; V, vacuole; T, tonoplast; N, nucleus; Chl, chloroplast; M, mitochondrion. Subchloroplast structures (thylakoids?) are marked by arrows.

tural preservation down to the mitochondrial and subchloroplast level. Very good preservation of the cytoplasm and tonoplast in root cells was recently documented by Storey and Walker.[13]

Our cryofracturing technique permits a high degree of reproducibility in structural preservation. Maize blade segments consistently fractured in a very reproducible manner and in a relatively smooth plane across the tissue despite the presence of considerable intercellular air spaces. A greater overall surface area of the tissue segments was found to be acceptable for X-ray microanalysis than in earlier studies. The quality of each fracture was judged by observing the specimen tilted at a 45° angle and low

[13] R. Storey and R. R. Walker, *J. Exp. Bot.* **38,** 1769 (1987).

magnification. Only samples which were found to be relatively flat and devoid of artifactual topographical features were used for analysis. Specimens prepared for analysis using this technique had a substantially greater proportion of their cells fractured open to expose intracellular organelles than conventionally fractured tissue samples. Similarly, fractures of root tissues showed significant structural improvement. The major structural differences occurring between stelar and cortical cells did not pose a substantial obstacle to obtaining smooth fracture faces, and root specimens did not fracture in a concave shape, as was the case in earlier investigations.

The specimen preparation method described here is a relatively simple technique which affords significantly greater control over the fracture plane than most techniques used for plant specimens (with the exception of that employed by Storey and Walker[13]). In addition, this method allows one to fracture consistently over large specimen surface areas and thus constitutes a substantial improvement over conventional cryofracturing techniques. A common characteristic of many cryofracturing systems is that they offer relatively little control over the fracturing process itself. To a large extent the fracture process is a somewhat random event, the tissues generally cleaving along the lines of least resistance, typically around rather than through individual cells.[6] This problem is especially evident in tissues with considerable intercellular spaces such as leaf mesophyll cell layers and cortical tissue in roots.

Figure 3 shows a typical energy-dispersive X-ray spectrum from a leaf cell of maize with normal background distribution and clearly distinguishable peaks of the elements Na, P, S, Cl, K, Ca, and Cr (the chromium peak is due to chromium coating of the specimen). Other spectra (not shown) revealed that the element Mg can also be determined in plant specimens using this technique. Although the described cryofracturing technique does not produce absolutely flat specimen surfaces which would be ideal for quantitative X-ray microanalysis,[5] it does permit easy identification of elements in contrast to rough specimen surfaces that cause difficulties in qualitative element identification. In addition, quantitative X-ray microanalysis is feasible.

Comments on Quantitative X-Ray Microanalysis of Frozen Hydrated Bulk Plant Specimens

In recent years several investigators have explored the potentials for deriving quantitative information from X-ray microanalysis of frozen hydrated bulk plant specimens. The first question is whether the background radiation should be accounted for by subtracting background from peak areas $(p - b)$ or calculating peak/background ratios (p/b). Boekestein et al.

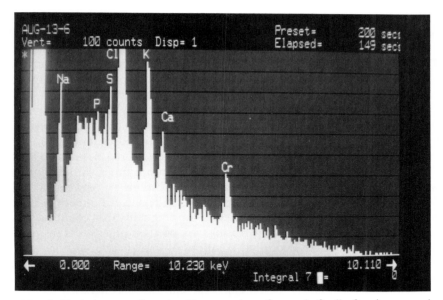

Fig. 3. Typical energy-dispersive X-ray spectrum from a leaf cell of maize grown in salinized nutrient solution. The Cr peak is due to chromium coating of the specimen.

provided evidence in support of the p/b ratio method for bulk specimens where the background intensity is calculated for the same energy region as for the peak.[14] Expressing data as p/b ratios gives relative quantitative information on elemental distribution in the specimen. The use of appropriate standards which resemble the biological specimens in chemical, physical, and structural properties can provide absolute quantitative data in units of concentration (amount of element per volume of specimen analyzed), if the relationship between p/b and concentration (preferably linear) allows accurate calibration.[15]

Standards consisting of frozen electrolyte solutions, sometimes in a loose cotton fiber matrix, do not resemble plant specimens enough to provide accurate absolute quantitative data.[2] These standards did allow, however, determination of the sensitivity of the analytical system for various elements. It was found that the particular energy-dispersive system used was about 4 to 5 times less sensitive for Na than for heavier elements but equally sensitive for K and Cl.[2] Whitecross et al. proposed the use of a

[14] A. Boekestein, F. Thiel, A. L. H. Strols, E. Bouw, and A. M. Stadhouders, J. Microsc. (Oxford) **134**, 327 (1984).

[15] D. M. R. Harvey, T. J. Flowers, and B. Kent, J. Microsc. (Oxford) **134**, 93 (1984).

colloidal graphite slurry in frozen electrolyte standard solutions for calibration of X-ray microanalysis data from biological specimens.[16] This proposal was experimentally tested by Treeby *et al.* on leaf specimens.[17] They found that a polynomial function described precisely the relationship between p/b and concentration, and they used this approach to estimate vacuolar phosphate concentrations in lupine leaf cells.[17]

Probably the most realistic approach to quantitative X-ray microanalysis of bulk frozen plant samples by means of standards has been described by Storey and Walker.[13] The latter authors established calibration curves using KCl standards prepared by extracting slices of red beet tissue with 80% boiling ethanol. The red beet was rinsed several times with deionized water and perfused overnight with standard KCl solutions between 50 and 300 mol m^{-3}. These standards were prepared and analyzed under the same conditions as the experimental material. Similar standards could be prepared for other elements by perfusing red beet or other suitable plant tissues with appropriate electrolyte solutions of known composition.

One can also estimate elemental concentrations in compartments of homogeneous cell aggregates or cell populations, provided that the concentrations in the total cell are known from conventional chemical analysis and the relative volumes of the major cell compartments are determined.[18,19] This was achieved for cells of the alga *Porphyra* with the volumes of the major cell compartments cytoplasm, plastid, and vacuoles determined by morphometric analysis.[18] Concentrations of the ions Na$^+$, Cl$^-$, and K$^+$ in these compartments were calculated using the following equation:

$$vol_{cyt}[\text{ion}]_{cyt} + vol_{vac}[\text{ion}]_{vac} + vol_{pl}[\text{ion}]_{pl} = [\text{ion}]_{cell}$$

where vol_{cyt}, vol_{vac}, and vol_{pl} are the volumes of cytoplasm, vacuoles, and plastid, respectively (expressed as percentage of protoplasmic volume), and $[\text{ion}]_{cyt}$, $[\text{ion}]_{vac}$, and $[\text{ion}]_{pl}$ denote the ion concentrations in the respective compartments. $[\text{ion}]_{cell}$ represents the ion concentration in the total cell determined by chemical analysis. A similar approach has been successfully applied to plant cell suspensions.[19]

[16] M. I. Whitecross, G. D. Price, and J. S. Preston, *J. Microsc. (Oxford)* **128**, RP3 (1982).
[17] M. T. Treeby, R. F. M. Van Steveninck, and H. M. DeVries, *Plant Physiol.* **85**, 331 (1987).
[18] C. Wiencke, R. Stelzer, and A. Läuchli, *Planta* **159**, 336 (1983).
[19] M. L. Binzel, F. D. Hess, R. A. Bressan, and P. M. Hasegawa, *Plant Physiol.* **86**, 607 (1988).

Application of X-Ray Microanalysis to Ion Compartmentation in Root Cells and Tissues

Recent examples of studies on ion compartmentation in root cells and tissues by means of X-ray microanalysis are given in Table I. With the exception of the study by Harvey, who used freeze-substituted specimens, all others were done on frozen hydrated bulk root samples.[20] In several of these studies the cytoplasm and vacuole compartments were analyzed separately for the ions Na^+, Cl^-, and K^+ and the elements Mg, P, and S. Ca was detectable in some cases. In most of these investigations quantitative information was limited to the calculation of element ratios. The only two reports on roots which provided quantitative information in units of concentration are the ones by Harvey[20] and by Storey and Walker.[13] Harvey presented quantitative data on the distribution of Na^+, K^+, and Cl^- in maize root cells for the compartments cytoplasm, vacuole, and cell wall.[20] However, since freeze substitution was used for specimen preparation, the possibility of internal redistribution of ions during the preparation cannot be ruled out. Quantitative vacuolar concentrations of K^+ and Cl^- in cells of frozen hydrated bulk root samples have recently been reported by Storey and Walker.[13]

[20] D. M. R. Harvey, *Planta* **165**, 242 (1985).

[20] Xylem Transport

By A. H. DE BOER

Introduction

It has long been believed that ion delivery to the xylem is simply a diffusion process, once ions have entered the root cortical cells. In light of the precise control that a normally functioning plant exerts over its internal ion fluxes these conclusions appeared questionable. Evidence has now accumulated that ion transport into and out of the xylem is not merely passive and is regulated.[1-5] The stumbling block for thorough study of the

[1] A. J. E. Van Bel and A. J. Van Erven, *Plant Sci. Lett.* **15**, 285 (1979).
[2] J. G. Johanson and J. M. Cheeseman, *Plant Physiol.* **73**, 153 (1983).
[3] D. T. Clarkson, W. L. Williams, and J. B. Hanson, *Planta* **162**, 361 (1984).
[4] A. H. De Boer, K. Katou, A. Mizuno, H. Kojima, and H. Okamoto, *Plant, Cell Environ.* **8**, 579 (1985).
[5] A. H. De Boer and H. B. A. Prins, *Plant, Cell Environ.* **8**, 587 (1985).

mechanism of ion transport along xylem vessels is their location deep inside the tissue. In this discussion I mainly refer to a technique, already used in 1929,[6] but which until recently had not been further developed, namely xylem perfusion.

Xylem perfusion is the use of pressure or suction to bring about the flow of an artificial solution through xylem vessels in an excised root or shoot segment. The advantages of a perfusion system are (1) control over the composition of the solution entering the xylem vessels, (2) ability to bring hormones, specific inhibitors, etc. into direct contact with parenchyma cells lining the vessels while avoiding direct contact with cortical parenchyma cells, (3) control over the flow rate of xylem sap, and (4) ability to study processes of ion uptake from the medium and ion release into and uptake from the xylem vessels separately.

A variety of different materials and different experimental systems have been used in perfusion experiments (Table I). In order to draw valid conclusions from perfusion experiments the following criteria should be met: (1) the tissue must not be irreversibly damaged by excision and mounting, (2) the perfusion solution must pass through the stele, preferably confined to the xylem vessels and not through the cortical apoplast, (3) the hydraulic resistance of the xylem vessels should be sufficiently low to allow a rapid flow rate at moderate applied pressure or suction, and (4) the energy supply of the excised segment should not become a limiting factor during the experiment. The aforementioned points can be further specified:

1. Excision and mounting causes depolarization of the membrane potential of cortical and stelar cells.[5,7] With aging and washing the membrane potential usually recovers, although this may not be always the case.

2. To avoid flow of the perfusion solution through the cortex, the cortex has to be stripped at the end where the perfusion solution enters the stele. After moisture removal, the exposed transverse cortical surface is sealed with wax[3] or cyanoacrylate glue.[5,8] The pathway of the perfusion solution can be followed with Tinopal CBS-X, 0.02%.[3] Tinopal binds to cellulose and produces a characteristic blue–white fluorescence. Neutral (0.1% Congo red[5,7]) or acid (acid fuchsin[9,10]) membrane-impermeable dyes have been used as well. Basic dyes cannot be used since they are immobile

[6] P. Frenzel, *Planta* **8,** 642 (1929).

[7] H. Okamoto, A. Mizuno, K. Katou, Y. Ono, Y. Matsumura, and H. Kojima, *Plant, Cell Environ.* **7,** 139 (1984).

[8] A. Mizuno, *Plant, Cell Environ.* **8,** 525 (1985).

[9] B. Jacoby, *Physiol. Plant.* **18,** 730 (1965).

[10] A. J. E. Van Bel, *Plant Physiol.* **57,** 911 (1976).

TABLE I
MATERIALS AND TECHNIQUES USED IN PERFUSION EXPERIMENTS

Plant	Segment	Force	Ref.
Allium cepa (onion)	Root	Suction	3
Plantago maritima (plantain)	Taproot	Air pressure	5
Vigna unguicalata L. (cowpea)	Hypocotyl	Pump pressure	4, 7, 8
Apple	Stem	Gravity	16
Phaseolus vulgaris L. (brittle wax bean)	Hypocotyl	Air pressure	9
Lycopersicum esculentum (tomato)	Stem internode	Gravity	1, 10

as a result of adsorption to the negatively charged vessel walls.[11] All these controls showed that flow was mainly limited to the xylem vessels, with no flow through the cortical apoplast.

3. Perfusion can be difficult in young tissue where the xylem conducting elements have not yet fully developed.[3] Large open vessels greatly facilitate perfusion as the resistance to flow in a vessel is inversely proportional to the fourth power of the effective vessel radius. High perfusion pressures may lead to radial water movement, thereby breaking insulating structures[7] and causing waterlogging of the cortex.[4,5] Rapid replacement of the total vessel volume is important to reduce measurement lag time and to expose the cells around the vessels to identical solutions. Assuming a vessel volume of $1-2\%$,[3,10] a flow rate of $10-20\mu l$/min/fresh weight will replace the xylem volume every minute. Care should be taken that the perfusion solution is filtered through a micropore filter to prevent blockage of xylem vessels with bacteria and suspended material. High concentrations of Ca^{2+} (>2 mM) should be avoided as this may precipitate with organic ions and/or phosphate.[12]

4. Excised plant tissue may exhaust its internal supply of sugar during aging. External supply of sugars can restore metabolic activity.[13] In hypocotyl segments of etiolated cowpea, addition of 1% sucrose to the perfusion solution was necessary to activate proton pumps located inside the stele (A. Mizuno, personal communication). However, the presence of sucrose may affect the exudate pH and its K^+ content during uptake by a proton-symport mechanism.[1] Not only a shortage of metabolic substrate but also a shortage of oxygen may easily prevail in the core of a segment.[14] The

[11] A. J. E. Van Bel, *Z. Pflanzenphysiol.* **89**, 313 (1978).
[12] D. G. Hill-Cottingham and C. P. Lloyd-Jones, *Physiol. Plant.* **28**, 443 (1973).
[13] P. H. Saglio and A. Pradet, *Plant Physiol.* **66**, 516 (1980).
[14] A. H. De Boer and H. B. A. Prins, *Plant Cell Physiol.* **25**, 643 (1984).

presence of air-filled intercellular spaces greatly facilitates radial oxygen diffusion. Care should be taken that water logging does not occur during perfusion.[14]

Xylem Vessel Properties

Many properties of xylem vessels are of importance for studies of xylem transport. In vessels, the primary cell wall is strengthened internally by the deposition of secondary wall (see Fig. 1). The secondary wall is composed of cellulose microfibrils and deposits of hemicellulose and lignin. The vessel wall is negatively charged, and, because of the large surface area, it acts as an effective exchange column.[12] The cation exchange capacity of the wall is estimated at $2-4$ mEq/g fresh weight.[15] Therefore, if a perfusion solution differs in ionic composition from that of the natural xylem sap the exudate may be enriched in ion X, simply by release of X adsorbed to the walls[4] or vice versa. The affinity of monovalent cations for exchange sites on the wall is very much less than that of divalent cations. Perfusion in the presence of $1-2$ mM $CaSO_4$ prior to and during the experiment should reduce the interference of monovalent cation exchange with the wall during the experiment. Another problem associated with the negatively charged walls is the immobility of organic cations during perfusion.[10,14] The effectiveness of inhibitors or hormones which are positively charged at pH $5-6$ in perfusion experiments is questionable as they may not reach the site of action; perfusion of strongly adsorbing cations may improve the mobility.[16]

Vessel walls seem to be permeable to water molecules, but the resistance to water flow in the walls is considerably higher than in the xylem vessels themselves.[17] Surprisingly, little is known about ion permeability of vessel walls with secondary thickening. The findings that xylem parenchyma cells are frequently transformed into transfer cells (Fig. 1) and that the phosphatase activity in the part of the plasmalemma of a xylem parenchyma cell which faces a pit in the vessel wall is higher than elsewhere[18] indicate that ion exchange occurs mainly through the pit membrane. However, it cannot be excluded that part of the ion exchange between stelar and vessel apoplast takes place by diffusion through the wall.

The question of ion permeability of vessel walls is of relevance to the

[15] S. C. Van de Geyn and C. M. Petit, *Plant Physiol.* **64**, 954 (1979).

[16] D. G. Hill-Cottingham and C. P. Lloyd-Jones, *Nature (London)* **220**, 389 (1968).

[17] E. I. Newman, *in* "The Plant Root and Its Environment" (E. W. Carson, ed.), p. 362. University of Virginia, Charlottesville, 1974.

[18] J. J. Sauter, *Z. Pflanzenphysiol.* **67**, 135 (1972).

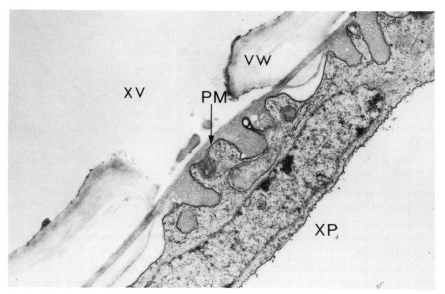

FIG. 1. Half-bordered pit in the wall of a xylem vessel in the stem of *Phaseolus vulgaris*. The plasma membrane of a bordering xylem parenchyma cell, with transfer-cell-like structures, is exposed to the vessel lumen. XV, Xylem vessel; XP, xylem parenchyma cell; PM, plasma membrane, VW, vessel wall. Micrograph courtesy of D. Kramer.

electrophysiology of the xylem. A considerable electrical potential difference is maintained between the xylem and medium[19,20] or the xylem and cortical apoplast.[21] This implies the presence of an insulating structure between vessels and cortical apoplast. The suberized Casparian strip in the endodermis may fulfill this task. In tissues where the strip is present "ion leakiness" of the vessel walls does not necessarily short-circuit the xylem potential. However, the endodermis plus Casparian strip disappear in tissues with secondary growth.[5] In that case it seems plausible that the vessel walls are lignified to such a degree that they form a high resistance to current flow. Radial current flow is now only possible through the vessel doors, namely, the pits, where the plasmalemma of the adjacent parenchyma cells forms the only barrier.

[19] J. Dunlop, *J. Exp. Bot.* **33,** 910 (1982).
[20] A. H. De Boer, H. B. A. Prins, and P. E. Zanstra, *Planta* **157,** 259 (1983).
[21] H. Okamoto, K. Katou, and K. Ichino, *Plant Cell Physiol.* **20,** 103 (1979).

Description of a Perfusion Apparatus

Clarkson *et al.*[3] give an elaborate description of a device to perfuse onion roots using suction. A technique to perfuse cowpea hypocotyl described by Okamoto *et al.*[7] was later modified slightly.[4,8] In case of very fast growing tissues the device as described by Okamoto *et al.*[7] is recommended. Here, I describe a perfusion apparatus, modified after Refs. 4 and 5, which is suitable for both root and shoot segments, to perform continuous measurements using ion-selective electrodes and an automized flow meter, or to do discontinuous measurements with isotopes.

The main chamber is constructed of acrylic plastic (Plexiglas) (Fig. 2A). Two tubes in the side wall, just above the chamber bottom, allow the flow through of solution (root) or gas (shoot). A reference electrode (Ag/AgCl) is mounted in the side wall. The lid ensures that an environment different from that outside can be maintained inside the chamber. A narrow slot in the lid allows entry of a microelectrode for intracellular measurements.

The plant segment is cut under water to prevent the entry of air into the vessels. The length of the segment should be 12 mm longer than the

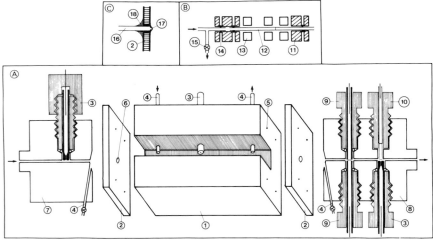

FIG. 2. Experimental setup for perfusion of root or shoot segments. (A) Main chamber plus inflow/outflow units with electrodes. (B) Flow meter. (C) Sealing of plant segment in flange. 1, Main chamber; 2, flange; 3, reference electrode (Ag/AgCl); 4, inlet/outlet; 5, screw hole; 6, hole for plant segment; 7, inflow unit; 8, outflow unit (7 and 8 shown in cross section); 9, liquid membrane ion-selective electrode; 10, glass membrane ion-selective electrode; 11, capillary support; 12, capillary; 13, slotted optical switch; 14, O ring; 15, valve; 16, plant segment; 17, stele; 18, cyanoacrylate glue.

chamber length. On each side of the segment, 3 mm of the cortex is stripped, leaving the stele exposed (Fig. 2C). The junction between the exposed stele and the cortex is dried and covered with a thin layer of nonpoisonous cyanoacrylate glue (super glue). This glue polymerizes when it contacts water and does not seem to harm the tissue. Both ends of the stele should be moist throughout.

Consequently, both ends are glued into a Perspex flange (3 mm), with cyanoacrylate glue. The best results are obtained if the segment fits precisely into the hole of the flange, allowing room to apply glue between the segment surface and the inside of the hole.

Next, the flanges are screwed onto each end of the chamber. The flow unit is attached after the protruding stele is excised with a sharp razor blade to make a fresh cut. The inflow unit has an outlet close to the segment end to allow a rapid change of perfusion solution. An Ag/AgCl reference electrode with ceramic tip (KCl or choline chloride filled) is mounted into the inflow tube to enable measurement of the xylem electrical potential at the inflow side. This electrode must be tightly sealed since there is pressure in the tube. To perfuse a solution, either a pump (which gives a constant flow rate) or gas pressure (which ensures constant pressure) may be used. A pressure transducer is mounted into the inflow tube to register the perfusion pressure.

The outflow unit is attached to the outflow flange. An outlet close to the end of the segment allows sampling of the exudate to measure radioactivity, osmotic potential, ion composition, etc. Ion-selective electrodes with their sensing part just protruding into the central capillary continuously register the ion activity of the exudate. An Ag/AgCl reference elelctrode (filled with choline chloride when K^+ is measured in the exudate) is mounted in the same way. Measurements with ion-selective electrodes are described elsewhere in this series.[22] Common ion-selective electrodes are either glass electrodes or liquid membrane electrodes. Both can be obtained through scientific supply houses. Microelectrodes Inc. (Londonderry, NH) provides glass microelectrodes that have a thin glass tip [external diameter (e.d.) 1.5 mm] connected to a metal shaft (e.d. 2 mm). For mounting convenience and because the glass tip is fragile, a hollow screw is glued around the metal shaft (Fig. 2A).

Liquid membranes which are selective for various ions can be obtained commercially or produced in the laboratory.[23] The electrode is made by gluing a 3-mm membrane punchd with a droplet of tetrahydrofuran (2 μl/3 mm membrane) to 3-mm poly(vinyl chloride) (PVC) tubing. The

[22] J. L. Walker, this series, Vol. 56, p. 359.
[23] H. Affolter and E. Sigel, *Anal. Biochem.* **97,** 315 (1979).

PVC tubing is attached to the membrane, and the tetrahydro-furan is evaporated. The PVC tubing with the ion-selective electrode is glued into a hollow screw to facilitate mounting. The electrode is filled according to factory specifications, and a reversible electrode (Ag/AgCl) is dipped into the filling solution to connect the electrode to the amplifier.

The principle of the automized flow meter is as follows. The central capillary of the outflow unit is connected to a calibrated microcapillary (Fig. 2B). The capillary is centered in the slot of three slotted optical switches (S.O.S.), which are connected to a microprocessor. The xylem exudate runs into the capillary, and when the meniscus breaks through the lightbeam of the second S.O.S. a counter is started. Counting stops when the meniscus breaks through the lightbeam of the third S.O.S. At the same moment the microprocessor opens the valve of a siphon and the liquid starts to withdraw. The valve is closed when the withdrawing meniscus passes the first S.O.S. The flow rate is calculated by the microprocessor as the capillary volume between the light-beams of second and third S.O.S., divided by the number of counts (sec). The frequency of measurements can be varied by varying the distance between the second and third S.O.S. or by varying the micropillary volume. Signals from both the electrodes and the flow meter are led into a channelyzer (Bailey: CHL-1) and recorded on a chart recorder every few seconds. Processing with a personal computer should be possible as well.

Electrophysiology

Cells of root and stem tissue are interconnected in the radial direction by numerous cytoplasmic bridges, the plasmodesmata. Thus, a cytoplasmic continuum, the symplast, is formed extending from the outermost epidermal cells to the innermost stelar cells which line the xylem vessels. The symplast forms an electrical continuum wherein no gradient in electrical potential exists.[24] Since there are few plasmodesmata in the longitudinal direction, a root or stem segment may be regarded as being composed of stacked, disklike parenchyma symplasts, penetrated by vascular bundles,[21] as diagramed in Fig. 3A. The electrical potential difference between the sap in the vessel lumen and the medium is the result of two potential jumps in series: the electrical potential difference across the plasmalemma of xylem parenchyma cells and that of cortical/epidermal parenchyma cells, respectively. Evidence has accumulated in recent years that in certain tissues both potential differences can be varied independently and that they are maintained by spatially separated proton pumps which work in oppo-

---

[24] K. Katou, *Plant Cell Physiol.* **19**, 523 (1978).

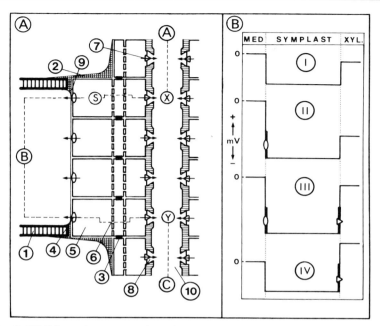

FIG. 3. (A) Schematic representation of the electrophysiological organization of a root or shoot segment. The outer pump brings protons into the cortical apoplast of medium, and the xylem pump brings protons into the stelar apoplast or directly into the vessel lumen. 1, Flange; 2, cyanoacrylate glue; 3, Casparian strip; 4, outer proton pump; 5, symplast; 6, plasmodesmata; 7, xylem proton pump; 8, half-bordered pit; 9, cell wall apoplast; 10, xylem vessel. (B) Theoretical radial potential profile in root or shoot segment. I, Both outer and xylem pump inactive, diffusion potentials only; II, outer pump active, xylem pump inactive; III, both pumps active; IV, xylem pumps active, outer pumps inactive.

site directions, namely, a "xylem pump" pumping protons from the symplast into the stelar apoplast or directly into the vessel lumen and an "outer pump" pumping protons from the symplast into the cortical apoplast or medium.[5,8,20,21] The apoplastic pathway between stelar and cortical apoplast is blocked by the suberized Casparian strip and/or by the lignified vessel walls.

The ion activity in the xylem is usually higher than in the medium. This results in a diffusion potential across the symplast, dominated by potassium. The diffusion potential is manifested under anoxic conditions, with the xylem as the more negative compartment[20] as symbolized in Fig. 3B. Depending on the electrogenicity of both pumps, the electrical potential difference across the symplast can range from negative to positive values (medium or cortical apoplast grounded).[20]

Trans-Root Potential

The trans-root potential (TRP) is the electrical potential difference (PD) between a measuring electrode in the xylem exudate and a reference electrode in the medium. Both electrodes are of the Ag/AgCl type, equipped with a ceramic tip and filled with 3 M KCl. For TRP measurements it is important to know that the root may be considered as a leaky cable:[25] the xylem resistance is not negligible compared to the radial resistance. This implies that the PD between points A and B (Fig. 3A) represents the PD across the symplast of a disk which is at a distinct point from the cut end. There is evidence that this point is not too far from the cut end.[14,26] This also explains why (1) TRPs measured from the inflow and outflow side of a perfused segment are not necessarily the same,[5] (2) the axial PD does not equal zero and responds to metabolic inhibitors,[5,24] and (3) the TRP of a segment is not dissipated until the root is reduced to a short segment.[26]

In studying the properties of the TRP in a perfused segment, using anoxia, one has to be aware of the fact that the perfusion solution may contain enough oxygen to keep the parenchyma cells which line the vessels aerobic, even though the medium is anaerobic. The advantage of TRP determinations is that measurements are stable and continuous. The disadvantage is that a precise localization is difficult.

Intravessel Measurements

The intravessel electrical potential is measured by bringing a microelectrode directly into the vessel lumen, with a reference electrode in the medium (PD between points X and B in Fig. 3A). Successful measurements of this type have been reported only by Dunlop.[19] There are two main problems with this technique. (1) It is difficult to localize the exact position of the electrode tip. If the anatomy of the tissue is regular, the position of the electrode tip can be inferred from the electrode response while penetrating the tissue.[19] It is difficult to transfix the thickened vessel walls without breaking or plugging the electrode tip. In order to combine a sharp tip with a reasonably low electrode resistance the tip can be beveled.

As with the TRP, an intravessel measurement gives the transsymplast PD, the latter having the advantage that the zone where the actual measurement takes place is better defined. An intravessel electrode in combi-

[25] H. Ginsburg, *J. Theor. Biol.* **37**, 389 (1972).
[26] H. Ginsburg and G. G. Laties, *J. Exp. Bot.* **24**, 1035 (1973).
[27] K. Ichino, K. Katou, and H. Okamoto, *Plant Cell Physiol.* **14**, 127 (1973).

nation with an intrasymplast electrode will enable a direct study of the xylem pumps, as this measures the PD between points X and S (Fig. 3A).

Intrasymplast Measurements

For the measurement of intrasymplast electrical PD, a microelectrode is brought into the symplast. With a reference electrode in the medium (or on the wet segment surface in the case of shoot tissue[21]) and a reference electrode in the exudate the properties of the outer pump (PD between points S and B, Fig. 3A) and the xylem pump (PD between points S and A) can be studied separately and simultaneously. Measurements of this kind can be stable for hours and provide a good tool to gain insight into the electrophysiological structure of a root or shoot segment.

Other Methods to Study Stelar Ion Transport Mechanisms

In certain plants, cortical and stelar tissue can be easily separated; maize roots and mesocotyls have been used most often.[28-31] This separation enables the study of the properties of isolated plasmalemma, tonoplast, soluble enzymes, etc. of stelar cells without interference of cortical cells. A number of reports on maize roots indicate that the physiology of stelar cells differs from that of cortical cells: (1) stelar cells contain several proteins which are absent from cortical cells,[28] (2) the specific activity of several enzymes, isolated from stelar cells, exceeds that of the same enzymes of cortical cells,[29] (3) cells of both cortex and stele possess cation ATPases at similar levels but not with identical properties,[30] and (4) the stele contains indolyl-3-acetic acid (IAA) at a level 5 times higher than the cortex.[31] The latter observation is of special interest in view of possible differences in control mechanisms over ion transports in stele and cortex. The reason why stelar tissue has as yet been studied in only a few cases is obvious: the isolation of sufficient material is laborious, 1 g of fresh maize roots giving approximately 0.15 – 0.20 g of stelar tissue.

Acknowledgments

This work was supported by National Science Foundation Grant DMB 8502021.

[28] E. E. Khavkin, E. Yu. Markov, and S. I. Misharin, *Planta* **148,** 116 (1980).
[29] I.V. Zeleneva, E. V. Savost'yanova, and E. E. Khavkin, *Biochem. Physiol. Pflanz.* **177,** 97 (1982).
[30] R. T. Leonard and C. W. Hotchkiss, *Plant Physiol.* **61,** 175 (1978).
[31] M. Saugy and P. E. Pilet, *Plant Sci. Lett.* **37,** 93 (1984).

[21] Phloem Transport

By GABRIELE ORLICH and EWALD KOMOR

The term phloem transport refers to the flow of assimilates from their site of synthesis or storage to their site of consumption, a process which, for methodological reasons, may be separated into three phases: phloem loading, long-distance transport, and phloem unloading. To gain a better understanding of phloem transport, a close relationship between structural and functional information is necessary. Therefore, methods for studying phloem transport should include anatomical, histological, and physiological approaches.

In this chapter a selection of techniques is presented with special emphasis on their methodological value, and open questions are briefly outlined on occasion. The different techniques are introduced by those who are working in the respective field to provide firsthand information.

Phloem Loading

General Comments

The term phloem loading as used in this chapter denotes the transfer of assimilates from the sites of CO_2 fixation (or the sites of mobilization of storage carbohydrates in stems or roots) to the phloem region via symplastic and/or apoplastic routes as well as the process of sucrose uptake into the sieve tube–companion cell complex as the final, most important step in this sequence, the mechanism of which may be termed either symplastic, if accomplished by plasmodesmatal connections between adjacent cells, or apoplastic, if mediated by a plasma membrane-bound carrier system. Symplastic versus apoplastic pathways through the tissue and symplastic versus apoplastic uptake of sucrose into the sieve tube–companion cell complex depend on species-specific properties of tissues, such as the arrangement of different cell types in mature source leaves, their enzyme and carrier composition, including cell wall-bound invertase (β-fructofuranosidase), and the frequency of plasmodesmata between different cell types.

There is a common consensus that sucrose is translocated within the sieve tubes. The mechanism of sucrose uptake into the sieve tubes, however, has been debated since the 1960s[1,2] for reasons which, as we see it,

---

[1] R. T. Giaquinta, *Annu. Rev. Plant Physiol.* **34**, 347 (1983).
[2] A. L. Kursanov, "Assimilate Transport in Plants," p. 81. Elsevier, Amsterdam, 1984.

must be ascribed to the experimental problem of either quantifying or eliminating the contribution of nonphloem cells to the available data on sucrose movement through leaf tissues. So, quite consistently, more recent conceptual assessments[3] (see also earlier discussions[4,5]) again concentrate on the fate of sucrose along its way to the phloem, in an attempt to elucidate its compartmentation at the cell type level and thereby trying to determine a potential regulatory role of assimilate transfer through the tissue in sucrose uptake at the sieve tube–companion cell complex.

The choice of the ideal object with which to investigate phloem loading, as well as the kind of pretreatment of a selected plant material, should depend on the question under study; thus, to trace the pathway(s) of assimilates (sucrose) from their site of synthesis or application to the site of sucrose uptake, strict methodology would dictate the use of intact, non-wounded tissue, whereas isolated, enriched or purified cells, protoplasts or membranes of a distinct cell type should be used to characterize mechanisms of carrier-mediated sugar uptake. Isolating viable phloem cells remains a technical problem; on the other hand, and more importantly, many of the efforts to analyze compartmentational and mechanistic aspects of phloem loading suffer from methodological restrictions and therefore often do not allow clear interpretation. In these introductory remarks we have tried to connect the pertinent questions with the available techniques and thus elucidate their methodological value.

Uptake of [¹⁴C]sucrose

The preferential labeling of veins in autoradiographs after application of [¹⁴C]sucrose to a pretreated mature green leaf has been taken as evidence that phloem loading has occurred; therefore [¹⁴C]sucrose uptake experiments have become a widely used technique to study compartmentational, kinetic, and energetic aspects of phloem loading. Interpretation of results from [¹⁴C]sucrose uptake experiments with leaf tissue is often impeded for the following reasons: (1) Sucrose applied via the apoplast may not follow a route in the tissue that is identical to the *in vivo* pathway after CO_2 fixation[5] (see also Giaquinta *et al.*[6] for the pathway of unloading). (2) The leaf tissue has to be made accessible for [¹⁴C]sucrose either by abrading the leaf surface or by cutting small leaf disks. These pretreatments, especially in

---

[3] W. J. Lucas, *in* "Regulation of Carbon Partitioning in Photosynthetic Tissue" (R. L. Heath and J. Preiss, eds.), p. 254. American Society of Plant Physiologists, Rockville, Maryland, 1985.

[4] B. R. Fondy and D. R. Geiger, *Plant Physiol.* **59**, 953 (1977).

[5] M. Madore and J. A. Webb, *Can. J. Bot.* **59**, 2550 (1981).

[6] R. T. Giaquinta, W. Lin, N. L. Sadler, and V. R. Franceschi, *Plant Physiol.* **72**, 362 (1983).

combination with prolonged incubation times to reduce internal sugar levels, may cause wound responses[7] such as ethylene-induced inactivation and formation of (new) carrier systems[8] or synthesis of (additional) invertase[9,10] thereby changing the relative contributions of sugar pathways to phloem supply in the tissue.[11]

For maize leaves, Heyser *et al.*[12] have developed a technique for supplying sucrose from the apoplast that avoids severe wounding by introducing sucrose into the xylem via the transpiration stream. The cotyledons of castor bean have the advantage of being accessible to externally supplied sucrose without abrasion of the epidermis since they take up sugars from the apoplast (endosperm) during germination.[13] However, because of the different physiological (heterotrophic) status in terms of simultaneous source and sink characteristics, the results obtained with the cotyledon may not adequately reflect compartmentational properties of phloem loading in a green source leaf.

(3) Sucrose is not exclusively taken up by phloem cells. Although autoradiographs show that label is preferentially accumulated in the veins of leaf tissue, Fondy and Geiger,[4] in an analysis of labeled products following [14C]sucrose uptake by sugar beet leaves and quantitative autoradiography, found that label was also present in the mesophyll cells, part of which had been identified as sucrose. This indicates that sucrose is not exclusively taken up by phloem cells. Alternatively, either mesophyll cells can take up sucrose as well or part of the sucrose is hydrolyzed by a cell wall-bound invertase and hexoses are taken up from which sucrose is resynthesized: in both cases [14C]sucrose uptake does not exclusively reflect "phloem loading" (in the traditional sense of uptake into the sieve tube–companion cell complex). Therefore, interpretation of kinetic data from sucrose uptake experiments with whole leaf tissue to characterize the presumed sucrose carrier in the sieve tube–companion cell complex is ambiguous.

Preferential labeling of veins in autoradiographs after [14C]sucrose application may indicate preferential sucrose uptake into the sieve tube–companion cell complex from the apoplast but would also be consistent with the interpretation that sucrose is taken up by mesophyll cells exclusively and subsequently loaded symplastically into the sieve tube–

[7] R. F. M. van Steveninck, *Annu. Rev. Plant Physiol.* **26**, 237 (1975).

[8] G. Abraham and L. Reinhold, *Planta* **150**, 380 (1980).

[9] M. Turkina and S. Sokolova, *Sov. Plant Physiol (Engl. Transl.)* **15**, 1 (1968).

[10] C. J. Pollock and E. J. Lloyd, *Z. Pflanzenphysiol.* **90**, 79 (1978).

[11] R. Lemoine, S. Delrot, and E. Auger, *Physiol. Plant.* **61**, 571 (1984).

[12] W. Heyser, O. Leonhard, R. Heyser, E. Fritz, and W. Eschrich, *Planta* **122**, 143 (1975).

[13] P. Kriedemann and H. Beevers, *Plant Physiol.* **42**, 161 (1967).

companion cell complex. If the sugar carrier composition of the different cell types are known, this would give a clear-cut answer in case no sucrose carrier is found in the plasmalemma of the sieve tube – companion cell complex: in this case a completely symplastic loading mechanism must be operative (see discussion in van Bel et al.[14]). However, the presence of a sucrose carrier in the plasmalemma of the sieve tube – companion cell complex cannot be taken as proof for an exclusively apoplastic final uptake step since the carrier may function as a retrieval mechanism for sucrose leaked out into the apoplast,[15] while the bulk of assimilates is supplied symplastically. Although there have been some attempts to isolate tissue enriched in phloem either by mechanical separation of veins from leaves[16] and petioles[15,17] or by enzymatic digestion of cell walls of leaves with subsequent separation of parenchyma and vein-type cells,[18,19] all procedures still yield a mixture of different cell types so that the requirements for an unequivocal localization of a sucrose carrier in the plasmalemma of the sieve tube – companion cell complex are not yet met.

As long as cell cultures containing pure phloem cells are not available, the only advantage of a differentiated cell culture as compared to leaf tissue may be that phloem cells can be more readily separated and isolated by using a phloem-enriched (maximum 10%) cell clump as the starting material.[20] It must also be taken into consideration that the composition of carriers and enzymes in a hormone-treated cell culture may not be identical to the protein pattern of the same cell type after differentiation in the intact plant tissue.

Isolation of mesophyll cells and comparison of their uptake kinetics with those of leaf disks is another approach to reveal the uptake characteristics of phloem cells and the involvement of mesophyll cells during phloem loading (van Bel et al.[14]; see also Ref. 21 for comparison of parenchymatous suspension cells and cotyledons of castor bean). If isolated cells or protoplasts of a single cell type are used to study uptake activities, it should be kept in mind that enzymatic digestion of cell walls may damage transport sites and modify membrane energetic parameters, or lead to loss of potential binding proteins. Therefore properties of iso-

[14] A. J. E. van Bel, A. Ammerlaan, and G. Blaauw-Jansen, J. Exp. Bot. 37, 1899 (1986).
[15] J. W. Maynard and W. J. Lucas, Plant Physiol. 70, 1436 (1982).
[16] M. J. Brovchenko, Sov. Plant Physiol. (Engl. Transl.) 12, 230 (1965).
[17] J. Daie, J. Am. Soc. Hortic. Sci. 111, 216 (1986).
[18] C. Wilson, J. W. Oross, and W. J. Lucas, Planta 164, 227 (1985).
[19] A. J. E. van Bel and A. J. Koops, Planta 164, 362 (1985).
[20] R. D. Sjölund and C. Y. Shih, J. Ultrastruct. Res. 82, 111 (1983).
[21] B.-H. Cho and E. Komor, J. Plant Physiol. 118, 381 (1985).

lated cells and protoplasts should be compared with those of intact tissue and isolated cells, respectively.[14,22-24]

To characterize the sugar carrier composition of a distinct cell type it has to be established whether the [^{14}C]sucrose applied is taken up unhydrolyzed or whether (part of) it is split to hexoses by a cell wall-bound invertase. As long as there is no specific inhibitor for a cell wall-bound invertase available, the uptake rates of differently labeled sucrose (glucose versus fructose versus uniformly labeled sucrose) can be compared in order to distinguish between sucrose and hexose uptake. If the sucrose is taken up without previous hydrolysis, the uptake rates of the three differently labeled sucrose molecules should be identical, whereas if the sucrose is hydrolyzed prior to uptake the rates will be different. A prerequisite for this type of experiment is, in case the cells can take up hexoses, too, that the uptake kinetics for glucose and fructose be different.

Another approach to reveal whether sucrose is hydrolyzed during [^{14}C]sucrose uptake is to supply asymmetrically labeled sucrose (labeled either in the glucose or fructose moiety) and subsequently analyze the distribution of the label in the sucrose recovered. Sucrose which is not hydrolyzed will remain asymmetrically labeled, whereas sucrose which is resynthesized from the products of [^{14}C]sucrose hydrolysis will be labeled more or less randomly [in case the activity of the cytoplasmic glucose-6-phosphate isomerase (phosphoglucoisomerase) is not rate-limiting]. However, if the asymmetry of the label is retained, it still cannot be concluded that no sucrose hydrolysis has occurred unless the sucrose recovered is approximately 100% of the incorporated label. For instance, if hexoses derived from sucrose hydrolysis are not entirely resynthesized to sucrose, this proportion of the label would be neglected, although it might contribute rather substantially to total sucrose uptake. (Giaquinta[25] showed that for sugar beet leaves the sucrose recovered retained asymmetry of the label to 95% but only amounted to about 60% of the total sucrose taken up.) By comparing uptake characteristics of sucrose and hexoses in isolated mesophyll cells and their protoplasts the involvement of cell wall-bound invertase during sucrose uptake will also show up (see Refs. 23 and 24 for this kind of experiment with heterotrophic tissues).

Recently the sucrose analog fluorosucrose has been introduced as a new tool to study sucrose uptake, without interference of invertase activity[26];

[22] R. J. Henry, A. Schibeci, and B. A. Stone, *Aust. J. Plant Physiol.* **11**, 119 (1984).
[23] W. Lin, M. R. Schmitt, W. D. Hitz, and R. T. Giaquinta, *Plant Physiol.* **75**, 936 (1984).
[24] M. Stanzel, R. D. Sjolund, and E. Komor, *Planta* **174**, 201 (1988).
[25] R. Giaquinta, *Plant Physiol.* **59**, 750 (1977).
[26] W. D. Hitz, M. R. Schmitt, P. J. Card, and R. T. Giaquinta, *Plant Physiol.* **77**, 291 (1985).

fluorosucrose is not hydrolyzed by invertase but is taken up by sucrose carrier systems. This compound will be very useful for further studies investigating the sugar carrier composition of different cell types, also in combination with microautoradiography.

Techniques of Loading Sucrose into Leaves*

A method for labeling and autoradiography of leaves is given that can be adapted with variation for different plant species. For sugar beet leaves,[27] several different methods of phloem-loading measurements are possible: reverse flap feeding, abraded epidermis feeding, or uptake by leaf disks. Only the latter two are described here.

A sugar beet leaf (usually kept dark overnight) is abraded on the upper surface over an area of 10 cm² by rubbing gently with 300-mesh ceroxide paste and rinsed with water, or the lower epidermis is removed by cracking the upper epidermis and leaf by bending the leaf past a 90° angle. The leaf piece is used as a tab to peel the lower epidermis from the other piece. With practice, areas several centimeters in length can be peeled in one motion. The leaf portion is then sealed with cord-type caulking compound into a Plexiglas incubation chamber, so that the lower surface of the leaf is aerated by an air stream whereas the upper surface is supplied with a solution of radioactive sucrose in 5 mM potassium phosphate, pH 6.5. The sucrose concentration, the specific radioactivity, and the duration of incubation are varied according to the experimental aims. A time course of approximately 30 min is appropriate for sugar beet. In sugar beet the uptake rates are of the order of 0.5 μmol sucrose min^{-1} dm^{-2} at 100 mM sucrose applied. After the radioactive incubation, the surface of the leaf is usually rinsed carefully with water or with the same solution without the radioactive sugar. However, the time needed for wash of free space must be determined by following the time course of exit of labeled material. In sugar beet, 10 min suffices.

Another method is incubation of leaf disks, e.g., of 1 cm², cut from the interveinal areas with a razor blade. The disks, after a short rinsing in buffer, are floated in 5 mM potassium-phosphate, pH 6.5, and labeled sucrose (ranging from 10 to 400 mM), e.g., 50 μl of labeled solution, on which one disk is allowed to float for 30 min. The experiment is terminated by removing the disk, rinsing it with water, and digesting it with H_2O_2 and perchloric acid or extracting it with boiling ethanol, etc., and the

---

* Section prepared by B. R. Fondy, D. R. Geiger, and S. Delrot.

[27] S. A. Sovonick, D. R. Geiger, and R. J. Fellows, *Plant Physiol.* **54**, 886 (1974).

radioactivity in the samples is determined. Analysis for either method can be by autoradiography or by chemical means.

Chemical Analysis of Metabolic Products of Sucrose Feeding. The tissue is quickly immersed in a small test tube with 20 ml of chloroform–methanol (1 : 4, v/v) at 65° for 10 min (refluxed in the tube with a stainless steel bolt to condense the solvent). The extract is collected, and the extraction is repeated with 80% methanol. The extracts are combined (rinse the tubes with chloroform) and may be used for thin-layer chromatography (TLC), enzymatic assay, or ion-exchange separation. For ion-exchange separation, chloroform is added to the extract to give two phases (after overnight in the refrigerator and centrifugation at 1000 g for 10 min). The chloroform layer is discarded, and the extract is dried on an evaporating block at 60° with a stream of dry nitrogen. The sample is then dissolved in 5 ml of 50% ethanol in water and subjected to ion-exchange Sepharose chromatography. The extract can be chromatographed on cellulose TLC plates (prepare 2 cm wide lanes with a razor blade and spot the extract, maximally 0.5 cm in diameter). Up to 80 μg of total sugar can be applied to 500-μm plates.

Lipids in the sample are moved to the solvent front by placing the plates in a tray of chloroform and allowing the front to advance 2–3 cm. This procedure is repeated twice, with evaporation of the solvent between runs. The plates are finally developed in ethyl acetate–acetic acid–formic acid–water (18 : 3 : 1 : 4) (for separation of mannitol, sucrose, glucose, and fructose) or in butanol–acetic acid–water (3 : 3 : 1) (for separation of stachyose, raffinose, sucrose, glucose, and fructose). The separation is improved by repeating the development 2 to 3 times in these solvent systems.

Labeled sugars are visualized with a fast-reacting X-ray film (e.g., Kodak Type AA) with a cumulative radiation of 4×10^6 counts/cm^2. The labeled spots are quantified by stripping off the labeled thin layer [apply a stripping mixture (7 g cellulose acetate, 3 g diethylene glycol, 2 g camphor, 25 ml *n*-propanol, 75 ml acetone) and the spots will curl after the mixture dries]. The bands can be mixed with the scintillation fluid and the radioactivity determined. Enzymatic analysis of the sugar in the extract can be performed by common methods, starch analysis by the method of Outlaw and Manchester.[28]

Microautoradiography

A technique for tracing the route of assimilates from mesophyll to phloem tissue at the cellular level is microautoradiography after feeding a

---

[28] W. H. Outlaw and J. Manchester, *Plant Physiol.* **64**, 79 (1979).

leaf with $^{14}CO_2$. In a pulse–chase experiment the cells involved in the pathway will show up by successive appearance of silver grains with increasing duration of the chase (see also Outlaw and Fisher[29] and Outlaw et al.[30] for details of this approach).

A cell type actively accumulating assimilates may be identified by the level of grain density, if certain considerations are noted. (1) A transient increase in grain density is expected in each cell type involved in the pathway during a pulse–chase experiment; therefore, the relative proportions of grain densities in the different cell types have to be analyzed over a time-course experiment. (2) The highest relative grain density always means that active assimilate uptake has occurred, but maximal relative grain density cannot be expected in each cell type actively accumulating labeled assimilates, because an active uptake site is masked if the rate-limiting step in the pathway is localized before an active uptake site, thus always generating a lower level of grain density in the following steps and cells, respectively. In this respect, experimental conditions to provide slow fluxes of label (extensive predarkening[31] or low-light intensity during $^{14}CO_2$ uptake[32]) would probably prevent rate-limiting steps from being detected (or rather avoid masking of actively accumulating cell types) from the outset, because fluxes higher than those determined by a rate-limiting step might thus be suppressed. (3) The rate-limiting step in a sequence usually is a target for regulation; therefore, varying the photosynthetic rate or translocation rate may change the pattern of grain density distribution in a way that could help elucidate this interesting aspect of phloem loading.

Although the cell types involved in the pathway of assimilates to the phloem can be identified by microautoradiography, the method does not reveal whether the assimilates are transported via the symplast or the apoplast. Geiger et al.[33] have tried to discriminate these routes by inhibition of fluxes of label through plasmodesmata using plasmolyzed leaf cells, but since disruption of cytoplasmic connections was not complete and photosynthetic activity was also impaired, this approach did not yield clear-cut results.

If application of [^{14}C]sucrose is used in microautoradiographic studies, the restrictions of interpretation as outlined in the section on [^{14}C]sucrose uptake should be noted. Comparing the labeling pattern after $^{14}CO_2$ feeding and [^{14}C]sucrose uptake, Fritz et al.[31] have produced results which

[29] W. H. Outlaw, Jr., and D. B. Fisher, *Plant Physiol.* **55**, 699 (1975).
[30] W. H. Outlaw, Jr., D. B. Fisher, and A. L. Christy, *Plant Physiol.* **55**, 704 (1975).
[31] E. Fritz, R. F. Evert, and W. Heyser, *Planta* **159**, 193 (1983).
[32] X.-D. Wang and M. J. Canny, *Plant, Cell Environ.* **8**, 669 (1985).
[33] D. R. Geiger, S. A. Sovonick, T. L. Shock, and R. J. Fellows, *Plant Physiol.* **54**, 892 (1974).

indicate that the pathway of [^{14}C]sucrose applied to a green leaf from the apoplast may be different from the pathway of assimilates after $^{14}CO_2$ fixation.

Technique of Microautoradiography. Microautoradiography of water-soluble compounds poses a serious technical problem since precise localization of the label on a cellular basis requires that neither loss nor dislocation of the water-soluble label must occur; that is, once the water is removed from the tissue either by freeze-drying or by freeze substitution, contact with water has to be avoided during the further processing of the tissue. A procedure for microautoradiography which uses freeze-drying has been described in detail by Fritz.[34] A detailed procedure which uses freeze substitution with special emphasis on the description of "dry" processing is given by Altus and Canny[35] and Wang and Canny.[32] The reader is also referred to the articles of Fisher[36] and Fisher and Housley[37] in which the advantages and drawbacks of the two procedures are evaluated.

In brief, the procedure comprises the following steps:

1. Application of label: either $^{14}CO_2$ fixation or uptake of [^{14}C]sucrose may be used (see previous and following sections).
2. Freezing: quick freezing in isopentane, liquid propane, or Freon 22 cooled by liquid nitrogen without cryoprotectant is usually sufficient for preserving the tissue structure at the light microscope level.
3. Removal of water and infiltration of organic solvent and plastic: two techniques can be used, either freeze-drying or freeze substitution.

FREEZE-DRYING. The frozen tissue is freeze-dried at low temperature and high vacuum. The dry tissue can be stored for several days without dislocation of the label, provided air humidity is kept low ($\leq 25\%$). The problems of infiltrating freeze-dried tissue are overcome by infiltrating the tissue with diethyl ether and plastic under high pressure.[34]

FREEZE SUBSTITUTION. The frozen tissue is transferred to a plastic miscible solvent (acetone, propylene oxide, or diethyl ether with 10% acrolein) containing a molecular sieve, cooled at $-70°$, and maintained at that temperature for several days. The volume ratio of solvent to tissue has to be high.[36] After slowly warming to room temperature, the tissue is (immediately) processed for embedding in plastic by conventional procedures with the added precaution that all solutions be dried over molecular sieves.

4. Cutting: sections of $1-2 \mu m$ are cut dry with a glass knife and transferred to microscopic slides coated with gelatin.

[34] E. Fritz, *Ber. Dtsch. Bot. Ges.* **93,** 109 (1980).
[35] D. P. Altus and M. J. Canny, *Plant, Cell Environ.* **8,** 275 (1985).
[36] D. B. Fisher, *Plant Physiol.* **49,** 161 (1972).
[37] D. B. Fisher and T. L. Housley, *Plant Physiol.* **49,** 166 (1972).

5. Processing for autoradiography: the slides are coated with nuclear emulsion (Ilford K5 or L4, or Kodak NTB2) by gently blowing a thin film of emulsion dried in a metal loop.[35] The slides are then kept in a dry box for several days or weeks, depending on the amount of radioactivity incorporated, and developed according to conventional procedures.

Electron Microscopy

As a structural prerequisite for symplastic movements of assimilates, the plasmodesmatal frequency between different cell types has to be sufficiently high to account for the observed translocation rates. Only a few quantitative electron microscopic studies are available to show that the number of plasmodesmata between mesophyll cells, bundle sheath cells, and phloem parenchyma allow for symplastic movement of assimilates along this route. The frequency of plasmodesmata between phloem (vascular) parenchyma and the sieve tube–companion cell complex is different in different plant species.[38-42] For instance, in maize[38] and *Amaranthus*[39] leaves the sieve tube–companion cell complex is virtually isolated symplastically from the surrounding parenchyma, a fact that can hardly be reconciled with symplastic loading, whereas in sugar beet[40] and *Cucurbita*[41] leaves plasmodesmata between phloem parenchyma and companion cells are abundant, i.e., an entirely symplastic pathway of assimilates could be assumed. On principle, it is premature to take the existence of cytoplasmic connections as proof for symplastic transport because our knowledge of the function and the regulation of plasmodesmata is still too scant.

The technique of injecting a membrane-impermeable fluorescent dye into the cytoplasm of a cell and following the spreading of the dye is a promising approach to study functional symplastic continuity. The dye may be injected either directly,[43] with the risk of vacuolar injection, or encapsulated into liposomes[44] which after injection into the vacuole can fuse with the tonoplast membrane and release the dye into the cytoplasm. By coinjection of compounds like calcium[43] and variation of environmental conditions more information on the regulation of the functional status of the plasmodesmata may be gained.

[38] R. F. Evert, W. Eschrich, and W. Heyser, *Planta* **138,** 279 (1978).
[39] D. G. Fisher and R. F. Evert, *Planta* **155,** 337 (1982).
[40] D. R. Geiger, R. T. Giaquinta, S. A. Sovonick, and R. J. Fellows, *Plant Physiol.* **52,** 585 (1973).
[41] R. Turgeon, J. A. Webb, and R. F. Evert, *Protoplasma* **83,** 217 (1975).
[42] D. G. Fisher, *Planta* **169,** 141 (1986).
[43] M. G. Erwee and P. B. Goodwin, *Planta* **158,** 320 (1983).
[44] M. A. Madore, J. W. Oross, and W. J. Lucas, *Plant Physiol.* **82,** 432 (1986).

pH Electrode and Membrane Potential Measurements

The mechanism of active sucrose uptake during phloem loading is postulated to be a sucrose–H^+ symport energized by the electrochemical gradient of protons across the plasma membrane of the sieve tube–companion cell complex. Therefore, a transient alkalinization of the medium is expected during sucrose uptake, and this has been shown in several cases with leaf tissue.[45-47] Again, this technique does not allow one to identify the cell type responsible for the pH change. This can be accomplished, however, by measuring membrane potential changes with a microelectrode. By putting the tip of a microelectrode into the sieve tube sap exuded from a cut aphid stylet, Wright and Fisher measured a decrease of the membrane potential after supplying sucrose to bark strips of willow, a finding consistent with a sucrose–H^+ symport mechanism for sucrose uptake into the sieve tube.[48] An advantage of this technique is the precise knowledge of the site of the electrode impalement; on the other hand no cell type but the sieve tube is accessible for such a measurement. Without the use of aphids, the cell type can be identified by labeling the cell with a (charged) fluorescent dye (e.g., Lucifer yellow) injected into the cell by a current pulse via the microelectrode after recording the membrane potential change.

Long-Distance Transport

General Comments

The translocation of substances within the sieve tubes from the phloem loading site to the sink region is covered under the term long-distance transport. Several features of long-distance transport can be determined, for example, the speed and mass transfer rate of transport; the direction of the flow of substances from a particular leaf, leaf part, or storage organ; the nature of substances which are transported and their concentrations; the factors (e.g., hormones, sink strength, temperature, water availability, and transpiration) which control or modulate transport speed and direction; and the specialization and diversity of sieve tubes within a bundle, or of bundles within an organ, with respect to transport speed, transport direction, and composition of the transported compounds.

[45] E. Komor, M. Rotter, and W. Tanner, *Plant Sci. Lett.* **9**, 153 (1977).
[46] W. Heyser, *Ber. Dtsch. Bot. Ges.* **93**, 221 (1980).
[47] S. Delrot, *Plant Physiol.* **68**, 706 (1981).
[48] J. P. Wright and D. B. Fisher, *Plant Physiol.* **67**, 845 (1981).

Two types of methods are described in detail: measurement of the translocation profile and sieve tube sap collection. The translocation profile of a particular transported compound can give information about the speed of transport and about its direction in case several plant parts or organs are monitored simultaneously. Measuring translocation profiles in plant parts either by external monitors or by extracting plant tissue will yield the translocation averaged over all bundles and sieve tubes. The collection of phloem sap by incision of only a small bark area or by an aphid which is feeding on one sieve tube, yields translocation profiles of particular bundles or sieve tubes.

Collection of phloem sap also allows analysis of the composition of phloem sap and the concentration of the transported compounds. The disadvantage of phloem sap collection is that it is an invasive technique involving opening of the sieve tube and allowing maximal outflow of sap, which probably changes several parameters of transport such as flow speed, sap concentration, and, perhaps, even shifting of the direction of phloem transport.

The direction of the flow of substances is also dependent on the vasculature of the plant and its organs and the place and frequency of connections between sieve tubes. Some morphological methods and results are described in Refs. 49–51.

Translocation Profile of Long-Lived Isotopes

The flow of assimilates can be followed in the conventional way by feeding $^{14}CO_2$ to a leaf, cutting the petiole and the stem in portions of equal length a certain time after feeding, and determining the radioactivity in the cut pieces. This method, used in steady-state labeling or pulse-labeling experiments, requires, however, an enormous amount of plant material in parallel sets to follow a detailed time course. The method of feeding the leaf is described by Geiger.[52] The extraction of tissue is principally the same as described before (chemical analysis). When high amounts of ^{14}C label are used, the translocation can also be followed with externally positioned detectors on the leaf blade.[53] Quantitative analysis of the radioactivity profile data can be complicated and is described elsewhere.[54,55]

[49] J. Stieber and H. Beringer, Bot. Gaz. (Chicago) 145, 465 (1984).
[50] T. C. Vogelmann, P. R. Larson, and R. E. Dickson, Planta 156, 345 (1982).
[51] J. T. Colbert and R. F. Evert, Planta 156, 136 (1982).
[52] D. R. Geiger, this series, Vol. 69, p. 561.
[53] D. R. Geiger and B. R. Fondy, Plant Physiol. 64, 361 (1979).
[54] P. E. H. Minchin and J. H. Troughton, Annu. Rev. Plant Physiol. 31, 191 (1980).
[55] P. Young, "Recursive Estimation and Time-Series Analysis: An Introduction." Springer-Verlag, Berlin and New York, 1984.

*Translocation Profile of Short-Lived Isotopes**

The application of short-lived isotopes with long-ranging radiation has the advantage that the tracer flow can be followed with monitors located along the outside of the plant shoot and can be repeated under slightly varied experimental conditions on the same plant, thus saving plant material and allowing excellent comparison of data.

Carbon-11 has a half-life of 20.4 min and decays emitting a β^+ (positron) which has a maximum path length in water of about 4 mm (mean of 1 mm) and is annihilated with a β^- (electron), producing a pair of antiparallel γ rays of 511 keV. Carbon-11, in the chemical form of $^{11}CO_2$, can readily be produced in quantities adequate for phloem research using a low-energy particle accelerator such as a 3-MeV Van de Graaff accelerator.[56]

Carbon-11 can be observed either by detecting the charged β^+ particle or by detecting the annihilation radiation. Geiger-Müller (GM) tubes are charged particle detectors and have been used with carbon-11.[57] The γ rays can also be detected with scintillation detectors, followed by pulse-height analyzers. This method gives a very large dynamic range that is extended by the use of attenuators which are sequentially removed as the tracer level decays. The main disadvantage of scintillation detection is that large quantities of lead shielding are needed to collimate the 511-keV γ rays. Coincidence counting, in which both of the annihilation γ rays are detected,[58] partially overcomes the difficulties of shielding but requires duplication of expensive detectors and their associated electronics.

The 511-keV γ radiation is not appreciably absorbed by plant material, soil, or air, so *in vivo* measurements are readily carried out. When following the tracer levels within phloem source or sink regions, it is possible to follow the carbon-11 for up to about 12 half-lives (i.e., about 250 min). When following the movement of ^{11}C-labeled photosynthate pulses along the phloem pathway, only 1 h or less of useful data can be obtained. The major disadvantage of ^{11}C is that it must be produced near the site of usage. Access to a particle accelerator is essential, but often small accelerators are

---

* Section prepared by P. E. H. Minchin and M. R. Thorpe.

[56] G. J. McCallum, G. S. McNaughton, P. E. H. Minchin, R. D. More, M. R. Presland, and J. D. Stout, *Nucl. Sci. Appl.* **1**, 163 (1981).
[57] D. S. Fensom, E. J. Williams, D. Aikman, J. E. Dale, J. Scobie, K. W. D. Ledingham, A. Drinkwater, and J. Moorby, *Can. J. Bot.* **55**, 1787 (1977).
[58] C. E. Magnuson, T. Fares, J. D. Goeschl, C. E. Nelson, B. R. Strain, C. H. Jaeger, and E. J. Bilpuch, *Radiat. Environ. Biophys.* **21**, 51 (1982).

available that are not being fully used as they have been superseded by larger machines.

The experimental method of observing the movement patterns of ^{11}C-labeled photosynthate is as follows. A clear Perspex chamber is sealed over part or all of the leaf, using a lanolin–calcium carbonate paste. Air is continuously drawn through the chamber using a suction pump vented well away from the experimental area. To load the leaf with [11]CO_2, the pump is turned off, and [11]CO_2 is injected into the chamber. After 1–10 min normal air flow is resumed, thus flushing unfixed [11]CO_2 from the chamber. The loading time is varied so as to obtain reasonable [11]CO_2 uptake. With maize 100% uptake is usual, while with wheat as little as 10% may be absorbed. Uptake depends on the area of leaf exposed to the [11]CO_2.

Radiation detectors are placed at various areas around the plant to monitor [11]CO_2 levels. A detector monitoring the loaded region will observe ^{11}C-labeled photosynthate prior to phloem loading as well as that just loaded into the sieve tubes. Since phloem transport occurs at a speed of about 1 cm min^{-1} in most species and the sieve tubes are very small, little of the observed tracer in the load zone will be within the phloem pathway. The shape of the tracer profiles seen downstream of the loaded zone is determined by the loading processing, while changes in profile shape seen between two pathway detectors depends on the phloem transport process.[59] A detector monitoring the entire plant downstream of any point will register tracer accumulation into this downstream sink. Many variations of this simple system are possible: for example, removal of the epidermis from a bean stem allows [^{11}C]sucrose to be exuded from the stem apoplast. Using surgically modified pea seeds still attached to the pod, phloem unloading into the seed coat can be followed.

Direct quantitative comparison between consecutive experiments is not simple. Since ^{11}C-labeling studies are necessarily short-term, and when pulse-labeling is used, ^{11}C levels and accumulation rates rarely have time to become constant. More complex, and more clear-cut, methods of analysis are needed.

Immediate changes in tracer profiles resulting from a treatment can be seen using a qualitative approach. Quantitative analysis based on simple heuristic measurements of tracer profile movement have been used and criticized.[59,60] Analysis based on the assumptions of compartmental analysis have been used and more recently model-independent methods have begun to be employed.[54,55]

[59] P. E. H. Minchin and J. H. Troughton, *Annu. Rev. Plant Physiol.* **31**, 191 (1980).
[60] M. J. Canny, *Biol. Rev. Cambridge Philos. Soc.* **35**, 507 (1960).

Collection of Sieve Tube Sap by Cutting *

The collection of phloem exudate (the mobile fluid of sieve tubes) is a process that is still not fully understood and must, for this reason, be regarded as something of an art. The following review of techniques gives guidance on the following components: (1) plant material, (2) wounding techniques, (3) sap collection methods, and (4) prolongation and control of exudate collection.

Not all plants exude phloem sap on wounding; indeed, most species generally exude only insignificant amounts of sap, regardless of how skillfully they are incised. The capacity to exude varies among species and even varieties. Nevertheless, a great many plants have yielded analyzable amounts of exudate. It is advisable to select species and varieties which have proven to yield consistent and reliable amount of phloem sap. A short list is given in Table I.

Material selection is important, and it is essential that it be in good physiological state. Generally, plants should be capable of vigorous sieve tube transport. Hence, all conditions which promote rapid growth are beneficial—good supplies of water, nutrients, light, and an appropriate temperature regime are all important factors. The rate of growth does not in itself appear to be the predominant factor, however, since many plants exude extremely well even after rapid growth has been suspended through the imposition of water stress.

Depending on the toughness and magnitude of the tissue involved, several methods are used to incise sieve tubes. Large trees, e.g., Manna ash and palms (e.g., coconut and *Arenga*), are cut with machetes, sickles, or strong knives.[61] Small plants, e.g., *Yucca,* or woody herbs, e.g., *Ricinus,* can be incised conveniently using razor blades.

Monocotyledonous tissues are cut transversely *in toto,* but dicotyledonous species are normally cut to the depth of the cambium so that the secondary phloem, but not the xylem, is severed. In all techniques a diagonal cut is made at 30° to 45° to the horizontal so that exudate will drain laterally to a collection point. It is essential that the blade severing the phloem be very sharp: experience indicates that slicing is more effective than a crushing action. Many tissues contain mineral crystals, e.g., oxalates and silica, which can blunt blades, and therefore their renewal or resharpening is very important. The slicing action should "skate" over the cambium and xylem quite delicately.

Recently, solid needles or syringe needles have been used successfully

---

* Section prepared by J. A. Milburn.

[61] J. Kallarackal, *Sci. Rep.* **12,** 172 (1975).

TABLE I
PLANT MATERIALS, REGIONS TAPPED, LOCATION OF INCISION, AND APPROXIMATE
MAXIMAL RATES OF FLOW

Category	Genus, species, and variety	Tissue tapped	Incisions	Vol. hr^{-1}, max.	Ref.
Monocots					
Palms	*Yucca, Agave*	Inflorescence	Whole	Several cm^3	a, b
	Arenga, Cocos, Phoenix	Inflorescence	Whole and side	Many cm^3	
Dicots					
Manna and American ashes	*Fraxinus ornus, F. americana*	Main trunk (bark)	5-cm cut	Up to 1 cm^3	c
Castor bean	*Ricinus communis* cv. *gibsonsii*	Stem, bark, petioles, peduncles	1-cm cut/ whole	Up to 1 cm^3	d
Squashes	*Cucurbita* spp.	Stems, petioles	Whole	Up to 1 cm^3	e
Legumes	*Lupinus alba*	Stems, petioles, pods	½-cm wound	Up to 10 mm^3	f

[a] J. Kallarackal, *Sci. Rep.* **12**, 172 (1975).
[b] J. Van Die and P. M. L. Tammes, *Encycl. Plant Physiol., New Ser.* **1**, 196 (1975).
[c] M. H. Zimmerman and H. Ziegler, *Encycl. Plant Physiol., New Ser.* **1**, 480 (1975).
[d] J. A. Milburn, *Planta* **117**, 303 (1974).
[e] A. S. Crafts, "Translocation in Plants." Holt, Rinehart & Winston, New York, 1961.
[f] J. A. Smith and J. A. Milburn, *Planta* **148**, 35 (1980).

to promote exudation. In the cryopuncture technique[62] a needle with a finely pointed tip and previously cooled in liquid nitrogen is used to puncture the tissue, incising the vascular trace, and is held in position for 5 sec or longer. Unfortunately, the cryopuncture method does not work with all species but is effective on cowpea (*Vigna unguiculata* L.).

Palms have been tapped since antiquity by placing the severed organ, often trained to incline downward over a period, into a hollow gourd or large pot. Manna ash is tapped by collecting the dried sap from either a collection bowl or the bark itself. Purer sap can be collected in volumetric glassware or plastic bags if the sap can be led from the collection point via some form of tubing.

In the laboratory, phloem sap is most conveniently and accurately collected into precision-fabricated microcapillaries. Surface tension draws the liquid horizontally from the collection point. If a scale is fitted along-

[62] J. S. Pate, M. B. Peoples, and C. A. Atkins, *Plant Physiol.* **74**, 499 (1984).

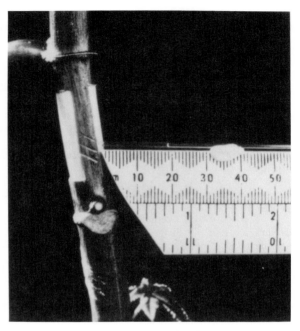

Fig. 1. Collection of phloem sap from bark incisions in a *Ricinus* internode. Sap is being collected from third incision in series. The stem has been partially ringed to isolate a longitudinal strip of bark. Fine divisions of the upper scale on the ruler are millimeters. (From Smith and Milburn.[63])

side the tube, the volume of exudate can be measured per unit time, hence exudation rates can be calculated (Fig. 1).[63]

It may be necessary to rely on the waxy surface on the cuticle or bark to prevent sap from escaping down the stem. Such losses will occur if exudation continues after the horizontal microcapillary becomes overfilled. One can tilt the capillary slightly and fit a collection bottle to the distal end for prolonged collections. However, the handling of microcapillaries is a delicate matter. Fine tubes hold sap quite well, whereas air bubbles rapidly displace the sap in wider tubes unless the tube is slightly inclined to the horizontal or one end is physically blocked.

Various methods can be used to enhance the duration of exudation. Plants cannot exude if the loading capacity for solutes is exhausted, and the composition of the sap may be altered. The availability of materials is

[63] J. A. C. Smith and J. A. Milburn, *Planta* **148**, 35 (1980).

dependent on stored materials from previous growth and also good growth conditions during the exudation process itself.

Cooling the severed flower stalk in an ice-filled Dewar flask has been found to prolong exudation in *Yucca,* presumably because the sealing processes are slowed at low temperature. The cryopuncture method presumably also stops immediate sealing responses because subsequent exudation can be prolonged for many hours or days, though the quantity of exudate is small. It has been reported that chelating agents can prolong exudation, but more recent evidence indicates that such "exudation" is seepage from phloem parenchyma rather than from sieve tubes themselves.

Several methods have been used to ensure that the cross section of sampled sieve tubes remains constant. One is a (previously performed) surgical treatment whereby strips of intervening bark are removed. This ensures that a blade *always* severs the same remaining cross section of bark through which longitudinal transport takes place.[63] Another technique which may be useful is that of bark compression applied some distance from the bleeding incision which can stop exudation instantly. The advantage of this technique is that, at least in *Ricinus* and *Cocos,* the compressed tissue regains its capacity to conduct longitudinally after a recovery period of 1 h or more.

A potential problem of phloem sap collection by exudation from incision is contamination with xylem sap. In case the plant is transpiring the negative pressure in the xylem vessels will prevent xylem sap from exuding (but will lead to some loss of phloem sap), but when a petiole or stem is cut or transpiration is low, root pressure will cause exudation of xylem sap at the wound. Since no unambiguous xylem sap marker can be recommended, the only way to estimate contamination by xylem sap is to stop phloem transport by compression of the bark or by treatment of the wound with steam.

Collection of Sieve Tube Sap by Aphid Stylets*

General Techniques with Aphids. The interest of plant physiologists in aphids is based on the fact that these insects feed on sieve tubes of angiosperms or on sieve cells of gymnosperms, pteridophytes, and even mosses. Most aphids prefer the vascular bundles of leaves and young shoots, but some species, the lachnids, feed on secondary phloem of twigs and branches. Most aphids can be maintained year-round on fresh host plants, which are more or less specific for each aphid species.[64] Some aphid species

---

* Section prepared by W. Eschrich.

[64] D. B. Fisher and J. M. Frame, *Planta* **161,** 385 (1984).

have been successfully kept on artificial diets for about 40 generations.[65] Their enemies are ladybugs, red spiders, and lacewings (aphid lions). Cages should be lined with nylon fabric of about 300 μm mesh width and irradiated continuously, but not exceeding 60 μEq m^{-2} sec^{-1}. Phloem-feeding insects other than aphids seldom are used.[64]

Honeydew. The excrement of aphids, honeydew, is secreted in fairly constant intervals. Honeydew is secreted only when the aphid has pierced a sieve element, thus, honeydew consists mainly of sieve tube sap. One droplet of honeydew contains the content of about 5000 sieve elements. Honeydew can be collected on a turntable attached to a 24-hr clock, and when the sieve tubes carry ^{14}C-labeled compounds, the resulting honeydew chronogram can be autoradiographed and their mobility, in principle, can be tested according to the appearance of labeled honeydew. However, phloem mobility can be altered, because the sucking of the aphid can attract the labeled compound not only from far away in a sieve tube, but also from the apoplast and adjacent parenchyma.

Aphid Stylets. When an aphid is severed from its mouth parts while sucking sieve tube sap, the high turgor of the sieve tube can cause the stylet stump to continue to produce sieve tube exudate. The exudate can be collected with a glass capillary, fastened to a micromanipulator, under the microscope. The advantages of obtaining stylet exudate are that (1) it is true sieve tube content, not contaminated with juices of the aphids intestines and not altered by the action of the aphids' enzymes, and (2) the stylet is buried in a canal of solidified saliva, thus preventing any contamination or dilution with apoplastic solutes of water.

The stylet exudate is the purest sieve tube sap which can be obtained. However, it must be considered that any turgor release in a sieve tube causes surging of its contents, and callose deposition. Surging can be accompanied with secretion of material from companion cells, especially enzymes and oligonucleotides.

Cutting Stylets. Aphid stylets are cut with (1) the microscope laser[64,66,67] or (2) the radio-frequency microcauter (RFM).[64,68,69] The aphids usually are directed to certain areas of the experimental plant in clip cages, made from a hair clip to which a suitable aerated cage is attached (Fig. 2). After a few hours or overnight, most aphids have settled and started to produce honeydew. Aphids may reproduce in such cages. They usually feed from

[65] P. Eberhardt, *Z. Vergl. Physiol.* **46**, 169 (1962).
[66] C. A. Barlow and M. E. McCully, *Can. J. Zool.* **50**, 1497 (1972).
[67] C. A. Barlow and P. A. Randolph, *Ann. Entomol. Soc. Am.* **71**, 46 (1978).
[68] N. Downing and D. M. Unwin, *Phys. Entomol.* **2**, 275 (1977).
[69] D. M. Unwin, *Phys. Entomol.* **3**, 71 (1978).

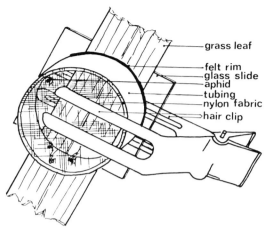

grass leaf

felt rim
glass slide
aphid
tubing
nylon fabric
hair clip

FIG. 2. Cage for directing aphids on a grass leaf. (Drawn by W. Eschrich.)

the lower side of a leaf, the phloem side. On branches, aphids may be encaged simply between two girdles of stiff grease.

When using the RFM, the microantenna may be touched to the aphid mouth part in any position. Some aphid species are very nervous, and they will retract the stylets if they are not narcotized before with a gentle stream of CO_2. [In some countries, RFM, which operates at citizen band (CB) frequencies, must have a permit from the post office.]

The microscope laser is preferentially mounted on a vibration-reduced table, on which also the plant with the aphids are placed. Either the microscope with the attached laser or the aphid must be movable in three dimensions. The microscope is usually mounted in a horizontal position and must have objectives with long operating distances (at least 6 mm). The outlet of the laser beam is attached to the tubes of a microscope camera. Convenient are lasers like neodymium–glass lasers which operate in the near-infrared (> 1000 nm) because this light can be absorbed by thick glass which protects the operator's eyes. The laser beam has to be centered on black target paper with a hairline across inserted in the eye piece. Microscope laser combinations are not yet available on the market. (Zeiss, Oberkochen, FRG, and Lasertechnik GmbH, 6056 Heusenstamm, FRG, produce such combinations on request.)

Stylet Exudate. Stylet exudate contains sucrose in fairly high concentrations (40% or more), but since collection in capillaries can take hours or even days, considerable evaporation of water occurs. For exact quantitative determinations, the exuding stylet stump can be wrapped in a drop of oil.[70] For determination of amino acids, hormones, and inorganic components in the sieve tube exudate, techniques of gas chromatography (GC), high-performance liquid chromatography (HPLC), and atomic absorption spectroscopy are indispensible. Stylet stumps differ greatly in their exuding capacities, and exudation can vary during the period of collection. Sucrose concentration of two stylet stumps positioned at the same leaf may differ considerably.

Phloem Unloading

General Comments

The term phloem unloading denotes the pathway of assimilates out of the sieve tube into heterotrophic, growing, or storage tissue like young (sink) leaves, stems, fruits, and roots. The same questions outlined for phloem loading concerning symplastic versus apoplastic pathways, localization of active transport sites, and involvement of cell wall-bound invertase also have to be solved at this stage of phloem transport. Different mechanisms of unloading appear to be operative within different sink tissues. Data from sink leaves (sugar beet[71]) and roots (pea[72] and corn[6,73]) have been interpreted to support the view of an entirely symplastic pathway of assimilates from the sieve tube to the final destination cells, whereas an apoplastic step seems to be involved during unloading in the stem (sugarcane[74] and bean[75]) and has to occur in fruits (maize,[76] soybean,[77] and beans[78,79]) because of the lack of symplastic connections between maternal and embryonic tissues. According to the frequency of plasmodesmata between sieve tube–companion cell complexes and parenchyma cells, the apoplastic step has been proposed to be localized in the parenchyma cell region, not at the sieve tube–companion cell complex itself.[76,77,79]

For symplastic unloading a sucrose gradient has to be maintained either by metabolic conversion of sucrose in the cytoplasm of the sink cells,

[70] S. Kawabe, T. Fukumorita, and M. Chino, *Plant Cell Physiol.* **21**, 1319 (1980).
[71] J. Gougler Schmalstig and D. R. Geiger, *Plant Physiol.* **79**, 237 (1985).
[72] P. S. Dick and T. Ap Rees, *J. Exp. Bot.* **26**, 305 (1975).
[73] R. D. Warmbrodt, *Bot. Gaz. (Chicago)* **146**, 169 (1985).
[74] K. T. Glasziou and K. Gayler, *Bot. Rev.* **38**, 471 (1972).
[75] P. E. H. Minchin and M. R. Thorpe, *J. Exp. Bot.* **35**, 538 (1984).
[76] F. C. Felker and J. C. Shannon, *Plant Physiol.* **65**, 864 (1980).
[77] J. H. Thorne, *Plant Physiol.* **67**, 1016 (1981).
[78] P. Wolswinkel and A. Ammerlaan, *Planta* **158**, 305 (1983).
[79] C. E. Offler and J. W. Patrick, *Aust. J. Plant Physiol.* **11**, 79 (1984).

promoted by neutral invertase or sucrose synthase, or by hydrolysis of sucrose in the vacuole by an acid invertase. The activity of at least one of these enzymes has to be high in tissues where symplastic unloading occurs.

Although reasonably postulated, the involvement of carrier proteins in apoplastic unloading has not yet been proved, nor has the requirement for metabolic energy been established as a general feature of assimilate release into the apoplast. Recent investigations using protein [p-chloromercuribenzenesulfonic acid (PCMBS)] and metabolic [KCN, carbonyl cyanide m-chlorophenylhydrazone (CCCP), azide] inhibitors have produced differing results.[78,80,81] Since it could be deduced from the direction of the concentration gradient at the apoplastic step whether an active transport mechanism has to be postulated or whether a facilitated diffusion mechanism would be sufficient, the sucrose concentrations in the apoplast and the adjacent parenchyma have to be determined. This has been done for the soybean seed coat,[82] and results demonstrate that sucrose release into the apoplast is an active transport process.

During apoplastic unloading a cell wall-bound invertase may[74,76] or may not[83] be involved, depending on the species. Therefore, its role as a "reflux valve," as suggested by Eschrich,[84] cannot be generalized. The presence of a cell wall-bound invertase, together with a very active hexose uptake system in tissues where symplastic unloading has been shown to be operative,[6] points to a possible additional pathway of sucrose unloading into the apoplast and retrieval as hexoses, which may undergo a different metabolic fate than incoming sucrose, thereby increasing sink strength and regulating partitioning of incoming assimilates into different metabolic pools in the absence of neutral invertase. In sink tissues as well as in source tissue the cell wall-bound invertase may be associated with a special cell type; this could be identified by immunohistochemical techniques.[85]

In principle, the same approaches as described for phloem loading can be followed for phloem unloading. These include radioactive tracer studies, microautoradiography, and electron microscopy. A technique for studying apoplastic unloading within legume seed coats is presented here. The method should also be applicable to other plant species with relatively large seeds.

[80] J. W. Patrick, Z. Pflanzenphysiol. **111**, 9 (1983).
[81] G. A. Porter, D. P. Knievel, and J. C. Shannon, Plant Physiol. **77**, 524 (1985).
[82] R. M. Gifford and J. H. Thorne, Plant Physiol. **77**, 863 (1985).
[83] J. H. Thorne, Plant Physiol. **65**, 975 (1980).
[84] W. Eschrich, Ber. Dtsch. Bot. Ges. **93**, 363 (1980).
[85] L. Faye and A. Ghorbel, Plant Sci. Lett. **29**, 33 (1983).

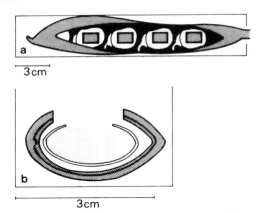

FIG. 3. Diagrammatic representation of the procedure for obtaining "empty" seed coats of *Vicia* and measuring the release of sugar and amino acids. (a) The pod is placed in a trough of metal in such a way that both halves of the pod are in a more or less horizontal plane. Subsequently, a window is made in the upper half of the pod wall and the four ovules are surgically treated. (b) As shown in cross section, an "empty" ovule is filled with a buffered solution. During the experiment the bottom of the trough is covered with a layer of tissue paper moistened with water and the trough is covered with aluminum foil to maintain a high humidity. (From Wolswinkel and Ammerlaan.[78])

Techniques of Measuring Phloem Unloading in Legume Seed Coats*

The development of legume seeds is characterized by a continuous transfer of assimilates from tissues of maternal origin (seed coat) to the embryonic tissues. Photosynthate imported from the pod wall must pass the seed coat before it is available for uptake by the embryo. For soybean[77] and *Phaseolus vulgaris*[79] data on phloem in the seed coat have been published. Recently a technique has been developed to measure unloading of assimilates from the seed coat of developing legume seeds. After removal of the embryo from a developing ovule, the "empty" ovule can be filled with a solution to measure assimilate release from the seed coat (Fig. 3). The sites of sucrose and amino acid unloading in the seed coat are accessible via the seed coat apoplast and can be challenged with inhibitors or other solutes to characterize the unloading process.

Seed Coat Preparation. The pod is placed in a small chamber (e.g., a trough) in such a position that both halves of the pod wall are in a more or less horizontal plane. First, an approximately rectangular window is made in the upper half of the pod wall. Care is taken to prevent damage to the

---

* Section prepared by P. Wolswinkel.

ventral or dorsal vascular bundles of the pod. Subsequently, in experiments with pea or broadbean, a more or less rectangular incision is made in the seed coat of several ovules. The part of the seed coat which is excised from each ovule during this procedure represents about 10% of the seed coat. After removal of the embryo from a number of developing seeds (during this procedure each embryo is cut into pieces), the pod is placed in a horizontal position in the "operating room," and the empty ovules are filled with a buffered solution [e.g., 2 or 10 mM 2-N-morpholinoethanesulfonic acid (MES) buffer, pH 5.5]. During the experiment, the bottom of the chamber is covered with a layer of tissue paper moistened with water, and the chamber is covered with aluminum foil to maintain a high humidity. The empty ovules (without embryo but filled with solution) remain attached to the pod vascular bundles via the funiculus. The formation of a funicular abscission layer in nearly mature seeds can add considerable difficulty to the surgical treatment, but until the stage of development in which the embryo has attained about one-half the final dry weight, this is not a serious problem in *Pisum sativum* and *Vicia faba*.

Pulse-Labeling Procedures. Several procedures have been described in recent reports.[78,80,86,87] In most cases, labeled compounds are administered to the plant after the surgical treatment. The exposure of a leaf or a leaf part to $^{14}CO_2$ in an assimilation chamber will produce a ^{14}C-labeled photosynthate pulse. Labeled compounds can also be applied to an abraded leaf. In double-label experiments with a mixture of ^3H and ^{14}C isotopes of sucrose and/or amino acids, label can also enter the plant via the petiole subtending the fruit used for the experiment. First, the leaflets of a compound leaf are removed and the rachis is severed about 1 cm below the site of attachment of the uppermost leaflets. Immediately after severing the midvein, it is placed in a narrow glass tube containing labeled solutes in water (e.g., 0.1 ml in the case of pea). This solution is rapidly taken up into the plant (in most cases within 15–30 min). The distribution pattern of ^3H and ^{14}C introduced in this way indicates a rapid entry from the xylem into phloem cells.

Exudate Collection. During the course of an experiment (10–12 hr), the solution filling an empty ovule is regularly collected for analysis and a fresh solution transferred into the seed coat cavity. Standard methods of radioactivity measurements or chemical analysis can be used for measurements of solutes present in the seed coat exudate or solutes present in seed coat extracts.

Comments. The rate of phloem transport of sucrose and amino acids

[86] J. H. Thorne and R. M. Rainbird, *Plant Physiol.* **72**, 268 (1983).
[87] P. Wolswinkel and A. Ammerlaan, *Physiol. Plant* **61**, 172 (1984).

into "empty" ovules is dependent on the solute concentration in the solution filling empty ovules. In experiments with pea, a solute (e.g., sucrose or mannitol) concentration of about 400 mM is optimal for sucrose and amino acid transport into the cavity within the empty ovules. The results were comparable to transport into intact ovules. When empty ovules are excised from the maternal plant by cutting the funiculus, the release of solutes from the seed coat is completely dependent on material present in the seed coat at the moment of cutting. By comparing solute release from excised seed coats with solute release from attached seed coats, it is possible to discriminate between effects on unloading from the seed coat, and effects on assimilate transport into the seed coat.

Acknowledgments

The authors are extremely grateful to Drs. W. Eschrich (Göttingen), P. E. H. Minchin and M. R. Thorpe (Lower Hut), B. R. Fondy, D. R. Geiger, and S. Delrot (Dayton), J. A. Milburn (Armidale), P. Wolswinkel (Utrecht), and D. B. Fisher (Pullman) for giving detailed descriptions of methods and comments on them.

[22] Patch Clamp Measurements on Isolated Guard Cell Protoplasts and Vacuoles

By KLAUS RASCHKE and RAINER HEDRICH

Guard cell protoplasts provide a suitable material for ion transport studies for three reasons. (1) The osmotic motor of the guard cells regulates gas exchange of leaves relatively rapidly. Large ion fluxes per unit area of plasma membrane are necessary to produce the volume changes in the guard cells that are required to open and close the stomatal pores. Channels and pumps in the plasmalemma and tonoplast of guard cells are expected to be numerous or of larger capacity than in other plant cells.[1] (2) Because, in the whole leaf, gas exchange has to be regulated in response to changes in external and internal factors,[2] mechanisms that control ion fluxes are presumably highly developed in guard cells. (3) Because ion

---

[1] K. Raschke, R. Hedrich, U. Reckmann, and J. I. Schroeder, *Bot. Acta* **101**, 283 (1988).
[2] K. Raschke, *Encycl. Plant Physiol., New Ser.* **7**, 382 (1979).

transport into guard cells requires energy, and because metabolism of organic anions is involved in stomatal movement, guard cells lend themselves to a study of the mutual relationships between ion transport, dissimilation of carbohydrates, and their interconversion with organic acids.

Guard cells can serve as models for other cells which export or import ions and break down carbohydrates in order to make organic acids for balancing cation fluxes. Such cells occur in roots, growing tissue, and in the pulvini of leaves of plant species displaying nastic leaf movements.

Guard cells are embedded in epidermal tissue. Investigations into processes occurring in these cells require an enzymatic isolation of wall-free protoplasts from the rest of the tissue. Removal of the cell walls is necessary for the elimination of the large exchange capacity for cations that resides in the walls and that would interfere with measurements of cation import and export.

Guard cell protoplasts were first prepared by Zeiger and Hepler in small numbers,[3] and in quantities suitable for biochemical analysis by Schnabl *et al.*[4] The glycolytic activity of guard cell protoplasts was shown to be high enough to account for the observed organic acid production.[5] Fitzsimons and Weyers presented evidence for comparable responses to CO_2 and light and comparable rates of ion transport of guard cells in complete tissue and in guard cell protoplasts.[6,7]

Preparation of Guard Cell Protoplasts and Vacuoles

Guard cells occupy not more than 0.1–1% of a thin leaf's volume. Therefore, the only way to prepare large numbers of viable guard cell protoplasts is to start from epidermal pieces. Epidermal tissue can be obtained either by peeling from leaves with detachable epidermis or by fractionation of leaf tissue after it has been minced in a blender.[8] The latter method allows guard cell protoplasts to be prepared from leaves of species hitherto not amenable to direct investigations into the ion transport and metabolism of their guard cells. We give instructions for the peeling method first and then list modifications to be followed for the application of the blender method.

[3] E. Zeiger and P. K. Hepler, *Plant Physiol.* **58**, 492 (1976).

[4] H. Schnabl, C. H. Bornman, and H. Ziegler, *Planta* **143**, 33 (1978).

[5] R. Hedrich, K. Raschke, and M. Stitt, *Plant Physiol.* **79**, 977 (1985).

[6] P. J. Fitzsimons and J. D. B. Weyers, *Physiol. Plant.* **66**, 463 (1986).

[7] P. J. Fitzsimons and J. D. B. Weyers, *Physiol. Plant.* **66**, 469 (1986).

[8] R. Hedrich, I. Baumann, and K. Raschke, *Plant Physiol.* Suppl., Abstr. No. 885 (1989).

Plants Suitable for Guard Cell Protoplast Preparation from Peeled Epidermis

There are only a few species which possess an easily detachable epidermis with guard cells of fair metabolic activity: *Allium cepa* L. (onion),[4] *Commelina communis* L. (day flower, spider wort),[9] *Pisum sativum* L. (garden pea) Argenteum mutant,[10] *Tulipa* hybrids (tulip), and *Vicia faba* L. (broad bean).[4] Growing conditions determine cell wall composition and amounts of secondary substances in the cells which in turn determine the concentration at which cell wall-digesting enzymes will have to be applied and the time of incubation. Standardized cultivation of plants is essential for successful preparation of protoplasts of predictable performance. The choice of a particular growing condition is, however, less important than strict adherence to one that has been established. Similarly, the choice of a particular cultivar of one of the species mentioned above does not appear to be critical, with the exception of the *Arg* mutant of pea. The information given in Table I on the cultivation of plants suitable for preparation of guard cell protoplasts is meant to facilitate a start in this field of work.

Procedure for Preparation of Guard Cell Protoplasts from Peeled Epidermis

General Procedure

1. Peel epidermis with Dumont No. 5 watchmakers' forceps. Cut strips in a puddle of water with razor blade into 10 mm² pieces. Collect the pieces in several dishes on a relatively large volume of solution I (e.g., 100 ml each in 95 mm diameter crystallization dishes) for dilution of inhibitory compounds released from damaged cells. The composition of solution I is given in the next section.

2. Let dishes stand on ice for 15 min (primarily for *V. faba*). Then sonicate to rupture epidermal cells and remove adhering mesophyll debris.

3. Preplasmolyze guard cells in solution II (necessary for some but not all of the suitable species).

4. Rinse epidermal pieces on a large (200 or 400 μm) mesh screen mounted in a frame of about 95 mm diameter (screens, for instance, from Heidland Technische Gewebe, Postfach 1921, 4830 Gütersloh 1, FRG). Use a Pasteur pipette for applying the rinsing solution. Pick up epidermal

---

[9] P. J. Fitzsimons and J. D. B. Weyers, *J. Exp. Bot.* **34,** 55 (1983).

[10] P. C. Jewer, L. D. Incoll, and J. Shaw, *Planta* **155,** 146 (1982).

[11] K. Gotow, K. Shimazaki, N. Kondo, and K. Syōno, *Plant Cell Physiol.* **25,** 671 (1984).

[12] E. J. Hewitt and T. A. Smith, "Plant Mineral Nutrition." English Universities Press, London, 1975.

TABLE I

PLANTS FOR PREPARATION OF GUARD CELL PROTOPLASTS AND CONDITIONS FOR CULTIVATION

Parameter	Vicia faba	Pisum sativum	Commelina communis	Allium cepa	Tulipa hybrids
Cultivars	Weisskeimige Hangdown or Long Pod	Argenteum mutant, originally from G. A. Marx, Cornell Univ., Ithaca, NY	Seeds originally from T. A. Mansfield, Univ. of Lancaster, U.K.	Gelbe Zittauer or Stuttgarter Riesen	Commercial hybrids
Germination in	Compost[a]	Vermiculite, compost for greenhouse cultivation	Vermiculite, compost for greenhouse cultivation	Vermiculite	Vernalize at 4° for 6–8 weeks, vermiculite
Cultivation in	Greenhouse	Growth chamber (or greenhouse)	Growth chamber (or greenhouse)	Greenhouse	Greenhouse
Substrate	Compost[a]	Clay pebbles for hydroponics[b]	Clay pebbles for hydroponics[b]	Vermiculite	Vermiculite
Quantum flux, μmol m^{-2} sec^{-1}	300 (metal halide)	170–250 (fluorescent tubes)	250 (fluorescent tubes)	100 (daylight and fluorescent tubes)	100 (daylight and fluorescent tubes)
Light period, hr	14	12	12	16	16
Temperature, day/night, °	20/14	20/14	20/14	20/17	20/17
Relative humidity, %	65–70	40	50/66	70–80	70–80
Leaves ready after	2–3 weeks	4–5 weeks	5–6 weeks	12 days	12–14 days
Leaves used	Youngest two fully expanded	Youngest two fully expanded	Youngest two fully expanded	Second and third if longer than 5 cm	First two before plant flowers
Alternate procedure	Ref. 11	Ref. 10	Ref. 8	Ref. 4	—

[a] Potting mixture, e.g., Fruhstorfer Einheitserde, Type A.
[b] Long Ashton nitrate-type nutrient solution.[12] Potting mixture is suitable for cultivation in the greenhouse.

strips with forceps and transfer them into several dishes with 25 ml each of the incubation medium (solution III).

5. Incubate in solution III in a shaking water bath (e.g., reciprocating at 28 excursions/min, amplitude 2 cm) at constant temperature. Interrupt incubation when the protoplasts of the ordinary epidermal cells have been released and their collection is desired (not suitable for *V. faba*).

6. Separate protoplasts from epidermal debris by passage through a large mesh nylon screen mounted in a frame of about 95 mm diameter. Rinse with solution IV until all guard cell protoplasts have been released; inspect a few epidermal strips under a microscope.

7. Pass the filtrate of step 6 (containing guard cell protoplasts) through a narrow mesh nylon screen mounted across the bottom part (~ 25 mm wide) of a powder funnel. Choose a mesh size that will just allow guard cell protoplasts to pass (as a rule 20 μm). Collect filtrate in 50-ml centrifuge tubes (round bottomed).

8. Centrifuge; remove the supernatant with a Pasteur pipette. Resuspend the sediment in a smaller volume of solution IV than in step 7.

9. Purify the protoplast suspension by centrifugation through a Percoll gradient, solution(s) V. Count the number of cells in a small sample, using a hemocytometer.

10. Finally centrifuge and resuspend the protoplasts. Choose a volume of final medium (in general solution IV) which will result in the desired number of cells per unit volume.

Note: Avoid too large a shearing stress on the protoplasts. Therefore, always accelerate the centrifuge *slowly* and decelerate *slowly;* do not use the brake. All centrifugations are to be done at 4°.

Solutions and Specific Instructions

First, we list the compositions of the solutions needed. Then we go through each specific procedure, following the sequence of the general description noted above.

All solutions required for the preparation of guard cell protoplasts are 1 mM with respect to CaCl$_2$. If Cl$^-$ is to be excluded (in order to stimulate maximal production of organic acids by the guard cells), calcium iminodiacetate [Ca(IDA)$_2$] can be used in place of CaCl$_2$. During the process of peeling, epidermal strips are collected on solution I. The following composition for solution I can be used for all species (prepare 400 ml): 10 mM sodium ascorbate and 1 mM CaCl$_2$ or Ca(IDA)$_2$. The pH of this solution will be about 6.4. The compositions of Solutions II, III, IV, and V are listed in Table II.

Note: D-Mannitol serves as an osmoticum in solutions II to V. Com-

TABLE II

SOLUTIONS II, III, IV, AND V FOR PREPLASMOLYSIS OF GUARD CELLS, CELL WALL DIGESTION, RINSING PROTOPLASTS, AND GRADIENT CENTRIFUGATION

Composition	V. faba[a]	P. sativum	C. communis	A. cepa IIa	A. cepa IIb	Tulipa
Solution II						
Prepare 100 ml						
D-Mannitol	—	—	300 mM	200 mM	400 mM	400 mM
Sodium ascorbate	—	—	10 mM	10 mM	10 mM	10 mM
$CaCl_2$	—	—	1 mM	1 mM	1 mM	1 mM
Solution III[b]						
Conveniently prepare	100 ml	200 ml	100 ml	100 ml		100 ml
D-Mannitol	400 mM	350 mM	300 mM	700 mM	400 mM	400 mM
Sodium ascorbate	10 mM	10 mM	10 mM	10 mM		10 mM
$CaCl_2$	1 mM	1 mM	1 mM	1 mM		1 mM
Cellulase (Onozuka RS)	1–4% (w/v)	1% (w/v)	2.25% (w/v)	1% (w/v)		2% (w/v)
Pectinase (z.B. Mazerozyme)	0.5–1%	0.5%	0.625%	0.25%		0.5%
BSA	0.5%	0.5%	0.75%	0.5%		0.5%
pH	5.5	5.5	5.3	5.3		5.3
Solution IV		IVa / IVb				
Prepare	100 ml	100 ml / 100 ml	100 ml	200 ml		200 ml
D-Mannitol	500 mM	350 mM / 400 mM	300 mM	700 mM		700 mM
$CaCl_2$	1 mM	1 mM / 1 mM	1 mM	1 mM		1 mM
pH 6.0						
Solution V						
Percoll	—	—	10 ml	1.2 ml		6 ml
D-Mannitol	—	—	—	—		0.765 g (solid)
$Ca(IDA)_2$, 50 mM	—	—	—	—		120 μl
Solution IV	—	—	—	4.8 ml		—

[a] Alternatively[11]: D-mannitol, 250 mM; cellulase (Onozuka RS, Yakult Honsha, Tokyo, Japan), 2%; Pectolyase Y-23 (Seishin Pharmaceutical, Tokyo, Japan), 0.02%; BSA, 0.2%; pepstatin A, 1 μg/ml; $CaCl_2$, 1 mM.

[b] For incubations of long duration (e.g., overnight), add antibiotic, like kanamycin sulfate, to give 0.1% (w/v).

mercial mannitol frequently contains the phytohormone abscisic acid, in quantities which result in final concentrations around 1 μM in the media. For removal of abscisic acid, add activated carbon to a stock solution of mannitol, shake for 1 min, pour through a prepleated filter, and centrifuge at 3500 g until clear. Osmotica other than mannitol can be used, for instance, sorbitol or glycine betaine. At equal molality, the osmolality of betaine is higher than that of mannitol. It is necessary to determine the osmolality of the solutions before use (e.g., in a Wescor 5100 vapor pressure osmometer).

Enzyme solutions must be centrifuged for removal of solid materials in commercial enzyme products: 3500 g (or more) for 5 to 10 min. As with the cultivation of the plants, the compositions of the wall-digesting solutions we give, as well as the durations of the various steps of the isolation procedures, may need modifications. For instance, plants grown in strong light may require a higher osmolality for plasmolysis of the guard cells than plants cultivated at a low light level, and we found that the activity of some of the enzymes used for cell wall digestion can vary from batch to batch by a factor of 4, in spite of comparable specifications (possibly because the activities of accompanying hemicellulases and other substances which contribute to the overall digestive activity vary among batches).

Procedures for the Various Plant Species

Vicia faba. The lower epidermis of 50 to 70 leaflets will yield 4 to 6 × 10⁶ guard cell protoplasts.

1. Collect epidermal peels on solution I.
2. Let dishes stand on ice for 15 min. Then sonicate for 30 to 60 sec in ice (sonicator: Branson B 15, large head, pulsed, 50% duty cycle, output control 7). Omit step 3.
4. Rinse on a 400 μm screen with deionized water.
5. Incubate strips in solution III at 26° for 2 to 3 hr (for a modification, see Ref. 11).
6. Pass through 200 μm screen, rinse with solution IV.
7. Pass through 20 μm screen.
8. Centrifuge at 100 g for 7 min. Remove supernatant. Resuspend in solution IV. Count cells in a sample. Omit step 9.
10. Centrifuge at 80 g for 7 min, remove supernatant. Resuspend in desired volume.

Pisum sativum argenteum Mutant. The upper and lower epidermis of 50 leaflets will yield 3 to 4 × 10⁶ guard cell protoplasts and 1 × 10⁶ protoplasts of the common epidermal cells.

1. Collect epidermal peels on solution I. Omit steps 2 and 3.

4. Rinse on a 200 μm screen with deionized water.

5. Incubate strips in solution III at 26°. After 20 to 30 min, proto-plasts of the common epidermal cells will be released. Filter epidermal protoplasts through a 200 μm screen, clean them by passage through a 50 μm screen. Rinse epidermal peels on the 200 μm screen with solution IVa. Transfer peels into fresh solution III and incubate for additional 1 to 1.5 hr until guard cell protoplasts have been released.

6. Pass through 200 μm screen. Rinse with solution IVb.

7. Pass through 14 μm screen.

8. Centrifuge at 100 g for 7 min. Resuspend in solution IVb; count cells. Omit step 9.

10. Centrifuge at 80 g for 7 min. Resuspend in desired volume of solution IVb.

Commelina communis. The lower epidermis of 75 leaves will yield 0.5 to 0.7 × 10⁶ protoplasts.

1. Collect epidermal peels in solution I. Omit step 2.

3. Preplasmolyze for 30 min on solution II.

4. Collect on 200 μm screen. Rinse with solution II.

5. Incubate in solution III (pH is important, adjust to 5.3) at 30°. After 2 to 2.5 hr, protoplasts of epidermal cells will be released. Pass them through a 200 μm net. Rinse with solution IV. Transfer peels into fresh solution III. Incubate in solution III for additional 2 to 3 hr.

6. Pass through 200 μm screen. Rinse with solution IV.

7. Pass through 50 μm screen.

8. Centrifuge at 110 g for 7 min.

9. Transfer pellet onto a Percoll gradient consisting of up to 5 layers containing between 10 and 90% (v/v) Percoll. Centrifuge at 200 g for 7 min.

10. Collect interphase containing protoplasts, suspend in solution IV. Centrifuge at 100 g for 7 min. Count cells. Repeat the centrifugation, if necessary.

Allium cepa. Epidermis of 100 leaves will yield 2 × 10⁶ guard cell protoplasts.

1. Collect epidermal strips on solution I. Omit step 2.

3. Preplasmolyze in solution IIa for maximally 10 min. Rinse on 200 μm screen with solution IIa. Preplasmolyze in solution IIb for maxi-mally 10 min.

4. Collect on 200 μm screen. Rinse with solution IIb.

5. Incubate strips in solution III at 30° for 2 hr.

6. Pass through 200 μm screen. Rinse with solution IV.

7. Pass through 20 μm screen.

8. Centrifuge at 100 g for 5 min.

9. Suspend pellet in a few milliliters of solution IV, layer on solution V, and centrifuge at 155 g for 5 min.

10. Remove supernatant with a pipette. Suspend pellet in solution IV, centrifuge at 75 g for 4 min in 1.5-ml Eppendorf cups. Resuspend in desired volume.

Tulipa Hybrids. The upper and lower epidermis of 20 to 30 leaves will yield 1 to 2 $\times$ 10^6 guard cell protoplasts.

1. Collect epidermal strips on solution I. Omit step 2.

3. Preplasmolyze for 10 min in solution II.

4. Collect on 200 μm screen. Rinse with solution II.

5. Incubate strips in solution III at 28° for 3 to 4 hr.

6. Pass through 200 μm screen; rinse with solution IV.

7. Pass through 40 μm screen.

8. Centrifuge at 100 g for 5 min. Resuspend pellet in solution IV.

9. Prepare the following three mixtures and layer 1 ml of each onto each other in 5 ml centrifuge tubes.

Solution IV	Solution V
70%	30% top
45%	55%
0%	100% bottom

Resuspend pellet from step 8 in solution IV; put about 2 ml of the solution on each of gradient tubes. Spin at 155 g for 5 min.

10. Remove protoplast-containing interface with Pasteur pipette into 1.5-ml Eppendorf cups and centrifuge at 75 g for 4 min. Resuspend in solution IV. Repeat centrifugation if necessary.

Procedure for Preparation of Guard Cell Protoplasts from Blender-Minced Leaf Tissue

Using the blender method it is possible to prepare guard cell protoplasts from leaves of various species [among them *Beta vulgaris, Commelina communis, Mesembryanthemum crystallinum* (C$_3$ and CAM), *Nicotiana tabacum, Pisum sativum, Vicia faba, Xanthium strumarium*]. Guard cell suspensions obtained by the blender method contain more debris (in particular xylem elements) and mesophyll protoplasts than preparations from isolated epidermal peels. These contaminations, however, do not present problems for patch clamp investigations which are conducted on individual selected cells. Additional purification steps (density gradients) will be required if batches of cells are to be subjected to biochemical analyses. Obviously, the blender method yields a mixture of guard cells from the upper and lower epidermis of the leaves used.

General Procedure

1. Remove major vein of the leaves with a razor blade. Collect the remaining parts of leaf lamina in a Waring Commercial Blendor on 150 ml of water kept at 0° by floating pieces of ice.

2. Blend for 15 sec, pour minced material on 200 μm mesh screen, and rinse with 500 ml cold water or buffer, squirted from a distance of 20 to 30 cm. Repeat blending 3 to 4 times with bursts of 10-sec duration each, with cold washes in between. The last run is followed by a wash with 500 mM sorbitol, 1 mM CaCl$_2$ (Pasteur pipette).

3. Invert the screen and wash the epidermal pieces into a crystallization dish containing 30–40 ml of the sorbitol solution. Guard cells should be plasmolyzed.

4. After 10 min, decant onto a 200 μm mesh screen mounted across the narrow opening of a powder funnel. Transfer the epidermal tissue by forceps to small petri dishes containing 2.5 ml each of solution III (with osmolality adjusted to produce incipient plasmolysis of the guard cells).

5. Incubate as described above, or overnight. In the case of an overnight incubation, no shaking is required, and temperature can be kept at about 20° (add antibiotic; see Table II). For further procedure see steps 6 through 10 above under General Procedure.

Preparation of Guard Cell Vacuoles

Suspensions of guard cell protoplasts are placed into measurement cuvettes, as described below. When the cells begin to stick to the glass bottom of the chambers, the incubation medium ($\pi \sim 500$ mosmol/kg) is to be replaced by the osmotic shock medium (100 mM KCl, 10 mM HEPES–KOH, pH 8, 2 mM EGTA, $\pi \sim 200$ mosmol/kg). The low osmolality, in combination with an absence of Ca^{2+}, causes the plasmalemma to rupture within 5–10 min. When about 50% of the protoplasts have released their vacuoles, the medium is replaced with the one to be subsequently used for the patch clamp experiments (e.g., 100–200 mM KCl, 10 mM HEPES–KOH, pH 7.5, 2 mM MgCl$_2$, 0–1 mM Ca^{2+}, sorbitol to make $\pi \sim 500$ mosmol/kg). This change in solutions will also stabilize the vacuoles.

Ion Transport through Membranes Measured on Individual
   Protoplasts

The patch clamp technique, invented by Neher and Sakmann,[13] has
been described in detail by Hamill et al.[14] and Sakmann and Neher.[15] It
allows separate recordings of the activities of the transport systems in the
plasmalemma and in the tonoplast. Furthermore, characterization becomes possible of electrogenic pumps, ion channels, and carriers of very
high turnover in any type of membrane. The basic requirement for the
application of the technique is an absolutely clean membrane surface. This
can be accomplished by using the isolation procedures for protoplasts and
vacuoles described in this chapter and elsewhere.[16,17]

Pipette Fabrication[14]

Pulling. Pull patch pipettes from glass capillaries (e.g., Kimax-51,
Owens-Illinois) in two stages using commercial pullers (e.g., List Electronic, Type 3P-A, Darmstadt, FRG). During the first pull, the glass capillary is thinned to a minimum diameter of about 200 μm. The capillary is
then recentered with respect to the heating coil. A second heating and
pulling thins the glass further and causes a separation of the capillary into
two identical pipettes. The exact pipette diameter (affecting its electrical
resistance) depends on the heating current during the final pull and the
following fire polishing (below). Tip diameters should range between 0.5
and 2 μm.

Coating and Polishing. Coat the pipette with silicon rubber (e.g., Sylgard, Dow-Corning, Midland, MI) within 50 μm from the pipette tip.
Apply the rubber to the pipette with a small metal hook while care is taken
that the very tip remains uncoated. The silicon compound is then cured by
a heated air stream. For this purpose, the pipette is mounted on an
appropriate setup (e.g., List Electronic, CPZ-101) which consists of a microforge attached to an inverted microscope (magnification $\sim 50\times$). The
hydrophobic surface will reduce the capacitance after the pipette has been
submersed in the bath.

---

[13] E. Neher and B. Sakmann, *Nature (London)* **260**, 779 (1976).
[14] O. P. Hamill, A. Marty, E. Neher, B. Sakmann, and F. J. Sigworth, *Pfluegers Arch.* **391**, 85
   (1981).
[15] B. Sakmann and E. Neher, eds., "Single-Channel Recording." Plenum, New York, 1983.
[16] B. P. Marin, "Biochemistry and Function of Vacuolar Adenosine-Triphosphatases in
   Fungi and Plants." Springer-Verlag, Berlin and New York, 1985.
[17] P.-E. Pilet, "The Physiological Properties of Plant Protoplasts." Springer-Verlag, Berlin and
   New York, 1985.

Polish the pipette shortly after coating in order to produce a smooth tip and to decontaminate the glass. Supply heat by a V-shaped platinum–iridium filament which carries a glass bead of about 0.5 μm diameter on its tip (to prevent the deposition of evaporated metal on the electrode). Heat the filament to a dull red glow, and direct a stream of air toward the glass droplet to ensure that heat transfer is mainly by radiation, that temperature fluctuations remain small, and that the tip is kept free from particles emitted by the filament. Move the tip of the pipette close to the filament (1 – 2 μm) until the rim of the pipette melts, the tip diameter decreases, and the glass surface is polished.

Cleaning. Immerse the micropipette in methanol while applying pressure to its interior using a 10-ml syringe. The compression of the syringe required for the emergence of air bubbles is an indication of tip diameter. For patch-clamp investigations of the plasmalemma of guard cells, pipettes should have a tip diameter of 0.5 – 2 μm and a resistance of 1 – 5 MΩ in 150 mM KCl. Covered electrodes can be stored for 1 day.

Filling. Pipettes are filled by sucking a small amount of solution into the tip and then filling the remainder of the pipette with a syringe through the wide end. Air bubbles remaining in the tip of the pipette are removed by tapping the capillary until all the bubbles have risen to the wide end.

Solutions

For the measurement of K$^+$-selective single channels in excised patches and whole cells, prepare the following solutions. Pass all solutions through a 0.2 μm filter before use. Bath solution: 225 mM KCl or NaCl, 1 mM CaCl$_2$, 3 mM KOH, 10 mM HEPES, pH 7.0. Add mannitol or sorbitol to make 500 mosmol/kg. Internal solution (= pipette solution): identical with bath solution but slightly hyposmolar (by about 10%). For whole-cell recordings, the pipette solution should be 1 mM with respect to EGTA in order to reduce the concentration of free Ca^{2+} to the micromolar level. This prevents resealing of the membrane across the pipette tip, and an access of low resistance to the cell's interior is maintained. Refer also to Fig. 4B.

For electrogenic H$^+$ pumps in the plasmalemma of whole cells, prepare bath solution: 50 mM N-methylglucamine-glutamate, 5 mM MgCl$_2$, 10 mM HEPES, pH 7.0. Add mannitol or sorbitol to make 500 mosmol/kg. Fusicoccin at 0.1 – 1 μM will stimulate H$^+$ expulsion. Internal solution: identical with the bath solution plus 1 – 10 mM Tris-ATP. Pass all solutions through a 0.2 μm filter before use.

Mounting Protoplasts

Preparation of Cuvettes. Sonicate Falcon dishes (with glass bottoms) in water (2 min). Remove water, add a few drops of dichromate-sulfuric acid (10 min). Rinse the dishes in the following sequence: deionized H_2O, 100% methanol, 70% ethanol, 100% methanol, then dry in a filtered stream of nitrogen. Dishes can be prepared in batches and stored under cover for about 1 day. A dish should be used only once after cleaning to ensure that protoplasts will stick to the hydrophobic glass surface.

Mount a dish on the inverted microscope (Fig. 1). Place a drop of protoplast suspension, containing 10^3 to 10^4 cells, into the dish. During the following 5–10 min the protoplasts will settle to the bottom of the chamber. Add between 300 and 1000 μl of the bath solution. Perfuse the bath at a rate of 1–5 ml/min in order to remove cell debris and protoplasts which do not stick to the glass bottom. Insert the patch pipette into the suction holder (Fig. 1). The electrode of the pipette holder, as well as the reference electrode, are connected to the preamplifier of the patch-clamp amplifier (List Electronic EPC-7, for the measurement of currents in voltage-clamp recordings or of membrane potentials in current-clamp recordings, and for the determination of membrane capacitances). Apply a slightly positive pressure (2–5 cm H_2O) to the interior of the pipette to ensure that nothing can get into the tip while the pipette is being lowered into the bath. Position the pipette (using micromanipulators) near a protoplast with healthy looking chloroplasts.

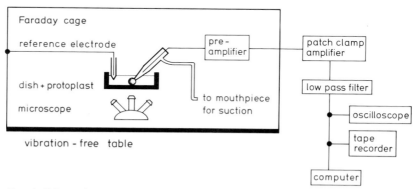

FIG. 1. Schematic representation of a patch clamp setup. Not shown is the patch-clamp tower (List Electronic) which carries an optical bench with fittings for the chamber holder, preamplifier, and suction hose, a stage for the glass dish, the perfusion equipment, the reference electrode, and the temperature control system.

Seal Formation and Recording Electrical Events in Various Patch-Clamp Configurations[14]

The pipette is pressed against the plasmalemma, which causes a seal to form with a resistance of at least 10 MΩ. Once the tip touches the membrane, the positive pressure is removed and suction is applied by mouth. Clamp the pipette potential at a value at which the current is zero. Apply test pulses (5–10 msec) increasing from 0.2 to 20 mV every 10 msec. An omega-shaped membrane patch forms inside the pipette. The leak current is followed on an oscilloscope connected to the amplifier. The disappearance of the leak current indicates the formation of a seal with a resistance ≫ 10 GΩ (Fig. 3). Resistances of up to 100 GΩ can be obtained if membrane surfaces and pipette tips are clean.

The current passing the membrane patch (of an area between 1 and 10 μm^2)[15] across the pipette tip is displayed on an oscilloscope, recorded on a PCM video recorder (VR 10 or VR 100, Adams and List Associates, Great Neck, NY), or, after passage of a low-pass filter (1–10 kHz; e.g., 902 LPF, Frequency Devices, Haverhill, MA), recorded by a microcomputer (e.g., PDP-11/73, Indec Inc., Sunnyvale, CA). For fitting and statistical analysis of the records, see Colquhoun and Sigworth in Ref. 15.

Currents through Single Channels

Cell-Attached Configuration. In the cell-attached configuration (Fig. 2) it is possible to record the activities of single channels as they depend on applied voltages and the ionic compositions on both sides of the membrane. In the cell-attached configuration the behavior of channels can be followed without interfering with intracellular processes (Fig. 3). The current I passing a single ionic channel is

$$I = g(V_{resting} - V_{holding})$$

where g equals the membrane conductance,

$$V_{resting} = \sum_{i=1}^{n} \frac{RT}{Fz_i} \ln(c_i^i/c_0^i) + \frac{I_j}{g_m}$$

and $V_{holding}$ is the voltage applied to the pipette; R is the gas constant, T absolute temperature, F the Faraday constant, z_i the valence of ion i, passing the channel, c_i^i the concentration of ion i in the cell, c_0^i the concentration of ion i in the pipette; I_j the (pump) current carried by actively transported ions, j, and g_m the effective conductance of the membrane.[18]

[18] R. M. Spanswick, *Annu. Rev. Plant Physiol.* **32**, 267 (1981).

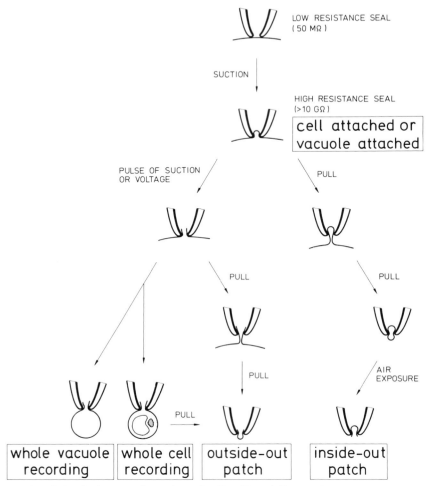

FIG. 2. Patch-clamp configurations after Hamill *et al.*,[14] as applied to plant cells.

Excised Membrane Patches: Inside-Out Patch. Pulling the pipette off the plasmalemma produces a closed vesicle in the tip (Fig. 2). Bringing the tip of the pipette to the surface of the bath causes the bath side of the vesicle to rupture. A membrane patch covers the opening of the pipette with the *cytoplasmic* side of the membrane facing the bath. Currents passing single channels can now be recognized in the recordings, and solutions can be changed while the experiment continues.

Excised Membrane Patches: Outside-Out Patch. After establishing a

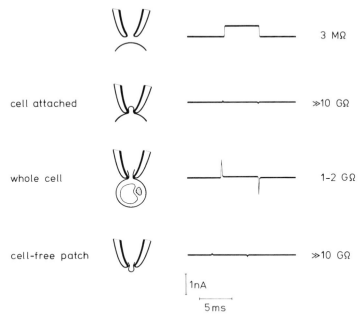

FIG. 3. Currents and resistances during a 20-mV pulse in various patch-clamp configurations of a guard cell protoplast.

whole-cell configuration, rupture by suction the membrane spanning the pipette tip. On pulling the pipette, a patch will be formed with its *extracellular* side facing the bath (Fig. 2). Sample records of the activity of K^+ channels in the plasmalemma of guard cell protoplasts are reproduced in Fig. 4A.[19] Opening and closing of channels can be recognized in a patch of an area of approximately 5 μm^2. The current shown was recorded while the voltage was clamped at various values for 50 msec each. Discrete increases and decreases in current indicate opening and closing of a single ion channel. The distribution of open and closed times and the amplitude of the stepwise changes in current depend on polarity and magnitude of the applied voltage and on ionic strength. The selectivity of a channel for particular ions can be determined after the establishment of known electrochemical gradients across the membrane and measuring the potentials at which single-channel currents reverse. Application of the Nernst and Goldman equations will yield relative permeabilities.[20]

[19] J. I. Schroeder, R. Hedrich, and J. M. Fernandez, *Nature (London)* **312,** 361 (1984).
[20] B. Hille, "Ionic Channels of Exitable Membranes." Sinauer Associates, Sunderland, Massachusetts, 1984.

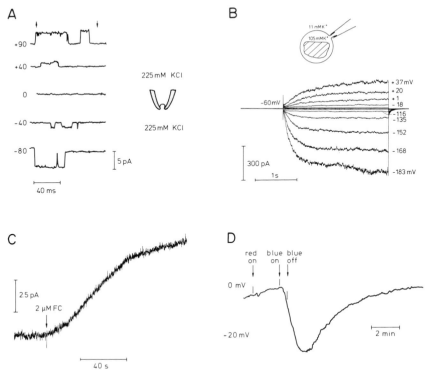

FIG. 4. Patch-clamp recordings of (A) single K$^+$-selective channels, (B) voltage-dependent K$^+$ currents, (C) fusicoccin-stimulated electrogenic pumps, and (D) blue light-activated pumps in the plasmalemma of guard cell protoplasts. (A) Recordings of K$^+$-selective channel currents in an inside-out membrane patch in symmetrical 225 mM KCl solutions (see inset). At positive membrane potentials (on the intracellular membrane side), upward deflections of the current trace are seen when channels open; they correspond to effluxes of K$^+$. At negative potentials, K$^+$ currents through single channels are inward (downward deflections of current traces). (Reprinted, with permission, from Schroeder et al.[19]) (B) Superposition of consecutive recordings of K$^+$ currents across the whole plasmalemma in response to various voltage pulses. The cytoplasm was loaded with 105 mM K$^+$ while the bath contained 11 mM K$^+$ (see inset). The membrane potential was held at -60 mV and stepped to various potentials. Depolarizations caused outward currents (upward deflections), and hyperpolarizations caused inward currents. (Reprinted, with permission, from Schroeder et al.[22]) (C) Record of the electrogenic pump current in the plasmalemma of a whole guard cell, stimulated by perfusion of the bath with 2 μM fusicoccin, with 5 mM MgATP in the pipette. (D) Record of a blue light-induced hyperpolarization of the plasmalemma of a guard cell protoplast in the whole-cell configuration. The membrane hyperpolarization was caused by the stimulation of an electrogenic pump by blue light of low fluence (50 μmol m^{-2} sec^{-1}). Nonpermeating ions were used on both sides of the plasmalemma, 50 mM N-methylglucamine glutamate and 5 mM MgATP in the pipette. [(C) and (D) reprinted, with permission, from Hedrich et al.[28]]

Whole-Cell Currents

The whole-cell configuration was developed by Marty and Neher in order to investigate effects of changes in the internal composition of the cells on ion transport.[21] Starting from a cell-attached configuration, the membrane patch covering the pipette tip can be ruptured while the seal between the pipette and the plasmalemma is maintained. Transient currents appearing in response to the voltage pulse indicate the charging of the plasmalemma as a capacitor (third trace from the top in Fig. 3). The contents of the pipette will diffuse into the cell within 15 sec to 3 min, depending on the solute and the size and structure of the cell.[23] Larger organelles, such as mitochondria, plastids, nuclei, and vacuoles, will remain in the cell. Potentials and currents recorded in this configuration will give information on the activity of all ion carriers, ion channels, and electrogenic pumps operating in the whole cell membrane. In voltage-clamp recordings, potentials must be corrected for the liquid–junction potential between the pipette and the bath. The voltage dependence of K^+-selective channels of guard cell protoplasts is shown in Fig. 4B.[22,24,25]

Whole-Vacuolar Currents

A whole-vacuole configuration (analogous to whole cell) can be established by rupturing the membrane patch covering the tip of the patch pipette (starting from a vacuole-attached configuration). Tonoplast patches can be broken by voltage pulses of $+$ and $-$ 500 to 600 mV amplitude and 1 to 3-msec duration each.[26] Vacuolar currents in response to depolarizing and hyperpolarizing voltages are shown in Fig. 5A.

Electrogenic Pumps

The current passing a single ionic channel is of the order of picoamperes which is equivalent to the flux of $10^6 - 10^7$ charges/sec. Because the resolution of the patch-clamp technique is about 0.1 pA, the behavior of ion channels can easily be studied in membrane patches. The transport activity of individual carriers or electrogenic pumps is with 10 to 10^4

[21] A. Marty and E. Neher, in "Single-Channel Recording" (B. Sakmann and E. Neher, eds.), p. 107. Plenum, New York, 1983.

[22] J. I. Schroeder, K. Raschke, and E. Neher, Proc. Natl. Acad. Sci. U.S.A. **84**, 4108 (1987).

[23] M. Püsch and E. Neher, Pfluegers Arch. **411**, 204 (1988).

[24] J. I. Schroeder, J. Gen. Physiol. **92**, 667 (1988).

[25] J. I. Schroeder, J. Membr. Biol. (in press).

[26] R. Hedrich, H. Barbier-Brygoo, H. Felle, U. I. Flügge, U. Lüttge, F. J. M. Maathuis, S. Marxs, H. B. A. Prins, K. Raschke, H. Schnabl, J. I. Schroeder, I. Struve, L. Taiz, and P. Ziegler, Bot. Acta **101**, 7 (1988).

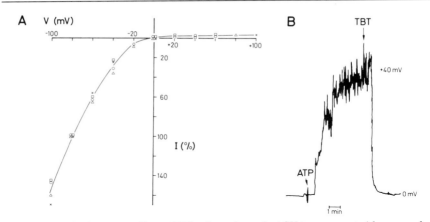

FIG. 5. Patch-clamp recordings of (A) voltage-dependent SV type currents (slow vacuolar type) and (B) ATP-dependent pump activity in the tonoplast of guard cells. (A) Current–voltage relationship of tonoplast currents. The vacuole was loaded with 100 mM KCl, 2 mM MgCl$_2$, 1 mM CaCl$_2$, pH 5.5, while the bath contained 100 mM KCl, 2 mM MgCl$_2$, 0.1 mM CaCl$_2$, pH 7.5. x, corn coleoptiles; O, broad bean guard cells; □, sugar beet taproot; △, suspension-cultured cells of *Chenopodium rubrum*. [Reprinted, with permission, from Hedrich *et al.*[26]] (B) Depolarization of the tonoplast on addition of 5 mM MgATP, indicating the presence of an inward H$^+$ pump. Application of the H$^+$-ATPase inhibitor tributyltin (TBT) abolished the depolarization totally. (Reprinted, with permission, from Raschke *et al.*[1])

charges/sec, much lower than that of channels,[19] but currents resulting from their combined simultaneous action can be recorded by the whole-cell method (Fig. 4C,D and Refs. 27 and 28).

For investigations into ATP-dependent H$^+$-pump activities, e.g., those stimulated by fusicoccin,[27,28] (Fig. 4C) or by red or blue light[27–29] (Fig. 4D), the interior of the cell is perfused with an ATP-containing solution. The solutions on both sides of the membrane should not contain any membrane-permeating ions except H$^+$ to ensure that there is no contribution made by ions other than H$^+$. The activity of electrogenic H$^+$-ATPases in the tonoplast can be studied in the same way (Fig. 5B).[26]

Acknowledgments

Our research was supported by grants from the Deutsche Forschungsgemeinschaft to K.R. and a fellowship to R.H. from the Max-Planck-Gesellschaft. We thank W. Lahr for determining abscisic acid contents of commercial osmotica. We are indebted to J. I. Schroeder for valuable suggestions and to U. Kirschner, A. Schön, and R. Seibert-Diaz for help during the preparation of the manuscript.

[27] S. M. Assmann, L. Simoncini, and J. I. Schroeder, *Nature (London)* **318,** 285 (1985).
[28] R. Hedrich, J. I. Schroeder, and J. M. Fernandez, *Trends Biosci.* **134,** 49 (1987).
[29] E. Serrano, E. Zeiger, and S. Hagiwara, *Proc. Natl. Acad. Sci. U.S.A.* **85,** 436 (1988).

[23] Intracellular and Intercellular pH Measurement with Microelectrodes

By D. J. F. BOWLING

Introduction

A number of methods have been used in the 1980s to measure intracellular pH. One method involves the use of weak acids labeled with ^{14}C to provide an indirect measurement of pH. Another uses nuclear magnetic resonance spectroscopy (^{31}P NMR). ^{31}P NMR has the advantage of being noninvasive and can provide simultaneous values for cytoplasmic and vacuolar pH in plant cells; however, the level of resolution is limited, and expensive equipment is required. A third method, the use of pH-sensitive microelectrodes, has a number of important advantages. Microelectrodes measure hydrogen ion activity directly, have rapid response times, and give immediate results. Disadvantages are that they have to be inserted into the cell without causing damage, and usually they have to be made by the experimenter, who therefore requires a certain level of manipulative skill.

Credit for the invention of the microelectrode is usually given to Ling and Gerard.[1] They pulled micropipettes by hand from glass capillaries and filled them with KCl. Electrical connection was made by introducing a silver wire coated with silver chloride into the wide end of the micropipette. Thus, the microelectrode is basically a Ag/AgCl electrode with a salt bridge pulled to a fine tip less than 1.0 μm in diameter. Caldwell[2] attempted to make pH microelectrodes by miniaturizing the conventional pH electrode, and he successfully used a semimicroelectrode with a tip diameter of about 100 μm to measure pH in large animal cells. The first true pH microelectrode with a tip of less than 1.0 μm was probably the one made by Thomas.[3] This was constructed by inserting the tip of a micropipette pulled from pH-sensitive glass into the tip of a conventional micropipette. Thomas has published a very clear account of how to make this type of microelectrode in his monograph.[4] He admits that it is rather difficult to make but states that it has proved to be reliable and long lasting.

An important step has been the development of liquid ion-exchanger type microelectrodes. Microelectrodes using liquid ion exchanger to mea-

[1] G. Ling and R. W. Gerard, *J. Cell Comp. Physiol.* **34**, 383 (1949).

[2] P. C. Caldwell, *J. Physiol. (London)* **126**, 169 (1954).

[3] R. C. Thomas, *J. Physiol. (London)* **238**, 159 (1974).

[4] R. C. Thomas, "Ion-Sensitive Intracellular Microelectrodes." Academic Press, London, 1978.

sure K$^+$ and Cl$^-$ were first described by Walker.[5] He introduced a small amount of commercially available ion exchanger (obtainable at that time from firms such as Orion and Corning) into the tip of a micropipette previously waterproofed by treatment with an organic silicon reagent. It thus became possible to make microelectrodes sensitive to a wide number of ions, depending on the availability of a suitable ion exchanger. A hydrogen ion-selective liquid membrane microelectrode was described by Ammann et al.[6] based on tri-n-dodecylamine as a neutral carrier for H$^+$. This microelectrode has a good performance with high selectivity for H$^+$ and a fast response time. It is relatively easy to make and has been used successfully to measure pH in higher plant cells.[7,8] What follows is an account of how to make this type of microelectrode and how to use it with cells of higher plants.

Construction of a pH Microelectrode

Take a length of borosilicate glass tubing of 2 mm outside diameter of the type containing an inner glass filament and cut it into 5 cm lengths. Use each length to pull two micropipettes on the microelectrode puller. The precise size of the tip of each micropipette can be controlled by adjustment of the heater and strength of pull settings on the puller. Stand the micropipettes, tips upward, in suitably sized holes drilled in an aluminum block. Stand the block in the bottom half of a petri dish and cover it with an inverted beaker (250 ml). Place the dish and its contents in an oven at 200° for about 1 hr in order to dehydrate the glass of the micropipettes. Then draw about 2 ml of dichlorodimethylsilane (2–4% v/v in carbon tetrachloride or trichlorethane) into a Pasteur pipette and release it under the lip of the inverted beaker. The silane will immediately vaporize and coat the micropipettes. Leave them in the oven for a further hour at 200° to allow the silane to bake onto the surface of the glass. On cooling, the silanized micropipettes can be stored indefinitely by standing them, tip upward, in holes drilled in a plastic block which should be placed in a closed container with silica gel to keep them dry.

A proton cocktail[6] can be made with the following: proton ionophore, tri-n-dodecylamine, 1.22 ml; solvent, 2-nitrophenol octyl ether, 5.93 ml; and stabilizer, sodium tetraphenylborate, 0.07 g. This mixture, which gives

[5] J. L. Walker, *Anal. Chem.* **43**, 89 (1971).

[6] D. Ammann, F. Lanter, R. A. Steiner, P. Schulthess, Y. Shijo, and W. Simon, *Anal. Chem.* **53**, 2267 (1981).

[7] A. Bertl and H. Felle, *J. Exp. Bot.* **36**, 1142 (1985).

[8] M. C. Edwards and D. J. F. Bowling, *Physiol. Plant Pathol.* **29**, 185 (1986).

approximately 10 g of proton cocktail, must be stored under CO_2 for 24 hr before use. A small volume of proton cocktail is introduced into the back of a silanized micropipette with a syringe. The filament should ensure that it moves down to the tip. The micropipette is then backfilled with KCl (3 M). The KCl should meet the ion exchanger and form a meniscus, and it is best to remove any large air bubbles which form with a length of fine wire, although the presence of small air bubbles will probably have no adverse effect on the performance of the microelectrode.

To complete the microelectrode a length of silver wire with its tip coated with silver chloride may be inserted into the back of the micropipette. However, it is preferable to insert the micropipette into a commercially available micropipette (or microelectrode) holder. The micropipette is pushed into the end of the holder which contains a sintered Ag/AgCl electrode and has either a male or female connector for a preamplifier at its other end (Fig. 1). A reference microelectrode is required to complete the electrical circuit. This is a conventional micropipette filled with KCl (3 M) inserted into a second micropipette holder.

It should be pointed out that more elaborate electrode arrangements are possible. Some workers have used a system in which the pH microelectrode and the reference microelectrode are joined together to form what is called a double-barreled microelectrode. This has the advantage that both electrodes are inserted into the cell at the same place. However, double-

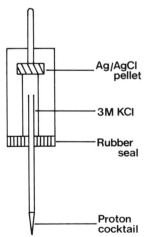

FIG. 1. Diagram of a micropipette containing proton cocktail in the tip, inserted into a micropipette holder containing a sintered Ag/AgCl electrode to form a complete H^+-selective microelectrode.

barreled microelectrodes are more difficult to make than individual microelectrodes, and more care is required with the electrical connections to the
preamplifier. They are not recommended for the beginner, but details on
how to make double-barreled pH microelectrodes are given by Reid and
Smith.[9]

Recording the Signal

The reference microelectrode will have an electrical resistance of 10^7–
10^8 Ω depending on the size of the tip. The pH microelectrode will have a
much higher resistance at 10^{10}–10^{11} Ω, so to obtain a faithful record of its
signal it must be connected to an electrometer with a high input resistance
such as, for example, Model FD 223 made by WP Instruments (New
Haven, CT) which has an input resistance of 10^{15} Ω. This high resistance
can cause difficulties such as high capacitance and noise in the cable
connecting the microelectrode to the electrometer. This allows interference
to the signal through physical movement by the operator or from neighboring electrical equipment. To reduce this a headstage or preamplifier is
introduced between the microelectrode and the electrometer. The microelectrode may be connected directly to the preamplifier which amplifies
the current in the signal but keeps the voltage the same. The preamplifier
may be quite small, about the size of a stubby pencil, and so can be brought
close to the experimental material. The reference microelectrode may be
connected to the ground terminal of the electrometer, but it is more
satisfactory to plug it into a second preamplifier and measure the difference
between the signals from the pH microelectrode and the reference microelectrode using a dual electrometer. Operating in this way means that
interference affecting both sides of the circuit is cancelled out (common
mode rejection). The electrometer will probably have a digital display of
the signal in millivolts, but it is almost essential to feed the signal to a chart
recorder to obtain a permanent copy of the data.

The microelectrode must be calibrated using a range of buffers of
known pH. A typical calibration curve for this type of microelectrode is
shown in Fig. 2. Once the slope has been determined it may be possible to
adjust the electrometer to display the signal directly in units of pH as with a
conventional pH meter. It is advisable to check the calibration immediately before and immediately after taking a series of measurements as
impalement may alter the characteristics of the microelectrode. A good
microelectrode will give 55–60 mV for each pH unit between pH 5.6 and

[9] R. J. Reid and F. A. Smith, *J. Exp. Bot.* **39**, 1421 (1988).

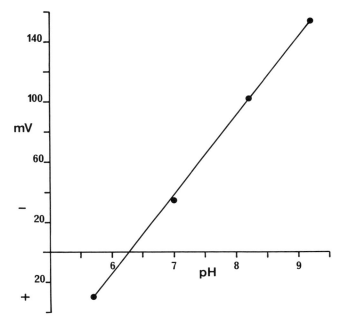

FIG. 2. Typical calibration curve for the ion-exchanger pH microelectrode.

9.0. If the response falls below this range after impalement, debris may have entered the tip of the microelectrode, and it should be discarded.

Impalement of Cells

To implant the microelectrodes in a cell under the microscope a micromanipulator is required. There are several types of micromanipulators, most of which work mechanically using levers or screw action, but at least one type works pneumatically and another by sliding one metal surface over another using thick grease as a lubricant. The aim is to reduce the movement of the operator's hand to microscopic proportions, and an instrument that can locate the tip of a microelectrode to 1 μm or less is required. The micromanipulator should have a carrier for two microelectrodes as both pH and reference microelectrodes must be inserted into the cell. A carrier which affords independent adjustment of the two microelectrodes is advantageous as it enables the tips of the microelectrodes to be brought together before impalement.

The tissue under investigation has to be held quite firmly to allow impalement by the microelectrodes. Roots are difficult to fix without damage to the tissues, but one method is to clamp a length of root between two coverslips glued together along one edge. The coverslips can then be mounted on top of a small chamber, made of, say, Plexiglas, on the microscope stage. A long working distance condenser may be required to illuminate the root in order to allow the use of all the microscope objectives. A suitable magnification for impalement of root cells is 500×. The microelectrodes can then be brought through a slit in the side of the chamber and between the open edges of the coverslips using the micromanipulator. An appropriate bathing solution can be circulated through the chamber with a peristaltic pump. Circulating fluids can cause a build up of charge on the apparatus so every piece of equipment must be grounded to the same point to eliminate ground loops.

Leaves are easier to mount than roots. A detached leaf or part of a leaf may be fixed to a slide with sealing compound or medical elastomer. The slide is then mounted at about 30° to the horizontal in a vessel with the base of the leaf dipping into water to prevent drying. The microelectrodes can then be inserted into the epidermal cells of the leaf in air using a stereomicroscope at a magnification of 200×.[10]

Unless the laboratory has a concrete floor and is situated on the ground floor of the building, vibration will be a problem. This can be overcome by using a vibration-free bench or by placing the micromanipulator and microscope on a metal plate or a slab of slate standing on four squash balls.

Some Results Using the Ion-Exchanger pH Microelectrode

Bertl and Felle[7] have used the ion-exchanger type of pH microelectrode to measure the cytoplasmic pH of root hair cells of *Sinapis alba.* They observed that these cells did not have a large central vacuole so they concluded that the tips of both microelectrodes, on entering the cells, resided in the cytoplasm. They found, however, that the turgor pressure of the cell forced the proton ion exchanger back from the tip when the microelectrode entered the cell. To prevent this, 0.1% poly(vinyl chloride) (PVC) dissolved in tetrahydrofuran was sucked into the tip before the micropipette was backfilled with ion exchanger. They found their microelectrode to have a resistance of $5-8 \times 10^{10}$ Ω and it displayed a slope of $55-58$ mV per pH unit between pH 4.0 and 9.0. This range of sensitivity was wider than that reported by Ammann *et al.*[6] who observed a marked decline in the slope below pH 5.6. Rise time was very fast at less than 1 sec over a change from pH 6.0 to 7.5.

[10] A. Edwards and D. J. F. Bowling, *J. Exp. Bot.* **35,** 562 (1984).

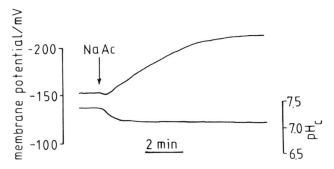

FIG. 3. Change in membrane potential and cytoplasmic pH of *Sinapis* root hair cells after the addition of 1.0 mol m^{-3} sodium acetate to the external medium (pH$_0$ = 5.0). (From Bertl and Felle.[7])

Bertl and Felle were able to follow the membrane potential and the pH of the cytoplasm simultaneously using a dual amplifier. They found the cytoplasmic pH of *Sinapis* root hairs to be 7.3 ± 0.2 at neutral external pH. Sodium acetate hyperpolarized the plasmalemma by about 60 mV and acidified the cytoplasm by 0.2–0.3 pH units (Fig. 3).

We have used the ion-exchanger microelectrode extracellularly to measure pH in the cell walls of the epidermis of leaves.[8] The pH and reference microelectrodes were simply placed on the surface of the abaxial side of the leaf mounted under the stereomicroscope as previously described. Figure 4 shows some results from *Commelina communis*. With the stoma open the pH of the cell wall was a little above 6.0 for all the cells of the stomatal complex, but when the stoma closed a marked gradient in pH developed.

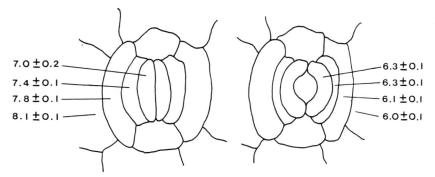

FIG. 4. Extracellular measurements of pH using the ion-exchanger pH microelectrode on the leaf surface of *Commelina communis*. (After Edwards and Bowling.[8])

We also measured the pH of the cytoplasm of these cells and found it to be
7.1 ± 0.5. Cytoplasmic pH did not vary between cells nor did it appear to
change with changes in stomatal aperture. For the intracellular measurements PVC was added to the proton cocktail using the following mixture:
0.17 g PVC, 0.4 g proton cocktail, and 7.0 ml tetrahydrofuran. This mixture was introduced into the tip, and the micropipette was left overnight to
allow it to set before backfilling with KCl ($3 M$).

With vacuolate cells such as those of leaf epidermis the cytoplasm
occupies only about 10% of the volume of the cell in a layer approximately
1 μm wide. A microelectrode entering the cell is therefore more likely to go
into the vacuole than the cytoplasm. The cytoplasm layer is too thin to
allow the microelectrode to be intentionally placed in the cytoplasm at
every impalement. Also it is not always possible to see where the tip is
located. Fortunately, the microelectrode tip does sometimes remain in the
cytoplasm after penetrating the plasmalemma but not the tonoplast. We
have found that in about 2 of every 10 impalements the tip is in the
cytoplasm. Any doubt about the location of the tip of the microelectrode
may be resolved by the pH reading obtained, because the pH of the vacuole
is always acid whereas the pH of the cytoplasm is always alkaline under
normal conditions.

[24] Water Relations of Plant Cells: Pressure Probe Technique

By U. Zimmermann

Introduction

Current interest in application of the pressure probe to osmoregulatory,
growth, and turgor pressure-dependent processes in single and individual
plant tissue cells results largely from the limitations of more traditional
procedures. The pressure probe technique provides direct access to many
biophysical parameters which control the short- and long-term responses
of plant cells to water and salt stress in the environment. These are the
turgor pressure and the internal osmotic pressure as well as the hydraulic
conductivity and the reflection coefficients of the cell membrane and the
elastic properties of the cell wall.

Knowledge of these parameters is required for the description of water
transport in plant cells and tissues on the basis of the phenomenological

equations of the thermodynamics of irreversible processes.[1-6] This theoretical concept takes into account the coupling of water transport with solute fluxes in a system which contains transport barriers (membranes) which are semipermeable or impermeable.[7-10] Water and solute transport as well as the associated changes in turgor and internal osmotic pressure can, therefore, only be properly dealt with by this theory regardless of the method used for the measurement of the relevant parameters. For this reason we first discuss the fundamental equations which are required for analysis of pressure probe data before we describe the method. The importance of this procedure will become evident when the various applications of the pressure probe are introduced in the following sections. In the experimental part an attempt is made to present guidelines for the accurate use of the pressure probe in the evaluation of the various water relation parameters. Although it is not possible to provide detailed protocols for each cell or tissue system, these guidelines should allow the experimenter to estimate the conditions which are optimal for the cell to be studied.

Theoretical Considerations

Phenomenological Equations

Application of thermodynamics of irreversible processes to discontinuous systems shows that forces and fluxes are interdependent, resulting in coupled movement of water and solutes through biological membranes. In the case of water flow this means that water can be driven not only by hydrostatic pressure differences (i.e., turgor pressure), but also by other forces such as osmotic pressure, temperature, and electrical potential differences as well as metabolic energy. A rigorous treatment of this subject matter is given elsewhere.[1-10] Here we discuss only those equations used in the analysis of pressure relaxation data.

[1] J. Dainty, *Adv. Bot. Res.* **1**, 279 (1963).

[2] J. Dainty, *Encycl. Plant Physiol., New Ser.* **2**, 12 (1976).

[3] U. Zimmermann, *in* "Integration of Activity in the Higher Plant" (D. Jennings, ed.), p. 561. Cambridge Univ. Press, 1977.

[4] U. Zimmermann, *Annu. Rev. Plant Physiol.* **29**, 121 (1978).

[5] U. Zimmermann and E. Steudle, *Adv. Bot. Res.* **6**, 45 (1978).

[6] U. Lüttge and N. Higinbotham, "Transport in Plants." Springer-Verlag, Berlin and New York, 1979.

[7] O. Kedem and A. Katchalsky, *Biochim. Biophys. Acta* **27**, 229 (1958).

[8] P. Läuger, *Angew. Chem.* **81**, 56 (1969).

[9] A. Katchalsky and P. F. Curran, "Nonequilibrium Thermodynamics in Biophysics." Harvard Univ. Press, Cambridge, Massachusetts, 1965.

[10] D. Woermann, *Encycl. Plant Physiol., New Ser.* **3**, 419 (1976).

For simplicity we assume a two-compartment system as shown in Fig. 1, in which a membrane separates two aqueous phases I (corresponding to the cell) and II (corresponding to the medium). The compartments have different hydrostatic pressures and concentrations of a nonelectrolyte. Chemical reactions, including active processes, temperature, and electrical potential differences are excluded. The two phases are of uniform composition, hence the stirring arrangements of Fig. 1. The volumes of both compartments are assumed to be large. Thus, the entire system is in a stationary state, that is, the forces (hydrostatic and osmotic pressure differences) and flows (solute and water fluxes) are independent of time.

The formalism of nonequilibrium thermodynamics shows that the water flow, J_W, in this system is given by

$$J_W = L_P(P^I - P^{II}) + L_{PD}(\pi_S^I - \pi_S^{II}) \tag{1}$$

where P^I and P^{II} are the hydrostatic pressures in compartments I and II. The difference is equal to the turgor pressure, P. π_S^I are π_S^{II} are the osmotic pressures of a solute, S, in the two compartments, L_P the hydraulic conductivity (water permeability) of the membrane, and L_{PD} the osmotic coeffi-

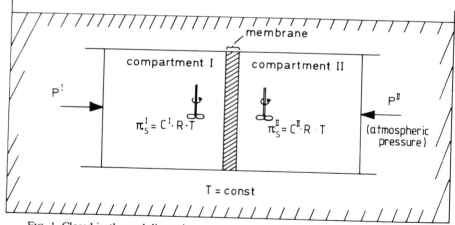

FIG. 1. Closed isothermal discontinuous system consisting of two aqueous, homogeneous compartments separated by a planar membrane. The two aqueous phases contain a solute at different concentrations resulting in different effective osmotic pressures, $\sigma\pi_S$. It is assumed that the hydrostatic pressure in compartment I, P^I, is larger than the atmospheric pressure in compartment II, P^{II}. Depending on the magnitude of the hydrostatic and effective osmotic pressure differences across the membrane, net water flow and solute flow will occur either from the left-hand to the right-hand side or vice versa.

cient of the solute. (In the presence of several solutes, average values of the respective osmotic pressures and of the osmotic coefficients are used in practice.) J_W has the dimensions cm^3/cm^2 sec and, therefore, corresponds to the velocity of water.

At equilibrium J_W vanishes, and Eq. (1) transforms into Eq. (2):

$$P_0 = \sigma(\pi_{S0}^I - \pi_{S0}^{II}) \qquad (2)$$

where π_{S0}^I and π_{S0}^{II} are the osmotic pressures of the solute S in compartments I and II and P_0 the corresponding turgor pressure at equilibrium. σ is defined by Eq. (3):

$$\sigma = -L_{PD}/L_P \qquad (3)$$

and is termed the reflection coefficient. This parameter is a measure for the coupling between flows of water and of solute (see below). If we include Eq. (3) defining σ in Eq. (1), we obtain the practical transport Eq. (4) for water flow.

$$J_W = L_P[P - \sigma(\pi_S^I - \pi_S^{II})] \qquad (4)$$

Equation (2) states that at equilibrium the turgor pressure is equal and opposite to the osmotic pressure difference only when $\sigma = 1$. The mathematical treatment shows that this is the case for a membrane which is impermeable to the solute. If σ is less than 1, the effective osmotic pressure, $\sigma\pi_S$, exerted by the osmotically active solute is smaller than the value calculated from the van't Hoff equation $\pi_S = cRT$. In this case the concentration, c, has to be increased until the effective osmotic pressure matches the turgor pressure. If $\sigma = 0$, the turgor pressure must be zero at $J_W = 0$, which means that no coupling exists between water and solute flow.

The coefficients L_P and σ can be easily determined from Eqs. (2) and (4) because the turgor pressure can be measured by means of the pressure probe and the osmotic pressure difference by cryoscopy of the concentration of the solute. The internal osmotic pressure, π_{S0}^I, can be also determined from the equilibrium condition [Eq. (2)] by means of the pressure probe, provided that an impermeable solute ($\sigma = 1$) of known concentration in the external medium is used.

Equation (4) is the fundamental equation for the analysis of water flow through biological membranes. However, the small size of the cell means that we cannot assume steady-state forces and fluxes. In contrast, a cell subjected to changes in turgor pressure or to osmotic pressure differences will swell or shrink, depending on the direction of net water flow. The turgor and the internal osmotic pressure and, in turn, the magnitude of the water flow will change as a result.

Pressure Relaxation

The time dependence of these interdependent parameters is obtained by integration of Eq. (4).[11-13] Integration can be carried out if we apply this Eq. (4) to time intervals where the deviations from the state of osmotic equilibrium can be considered to be small:

$$J_W = L_P\{P(t) - \sigma[\pi_S^I(t) - \pi_S^{II}]\} \tag{5}$$

It is assumed that the volume of the medium is large compared to the cell volume so that $\pi_S^{II} = \pi_{S0}^{II}$.

Equation (5) can be integrated if we take into account that changes in turgor pressure and volume changes during shrinking or swelling are interdependent. If it is assumed that the cell behaves as a classical osmometer and that the cell wall has only elastic (but not plastic) properties, the relationship between changes in turgor pressure and cell volume are given by Eq. (6):

$$dP = \epsilon(dV/V_0) \tag{6}$$

where dP and dV are differential changes in turgor pressure and volume, respectively, and V_0 the volume at the original equilibrium state.

ϵ is the so-called volumetric elastic modulus.[1-6] Its value determines the magnitude of the change in cell volume arising from changes of the turgor pressure during the shrinking or swelling process. Cells with ϵ values of 10 MPa or more exhibit smaller volume changes than cells with lower ϵ values if a given change in turgor pressure is induced.

Substituting Eq. (6) into Eq. (5) and further assuming that the internal osmotic pressure, π_S^I, at a given time, t, is given by Eq. (7):

$$\pi_S^I(t) = \frac{\pi_{S0}^I V_0}{V(t)} \tag{7}$$

then Eq. (5) can be integrated. Integration shows that turgor pressure (as well as cell volume and water flow) change exponentially, if the equilibrium is osmotically disturbed:

$$P - P_f = (P_0 - P_f) \exp(-kt) \tag{8}$$

where P is the pressure at time t, P_0 the initial equilibrium pressure, and P_f the final pressure. The rate constant k is equal to $\ln 2/T_{1/2}^P$.

[11] U. Zimmermann and E. Steudle, *Z. Naturforsch.*, **25B**, 500 (1970).
[12] U. Zimmermann and E. Steudle, *J. Membr. Biol.* **16**, 331 (1974).
[13] E. Steudle and U. Zimmermann, *Biochim. Biophys. Acta* **332**, 339 (1974).

The initial slope of the pressure relaxation, S_P, is given by Eq. (9):

$$S_P = -\frac{A\epsilon L_P \Delta P}{V_0} \tag{9}$$

and the half-time of the turgor pressure relaxation, the so-called half-time of water exchange, $T^P_{1/2}$, is given by Eq. (10):

$$T^P_{1/2} = \frac{\ln 2 V_0}{A L_P(\epsilon + \pi^I_{S0})} \tag{10}$$

Equations (9) and (10) are the key equations in the analysis of pressure relaxation curves. It is evident that the hydraulic conductivity can be determined both from the initial slope, S_P, and from the half-time of the pressure relaxation, $T^P_{1/2}$, if the other parameters are known.

As shown below, the pressure probe allows the measurement of the elastic modulus, ϵ, the half-time of water exchange, $T^P_{1/2}$, and the internal initial osmotic pressure, π^I_{S0} [because Eq. (2) still holds]. If the reflection coefficients of the internal osmotically active solutes are not equal to 1, π^I_{S0} has to be replaced by the effective osmotic pressure, $\sigma \pi^I_{S0}$. The volume-to-area ratio (V_0/A) can be determined by measurements of the cell dimensions under the microscope.

Equation (10) shows that the time to reach the new equilibrium state is independent of the reflection coefficient of the external osmoticum used for the initiation of the pressure relaxation. However, the value of the turgor pressure change and, in turn, of the cell volume change between the final and initial states depends on the effective external osmotic pressure. This pressure depends on both the reflection coefficients and the osmotic pressure calculated from the van't Hoff equation.

Pressure Clamp

Integration of Eq. (5) leads to different results when the osmotically or hydrostatically induced initial change in turgor pressure is clamped, i.e., is maintained constant.[14,15] This can be easily achieved by the pressure probe as shown below. Under these conditions the relaxation of the cell volume is measured instead of the turgor relaxation. Assuming pressure clamp, integration of Eq. (5) shows that volume relaxation occurs also exponentially; however, the initial slope, S_V, and the half-time of volume relaxation, $T^V_{1/2}$,

[14] S. Wendler and U. Zimmermann, *Plant Physiol.* **69**, 998 (1982).
[15] S. Wendler and U. Zimmermann, *Planta* **164**, 241 (1985).

are given by Eqs. (11) and (12), respectively:

$$S_V = -AL_P\Delta P \tag{11}$$

$$T^V_{1/2} = \frac{ln\, 2V_0}{AL_P\pi^I_{S0}} \tag{12}$$

It is interesting to note that the hydraulic conductivity can be calculated from either Eq. (11) or Eq. (12) without knowledge of ϵ and of the initial volume (if only the slope is analyzed). Clearly, at maintenance of a constant (nonequilibrium) turgor pressure the volume change is decoupled from turgor pressure, and, therefore, the elastic properties of the cell wall do not play any role. A new equilibrium is reached if the difference between the internal and external osmotic pressure matches the value of the turgor pressure. The internal osmotic pressure will indeed change during pressure clamp because of the induced volume relaxation.

Comparison of Eqs. (10) and (12) also shows that the half-time of volume relaxation at constant pressure is longer than that of a turgor pressure relaxation because of the term $(1 + \epsilon/\sigma\pi^I_{S0})$. Pressure clamp experiments are, therefore, advantageous in cases in which the half-time of pressure relaxations are too fast to be resolved accurately.

Pressure clamp experiments have the further advantage that they allow the accurate determination of cell volume, which is of importance if the cells under investigation are in the interior layers of tissues. Under these conditions geometric measurements of the cell volume are subject to large errors. This can be easily shown by comparison of the equilibrium equations for the turgor and osmotic pressure differences at the original and final equilibria. (The final equilibrium is reached after some time under clamped pressure conditions.) The mathematical treatment shows that Eq. (13) holds:

$$V_0 = \frac{\Delta V}{\Delta P}(\pi^I_{S0} + \Delta P) \tag{13}$$

where ΔP is the induced change in turgor pressure, given by $\Delta P = P_f - P_0$ where P_0 is the initial equilibrium turgor pressure and P_f the final, clamped pressure, and ΔV is the corresponding change in volume.

Principle and Procedure

The pressure probe consists of a microcapillary connected via pressure-tight seals to a Plexiglas chamber in which a pressure transducer is mounted (Fig. 2). The system is filled with oil of low viscosity or water

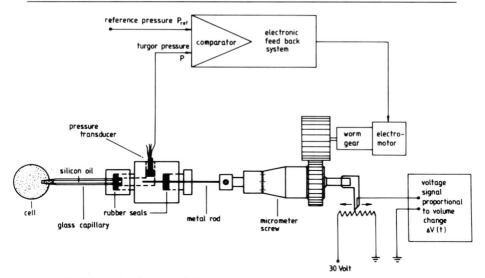

FIG. 2. Schematic diagram of the pressure probe. The turgor pressure, P (hydrostatic pressure inside the cell, compartment I, in reference to the atmospheric pressure in the medium, compartment II), is transferred via the cell sap–oil boundary into the probe chamber and converted to a proportional voltage signal by a pressure transducer. In pressure relaxation experiments, an electronic feedback system keeps the position of the cell sap–oil boundary constant via appropriate displacement of a metal rod. In pressure clamp experiments, the turgor pressure is maintained constant by a continuous shift of the metal rod in the pressure probe, thus compensating for the water flow through the cell membrane. In this case, movement of the rod is recorded by monitoring a position-dependent voltage and is then related to the volume change of the cell. For further details, see text.

depending on the pressure transducer used.[16,17] The pressure transducers measure pressure relative to atmospheric pressure. For turgor pressure measurements the microcapillary is introduced into the cell. The turgor pressure in the cell (which normally exceeds atmospheric pressure) pushes cell sap into the microcapillary until an equilibrium state is reached in which the pressure built up in the oil (or in the water) phase of the pressure probe equals the turgor pressure of the cell. This occurs in a fraction of a second. The pressure transducer transforms this equilibrium pressure into a proportional voltage which is displayed by an oscilloscope or by a recorder.

For measurements of changes in turgor pressure (induced, e.g., by

[16] U. Zimmermann, H. Räde, and E. Steudle, *Naturwissenschaften* **56,** 634 (1969).
[17] A. Balling, U. Zimmermann, K.-H. Büchner, and O. Lange, *Naturwissenschaften* **75,** 409 (1988).

osmotic means) the position of the boundary between cell sap and oil (or water) in the microcapillary is kept constant during the entire turgor pressure regulation. In this case, the pressure probe is effectively made a compensation method. This is performed by the displacement of a metal rod introduced via pressure-tight seals through the opposite wall of the plexiglass chamber.

In the conventional type of pressure probe, displacement of the metal rod is achieved manually by turning a micrometer screw. The direction of screw movement determines whether the boundary between cell sap and oil (or water) is shifted in the direction of the cell or in the opposite direction. Manual adjustment of the boundary position is sufficient in short-term turgor pressure experiments, provided that the half-time of the turgor pressure relaxation of the cell is not less than 1 sec. However for long-term measurements of turgor pressures in plant cells (e.g., during light/dark regimes as shown in Fig. 3[18,19]) an automatic adjustment of the boundary position is preferred. Different experimental approaches for automatic recording of turgor pressures in large as opposed to small cells have been published.[20-22]

The most simple and promising automatically regulating pressure probe (shown in Fig. 4) is based on an electronic control that is optically coupled to the movement of the highly refractive interface between oil and cell sap.[22] The position of the oil cell sap boundary is sensed by its movement along the image of the capillary on a focusing screen which is attached to the microscope. The movement of the boundary is monitored by means of a small sensor containing eight phototransistors in a row which is positioned on the surface of the screen along the image of the oil-filled capillary by means of an adjustable support. The eight phototransistors are used to detect the position, direction, and speed of movement of the boundary as well as the basic luminance of the focusing screen. These signals are fed into an electronic circuit connected to a small motor, which regulates the position of the boundary in the capillary by displacement of the metal rod in the plexiglass chamber. Depending on the position of the boundary, flip flops are set when the appropriate sensor element recognizes the movement of the boundary. As each of the flip flops represents different speed steps or directions for the motor regulation, the speed of the motor increases the more the boundary drifts away from its nominal position. When the phototransistor at the nominal position is crossed by

[18] J. Rygol, K. H. Büchner, K. Winter, and U. Zimmermann, *Oecologia* **69**, 171 (1986).
[19] J. Rygol, U. Zimmermann, and A. Balling, *J. Membr. Biol.* **107**, 203 (1989).
[20] D. Hüsken, E. Steudle, and U. Zimmermann, *Plant Physiol.* **61**, 158 (1978).
[21] D. J. Cosgrove and D. M. Durachko, *Rev. Sci. Instrum.* **57**, 2614 (1986).
[22] K.-H. Büchner, G. Wehner, W. Virsik, and U. Zimmermann, *Z. Naturforsch. C: Biosci.* **42C**, 1143 (1987).

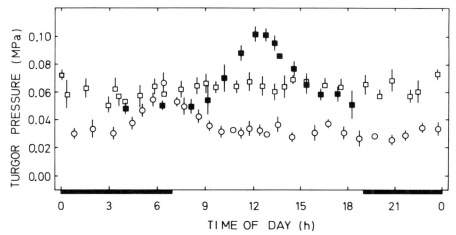

FIG. 3. Measurements of the steady-state turgor pressure in individual cells of leaves of *Mesembryanthemum crystallinum* as a function of a light/dark regime by means of the pressure probe. The plants used were grown under conditions of high salinity (400 mM NaCl) in order to induce crassulacean acid metabolism.[18,19] The symbols open circles, filled squares, and open squares denote turgor pressure values measured in the mesophyll and in the bladder cells of the upper and lower epidermis, respectively. It is evident that the pressure probe can record long-term changes in steady-state turgor pressures in individual cells of higher plant tissues, even though these changes may be quite different in the various cell types. In this case, a maximum turgor pressure is measured for the mesophyll cells at the beginning of the light period followed 6 hr later by a pressure maximum in the bladder cells of the upper epidermis. In contrast, the turgor pressure in the bladder cells of the lower epidermis remains constant during the light/dark regime. This turgor pressure pattern results from malate synthesis and subsequent degradation and the associated asymmetric water transport in the leaf. The data points represent the means of three or four measurements, the bars the standard deviations. (Data redrawn from Refs. 18 and 19.)

the boundary image it resets all flip flops so that the motor movement comes to a stop instead of constantly regulating to and from. In this way the position of the boundary can be kept constant up to a maximum deviation of 6 μm corresponding to a volume change of little more than 0.05 pl.

Figure 5 shows a typical turgor pressure recording from a cell of the peat-bog alga *Eremosphaera viridis.* Water efflux was induced by addition of sucrose ($\sigma = 1$) to the pond water. From the straight line in the semilogarithmic plot the half-time of water flow can be calculated to be approximately 9 sec [using Eq. (10)]. The automatic control system is fast enough to cope with the speed of the boundary movement at still shorter half-times.

The application of the (manual or automatic regulating) pressure probe is not restricted to measurements of steady-state turgor pressures or osmot-

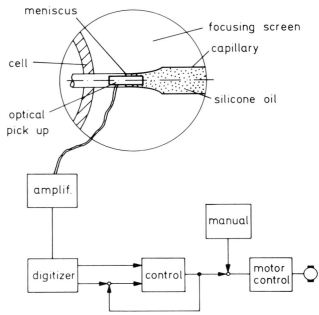

Fig. 4. An optical sensor, mounted on a flat screen along the image of the oil-filled capillary or the pressure probe, is used to monitor the movement of the cell sap–oil boundary. The signal from the sensor is fed into an electronic circuit which regulates the position of the boundary by driving the metal rod into or out of the probe chamber (see Fig. 2). A detailed circuit diagram is available on request from the author. (Redrawn from Ref. 22.)

ically induced pressure changes. The possibility of moving the oil–cell sap boundary by displacement of the metal rod allows changes of turgor pressure by injection of negative or positive pressure pulses into the cell. Thus, pressure relaxations can be hydrostatically instead of osmotically induced. The magnitude of the pressure change depends on the displacement of the metal rod. After an appropriate displacement, the boundary is again manually or automatically fixed in the new position, and the turgor pressure relaxation is recorded in the manner described above.

The possibility of changing the turgor pressure hydrostatically enables accurate turgor pressure measurements in individual cells located in internal layers of higher plant tissues. Changes of turgor pressure in such cells induced by osmotic changes in the external medium always lead to erroneous results, owing to problems arising from diffusion restriction and unstirred layers in the tissue. These will lead to (normally unknown) changes of the actual osmotic pressure in the close neighborhood of the cell under investigation, which may be quite different to the osmotic pressure in the bulk medium.

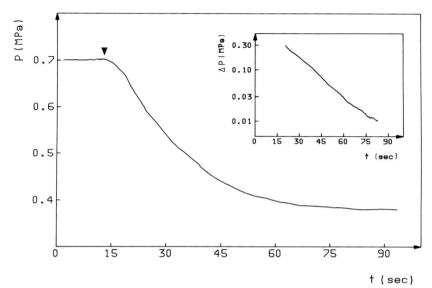

FIG. 5. Typical turgor pressure relaxation induced in a cell of *Eremosphaera viridis* and recorded by the automatic pressure probe shown in Fig. 4. The initial steady-state turgor pressure of the cell was 0.7 MPa when the cell was bathed in artificial pond water. After addition of a solution containing 127 mosmol sucrose (reflection coefficient, $\sigma = 1$) the pressure dropped exponentially and achieved a final value of 0.4 MPa. As can be seen from the semilogarithmic plot (inset), the pressure decay was purely exponential, exhibiting a half-time of 9 sec [Eq. (10)]. (Redrawn from Ref. 22.)

Measurements on algal cells immobilized in an alginate matrix cross-linked by bivalent ions have clearly shown that the analysis of the turgor pressure relaxation curves can only be applied to individual cells of higher plants (buried deeply in the tissue) provided that the turgor pressure is changed hydrostatically.[23] The elucidation of the water relations parameters of cells in higher plant tissues is therefore at present only feasible with the pressure probe technique, because other methods (such as the plasmolytic procedure) require osmotic pressure changes in the bathing medium of the tissue.

Determination of Reflection Coefficients

Experimentally, the pressure probe can be used for determination of reflection coefficients of giant algal and of individual cells of higher plant tissues. Knowledge of the reflection coefficients of osmotically active sub-

[23] K.-H. Büchner and U. Zimmermann, *Planta* **154**, 318 (1982).

stances is required because osmotically driven water transport through biological membranes is not controlled by the osmotic pressure calculated from the van't Hoff equation (assuming that the membrane has impermeable properties) but rather by the effective osmotic pressure, i.e., by the product $\sigma\pi_s$ [see Eq. (4)].

There are two different methods which enable accurate measurements of reflection coefficients of various solutes for a given cell system.[24-26] In the zero-flow method[24] the concentration of the given (permeable) osmoticum in the medium, c_2^{II}, is determined that matches the turgor pressure. The reflection coefficient of the permeable solute ($\sigma < 0$) can be obtained easily if a parallel experiment is performed in which the concentration of a solute to which the membrane is strictly impermeable ($\sigma = 1$, c_1^{II}) is determined for $J_w = 0$ and $P_0 = $ constant. Under this condition Eq. (14) holds according to Eq. (2):

$$\sigma = C_2^{II}/C_1^{II} \tag{14}$$

It is obvious that the determination of the reflection coefficients is traced back to a concentration method. The method assumes that changes in the intracellular pool of solutes do not occur during the procedure. The following protocol is normally used.

The microcapillary is introduced into the vacuole of the given single cell, and the equilibrium pressure is established by displacement of the cell sap–oil boundary as described above. The incubation (nutrition) medium is changed to one of the test solutions containing various concentrations of the compound under investigation (e.g., glucose, urea, glycerol, acetamide, ethanol, or salts). If the effective osmotic pressure of these solutions does not match the turgor pressure a fall or increase in pressure with time (dP/dt) is recorded (as shown in Fig. 6). Depending on the direction of the change turgor in pressure, the concentration of the nonelectrolyte is increased or decreased (Fig. 6) until the turgor pressure does not change.

This procedure is performed both for the (permeable) substance under investigation and for the impermeable osmoticum. Normally sucrose is used as the reference osmoticum because most biological membranes are impermeable to this compound. From the two concentration values σ is calculated according to Eq. (14).

Complications such as membrane disintegration may arise if the incubation medium of the cells is replaced by test solutions containing no electrolyte. If this is the case, certain amounts of electrolyte must be

[24] U. Zimmermann and E. Steudle, *Z. Naturforsch.*, **25B**, 500 (1970).
[25] E. Steudle and S. D. Tyerman, *J. Membr. Biol.* **75**, 85 (1983).
[26] S. D. Tyerman and E. Steudle, *Plant Physiol.* **74**, 464 (1984).

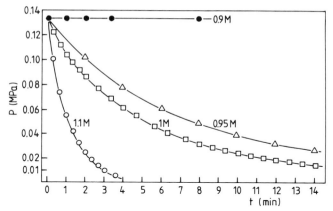

FIG. 6. Plot of turgor pressure as a function of time recorded in the Mediterranean alga *Valonia utricularis* after replacement of the seawater by solutions containing decreasing concentrations of glucose. For membrane stabilization a certain amount of electrolyte was added to the glucose solutions (50 mM NaCl, 5 mM KCl, 1 mM CaCl$_2$, and 2.4 mM NaHCO$_3$). It is evident that "zero flow" is observed at a glucose concentration of 0.9 M. Similar experiments with sucrose ($\sigma = 1$) showed that the reflection coefficient of glucose was 0.95. (Redrawn from Ref. 11.)

included in the test solutions, and the reflection coefficients of nonelectrolytes are determined in reference to this solution.

The second method for determination of reflection coefficients is based on measurement of the biphasic course of turgor pressure after osmotic disturbance of the equilibrium pressure caused by replacement of the incubation medium by a solution of a nonelectrolyte or salt.[25,26] If the effective osmotic pressure of this solution exceeds that of the incubation medium a drop in turgor pressure is recorded due to water efflux. After a certain time, a minimum value in turgor pressure is reached (Fig. 7). The magnitude of the entire pressure change depends on the effective osmotic pressure and, in turn, on the reflection coefficient as mentioned under Theoretical Considerations. For osmotica exhibiting σ values below 1 the turgor pressure subsequently increases again because the given osmoticum ($\sigma < 1$) enters the cell. Water follows owing to the change in the net driving force (Fig. 7). The initial turgor pressure is reached when the osmoticum is equilibrated between the internal and external phases. The reflection coefficient can be calculated from the time course of the turgor pressure as shown by Eq. (15) [derivation of this equation from Eqs. (2) and (4) is given in Ref. 26]:

$$\sigma = \frac{P_0 - P_{\min}}{\Delta \pi_S^{II}} \frac{\epsilon + \pi_{S0}^{I}}{\epsilon} \exp(k_S t_{\min}) \tag{15}$$

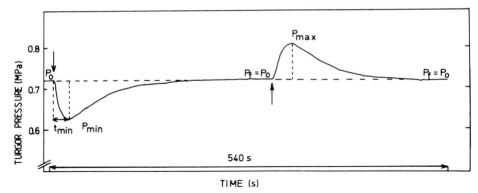

FIG. 7. Response in cell turgor pressure of an internodal cell of *Chara corallina* to an increase of the osmotic pressure in the incubation medium (artificial pond water). The increase of the external osmotic pressure was achieved by addition of *n*-propanol. It is obvious that the turgor pressure first decreased (owing to outwardly directed water flow) and reached a minimum value of (P_{min}). Then the turgor pressure increased again owing to solute influx until the original equilibrium pressure, $P_f = P_0$, was reached. Replacement of this solution by the original incubation medium results in the opposite course of turgor pressure with time, that is, a maximum in turgor pressure, P_{max}, occurred. (From J. Rygol and U. Zimmermann, unpublished data.)

whereby P_0 is the initial equilibrium turgor pressure and P_{min} the minimum turgor pressure. $\Delta\pi_S^{II}$ is the osmotic pressure difference between incubation medium and the test solution, k_S is the rate constant for the solute flow, and t_{min} is the time required to reach the minimum turgor pressure value, P_{min}.

The difference in turgor pressure $P_0 - P_{min}$ induced by the osmotic pressure difference, $\Delta\pi_S^{II}$, between nonelectrolyte solution and incubation medium is measured by the pressure probe. The rate constant, k_S, of solute influx is determined from a logarithmic plot of the second phase of the biphasic time course of turgor pressure. The elastic modulus of the cell wall [defined by Eq. (6)] can be determined by independent experiments as described below.

Comparison of Eq. (15) with Eq. (2) shows that the term $\epsilon + \pi_{S0}^I/\epsilon$ corrects the concentrations within the cell to take into account the decrease in cell volume that occurs during the experiments because of the limited size of the cell. On the other hand, the term $k_S t_{min}$ represents the correction that is necessary for the concentration changes that are due to the influx of osmoticum during the measuring period. It is obvious that Eq. (15) can be also used for calculation of the reflection coefficient from the reverse experiment (Fig. 7), that is, if the nonelectrolyte medium is changed back to the original incubation medium. In this case, a biphasic turgor pressure response with a pressure maximum is obtained, since $\Delta \pi_S^{II} < 0$.

Determination of Elastic Modulus of Cell Wall

The pressure probe also enables the independent determination of the elastic modulus of the cell wall of individual isolated or tissue cells. Analysis of the swelling and shrinking kinetics of a plant cell in response to external osmotic or water stress shows that this parameter, as much as the hydraulic conductivity of the membranes and the geometry of the cells, influences water exchange between the cell and its surroundings [Eq. (10)].

The experimental determination of ϵ requires the accurate measurement of instantaneous changes in cell turgor pressure, ΔP, and cell volume, ΔV [see Eq. (6)]. Otherwise, irreversible (plastic) deformation, hysteresis effects, and changes in cell volume and pressure owing to water flow have to be taken into account if tension is applied to the cell wall for a prolonged period of time. The conditions for ϵ measurements of individual isolated and tissue cells are as follows: (1) The pressure increments should not exceed 0.1 MPa because of the pressure dependence of ϵ.[2-5] Thus, small changes of pressure of the order of 0.001 MPa must be accurately measured. (2) Accurate determination of the corresponding instantaneous volume changes is required even though they are very small. The relative volume changes in full turgid plant cells are only of the order of 0.1 – 10% per 0.1 MPa (depending on the magnitude of the elastic modulus of the cell wall). This means that ΔV is somewhere between 0.01 to 0.10 μl per 0.1 MPa (giant algal cells) and 5×10^{-4} to 5×10^{-1} nl per 0.1 MPa (tissue cells). The anisotropic properties of the cell, the required resolution, and high speed of measurement render the accurate evaluation of ΔV by optical means difficult or even impossible.

The pressure probe satisfies both the above requirements. Pressure increments (induced hydrostatically by the displacement of the metal rod) of the order of 10^{-3} can be measured, and the corresponding volume changes can be readily calculated from the displacement and the diameter of the metal rod. Calibration of a given pressure probe therefore permits the immediate calculation of pressure and the corresponding volume changes.

The following experimental procedure is recommended for the determination of ϵ. Pressure increments, ΔP, of positive (and/or negative) signs and of increasing amplitude are injected into the cell before, during, or at the end of pressure relaxation experiments, and the corresponding ΔV values are calculated (Fig. 8). If ΔP is plotted versus ΔV, a straight line is normally obtained (Fig. 9A). The slope yields the elastic modulus provided

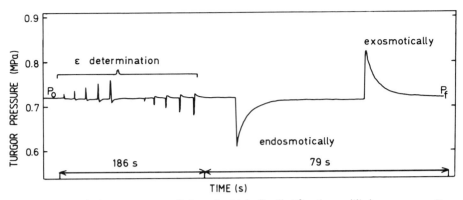

FIG. 8. Typical measurement of ϵ in a plant (algal) cell. After the equilibrium pressure, P_0, had been established, positive (and subsequently negative) pressure pulses of increasing amplitude were injected into the cell by appropriate displacement of the metal rod. After these measurements, turgor pressure relaxations were hydrostatically induced by decrease or increase of the effective osmotic pressure in the solution (endosmosis or exosmosis).

that the original volume of the cell was (microscopically) determined. A straight line indicates that ϵ can be assumed to be constant in the investigated pressure range and that water flow through the membrane (induced by disturbance of the steady state) can be neglected.

If these conditions are not fulfilled, bending of the curves occurs in the range of higher negative or positive pressure pulses, as indicated in Fig. 9B. Bending of the curves arising from the pressure dependence of the elastic modulus can be sometimes avoided by using smaller pressure amplitudes. Deviations from the straight line resulting from very rapid cellular water exchange times cannot be experimentally eliminated. In this rare case, an analysis of the data should be performed as described by Tomos et al.[27]

Determination of Hydraulic Conductivity

The hydraulic conductivity is experimentally determined by the following protocol: (1) A turgor pressure relaxation is either osmotically or hydrostatically induced and recorded (Fig. 5). (2) From a semilogarithmic plot of turgor pressure as a function of time (Figs. 5 and 6) the half-time, $T_{1/2}^P$, is calculated. (3) The volumetric elastic modulus is determined in an independent experiment as described above (Figs. 8 and 9). (4) The internal osmotic pressure is determined from the equilibrium state [Eq. (2)] by

[27] A. D. Tomos, E. Steudle, U. Zimmermann, and E.-D. Schulze, *Plant Physiol.* **68**, 1135 (1981).

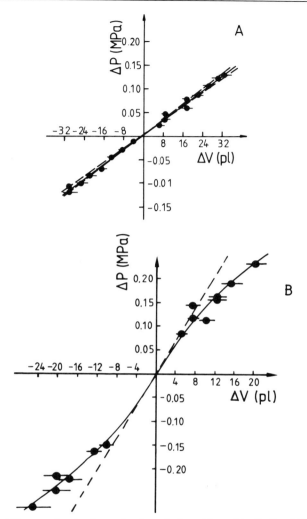

FIG. 9. $\Delta P - \Delta V$ curves for the determination of ϵ in epidermal leaf cells of *Tradescantia virginiana*. The plots are obtained by injection of pressure of increasing amplitude to the cell as shown in Fig. 8 and by simultaneous determination of the corresponding changes in volume from the displacement of the metal rod. For one type of cells a straight line was obtained (A). Owing to rapid water flow in the other type of cell investigated (B) the curves were bent. The sigmoid bending indicates that the elastic modulus can be assumed to be constant in the pressure range investigated. In the former case (A) the rate of water flow was apparently lower than in the other experiment (B). The bars indicate the pressure interval in which ϵ was determined. (Redrawn from Ref. 27.)

direct measurement of the stationary (original) turgor pressure and by cryoscopic determination of the external osmotic pressure. In the case of solutes, which exhibit reflection coefficients below 1, the reflection coefficient σ has to be measured in an independent experiment as described above. (5) The shape (spherical, ellipsoidal, etc.) of the cell and the corresponding geometric parameters are measured under the microscope, and the volume-to-area ratio is calculated using the appropriate equations for spheres, ellipsoids, etc. (6) The hydraulic conductivity is then calculated from Eq. (10).

Constraints of Pressure and Volume Measurements

The volume and compressibility of the pressure probe must be small in relation to cell volume changes and cell compressibility. After insertion of the microcapillary of the pressure probe into the cell, turgor pressure may compress the oil and/or other compressible parts of the plexiglass chamber (e.g. rubber seals) to some extent. Thus, the pressure recorded by the transducer is reduced or even zero. This effect can be ignored if volume changes arising from compression are negligible compared to the hydrostatically or osmotically induced changes in cell volume, i.e., if the volume of the apparatus is small. This condition is easily fulfilled for giant algal cells, even if the oil–cell sap boundary is not kept in exact position during turgor pressure regulation. Temperature variations (which change the compressibility of the system) do not interfere with the measurements either, provided they are not too large. For illumination of the cells and of the experimental setup, however, a cold light source or heat filters in the light pathways of the microscope should be used.

When investigating very small tissue cells of higher plants, interference effects resulting from temperature, from compressibility of the apparatus, and from capillary forces must be carefully excluded. This is achieved by the use of plexiglass chambers of very small volume and careful control of the position of the oil–cell sap boundary. By manual or automatic stabilization of the boundary position the effective volume of the pressure probe is reduced to the volume of the sap in the very tip of the microcapillary.

The compressibility of the probe also plays an important role in the determination of ϵ of very small tissue cells. If the pressure is increased within the cell by movement of the metal rod, the total volume change, ΔV, of the system (i.e., the parameter which is calculated from the displacement of the metal rod) will be equal to the sum of the decrease of the oil volume, ΔV_{Ch} (owing to compression), and the expansion of the cell, ΔV_C. Assuming that the cell sap such as water has negligible compressibility over the pressure range used in such experiments, Eq. (16a) holds:

$$\frac{\Delta V}{\Delta P} = -\frac{\Delta V_{Ch}}{\Delta P} + \frac{\Delta V_C}{\Delta P} \tag{16a}$$

Substitution of $\Delta V_{Ch}/\Delta P$ by the compressibility, K_{Ch}, of the probe and of $\Delta V/\Delta P$ by V_0/ϵ yields:

$$\frac{1}{V_0}\frac{\Delta V}{\Delta P} = \frac{V_{Ch}}{V_0} K_{Ch} + 1/\epsilon \tag{16b}$$

Equation (16b) gives the limits for the accurate determination of ϵ and, in turn, the conditions for constructing the pressure probe. If $\epsilon < V_{Ch}/V_0 K_{Ch}$, measurements will yield the correct value of the elastic modulus of the cell. On the other hand, if the ratio V_0/V_{Ch} is very large (as is the case for small cells) the first term on the right-hand side of Eq. (16) will be of the same order as $1/\epsilon$. In this case the error in the determination of ϵ according to Eq. (6) will be large. The same is true if the compressibility of the probe is large because of poor performance. Therefore, it is recommended that the compressibility of the pressure probe be measured before use. This can be achieved by simply blocking the tip with glue followed by displacement of the metal rod. Measurement of the pressure increase in the chamber and calculation of the corresponding volume change in the sealed probe yield the elastic modulus of the probe (or the reciprocal value of the compressibility coefficient, K_{Ch}). Only those pressure probes which exhibit an elastic modulus which is much lower than that of the cell should be used. If the effective volume is restricted to the very tip of the microcapillary by stabilization of the boundary, accurate measurements can be performed on very small cells as shown by the following example.

For typical pressure probes applied to higher plant cells the tip volume is 10^{-6} to 10^{-4} nl and K_{Ch} is approximately that of water (about 10^{-5} MPa). If V is 10 to 10 nl (corresponding to a cell diameter of 25 to 250 μm) and the elastic modulus of the cell wall of the order of 1 to 10 MPa, then $V_{Ch}/V_0 K_{Ch}$ will be 10^{-6} MPa or less and thus negligible compared to the term $1/\epsilon$ (10^{-2} to 10^{-3} MPa).

Another problem in pressure measurements is the occurrence of leakages around the microcapillary. During insertion leakage of water or cytoplasm into the medium will occur (assuming that the turgor pressure exceeds atmospheric pressure), because resealing and restoration of the original steady state need some time. Experience shows that at least 15 min is required before measurements can begin. Irreversible leakages can also occur during measurement or can be triggered by injection of positive pressure pulses of critical amplitude. Most of the leakages which occasionally occur over longer measuring periods are due to vibration of the experimental setup. This can be avoided by fixing the plant material in the Perspex chamber with a clamp having a small hole through which the probe may be introduced.

Leakages within the membrane are picked up by the pressure probe with high sensitivity. The pressure will drop immediately to zero and can be easily distinguished from a pressure relaxation curve with a very short half-time. Comparative measurements with microelectrodes and the pressure probe have shown that the pressure probe is much more sensitive in the recording of leakages than microelectrodes which measure membrane potential and resistance.[12]

Measurement of Negative Pressures

The pressure probe can also be used for direct determination of the negative pressures that have been proposed to exist in the water-conducting vessels (xylem) of cormophytes.[28] To this end, it is necessary to show that the pressure transducer can be used for accurate measurements of negative pressures.[17] Only those pressure transducers which can be wetted by water can be used. In the case of oil, gas bubble formation could occur at the oil–cell sap interface once the pressure falls below the partial pressure of water vapor.

For testing and calibration an apparatus based on a Hepp-type osmometer developed by Hansen can be used.[29] A water-filled glass capillary of about 20 mm length and 0.5 mm diameter is mounted vertically in a rectangular block of methacrylate (Fig. 10). At its upper end it is sealed to a dialysis membrane by an O ring. The membrane must be impermeable to polyethylene glycol (PEG, MW 6000) used as osmoticum for the generation of negative pressures in the glass capillary. For mechanical support of the dialysis membrane a porous copper grid is further introduced between membrane and glass capillary, as indicated in Fig. 10. This arrangement prevents a bending of the membrane owing to pressure gradients built up between the water-filled glass capillary and the PEG solution in the reservoir on top of the methacrylate block. The PEG solution in the chamber can be easily replaced so that different concentrations of PEG can be applied. Because of the nonideal behavior of PEG solutions, the osmotic pressure of the solutions has to be determined by psychometry.[17] The pressure transducer is sealed to the lower end of the glass capillary so that

[28] M. H. Zimmermann, "Xylem Structure and the Ascent of Sap." Springer-Verlag, Berlin and New York, 1983.
[29] A. T. Hansen, *Acta Physiol. Scand.* **53**, 197 (1961).

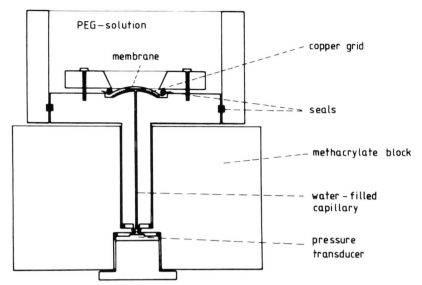

FIG. 10. Schematic diagram of the pressure probe–Hepp type osmometer. The water-filled glass capillary in the center of the methacrylate block was sealed at the upper end by a dialysis membrane (impermeable to PEG 6000), which was reinforced with a copper grid. A pressure transducer was bolted directly to the lower end of the capillary. Negative pressures in the glass capillary were built up by appropriate PEG concentrations in the reservoir above the copper grid.

the liquid in the capillary is constrained and negative pressures are directly recorded by means of the pressure probe with reference to atmospheric pressure.

It has to be emphasized that the glass capillary must be thoroughly cleaned to promote complete wetting of the walls and that the transitions between the different components of this apparatus must be smooth. These precautions minimize cavitation; for the same reason the water must be degassed for 12 to 24 hr in vacuum. The tracings in Fig. 11 show the pressure of the water in the glass capillary as a function of time. Pressure is decreased by stepwise increasing the concentration of PEG in the reservoir. At concentrations corresponding to an osmotic pressure greater than 0.1 MPa, negative pressures can be recorded. The replacement of the PEG solutions by water increases the pressure to zero, indicating that the procedure is reversible. Figure 12 shows a plot of the osmotic pressure of the PEG solutions versus the negative pressure within the capillary as measured by the pressure transducer. It is obvious that there is a 1 : 1 relationship between the changes in negative pressure and osmotic pressure.

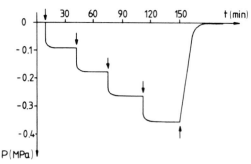

FIG. 11. Records of the pressure, P, within the water-filled glass capillary recorded by the pressure transducer (Fig. 10) with reference to atmospheric pressure, as a function of time. Addition of increasing concentrations of PEG (downward arrows, equal to pressures of 0.08, 0.17, 0.28, and 0.36 MPa) resulted in a corresponding decrease of the pressure in the water phase of the capillary. Replacement of PEG by water (upward arrow) led to restoration of the original atmospheric pressure. (Redrawn from Ref. 17.)

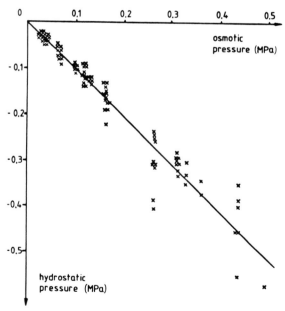

FIG. 12. Plot of the pressure in the water phase of the glass capillary of the pressure probe–Hepp osmometer (Figs. 10 and 11) versus the osmotic pressure of the PEG solution. Data points were obtained from 24 independent measurements. The straight line was fitted to the data by the method of least squares. (Redrawn from Ref. 17.)

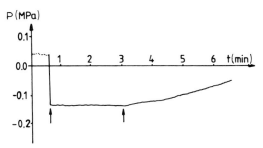

FIG. 13. In contrast to Figs. 10 and 11 the lower end of the glass capillary of the modified Hepp osmometer was sealed to an isolated vascular bundle of a leaf of *Plantago major*. One xylem tube of this bundle was penetrated by the pressure probe about 2 cm further down. Before insertion of the probe, the pressure in the osmometer had been adjusted to −0.13 MPa by addition of a defined PEG solution to the reservoir. During insertion of the microcapillary of the pressure probe into the tissue, the probe recorded the slight positive pressure in the parenchymal cells of the stem (dashed line). Entry into the xylem (left arrow) led to a sudden drop in pressure toward negative values; in this experiment a constant value of −0.13 MPa was obtained. Replacement of the PEG solution by water (right arrow) resulted in an increase in pressure. This happened more slowly than in Fig. 11, presumably because of the larger volume of the compartment in which negative pressure developed. Apparently, the system needed a much longer time to equilibrate. (Redrawn from Ref. 17.)

After this test the pressure probe can be used for measurements in the xylem tubes of plants. Successful insertion of the pressure probe into the xylem is indicated by pressure profile such as that shown in Fig. 13. During the very slow penetration of the microcapillary through the tissue a positive turgor pressure is recorded because the microcapillary of the pressure probe will pass through cells. Insertion of the microcapillary is then indicated by a sudden drop of the pressure to negative values or to values below atmospheric pressure. As soon as the boundary between xylem sap and water in the microcapillary is fixed the negative pressure can be recorded. Leakages occurring around the penetration area of the xylem are easily detected by a sudden increase in pressure to zero or to positive values.

In addition, it is useful to use dye (e.g., eosin) solutions instead of water in the microcapillary of the pressure probe in order to inject the dye into the xylem after the end of the experiment. Under the microscope, thin cross sections of the stem tissue show the location of the dye in the tissue. In the absence of leaks of dye will be the water-conducting vessel above the insertion area.

So far the pressure probe has been applied only for pressure measurements in the xylem of *Plantago major* and *Nicotiana rustica*. The average pressures measured in the petioles of well-watered *N. rusticana* plants are in the range of −0.06 to −0.2 MPa. These values, which can be recorded continuously for several hours, are only slightly higher than those obtained in parallel experiments using the conventional Scholander bomb.[30] However, in plants that have not been watered for 3 days no values lower than -0.35 MPa are recorded with the pressure probe, whereas parallel measurements with the bomb obtain values down to -1.4 MPa. This discrepancy in the pressure values is not yet understood.

Pressure Clamp

The pressure probe can also be used for pressure clamp experiments. By analogy with voltage clamp experiments, the turgor pressure is maintained at a constant level.[14,15] This nonequilibrium turgor pressure is maintained by appropriate displacement of the oil–cell sap boundary. The volume relaxation (which is equal to the volume change caused by water flow) can be measured from the displacement. Pressure clamp experiments have some advantages over turgor pressure relaxation experiments. As shown in the section on Theoretical Considerations, the cell volume can be accurately determined [Eq. (13)], which allows the volumetric elastic modulus to be calculated with greater accuracy. This is of particular importance for tissue cells. Measurements of the geometric dimensions of such cells, particularly if they are located in interior layers, is normally subject to large errors. The half-time of a volume relaxation can be considerably longer than that of the corresponding turgor pressure relaxation, and this may be advantageous for plant cells which exhibit very short water exchange times.

In practice, large volume changes occur in pressure clamp experiments. This implies that a considerable amount of cell sap must be pushed through the tip of the microcapillary. Under these conditions the assumptions made in the derivation of Eqs. (11) and (12) no longer hold. Therefore the following experimental conditions are normally used.[14,15]

In the first part of a pressure clamp experiment a (positive) pressure pulse of suitable amplitude is injected into the cell (Fig. 14). The final pressure $P_f = P_0 + \Delta P$ is held constant for about 60 to 90 sec. This time interval is sufficiently long to measure accurately the corresponding volume changes during the pressure clamp. From the initial slope of the volume relaxation the hydraulic conductivity is calculated according to Eq. (11). After this the cell is allowed to reach a new water equilibrium by a normal turgor pressure relaxation, in which the oil–cell sap boundary is stabilized as described above. The hydraulic conductivity can also be calculated from the half-time of the turgor pressure relaxation [Eq. (10)]

[30] P. F. Scholander, H. T. Hammel, E. A. Hemmingsen, and E. D. Bradstreet, *Proc. Natl. Acad. Sci. U.S.A.* **52**, 119 (1964).

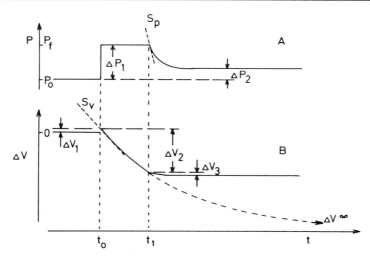

Fig. 14. Schematic diagram of a combined pressure clamp–turgor pressure relaxation experiment. (A) Turgor pressure versus time; (B) volume versus time. The volume relaxation is generated at a pressure $P_f = P_0 + \Delta P_1$ clamped for the time interval $t_1 - t_0$ (A). Note that the pressure step, ΔP_1, at t_0 is achieved by pushing a volume ΔV_1 through the tip of the pressure probe into the cell (B). The corresponding volume during the pressure clamp is plotted in B, the total amount being denoted by ΔV_2. At time t_1, the cell is allowed to reach a new equilibrium state by turgor pressure relaxation brought about by a relatively small, outwardly directed water flow. The water flow is associated with a change in cell volume denoted by ΔV_3. Although the ΔV values refer to different volumes (see text), they are here plotted on the same axis. $\Delta V' = \Delta V_1 + \Delta V_2$ is the experimentally determined volume change. The value of the hydraulic conductivity of the cell membrane is calculated either from the initial slope of the volume relaxation, S_V, or from the half-time of the pressure relaxation, according to Eqs. (11) and (10), respectively. For calculations of the cell volume from these combined pressure clamp–relaxation experiments, see text. The dashed line in B denotes the theoretically expected continuation of the volume regulation, provided that the pressure clamp could be maintained until a new equilibrium state is reached.

provided that the elastic modulus is determined in an independent experiment. It is, therefore, possible to compare the accuracy of the two methods. The cell volume can still be calculated from Eq. (13). However, it has to be noted that ΔP in Eq. (13) has to be replaced by ΔP_2 which is the pressure difference between the initial and final states of water equilibrium (Fig. 14).

As indicated in Fig. 14 the total change in cell volume can be subdivided into three parts: ΔV_1, ΔV_2, and ΔV_3. ΔV_1 represents the change in cell volume associated with the change in turgor pressure at the beginning of the pressure clamp experiment. This volume passes through the tip of the microcapillary of the pressure probe but does not cross the membrane as in turgor pressure relaxation experiments. ΔV_2 and ΔV_3 correspond to the volume changes resulting from the water flow through the membrane during pressure clamp and the subsequent turgor pressure relaxation, respectively. ΔV_3 is not recorded because of the fixation of the oil–cell sap boundary. Therefore the displacement of the metal rod in the pressure probe is equal to the sum $\Delta V' = \Delta V_1 + \Delta V_2$. The total volume change, ΔV_m, arising from the water flow through the membrane is, therefore, given by

$$\Delta V_m = \Delta V_2 + \Delta V_3 = \Delta V' - \Delta V_1 + \Delta V_3 \qquad (17)$$

The two correction terms, ΔV_1 and ΔV_3 in Eq. (17), can be rewritten as $\Delta V_1 = dV/dP \, \Delta P_1$ and $\Delta V_3 = dV/dP \Delta P - \Delta P_2$), which yields

$$\Delta V_m = \Delta V' - \Delta P_2 \frac{dV}{dP} \qquad (18)$$

According to Eq. (6) dV/dP equals $(\epsilon/V_0)^{-1}$ and can be determined in an independent experiment (see above).

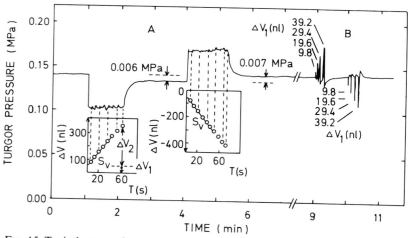

FIG. 15. Typical course of a combined pressure clamp–pressure relaxation experiment on an internode of *Chara corallina*. (A) A turgor pressure of 1.4 bar was established by adding mannitol to the artificial pond water. In the first run, the cell pressure was reduced by means of the pressure probe and clamped for 64 sec. After this the cell was allowed to reach a new water equilibrium state by a turgor pressure relaxation. In a second run, the same experiment was repeated with a pressure step of opposite sign. The corresponding volume changes occurring during the pressure clamp are given in the insets (for details, see text). (B) Volume changes of increasing amplitude in both directions were applied to the cell with the pressure probe in order to determine the elastic modulus of the cell wall.

From the foregoing considerations, it is evident that, for the combined pressure clamp–pressure relaxation experiment, the total water flow, ΔV_m, has to be used in Eq. (13) instead of ΔV.

A typical time course of a combined pressure clamp–pressure relaxation experiment in an internode of *Chara corallina* is shown in Fig. 15. The corresponding changes in volume with time during the pressure clamp are presented in the insets. The duration of this experiment was 100 sec. The volume change, ΔV, was generally only 80 to 900 nl. It can be shown that the values of the hydraulic conductivities deduced from pressure clamp and pressure relaxation are identical in the limits of accuracy.[14,15]

The cell volume determined by pressure clamp according to Eq. (13) is 7–20% smaller than the volume calculated by optical means.[14,15] This discrepancy is probably due to the thickness of the cell wall, since only the external diameter of the cell can be determined by optical means. So far the pressure clamp technique has been applied only to giant algal cells. Further miniaturization is required to apply this method to the tiny cells of higher plants.

Conclusions

The pressure probe technique is a powerful tool for the determination of water relation parameters of single and tissue plant cells. The success of the method depends on several factors. An understanding of the theoretical concepts underlying the analysis of coupled transport processes and knowledge of the sources of the artifacts involved in pressure relaxation and pressure clamp experiments are prerequisites for the development of the full potential of this technique.

Insertion of several (automatic recording) pressure probes in individual cells of different tissues (roots, leaves, water-conducting vessels, etc.) of an intact plant may help to elucidate the complicated regulatory processes involved in the water supply of tissue cells under water and salt stress. Coupling of the pressure probe to complementary techniques may further increase the possibility of exploring the water relations of plants on a molecular and cellular level. One conceivable scenario for future experiments is the combined application of the pressure probe and noninvasive NMR tomograph techniques to individual plant tissue cells. High-resolution NMR tomography would allow the determination of the local water distribution in tissues and would also show whether water is immobilized differently in the cells of various tissues.[31] This information, which is required for the correct analysis of the pressure measurements in terms of

[31] A. Haase, A. Balling, L. Walther, and U. Zimmermann, *Planta* (in press).

membrane and cell wall parameters, is not accessible by the pressure probe technique.

Acknowledgments

This work was supported by funds from the Deutsche Forschungsgemeinschaft (Forschergruppe Ökophysiologie and SFB 176). The author gratefully acknowledges Dr. W. M. Arnold for reading the manuscript and Dr. J. Rygol for drawing some of the artwork.

[25] Transport Systems in Algae and Bryophytes: An Overview

By J. A. Raven

Introduction

The organisms considered in Section II,B of this volume are lower photosynthetic plants. This statement means that we are dealing with organisms which grow photolithotrophically, and which lack the vegetative and reproductive characteristics of vascular plants (tracheophytes), namely, a diploid sporophyte generation with xylem, phloem, cuticle, stomata, and intercellular gas spaces which eventually grow independently after its initial nutritional dependence on the haploid gametophyte generation. To set the scene for detailed discussion of specific transport systems in [26]–[30] of this volume, we outline the special features of the use of lower phototrophs for transport studies.

General Characteristics of Algae

As befits the eukaryotic part of the title of this volume, and also in conformity with my own predelictions, the algae may be defined as eukaryotic, O_2-evolving photolithotrophs (and their immediate chemoorganotrophic relatives) which lack the special reproductive features of the archegoniates (Bryophyta and Tracheophyta). This definition excludes the O_2-evolving prokaryotic Cyanobacteria (alias Cyanophyceae) and Chloroxybacteria (alias Prochlorophyta).[1-3]

[1] L. Margulis, "Symbiosis in Cell Evolution." Freeman, San Francisco, 1981.
[2] J. A. Raven, *Prog. Phycol Res.* **5**, 1 (1986).
[3] T. Burger-Wiersma, M. Veerhuis, H. J. Korthals, C. C. M. Van de Wel, and L. R. Mur, *Nature (London)* **320**, 262 (1986).

As defined in this rather negative way at least 17 classes of algae in at least 12 divisions can be recognized on the basis of ultrastructural and biochemical features[2,4,5] (Table I). Many structural and reproductive features attributable to the occurrence of parallel evolution recur in various of these classes to yield organisms with similar macroscopic reproductive and vegetative features but with dissimilar chemical and ultrastructural features. The organisms considered in [26]–[30] in this volume all belong to a single division, the Chlorophyta, whose members are characterized by possessing chlorophyll *b* as well as chlorophyll *a* and by storing a 1,4-α-glucan (starch) *inside* plastids[2,4,5]; these attributes, and various ultrastructural features, show that Chlorophyta is the algal division most closely related to the Bryophyta (see below) and the Tracheophyta (vascular plants; Section II,A). Indeed, strict application of the criteria by which algae are assigned to divisions to the totality of O_2-evolving eukaryotes would subsume the Bryophyta and Tracheophyta into the Chlorophyta.[2,5] Accordingly, most of the biochemical and electrophysiological work on algal membrane transport,[2,5-7] and almost all of that discussed in [26]–[30] in this volume was conducted on these algae which are most closely related to the higher plants, and it may be unwise to generalize too far to the other major clades (divisions and classes) at the algal level of organization from results obtained on the Chlorophyta. Indeed, generalization of results obtained *within* the Chlorophyta must be approached with caution since there is at least as much diversity of plasmalemma and tonoplast transport systems in the Chlorophyta as has been discovered so far in the Tracheophyta.[2,5-7]

The Chlorophytan class which, on ultrastructural and biochemical criteria, is most closely related to the Bryophyta and Tracheophyta is the Charophyceae, two of whose genera (*Chara* and *Nitella*) are discussed in [28] and [30] in this volume; these genera belong to the family Characeae whose members have been very widely used in the work described in [27] in this volume. *Chlorella* ([26], this volume) is in the class Chlorophyceae, while *Acetabularia* ([30], this volume) belongs to the predominantly marine class Ulvophyceae.[2,5-8]

[4] H. C. Bold and M. J. Wynne, "Introduction to the Algae: Structure, Function and Reproduction," 2nd ed. Prentice-Hall, Englewood Cliffs, New Jersey, 1985.

[5] J. A. Raven, "Energetics and Transport in Aquatic Plants." Liss, New York, 1984.

[6] J. A. Raven, *in* "Solute Transport in Plant Cells and Tissues" (D. A. Baker and J. L. Hall, eds.), p. 166. Longmans, London, 1988.

[7] J. A. Raven, *Symp. Soc. Exp. Biol.* **40**, 347 (1986).

[8] D. E. G. Irvine and D. M. Johns, eds., "Systematics of the Green Algae." Academic Press, London, 1984.

TABLE I

Division (Suffix -phyta) and Classes (Suffix -phyceae) of Algae, with Organisms
Used for Transport Studies and Others Mentioned in This Chapter[a]

Division, class	Organism	Material used in transport studies
Chlorophyta: Chlorophyceae	*Chlorella, Eremosphaera*	Vegetative nonmotile unicell
	Chlamydomonas	Vegetative flagellate unicell (effectively wall-less; cannot develop turgor)
	Dunaliella	Vegetative flagellate unicell (wall-less)
	Hydrodictyon	Vegetative nonmotile giant cell from colony (not part of a symplast)
Chlorophyta: Charophyceae	*Spirogyra, Mougeottia*	Vegetative nonmotile cells in filament (not part of a symplast)
	Chara, Lamprethamnium, Nitella, Nitellopsis	Vegetative giant internodal and "leaf" cells of multicellular plant (part of a symplast)
Chlorophyta: Ulvophyceae	*Ulva, Enteromorpha*	Vegetative cells in foliose thallus (not part of a symplast)
	Cladophora, Chaetomorpha	Vegetative large (giant) cells in filaments (not part of a symplast)
	Acetabularia, Codium, Bryopsis, Halicystis, Valonia, Caulerpa	Vegetative giant cells (whole organism)
Euglenophyta: Euglenophyceae	*Euglena*	Vegetative flagellate unicell (wall-less)
Rhodophyta: Cyanidiophyceae	*Cyanidium*	Vegetative nonmotile walled unicell
Rhodophyta: Bangiophyceae	*Porphyra*	Vegetative cells in foliose thallus of "Porphyra" phase (not part of a symplast)
Rhodophyta: Floridiophyceae	*Gracilaria, Palmaria, Lemanea, Griffithsia*	Vegetative cells in (or giant cells from, *Griffithsia*) multicellular plant (part of a symplast)

TABLE I (*continued*)

Division, class	Organism	Material used in transport studies
Chrysophyta: Chrysophyceae	*Ochromonas* (= *Poteriochromonas*)	Vegetative flagellate unicell (wall-less)
Bacillariophyta: Bacillariophyceae	*Phaeoductylum Cyclotella, Ditylum, Fragilaria, Stephanodiscus*	Vegetative nonmotile unicell
Tribophyta: Tribophyceae	*Vaucheria*	Vegetative giant cells (whole organism)
Phaeophyta: Phaeophyceae	*Fucus, Pelvetia, Ascophyllum, Laminaria Macrocystis*	Vegetative cells in parenchymatous thallus (part of a symplast); also released eggs and zygotes of *Fucus, Pelvetia,* and *Ascophyllum*

[a] Not all algal divisions and classes are listed here; see Bold and Wynne[4] and Raven.[2,5,6]

Opportunities and Limitations in Use of Algae in Transport Studies

Algae have cell volumes (Table II) ranging from $\sim 10^{-18}$ m^3 for some of the picoplankton up to values in excess of 10^{-6} m^3 for "giant cells" in some cellular or multicellular algae.[2,5-7,9-17] Corresponding values for plasmalemma area are 10^{-12}–10^{-3} m^2. Such size differences have important repercussions for the applicability of techniques for the study of transport at the plasmalemma. Intracellular microelectrodes for the measurement of electrical potential differences, conductance at tonoplast and plasma-

[9] J. A. Raven, *Encycl. Plant Physiol., New Ser.* **3,** 129 (1976).

[10] G. Wagner and E. Bellini, *Z. Pflanzenphysiol.* **79,** 283 (1976).

[11] G. Wagner, *Planta* **118,** 145 (1974).

[12] A. J. Young, J. C. Collins, and G. Russell, *in* "Systematics of the Green Algae" (D. E. G. Irvine and D. M. John, eds.), p. 343. Academic Press, London, 1984.

[13] R. H. Reed, J. C. Collins, and G. Russell, *J. Exp. Bot.* **31,** 1521 (1980).

[14] H. Kauss, *in* "Membrane Transport in Plants" (U. Zimmermann and J. Dainty, eds.), p. 90. Springer-Verlag, Berlin and New York, 1974.

[15] B. Schobert, E. Untner, and H. Kauss, *Z. Pflanzenphysiol.* **67,** 385.

[16] R. W. Eppley and J. Rogers, *J. Phycol.* **6,** 344 (1970).

[17] J. A. Raven, *in* "Physiological Ecology of Photosynthetic Picoplankton in the Ocean" (T. R. Platt and W. K. Li, eds.), p. 1. Can. Bull. Fish. Aquat. Sci., 1986.

TABLE II

Protoplast Volume, Plasmalemma Area, and Fraction of Protoplast Occupied by Vacuole in Various Algal Cells[a]

Organism	Cell type	Protoplast volume (m³)	Plasmalemma area (m²)	Vacuole(s) as fraction of protoplast volume	Ref.
Chlamydomonas reinhardi	Vegetative cell	$63–217 \times 10^{-18}$	$77–175 \times 10^{-12}$	0.08	
Chlorella emersonii	Vegetative cell	23×10^{-18}	39×10^{-12}	0.07–0.14	
Chlorella fusca	Vegetative cell	$33–188 \times 10^{-18}$	$50–159 \times 10^{-12}$	0.08–0.13	
Eremosphaera viridis	Vegetative cell	1.66×10^{-12}	67.0×10^{-9}	?	
Hydrodictyon africanum	Vegetative cell (spherical)	14×10^{-9}	28×10^{-6}	0.98	
Chara corallina	Vegetative cell internodal cylindrical	10×10^{-9}	60×10^{-6}	0.95	
		100×10^{-9}	$300 \times 10{-6}$	0.97	
Mougeottia sp.	Vegetative cell (cylindrical)	20.7×10^{-15}	4.89×10^{-9}	0.80	10, 11
Spirogyra sp.	Vegetative cell (cylindrical)	0.25×10^{-12}	10×10^{-9}	0.9?	
Enteromorpha intestinalis	Vegetative cell (squat cylinder)	$407–877 \times 10^{-18}$	$343–524 \times 10^{-12}$	$<0.05–<0.80$	12
Valonia ventricosa	Vegetative cell (spherical)	$0.40–25 \times 10^{-6}$	$0.263–4.13 \times 10^{-3}$	0.995–0.997	
Euglena viridis	Vegetative cell	2.92×10^{-18}	0.987×10^{-15}	0.17	
Porphyra purpurea/P. umbilicalis	Vegetative cell (squat cylinder)	7×10^{-15} (*P. purpurea*)	2×10^{-9} (*P. purpurea*)	0.05–0.20 (*P. umbilicalis*)	13
Griffithsia monile	Vegetative cell (spheroid)	10^{-9}	6×10^{-6}	0.97–0.98	
Ochromonas (= Poteriochromonas) malhamensis	Vegetative cell	0.30×10^{-15}	0.25×10^{-9}	?	14, 15
Ditylum brightwellii	Vegetative cell	54×10^{-15}	6.9×10^{-9}	0.80–0.90	16
Fragilaria capucina	Vegetative cell	0.319×10^{-15}	0.226×10^{-9}	0.45	
Stephanodiscus biradians	Vegetative cell	0.776×10^{-15}	0.408×10^{-9}	0.66	
Pelvetia canaliculata	Ovum or uncleared zygote	5×10^{-12}	0.14×10^{-6}	<0.05?	

[a] From Raven,[5,9] unless otherwise stated.

lemma, and cytosol or vacuole ion activities have been used on cells as small as *Chlorella* sp. (plasmalemma area $\sim 50 \times 10^{-12}$ m^2), but are more reliably applied to cells such as *Eremosphaera viridis* and *Spirogyra* sp. (plasmalemma area $10-100 \times 10^{-9}$ m^2) (see Ref. 5, pp. 327, 328). The problems with the use of "lipid soluble cations" as probes of transplasmalemma potential difference, and of such "activity" probes as Quin 2 (for Ca^{2+}), in plant cells[18-23] mean that the transplasmalemma and transtonoplast passive driving forces on ions in small cells are not readily measured. Larger cells can, of course, also present problems in electrophysiological studies. Large cylindrical cells, to which cable theory has been applied to produce "real" nonlinear I/V relationships from "measured" nonlinear I/V values, are probably not completely "space-clamped" by the techniques usually applied[24]; this problem is exacerbated if, as is often the case for *Chara* internodal cells and *Acetabularia* cells (and in smaller algal cells), there are spatial variations of electrical properties over the cell surface.[2,5-7,24]

A further problem which is encountered with any cells in multicellular differentiated organisms with connections between cells via plasmodesmata (some Chlorophyceae, Charophyceae, and Pleurastrophyceae; many Phaeophyceae) or pit connections (many Rhodophyceae).[2,5-7,9,25,26] Where electrical measurements have been made (in membranes of the Characeae, and in rhodophyceaen *Griffithsia pacifica*) it has been found that the intercellular connections constitute current leaks in parallel with the plasmalemma or pit connections in dissipating charge injected into a single giant cell[5-7,26-30] which is still part of the symplastic continuum. The quantitation of the conductance of intercellular connections is of great interest in relation to the significance of plasmodesmata in the transmission of morphogenetic information (low M_r messenger solutes and electri-

---

[18] H. Gimmler and H. Greenway, *Plant, Cell Environ.* **6**, 739 (1983).

[19] R. J. Ritchie, *J. Exp. Bot.* **35**, 699 (1984).

[20] R. J. Ritchie, *Prog. Biophys. Mol. Biol.* **43**, 1 (1984).

[21] M. C. Astle and P. H. Rubery, *Plant Sci. Lett.* **39**, 43 (1984).

[22] R. J. Cork, *Plant, Cell Environ.* **9**, 157 (1986).

[23] C. Brownlee and J. W. Wood, *Nature (London)* **320**, 624 (1986).

[24] J. R. Smith, *Aust. J. Plant Physiol.* **11**, 211 (1984).

[25] H. J. Marchant, *in* "Intercellular Communication in Plants: Studies on Plasmodesmata" (B. E. S. Gunning and A. W. Robards, eds.), p. 59. Springer-Verlag, Berlin and New York, 1976.

[26] R. W. Bauman, Jr., and B. R. Jones, *J. Phycol.* **22**, 49 (1986).

[27] T. M. Goldsmith and M. H. M. Goldsmith, *Planta* **143**, 167 (1978).

[28] R. M. Spanswick and J. B. Costerton, *J. Cell Sci.* **2**, 451 (1967).

[29] U. Zimmermann and F. Beckers, *Planta* **138**, 173 (1978).

[30] T. Siboaka and T. Tabata, *Plant Cell Physiol.* **22**, 397 (1981).

cal currents) which may be significant in *directing* differentiation as well as in the movement of materials between sources and sinks produced as a *result* of differentiation.[2,5-7,9,31-34] However, surgical isolation of single giant cells (of *Griffithsia*) and of single ecorticate internodal cells *plus* their associated nodal cells at each end (of various Characeans) leads to "sealing" of the intercellular connections,[35,36] so that the isolated giant cells (with associated small nodal cells in the Characeans) become single cells surrounded by plasmalemma. It is important to recognize these systems as being derived surgically from a symplastic network rather than being analogous to *unicellular* (acellular) giant-celled organisms such as vegetative plants of *Acetabularia, Halicystis,* and *Valonia,* or unicells derived from *colonial* giant-celled organisms such as *Hydrodictyon.*

The measurement of intracellular compartmentation of solutes, and of fluxes between compartments, is also greatly influenced by the size and structure of the cells and whether the cells are part of a colony or of a differentiated organism. Small algal cells may have only a few percent of their total volume occupied by vacuoles, whereas the fraction may be as high as 99% in the largest cells of *Valonia.*[5] The large fraction of vacuoles and the large absolute cell size mean that characterization of vacuolar chemistry (for solute concentration) and isotope chemistry (following supply of label to the bathing medium, as a means of solute flux estimation) can be conducted directly on uncontaminated vacuolar sap samples. These large cells also allow vacuolar perfusion, which permits a number of important experiments to be performed. Varying the composition of the solution bathing the inner surface of the tonoplast permits better characterization of transtonoplast fluxes, their mechanism, energization, and regulation. The capacity to replace the vacuolar fluid with, at least in marine organisms, a solution essentially identical to the external solution permits (with a suitable voltage clamp) the measurement of short-circuit currents and their comparison with net charge carriage by particular ionic species as estimated from bidirectional tracer fluxes between bathing medium and vacuole. Such measurements can, with reservations, be used to suggest the location, direction, and magnitude of active (electrogenic) ion transport. These vacuole manipulations recall the possibilities for the use of meta-

[31] B. E. S. Gunning and A. W. Robards, eds., "Intercellular Communication in Plants: Studies on Plasmodesmata." Springer-Verlag, Berlin and New York, 1976.

[32] J. A. Raven, *Adv. Bot. Res.* **9**, 153 (1977).

[33] M. Andrews, R. Box, A. Fyson, and J. A. Raven, *Plant, Cell Environ.* **7**, 683 (1984).

[34] R. Box, M. Andrews, and J. A. Raven, *J. Exp. Bot.* **35**, 1016 (1984).

[35] S. B. Waaland and R. E. Cleland, *Protoplasma* **79**, 185 (1974).

[36] A. B. Hope and N. A. Walker, "The Physiology of Giant Algal Cells." Cambridge Univ. Press, London and New York, 1975.

zoan epithelia in transport studies, with the cytoplasm (cytosol *plus* organelles other than the central vacuole) bounded by the plasmalemma on the outside and the tonoplast within as a "syncytial" analog of the epithelial cells with their bounding mucosal and serosal plasmalemmas. This point will not be further pursued (except to mention that the giant-celled algae have been much less thoroughly investigated than have the animal epithelia) lest the lower plants are banished to the volume (Vol. 175) on epithelia (i.e., the approach of Giebisch *et al.*[37])!

Returning to a consideration of the smaller-celled algae, we have already seen that intact cells with an equivalent spherical radius of 2 μm present difficulties in the determination of plasmalemma electrical properties. However, small unicells or undifferentiated assemblies (colonies) of small cells can have very substantial advantages in transport studies, such as rapidity of growth under well-defined conditions, relative ease of axenic culture, and possibility of haploid genetic studies. These properties, which are shared by many other microorganisms, permit the relatively rapid production of the required amount of genotypically and phenotypically uniform material. It is significant that photolithotrophic growth is intrinsically slower than that of chemoorganotrophic.[17,38] For a cell of a given size growing at its maximum specific growth rate, there is necessarily a diversion to catalysis of phototrophy (photons $\rightarrow$ sugars) of cell components which would be used, in chemoorganotrophs, to produce additional catalysts of the later stages of metabolism (sugar $\rightarrow$ all other cells materials) which are common to both photolithotrophs and chemoorganotrophs.[38] However, this liability in terms of growth is observed to be (as predicted from computations based on catalyst size and specific reaction rate) only about 2-fold for cells of any given size.[38] At least as serious from the practical point of view for the rapid production of large quantities of material for experimentation is the problem of self-shading in large culture vessels with reasonable densities of phototrophic cells per unit volume.[38]

An important experimental system for transport studies is a unicellular stage in the life cycle of a multicellular plant, i.e., the eggs of members of the Fucales (Phaeophyta). These have been used extensively to study transport phenomena related to fertilization and subsequent differentiation, and they have recently been used in estimates of free cytosol [Ca^{2+}].[23,39]

[37] G. Giebesch, D. C. Tosteson, and H. H. Ussing, eds., "Membrane Transport in Biology," Vol. 3. Springer-Verlag, Berlin and New York, 1978.

[38] J. A. Raven, *in* "Microalgal Biotechnology" (M. A. Borowitzka and L. J. Borowitzka, eds.), p. 331. Cambridge Univ. Press, London and New York, 1987.

[39] D. Sanders and A. J. Miller, *in* "Molecular and Cellular Aspects of Calcium in Plant Development" (A. J. Trewavas, ed.), p. 149. Plenum, New York, 1986.

A final point about many of the algal cells which are employed in transport studies (and which is shared by other photosynthetically competent cells such as many bryophyte and some tracheophyte cells used in transport studies) is the presence of photosynthetically competent plastids. This complicates interpretation of the energetics, and regulation, of transport at the plasmalemma and tonoplast of such cells.[9,40-44] Giant-celled algae (freshwater Characeae), can be perfused and even have their "native" plastids replaced by intact plastids isolated from vascular plants, thus giving valuable information as to the taxonomic specificity of the energetic and informational interactions between plastids and plasmalemma. Further phototrophy-related considerations for plasmalemma transport in many phototrophic cells involve the implications for mediated and lipid–solution fluxes at the plasmalemma connected with acquisition of inorganic carbon for photosynthesis other than by diffusive entry of CO_2 from a periplasm in which the $HCO_3^- - CO_2$ equilibrium has not been altered by "specific" plasmalemma H^+ and OH^- fluxes. The multifarious mechanisms involved are best investigated in algae, but some also occur in some aquatic flowering plants.[2,5-7,41] In some cases the net mediated H^+/OH^- and/or inorganic carbon fluxes which are implicated in these "CO_2 accumulation mechanisms" are quantitatively among the most significant in the illuminated plasmalemma, and they have implications for other fluxes in terms of preempting a share of integral proteins of the plasmalemma, the energy supply for their operation, and their regulation.

Results from Transport Studies on Algae

Plasmalemma: Primary Active Transport

Much of the algal work has involved freshwater members of the Chlorophyceae and freshwater and euryhaline Charophyceae (Chlorophyta); this work has generally supported the occurrence of an MgATP-driven, vanadate-sensitive electrogenic H^+ efflux pump in the plasmalemma. This pump can, at least under some circumstances, operate with a 1 H^+ : 1 ATP stoichiometry, can consequently generate a $\Delta\bar{\mu}_{H^+}$ of up to 30 kJ (mol $H^+)^{-1}$, and has a site density-specific reaction rate combination which yields an active efflux of up to 1 μmol H^+ efflux m^{-2} sec^{-1}.[2,5,6,45] Much of

[40] J. A. Raven, in "The Intact Chloroplast" (J. Barber, ed.), p. 403. Elsevier, Amsterdam, 1976.

[41] J. A. Raven, New Phytol. **76**, 205 (1976).

[42] J. A. Raven, B. A. Osborne, and A. M. Johnston, Plant, Cell Environ. **8**, 417 (1985).

[43] T. Mimura and M. Tazawa, Plant Cell Physiol. **27**, 319 (1986).

[44] K. Takeshige, T. Shimmen, and M. Tazawa, Plant Cell Physiol. **27**, 237 (1986).

[45] M. Tazawa, T. Shimmen, and T. Mimura, Annu. Rev. Plant Physiol. **38**, 95 (1987).

the most convincing data arises from electrical and net H^+ flux work on characean internodal cells, either intact or with the tonoplast and much of the cytoplasm perfused away.[2,5,6,42,44] Work using isolated plasmalemma vesicles, or more completely resolved systems, is less advanced with algae than with certain fungi (Section III in this volume) and higher plants (Section II,A, this volume).[46] Another major thrust in the study of primary active transport at the plasmalemma of algal cells involves the marine *Acetabularia* (Ulvophyceae: Chlorophyta). Here a 2 Cl^- : 1 ATP pump occurs; this mechanism can produce a Cl^- active influx of up to 10 μmol m^2 sec^{-1} and a $\Delta\bar{\mu}_{Cl^-}$ of up to 20 kJ (mol $Cl^-)^{-1}$. Such a mechanism may also operate in other members of the Ulvophyceae.[2,5,6] However, in many marine members of the Ulvophyceae, and in members of the Phaeophyta and Rhodophyta, the identity of the major primary active transport mechanism(s) at the plasmalemma is unclear. In any case this latter-mentioned category has no readily detectable [by ψ_{co} (transplasmalemma electrical potential difference) measurements with microelectrodes] electrogenic pump, in that the value of ψ_{co} is never more negative than the most negative value allowed by diffusion potentials, i.e., ψ_{K^+}.[2,5,6]

Evidence for primary active transport of other solutes at the algal plasmalemma is less complete; some evidence assigns Ca^{2+}-ATPase of algae to the plasmalemma, although evidence from *Chara* "tonoplast-less" preparations suggests that active Ca^{2+} efflux at the plasmalemma is by secondary active transport.[2,5,6]

Plasmalemma: Secondary Active Transport

For the freshwater Chlorophyta (Chlorophyceae and Charophyceae), the majority of evidence is consistent with the occurrence of H^+ transport for the active influx of glucose in *Chlorella vulgaris* (1 H^+ : 1 glucose) and of chloride in *Chara corallina* (2 H^+ : 1 Cl^-) (see [27] and [28] in this volume and Raven[2,5,6]). Less complete data [i.e., not including all of the required demonstrations of reciprocal stimulation of H^+ influx by the symported solute, and of the symported solute by H^+, and measurements of stoichiometry of these fluxes (n H^+ : 1 symported solute), of the electrochemical potential differences per mol for H^+ and the symported solute, and the excess of inwardly directed driving force on n mol H^+ over that needed to move in 1 mol of symported solute] are available for other solutes.[5] This is also true of the possible Na^+ symport of many solutes in marine algae (mainly diatoms, see Raven[2,5,6]), for which the main evidence involves a dependence of the solute influx on, or a substantial stimulation

[46] K. J. Flynn, H. Opik, and P. J. Syrett, *J. Gen. Microbiol.* **133**, 93 (1987).

of the solute influx by, external Na^+. There is no clear evidence as to how the marine algae generate the Na^+ electrochemical potential difference (cytosol low relative to the medium), which has been demonstrated in many cases (rarely those in which Na^+ cotransport has been suggested!) and which would be the driving force for the suggested Na^+-coupled symports. By contrast, the mechanism of generation of the cotransport-powering $\Delta\bar{\mu}_{H^+}$ at the plasmalemma of the freshwater green (Chlorophycean, Charophycean) algae is rather better understood (see above).

While some cation antiport systems have been proposed at the algal plasmalemma (mainly involving H^+ and Na^+), the evidence on which these suggestions are based needs (as for Na^+ symport) extension.[2,5,6] Furthermore, the connection (if any) between the generation of a $\Delta\bar{\mu}_{Cl^-}$ by plasmalemma primary active Cl^- influx in *Acetabularia* (and its friends) and secondary active transport appears to be unknown.

Plasmalemma: Mediated Passive Uniport

Characterization of thermodynamically (hence passive) and kinetically (hence mediated) defined transport at the algal plasmalemma has not proceeded to the purification and reconstitution of the porters responsible for the transport processes. Probably the best joint electrical and (radio-) chemical analysis of plasmalemma mediated passive uniport in an alga is the Characean (Charophycean) ammonium/methylammonium porter.[47,48] Other uniporters which have been electrically characterized in Characean cells are for potassium, calcium, and chloride; the voltage dependence of the catalytic activity of these porters is involved in, *inter alia,* the excitability of the Characean plasmalemma.[2,5,6] The increased catalytic activity of an H^+ passive uniporters at high extracellular pH values is important in the "banding" phenomenon in the cells of certain ecorticate characeans[2,5,6]; the relative catalytic activities of the primary active H^+ efflux pump, the K^+ passive uniporter, and the H^+ passive uniporter are important in determining the three "states" of the characean plasmalemma.[2,5,6,47,49,50] The characean plasmalemma also appears to mediate passive transport of water.[5,51]

Even without detailed electrical (voltage-clamp) or (radio-) chemical

[47] N. A. Walker, M. Beilby, and F. A. Smith, *J. Membr. Biol.* **49**, 21 (1979).
[48] N. A. Walker, F. A. Smith, and M. Beilby, *J. Membr. Biol.* **49**, 283 (1979).
[49] N. A. Walker, *in* "Membrane Transport in Plants: Current Conceptual Issues" (R. M. Spanswick, W. J. Lucas, and J. Dainty, eds.), p. 287. Elsevier, Amsterdam, 1980.
[50] M. J. Beilby, *J. Membr. Biol.* **89**, 241 (1986).
[51] J. A. Raven, *in* "Plant Membranes: Structure, Assembly and Function" (J. C. Harwood and T. J. Walter, eds.), p. 239. Biochemical Society, London, 1988.

data on downhill solute fluxes it can be deduced that fluxes which correspond to permeability coefficients in excess of those found in lipid bilayers must involve mediated transport (possible uniport, or cotransport). Investigation of high-capacity passive uniport processes (e.g., the K^+, Ca^{2+}, Cl^-, and H^+ porters mentioned above) is clearly easier and is important in relation to algal functioning; however, in terms of maintenance energy costs, it is important to decide if the lowest downhill fluxes of solutes correspond to "lipid–solution" uniport or to some additional mediated process.[2,5,6]

Plasmalemma: Passive Nonmediated Uniport

Uncatalyzed transport at the plasmalemma of a number of solutes, and of water, is important in a variety of algal processes. In a number of cases, the physiological and ecological "needs" of the organism are apparently met by a "lipid–solution" flux driven by the prevailing driving force of a transmembrane concentration difference, granted the measured permeability coefficient for the solute in lipid bilayers with a composition similar to that of the lipid portion of the plasmalemma. Examples are the entry of CO_2 in photosynthesis for a number of algae grown under conditions that repress any "CO_2 concentrating mechanism" which they can express; the entry of NH_3 as nitrogen source in algae grown under conditions (e.g., high pH or relatively high external NH_3 concentration) that repress NH_4^+ uniport; O_2 entry for respiration; and H_2O fluxes in many microalgae (see Raven[2,5,6]).

In a number of cases, however, the measured permeability coefficients *in vivo* for algal plasmalemma are very different from those found for *in vitro* membrane bilayers. Where the *in vivo* permeability coefficient is higher than that found *in vitro* (e.g., for H_2O in characean internodes) there is ready "explanation" in terms of mediated uniport supplementing "lipid solution"; the quotation marks around explanation can be, at least partially, removed by consideration of the effects of protein modifiers on water permeability (see above). However, a number of cases involve an *in vivo* permeability coefficient which is much lower than that found *in vitro*. One is P_{CO_2} in at least some microalgal cells with a "CO_2 concentrating mechanism"; its value can be one-hundredth that of model membrane systems *in vitro*. This is perplexing in view of the same genotype, under other growth conditions (which repress the "CO_2 concentrating mechanism") exhibiting a plasmalemma P_{CO_2} value closer to that of model membrane systems (see above, and Raven[2,5,6,51]); a phenotypic change in P_{CO_2} is needed. In *Dunaliella* there is an apparently constitutive low $P_{glycerol}$ (10^{-3} – 10^{-4} that of characean plasmalemma, red blood cell plasmalemma, or artificial lipid bilayers); although no phenotypic change appears to be involved here, the discrepancy between observation and expectation is more dramatic than for CO_2.[2,5,6,51,52] Clear selective advantages (energy costs of CO_2 accumulation in cells and of "osmoregulation" with glycerol as compatible solute) may be seen, but no obvious mechanism is at hand.[2,5,6]

Plasmalemma: Summary and Prospects

Most of the progress that has been made in the analysis of the mechanism of solute transport at the plasmalemma of algal cells has involved relatively few species and frequently, for a given species, has concentrated on relatively few transport systems. Investigation of ecorticate members of the Characeae and on *Acetabularia* have yielded the most detailed information on a range of transport processes and their interactions. It is to be hoped that the exemplary biophysical work on the plasmalemma of intact cells and on simplified (perfused) plasmalemma systems can be extended to more biochemical approaches of purification and reconstitution in the near future. When this is achieved for individual transport systems there is still a great deal to be done in putting together the porters (plus nonmediated passive uniport) to "explain" the integrated plasmalemma fluxes in algae under various conditions. An obvious regulatory problem relates, in an organism with primary active transport of an ion $j^\pm$, to limiting the achieved rate of *consumption* of $\Delta\bar{\mu}_{j\pm}$ in cotransport and uniport processes so as not to exceed the achieved rate of *production* of $\Delta\bar{\mu}_{j\pm}$ in primary active transport when production and consumption are averaged in space (e.g., in a Characean internodal cell with "acid" and "alkaline" bands) and time (e.g., in Characeans showing action potentials and episodic nutrient availability) (see Raven[2,5,6]).

Tonoplast Transport

Much less is known about transport at the tonoplast than at the plasmalemma in algae. Categorization of transport processes as primary or secondary active transport, or as passive uniport, is less advanced even in the relatively well-investigated Characeae and various giant-celled marine algae (Ulvophyceae: Chlorophyta; *Griffithsia:* Rhodophyta). Raven[2,5,6] summarizes these data and suggests that the evidence is consistent with the occurrence of ATP-dependent, electrogenic primary active H^+ transport from the cytosol of *Chara* to the vacuole, while there could be primary active K^+ transport from cytosol to vacuole in *Valonia* (Ulvophyceae).

---

[52] H. Gimmler and W. Hartung, *J. Plant Physiol.* **153**, 165 (1988).

Recent data show that the tonoplast membranes of both *Nitella* and *Chara* have separate porters for ATP-powered and for PP_i-powered active transport of H^+ from cytosol to vacuole.[53,54] Evidence for *Griffithsia*[26] presents a contrast to earlier work on this genus with respect to ψ_{vc} (see Raven[2,5,6]). The secondary active transport, mediated passive uniport, and primary active transport processes other than of H^+ or K^+ are less well investigated for the algal tonoplast than for the plasmalemma.[54a,55] Mass production of isolated algal vacuoles and of algal tonoplast vesicles is not yet very far advanced.[46]

General Characteristic of Bryophytes

The Bryophytes are a more homogeneous group than are the algae (although both groups represent grades of organization and are not monophyletic[56]), as far as the features which are used to define divisions and classes of algae (i.e., ultrastructure, photosynthetic pigment chemistry, and nature of the storage products). Bryophytes also show much less diversity in their habitats, life cycles, and life forms.[2,57–60] As indicated above the bryophytes (mosses, liverworts, and hornworts: see Table III), like the vascular plants, are closely related to the class Charophyceae in the division Chlorophyta.[2,4,8,56] Bryophytes are never fully marine or even very halophytic, the ecological limit for continuous submersion apparently being a salinity of about one-fifth that of normal seawater[61]; many occur submerged in freshwater environments, while a large number are more or less successful in growing in terrestrial environments. The degree of vegetative complexity of most bryophytes is greater than that of the unicellular and colonial algae but does not generally much surpass that of the most complex multicellar macroalgae (e.g., the orders Laminarialas and Fucales of the class Phaeophyceae). The three major groups of Bryophyta (the classes Hepaticae, or liverworts; Anthocerotae, or hornworts; and Musci, or

[53] T. Shimmen and E. A. C. MacRobbie, *Protoplasma* **136**, 205 (1987).
[54] K. Takeshige, M. Tazawa, and A. Hager, *Plant Physiol.* **86**, 1168 (1988).
[55] M. Tester, M. J. Beilby, and T. Shimmen, *Plant Cell Physiol.* **28**, 1555 (1987).
[56] B. D. Mischler, *J. Bryol.* **14**, 71 (1986).
[57] J. A. Raven, *in* "On the Economy of Plant Form and Function" (T. Givnish, ed.), p. 421. Cambridge Univ. Press, London and New York, 1986.
[58] E. V. Watson, "The Structure and Life of Bryophytes," 3rd ed. Hutchinson, London, 1971.
[59] N. S. Parihar, "An Introduction to the Embryophyta. I. Bryophyta," 5th ed. Central Book Depot, Allahabad, India, 1965.
[60] D. H. S. Richardson, "The Biology of Mosses." Blackwell, Oxford, 1979.
[61] M. Waern, *Acta Phytogeogr. Suecica* **30**, 1 (1952).

TABLE III

CLASSES OF THE DIVISION BRYOPHYTA, WITH ORGANISMS USED FOR TRANSPORT
STUDIES AND OTHERS MENTIONED IN THIS CHAPTER[a]

Class	Organism	Gametophyte organization	Ref. for transport work
Hepaticae (liverworts; gametophytes thalloid or leafy)	*Riccia fluitans*	Thalloid	62–74
	Marchantia sp.	Thalloid	75
	Conocephalum conicum	Thalloid	76, 77
	Pellia sp.	Thalloid	78
	Cephalozia connivens	Leafy	79
	Leiocolea turbinata	Leafy	79
Anthocerotae (hornworts; gametophytes all thalloid)	*Phaeoceros laevis*	Thalloid	80, 81
	Anthoceros punctatus, Anthoceros sp.	Thalloid	82–85
Musci (mosses; gametophytes all leafy)	*Funaria hygrometrica*	Leafy	86–92
	Hookeria lucens	Leafy	93–95
	Mnium cuspidatum (= *Plagiomnium cuspidatum*)	Leafy	96–99
	Fontinalis antipyretica	Leafy	100–102
	Polytrichum commune	Leafy	101, 103
	Polytrichum formosum	Leafy	104, 105
	Tortella tortuosa	Leafy	106

[a] All of these organisms have filamentous protonemata.

mosses) have all been used for detailed studies of transport at the cell level
(Table III).[62–105]

[62] H. Felle, *Biochim. Biophys. Acta* **602,** 181 (1980).

[63] H. Felle, *Biochim. Biophys. Acta* **646,** 151 (1981).

[64] H. Felle, *Planta* **152,** 505 (1981).

[65] H. Felle, *Biochim. Biophys. Acta* **730,** 342 (1983).

[66] H. Felle, *Biochim. Biophys. Acta* **772,** 307 (1984).

[67] H. Felle and F. W. Bentrup, *Ber. Dtsch. Bot. Ges.* **87,** 223 (1974).

[68] H. Felle and F. W. Bentrup, *in* "Membrane Transport in Plants" (U. Zimmermann and J. Dainty, eds.), p. 167. Springer-Verlag, Berlin and New York, 1974.

[69] H. Felle and F. W. Bentrup, *J. Membr. Biol.* **27,** 153 (1976).

[70] H. Felle and F. W. Bentrup, *in* "Echanges ioniques transmembranaires chez les végétaux" (M. Thellier, A. Monier, M. Demarty, and J. Dainty, eds.), p. 193. CNRS, Paris, 1977.

[71] H. Felle and F. W. Bentrup, *Biochim. Biophys. Acta* **464,** 179 (1977).

[72] H. Felle and F. W. Bentrup, *Planta* **147,** 471 (1980).

[73] J. Bertl, H. Felle, and F. W. Bentrup, *in* "Membrane Transport in Plants" (W. J. Cram, K. Janacek, R. Rybova, and K. Sigler, eds.), p. 205, Akademia-Verlag, Prague, 1984.

[74] J. Bertl, H. Felle, and F. W. Bentrup, *Plant Physiol.* **76,** 75 (1984).

[75] E. Marré, *Biol. Cell.* **32,** 19 (1978).

Opportunities and Limitations in Use of Bryophytes in Transport Studies

Bryophytes have a much smaller range of cell volumes than do algae, i.e., $10^{-16}-10^{-12}$ m^3 (Table IV)[106]; accordingly many of the cells are amenable to study by electrophysiological techniques. Essentially all of the

[76] T. Zawadzki and T. Trębacz, *Physiol. Plant.* **64,** 477 (1985).

[77] K. Trębacz, and T. Zawadzki, *Physiol. Plant.* **64,** 482 (1985).

[78] J. G. Ellis and R. J. Thomas, *J. Plant Physiol.* **121,** 259 (1985).

[79] R. L. Jefferies, D. Laycock, G. R. Stewart, and A. P. Sims, *in* "Ecological Aspects of the Mineral Nutrition of Plants" (I. H. Rorison, ed.), p. 281. Blackwell, Oxford, 1969.

[80] R. F. Davis, *in* "Membrane Transport in Plants" (U. Zimmermann and J. Dainty, eds.), p. 197. Springer-Verlag, Berlin and New York, 1974.

[81] P. J. Contardi and R. F. Davis, *Plant Physiol.* **61,** 164 (1978).

[82] A. A. Bulychev, V. K. Andrianov, and G. A. Kurella, *Biochim. Biophys. Acta* **590,** 300 (1980).

[83] A. A. Bulychev, *Biochim. Biophys. Acta* **766,** 647 (1984).

[84] A. A. Bulychev, M. M. Niyazova, and V. B. Turovetsky, *Biochim. Biophys. Acta* **808,** 186 (1985).

[85] A. A. Bulychev, M. M. Niyazova, and V. B. Turovetsky, *Biochim. Biophys. Acta* **850,** 218 (1986).

[86] A. J. Browning and B. E. S. Gunning, *J. Exp. Bot.* **30,** 1233 (1979).

[87] A. J. Browning and B. E. S. Gunning, *J. Exp. Bot.* **30,** 1247 (1979).

[88] A. J. Browning and B. E. S. Gunning, *J. Exp. Bot.* **30,** 1265 (1979).

[89] S. Rose, P. H. Rubery, and M. Bopp, *Physiol. Plant.* **58,** 52 (1983).

[90] S. Rose and M. Bopp, *Physiol. Plant.* **58,** 57 (1983).

[91] J. Erichsen, B. Knoop, and M. Bopp, *Plant Cell Physiol.* **19,** 839 (1978).

[92] M. J. Saunders, *Planta* **167,** 402 (1986).

[93] J. Sinclair, Ph.D. Thesis, University of East Anglia (1965).

[94] J. Sinclair, *J. Exp. Bot.* **19,** 254 (1968).

[95] J. Sinclair, *J. Exp. Bot.* **18,** 594 (1967).

[96] U. Lüttge and K. Bauer, *Planta* **78,** 310 (1968).

[97] U. Lüttge and G. Zirke, *J. Membr. Biol.* **18,** 305 (1974).

[98] U. Lüttge, N. Higinbotham, and C. K. Pallaghy, *Z. Naturforsch., B: Anorg. Chem., Org. Chem., Biochem., Biophys., Biol.* **27B,** 1239 (1972).

[99] E. Fischer, U. Lüttge and N. Higinbotham, *Plant Physiol.* **58,** 240 (1976).

[100] D. H. Brown, *in* "The Experimental Biology of Bryophytes" (A. F. Dyer and J. G. Duckett, eds.), p. 229. Academic Press, London, 1984.

[101] M. C. F. Proctor, *in* "The Experimental Biology of Bryophytes" (A. F. Dyer and J. G. Duckett, eds.), p. 9. Academic Press, London, 1984.

[102] J. A. Raven, J. J. MacFarlane, and H. Griffiths, *in* "Plant Life in Aquatic and Amphibious Habitats" (R. M. M. Crawford, ed.), p. 129. Blackwell, Oxford, 1987.

[103] P. J. Grubb, Ph.D. Thesis, University of Cambridge (1961).

[104] D. Pichelin, G. Mounoury, S. Delrot, and J.-L. Bonnemain, *C. R. Acad. Sci. Paris* **298 III,** 439 (1984).

[105] J.-L. Bonnemain, D. Pichelin, C. Caussin, and S. Delrot, *in* "Phloem Transport" (J. Cronshaw, W. J. Lucas, and T. D. Steinke, eds.), p. 181. Alan R. Liss, New York, 1986.

[106] E. V. Watson, "British Mosses and Liverworts," 3rd ed. Cambridge Univ. Press, London and New York, 1981.

TABLE IV
PROTOPLAST VOLUME, PLASMALEMMA AREA, AND FRACTION OF PROTOPLAST VOLUME OCCUPIED BY VACUOLE IN VARIOUS BRYOPHYTE CELLS

Organism	Cell type	Protoplast volume (m^3)	Plasmalemma area (m^2)	Vacuole as fraction of protoplast	Ref.
Riccia fluitans[a]	Rhizoid cell	2×10^{-13}–2×10^{-12}	3×10^{-8}–3×10^{-7}	?	66–68
	Thallus cell	1–4×10^{-14}	2–8×10^{-9}	~0.8	63, 64, 106
Phaeoceros laevis[a]	Thallus cell	10^{-14}–3×10^{-13}	2×10^{-9}–2×10^{-8}	?	106
Mnium cuspidatum[a]	Leaf cell	4×10^{-15}–4×10^{-14}	1–6×10^{-9}	>0.5	97, 106
Hookeria lucens[a]	Leaf cell	2×10^{-13}–2×10^{-12}	2–8×10^{-8}	>0.7	93, 106
Polytrichum commune	Leaf lamella cell	5×10^{-17}	8×10^{-11}	?	106
Tortella tortuosa	(Green) leaf cells	6×10^{-17}	8×10^{-11}	?	102

[a] Organism used in membrane transport studies.

cells of bryophytes (at least within a single generation of the life cycle) are in symplastic contact, with the implications for electrophysiological measurements indicated above in connection with measurements on algae. The range, within Bryophytes, of the fraction of cell volume occupied by vacuoles is smaller than is the case in the algae (Table IV) as befits the absence of very small and very large bryophyte cells. These various considerations limit the utility of bryophytes (as a division) in solute transport studies relative to that of the algae.

Nevertheless, a number of significant data sets have been assembled on bryophytes, e.g., on plastid electrophysiology of the large plastid of the monoplastidic cells of the hornworts (Anthocerotae: *Anthoceros, Phaeoceros*), investigated by Davis,[80] Contardi and Davis,[81] and Bulychev and co-workers,[82-85] and voltage-clamp and tracer flux studies on the rhizoids of the aquatic (pleustophytic) liverwort *Riccia* (Hepaticae), used extensively by Felle and co-workers[62-66,73,74,107-111] in studies of the primary active transport and various passive uniports and cotransports at the plasmalemma, and for cytoplasmic pH estimation using a microelectrode. The cells of at least some members of each of the three main groups (classes) of bryophytes are such (Table IV) that the electrical characteristics of the plasmalemma and of the tonoplast can be measured separately, e.g., in the mosses *Hookeria*[93-95] and *Mnium,*[97-99] the hornwort *Phaeoceros,*[80] and the liverwort *Riccia.*[67,68] However, interpretation of the data in terms of ψ_{co} and ψ_{cv} is not always clear cut.[93,94,97] A complicating feature of bryophytes, relative to at least some (unicellular, colonial) algae, is the range of cell types present, which renders the interpretation of experiments on electrochemical potential differences or on solute fluxes less unambiguous than is the case for data obtained on more homogeneous (microalgal) cell populations. It would appear that bryophytes generally lack the "CO_2 concentrating mechanism(s)" found in many algae, thus simplifying the analysis of plasmalemma fluxes.[5,42]

Results from Transport Studies on Bryophytes

Electrophysiological studies on gametophytes of bryophytes from all three classes (liverworts, e.g., *Riccia;* hornworts, e.g., *Phaeoceros;* and mosses, e.g., *Mnium*) indicate that there is a significant electrogenic component of the transplasmalemma electrical potential difference; ion substi-

[107] H. Felle and A. Bertl, *J. Exp. Bot.* **37,** 1416 (1986).
[108] H. Felle and A. Bertl, *Biochim. Biophys. Acta* **848,** 176 (1986).
[109] E. Johannes and H. Felle, *Planta* **166,** 244 (1986).
[110] E. Johannes and H. Felle, *Planta* **172,** 53 (1987).
[111] I. Dahse, H. Felle, F. W. Bentrup, and B. Liebermann, *J. Plant Physiol.* **124,** 87 (1986).

tution effects and other considerations favor H^+ as the ion responsible. The thallus and rhizoid cells of the aquatic (pleustophytic) liverwort *Riccia fluitans* have ψ_{co} values, at least in the light, of -215 mV in 100 mmol K^+ m^{-3} when intracellular K^+ concentrations give a ψ_{K^+} value equal to -165 mV; an even more dramatic difference is found in 10 mol K^+ m^{-3} medium, when the ψ_{K^+} value of -50 mV is much less negative than is the ψ_{co} of -185 mV.[67-69]

The ion involved in this electrogenic process seems to be H^+.[63,67-72] With a measured cytoplasmic pH of 7.4[73,74] an external pH of 5.5, and a ψ_{co} (at 100 mmol K^+ m^{-3}) of -215 mV, $\Delta\bar{\mu}_{H^+}$ at the plasmalemma is 30 kJ (mol H^+)$^{-1}$, inside negative. Davis[80] and Contardi and Davis[81] have investigated the terrestrial (rhizophytic) hornwort *Phaeoceros laevis*. Here ψ_{co} is rather more negative in the dark than in the light; at an external pH of 5.5–5.8, ψ_{co} is -176 mV when ψ_{K^+} is -165 mV at an external $[K^+]$ of 100 mmol m^{-3}; when $[K^+]$ is 10 mol m^{-3}, ψ_{K^+} is only -49 mV, and ψ_{co} is -138 mV. ψ_{co} is -181 mV at an external pH of 4.2; for a cytoplasmic pH measured with antimony or stainless steel microelectrodes of pH 6.7, $\Delta\bar{\mu}_{H^+}$ at the plasmalemma is 33 kJ (mol H^+)$^{-1}$, inside negative. Spanswick and Miller[112] found that in characean internodal cells the cytoplasmic pH measured with antimony microelectrodes (~ 6.7) was lower than that measured by weak electrolyte (DMO: 2,4-dimethyloxazolidine-2,4-dione) distribution or by glass (pH-sensitive) microelectrodes, both of which gave a value of 7.4, a more usual value for cytoplasmic pH.[106] Furthermore, intracellular microelectrode studies on the liverwort *Riccia fluitans* also suggest a cytoplasmic pH of 7.3–7.4[72,73,107,108]; thus, $\Delta\bar{\mu}_{H^+}$ of the *Phaeoceros* plasmalemma may be greater (more inside negative) than the 33 kJ (mol H^+)$^{-1}$ quoted above. Lüttge and co-workers[97-99,113,114] have investigated leaf cells of the terrestrial moss *Mnium cuspidatum*. At an external pH of 7.0, ψ_{co} can be as negative as -300 mV in 0.1 mol K^+ m^{-3} (ψ_{co} is far more negative than ψ_{K^+}), so that $\Delta\bar{\mu}_{H^+}$ can (assuming cytoplasmic pH to be more alkaline than 7.0) be 30 kJ (mol H^+)$^{-1}$, inside negative.

In these three cases, then, it appears that electrogenic H^+ efflux is a major determinant of ψ_{co}, at least under some conditions, in representatives of all three bryophyte classes. Moreover, if the H^+ pump is ATP-dependent, in all three classes $\Delta\bar{\mu}H^+$ can be so high, relative to the likely *in vivo* free energy of hydrolysis of ATP, that only 1 mol H^+ can be actively transport per mole ATP converted to ADP and P_i. We note that, as with

[112] R. M. Spanswick and A. G. Miller, *Plant Physiol.* **59**, 664 (1977).

[113] F. A. Smith and J. A. Raven, *Annu. Rev. Plant Physiol.* **30**, 289 (1979).

[114] U. Lüttge and N. Higinbotham, "Transport in Plants." Springer-Verlag, Berlin and New York, 1979.

other plant groups, not all measurements of ψ_{co} in bryophyte cells have yielded values more negative than ψ_{K^+}.

Voltage-clamp studies on the rhizoid cells of *Riccia fluitans*[62-72] have quantified the conductance associated with the electrogenic active H$^+$ efflux. The capacity for active H$^+$ efflux revealed in these experiment is ~ 1 μmol H$^+$ m^{-2} sec^{-1} in "normal" cells and ~ 2 μmol H$^+$ m^{-2} sec^{-1} in CCCP-treated cells [carbonyl cyanide *m*-chlorophenylhydrazone (CCCP) treatment will reduce $\Delta\bar{\mu}_{H^+}$ and may double the H$^+$ efflux per mole ATP used while maintaining the rate of ATP use at 1 μmol m^{-2} sec^{-1}]. Alas, we have no comparative data on the H$^+$ efflux capacity based on net H$^+$ efflux of the sort available for some ecorticate internodal cells of the Characeae (see above). The voltage-clamp studies have also quantified change transfer associated with influx of solutes such as various hexoses (up to 0.2 μmol m^{-2} sec^{-1}),[72] neutral amino acids (up to 0.2 μmol m^{-2} sec^{-1}),[64-66] and methylammonium and ammonium (up to 0.7 μmol m^{-2} sec^{-1}).[62] These data show that the influx of all of these solutes is accompanied by positive charge influx. The electrical measurements have been supplemented by tracer flux studies with ^{14}C-labeled hexoses, neutral amino acids, and methylammonium.

Since the voltage-clamp studies relate only to the impaled rhizoid cell and the tracer studies involve the whole thallus, the data from the two techniques cannot be so directly compared as in the investigations of (methyl-) ammonium transport system at the plasmalemma of internodal cells of ecorticate characeans.[47,48] Indeed, tracer fluxes in whole thalli (assuming 0.1 m^2 plasmalemma area per gram dry weight)[63] are consistently lower (by 5- to 10-fold) at substrate saturation than are fluxes derived from voltage-clamp studies assuming one positive charge enters per methylammonium, hexose, or neutral amino acid. "Leakage" by nonchargetransporting processes could, at least in part, account for this discrepancy.[72] Overall the data are consistent with mediated passive uniport of NH$_4^+$ and CH$_3$NH$_3^+$, and mediated secondary active transport of hexose and neutral amino acids with H$^+$, with a 1 H$^+$:1 hexose or a 1 H$^+$:1 amino acid stoichiometry. The data also permit conclusions to be drawn as to the mechanism of the cotransport reactions (favoring the concept of a neutral H$^+$–solute–carrier ternary complex and a negatively charged unloaded carrier for the secondary active influxes) and the significance of "leakage" of the cotransported solute in making use of the maximum measured accumulation ratio of the symported solute, via thermodynamic reasoning, an unreliable indication to the stoichiometry of the cotransport reaction.

Other data on bryophytes are consistent with the requirements for active transport at the plasmalemma being essentially as in other phototrophs,[5,6,113,114] although data are scanty, particularly for inorganic ions other than K^+, Na^+, Cl^- and NH_4^+. Recent work[76,77,115,116] on plasma-lemma electrical properties of the liverwort *Conocephalum conicum* shows (1) that the resting potential has a component which probably relates to electrogenic H^+ efflux, (2) that subthreshold light (less than 1 μmol photons m^{-2} sec^{-1}, 400–700 nm, absorbed by photosynthetic pigments) causes a transient depolarization of the resting potential while (3) higher photon flux densities, or electrical stimulation, initiate an action potential whose (4) requirement for metabolism is as yet unexplained in terms of the occurrence of a "metabolic" cation potential of the type found in certain algae rather than a continuous need for pumping of ions leaked during a "classic" action potential.

Work on transport properties of the bryophyte tonoplast is relatively scarce. It would seem that the vacuole (as is almost universally the case in plants) has a higher H^+ activity than does the cytoplasm; vacuolar pH (antimony or steel electrodes) is about 1 unit lower than is cytoplasmic pH in *Phaeoceros laevis*[78] while a vacuole-acid pH difference of up to 3 units between cytoplasm and vacuole was shown, using liquid ion-exchanger microelectrodes, in *Riccia fluitans*.[73,74] The electrical potential difference across the tonoplast (ψ_{vc}) is generally held to be slightly vacuole-positive and small relative to that at the plasmalemma in *Mnium cuspidatum*,[112] *Riccia fluitans*,[66,67] and *Phaeoceros laevis*,[80] although the data for *M. cuspidatum* can be interpreted in terms of a ψ_{vc} of $+60$ to $+65$ mV,[97] while Sinclair[92,94] suggested a ψ_{vc} of -50 mV in *Hookeria lucens*. If this latter value is correct, the *H. lucens* vacuole would have to be at least one pH unit more acid than the cytoplasm before the "usual" active transport of H^+ from cytoplasm to vacuole would need to be invoked; for the other ΔpH and ψ_{vc} values, active H^+ influx at the plasmalemma would be needed.[113]

Raven *et al.*[117] and Raven[5] have pointed out that the Bryophyta in general resemble the Tracheophyta rather than algae in terms of their higher ratio of organic to inorganic low-M_r anions in vacuolate cells. This is presumably a reflection of different interactions of metabolism and tonoplast transport in algae from what occurs in the Bryophyta and Tracheophyta.[118]

[115] K. Trębacz, R. Tarneck, and T. Zawarski, *Physiol. Plant.* **75**, 20 (1989).

[116] K. Trębacz, R. Tarneck, and T. Zawarski, *Physiol. Plant.* **75**, 24 (1989).

[117] J. A. Raven, F. A. Smith, and S. E. Smith, *in* "Genetic Engineering of Osmoregulation: Impact on Plant Productivity for Food, Chemicals and Energy" (D. W. Rains, R. C. Valentine, and A. Hollaender, eds.), p. 101. Plenum, New York, 1980.

[118] D. J. Cove and N. W. Ashton, *in* "The Experimental Biology of Bryophytes" (A. F. Dyer and J. G. Duckett, eds.), p. 177. Academic Press, London, 1984.

While transport in organelles of Bryophyta may belong, more properly, in Section II,C of this volume, I cannot forbear mentioning the work[80] on *in situ* giant plastids of *Phaeoceros laevis* using microelectrodes measuring electrical potential differences between the cytosol and "plastid" and pH antimony or stainless steel electrodes. These data, showing light/dark differences, are not, alas, readily interpreted in terms of specific compartments in the plastid. Work[82-85] with microelectrodes measuring electrical potential differences in giant plastids of *Anthoceros* spp. at the onset of illumination have been interpreted, as have similar phenomena in giant plastids of the angiosperm *Peperomia metallica,* in terms of transthylakoid electrical potentials related to H^+ fluxes via photoredox catalysts and the CF_0-CF_1 ATP synthase.

Other transport work on bryophytes has involved a study of the apoplastic transfer of glucose and sucrose from the gametophyte to the (partially heterotrophic) generation of *Funaria hygrometrica* via the plasmalemma of the "transfer cells" of the haustorium of the sporophyte.[86-88] The measured influx of glucose at a saturating [glucose] in the sporophyte would involve a flux of 0.66 μmol glucose m^2 sec^{-1} if the transport was across the plasmalemma at the haustorial surface with no "transfer cell" amplification of plasmalemma area but only 0.073 μmol glucose m^{-2} sec^{-1} when the observed amplification of membrane area is taken into account.[87] For comparison, the maximum hexose influx at *Riccia fluitans* rhizoid plasmalemma is 0.2 μmol m^{-2} sec^{-1} based on current–voltage relationships [we assume that the lower ^{14}C tracer fluxes (Table V) relate to leakage short-circuit the cotransport influx]. The transfer cell amplification of plasmalemma area may indeed be useful at the gametophyte–sporophyte junction in enhancing the flux of hexose per area of interface surface, provided the membrane area-based capacity for hexose influx cannot exceed 0.2 μmol m^{-2} sec^{-1} in bryophytes[5] limited by either the specific reaction rate and areal density of H^+–sugar symporters or the primary active H^+ porters which lower the symport. Work[104,105] on haustoria of the sporophytes of *Polytrichum formosum* shows that the transplasmalemma electrical potential difference ψ_{co} of -160 mV has a large electrogenic component, and that the haustorium, unlike the rest of the sporophyte, shows a net H^+ efflux. Haustoria take up supplied glycine and α-aminobutyric acid by a saturable process which is inhibited by protonophores and inhibitors of the plasmalemma ATPase; transport of the amino acids is paralleled by a depolarization of ψ_{CO}. These data are consistent with H^+-coupled active uptake of amino acids by haustorial cells.

Other transport work on bryophytes has been related to development of the gametophyte generation. This has involved investigation of the transplasmalemma movement at the single cell level, the apoplastic and/or symplastic transport between cells, growth regulatory substances, the role

TABLE V

MAXIMUM CAPACITY FOR VARIOUS MEDIATED TRANSPORT PROCESSES AT THE PLASMALEMMA OF *Riccia fluitans*[a]

Transport process	Method	Maximum capacity (μmol m^{-2} sec^{-1})	Ref.
Primary active efflux of H$^+$ in control cells: probably 1 mol H$^+$ transported per mole ATP used	Voltage clamp	~ 1.0	63
Primary active efflux of H$^+$ in cells treated with CCCP; CCCP increased P_{H^+}/P_{K^+} from ~ 10 to ~ 1000 (if P_{K^+} is $\sim 0.5 \times 10^{-8}$ m sec^{-1}, from tracer flux measurements, P_{H^+} is increased from $\sim 5 \times 10^{-3}$ to 5×10^{-6} m sec^{-1}); the reduced $\Delta\mu_{H^+}$ produced by the higher P_{H^+} may alter the stoichiometry to 2 mol H$^+$ transported per mole ATP used	Voltage clamp	~ 2.0	63, 70, 71
Mediated passive uniport influx of K$^+$ (^{86}Rb$^+$) at a [K$^+$]$_o$ of 10 mol m^{-3}	Tracers	~ 0.10 (light) ~ 0.05 (dark)	69
Mediated passive uniport of K$^+$ at a [K$^+$]$_o$ of 10 mol m^{-3} (5°)	Voltage clamp	~ 0.20	63, 69
Mediated passive uniport of CH$_3$NH$_3^+$ or NH$_4^+$	Voltage clamp	~ 0.70	60
Mediated passive uniport of ^{14}CH$_3$NH$_3^+$	Tracers	~ 0.12	60
Active ^{36}Cl$^-$ influx at a [Cl$^-$]$_o$ of 10 mol m^{-3}	Tracers	~ 0.05	69
Hexose (glucose or 3-*O*-methylglucose)–H$^+$ cotransport.	Voltage clamp	0.20	71
Hexose (glucose or 3-*O*-methylglucose)–H$^+$ cotransport.	Tracers (assuming 0.1 m^2 plasmalemma area per gram dry weight of *Riccia*)	0.01	71
Amino acid (α-aminoisobutyric acid)–H$^+$ cotransport	Voltage clamp	0.20	64–66
Amino acid (α-aminoisobutyric acid)–H$^+$ cotransport	Tracers (assuming 0.1 m^2 plasmalemma area per gram dry weight of *Riccia*)	0.20	64–66

[a] Deduced from electrical (voltage-clamp currents) and tracer flux data obtained at 22–26° in saturating light at pH 5.5–5.8.

of circulating electrical currents in developmental processes, and the possible involvement of H^+ fluxes at the plasmalemma in the phototropism of the setae of *Pellia* sporophytes.[78] Rose and co-workers[89,90] have investigated indole-3-acetic acid influx and polar cell-to-cell transport in protonemata and detached rhizoids of the moss *Funaria hygrometrica;* evidence consistent with nonmediated indole-3-acetic acid (IAA) transport, "saturable" influx (mediated H^+ : IAA^- symport?), triiodobenzoic acid-inhibited efflux (mediated IAA^- uniport), and polar ("chemiosmotic"?) nonsymplastic cell-to-cell transport (only investigated in rhizoids) was obtained. By contrast, cell-to-cell transport of an exogenously supplied cytokinin was inhibited,[91] suggesting an involvement of symplastic transport. Such transport *via* plasmodesmata is consistent with electrical coupling demonstrated in bryophytes by *(inter alia)* Sinclair[93] for *Hookeria lucens* and Felle and Bentrup[67-69] for *Riccia fluitans,* and is consistent with the involvement of plasmodesmata as a component of internal current flow for the external current measured by Saunders[92] for *Funaria hygrometrica* protonemata (caulonema stages), bearing in mind that the currents span several cells.

It is of interest that developmental mutants of bryophytes[118] include some that are apparently deficient in net auxin uptake so that genetic approaches to these, and other, transport phenomena in bryophytes should be possible. The haploid nature of the gametophyte generation should (polyploidy permitting[119]) be of use in this regard. Furthermore, there is much stronger evidence for bryophytes than for algae[120] (cf. Ref. 121) that "higher plant" growth substances, especially IAA and cytokinins, are naturally present in bryophytes and are involved in regulation in those organisms[122]; indeed, the work of Raven[122] on IAA transport in the alga *Hydrodictyon africanum* was predicated on its not being a known natural growth substance for algae.

Conclusions: Past, Present, and Future Contributions of "Lower Plants" to Membrane Transport Studies

Giant-celled algae were of great importance in developing many techniques for the study of (and hypotheses of the mechanism of) membrane transport in plants and, indeed, in biological material in general.[36] They, as well as smaller-celled algae and bryophytes, will prove useful in analyses of

[119] M. E. Newton, *in* "The Experimental Biology of Bryophytes" (A. F. Dyer and J. G. Duckett, eds.) p. 65. Academic Press, London, 1984.

[120] R. G. Buggeln, *in* "The Biology of Seaweeds" (C. S. Lobban and M. J. Wynne, eds.), p. 627. Blackwell, Oxford, 1981.

[121] P. A. Mooney and J. van Staden, *J. Plant Physiol.* **123,** 1 (1986).

[122] J. A. Raven, *New Phytol.* **74,** 163 (1975).

general membrane transport phenomena as well as of specifically "plant" processes (e.g., the active transport of one or more inorganic carbon species related to the occurrence of a "CO_2-concentrating mechanism" in photosynthesis). Further advances in the use of algae and bryophytes in general membrane transport studies require that their demonstrated utility in relation to perfusion, voltage-clamp, and (radio) chemical tracer studies be integrated with the genetic, biochemical, and patch-clamp techniques developed for other groups of organisms. It must not, of course, be forgotten that such integration would also greatly benefit studies of membrane transport in algae and bryophytes in relation to studies of the structure, physiology, and ecology of the organisms.

Acknowledgments

Research grants from NERC and SERC are gratefully acknowledged.

[26] Uptake of Sugars and Amino Acids by Chlorella

By W. TANNER and N. SAUER

Chlorella is a unicellular eukaryotic green alga, a genus of the nonmotile Chlorococcales within the Chlorophyta. It has been one of the green cells most widely used for biochemical investigations, for example, to study the process of photosynthesis. Although *Chlorella* can grow with CO_2 as the sole carbon source, it has been known that these cells are able in addition to assimilate organic molecules like sugars.[1] In some *Chlorella* species this capability to take up organic molecules is inducible.[2-4] It has been mainly for the latter reason that these species have been used to study transport properties in greater detail.

Growth of Chlorella kessleri

Chlorella kessleri (available from Sammlung von Algenkulturen, Göttingen, FRG; SAG 27.87) is grown with 2% CO_2 as the sole carbon source at a light intensity of $5-10 \times 10^3$ lux in a mineral medium containing the

[1] R. Emerson, *J. Gen. Physiol.* **10**, 469 (1927); J. Myers, *ibid.* **30**, 217 (1947).
[2] W. Tanner and O. Kandler, *Z. Pflanzenphysiol.* **58**, 24 (1967).
[3] W. Tanner, *Biochem. Biophys. Res. Commun.* **36**, 278 (1969).
[4] D. Haass and W. Tanner, *Plant Physiol.* **53**, 14 (1974).

following (per liter): 0.4 g KNO_3, 0.1 g $Ca(NO_3)_2 \cdot 4H_2O$, 0.1 g $MgSO_4 \cdot 7H_2O$, 0.1 g KCl, 0.1 g $KH_2PO_4 \cdot H_2O$, 2 mg $FeCl_3$, 5 ml saturated EDTA solution (free acid), and 1 ml Hoagland's A–Z solution [containing (mg per liter) $Al_2(SO_4)_3$ 55, KI 28, KBr 28, TiO_2 55, $SNCl_2 \cdot 2H_2O$ 28, LiCl 28, $MnCl_2 \cdot 4H_2O$ 389, $B(OH)_3$ 614, $ZnSO_4$ 55, $CuSO_4 \cdot 5H_2O$ 55, $NiSO_4 \cdot 7H_2O$ 59, and $CO(NO_3)_2 \cdot 6H_2O$ 55].

Inducibility and Specificity of Sugar and Amino Acid Uptake Systems in *Chlorella kessleri*

Hexose Uptake

Inducibility. The inducibility of autotrophically grown *Chlorella* to take up hexoses is species and strain specific. Although in all *Chlorella* strains and species tested the rate of 3-*O*-methylglucose uptake increased when the cells were preincubated with D-glucose, this increase generally was only 2- to 5-fold.[4] Only in two strains, *Chlorella vulgaris* K (a strain isolated by Otto Kandler and later identified as *C. kessleri*) and *C. vulgaris* 211-11h/P, did the increase amount to 100-fold and more.[4] Various species of *Scenedesmus* and *Ankistrodesmus* increased their rates by less than a factor of 2.[4] All further studies reported here have been carried out with *Chlorella kessleri* (SAG 27.87).

Incubation for Induction. Cells are harvested by low-speed centrifugation and washed once with distilled water, and 50 μl packed cells/ml is incubated in 25 mM sodium phosphate buffer (pH 6.0) at 27° in the presence of 15 mM glucose. The cell suspension is rapidly shaken in the dark. After 60 to 80 min the cells are fully induced for uptake. To ensure that no residual glucose is left in the buffer, cells are washed once with 25 mM sodium phosphate buffer, pH 6.0, and resuspended at the cell density desired.

Turnover. Fully induced cells lose their uptake activity at 27° with a $t_{1/2}$ of about 6 hr; at 0° the induced uptake system is stable for at least 24 hr.[4] The minimum time required to see an increased rate of uptake after the addition of inducer is 15 to 18 min.[4]

Specificity. From coinduction, countertransport, and inhibition experiments[5] it is clear that the uptake system of *Chlorella* transports nearly all physiological hexoses including a large variety of glucose analogs (like 3-*O*-methylglucose, 6-deoxyglucose, 1-deoxyglucose) as well as D-xylose.

[5] W. Tanner, R. Grünes, and O. Kandler, *Z. Pflanzenphysiol.* **62,** 376 (1970); E. Komor and W. Tanner, *Biochim. Biophys. Acta* **241,** 170 (1971).

α-Methylglucoside, as well as any disaccharide, is not transported; this is also true for hexitols, ribose, L-glucose, and L-rhamnose.

Uptake of Amino Acids

Inducibility and Specificity. Surprisingly, pretreatment of *Chlorella kessleri* with D-glucose or a nonmetabolizable D-glucose analog not only induces hexose uptake but also increases the rates of two amino acid transport systems up to 100-fold.[6] When this behavior was investigated systematically, it turned out that in *Chlorella* three amino acid transport systems can be significantly increased in rate (Table I). Whereas an arginine (basic amino acids) and proline system can also be induced by lack of an inorganic nitrogen source in the medium,[7] an additional general amino acid transport system[8] is induced by pretreatment with glucose plus NH_4^+ (or NO_3^-). Finally, *Chlorella* possesses additional amino acid transport systems; they transport various amino acids with low rates and have so far not been found to be inducible.[9] Growth of the algae and incubation for induction is as given above or as indicated in Table I. For nitrogen-depleted media KNO_3 and $Ca(NO_3)_2$ were omitted from the medium described above; 0.1 g $CaCl_2$ was added instead.

Turnover. Cells induced for amino acid uptake with D-glucose lose the uptake activity of the proline system with a half-life identical to that of the hexose transport system ($t_{1/2} \sim 6$ hr), whereas the arginine system with a $t_{1/2}$ of 25 hr is much more stable.[6]

Influx and Efflux Measurements

General Methodology

If influx of substrate into cells, or efflux of substrate out of cells, is to be determined, one must in principle study concentration changes in a defined compartment. One can chose between the two compartments for these measurements, medium or cell.

Concentrations of sugars or amino acids in solution can be determined photometrically, for example, by the procedures of Folin and Wu[10] and

[6] B.-H. Cho, N. Sauer, E. Komor, and W. Tanner, *Proc. Natl. Acad. Sci. U.S.A.* **78**, 3591 (1981).

[7] N. Sauer, E. Komor, and W. Tanner, *Planta* **159**, 404 (1983).

[8] N. Sauer, *Planta* **161**, 425 (1984).

[9] B.-H. Cho and E. Komer, *Biochim. Biophys. Acta* **821**, 384 (1985).

[10] O. Folin and H. Wu, *J. Biol. Chem.* **82**, 83 (1929).

TABLE I
INDUCIBLE AMINO ACID TRANSPORT SYSTEMS IN *Chlorella kessleri*

System characteristic	Arginine system	Proline system	General system
Amino acids transported	L-Arg, L-Lys, L-His	L-Pro, L-Ala, L-Ser, Gly	L-Gln, L-Glu, L-Ala, L-Cys, L-Ser, L-His, Gly, L-Met, L-Asn, L-Leu
Method for induction of system	Preincubation in the presence of glucose or growth in nitrogen depleted medium		Preincubation in the presence of glucose and relatively high concentrations of NH_4^+ or NO_3^- (10 mM)

Yemm and Cocking.[11] With such methods, however, only changes in substrate concentrations in the medium, and not inside the cells, can be studied, since other internal sugars and amino acids interfere with these rather nonspecific tests. The substrate may also be converted to other compounds which are no longer accessible to measurement. These difficulties can be avoided by the use of radiolabeled substrates, which are a convenient tool for the determination of fluxes in either direction.[3,12]

Efficient and rapid separation of cells and medium is an important prerequisite for any uptake experiment. Cell-free supernatants of *Chlorella* cell suspensions can be obtained by spinning 200-μl samples for 15 sec in an Eppendorf centrifuge (at 8,000 to 10,000 g). The shortest intervals between two measurements are about 25 to 30 sec.

Intervals of 10 sec or less can be reached by filtering the cells on nitrocellulose filters (pore size 0.8 μm). Such experiments are performed using "filtering benches" consisting of a chamber with six or more filtering units and connected to an aspirator. Reduced pressure can be applied to each filtering unit individually by means of opening a tap. Before the experiment starts, a filter is placed in each unit, and 2 ml of ice-cold incubation buffer is pipetted into the chamber above the filter. Aliquots of cell suspensions are injected into the buffer, immediately filtered, and washed with 10 ml of ice-cold buffer.

When the experiment is performed with a ^{14}C-labeled substrate, radioactivity in the cells can be determined directly by immersing the cells (together with the filter) in scintillation cocktail. Use of ^{3}H-labeled substrates makes it necessary either to bleach the chlorophyll chemically or, if the substrate is not metabolized, to extract the radiolabel from the cells.

[11] E. W. Yemm and E. C. Cocking, *Analyst* **80**, 209 (1955).
[12] E. Loos, *Anal. Biochem.* **47**, 90 (1972).

For the latter purpose filter plus cells are boiled for 10 min in 1 ml 0.1 N HCl (close test tube with glass bead to keep volume constant or use screw-top tubes). An aliquot of the clear supernatant is counted directly in scintillation cocktail.

Net and Exchange Uptake

Net uptake and substrates can be measured when a substrate analog is available and its influx is followed for short times (2–3 min). Although cells obviously will not contain such an analog and thus exchange flux seems excluded, this is not necessarily true if, for example, physiological substrates using the same carrier will exchange for the external substrate analog. For *Chlorella* under physiological conditions this does not seem to be the case: the inside concentration of free hexoses is less than 10^{-5} M and thus far below the K_m for efflux (see below). For amino acids an exchange flux cannot be measured even with preloaded cells.[7]

The disappearance of substrate from the medium within the first few minutes can, therefore, be taken as the true rate of net uptake. If sugar analogs accumulate in the cell, however, flux rates across the plasma membrane change considerably: the rate of net uptake decreases, whereas the rate of unidirectional influx (or "gross" influx) under exchange conditions increases significantly. The same is true for the unidirectional efflux, until under steady-state conditions the two fluxes in opposite directions constitute a 1-to-1 exchange transport. For *Chlorella* the various rates have been determined for 6-deoxyglucose and are summarized in Table II.[13,14] In the footnotes to Table II the various K_m values for 6-deoxyglucose influx and efflux are given. The K_m value estimated for net efflux[13] corresponds nicely to the K_m value determined for exchange efflux according to Fig. 1. In this experiment the increase in the rate of unidirectional influx with increasing inside concentrations (achieved by preloading) has been determined and plotted.

The effect of preloading observed for influx is even more pronounced in the case of efflux: net efflux is very slow whereas exchange efflux is higher by a factor of 80 (Table II). This led to the conclusion that the rate-limiting step for net efflux is the "returning" of the "empty" carrier to the inside.[13]

For amino acids the situation is different. With rising inside concentration the net as well as the unidirectional influxes approach zero, which is most easily explained by assuming a feedback inhibition (or inactivation)

[13] E. Komor, D. Haass, B. Komor, and W. Tanner, *Eur. J. Biochem.* **39**, 193 (1973).
[14] M. Decker and W. Tanner, *Biochim. Biophys. Acta* **266**, 661 (1972).

TABLE II
INFLUX AND EFFLUX OF 6-DEOXYGLUCOSE:
NET AND EXCHANGE FLUXES

| | Flux | | | |
Transport	Net (μmol/ml)	Exchange (packed cells/hr)	Ratio net/ exchange	Ref.
Influx[a]	280	570	2.0	13, 14
Efflux[b]	< 10	570	> 57	13, 14

[a] Outside concentration 10^{-2} M, inside concentration 0 and 10^{-1} M, respectively. The K_m for net influx is 2×10^{-4} M; the K_m for exchange influx 3×10^{-4} M.[13,14]

[b] Outside concentration 0 and 6×10^{-3} M, respectively; inside concentration 10^{-1} M. The K_m for net efflux is 2.1×10^{-2} M; the K_m for exchange efflux 3.2×10^{-2} M (see Fig. 1 and Refs. 13 and 14).

of the transport protein.[7] For amino acids neither a significant net nor gross efflux can be measured; the cells seem to be "tight," whatever the reason for this asymmetric behavior may be.

Thus, different "trans effects," i.e., the effect of transport substrate on the corresponding "other" side of the membrane to which the flux of the substrate in question is followed, are found with different transport systems. They depend on the substrate (sugar or amino acid) and on which side is the "trans side."

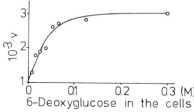

FIG. 1. Initial velocities of 6-deoxyglucose influx into cells preloaded with different amounts of 6-deoxyglucose. The external sugar concentration was 10 mM in each case. The velocity v is expressed as counts/min^2 in 5 μl packed cells.

Energization of Transport

Respiratory Increase

Since *Chlorella* cells induced for transport show a respiratory increase when hexoses or hexose analogs are supplied[14] the energy requirement for the transport reaction can be estimated. For 6-deoxyglucose a value of 1 ATP per deoxyglucose taken up has been determined.[14] The O_2-uptake measurements were carried out by Warburg manometry or by following the disappearance of O_2 in solution polarographically.[14] The experimental conditions were as follows.

The O_2 uptake of 60 μl packed cells of induced algae in 2 ml 25 mM sodium phosphate buffer, pH 6.0, is followed by conventional manometric techniques for 30 min, then 0.5 ml of D-glucose or an analog is added from the side arm to give a final concentration of 100 mM. Two controls with either mannitol or just water are included. The uptake of radiolabeled sugars is measured in parallel experiments under identical conditions.

Proton Symport

From unbuffered solutions the disappearance of protons from the medium after addition of sugars or sugar analogs can be followed by a sensitive pH meter.[15,16] Under the following conditions a transient change of pH of 0.07 pH units in 10 sec can be observed in the medium.

Packed cells (300–600 μl) of *Chlorella* induced for glucose transport are incubated aerobically in 5.5 ml distilled water, which is stirred vigorously with a glass stirrer. A glass electrode (Ingold lot 205 M5) is inserted into the suspension, and a thin plastic tube, filled with 1 mM KCl, connects the reaction vessel with the vessel of the reference electrode.[15,16]

Electrogenic Transport

Proton uptake is accompanied by an almost stoichiometric release of K^+.[17] That protons are the primary ion moved together with the sugar and that the transport, therefore, is electrogenic can be shown by measuring changes in the membrane potential $\Delta\psi$ caused by the addition of the substrate to be transported. Although membrane potentials for *Chlorella* have been measured with electrodes,[18,19] because of the small size of the cells (< 10 μm) and the tough cell wall, in our opinion the method using

[15] E. Komor, *FEBS Lett.* **38**, 16 (1973).
[16] E. Komor and W. Tanner, *Eur. J. Biochem.* **44**, 219 (1974).
[17] E. Komor, unpublished results.
[18] J. Barber, *Biochim. Biophys. Acta* **150**, 618 (1968).
[19] G. Langmüller and H. Springer-Lederer, *Planta* **120**, 189 (1974).

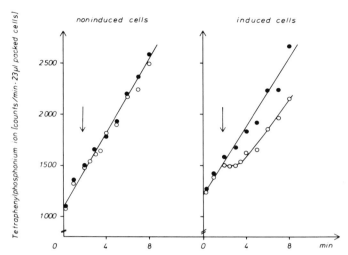

FIG. 2. Time course of [³H]tetraphenylphosphonium (15 μM; specific activity 1 Ci/mol) uptake and the effect of subsequent addition of sugar. The final concentration of sugar was 10 mM. Radioactivity is expressed for 23 μl of packed cells. Filled circles, control; open circles, batch to which sugar was added 2 min after start.

lipophilic cations[20] seems to be the more reliable one for *Chlorella*.[21] The best experience was made using tritium-labeled tetraphenylphosphonium (TPP⁺), introduced first by Heinz *et al.*[22]

A problem arises, however, if one wishes to measure rapid membrane potential changes, since it takes more than 1 hr before TPP⁺ equilibrates across the plasma membrane of *Chlorella*.[21] This problem can be overcome if the *rate* of TPP⁺ uptake into *Chlorella* is followed, which itself also is a function of $\Delta\psi$. A change in rate, which can be determined within seconds, indicates a change in $\Delta\psi$. In Fig. 2 the decrease in the rate of TPP⁺ uptake reflects a depolarization of $\Delta\psi$.

For the experiment, 70 μl packed cells/nl in 25 mM sodium phosphate buffer, pH 6.0, are incubated with [³H]tetraphenylphosphonium (15 μM, specific activity 1 Ci/mol). Aliquots are withdrawn and put on top of a 1-cm layer of silicone oil (S 20 : S 200 from Wacker Chemie, Munich, in a ratio of 1 : 3) in an Eppendorf centrifuge tube. The cells are sedimented rapidly through the silicone layer, and after removal of the medium plus the silicone oil the lower part of the centrifuge tube is cut off, combusted by

[20] L. L. Grinius, A. A. Jasaitis, Y. P. Kadziauskas, E. A. Liberman, V. P. Skulachev, V. P. Topali, L. M. Trofina, and M. A. Vladimirova, *Biochim. Biophys. Acta* **216,** 1 (1970).
[21] E. Komor and W. Tanner, *Eur. J. Biochem.* **70,** 197 (1976).
[22] E. Heinz, P. Geck, and C. Pietrzyk, *Ann. N.Y. Acad. Sci.* **264,** 428 (1976).

an Intertechnique liquid scintillation sample oxidizer, the tritium collected as labeled water in scintillation fluid, and the radioactivity determined. This separation method is chosen since the tetraphenylphosphonium ion adsorbs onto membrane filters to a significant extent.

If the rates of TPP$^+$ uptake are calibrated by correlating them with a TPP$^+$ equilibrium potential obtained under comparable conditions,[21] the changes in $\Delta\psi$ can be estimated from the relative changes in the rate of TPP$^+$ uptake. In the case of Fig. 2, a depolarization of 70 mV has been estimated.[21]

Measurements of Intracellular pH

Using weak acids, measurements of intracellular pH can, in principle, be carried out as described.[23,24]

Effects of pH and $\Delta\psi$ on V_{max} and K_m

For sugar transport the effects of pH and $\Delta\psi$ on V_{max} and K_m have been carried out[23,25,26]; however, they are beyond the scope of this chapter. Only one point is stressed here: at alkaline pH values of the medium (8.0) the K_m for 6-deoxyglucose *influx* is 50 mM and thus very close to the *efflux* K_m normally observed (see above).[23]

Attempts at in Vitro Studies

All attempts to break *Chlorella* cells and to obtain actively transporting membrane vesicles have failed so far. This failure most likely is mainly due to the fact that during cell breakage almost instantaneously large amounts of free fatty acids are set free, which uncouple any proton gradient-driven transport.[27] The same observation has been made with other eukaryotic cells.[27]

Isolation of Transport Mutants

For further investigations it is helpful to have mutant strains available that are deficient in the transport activities in question. Such mutants were obtained from *Chlorella* using acridine orange or ethidium bromide as mutagenizing agents at final concentrations of 0.1 mg/ml for 3 hr at 27°.

[23] E. Komor and W. Tanner, *J. Gen. Physiol.* **64**, 568 (1974).
[24] H. W. Heldt, K. Werdan, M. Milocancev, and G. Geller, *Biochim. Biophys. Acta* **314**, 224 (1983).
[25] W. G. W. Schwab and E. Komor, *FEBS Lett.* **87**, 157 (1978).
[26] E. Komor, W. G. W. Schwab, and W. Tanner, *Biochim. Biophys. Acta* **555**, 524 (1979).
[27] M. Decker and W. Tanner, *FEBS Lett.* **60**, 346 (1975).

For detection and isolation of the mutants it is advantageous that *Chlorella* cells can be handled almost like bacteria. Cells can be plated on petri dishes containing either the same inorganic medium as for liquid culture supplemented with 1.8% agar (minimal medium) or any other solid media, as, for example, YEPD medium (1% yeast extract, 2% bactopeptone, 2% D-glucose, 2% agar) which is commonly used for growth of *Saccharomyces cerevisiae* or *Escherichia coli*. On such plates single colonies can be isolated by the usual procedure, cells can be replica plated, and so on.

The selection methods used for identification of mutants are in principle identical to those described for prokaryotic or other unicellular organisms.[28] The easiest way is to select for growth in the presence of a substrate that is translocated by the carrier protein under investigation but is toxic for the cell. Such substrates are L-canavanine for the arginine transport system,[29] L-azetidine-2-carboxylic acid for the proline transport system,[30] and 2-deoxy-D-glucose for the hexose transport system.[31]

Mutagenized cells are washed several times and plated on minimal medium (about 10^7 cells/plate). These plates are supplemented either with 100 mM 2-deoxy-D-glucose or with 0.3 mM of one of the two amino acid analogs. In the latter case D-glucose is also added (10 mM) for the induction of the amino acid transport systems (see above). Cells are incubated for growth at room temperature in daylight, and after 10 to 20 days mutants are observed as single colonies either on a clear (amino acid analogs) or on a yellowish green background (2-deoxy-D-glucose). For further purification these colonies are streaked on plates with the same composition. Purified single colonies can then be transferred either to YEPD plates and stored in a refrigerator up to 3 months (cells should then be replated) or to liquid medium if used right away for further studies.

Table III lists all *Chlorella* mutants obtained by this method. Their transport activities following the two induction procedures (see above) are compared. Mutants defective in the uptake system for basic amino acids are called AUP⁻ (for arginine uptake negative), mutants defective in the proline uptake system are called PUP⁻, and mutants defective in the hexose uptake system are called HUP⁻.

Identification of Transport Proteins

For the characterization of transport proteins *Chlorella* exhibits the advantage that transport activities are inducible. Since this induction is due

[28] C. W. Slayman, *Curr. Top. Membr. Transport.* **4,** 1 (1973).
[29] J. H. Schwartz, W. K. Maas, and E. J. Simon, *Biochim. Biophys. Acta* **32,** 582 (1959).
[30] H. Tristam and S. Neale, *J. Gen. Microbiol.* **50,** 121 (1968).
[31] N. Sauer, *Planta* **168,** 139 (1986).

TABLE III
UPTAKE ACTIVITIES OF VARIOUS *Chlorella* MUTANTS

| | Uptake[a] following pretreatment of cells | | | | | |
| | Addition of D-glucose | | | Starved for nitrogen | | |
Strain	Hex	Arg	Pro	Hex	Arg	Pro
Wild type (WT)	+	+	+	−	+	+
HUP⁻	−	−	−	−	+	+
AUP⁻	+	−	+	−	−	+
PUP⁻	+	+	−	−	+	−
AUP⁻ PUP⁻	+	−	−	−	−	−

[a] +, Uptake; −, no uptake.

to *de novo* synthesis of transport proteins[32] *Chlorella* cells induced for transport by treatment with D-glucose should have additional proteins inserted into their plasma membranes.

These additional proteins can be studied by comparing detergent extracts of induced versus noninduced cells on sodium dodecyl sulfate (SDS)–polyacrylaminde gels. For this purpose pure plasma membranes would be the best starting material, but so far it has not been possible to obtain such material. A so-called plasma membrane-enriched fraction, however, can be prepared by collecting the cell walls of homogenized cells (French press, 20,000 psi) by low-speed centrifugation (3000 g for 10 min) and washing the pellet with 25 mM sodium phosphate buffer, pH 6.0. Besides some starch grains and very few unbroken cells, this fraction consists mainly of cell wall pieces with adhering patches of plasma membrane.[32]

Solubilization and separation of proteins from such fractions on SDS–polyacrylamide gels does not result, however, in any detectable differences in protein pattern between induced and noninduced cells when stained with Coomassie blue or silver stain.[31] A further increase in sensitivity was obtained by introducing a radiolabel into the newly synthesized proteins. [¹⁴C]Phenylalanine was chosen as radioactive precursor for two reasons: first, as a lipophilic amino acid it might be used to a greater extent for the synthesis of membrane proteins, and, second, it is not a substrate for any of the known inducible amino acid transport systems in *Chlorella* (see above!). A typical labeling experiment is carried out in the following way (Table IV).

[32] F. Fenzl, M. Decker, D. Haass, and W. Tanner, *Eur. J. Biochem.* **72**, 509 (1977).

TABLE IV
RADIOLABELING OF INDUCIBLE PROTEINS

Component of reaction mixture	Induced sample	Control
Algae (50 μl packed cells/mi in 25 mM sodium phosphate buffer, pH 6.0)	2 ml	2 ml
D-Glucose (28 mg/ml)	0.2 ml	—
H$_2$O	—	0.2 ml
L-[U-^{14}C]Phenylalanine (specific activity 495 Ci/mol)	8 μCi	8 μCi

The algae are shaken at 27° for 2 hr. After this time, when maximal transport activity is reached, cells are collected by centrifugation, washed, and disintegrated with a French press.

If the plasma membrane-enriched fractions prepared from these cells are compared on fluorograms of SDS–polyacrylamide gels,[31] differences become visible (Fig. 3). Two zones of radioactivity obtained with the extract of induced cells are not present in noninduced cells: a protein band with an apparent molecular weight of 47 K and a more diffuse region in the molecular weight range of 34 K to 41 K. Within this diffuse zone, there is one band with an apparent molecular weight of 38 K showing up more distinctly. It seems likely that these various proteins are responsible for the different transport activities.

With the help of the mutants described above these various bands may be further characterized. The protein with the apparent molecular weight of 38 K is lacking, for example, in all AUP$^-$ mutants that are defective in the transport of basic amino acids.[31]

Identification of a cDNA Clone Encoding the Glucose–H$^+$ Cotransporter

Inducibility of transport activities in *Chlorella* was also a big advantage in attempts to obtain cDNA clones of the transport proteins described so far. Poly(A)$^+$ RNA was isolated from induced cells[33,34] and the corresponding cDNA was cloned into λgt10 arms. The resulting library was screened differentially with ^{32}P-labeled cDNA from either induced or noninduced *Chlorella* cells. Out of about 20 clones, which lit up only when

[33] J. N. Bell, T. B. Ryder, V. P. M. Wingate, J. A. Bailey, and C. J. Lamb, *Mol. Cell. Biol.* **6**, 1615 (1986).
[34] T. Maniatis, E. F. Fritsch, and J. Sambrook, "Molecular Cloning: A Laboratory Manual." Cold Spring Harbor Laboratories, Cold Spring Harbor, New York, 1982.

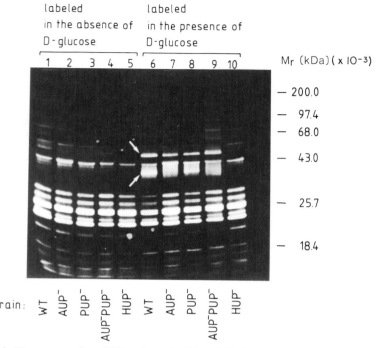

labeled
in the absence of
D-glucose

labeled
in the presence of
D-glucose

1 2 3 4 5   6 7 8 9 10

Mr (kDa)(x 10⁻³)

— 200.0
— 97.4
— 68.0
— 43.0
— 25.7
— 18.4

strain: WT AUP⁻ PUP⁻ AUP⁻PUP⁻ HUP⁻ WT AUP⁻ PUP⁻ AUP⁻PUP⁻ HUP⁻

FIG. 3. Fluorogram of an SDS–polyacrylamide gel of plasma membrane-enriched fractions of wild-type and mutant cells of *Chlorella*. Cells (induced or not induced) were radiolabeled as described in the text.

screened with labeled cDNA from induced cells, one clone has been identified as the cDNA of the glucose carrier mRNA. This clone is homologous to a group of other sugar translocators (e.g., the human hepatoma glucose transport protein[35] or the arabinose carrier of *E. coli*[36]). However, it exhibits no homology to the *lac* permease of *E. coli*[37] or the human Na⁺–glucose cotransporter.[38]

[35] M. Mueckler, C. Caruso, S. A. Baldwin, M. Panico, J. Blench, H. R. Morris, W. J. Allard, G. E. Lienhard, and H. F. Lodish, *Science* **229**, 941 (1985).
[36] M. C. J. Maiden, E. O. Davis, S. A. Baldwin, D. C. M. Moore, and P. J. F. Henderson, *Nature (London)* **325**, 641 (1987).
[37] D. E. Büchel, B. Gronnenborn, and B. Müller-Hill, *Nature (London)* **283**, 541 (1980).
[38] M. A. Hediger, M. J. Coady, T. S. Ikeda, and E. M. Wright, *Nature (London)* **330**, 379 (1987).

[27] Electrophysiology of Giant Algal Cells

By M. J. BEILBY

Electrical Measurements

Making Contact with the Cell

The giant algae provide a unique system suitable for the study of a single plasma membrane *in vivo.* The large cell sizes (0.5–1.0 mm in diameter by several centimeters in length for freshwater algae, even larger for some marine algae such as *Valonia ventricosa*) minimize damage on making electrical contact. Two methods are employed: the microelectrodes pioneered by Walker[1] and K^+ anesthesia explored in early experiments by Osterhout and Harris[2] and reintroduced by Shimmen *et al.*[3] The methods described in this chapter are somewhat biased toward freshwater giant algae, as my experience lies in this area. Most of these techniques can be extended to giant-celled marine algae. Some aspects of the electrophysiology of *Acetabularia* are discussed by Gradmann [30] in this volume.

Microelectrodes. The electrode method is more accurate and versatile than K^+ anesthesia. The microelectrodes are manufactured from borosilicate glass tubing in an electrode puller. The diameter of glass blanks varies, the upper limit is probably 5 mm outer diameter (o.d.) (used in biophysics laboratory, University of NSW). However, large micropipettes make manipulations difficult in the limited space around the cell holder. Tubing of 2 mm o.d. is satisfactory, but, presumably, smaller diameter tubing can be used if the glass is thick enough for the micropipette to pierce the cell wall. The glass blanks are softened in a high-resistance filament heated by passage of electrical current. The ends are then drawn apart, usually in two stages. The force is provided by solenoids, but in a vertical puller the gradual pull is due to gravity. The horizontal puller is more useful for complex electrodes (e.g., multiple barrels), as a greater degree of control can be exercised over the pulling forces. However, for work with giant algae, the vertical puller (e.g., Narishige, Japan) is quite sufficient. The amplitude of the current is the major controlling factor in the shape of the micropipette tip. The taper is very important: if it is too gradual, the microelectrode will be too flexible, if too short, it will not be flexible

[1] N. A. Walker, *Aust. J. Biol. Sci.* **8**, 476 (1955).
[2] W. J. Osterhout and E. S. Harris, *J. Gen. Physiol.* **12**, 761 (1929).
[3] T. Shimmen, M. Kikuyama, and M. Tazawa, *J. Membr. Biol.* **30**, 249 (1976).

enough. In either case the tip will break upon insertion, causing greater damage. The micropipette tips range from 0.2 to several microns (see Fig. 2A).

To make the micropipettes conductive, they are filled with KCl. This task used to be very tedious, until the thin fiber was included to allow the liquid to penetrate the tips (Kwik-Fil single capillaries with inner filament, Clarke Electromedical Instruments). The concentrations vary. The upper limit of 3 M KCl originally derived from a need to lower the microelectrode resistances and time constant.[4,5] Lower concentrations (below 100 mM) were used for small cells to avoid contamination by Cl^- leaking out of the electrode tip.[6] Cytoplasmic impalements in giant algae, however, are better done with microelectrodes filled with KCl more concentrated than 1 M, as the Donnan potential arising from the negative fixed charges in the cytoplasm will add to the membrane potential difference (p.d.).[7] In any case, the Cl^- contamination is not a problem, as it is possible to starve the cytoplasm for chloride over several hours with the electrode inserted.[8] To connect the microelectrodes to the circuitry, agar bridges and calomel half-cells or silver wire coated with AgCl are traditionally used. Ag/AgCl half-cells can be prepared in the laboratory and are also now available from WP Instruments (New Haven, CT). These are very convenient, as they need no repeated coating. While they are not as stable as calomel half-cells, they fit snugly over the glass micropipette ends and plug straight into the instrumentation operational amplifier.

K^+ *Anesthesia.* In the K^+ anesthesia method the cell is placed into a chamber divided into two compartments (electrically insulated by silicone grease), one of which is filled with ~100 mM KCl (see Fig. 1A). This medium reduces the resting p.d. and resistance of the exposed membrane to negligible values. Placing an electrode into each compartment effectively measures the p.d. across the membrane in the low-K^+ compartment. The cells survive for hours, so the treatment is not detrimental like the usage of chloroform in the early experiments.[2] The large conductance of the membrane in the high-K^+ compartment is probably brought about by the opening of the $[K^+]_o$-activated K^+ channels at the plasmalemma of many giant algae,[9-11] This technique is particularly suitable for experiments

[4] G. Ling and R. Gerard, *J. Cell. Comp. Physiol.* **34**, 383 (1949).

[5] W. Nastuk and A. Hodgkin, *J. Cell. Comp. Physiol.* **35**, 39 (1950).

[6] M. R. Blatt and C. L. Slayman, *J. Membr. Biol.* **72**, 223 (1983).

[7] H. G. L. Coster, E. P. George, and V. Rendle, *Aust. J. Plant. Physiol.* **1**, 459 (1974).

[8] M. J. Beilby, *J. Membr. Biol.* **62**, 207 (1981).

[9] A. I. Sokolik and V. M. Yurin, *Sov. Plant Physiol. (Engl. Transl.)* **28**, 206 (1981).

[10] G. P. Findlay and H. A. Coleman, *J. Membr. Biol.* **75**, 241 (1983).

[11] M. J. Beilby, *J. Exp. Bot.* **36**, 228 (1985).

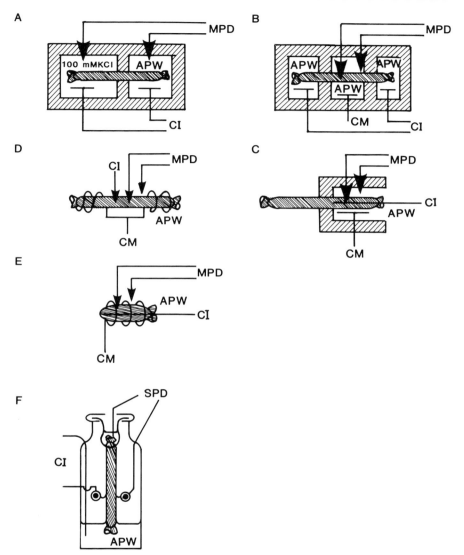

FIG. 1. Types of cell holders. The cells are shown with uneven hatching. (A) K$^+$ anesthesia as used by Shimmen *et al.*[3] The chambers of the holder are separated by a greased partition. One set of electrodes is used to measure the membrane potential (MPD), the other for current injection (CI). [The current measurement (CM) is probably effected between the artificial pond water (APW) compartment and ground.] (B) Three-compartment holder used by V. Z. Lunevsky *et al.* [*J. Membr. Biol.* **72**, 43 (1983)]. The MPD was measured with glass micro-electrodes, and the CI was effected by Ag/AgCl wires placed in the lateral compartments. Current measurement (CM) was done between the holder and ground. (The reader is re-

where the cells have been incubated in radioactive tracers and electrode insertion might cause a loss of the tracer. It is also a good method to be used for practical classes, as the degree of success is much higher for the unpracticed hand than electrode insertion. Problems can arise from leakage of the electrolyte around the grease, thus short-circuiting the cell p.d. Another shortcoming is that the electrical properties of the plasmalemma and the tonoplast cannot be separated.

Measuring Membrane p.d.

Preparing Cell for Impalement. To insert the microelectrodes, the cell must be fixed in a holder. Several cell holders are displayed in Fig. 1, and they are discussed in greater detail in the next section, when the geometries become important. The cells are immobilized by greased partitions or in a silver coil coated with AgCl. A dissection microscope and micromanipulators are used to guide the microelectrodes into the cell. Magnification of up to $100 \times$ is usually required.

At the time of the experiment, the cell is bathed in some medium. While various workers have their special recipes (see Hope and Walker,[12] for review), most of the media for freshwater algae are variations of artificial pond water (APW), which approximates the composition of pond water in the field (APW = 1.0 mM NaCl, 0.1 mM KCl, 0.5 mM CaCl$_2$, 1–5 mM zwitterionic buffer, and NaOH to adjust to desired pH). Fresh

---

[12] A. B. Hope and N. A. Walker, "The Physiology of Giant Algal Cells." Cambridge Univ. Press, London and New York, 1975.

---

minded that no current can be measured unless some is injected. By definition the sum of all currents in the resting state is 0.) Using greased partitions may lead to some error in estimates of area-specific conductance, as the cell wall will make such insulation imperfect [see J. R. Smith, *Aust. J. Plant Phys.* **12**, 403 (1985)]. (C) Combination of space clamp and use of compartmented holder, from U. Kishimoto [*Plant Cell Physiol.* **9**, 539 (1968)]. The compartment makes it possible to effectively space clamp long internodal cells. The internal current electrode is made from tungsten wire coated with platinum black. (D) Point clamp arrangement used by N. A. Walker *et al.*[14] Both the MPD measurement and the CI are done via glass micropipettes. (E) Space clamp setup originated by G. P. Findlay and A. B. Hope.[16] The Ag/AgCl coil is the low-current electrode and also holds the cell fixed. This arrangement is mainly suitable for very young leaf cells. The axial electrode is Pt/Ir wire. (F) The "baby bottle" method of K. Ogata.[120] Long internodal cells are suspended from an inverted teat, filled with APW. The other end of the cell is also submerged in APW, and the air in the bottle is humidified to prevent the cell from drying out. The Ag/AgCl coil can be moved along the length of the cell. Coating it with agar ensures a water film between it and the cell making electrical contact with the cell surface, measuring the surface membrane p.d. (SPD). The CI is also effected between the coil and Ag/AgCl wire in the bath at the bottom of the bottle.

and brackish water algae live in an enormous range of media. Greater uniformity of cell behavior can thus be achieved if the cells are precut a few days prior to the experiment (to produce closure of the plasmodesmata, which appear between adjoining cells in the charophytes[13]) and stored in APW. Marine algae live in comparatively more constant compositions, the sea environment containing high NaCl (~ 0.5 M) and considerably lower KCl.

To prevent any modification to the APW composition by an efflux from the cell, a fast flow past the cell is advisable. Fast flow rates ($5-10$ mm/sec) are also important to minimize unstirred layers when the cell response to a substance is studied.[14]

Electrode Tip p.d.'s. The reference electrode for the measurement of the transmembrane p.d. can be provided by another microelectrode identical to that inside the cell. Some workers prefer to break the tip, or to use an agar bridge [Polythene tubing filled by the same electrolyte as in the high electrode with about 3% agar (w/v)]. The first alternative offers two similar input resistances to the instrumentation operational amplifier, and such a symmetrical situation should be easier for the electronics to handle. The p.d. between the two electrodes in the medium should be below 10 mV. Backing off the electrode p.d. is not advisable based on the work of Blatt and Slayman,[6] who found that the tip p.d.'s are likely to decrease once in a cell.

There are various techniques of minimizing the tip p.d.,[15] such as decreasing the pH of the electrolyte or buffering the pH to 8.5 with glycylglycine, boiling the electrodes prior to filling, or using glass blanks of triangular cross section. Most freshwater algae have transmembrane p.d.'s between -100 and -200 mV, and a few millivolts are probably not of vital importance. However, those studying the tonoplast or some marine algae might find the tip p.d.'s more of a problem.

Noise Immunity. The microelectrode resistance is typically $10-15$ megohms ($M\Omega$), and in such a system the noise pickup is high. To avoid noisy readings, the cell holder is placed in an electrostatic (Faraday) cage. The medium flowing over the cell can often be a source of noise, and the reservoir is best placed inside the cage. Likewise, the outflow of the solution can be shielded by grounded aluminum foil. It is also necessary to have a single ground point for all the circuitry, as ground loops introduce noise. To minimize the transmission distance of the high-impedance sig-

[13] J. R. Smith, *Aust. J. Plant Physiol.* **11**, 211 (1984).
[14] N. A. Walker, M. J. Beilby, and F. A. Smith, *J. Membr. Biol.* **49**, 21 (1979).
[15] R. C. Thomas, "Ion-Sensitive Intracellular Microelectrodes." Academic Press, London, 1978.

nal, field effect transistor (FET) input operational amplifiers can be placed adjacent to Ag/AgCl half-cells (this feature is now basic to design of any electrophysiological equipment). Cleanliness in the experimental setup is strongly advised. Highly conductive pools of concentrated KCl solution spilled from the electrodes often cause 50-Hz noise to appear in the signals, probably by producing ground loops. Unfortunately, noise immunity is more of an art than a science and requires all types of adaptations.

Signal Recording. A range of operational amplifiers (we used 741) is now available to process the signals from the FET's, which can then boost the cell membrane p.d. Fast chart recorders are now available with time resolution sufficient to record accurately any plant membrane phenomenon. However, these also condemn the experimenter to sorting out kilometers of chart paper. Cathode ray oscilloscopes (CRO) offer accurate display of the fast transients, but records obtained by photographing the screen are often difficult to analyze. Fortunately, with newer technology the signal can now be stored on magnetic tape, in a digital storage CRO, or datalogged directly by a computer. In each case it is possible to perform complex mathematical manipulations, such as statistical analysis, Fourier analysis, or fitting of various models to the data. Thus, one can obtain feedback *at the time* of the experiment. The chart recorder does have its place even in the "high-tech" laboratory. It can provide a real time recording of the whole experiment to monitor the long-term cell responses to whatever challenges one presents to it.

Placement of the p.d.-Measuring Electrode. Impalement into the vacuole is easy and the electrode usually does not block. As the tonoplast is believed to be substantially more conductive than the plasmalemma,[16-21] the two membranes in series are therefore thought to reflect primarily the properties of the plasmalemma. However, there is a range of conditions where the plasmalemma resistance may become comparable to that of the tonoplast (such as the action potential,[16,22] K^+ state,[9,11] or the proton-conductive state at high pH_o[23,24]). Further, the work with isolated vacuoles[25]

[16] G. P. Findlay and A. B. Hope, *Aust. J. Biol. Sci.* **17**, 62 (1964).
[17] R. M. Spanswick and J. W. F. Costerton, *J. Cell Sci.* **2**, 451 (1967).
[18] G. P. Findlay, *Aust. J. Biol. Sci.* **23**, 1033 (1979).
[19] R. M. Spanswick, *J. Exp. Bot.* **21**, 617 (1970).
[20] H. G. L. Coster and J. R. Smith, *Aust. J. Plant. Physiol.* **4**, 667 (1977).
[21] J. R. Smith, *J. Exp. Bot.* **34**, 120 (1983).
[22] M. J. Beilby and H. G. L. Coster, *Aust. J. Plant Physiol.* **6**, 337 (1979).
[23] M. A. Bisson and N. A. Walker, *J. Membr. Biol.* **56**, 1 (1980).
[24] M. J. Beilby, *J. Membr. Biol.* **93**, 187 (1986).
[25] B. P. Marin, ed., "Biochemistry and Function of Vacuolar Adenosine-Triphosphatase in Fungi and Plants." Springer-Verlag, Berlin and New York, 1984.

shows that this membrane emerges with substantial differences and thus lumping it together with the plasmalemma will make data difficult to interpret. The cytoplasmic electrode or one of the tonoplastless systems (see section on controlling the cell compartments) is probably the best route to take in future investigations.

The electrodes for insertion into cytoplasm are best suited to this task if the taper is steep (see Fig. 2A). Such electrodes are obtained by turning down the current heating the wire coil of the puller. Further hints for successful cytoplasmic insertion are the use of young cells, positioning of the cell with the electrode at right angles to the cell surface, and the firm fastening of the cell in a coil. Watching the chart recorder for the sudden p.d. rise is also useful. An audio monitor (voltage-controlled oscillator providing a tone with frequency proportional to the cell p.d.) can also be very convenient. Once in the cytoplasm, the electrode may block within minutes, or it may show a steady reading for hours. Using a piezoelectric inchworm to insert the tip into the cytoplasm makes it very easy, but unfortunately the electrodes then block very quickly. The blockage is thought to be a new membrane which excludes the electrodes out of the cell. The process is not well understood. The longer survival time of hand-inserted microelectrode before blocking is probably due to greater disturbance of the immediate surroundings.

There are many criteria to confirm the position of the electrode in the cytoplasm, such as lack of gushing, absence of action potential following the insertion, and the shape of the action potential, when induced. In my experience the most reliable is the short-circuit current discussed in the next section.

Resting p.d. in Giant Algae. When the electrode is placed in the cell, what are we to expect? This depends greatly on the outside conditions. Most freshwater algae exhibit several distinct "states," in which cell properties are drastically different.[23,24] These states are mainly ascribed to the plasmalemma membrane.

Under "normal conditions" for freshwater algae (low $[K^+]_o$ less than 1.0 mM, some Ca^{2+} 0.5 mM, slightly acid to neutral pH_o, and light), the membrane p.d.'s are typically hyperpolarized to vicinity of -200 mV, which is dependent on pH_o, only weakly on $[K^+]_o$, and susceptible to a range of metabolic inhibitors, CO_2, and darkness in *Nitella*.[12,26,27] This state is believed to be dominated by an electrogenic proton pump that moves protons out of the cytoplasm thereby creating an electrochemical gradient for protons and perhaps regulating cytoplasmic pH. The pH_o

[26] R. M. Spanswick, *Biochim. Biophys. Acta* **288**, 73 (1972).
[27] R. M. Spanswick, *Biochim. Biophys. Acta* **332**, 387 (1974).

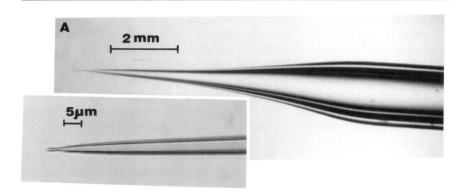

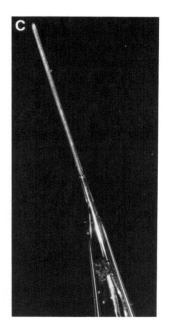

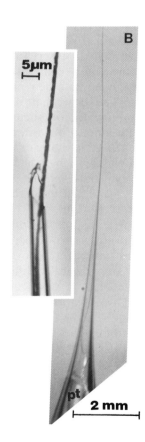

dependence of the proton pump places the maximum resting p.d. between pH_o 7 and 8. The resting p.d. depolarizes considerably at pH_o 4.5, but the membrane conductance is not diminished.[28,29] This behavior is thought to be due to a change in pump kinetics[29] (see Fig. 3). Some brackish water algae also seem to have a proton pump at their plasmalemma.[30] The marine algae such as *Acetabularia* or *Halicystis* do possess an electrogenic pump, but the ion transported is Cl^-[31,32] (also see Gradmann [30], this volume).

The K^+ state is initiated by high $[K^+]_o$ of greater than ~ 1.0 mM. The resting p.d. is then low and close to the estimated E_k (see Fig. 3). The conductance in this state increases with $[K^+]_o$.[11] Such behavior is observed in many freshwater, brackish, and marine algae.[9-11,18,30,33-35] At very high pH_o the proton-permeable state takes over the membrane properties.[23] The membrane p.d. then behaves as a pH electrode (Fig. 3).

Under stable conditions and flowing APW, the resting p.d. recording tends to be steady for many hours (longest time in our experience, 2 days), except for blocking of the p.d.-measuring electrode. In some cases, however, the p.d. can be seen to oscillate with an amplitude of ~ 20 mV. These oscillations have been classified[36] into 15-min and 1-hr oscillations. Simultaneous measurements of the membrane resistance seem to point to the fact that the former are associated with the proton pump, whereas the latter are ascribed to K^+ channels. Spatially the potential is also not uniform in many freshwater algae; this effect is discussed in the section on banding.

[28] D. W. Keifer and R. M. Spanswick, *Plant Physiol.* **62**, 653 (1978).
[29] M. J. Beilby, *J. Membr. Biol.* **81**, 113 (1984).
[30] M. A. Bisson and G. O. Kirst, *J. Exp. Bot.* **31**, 1223 (1980).
[31] D. Gradmann and F. W. Bentrup, *Naturwissenschaften* **57**, 46 (1970).
[32] J. S. Graves and J. Gutknecht, *J. Membr. Biol.* **36**, 65 (1977).
[33] K. Oda, *Sci. Rep. Tohoku Univ., Ser. 4* **28**, 1 (1962).
[34] G. P. Findlay, A. B. Hope, and E. J. Williams, *Aust. J. Biol. Sci.* **22**, 1163 (1969).
[35] H. W. D. Saddler, *J. Gen. Physiol.* **55**, 802 (1970).
[36] J. Fisahn, E. Mikschl, and U.-P. Hansen, *J. Exp. Bot.* **37**, 34 (1986).

FIG. 2. (A) Glass micropipette with the taper used for cytoplasmic insertion (o.d. of the glass blank 2 mm). The electrode was pulled on the Narishige vertical electrode puller with the current setting at 12. The inset shows a greater magnification of the tip. The filling filament cannot be seen. (B) Pt/Ir wire etched in KCN, passed through a glass micropipette, and ready to insert into the node of the cell to be used for current injection. The wire is threaded through plastic tubing (pt) for insulation and easier handling. The inset shows the glass micropipette tip, which was broken. Here the wire has been pulled back so that only the tip is protruding. When the whole length of the etched wire is pushed out, it gives a snug fit (thus minimizing loss of contents from the cell when the electrode is inserted). The jagged edge of the glass helps to penetrate the tough cell wall. (C) Alternative technique for making the current-injecting electrode using the "thick" wire. Note the smooth finish of the glass forged onto the wire.

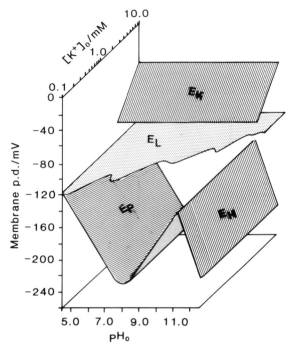

FIG. 3. Plasmalemma p.d. as function of pH_o and $[K^+]_o$. The leak p.d. (E_L) probably coexists with all the other states. It is not known to what extent it is affected by pH_o or $[K^+]_o$, so the E_L plane is an approximation. The pump state (E_P), the proton-permeable state (E_H), and the K^+ state (E_K) do not coexist. In the proton-permeable state and the K^+ state the resting p.d. follows E_H and E_K, respectively, and can be approximated by planes [the K^+ conductance is affected by pH_o but not the resting p.d., while the H state is insensitive to $[K^+]_o$ as long as the membrane is not depolarized; see, for instance, M. J. Beilby.[96] The E_P surface is not entirely independent of $[K^+]_o$ (as shown here for simplicity), but the changes are small. The surface probably extends further than 1 mM K^+, but at $[K^+]_o$ greater than this concentration the transition to the K^+ state is sharp both temporally and with respect to membrane properties. The transition is effected by depolarization.

It is advisable to build some temperature control into the cell holder as temperature does strongly affect the membrane p.d.'s. The most hyperpolarized values are found between 20° and 35°, and depolarization is observed at temperatures above and below this window.[37–40] In the 1980s

[37] J. Hogg, E. J. Williams, and R. J. Johnston, *Biochim. Biophys. Acta* **150**, 518 (1968).

[38] A. B. Hope and P. A. Aschberger, *Aust. J. Biol. Sci.* **23**, 1047(1970).

[39] F. J. Blatt, *Biochim. Biophys. Acta* **339**, 382 (1974).

[40] M. J. Beilby and H. G. L. Coster, *Aust. J. Plant Physiol.* **3**, 275 (1976).

temperature studies seemed to have become unfashionable, but interest has been recently revived by an elegant study by Fisahn and Hansen.[41] Another important factor determining the resting p.d. in some algae is illumination. Here the reader is directed to [29] by Spanswick in this volume.

Controlling Transmembrane p.d.

The easiest way to change the membrane p.d. is by passing a current across it. The direction of the current will cause either hyperpolarization or depolarization. While the measurement of the p.d. involves neither the size of the cell nor its geometry, the passing of the current and measurement of area-specific conductance introduce these variables with concomitant uncertainties.

Action Potential. One of the simplest uses of the current injection is the stimulation of an action potential. The depolarizing current is applied as a series of increasing steps or as a ramp until threshold for excitation is reached. The membrane p.d. then continues to depolarize spontaneously, reaches a peak close to, but usually negative of, the zero p.d., and then repolarizes to the resting level. Researchers fascinated by the threshold for excitation will find this elusive, as it changes with the slope of the current ramp or length of the pulse.

The action potential can be observed in most freshwater algae, some brackish algae, and some marine algae.[16,18,33,42-45] The depolarization phase of the action potential is brought about by a transient opening of the Cl^- channels. $Cl^{-43,46}$ then flows out of the cell, down its electrochemical gradient. Involvement of Ca^{2+} is strongly indicated.[47,48] Recent experiments with perfused cells of *Nitellopsis obtusa* indicate that opening of the Cl^- channels is activated by the transient rise in cytoplasmic Ca^{2+} [49,50] mainly due to an inflow from the outside medium, as the plasmalemma Ca^{2+} channels open in response to depolarization[51] (see also the section on controlling the cell compartments). The repolarizing current was assumed

[41] J. Fisahn and U.-P. Hansen, *J. Exp. Bot.* **37**, 177 (1986).
[42] C. T. Gaffey and L. J. Mullins, *J. Physiol. (London)* **144**, 505 (1958).
[43] G. P. Findlay, *Aust. J. Biol. Sci.* **12**, 412 (1959).
[44] U. Kishimoto and M. Tazawa, *Plant Cell Physiol.* **6**, 529 (1965).
[45] D. Gradmann, *Planta* **93**, 323 (1970).
[46] A. B. Hope and G. P. Findlay, *Plant Cell Physiol.* **5**, 377 (1964).
[47] M. J. Beilby, *Plant, Cell Environ.* **7**, 415 (1984).
[48] I. Tsutsui, T. Ohkawa, R. Nagai, and U. Kishimoto, *J. Membr. Biol.* **96**, 65 (1987).
[49] T. Shiina and M. Tazawa, *J. Membr. Biol.* **99**, 137 (1987).
[50] T. Shiina and M. Tazawa, *J. Membr. Biol.* **106**, 135 (1988).
[51] T. Shiina and M. Tazawa, *J. Membr. Biol.* **96**, 263 (1987).

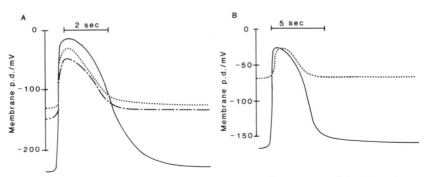

FIG. 4. (A) Action potential of *Chara corallina* in different states of the plasmalemma: —, pump state, pH_o 7.5;---, pH_o 4.5; ·····, proton-permeable state, pH_o 11. All records are from one cell. In the pump state the membrane is probably repolarized by the pump current. At low pH_o the resting pump conductance seems to increase,[29] and this might account for the less depolarized action potential peak. Similarly, in the proton-permeable state, where the membrane conductance increases further, the peak shifts to even more negative p.d.'s. $[K^+]_o$ was 0.1 mM throughout. (B) Comparison of the action potential shape in the pump state, pH_o 7.5 —, and the K^+ state in $[K^+]_o$ of 5 mM (---). (From Beilby.[11]) A different cell from that in A was used. As $[K^+]_o$ increases further, cells tend to become inexcitable, probably owing to lack of an electrochemical gradient for Cl^-.

to be due to outflow of K^+ as in nerves. An increase in K^+ efflux is, indeed, observed in some species.[43,52,53] However, it may be important which state the cell membrane is in. Figure 4 shows the shape of the action potential at the plasmalemma in the different states of *Chara*. It seems that different mechanisms are responsible for the return to resting p.d.

At 20°, the action potential across the plasmalemma is completed within 5 sec.[16,22] (The duration of the excitation increases with decreasing temperature.[39,40]) However, if the resting p.d. and conductance are monitored, they undergo a decrease, reaching a minimum in several minutes.[54] The reason for this behavior is not known, but it may be due to temporary inhibition of the proton pump following the increase in cytoplasmic Ca^{2+}. In any case, it is advisable to wait about 30 min between excitations, so that the resting p.d. and conductance can recover to preexcitation level.

Both membranes are excitable, the tonoplast with much smaller amplitude but longer time constant. When measuring across both membranes the two responses are additive.[16,18]

[52] K. Oda, *Plant Cell Physiol.* **47**, 1085 (1976).
[53] T. Shimmen and M. Tazawa, *Plant Cell Physiol.* **24**, 1511 (1983).
[54] J. R. Smith and M. J. Beilby, *J. Membr. Biol.* **71**, 131 (1983).

Need for Space Clamp. Another complication to consider is that the action potential is conducted away from the point where it was first initiated. To simplify data analysis, the excitation can be stimulated along the whole cell at once. This technique, called the space clamp, is effected by a thin wire passed along the whole length of the cell. This approach was pioneered by Findlay and Hope,[16,55] who electropolished Pt/Ir wire but left it thick enough to puncture the cell wall. The insulation was provided by glass, which was melted onto the wire, leaving only ~7 mm exposed.

We have employed two different versions of this method in our experiments. One procedure starts with a Pt/Ir wire of 25 μm (now available from Goodfellow Metals, Cambridge, England), etching it in ~2 M KCN or NaCN solution (experimenters are cautioned to include some hydroxide when making the cyanide solution, otherwise cyanide gas may be evolved) to ~5 μm in diameter after it had been straightened by annealing. This is achieved by passing a current through a 10-cm length of the wire until it glows red hot, while stretching it. The straightened wire is threaded through thin plastic tubing with both ends protruding. At one end the wire is wrapped around the tubing (a connection to the apparatus can then be made via alligator clip), while at the other end about 3 cm is left exposed. Attaching the wire to a 6-V source (ac or dc) and dipping into the KCN solution completes the preparation. The etching takes only few minutes, and one learns by experience to estimate the thickness of the wire by its flexing, as it is moved in the etching medium (leaving the wire motionlessly suspended leads to the etched portion breaking off at the liquid – air interface). The final etched wire often has a very fine tip (see Fig. 2B) and grows thicker. It is too soft to pierce the cell wall and must be threaded into a glass micropipette, which then passes through the cell wall and also provides insulation. The threading of the wire into the micropipette is a procedure requiring infinite patience. To locate the tip of the wire, reflected light can be helpful. Freshly pulled micropipettes without filling filament must be used, and it is advisable to break the tip only after the wire has been pushed as far as it can go (see Figs. 2B and 5). The micropipette and the wire are mounted on separate micromanipulators. The wire is inserted immediately after the glass micropipette and passed throughout the whole length of the cell, taking care not to scrape the cell sides. Usually the cell p.d. recovers quickly (30 min), although achieving true stability of electrical properties requires a few hours.

[55] G. P. Findlay and A. B. Hope, *in* "Encyclopedia of Plant Physiology, Transport in Plants II, Part A Cells" (U. Lüttge and M. G. Pitman, eds.), p. 53. Springer-Verlag, Berlin and New York, 1976.

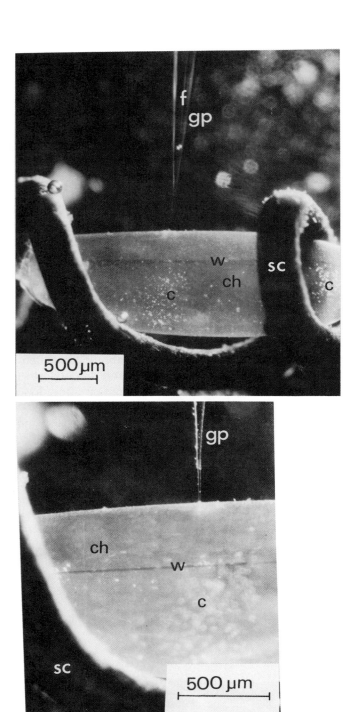

The other method of making the wire electrode was developed in 1986.[56] Pt/Ir wire of 50 μm is used. (Thinner wire, such as 25 μm, is not capable of penetrating the wall of most cells.) As before, the wire is soldered to a thicker wire and annealed. While it glows red hot it is stretched until a small portion (1 to 2 cm) breaks off. Under such conditions a sharp point forms at the end. The wire is then passed through a broken micropipette, and the glass is forged on in an electrode puller (David Kopf Model 720) (see Fig. 2C). The advantages of this wire electrode are (1) only one micromanipulator is necessary for insertion, (2) no etching is required, and (3) the electrode lifetime is long. (One slight disadvantage is the fixed length of the wire, so that cells of particular dimensions have to be selected.) The wire electrode produced by this second method seems gross compared to that shown in Fig. 2B. However, the cells seem to survive the insertion of the thicker wire better than the two-stage penetration by the hollow micropipette followed by the thinner wire, which is usually accompanied by gushing of the cell contents into the micropipette.

Current Clamp. The action potential is effectively held (clamped) at total membrane current = 0. To clamp the current to other levels a resistance much greater than that of the membrane must be placed in series with the current source (see Fig. 6). A range of integrated circuits can be used to obtain a programmable constant current source, precluding the necessity for high voltages. In general, the current clamp is not widely used, as the results are difficult to interpret, some exceptions being studies of K^+ channels in *Hydrodictyon*[10] and *Chara*.[57,58] The current clamp is also at serious disadvantage when the membrane current becomes a multivalued function of the membrane p.d., as can happen in regions of negative conductance.

Conductance Measurement. The main purpose of passing a current across the membrane is finding the resistance R or conductance $G (= 1/R)$. The knowledge of membrane conductance is very valuable, as upon changes of resting p.d. the G value can distinguish whether these are due to inhibition of electrogenic pumps (decrease in conductance) or opening of channels (increase in conductance). The resting conductance is usually measured by the perturbation method: injecting a small sinusoidal wave-

[56] M. J. Beilby and V. A. Shepherd, *Protoplasma* **148**, 150 (1989).
[57] F. Homble and A. Jennard, *J. Exp. Bot.* **35**, 1309 (1984).
[58] F. Homble, *J. Exp. Bot.* **36**, 1603 (1985).

FIG. 5. Young leaf cell of *Chara* (ch) in the Ag/AgCl coil (sc) with longitudinal current-injecting electrode (w) and p.d. measuring glass micropipette (gp) inserted. Note the filling filament (f). The reference micropipette is not shown. Although the cell is very young, some $CaCO_3$ crystals (c) can be seen, indicating the possible presence of banding. The bottom micrograph shows a greater magnification.

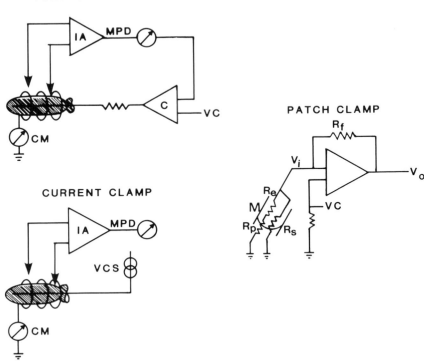

FIG. 6. Clamp circuits. In voltage clamping the membrane p.d. (MPD) is measured by an instrumentation amplifier (IA), and its output is fed into one side of a comparator operational amplifier (C) and compared to voltage commands (VC). If the two are different, current of appropriate direction is pumped into the cell via the current electrode until the membrane p.d. reaches the desired level. The current is measured (CM) after it leaves the cell on the way to ground. The current clamp is rather simpler. The current is provided by a constant voltage source (VCS), which is insensitive to changes in membrane resistance. Both the current and membrane p.d. are measured as above. The patch clamp is used to clamp a small patch of membrane "gigasealed" to the micropipette (M). The operational amplifier is used in the current-to-voltage configuration. The input resistance is a combination of the electrode resistance (R_e), the patch resistance (R_p), and the seal resistance (R_s). The current flowing through the patch is determined as the ratio of the output p.d. (V_o) to the feedback resistor (R_f).

form and observing the response of the p.d.[12] The ratio of the amplitudes ($\Delta I/\Delta V$) gives the conductance according to Ohm's law.

There are some complexities to be considered. If the current is injected into a large cylindrical cell through a microelectrode, the cable properties may under some conditions make the effective surface area considerably smaller than the total surface area of the cell,[59-62] that is, the measured and the area-specific conductance are no longer linearly related, the area being the constant of proportionality. While various mathematical corrections can be used (the latest being a very thorough series by Smith,[62-65] one of the practical remedies for this is space clamping. A possible problem with this method is that the wire can be etched in short lengths only (up to ~ 1 cm) and very young leaf cells are the only ones suitable for such treatment. In such cells the node constitutes a large percentage of the total area and might provide large conductances owing to plasmodesmata.[13] Possible remedies to this are discussed in the section on controlling the cell compartments.

Membrane Property of Reactance. Proper measurement of the conductance must take the time dependence of membrane transport into consideration. This dimension becomes apparent if the current is changed stepwise. The response of the p.d. can be an exponential growth to the final new level or an overshoot. The first type of behavior is phenomenologically identical to a response of a resistor and capacitor in parallel; the second can be approximated by an inductance and resistance network. In the cell these are brought about by rearrangement of ionic profiles at membrane–cell wall interfaces, opening/closing of channels, electroosmosis, or changes in the electrogenic pumping rate. Unstirred layers can also contribute.

The presence of these elements necessitates measuring the phase difference of the two sinusoids, as well as their amplitude. The most elegant mathematical approach is to attribute to the membrane the property of complex impedance, the conductance being the real part, the reactance (capacitance or inductance) the imaginary part. Attempts to estimate the reactance elements in giant algae have been made since 1938[66] using ac bridges, but the results only became readily accessible and reliable with the advent of digital computers. In plants, such a system was pioneered by Bell *et al.*[67] The current injected into the longitudinal wire electrode is synthesized by a PDP 11/20 computer, and the response of the membrane p.d. is digitized and stored in the memory. The two sine waveforms are then

[59] N. A. Walker, *Aust. J. Biol. Sci.* **13**, 468 (1960).
[60] E. J. Williams and J. Bradley, *Biochim. Biophys. Acta* **150**, 626 (1968).
[61] J. Hogg, E. J. Williams, and R. J. Johnston, *J. Theor. Biol.* **24**, 317 (1969).
[62] J. R. Smith, *Aust. J. Plant Physiol.* **10**, 329 (1983).
[63] J. R. Smith, *Aust. J. Plant Physiol.* **12**, 403 (1985).
[64] J. R. Smith, *Aust. J. Plant Physiol.* **12**, 413 (1985).
[65] J. R. Smith, *Aust. J. Plant Physiol.* **12**, 423 (1985).
[66] K. S. Cole and H. J. Curtis, *J. Gen. Physiol.* **22**, 37 (1938).
[67] D. J. Bell, H. G. L. Coster, and J. R. Smith, *J. Phys. E* **8**, 66 (1975).

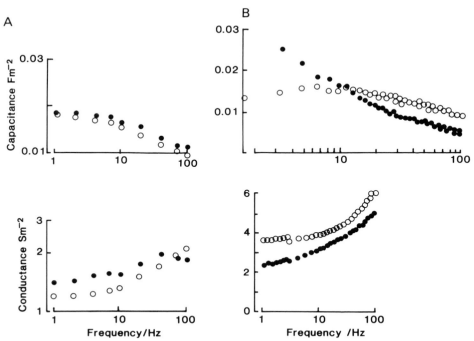

FIG. 7. Dispersion of the capacitance and conductance with frequency in *Chara*. (A) In this set of data (from Coster and Smith[20]), the dispersion of conductance and capacitance of plasmalemma and tonoplast in series (O) is compared to dispersion of the plasmalemma only (●). The results were obtained by injection of a small sinusoidal signal of varying frequency. The pH_o was ~5.5, and the cells were probably in the pump state. Under these conditions the combination of the plasmalemma and tonoplast in series is dominated by the electrical characteristics of the plasmalemma. (B) Using the data of Ross *et al.*,[68] this time the dispersion of the conductance and capacitance of both membranes in series are shown in the pump state, pH_o 7.5 (●) and the proton-permeable state (O). The results were obtained by Fourier analysis of the white noise signal injected into the cell. The data reflect the high conductance of the proton-permeable state.

compared, and the phase difference is calculated. This can be repeated at frequencies from 0.1 to 1000 Hz. At very high frequencies the limit is the speed of the analog-to-digital conversion. The accurate membrane conductance is thus obtained with reactance added into the bargain. The change in frequency revealed that both capacitance and reactance disperse with frequency: capacitance decreases with increasing frequency, apart from some cells which swing inductive at low frequencies below 1 Hz; conductance increases with frequency (see Fig. 7A).

The approach of Ross et al.[68,69] introduces further technological escalation, where the dispersion is obtained by recording the p.d. response to white noise (all frequencies included). The Fourier analysis can isolate the dispersion rather faster than the previous method. Their results are similar to those of Smith and Coster (see Fig. 7B), but the reactance always swings negative below 2 Hz. Ross et al. consider the reasons for the rise of the negative capacitance at low frequencies in greater detail and suggest that it may be caused by unstirred layers, when the conductance of the membrane increases. Some support to this idea is lent by the finding of Beilby and Beilby,[70] that reactance becomes negative whenever the conductance rises, such as at the time of action potential or punch through.

The reactance data no doubt contain valuable information about the membrane, but at present this is not easily accessible. The dispersion can be modeled by addition of several elements consisting of parallel resistance and capacitance (while each element is frequency independent, the combination, if their values are sufficiently different, will show so-called Maxwell–Wagner dispersion). The lipid bilayers, which are much simpler in structure, can indeed yield much information, when treated in this way.[71]

The measurement of the area-specific capacitance could throw some light on the variation in the area of the plasmalemma or tonoplast. At present the area is usually approximated as a cylinder (or cylinder and a cone in very young leaf cells). Electron micrographs indicate, however, that there are convolutions known as charasomes. These are found in cells of high pH_o and might considerably increase the membrane surface area.[72] The experiment is waiting to be performed.

Limits of Current Injection. How large a current can be injected into the cells without damage? Such limits have been explored in dielectric breakdown studies, where current pulses of increasing amplitude (but short duration of 0.5–1.0 msec) are injected via the longitudinal wire electrode.[73,74] The membrane p.d. in *Valonia utricularis*[73] grows more positive (or negative, as the effect shows symmetry about the zero p.d.) in response to the rising current until a critical p.d. (of ~700 mV at 10°) is reached, where no further change in p.d. occurs, despite an increase in current. The critical p.d. is strongly temperature dependent. This phenomenon is called

[68] S. M. Ross, J. M. Ferrier, and J. Dainty, *J. Membr. Biol.* **85**, 233 (1985).

[69] J. M. Ferrier, J. Dainty, and S. M. Ross, *J. Membr. Biol.* **85**, 245 (1985).

[70] M. J. Beilby and B. N. Beilby, *J. Membr. Biol.* **74**, 229 (1983).

[71] R. G. Ashcroft, K. R. Thulborn, J. R. Smith, H. G. L. Coster, and W. H. Sawyer, *Biochim. Biophys. Acta* **604**, 299 (1980).

[72] W. J. Lucas, D. W. Keifer, and T. C. Pesacreta, *Protoplasma* **130**, 5 (1986).

[73] H. G. L. Coster and U. Zimmermann, *J. Membr. Biol.* **22**, 73 (1975).

the dielectric breakdown and is thought to arise by a compression of the membrane owing to stresses set up by the electric field (electrostriction). Coster and Smith[74] speculate that the membrane proteins are compressed rather than the lipid matrix. Resultant hydrophobic interactions may lead to rotation of the protein molecules, thus opening new high-conductance channels. These reseal again once the current pulse ceases.

The shortcoming of current injection is that the membrane p.d. can be controlled only to a limited extent. The p.d. changes as membrane resistance varies. As most transport systems in the membrane exhibit some p.d. dependence, current clamping might activate/inactivate a range of different systems, and the results are then difficult to interpret. (In the case of impedance measurement the injected waveforms are kept within an amplitude of ~ 10 mV, to prevent distortion to the sinewave or departure from linearity, and so only the resting p.d. or the action potential can be measured.)

Voltage Clamp. To command total control of the membrane p.d., the voltage clamp is employed. The voltage clamp method was introduced to animal biophysics by Marmont.[75] Its application to the excitation in giant axons was the basis for the Hodgkin–Huxley equations, describing the dynamics of the Na^+ and K^+ channels.[76] In voltage-clamp experiments, the membrane p.d. is fed into one side of a comparator operational amplifier while the other input receives a command voltage which can be set to a single level or any complex function. The output of the comparator is connected to a current-injecting electrode (see Fig. 6). As long as the membrane p.d. is different from the command, the comparator provides current, I, in the appropriate direction for the IR drop to make this difference smaller. In a good clamp circuit the command and the cell p.d. become equal in ~ 1 msec.

In plants this technique was adopted independently by Findlay[77] and Kishimoto.[78] As plant cell conductances are about an order of magnitude smaller than those in animal tissues, initially the clamp was not very efficient. However, technology has brought improvements. If the wire electrode is used for current injection, the resistance is only between 300 and 500 kΩ, and a range of integrated circuits is now available for efficient clamping. We used the ubiquitous 741, which proved to be stable and

[74] H. G. L. Coster and J. R. Smith, *in* "Plant Membrane Transport: Current Conceptual Issues" (R. M. Spanswick, W. J. Lucas, and J. Dainty, eds.), p. 607. Elsevier/North-Holland Biomedical Press, Amsterdam, 1980.

[75] G. Marmont, *J. Cell. Comp. Physiol.* **34**, 351 (1949).

[76] A. L. Hodgkin and A. F. Huxley, *J. Physiol. (London)* **117**, 500 (1952).

[77] G. P. Findlay, *Nature (London)* **191**, 812 (1961).

[78] U. Kishimoto, *Biol. Bull. (Woods Hole, Mass.)* **121**, 270 (1960).

sufficiently fast for plant transport phenomena.[70] The clamp current is measured on its way to ground. This can be done by introducing a dropping resistor. Such an arrangement, however, causes the cell bath to raise to some p.d. above ground when current is flowing through it. Such changes may introduce oscillations in the cell p.d. measurement, if the p.d.-measuring amplifier does not have a good common-mode rejection ratio. A more sophisticated method is to use a operational amplifier to provide a virtual ground and to measure the current drop. As with the current clamp, computerization also revolutionized voltage clamping. The command p.d. can be synthesized digitally, which provides any complicated waveform.[70]

The voltage clamp has been put to many various uses. One of the simplest but most effective techniques is the possibility of isolating effects from other membrane events. For instance, cell exposure to amines or to Cl^- (especially after Cl^- starvation[79]) causes the membrane p.d. to depolarize.[80-82] If the membrane is clamped to the preexposure resting level, the current subsequently measured is due only to the transport phenomenon under study, rather than to others activated by change in membrane p.d.[83] Coupled with radioactive tracer analysis (in the same cell), this is a powerful tool in determining stoichiometries.[84]

As in the case of axons, transient currents flowing on clamping the membrane p.d. at a level more depolarized than the excitation threshold was examined in *Chara* by Findlay and Hope[16] and later by Beilby and Coster.[22,40,85] It is possible to adapt the Hodgkin–Huxley equations[76] to excitation in *Chara*.[85] The transients in *Nitella axillaris*[86,87] and *Nitellopsis*[88] were also analyzed. Some examples of excitation currents are shown in Fig. 8. In *Chara* the short-circuit currents are strikingly different across plasmalemma and both membranes in series and provide a convenient check on the electrode position.[29]

I/V Analysis. As the excitation transients die down (in ~5 sec at 20°), the current settles to a steady value. This steady-state current can thus be recorded at a range of membrane p.d.'s.[88,90] This time-consuming method

[79] D. Sanders, *J. Membr. Biol.* **52**, 51 (1980).

[80] D. G. Spear, J. K. Barr, and C. E. Barr, *J. Gen. Physiol.* **54**, 397 (1969).

[81] F. A. Smith, *New Phytol.* **69**, 903 (1970).

[82] F. A. Smith, N. A. Walker, and J. A. Raven, *Colloq. Int. C.N.R.S.* **258**, 233 (1977).

[83] M. J. Beilby and N. A. Walker, *J. Exp. Bot.* **32**, 43 (1981).

[84] N. A. Walker, F. A. Smith, and M. J. Beilby, *J. Membr. Biol.* **49**, 283 (1979).

[85] M. J. Beilby and H. G. L. Coster, *Aust. J. Plant Physiol.* **6**, 337 (1979).

[86] U. Kishimoto, *Jpn. J. Physiol.* **14**, 515 (1964).

[87] C. Hirono and T. Mitsui, *Plant Cell Physiol.* **24**, 289 (1983).

[88] V. Z. Lunevsky, O. M. Zherelova, I. Y. Vostrikov, and G. N. Berestovskij, *J. Membr. Biol.* **72**, 43 (1983).

[89] G. P. Findlay and A. B. Hope, *Aust. J. Biol. Sci.* **17**, 388 (1964).

[90] U. Kishimoto, *Plant Cell Physiol.* **9**, 539 (1968).

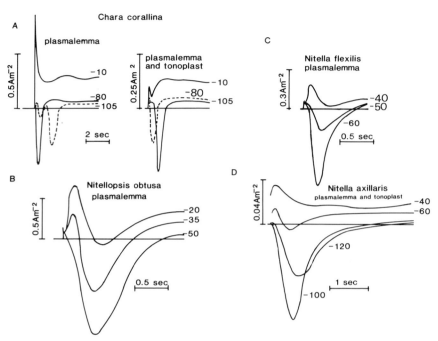

FIG. 8. Examples of transient excitation currents elicited on clamping the membrane p.d. to a level more depolarized than the threshold p.d. (A) Comparison between excitation currents across plasmalemma and plasmalemma and tonoplast in series in *Chara corallina* [M. J. Beilby and Coster, *Aust. J. Plant Physiol.* **6**, 323 (1979)]. Note the double peak at −105 mV and the sharp positive spike at −10 mV, which are found when plasmalemma only is clamped. The plasmalemma of *Nitellopsis obtusa* (B) and *Nitella flexilis* (C) both show prompt positive currents close to 0 p.d. (From Lunevsky et al.[88].) The vacuolar currents in *Nitella axillaris* (D) are rather smaller in amplitude compared to other giant algae [data from U. Kishimoto, *Plant Cell Physiol.* **7**, 559 (1966)].

of obtaining the steady-state current–voltage characteristics was later replaced by clamping the membrane to a ramp command p.d.[91] The problem with this approach is that the electrical characteristics show a hysteresis, probably because of rearrangement of ionic profiles at the cell wall–membrane interfaces due to prolonged flow of large currents. The idea of the bipolar staircase was developed by Slayman and colleagues.[92] In this method the command p.d. is stepped from the resting level alternately in the depolarizing and hyperpolarizing direction (see Fig. 9). The width of

---

[91] U. Kishimoto, *Adv. Biophys.* **3**, 199 (1972).

[92] D. Gradmann, U.-P. Hansen, W. S. Long, C. L. Slayman, and J. Warncke, *J. Membr. Biol.* **39**, 333 (1978).

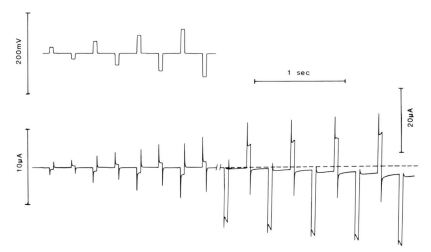

FIG. 9. Making of the I/V curve. The bipolar staircase clamp command is depicted in the upper curve. The lower curve shows the *Chara* plasmalemma current, which is at first well behaved: the current returns to 0 after each pulse. At the time of each pulse the current becomes steady after the initial spike (the current value for each p.d. step is then taken). Under some conditions, such as initiation of an action potential, these criteria no longer apply and the I/V curve loses reproducibility (see right-hand side of current curve).

the staircase pulses varies from tens to hundreds of milliseconds, with intervening times at resting level of hundreds of milliseconds. This method certainly yields more reproducible data; however, as many transport phenomena are involved in making up the whole profile, the I/V curve is inherently shifty, and usually profiles obtained in short time spans are compared. Caution must be exercised in the interpretation of results, and the data need to be used in conjunction with time-dependence studies of the currents at various clamp levels. For more details, see Beilby.[93]

Space clamping again turns out to be a very useful technique, as distortion arising from to the cable properties of the cell (if the cell is point clamped) tends to linearize the I/V curves.[13] This is particularly relevant to the K^+ state, where the conductance is strongly p.d. dependent.[11]

Limits of I/V Curve. I/V analysis is a valuable tool in characterizing the various states of the plasmalemma[9,24] (also see Fig. 10). In the pump state the I/V profile can be fitted by an enzyme kinetic model.[29,94,95] In the K^+

[93] M. J. Beilby, *J. Exp. Bot.* (in press) (1989).
[94] U.-P. Hansen, D. Gradmann, D. Sanders, and C. L. Slayman, *J. Membr. Biol.* **63**, 165 (1981).
[95] Y. Takeuchi, U. Kishimoto, T. Ohkawa, and N. Kami-ike, *J. Membr. Biol.* **86**, 17 (1985).

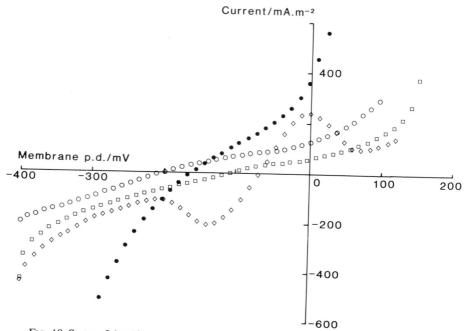

FIG. 10. States of the *Chara* plasmalemma as seen by the I/V technique. o, Pump state; •, proton-permeable state; ◇, K^+ state; □, leak state. All results are from one cell.

state the I/V profile reveals the unusual p.d. dependence of the K^+ channels, their closure with tetraethylammonium (TEA), and low $[K^+]_o$.[11,96] In the high-pH proton-permeable state the I/V profile is close to linear, indicating passive conductance.[24]

The I/V analysis becomes useful only when it is recorded over a large range of membrane p.d. This became obvious from the recent analysis of I/V relations based on kinetic modeling by Blatt.[97] What constitutes the limits of an I/V curve? At negative p.d.'s the currents become large and eventually fatal to the cell. This effect used to be called punch through and was explained by the double fixed-charge membrane theory, which models the membrane as two regions of opposite fixed charge.[98] The current was found to be Cl^- outflow.[99] It now seems certain, however, that this phe-

[96] M. J. Beilby, *J. Membr. Biol.* **89**, 241 (1986).
[97] M. R. Blatt, *J. Membr. Biol.* **92**, 91 (1986).
[98] H. G. L. Coster, *Biophys. J.* **5**, 669 (1965).
[99] H. G. L. Coster and A. B. Hope, *Aust. J. Biol. Sci.* **21**, 243 (1968).

nomenon is due to hyperpolarization-activated Cl⁻ channels,[100,101] and it is now referred to as inward rectifier.

Going in the positive direction excitation is encountered which makes the I/V curve very difficult to interpret (see Fig. 9). The excitation can be blocked (at least in *Chara*) by treating the cells with 0.1 mM LaCl$_3$ for several hours.[29,93] The excitation remains inhibited even on prolonged return into normal APW. This treatment extends the I/V analysis for up to 300 mV and reveals interesting features such as the K$^+$ channel closure at positive p.d.'s.[96] Eventually, the currents once again become very large (positive). This rectifying feature was thought to be carried by K$^+$ and Na$^-$ currents.[102] However, it is not blocked by TEA (which blocks the K$^+$ channels[9,10,96,100]), and changing [Na$^+$]$_o$ by an order of magnitude does not change the rectification feature.[96] The rectification can be diminished by high [Ca^{2+}]$_o$ and inhibitors.[96] In most studies +100 mV is the limiting positive p.d. under normal circumstances.

Voltage Clamp and Impedance Measurement. Impedance measurements can also be performed using the voltage clamp. This method has the advantage that the membrane p.d. can be controlled at the same time. Measurements at 5 Hz indicate increases in conductance at the depolarized and hyperpolarized extremes of the p.d. range, with a small conductance maximum near the resting p.d. (as shown by the I/V analysis). In the steady state the capacitance stays unchanged throughout the p.d. range except for the punch through where it turns inductive.[70] Possible reasons are electroosmosis or unstirred layer interference.[69] The conductance at the time of excitation is close to that obtained by simpler methods,[16,70] but the capacitance undergoes a decrease followed by an increase. The decrease may be due to reactive elements associated with the activation and inactivation processes, as expressed by mathematical analysis linearizing the Hodgkin–Huxley equations.[103] The capacitance increase may be due to transport number effects,[104] which should disappear at higher frequencies. This was, indeed, observed.[105] Dispersion measurements over a range of membrane p.d.'s have yet to be performed.

Banding. As if electrophysiology were not complex enough, some freshwater algae display the property of banding.[80,106] Passing a small pH elec-

[100] S. D. Tyerman, G. P. Findlay, and G. J. Paterson, *J. Membr. Biol.* **89**, 89 (1986).

[101] H. A. Coleman, *J. Membr. Biol.* **93**, 55 (1986).

[102] D. E. Goldman, *J. Gen. Physiol.* **27**, 37 (1943).

[103] W. K. Chandler, R. FitzHugh, and K. C. Cole, *Biophys. J.* **2**, 105 (1962).

[104] P. H. Barry and A. B. Hope, *Biophys. J.* **9**, 700 (1969).

[105] M. J. Beilby and B. N. Beilby, *in* "Membrane Transport in Plants" (W. J. Cram, K. Janacek, R. Rybova, and K. Sigler, eds.), p. 203. Akademia-Verlag, Prague, 1984.

[106] W. J. Lucas and F. A. Smith, *J. Exp. Bot.* **24**, 1 (1974).

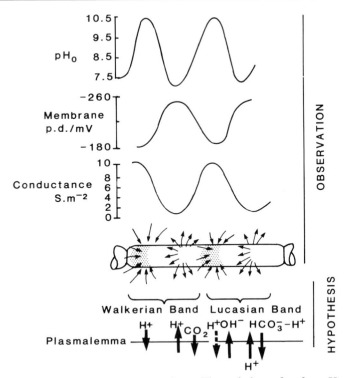

FIG. 11. Banding: observation and hypotheses. The variations of surface pH_o, p.d., and conductance when banding occurs are graphed. The arrows show the entry and departure of the circulating currents at the cell surface. $CaCO_3$ crystals in the alkaline zones are shown as dots. Two different schemes for cell carbon assimilation are displayed. See text for details.

trode along the surface reveals variation in pH_o of up to 3 pH units[106,107] (see Fig. 11). This effect persists even in strongly buffered APW between pH 5 and 10.[108] The pH banding is maintained by currents circulating between the zones. These can be calculated from the resulting p.d. profiles along the cell surface[109] or measured directly by the vibrating probe.[110] The type employed by Lucas and Nuccitelli[110] consists of a small sphere coated with platinum black and vibrated over several microns. This technique might interfere by stirring the solution near the cell surface. The currents

107 J. R. Smith and N. A. Walker, *J. Membr. Biol.* **83**, 193 (1985).
108 W. J. Lucas, J. M. Ferrier, and J. Dainty, *J. Membr. Biol.* **32**, 49 (1977).
109 N. A. Walker and F. A. Smith, *J. Exp. Bot.* **28**, 1190 (1977).
110 W. J. Lucas and R. Nuccitelli, *Planta* **150**, 120 (1980).

indicated are rather high (up to 800 mA m^{-2}). The conductance also varies in the acid and alkaline bands, giving low values of ~ 1.9 siemens (S) m^{-2} in the former, higher values of up to 8 S m^{-2} in the latter.[111] Similar values are measured by a rather different technique by Chilcott et al.[112,113] (see next paragraph). The conductance in the alkaline bands matches that observed when the whole cell is bathed in media with a pH greater than 10,[10,23] suggesting that under those circumstances, the whole surface becomes an alkaline band.

Methods to Investigate Banding. Various setups have been devised to investigate the banding phenomena. An air gap between two agar blocks embedding the alkaline and acid bands provided a means for measuring the short-circuit current between the zones.[109] The p.d. and pH profiles were measured by moving an electrode (an ordinary glass micropipette filled with KCl and agar or a pH-sensitive electrode) along the cell surface with the aid of micromanipulators in some cases driven by stepping motors and under computer control.[109,111] In other experiments, the bands were isolated by means of greased partitions in the holder (similar to some of the designs of Fig. 1), and the current was passed through external electrodes in different partitions.[111] Ogata and co-workers[112,113] developed another method for measurement with the cell mounted vertically in water-saturated air with a wire loop that moves along the cell length (contact made through water film), thus enabling measurements of electrical properties of different parts of the cell (see Fig. 1).

Why Do the Cells Band? The banding system is a strategy of providing the cell with carbon at high pH$_o$ (it is necessary for photosynthesis, and, indeed, banding is only observed in the light). However, the pH dependence of the very fast reaction rates between different species in the $CO_2/H_2CO_3/HCO_3^-/CO_3^{2-}$ system and the lack of knowledge of the exact pH$_o$ values close to the plasmalemma make the exact mechanism difficult to discern.[114] Lucas[115] suggested that in the acid zones the bicarbonate is imported into the cell, while OH$^-$ is exported at the alkaline regions. The hypothesis of Walker et al.[116] states that the acid regions are created by the proton pump (H$^+$ outflow) and that the protons pour back in the alkaline regions. In the acidified regions the bicarbonate ion dissociates and result-

[111] J. R. Smith and N. A. Walker, *J. Membr. Biol.* **73**, 193 (1983).

[112] T. C. Chilcott, H. G. L. Coster, K. Ogata, and J. R. Smith, *Aust. J. Plant Physiol.* **10**, 353 (1983).

[113] K. Ogata, T. C. Chilcott, and H. G. L. Coster, *Aust. J. Plant Physiol.* **10**, 339 (1983).

[114] N. A. Walker, *in* "Membranes and Transport" (A. N. Martonosi, ed.), p. 645. Plenum, New York, 1982.

[115] W. J. Lucas, *J. Exp. Bot.* **27**, 19 (1976).

[116] N. A. Walker, F. A. Smith, and I. Cathers, *J. Membr. Biol.* **57**, 51 (1980).

ant CO_2 crosses the membrane. Later Lucas discovered that the circulating currents can be observed *in absence of bicarbonate*.[117,118] Under some conditions, the membrane shows large hyperpolarizations up to -350 mV. (These hyperpolarizations can be observed only if the cell is not perturbed by electrode insertion.) Lucas concluded that the bicarbonate is cotransported with protons (this being electrically neutral) and that the proton pump is responsible for the acid bands (as in the Walker model) and also for the transient hyperpolarization. Photosynthesis-produced CO_2 and OH^- flow out, the latter constituting the alkaline zone. Details of this model can be found in [28] by Lucas and Sanders in this volume.

Impact of Banding on Methods. What are the implications of the inhomogeneity of the many types of giant algal cells for measurements? Area-specific conductances measured at a point with unknown band properties are likely to result in serious errors. In space-clamped cells, the average value over the whole surface is obtained. Thus space clamp experiments should yield a higher conductance than point clamps (survey of literature seems to support this), although the matter is complicated by comparing measurements in different species.[111,119] Further, there are some differences in observations of incidence of banding between the Walker and Lucas laboratories. While the former found that nothing except prolonged darkness or pH below 5.0 will abolish the banding,[107] the latter claim that even slight mechanical manipulations diminish the currents.[110]

Our experiments seem to indicate that the banding may not be as omnipresent as thought by Walker and Smith.[107,111] For instance, if the high conductive alkaline bands were present at pH_o above 5, the membrane conductance measured by space clamp should drop below this pH_o value. Instead, it remains at the same level, or grows larger.[29] The I/V characteristics of the cells at $pH_o > 10$ (i.e., in an alkaline band) are very different from those at $pH_o < 10$[24] (see Fig. 10). If the alkaline bands contributed to the total I/V characteristics, the I/V profile would be expected to change drastically below pH_o 5. Again this is not seen; only a continuous shift of the sigmoid curve to more depolarized p.d. occurs as pH_o decreases from 8 to 4.5.[29] (This change is thought to be due to the pump kinetics, but may, of course, be due to changes in banding.) My experiments have been done in light, but in the absence of any added bicarbonate and with freshly made, buffered APW. The length of the cells is between 3 and 5 mm and only rarely are there any calcifications (these

[117] W. J. Lucas, *Planta* **156**, 181 (1982).
[118] W. J. Lucas, *in* "Membrane Transport in Plants" (W. J. Cram, K. Janacek, R. Rybova, and K. Sigler, eds.), p. 459. Akademia-Verlag, Prague, 1984.
[119] J. R. Smith, *J. Membr. Biol.* **73**, 185 (1983).

form in the alkaline zones). Such cells are thought to flicker from acid to alkaline behavior with concomitant changes in conductance,[111,112] but after initial large conductances (perhaps due to some damage) the cells settle to a fairly steady conductance and I/V profile.

Membrane Noise. The Ogata "baby bottle" method[120] allows the measurement of membrane noise in the acid and alkaline zones. The loop (Fig. 1) is positioned over a zone, and the p.d. is continually sampled for long periods (1 hr). The results are Fourier analyzed, and the power spectra and phase correlation are obtained. The power spectra are similar in both bands, but the phase shows a strong correlation above 0.1 Hz in the acid zone.[121] This might be the feedback mechanism of the individual proton ATPases. Injection of current of frequency below 0.1 gives oscillatory or negative capacitances, suggesting interference coupling with some internal electrogenic transport mechanism.

The vibrating probe experiments, on the other hand, demonstrated a low frequency noise (period of about 1 min) associated with the alkaline zones.[110] Lucas and Nuccitelli postulated that these are associated with the synchronization of the OH^- and HCO_3^- transport.

Patch Clamp. As the study of "simple" giant algal cells grows steadily more complex, a new technique of patch clamping promises to solve some of our problems. The idea of isolating small patches of membrane for study is not new,[122] but the technique gathered momentum only when the gigaseal was discovered and low leakage current and noise amplifiers became available. The gigaseal is a high resistance ($> 10^9$ ohm) seal between the inside and outside of the pipette, as the membrane adheres to the tip on applying negative pressure by mouth suction. The pipettes are pulled on electrode puller. A hydrophobic coating (such as Sylgard 184, Dow Corning) is applied to the outside to diminish pipette capacitance and to increase the shunt resistance. The tips are then fire polished with fine Pt wire on a microforge. The tips are typically $1-2$ μm in diameter. There are several configurations, after the gigaseal has been formed. The simplest is leaving the cell attached. On withdrawing the pipette, the patch becomes cut out, forming an inside-out patch. Another choice is to blow out the patch without losing the gigaseal, thus getting into the cell interior (important for small cells damaged by ordinary electrode insertion). Withdrawing the electrode often leads to formation of an outside-out membrane across the tip.

[120] K. Ogata, *Plant Cell Physiol.* **24**, 695 (1983).
[121] H. G. L. Coster, T. C. Chilcott, and K. Ogata, *in* "Membrane Transport in Plants" (W. J. Cram, K. Janacek, R. Rybova, and K. Sigler, eds.), p. 189. Akademia-Verlag, Prague, 1984.
[122] E. Neher and B. Sakmann, *Nature (London)* **260**, 779 (1976).

The simplified patch clamp circuit is shown in Fig. 6. The amplifier supplies the current to keep the patch at predetermined p.d. The current flowing across the patch (and hopefully through one channel) is given by the values of the output p.d. and feedback resistor. Matters are complicated by the fact that the input resistance (consisting of electrode resistance, seal resistance, patch resistance) determines the amplifier gain, and if it is smaller than the feedback resistance (e.g., before seal formation), the gain is large and the output becomes noisy. Only after the gigaseal is formed does the signal becomes less noisy as the input resistance rises. For further details of the technique, see the reviews by Sakmann and Neher,[123] Auerbach and Sachs,[124] and Takeda et al.[125]

In animal tissues the ion channels are well characterized and amenable to pharmacological dissection. In plants the electrical characteristics are known in some detail only in giant algae. However, protoplasts, whose electrical properties are not well explored, were the first to be "patched."[126,127] The main problem with plant cells is the cell wall, which prevents access to the cell membrane and seal formation.[128] Even protoplasts are difficult to patch, and this could be due to cell wall regeneration.[129] In giant cells two approaches have been adopted: separating the plasma membrane from the wall of very young whorl cells by increasing the osmotic pressure of the medium or generating cytoplasmic drops. In the first case the cell wall can be cut with scissors, and the plasma membrane is then accessible to the patch pipette.[101] In the second case a membrane forms around the cytoplasmic droplets, which are obtained by cutting the cell open in a medium isotonic with the cell sap. This membrane is thought to be tonoplast derived.[130] The droplets provide a wealth of information about K^+ channels.[131-134] It is somewhat puzzling, however, that the properties of this supposed tonoplast K^+ channel are very similar to those obtained by whole-cell electrophysiology from the plasmalemma of the giant cells.[11,24,96,135] Thus there may be some doubt about the origin

[123] B. Sakmann and E. Neher, eds., "Single-Channel Recording." Plenum, New York, 1983.
[124] A. Auerbach and F. Sachs, Annu. Rev. Biophys. Bioeng. 13, 269 (1984).
[125] K. Takeda, A. C. Kurkdjian, and R. T. Kado, Protoplasma 127, 147 (1985).
[126] N. Moran, G. Ehrenstein, K. Iwasa, C. Bare, and C. Mischke, Science 226, 935 (1984).
[127] J. I. Schroeder, R. Hedrich, and J. M. Fernandez, Nature (London) 312, 361 (1984).
[128] E. A. C. MacRobbie, Nature (London) 313, 529 (1985).
[129] J. Burgess, Intl. Rev. Cytol. Suppl. 16, 55 (1983).
[130] K. Sakano and M. Tazawa, Protoplasma 131, 247 (1986).
[131] H. Luhring, Protoplasma 133, 19 (1986).
[132] D. R. Laver and N. A. Walker, J. Membr. Biol. 100, 31 (1987).
[133] D. R. Laver, K. A. Fairley, and N. A. Walker, J. Membr. Biol. 108, 153 (1989).
[134] S. D. Tyerman and G. P. Findlay, J. Exp. Bot. 40, 105 (1989).
[135] M. Tester, J. Membr. Biol. 103, 159 (1988).

of the membrane bounding the droplets, especially since the study with permeabilized cells did not show K^+ currents across the tonoplast.[136] Plasmalemma patch-clamp experiments, on the other hand, reveal channels predominantly permeable to Cl^-.[101,137]

The patch clamp technique opened new horizons in plant electrophysiology (for review of perspectives and results in higher plant tissues, see Hedrich *et al.*[138]). To assess the relevance of patch-clamp data, whole-cell electrophysiology is becoming even more important. Further, recent electrophysiological studies of higher plant cells[139,140] indicate that the giant algal cells show fairly typical electrophysiological behavior representative of many plant cells.

Controlling Cell Compartments

Even after we have measured the resting p.d. and the conductances of the plasmalemma and tonoplast, we do not have all the information necessary to understand the workings of the cell. The phenomena we are observing are electrochemical in nature, and we need to know the concentrations of the most abundant ions in each compartment. Measurements of the vacuolar concentrations are relatively simple, as the vacuole comprises 90–95% of the total cell volume and contamination by the viscous cytoplasm is not likely. If a mature internodal cell is cut, enough sap drains out to use conventional methods, such as flame photometry.[141,142] However, as it is the plasmalemma that dominates the electrical characteristics of the cells, the measurement of principal ions in the cytoplasm is more important. A sample from the cytoplasm is more difficult to obtain. Gentle centrifugation can be employed to move the cytoplasm to one end of the cell.[141–143] An estimate of the cytoplasmic volume can be provided by measuring the length of the cytoplasmic plug at the end of the centrifugation.[143,144] However, the results have become reliable only since the method of vacuolar perfusion described below has been employed. (The vacuole is perfused by ions other than the ones measured.)

The use of ion-sensitive electrodes also presents some difficulties be-

[136] M. Tester, M. J. Beilby, and T. Shimmen, *Plant Cell Physiol.* **28**, 1555 (1987).
[137] D. Laver, in preparation.
[138] R. Hedrich, J. I. Schroeder, and J. M. Fernandez, *Trends Biol. Sci.* **12**, 49 (1987).
[139] M. R. Blatt, *J. Membr. Biol.* **98**, 257 (1987).
[140] M. R. Blatt, *J. Membr. Biol.* **102**, 235 (1988).
[141] R. M. Spanswick and E. J. Williams, *J. Exp. Bot.* **15**, 193 (1964).
[142] E. A. C. MacRobbie, *J. Gen. Physiol.* **45**, 861 (1962).
[143] U. Kishimoto and M. Tazawa, *Plant Cell Physiol.* **6**, 507 (1965).
[144] M. Tazawa, U. Kishimoto, and M. Kikuyama, *Plant Cell Physiol.* **15**, 103 (1974).

cause reference as well as ion-sensitive electrodes need to be inserted into the thin cytoplasmic layer. Only Cl⁻ and K⁺ concentrations have been measured via this method.[145-147] Even when the problem of multiple electrodes is overcome (see section on single membrane samples), the turgor pressure makes the use of some types of ion-sensitive electrodes difficult.[147] While technology is providing easier procedures for manufacturing ion-sensitive electrodes, there is still a great impetus for researchers to control the ionic concentrations inside cells and to simplify the compartmentation.

Vacuolar Perfusion. Replacing the cell sap has been realized on the big, sturdy, spherical marine algae *Valonia*[148] and *Halicystis.*[149] In these cells, it is possible to insert two capillaries and slowly exchange the cell sap for artificial medium. In Characeae the perfusion was introduced by Tazawa in 1964.[150] A long internodal cell is placed on cell holder with three pools (see Fig. 12A). The two outside pools are filled with perfusion medium, and to ensure that the level is the same in both they are connected by rubber tubing. The middle pool is left empty, and the "creep" of liquid along the surfaces is prevented by application of grease between the pools. Most of the cell is left in air until enough water evaporates from the cell for loss of turgor to occur. (If the turgor is diminished by hypertonic solution, the cell plasmolyzes and the membrane becomes separated from the wall and damaged.) Once the cell loses turgor, the ends are amputated with sharp scissors. The rubber tubing is then pinched off, and a small pressure head is introduced by a slight tilt, an addition of more perfusion medium to one pool or removal of the perfusion medium from one pool by dipping in filter paper. The progress of the perfusion medium through the cell can be monitored by staining the cell with neutral red (which turns red in the acidic vacuole). As the artificial medium flows through the cell, the green color returns. When the perfusion is complete, the cell is ligated with silk or polyester thread. Two or three ligations are suggested, as a partially turgid cell (irrigated by APW about 100 mM hypotonic to the cell sap) is less liable to damage in the process of ligation. The inner segment is then cut out and can be used as an intact cell, and electrodes can be inserted.

The perfused cells show the best rate of survival if the perfusion medium is close to the ionic composition of the original sap,[144] and such cells continue to stream for up to 1 month. (The K⁺ concentration in the cytoplasm tends to be slightly higher than in the vacuole, but Na⁺ and Cl⁻

[145] L. Vorobiev, *Nature (London)* **216**, 1325 (1967).
[146] H. G. L. Coster, *Aust. J. Plant Physiol.* **19**, 545 (1966).
[147] M. J. Beilby and M. R. Blatt, *Plant Physiol.* **82**, 417 (1986).
[148] E. B. Damon, *J. Gen. Physiol.* **13**, 207 (1929).
[149] L. R. Blinks, *J. Gen. Physiol.* **13**, 223 (1929).
[150] M. Tazawa, *Plant Cell Physiol.* **5**, 33 (1964).

TABLE I
PERFUSION (PM) AND PERMEABILIZING (PerM) MEDIA[a]

Component	Vacuolar PM[b]		Cytoplasmic PM[c]	PerM[d]
	Chara	Nitella		
EGTA	—	—	5.0	0.1–5.0
MgCl$_2$	—	10.0	6.0	2.0
NaCl	30.0	40.0	—	2.0
KCl	110.0	92.0	—	—
CaCl$_2$	1.0	15.0	—	—
KOH	—	—	17.0–71.0	90.0
Buffer	—	—	5.0–30.0 (Tris, HEPES, or PIPES)	HEPES
pH	—	—	7.0	7.5
Sorbitol	70.0	—	Adjust to 330 mM	Adjust to 320 mM
Ficoll[e]	—	—	5% (w/v)	—
ATP	—	—	2.0–5.0	1.0
Alternatives[f]				
Hexokinase	—	—	1 mg/ml	—
CyDTA	—	—	5	—
Glucose	—	—	5	—

[a] Unless specified otherwise, concentrations are in mM. EGTA, ethylene glycol bis(β-aminoethyl ether)-N,N, N′,N′-tetraacetic acid; CyDTA, 1,2-cyclohexanediamine-N,N, N′,N′-tetraacetic acid ATP, adenosine 5-triphosphate.

[b] Vacuolar media of by Tazawa et al.[151]

[c] Cytoplasmic perfusion medium from Tazawa et al. [Curr. Top. Membr. Transp. 16, 49 (1982)].

[d] Permeabilization medium used in the Cambridge laboratory by T. Shimmen, M. Tester, and M. J. Beilby.[136]

[e] Ficoll was introduced by Lucas and Shimmen[166] for greater stability of membrane p.d.

[f] Hexokinase is used in conjunction with glucose to remove ATP from the cell (replaces ATP in the medium) and CyDTA removes Mg^{2+} from the cell (replaces Mg^{2+}).

are significantly lower in the cytoplasm than in the vacuole.) The cells tolerate some changes, but survive for only a few hours. For the composition of perfusion media, see Table I.

Open Vacuole Method. If the ligated perfused cell survives, the composition of the artificial medium will be modified by the fluxes across the tonoplast. To have total control of the vacuole composition the cells are left open, and the perfusion becomes continuous.[151] The added advantage

[151] M. Tazawa, M. Kikuyama, and S. Nakagawa, *Plant Cell Physiol.* 16, 611 (1975).

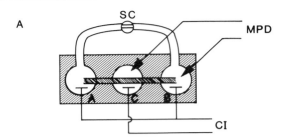

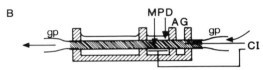

Fig. 12. Holders for perfusion. (A) Holder originally used for vacuolar perfusion by Tazawa.[150] The cell is shown with uneven hatching. Pools A and B are filled with perfusion medium, and pool C is left empty. The stopcock (SC) is open to ensure that there is no flow between pools A and B. When the cell loses turgor in air, the ends of the cell in pools A and B are amputated. The stopcock is then closed, a small pressure head is introduced (see text), and after the cell sap is replaced by perfusion medium the cell is ligated (the ligations are prepared while the cell is still intact, between pools A and C and B and C). After the first ligation pool C is filled with APW with ~ 100 mM sucrose and ligated again. The inner segment can then be cut off and used for electrode insertion. Alternatively, the cell can be used in the "open vacuole" configuration.[151] The p.d.-measuring electrodes (MPD) (plastic tubing filled with agar and concentrated KCl) are then simply added to pools A and C. Current is injected (CI) by silver wires placed in the different compartments. The same holder can be used for cytoplasmic perfusion. In this case the cells can also be ligated or left *in situ* for continuous perfusion. (B) Vessel used for continuous vacuolar perfusion by Strunk.[155] The most interesting features are the seals of the amputated cell ends onto the glass pipettes (gp) (see text), which allow the cell (irregular hatching) to remain turgid after perfusion. The flow of the perfusion medium is indicated by arrows. The membrane p.d. is measured by an electrode inserted in the central compartment. The current is injected via wire passed through one of the pipettes.

of this approach is that no electrodes are necessary for p.d. measurement and current injection (see Fig. 12). In this method much greater latitude is possible in the composition of the perfusion medium. (It seems, for instance, that low Ca^{2+} makes recovery from injury due to electrode insertion more unlikely.) Tazawa *et al.*[151] compared fragments of perfused cells that have been ligated and impaled with others measured by the open vacuole method. The latter showed slightly more negative resting p.d.'s and significantly higher resistances in *Nitella flexilis* but not in *Chara australis*.

It is possible that the low resistance may have been caused by the leakage through the ligations.

The open vacuole method can be used for exploring the properties of the tonoplast.[152] The change in p.d. across both membranes in response to changes in outside medium was assumed to be due to the plasmalemma processes, while the changes in the perfusion medium were thought to affect only the tonoplast. The tonoplast of *Nitella pulchella* is comparatively insensitive to changes in $[K^+]_v$, but tonoplast p.d. changes with $[Cl^-]_v$ in a direction opposite to what would be expected if the response was passive. Extremes of pH_v also seem to affect the tonoplast p.d. The action potential across the tonoplast was also studied using the open vacuole method. In *N. pulchella* the transient p.d. change reverses sign if the $[Cl^-]_v$ is greatly lowered, indicating that this ion is responsible for the excitation.[153] Interestingly, this is not observed in *Chara corallina.* Recent studies[154] suggest that the tonoplast action potential is due to Cl^- channels opening in response to increases in $[Ca^{2+}]_c$.

The open vacuole method described above has the disadvantage that the cell is totally turgorless. It is not known at present what effect turgor pressure has on ions transport across the membrane, but it is conceivable that stretching the membrane under turgor might alter its properties. To study the cells under turgor, a very ingenious continuous perfusion method was introduced by Strunk[155] (see Fig. 12B), who dried the ends of *Nitella* cells (after amputation of the nodes) onto glass pipettes, forming a watertight seal. Such cells can be continuously perfused under pressure. Unfortunately, the experiments seemed to have faded out after establishing the method. A similar method for continuous pressurized perfusion was introduced by Yurin and Plaks,[156] who inserted glass capillaries into both ends of *Nitella* cells.

Cytoplasmic Perfusion. Vacuolar perfusion gave information about the tonoplast, made measurements of the cytoplasmic concentrations more accurate, and became an important stepping stone to the tonoplastless system (see Fig. 13). Either rapid perfusion or inclusion of EGTA (strong chelator of Ca^{2+}) in the perfusion medium leads to the disintegration of the tonoplast.[157,158] The tonoplast is thus mechanically brittle, and Ca^{2+} on the

[152] M. Kikuyama and M. Tazawa, *J. Membr. Biol.* **30**, 225 (1976).
[153] M. Kikuyama and M. Tazawa, *J. Membr. Biol.* **29**, 95 (1976).
[154] T. Shimmen and S. Nishikawa, *J. Membr. Biol.* **101**, 133 (1988).
[155] T. H. Strunk, *Science* **169**, 84 (1970).
[156] V. M. Yurin and A. V. Plaks, *Sov. Plant Physiol. (Engl. Transl.)* **22**, 953 (1976).
[157] R. E. Williamson, *J. Cell Sci.* **17**, 655 (1975).
[158] M. Tazawa, M. Kikuyama, and T. Shimmen, *Cell Struct. Funct.* **1**, 165 (1976).

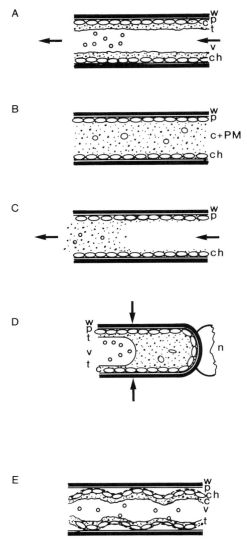

FIG. 13. Schematic diagrams of various systems simplifying cell compartmentation. The cell wall (w), cytoplasm (c), plasmalemma (p), tonoplast (t), vacuole (v), and chloroplasts (ch) are shown. (A) In the vacuolar perfusion, the sap is replaced by artificial medium. If the above medium contains EGTA and the cell is ligated, the tonoplast disintegrates, and the cell then contains mixture of perfusion medium (PM) and cytoplasm, as shown in B. If the cell is not ligated, the interior can be perfused again (C) to control the inside composition. (D) In the single membrane sample, the cytoplasm moves to one end of the cell after 30 min of gentle centrifugation. The node (n) can be either left on or removed by ligation. The single membrane sample is ligated at the point shown by arrows. (E) In the permeabilized cell, the Ca^{2+} from the outside is removed with EGTA. The plasmalemma disintegrates, and the flowing cytoplasm is washed away. The chloroplasts are disrupted by plasmolysis. The tonoplast and the vacuolar compartment are undisturbed.

vacuolar side is necessary for its stability. The tonoplastless cells can be either ligated[158] or continuously perfused.[159]

While the perfusion does not sound difficult, beginners may find that many of their cells are not viable. Meticulous mixing of solutions is required, and deft fingers are necessary for manipulating the cells.

Properties of the Tonoplastless System. Tonoplastless cells display the hyperpolarized p.d.'s of cells in the pump state only if ATP and several millimoles Mg^{2+} are included. Under such conditions the p.d. dependence on pH_o is similar to intact cells[160,161] (see Table I and Fig. 14), and H^+ efflux is observed.[162] The electrogenic pump at the plasmalemma is thus identified unequivocally as an ATPase pumping H^+ out of the cell. For details on the dependence of the resting p.d. of various freshwater algae on ATP concentrations, see the review by Tazawa and Shimmen.[163] The cells without ATP seem to enter a K^+ state similar to intact cells.[160,164] A depolarization by a current pulse transfers cells perfused with ATP to the low state (see Fig. 14). The excitation also requires ATP and Mg^{2+}. It takes a somewhat different form than that in intact cells. The action potentials are rectangular in shape, and the p.d. seems to move between two levels. The peak p.d. of the excitation is very close to the resting p.d. of cells without ATP.[165] Under some conditions (such as 2 mM K_2SO_4 in the medium), the p.d. stays at each level until removed to the other by a current pulse.[158] The transition from the high state to the low state can be induced by higher concentrations of K^+ and reversed by increasing concentrations of Ca^{2+}. The peak of the excitation response follows E_K above 1.0 mM K^+. Further, there is no refractory period, the peak of the action potential is independent of the Cl^- concentration in the perfusion medium (unlike intact cells) and the total replacement of Cl^- by a wide range of anions did not alter the excitation response.[165]

It is now thought that only Ca^{2+} channels open in perfused cells in response to depolarization.[51] The two stable states may thus be characterized as the closed and the opened states of the Ca^{2+} channel. The main reason for not seeing the Cl^- channel in tonoplastless cells is the use of EGTA in the perfusion medium (see Table I), which prevents the internal

[159] P. T. Smith and N. A. Walker, *J. Membr. Biol.* **60**, 223 (1981).
[160] T. Shimmen and M. Tazawa, *J. Membr. Biol.* **37**, 167 (1977).
[161] G. Kawamura, T. Shimmen, and M. Tazawa, *Planta* **149**, 213 (1980).
[162] T. Shimmen and M. Tazawa, *Plant Cell Physiol.* **21**, 1007 (1980).
[163] M. Tazawa and T. Shimmen, *Int. Rev. Cytol.* **109**, 259 (1987).
[164] P. T. Smith, *Aust. J. Plant Physiol.* **11**, 303 (1984).
[165] M. Tazawa and T. Shimmen, in "Plant Membrane Transport: Current Conceptual Issues" (R. M. Spanswick, W. J. Lucas, and J. Dainty, eds.), p. 349. Elsevier/North-Holland Biomedical Press, Amsterdam, 1980.

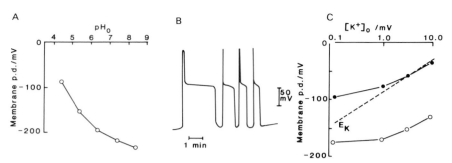

FIG. 14. Behavior of tonoplastless cells. (A) Resting p.d. as a function of pH_o. This pH_o dependence is observed only when the perfusion medium contains Mg^{2+} and ATP and the cells are illuminated.[161] (B) Typical action potential in perfused *Chara* cells. The initial spike is due to the stimulation pulse. The duration decreases if a train of action potentials is elicited. Compare the duration and shape to those of intact cells in Fig. 4. (From Shimmen *et al.*[3]) (C) Dependence of membrane p.d. of perfused *Chara* on $[K^+]_o$. The perfusion medium was with (o) and without ATP (●). When the cells are in the hyperpolarized state, the peak of the action potential is close to the p.d. level attained by cells perfused with medium lacking ATP.[160]

rise in Ca^{2+} [50] (which normally activates the Cl^- channels). Consequently, a new picture of excitation is emerging: the Ca^{2+} channels are activated by depolarization and spontaneously close in intact cells. (It is not clear why the inactivation is disturbed in the tonoplastless cells.) The transient opening of the Cl^- channels is conferred by the transient rise in Ca^{2+} concentration.

One disappointing finding is that the tonoplastless cells do not band. The disappearance of banding seems to be associated with some organelle in the flowing cytoplasm.[166]

The tonoplast-free cells can be further simplified by centrifuging at 7,000–20,000 *g*. The chloroplasts collect in the centrifugal end and can be separated by ligation. Chloroplasts from other plants can be substituted, and the cells recover their photosynthetic function.[163]

Permeabilization of the Plasmalemma. The plasmalemma permeabilization system has been devised to obtain better access to the cytoplasm and to study the tonoplast. The method has evolved somewhat,[167,168] but at present essentially depends on the removal of extracellular Ca^{2+} by EGTA. The medium is tailored to be close to the cytoplasm in ionic composition and osmolarity[167] and is chilled before cells are exposed to it. (For an example of a permeabilizing medium, see Table I.) The lack of calcium

[166] W. J. Lucas and T. Shimmen, *J. Membr. Biol.* **58**, 227 (1981).
[167] T. Shimmen and M. Tazawa, *Protoplasma* **113**, 127 (1982).
[168] T. Shimmen and M. Tazawa, *Protoplasma* **116**, 75 (1983).

seems to destabilize the plasmalemma but leaves the cytoplasmic gel in place. The tonoplast is accustomed to very low calcium levels and thus should also remain unmodified. As the plasmalemma disintegrates, cytoplasmic streaming stops with the washing away of ATP, but it can be restarted by supplying ATP to the outside medium. The chloroplast arrangement becomes disordered as the cell is plasmolyzed by slightly hypertonic medium.

Current voltage analysis shows that permeabilized cells of *Chara* develop two regions of negative conductance on addition of external (in this case cytoplasmic) Cl^-,[136] suggesting presence of p.d.-dependent Cl^- channels. Changes in K^+ concentration, on the other hand, have little effect on the current–voltage profile.

Cytoplasm-Enriched Fragments. The treatments described above are becoming increasingly drastic, and perhaps it is time to move in another direction. Cytoplasm-enriched fragments (also called single membrane samples) have been used for a long time to obtain samples of cytoplasm,[143] but only recently have their electrical characteristics been explored.[87,169] For this method long, young internodal cells are gently centrifuged at ~2 *g* for 30 min. The cytoplasmic plug is then clearly visible against a dark background and can be tied off by silk thread, after the cell has been allowed to lose turgor in air. These fragments live and stream for weeks. Small vacuoles are formed after several days. The fragments exhibit all the cell states,[56,147] but the pump state is weak and the excitation is often absent, although normal excitation has been observed in some fragments. This system is suitable for measurement with ion-sensitive electrodes, as it eliminates the problems of an inaccessible cytoplasmic layer. Several fragments can be seen in Fig. 15.

When the fragments were stained with neutral red and fluorescent dyes, many membrane-bound vesicles ranging in size from few to several hundred microns[56] were observed (thus the name single membrane samples was deemed inaccurate). The presence and origin of the vesicles are interesting in themselves and the system might provide some data on vacuole formation. Despite the fact that the structure of the fragments is more complex than originally expected, they still constitute a convenient system for the study of the plasmalemma.

Conclusion

The study of giant algae spans several disciplines, which makes it a more interesting but also a more demanding field of research. Knowledge

---

[169] C. Hirono and T. Mitsui, *in* "Nerve Membrane" (G. Matsumoto and K. Kotani, eds.), p. 135. Univ. of Tokyo Press, Tokyo, 1981.

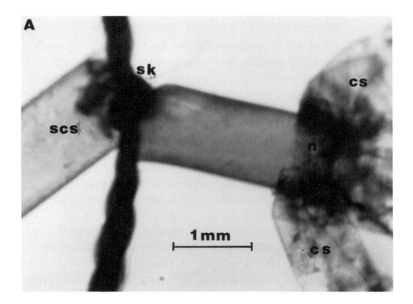

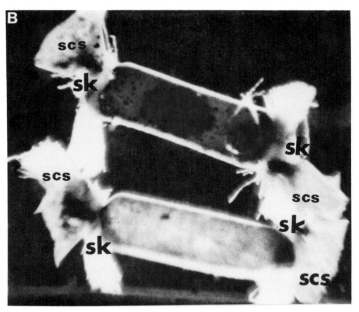

FIG. 15. (A) Cytoplasm-enriched fragment from a long internodal cell of *Chara*. The cytoplasmic plug obtained by centrifugation is ligated by a silk thread knot (sk), and part of the same cell sleeve (scs) is usually left attached. The node (n) and remnants of the neighboring internodal cell and leaf cells can also be seen (cell sleeves, cs). (B) Fragments without nodal cells can be made by tying the first node off prior to centrifugation. Such system may be useful for radioactive tracer studies, as no tracer is taken up (or released) by the nodal cells.

of computing and electronics will be a great advantage. Some of the techniques defy detailed scientific description (such as making the setup noise-free or performing cytoplasmic perfusion successfully), and new researchers in this area are advised to spend some time in a laboratory of someone well versed in the methods.

Acknowledgments

I would like to thank Professor E. A. C. MacRobbie, and Drs. M. R. Blatt, T. Shimmen, and M. Tester for critical reading of the manuscript and helpful discussion. I am also grateful to Bruce Beilby for constructive arguments and support at the time of writing the review.

[28] Ion Transport in *Chara* Cells

By WILLIAM J. LUCAS and DALE SANDERS

Introduction

Algal cells of the family Characeae have, for many decades, provided plant cell biologists with an ideal material for studies of ion transport.[1] The size and robustness of the ecorticate giant internodal cells (diameter typically 0.4–1.0 mm, length up to several centimeters) have enabled measurement of ion fluxes on single cells without the interpretational difficulties which relate to cellular heterogeneity of most multicellular higher plant tissues. The electrochemical driving forces across the plasma membrane and tonoplast for a variety of ions have been determined by direct measurement of the relevant electrical and chemical components; thus, by application of the Nernst equation, it has been possible to state which ions are actively transported, in which direction, and at which membrane. In addition, single internodal cells can be perfused internally,[2] thereby facilitating studies on intracellular control of ion transport.

Ion transport has been most extensively investigated in the genus *Chara*, although *Nitella*, *Nitellopsis*, and *Lamprothamnium* are also commonly used. Within the genus *Chara*, *Chara corallina* has been the species

---

[1] A. B. Hope and N. A. Walker, "The Physiology of Giant Algal Cells." Cambridge Univ. Press, London and New York, 1975.

[2] M. J. Beilby, this volume [27].

METHODS IN ENZYMOLOGY, VOL. 174

of choice for most studies: it is easily cultured in the laboratory by rooting the thalli in river mud which is overlayed with a simple salt solution (usually termed *Chara*. pond water, CPW). [By contrast, culture in axenic media of defined composition is rarely attempted, though success is possible using the formulation of Forsberg.[3,4]] Other species (e.g., *Chara inflata*) have also been used occasionally for transport studies.[5] Substantial interspecific variation in the electrophysiological properties of *Chara* implies that the results of transport studies should only be extrapolated from one species to another with extreme caution.[6]

Despite the simple geometry of the internodal cells, the plasma membrane of *Chara* cannot be viewed as homogeneous with respect to its ion-transporting properties. Zones capable of selectively acidifying or alkalinizing the medium are usually present. While the immediate effects of localized H^+ transport can be largely abolished with strongly buffered media, the implications for spatial variation in the transport of other ions are not clear. Flux studies are generally expressed on the basis of cell surface area, though the possibility that this represents some sort of weighted average for the whole plasma membrane should not be overlooked. Furthermore, *Chara* cells cultured in alkaline conditions possess significant membrane infoldings (charasomes[7] or plasmalemmasomes[8]). This means that the measured ion fluxes should be considered as a relative guide to the transporting capabilities of cells from a given culture rather than as a physically interpretable and quantitative reflection of transport activity across the membrane as a whole.

Cell Surgery and Ion Transport

Ion transport studies are generally conducted with excised *Chara* cells bathed in CPW. One suitable formulation for CPW is as follows: K_2SO_4, 0.2 mM; NaCl, 1.0 mM; $CaSO_4$, 1 mM; pH adjustment via an appropriate zwitterionic buffer, 5 mM. *Chara* has a very high-affinity electrophoretic amine uptake system in the plasma membrane,[9] and in order to avoid interference from this system, via its effect on membrane potential, it is important to maintain contaminating NH_4^+ as low as possible. Analytical

[3] C. Forsberg, *Physiol. Plant.* **18**, 275 (1965).
[4] D. Sanders, *Plant Physiol.* **68**, 401 (1981).
[5] S. D. Tyerman, G. P. Findlay, and G. J. Paterson, *J. Membr. Biol.* **89**, 139 (1986).
[6] S. D. Tyerman, G. P. Findlay, and G. J. Paterson, *J. Membr. Biol.* **89**, 153 (1986).
[7] V. R. Franceschi and W. J. Lucas, *Protoplasma* **104**, 253 (1980).
[8] G. D. Price and M. I. Whitecross, *Micron* **13**, 309 (1982).
[9] N. A. Walker, M. J. Beilby, and F. A. Smith, *J. Membr. Biol.* **49**, 21 (1979).

grade reagents should be used, as well as a reliable source of H_2O (either freshly glass-distilled or passed through a cation-exchange column).

Obtaining Internodal Cells for Flux Measurements

The *Chara* thallus assumes an upright growth habit, attaining a height of a meter or more, depending on the depth of the culture solution. The nodes which are present at the ends of the large internodal cells consist of clusters of small cells, and lateral internodes emerge from each node. For studies of ion transport at the single cell level, some degree of cell surgery is required.

Normally, the thalli are harvested from the culture by cutting with a pair of sharp dissecting scissors just above the level of the river mud. The thalli are then transferred to a large bowl of water or CPW to enable easy inspection of suitable cells for flux work. For most studies, mature cells of the third or fourth internodes are suitable. These are commonly between 20 and 60 mm in length, which facilitates uptake of sufficient radioactivity in short periods even for ions available only at low specific activity. The adjacent internodal cells are cut away, and the lateral internodes carefully trimmed as close as possible to the node. The main axis internodes are then washed free of cell debris in CPW, placed in fresh CPW, and divided into appropriate batches for the particular flux experiment. At this time, or at the conclusion of an experiment, the surface area of the cells must be determined. This is done by measuring the diameter of the cell ($\pm 10 \mu m$), using a binocular microscope, and the length by means of tracing the shape of the cell onto paper. After these measurements, cells are placed in a large, covered petri dish (10 ml/cell is appropriate) and left to recover overnight. Incubation periods exceeding 1 day should not be used, since there is a marked tendency for time-dependent reduction in transport capacity.[10]

Sampling Vacuolar Contents

Direct sampling of the vacuolar sap is necessary for measurement of intracellular pH with weak acids and for compartmental analysis in which the tonoplast fluxes are estimated (see below). An isolated internodal cell (and its associated nodes) is blotted dry with a tissue and allowed to wilt slightly in air before one of the nodes is gently removed with a pair of *very* sharp scissors. The wilting process is essential before cutting in order to avoid explosion of the cell contents under the influence of cell turgor. The loss in cell turgor is accomplished with minimal change in cell volume — and hence minimal change in vacuolar solute concentration — because the

[10] H. G. L. Coster and A. B. Hope, *Aust. J. Biol. Sci.* **21**, 243 (1968).

volumetric elastic modulus of *Chara* is high.[11] A 100-μl Microcap (Drummond Scientific Company, Broomall, PA) drawn to a fine tip in a flame is inserted into the cut end of the cell as it lies horizontally in the hand. The cell is then moved to a vertical position with the capillary uppermost. The remaining node is cut, and the vacuolar sap can be induced to emerge from the lower end by blowing gently into the capillary. The sample should appear clear: contamination by cytoplasm can be easily visualized as cloudiness or, in extreme cases, the appearance of chloroplasts. If the sample is collected on a preweighed planchet, the volume of sap can be estimated from its weight, assuming a density of unity.

The sap collected (10–20 μl) is sufficient for determination of radioactivity by scintillation counting, of osmotic pressure using an osmometer (e.g., Model 5100C, Wescor Inc., Logan, UT), or of concentration of the major inorganic ions, each measurement being possible on a single cell. The chloride concentration can be determined by electrometric titration,[12] while the Na^+, K^+, and Ca^{2+} concentrations can be measured by atomic absorption spectroscopy (e.g., aa/ae spectrophotometer, Advanced Instrumentation Inc., Warrington, Cheshire, England).

Intracellular Perfusion

The possibility of controlling the internal milieu is one of the fortes of using giant internodal cells for transport work.[13–16] The general technique described below is based on the work of Tazawa and co-workers for selective perfusion of the vacuolar sap,[17–19] as well as for perfusion of the sap and flowing cytoplasm.[20]

In order to preserve the tonoplast, it is necessary to include Ca^{2+} (10 mM) in the perfusion solution; conversely, the tonoplast can be destroyed by reducing free Ca^{2+} with chelators. Two sample perfusion media for vacuolar and cytoplasmic perfusion, respectively, are given in Table I. In each case it is necessary first to measure the osmotic pressure of the intact cell by incipient plasmolysis and to match the perfusion medium accordingly, since cell-to-cell variation in osmotic pressure can be consid-

[11] U. Zimmermann and E. Steudle, *Aust. J. Plant Physiol.* **2**, 1 (1975).

[12] J. A. Ramsay, R. H. J. Brown, and P. C. Croghan, *J. Exp. Biol.* **32**, 822 (1955).

[13] D. Sanders, *J. Membr. Biol.* **52**, 51 (1980).

[14] D. Sanders and U.-P. Hansen, *J. Membr. Biol.* **58**, 139 (1981).

[15] R. J. Reid and N. A. Walker, *J. Membr. Biol.* **78**, 157 (1984).

[16] G. M. Clint and E. A. C. MacRobbie, *Planta* **171**, 247 (1987).

[17] M. Tazawa, *Plant Cell Physiol.* **5**, 33 (1964).

[18] U. Kishimoto and M. Tazawa, *Plant Cell Physiol.* **6**, 529 (1965).

[19] M. Tazawa, U. Kishimoto, and M. Kikuyama, *Plant Cell Physiol.* **15**, 103 (1974).

[20] M. Tazawa, M. Kikuyama, and T. Shimmen, *Cell Struct. Funct.* **1**, 165 (1976).

TABLE I
MEDIA FOR VACUOLAR AND CYTOPLASMIC PERFUSION

Component (mM)	Vacuolar medium[a]	Cytoplasmic medium[b]
KCl	150	—
NaCl	10	—
CaCl$_2$	10	—
HEPES	—	100
ATP	—	2
KOH	—	78
MgCl$_2$	—	6
EGTA	—	5
Sorbitol	—	50
Ficoll-70 (%, w/v)	—	5
pH	5.5	7.5
Total osmotic pressure (kPa)	793	545

[a] After Tazawa.[17]
[b] After Lucas and Shimmen.[21]

erable. Apart from the presence of Ca^{2+}, the actual composition of the vacuolar medium is not critical, and, indeed, all other ions can be replaced with a sugar or sugar alcohol osmoticum. In contrast, the cytoplasmic perfusion medium composition is more critical to the survival of the cell. Lucas and Shimmen[21] found that incorporation of 5% Ficoll in the cytoplasmic perfusion medium enhanced the survival rate. Other recipes for cytoplasmic perfusion media have also been reported.[13,15,16,20,22,23]

Figure 1 shows an apparatus of the type commonly used for both vacuolar and cytoplasmic perfusion. Apeizon L grease is applied liberally to the Plexiglas base plate at the prospective contact points of the three chambers. A straight internodal cell (diameter 0.7–0.9 mm, length 40–70 mm) is selected, blotted dry, and six ligatures are tied very loosely around it. The ligatures are of waxed dental floss which has been cut to a length of about 100 mm and divided longitudinally into about four or five equal strands. The cell is placed on the elevated section of the base plate such that three of the ligatures reside between the eventual positions of each of the end chambers and the central chamber. The elevated section of the base plate enables clear viewing of the cell with a binocular microscope through the coverslip which forms one side of the central chamber. The

[21] W. J. Lucas and T. Shimmen, *J. Membr. Biol.* **58**, 227 (1981).
[22] P. T. Smith and N. A. Walker, *J. Membr. Biol.* **60**, 221 (1981).
[23] R. E. Williamson, *J. Cell Sci.* **17**, 655 (1975).

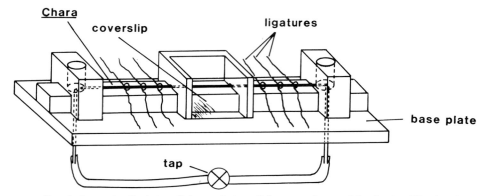

Fig. 1. Apparatus for intracellular perfusion of characean internodal cells. A cell is placed on the elevated section of the base and is perfused with isotonic medium introduced to the two end chambers. Influx of isotope takes place in the central chamber after ligation of the cell in the two interchamber regions. The height of the end chambers is 5 mm, and the length of the central chamber is 20 mm.

elevated section also allows easy access to the ends of the ligatures which hang over each side.

The three chambers, all pregreased at the contact points, are then placed over the cell as indicated in Fig. 1 before turgor is lost. The chambers are each grooved (1 mm depth) at points where contact is to be made with the cell. After positioning of the chambers, CPW can be added to the central chamber in order to restore turgor, pending further manipulation of the cell. To begin perfusion, the CPW in the central chamber is removed with a Pasteur pipette and the surface film blotted with a tissue. The two end chambers are then filled with an isotonic, artificial vacuolar sap. After loss of turgor, the nodes are removed with sharp dissecting scissors.

In our experience, the critical point of the operation concerns the point at which turgor is judged to have been lost sufficiently to prevent forced expulsion of the cell contents on cutting, yet insufficiently to result in physical distortion of the cell surface. If either of these conditions is exceeded, the selective permeability properties of the cell are lost. One convenient method for assessment of the time at which the nodes can be excised is to apply very gentle pressure to one end of the cell with the dissecting scissors: if a slight concavity remains in the surface, the cell can be cut without detrimental effect.

During removal of the nodes, the tap (Fig. 1) connecting the end chambers is open, thereby prohibiting the establishment of a hydrostatic pressure difference across the ends of the cell. However, as soon as excision is complete, the tap is closed, and medium is removed from one of the end chambers with a Pasteur pipette. Simultaneously, the cell is observed, with the microscope focused on the vacuolar contents. For vacuolar perfusion, the flow rate is adjusted empirically to around 1000 μm/sec. For cytoplasmic perfusion, the flow rate can be considerably faster, the sheer forces aiding the removal of the tonoplast. Actually, perfusion can be allowed to proceed driven by a 5 mm hydrostatic head. After 1–2 min of perfusion, the flow is stopped by opening the tap. The outermost pair of ligatures is tied, taking great care not to bend the cell. CPW plus 100 mM sorbitol is then added to the central chamber, and the cell partially restores its turgor pressure within seconds. The middle pair of ligatures is tied, and the experimenter should now feel considerable resistance from the cell turgor pressure. Finally, to ensure tight sealing of the perfused cell, the cell is ligated again, this time as close as possible to the central chamber. After this, full osmotic pressure can be restored by replacement of the medium in the central chamber with CPW.

There are several variations on the above theme.[24,25] The simplest is to omit ligation altogether, in which case isotonic CPW must be added to the central chamber at the end of perfusion and the exposed cell segments covered with liquid paraffin to avoid evaporative losses. This "open vacuole" method has seen extensive use for electrophysiological studies,[2] but, despite the clear advantages of obtaining more than one flux measurement on a single cell, it has not yet been applied for straightforward transport work.

For influx studies, isotope can be added to the central chamber after allowing a period of 5 min for recovery of the cell and inspection for signs of gross physical damage. An influx period of 5 min or less enables sufficient radioactivity to enter the cell, after which time the solution is removed and replaced by an identical nonradioactive solution. Cell wall exchange of radioisotopic anions is complete within 2 min (see below). The cell can then be briefly rewashed before placing in a scintillation vial for counting. Ion efflux is measured simply by sampling the external medium, which can be stirred periodically by solution withdrawal and expulsion with a Pasteur pipette.

[24] T. H. Strunk, *Science* **169**, 84 (1970).
[25] C. S. Spyropoulos, *J. Membr. Biol.* **76**, 17 (1983).

Membrane Fluxes

Isotopic Influx at Plasma Membrane

Anions. CHLORIDE. Chloride is the major anionic constituent of the *Chara* vacuole, and as such its transport has been more intensively studied than that of other anions. Thus, we shall illustrate anion influx measurements using Cl^- as the case ion. The special case of HCO_3^- is dealt with below.

After overnight recovery from excision (see above), and 2 hr before the start of $^{36}Cl^-$ influx measurement, the internodal cells are collected into batches of 10 cells, which are held together very loosely at one end with cotton thread. One strand of the cotton is arranged to extend beyond the end of the batch of cells to enable subsequent manipulations with forceps. At this stage, the cells can be transferred to fresh CPW or to some other pretreatment solution.

If influx of Cl^- is to be studied in the absence of another constituent of CPW, the cells are subjected to a 5-sec wash in a nonradioactive solution identical with the influx solution: this prevents contamination of the influx solution with the absent ion. Since radioactive $^{36}Cl^-$ is available only at rather low specific activity ($\sim 10^7$ Bq mmol^{-1}), it is normal to use the radioisotope solution in the undiluted form. The isotopic influx solutions are made up in CPW, and cells are exposed to these solutions for 5 min. Normally, influx is measured with illumination provided by two 40-W "daylight" fluorescent tubes positioned about 50 cm above the surface of the water bath: this is sufficient to provide an intensity of photosynthetically active radiation equal to 12 W m^{-2}. If steady-state dark influx is to be measured, the cells must be preincubated in darkness for at least 20 min before contact with the isotope.[26]

To terminate influx of radioisotope, the cells are removed and placed for 200 sec in at least 200 ml of a solution identical in all other respects to the influx solution. The cotton is then removed using forceps, and the cells are blotted lightly with tissue and placed on a sheet of paper. After the cells have wilted slightly, the nodes must be removed using sharp dissecting scissors. During this operation the cell is held over a scintillation vial so that all the cell sap is retained for analysis.

A small volume of H_2O (100 μl) is added to each of the vials, which are capped after addition of scintillation cocktail. Aliquots (e.g., 50 μl) of the influx solution are also placed in vials for estimation of specific activity. [This, of course, should be identical with the specific activity of the stock

[26] D. Sanders, *J. Exp. Bot.* **31**, 105 (1980).

radioisotope solution as no unlabeled Cl⁻ is added, and any variation will therefore more likely reflect small concentration differences resulting from pipetting errors.] Counting efficiency is generally about 95% and can be estimated from the channels ratio on the scintillation counter after preparation of quench curves by addition of *Chara* cells to radioactive standards.

To calculate the flux on a surface area basis, the cell is assumed to be cylindrical in shape. In general, fluxes of the order of 10 nmol m⁻² sec⁻¹ are measured, which represents between 100 and 200 cpm/cell for the 5-min uptake period.

The cytosplasmic Cl⁻ content has been estimated as 65 μmol m⁻² cell surface (see section on tonoplast fluxes). This implies that the cytoplasmic specific activity will rise to a maximum of 5% of the external specific activity during the influx period. Thus, the errors due to effluxing of isotope during the uptake period and the subsequent wash will be small and are insignificant in comparison with cell-to-cell variation of Cl⁻ influx. In practice, a sizable proportion of the influxed label is also transferred to an even larger pool of Cl⁻ in the vacuole, so that Cl⁻ uptake proceeds at an apparently constant rate for influx periods of at least 15 min.[27] However, maintenance of a shorter influx period is generally desirable for investigation of non-steady-state events such as the effects of inhibitors, since the cell is able to compensate for modulation of influx via the resulting cytosolic Cl⁻ concentration.

If very short uptake periods are required and the expected Cl⁻ flux is low, then it may not be possible to obtain sufficient radioactivity inside the cell for unambiguous determination. In these cases, it is more convenient to use ⁸²Br⁻ as an analog for Cl⁻. The influx system appears to have similar K_m values for Br⁻ and Cl⁻,[14] although the maximal influx is higher by about 30% in the case of Cl⁻.

H¹⁴CO₃⁻. Experiments involving the use of H¹⁴CO₃⁻ require special attention because the ¹⁴C in solution can exchange readily with ¹²CO₂ in the atmosphere. Sterile NaH¹⁴CO₃ solutions can be purchased having a specific activity of approximately 1.5×10^9 Bq mmol⁻¹; once opened, these solutions must be stored (4°c) in sealed containers to reduce the loss of radioactivity. All stock and experimental solutions required for H¹⁴CO₃⁻ experiments are prepared just prior to their use. CPW is scrubbed with CO₂-free air for 1.5 hr to remove dissolved CO₂. A stock solution of 20 mM Na₂CO₃ is then prepared and kept in a sealed container. Prior to the actual experiment, a test series of solutions is prepared in which the HCO₃⁻ concentration range and experimental pH are *established* using aliquots of the stock 20 mM Na₂CO₃ and a stock 100 mM H₂SO₄ solution.

[27] W. J. Lucas and F. A. Smith, *Aust. J. Plant Physiol.* **3**, 443 (1976).

These solutions are used to pretreat the *Chara* cells prior to the introduction of ^{14}C.

During all treatments, cells are held in gas-tight glass vials and placed in a 25° water bath under 20 W m^{-2} illumination. A few minutes before the end of the pretreatment period (usually 30 min duration) a second set of HCO_3^- solutions is prepared. An aliquot of the $NaH^{14}CO_3$ stock is added to each of these solutions and, after mixing, two 100-μl samples are withdrawn from *each* solution to determine its specific activity. Although the $NaH^{14}CO_3$ solution is approximately 25 mM, its specific activity is so high that only very small aliquots are required, and thus no correction is needed to the HCO_3^- concentrations. (For example, addition of 10 μl of 25 mM $NaH^{14}CO_3$ to 100 ml of a 100 μM $NaHCO_3$ solution would raise the HCO_3^- level to 102.5 μM.) The HCO_3^- transport system of *Chara* is quite labile (easily inactivated), and so the change of solutions (removal of pretreatment and introduction of $H^{14}CO_3^-$) must be carefully done. At the conclusion of the $H^{14}CO_3^-$ exposure (usually 30 min), the radioactive solutions are decanted and again sampled for specific activity as well as a final pH measurement.

After a brief wash in nonradioactive pretreatment solution to remove the excess surface radioactivity, cells are prepared for scintillation counting as described above for $^{36}Cl^-$. However, a 60-min treatment with 0.1 ml of 30% H_2O_2 and 0.1 ml of 7% perchloric acid is required to remove unfixed, precipitated ^{14}C and to digest the tissue to prevent chlorophyll quenching during scintillation counting. The digested tissue is suspended in an appropriate cocktail, and radioactivity is detected using a scintillation spectrometer. By knowing the specific activity of the solution, the efficiency of the scintillation counter, and the cell surface area, the net flux of $H^{14}CO_3^-$ into the cell can be computed.

Cations. Influx of monovalent cations can be studied essentially as detailed for Cl^-, with the exception that precautions must be taken to ensure that all exogenous label is removed from the negatively charged cell wall during the wash period.[28]

Isotopic Efflux at Plasma Membrane

Most experimental work has focused on Cl^- efflux, and this case is therefore treated in detail. In principle, the measurement of plasma membrane efflux is quite simple, but the interpretation of data is complicated considerably by the presence of more than one intracellular compartment

---

[28] E. A. C. MacRobbie, *Aust. J. Biol. Sci.* **19**, 371 (1966).

(especially the vacuole). The experimental methods are outlined below, but discussion of data analysis is reserved for the section on tonoplast fluxes (see below).

The vacuole of *Chara* has a very small rate constant for turnover of Cl^-, and cells therefore have to be loaded with isotope for several weeks to attain a specific activity of Cl^- which approaches that of the labeling solution. The cells are loaded at room temperature in CPW containing $^{36}Cl^-$, and the solution is renewed every week to return the pH to lower levels at which Cl^- influx is optimal. The light regime is 16L (light)–8D (dark), as constant illumination damages the chloroplasts. (If complementary influx data are required, it is important to treat unlabeled cells similarly.[29]) Prior to the efflux experiment, the cells are bathed in nonradioactive CPW overnight, to allow the cytoplasm and its constituents to come to a constant specific activity.

Each efflux experiment is conducted with a batch of about 5 cells loosely held together with cotton thread. The cells are inserted into glass tubing [5 mm internal diameter (i.d.)] which is sealed at each end with a flexible compound rubber cap. Narrow-bore (1 mm diameter) nylon tubing is inserted into these caps. The free end of one piece of tubing is connected to a 50-ml burette, and the other free end is positioned over a 20-ml scintillation vial. For temperature control, the efflux apparatus is immersed in a water bath.

The burette acts as a gravity-feed reservoir from which a measured volume of solution can be flowed past the cells and collected with effluxed tracer. Normally, 10 ml is collected over 30 sec at the end of each 10-min efflux period. A further 30 ml over 90 sec is then flushed through the apparatus and discarded. For a total chamber volume and downstream dead space of 1.5 ml, 99.7% of the extracellular activity is flushed from the chamber by the first 10 ml: from the point of view of contamination of the succeeding samples, the subsequent wash is therefore unnecessary. However, in many experiments the effect of rapid removal of extracellular Cl^- is investigated, and the second wash is necessary to remove remaining micromolar levels of the ion.

Each 10-ml sample is dried in an oven at 95° overnight, rehydrated with 1.5 ml H_2O, and counted by liquid scintillation. At the end of the efflux experiment, the cells are removed from the chamber, and vacuolar samples are obtained in order for the Cl^- concentrations to be measured by electrometric titration (see above). The precipitated chloride (as AgCl) can be dissolved with a drop of NH_3 and each sample washed into a separate scintillation vial. After counting, the specific activity of the sample can be

[29] F. A. Smith and N. A. Walker, *J. Exp. Bot.* **27**, 451 (1976).

calculated. These specific activities must be weighted in proportion to the original surface area of the cell to enable the effective overall specific activity of the sap of the five cells to be calculated with respect to surface area. Values of about 3 nmol m^{-2} sec^{-1} are commonly observed in CPW with illumination, rising to 13 nmol m^{-2} sec^{-1} in darkness.[26]

Several factors must be considered in assessing the accuracy of efflux estimates. First, the presence of the nodes must be taken into consideration, since these will also efflux label to the external medium; their contribution can be as high as 40%. Nodal fluxes can be eliminated quite simply by ligation and excision, to leave a large fragment of the internodal cell.

Second, in Cl^{-}-free media, it is possible that a significant proportion of effluxed label is reabsorbed before it diffuses through the cell wall. Tyree[30] has estimated the wall permeability to Cl^{-} as 8×10^{-10} m^2 sec^{-1}, and, even given the higher estimates of efflux, it is unlikely that the concentration of Cl^{-} adjacent to the membrane would exceed 1 μM, which is well below the measured K_m for the uptake system. However, an alternative estimate of wall permeability is given by Mailman and Mullins[31] as 5×10^{-12} m^2 sec^{-1}, which could result in a Cl^{-} concentration as high as 20 μM (or about one-half the measured K_m) at the external surface of the membrane.

Third, to calculate efflux it is normally assumed that the cytosolic specific activity is equal to that of the vacuole.[26,32,33] However, the true unidirectional efflux is given as:

$$\phi_{co} = \frac{r}{s_c} = \frac{s_v \phi_{vc} - s_c \phi_{cv}}{s_c} \qquad (1)$$

where the subscripts o, c, and v refer to the external, cytoplasmic, and vacuolar compartments, respectively, ϕ_{ij} is the unidirectional flux from compartment i to compartment j, r is the rate of emergence of radioactivity from the cell (units: cpm m^{-2} sec^{-1}), and s is specific activity (units: cpm nmol^{-1}). Rearrangements of Eq. (1) shows that the true efflux, ϕ_{co}, will be higher than the apparent efflux, (r/s_v) by the factor $(\phi_{co} + \phi_{cv})/\phi_{vc}$. Thus, a crucial question concerns the magnitudes of the tonoplast fluxes. Only if these are considerably larger than the plasma membrane fluxes is an accurate estimate of efflux obtained.

[30] M. T. Tyree, *Can. J. Bot.* **46**, 317 (1968).
[31] D. S. Mailman and L. J. Mullins, *Aust. J. Biol. Sci.* **19**, 385 (1966).
[32] A. B. Hope, A. Simpson, and N. A. Walker, *Aust. J. Biol. Sci.* **19**, 355 (1966).
[33] G. P. Findlay, A. B. Hope, M. G. Pitman, F. A. Smith, and N. A. Walker, *Biochim. Biophys. Acta* **183**, 565 (1969).

Tonoplast Fluxes

Two general strategies have been used in attempts to determine ion fluxes across the tonoplast of characean cells. In the first, the vacuolar specific activity is sampled at defined times after the onset of isotopic influx, and the tonoplast influx is derived from the calculated rate constant for turnover of the ion in the cytoplasm (k_c). In the second, the cell is loaded with isotope for a considerable time (at least $5k_c$), and k_c is calculated from the washout kinetics. In both cases, it is strictly possible to carry out the analysis only for steady-state conditions in which the cytoplasmic and vacuolar contents of the ion are constant during the measurement period. The theory underlying both methods is comprehensively discussed by Walker and Pitman.[34] As in the case of ion effluxes, Cl⁻ has received most attention, and the accounts below treat Cl⁻ as a case study. The same techniques are applicable to other ions.

Vacuolar Fraction Method. Since its initiation in the late 1950s the vacuolar fraction method[35] has been used by a number of workers.[10,28,36-40] A premeasured internodal cell is incubated in ^{36}Cl⁻ for a period during which s_v is insignificant compared with s_c. [MacRobbie [28] set an arbitrary limit at $s_v < 0.1s_c$, which is suggested to occur at $k_ct = 2.2$. However, this criterion is based on a high estimate of cytoplasmic Cl⁻ concentration (see below), and for the lower estimate obtained using solid-state Cl⁻ microelectrodes[41] of 10 mM, or 0.065 mmol m⁻², an incubation corresponding to $k_ct = 10$ can legitimately be used *before* isotope backflux from the vacuole becomes problematic.]

After incubation, the cell is washed free of isotope for 100 sec, blotted, rapidly wilted under the heat of a light bulb, and pinched firmly with forceps 15–20 mm from one node. This portion of the cell is removed and its length measured before trimming the node and measuring radioactive uptake, which provides an estimate of radioactive uptake by the whole cell (Q_T^*). A vacuolar sample is taken from the other portion of the cell, and the radioactivity is counted in a measured aliquot. Assuming the tonoplast diameter to be 50 μm less than that of the whole cell, the fraction of

[34] N. A. Walker and M. G. Pitman, *Encycl. Plant Physiol., New Ser.* **2**, Part A, 93 (1976).
[35] J. M. Diamond and A. K. Solomon, *J. Gen. Physiol.* **42**, 1105 (1959).
[36] E. A. C. MacRobbie, *J. Gen. Physiol.* **45**, 861 (1962).
[37] E. A. C. MacRobbie, *J. Gen. Physiol.* **47**, 859 (1964).
[38] E. A. C. MacRobbie, *J. Exp. Bot.* **20**, 236 (1969).
[39] E. A. C. MacRobbie, *J. Exp. Bot.* **22**, 487 (1971).
[40] G. P. Findlay, A. B. Hope, and N. A. Walker, *Biochim. Biophys. Acta* **233**, 155 (1971).
[41] H. G. L. Coster, *Aust. J. Biol. Sci.* **19**, 546 (1966).

intracellular isotope which has entered the vacuole (F_v^*) can then be established.

If the influx period is in excess of about 15 min, the cytoplasm behaves as a single kinetic compartment.[26] However, for shorter influx times, a smaller component of vacuolar transfer is apparent. Although the physical identity of this component was the subject of some controversy,[38,42,43] it now appears likely that the fast component is an artifact introduced by cutting the cell.[42] The errors arising from the presence of the fast phase can be corrected for, and k_c estimated, as follows. The apparent fraction of tracer observed in the vacuole (F_v^{*A}) will differ from that which describes the true *in vivo* transfer (F_v^{*T}) according to the relationship

$$F_v^{*A} = F_v^{*T} + PF_c^* \tag{2}$$

where P is the proportion of cytoplasmic isotope transferred as a result of cutting and F_c^* the fraction of isotope in the cytoplasm. Thus,

$$F_v^{*A} = F_v^{*T} + P(1 - F_v^{*T}) \tag{3}$$

With the cytoplasm acting as a single kinetic phase, the vacuolar fraction is given as

$$F_v^{*T} = 1 - \left[\frac{1 - \exp(-k_c t)}{k_c t} \right] \tag{4}$$

Values of F_v^{*A} measured over a range of influx times (5–75 min) are plotted as a function of time, and the intercept at $t = 0$ ($F_v^{*A} = P$) is estimated assuming a smooth function. By substituting this value of P into Eq. (3) it is possible to calculate a value of k_c for each point (from a graph of $\{1 - [1 - \exp(-k_c t)]\}/k_c t$ versus $k_c t$). The values of k_c should be time independent.

Using this method, k_c for Cl^- in *Chara* has been estimated as 2.9×10^{-4} sec^{-1} (time constant = 57.5 min). The tonoplast flux can be calculated from the relationship

$$\phi_{cv} = Q_c k_c - \phi_{co} \tag{5}$$

where Q_c is the Cl^- content of the cytoplasm (units: mmol m^{-2}). Q_c is obtained from separate measurements of the cytoplasmic Cl^- concentration and the volume of cytoplasm per unit cell surface area. Ion-selective electrode measurements suggest that cytoplasmic Cl^- concentration is 10 mM,[41] and from studies in which the cytoplasm is centrifuged to one

[42] E. A. C. MacRobbie, *J. Exp. Bot.* **26**, 489 (1975).
[43] N. A. Walker, *in* "Membrane Transport in Plants" (U. Zimmermann and J. Dainty, eds.), p. 173. Springer-Verlag, Berlin and New York, 1974.

end of the cell, the volume-to-surface ratio is 6.5 ml m^{-2} (which corresponds to a cytoplasmic thickness of 7 μm[10]).

Thus, for $Q_c = 0.065$ mmol m^{-2} and $\phi_{co} = 3$ nmol m^{-2} sec^{-1} (see above), ϕ_{cv} is 15.9 nmol m^{-2} sec^{-1}. However, it must be considered that ϕ_{co} itself will be underestimated if the tonoplast and plasma membrane fluxes are comparable [see Eq. (1) and discussion] so that the true fluxes by this method are more likely to be in the region $\phi_{co} \simeq 4$ nmol m^{-2} sec^{-1} and $\phi_{cv} \simeq 15$ nmol m^{-2} sec^{-1}.

There are several potential sources of error in the vacuolar fraction method which must be considered. One arises from the assumption,[28] implicit in Eq. (4), that the total measured isotope entry (Q_T^*) is given by $\phi_{oc}t$. However, the full expression is

$$Q_T^* = \phi_{oc}t - \phi_{co} \int_0^t S_c \, dt \tag{6}$$

where the cytoplasmic specific activity as a fraction of that in the loading solution is defined as

$$S_c = \frac{\phi_{oc}}{\phi_{co} + \phi_{cv}} [1 - \exp(-k_c t)] \tag{7}$$

Integration of Eq. (7) yields

$$Q_T^* = \phi_{oc}t - \frac{\phi_{co}\phi_{oc}}{\phi_{cv} + \phi_{co}} \left\{ t - \frac{1}{k_c} [1 - \exp(-k_c t)] \right\} \tag{8}$$

The assumption is therefore valid if ϕ_{cv} is much greater than ϕ_{co}. Since that does not appear to be the case, the rise of S_c with time is more likely to set an upper limit on the length of influx period which can be used than is the rise in S_v. In practice, the time limit for loading, given by the condition $\phi_{co} \int S_c \, dt < 0.1\phi_{oc}t$, for Cl$^-$ in *Chara* is equal to about 75 min.

The other uncertainties in the estimate cannot be so easily compensated. P itself contributes significantly to F_v^{*A} at early times, and, with a value of 0.3, it accounts for around one-half of F_v^{*A} even after 1 hr of loading. Thus, the $(1 - \exp)$ rise in F_v^* has to be detected against considerable system noise.

Finally, the value of Q_c has a considerable influence on the estimate of the tonoplast flux. Early measurements of Q_c (obtained by physical separation of the cytoplasm and vacuole[18,32,37,44] are almost certainly erroneously high owing to contamination of the cytoplasm with vacuolar sap. Estimates of Q_c are therefore much more reliably obtained by the solid-state

[44] R. M. Spanswick and E. J. Williams, *J. Exp. Bot.* **15**, 193 (1964).

Cl^- microelectrode method.[41] The effect of a high value of Q_c on ϕ_{cv} can most simply be seen with reference to Eq. (5). Furthermore, the estimate of ϕ_{co} itself normally rests on the assumption that $S_c = S_v$ during efflux experiments: if ϕ_{cv} is comparable to ϕ_{co} this assumption will not hold, and the value of ϕ_{co} in Eq. (5) is correspondingly underestimated. Both these errors result in considerable amplification of the estimate of ϕ_{cv}, and most published values using the vacuolar fraction method have been of the order $500-1000$ nmol m^{-2} sec^{-1}.[10,28,37] Given these difficulties, it seems more reliable to estimate tonoplast flux and Q_c by a noninvasive technique such as compartmental analysis of isotope efflux.

Compartmental Analysis. The technique of compartmental analysis was first applied to giant-celled algae by MacRobbie and Dainty,[45] and it has most recently been used to estimate steady-state Cl^- fluxes in *Chara*.[46] The theory and derivation of the rate equations are dealt with elsewhere.[34,37]

To prevent interference in the subsequent efflux studies, the nodes are either removed or coated with silicone grease. A single internodal cell is then placed in the same apparatus as that used for plasma membrane efflux measurements (see above) the one exception being that the reservoir end of the chamber is connected to a 10-ml syringe which enables faster flow rates to be established. Isotopic CPW is then introduced via the syringe for the required loading period, after which the chamber is flushed through with 10-ml unlabeled CPW. Three further aliquots of CPW are passed through the system before the first 10-ml sample is collected for counting at 140 sec after the end of the loading period. Further samples are taken at 50 sec intervals, decreasing in frequency to 1 sample/100 sec at 500 sec, 1/0.3 ksec at 0.9 ksec, 1/0.5 ksec at 1.5 ksec, 1/ksec at 4 ksec, 1/2 ksec at 8 ksec, and 1/4 ksec at 12 ksec. This series of frequencies allows radioactivity to emerge at greater than 3 × (background) for each sample, given a loading solution specific activity of around 16.7 Bq nmol^{-1}. If a separate estimate of the rate constant for vacuolar turnover is required, sampling must be continued for many days, with correspondingly reduced sampling frequencies.

The samples are dried and counted as described for plasma membrane efflux above. At the end of the experiment, the cell dimensions are measured and estimates of whole-cell and vacuolar isotope contents obtained as described above. The vacuolar Cl^- concentration can also be measured

[45] E. A. C. MacRobbie and J. Dainty, *J. Gen. Physiol.* **42**, 335 (1958).
[46] S. Jones and N. A. Walker, *in* "Plant Membrane Transport: Current Conceptual Issues" (R. M. Spanswick, W. J. Lucas, and J. Dainty, eds.), p. 583. Elsevier/North-Holland, Amsterdam, 1980.

prior to determination of radioactivity (see section on sampling vacuolar contents above).

Data are analyzed as follows. The cell content of isotope (Q_T^*: units, dpm cm^{-2}) is plotted for each collection period logarithmically as a function of time. Isotope efflux from the cell wall dominates the exchange kinetics for the first 500 sec, but at times in excess of this, it may be possible to fit the decay in isotope efflux as the sum of two exponentials.[46] Ideally, a least-squares fitting routine[47] should be used to facilitate an unbiased estimate of where the "break point" in the two exponentials occurs. For the steady-state condition, the fluxes and contents of the two compartments can be estimated from the rate constants (k_1, k_2) and zero-time intercepts (I_1, I_2) of the two exponentials and from the loading time (T) and specific activity of the loading solution (s_o) as follows[48]:

$$\phi_{oc} = \frac{I_1 k_1 [1 - \exp(-k_1 T)] + I_2 k_2 [1 - \exp(-k_2 T)]}{s_o} \tag{9}$$

$$Q_T = \frac{I_1 [1 - \exp(-k_1 T) + I_2 [1 - \exp(-k_2 T)]}{s_o} \tag{10}$$

$$Q_c = \frac{\{I_1 k_1 [1 - \exp(-k_1 T)] + I_2 k_2 [1 - \exp(-k_2 T)]\}^2}{s_o \{I_1 k_1^2 [1 - \exp(-k_1 T)] + I_2 k_2^2 [1 - \exp(-k_2 T)]\}} \tag{11}$$

$$Q_v = Q_T - Q_c \tag{12}$$

$$\phi_{vc} = \frac{k_1 k_2 Q_1 Q_2}{\phi_{oc}} \tag{13}$$

The estimates of content can then be expressed relative to volume, rather than surface area, if cytoplasmic volume is calculated from centrifugation experiments. The supplementary sampling of vacuolar Cl$^-$ concentration allows an independent verification of the estimate of Q_v. Jones and Walker[46] have derived the following values for Cl$^-$ fluxes and concentrations in *Chara*: $\phi_{oc} = 10.6$ nmol m^{-2} sec^{-1}; $\phi_{vc} = 14.3$ nmol m^{-2} sec^{-1}; [Cl$^-$]$_c$ = 1.8 mM; [Cl$^-$]$_v$ = 134 mM. The respective values of k_c and k_v were 8.5×10^{-4} and 1.3×10^{-7} sec^{-1}.

Net Fluxes at Plasma Membrane

In the previous sections we have outlined the methodology involved in the use of radioisotopes to obtain unidirectional fluxes crossing both the

[47] D. W. Marquardt, *J. Soc. Ind. Appl. Math.* **11**, 431 (1963).
[48] C. K. Pallaghy and B. I. H. Scott, *Aust. J. Biol. Sci.* **22**, 585 (1969).

plasmalemma and the tonoplast. For influx determinations it is necessary to destroy the *Chara* cells in order to measure the radioactivity. As pointed out above, a second limitation associated with this method of flux analysis is that fluxes are usually expressed on the basis of *total* cell surface area; i.e., it is generally not possible to obtain data on localized flux densities. These limitations can be overcome by using either ion-selective electrodes[49,50] or the extracellular vibrating probe technique.[51,52]

Ion-Selective Electrodes

In many physiological studies net H^+ fluxes are determined by back-titration of the medium bathing the experimental tissue. These experiments can be performed with almost any conventional glass pH electrode system. However, when information on the spatial aspect of the H^+, OH^-, K^+, or Ca^{2+} flux is required, miniature or ion-specific microelectrodes must be employed.

In ion-specific microelectrodes, an organic solution of a neutral (or lipophilic electrolyte), ion-complexing agent(s) [ionophore(s)] acts as the ion-selective liquid "membrane." Construction and calibration of these microelectrodes are straightforward and have been discussed in detail elsewhere.[53-55] The only points we wish to stress concerning the use of these microelectrodes is that they be calibrated before and after each experimental sequence, until one becomes familiar with the particular stability features of the ion-specific electrode being employed. Second, it is our experience that most problems in the use of these electrodes can usually be traced to the reference electrode, and so it is wise to take care in the preparation of this part of the circuit.

Application to Diffusion Analysis. In an aqueous bathing medium, the flux vector of an ionic species like OH^- (or H^+) can be expressed as the sum of three terms, involving the activity and electric potential gradients and the convection velocity:

$$J_{OH^-} = -D_{OH} \nabla C_{OH} - (D_{OH} C_{OH} z_{OH} F/RT) \nabla E + v C_{OH} \qquad (14)$$

[49] W. J. Lucas, *J. Exp. Bot.* **26**, 271 (1975).

[50] W. J. Lucas, J. M. Ferrier, and J. Dainty, *J. Membr. Biol.* **32**, 49 (1977).

[51] W. J. Lucas and R. Nuccitelli, *Planta* **150**, 120 (1980).

[52] W. J. Lucas, *Planta* **156**, 181 (1982).

[53] W. J. Lucas and L. V. Kochian, *in* "Advanced Agricultural Instrumentation: Design and Use" (W. G. Gensler, ed.), p. 402. Nijhoff, Dordrecht, 1986.

[54] D. Ammann, F. Lanter, R. A. Steiner, P. Schulthess, Y. Shijo, and W. Simon, *Anal. Chem.* **53**, 2267 (1981).

[55] P. C. Meier, D. Ammann, W. E. Morf, and W. Simon, *in* "Medical and Biological Applications of Electrochemical Devices" (J. Koryta, ed.), p. 13. Wiley, New York, 1980.

where D_{OH} is the diffusion coefficient for OH^-, C_{OH} the OH^- concentration or activity, E the electric potential, ∇ the gradient operator, and v the convection velocity; R, T, z, and F have their usual meanings.

By placing the *Chara* cell into an experimental medium that contains agar, convection is damped [49] and Eq. (14) reduces to

$$J_{OH^-} = -D_{OH} \nabla C_{OH} - (D_{OH} C_{OH} z_{OH} F/RT) \nabla E \qquad (15)$$

Under these convection-damped conditions, the light-dependent pH and electric potential profiles that develop along the surface of *Chara* internodal cells can be mapped. The pH values can be measured using either a commercial miniature electrode (Model MI-405; Microelectrodes Inc., Londonderry, NH) in conjunction with a standard pH meter (Model 4500; Beckman, Palo Alto, CA) or an ion-specific pH microelectrode system.[56] Electric potential (external) measurements can be made using a differential amplifier (Model 750; WPI, New Haven, CT). For optimal measurements, glass microelectrodes having a tip diameter of $50 \, \mu m$ are filled with a solution containing 0.7% (w/v) agar and 10 mM KCl. If the system is to be thermostatted by means of a bath, spurious electrical noise can be significantly reduced by using high-purity, low-viscosity, liquid paraffin as the thermostatting fluid.[57]

A high-resolution micromanipulator must be used to move the pH and electric potential electrodes to the required position. It is relatively easy to manufacture a small holder that allows both electrodes to be held in an orientation such that the two sensing tips are very close together. This arrangement enables one to measure simultaneously the pH and electric potential profiles that develop on the surface of *Chara* internodal cells during photosynthetic assimilation of exogenous HCO_3^-.

Although the external electric potential is readily detectable, it is generally the case that $\nabla E \ll \nabla C_{OH}$, and so Eq. (15) simplifies to

$$J_{OH^-} = -D_{OH} \nabla C_{OH} \qquad (16)$$

which is Fick's first law of diffusion. By establishing the *symmetry* of the diffusion pattern, produced by the transplasmalemma OH^- transport system, Eq. (16) can be used to give Fick's second law of diffusion, which can be solved to give an explicit expression for J_{OH^-} as a function of C_{OH} and the spatial coordinates. Since the outer surface of the *Chara* cell acts as a boundary between the site of transport and the diffusion system, the appropriate mathematical solution usually involves "hollow" cylindrical or spherical coordinates. Lucas[49] demonstrated that the OH^- or alkaline

[56] W. J. Lucas, D. W. Keifer, and D. Sanders, *J. Membr. Biol.* **73**, 263 (1983).
[57] J. M. Ferrier and W. J. Lucas, *J. Exp. Bot.* **30**, 705 (1979).

bands of *Chara* form spherical diffusion profiles. Thus, under steady-state conditions, the flux leaving a particular OH⁻ band can be determined using the following:

$$J_{OH^-} = 4\pi D_{OH}r_1 r_2[C_{OH(1)} - C_{OH(2)}]/r_2 - r_1 \qquad (17)$$

where $C_{OH(1)}$ and $C_{OH(2)}$ are the OH⁻ activities at the radial positions r_1 and r_2, respectively. Since the OH⁻ bands exhibit a reasonable degree of symmetry,[49] the flux associated with each band can be determined from a single radial scan through its center of symmetry.

The accuracy of these flux determinations is high, provided the radial values employed (r_1, r_2) are larger than the physical dimensions of the miniature pH electrode. For studies on the diffusion characteristics within the region close to the *Chara* cell wall, where the unstirred layer (200 μm) restricts convective mixing, ion-specific microelectrodes must be employed.[56]

PVC-Gelled Ion-Specific Electrodes. Intracellular ion-specific microelectrodes are not used to measure actual fluxes crossing the plasmalemma, but rather to monitor cytoplasmic (or vacuolar) pH, pCa, etc., parameters having essential roles with respect to regulation of transmembrane fluxes. Because the plant cell is under high hydrostatic pressure (turgor), insertion into the cytosol or vacuole tends to displace the liquid membrane back from the tip of the microelectrode. This introduces the possibility of an electrical shunt between the liquid membrane and the glass surface. This problem can be overcome by mixing the liquid membrane with poly(vinyl chloride) (PVC) in order for the sensor to be gelled into the tip of the microelectrode. The basic protocol for production of these gelled tips is described below for the Ca²⁺ sensor.[58]

A small quantity of PVC (1 mg, MW 200,000; BDH Chemicals Ltd., Dorset, England) is dissolved in tetrahydrofuran (50 μl; Sigma, St. Louis, MO) held in a Reactivial. The solution is kept on ice to prevent solvent evaporation. Fluka Ca²⁺ cocktail [89% (w/w) 2-nitrophenyl octyl ether (solvent), 10% ETH 1001, and 1% sodium tetraphenylborate (functions as lipophilic anion to limit interference from anion-generated diffusion potentials)] is then mixed with the PVC solution to give a final dilution of 10% (w/w). A small quantity of this solution is back-filled into the tip region of the micropipette. A cat's whisker (or similar object) is then used to move the sensor into the tip and to remove any air bubbles that may have formed. Filled micropipettes should be stored, tips downward, in a desiccator.

---

[58] D. Sanders and A. J. Miller, *in* "Molecular and Cellular Aspects of Calcium in Plant Development" (A. J. Trewavas, ed.), p. 149. Plenum, New York, 1986.

Once the PVC–Ca^{2+} cocktail has gelled (leave for about 12 hr) the electrode reference solution (0.1 μM $CaCl_2$) can be back-filled and air bubbles between the gelled membrane–reference solution interface removed using a cat's whisker. A reference electrode, having a tip diameter similar to the Ca^{2+} microelectrode, is pulled using microfiber-borosilicate glass and is filled with 100 mM KCl. After the electrodes have been connected to the high-impedance amplifier, they should be given a 10–30 min preconditioning before calibration against a series of standard solutions (pCa 3 to 8). Interference characteristics for Mg^+, K^+, and H^+ can be determined using the separate solution method.[53]

Flux Measurements Using the Vibrating Probe

One-Dimensional Vibrating Probe. Numerous biological systems appear to generate electric fields/currents, but in many situations these fields are very small. This meant that traditional KCl-filled microelectrodes were of limited value in studying these phenomena, since the noise associated with these electrodes limits their resolution to about 10 μV. It was not until Jaffe and Nuccitelli[59] developed an innovative technique, called the vibrating probe, that sophisticated research became possible on the nature of these fields. Basically, the enhanced resolution of this system was achieved by lowering the electrode resistance by a factor of 1,000. This was done by filling the glass micropipette (microelectrode) with solder; the new level of resolution was 10 nV.

At present, several types of vibrating probes are in use. We first describe the theory and operation of the platinum black-tipped glass micropipette probe, since it was the first system introduced for commercial use. This will be followed by a brief description of the very recent two-dimensional (2-D) and 3-D probes.

The important details of the 1-D probe and its housing are illustrated in Fig. 2. The small platinum black electrode (tip diameter 10–30 μm) can be manufactured in the laboratory or purchased commercially (Vibrating Probe Company, Davis, CA). A special alignment tool is used to hold the platinum black electrode during its insertion into the Lucite boat (see Fig. 2). A coaxial cable is then connected to the upper end of the glass micropipette. The outer portion of the cable connects to the reference on the shank of the electrode housing and is grounded in the measuring amplifier circuit. The inner portion of the cable is used to transmit the electrical signal to a phase-sensitive lock-in amplifier (Model 5101, EG & G Princeton Applied Research, Princeton, NJ).

[59] L. J. Jaffe and R. Nuccitelli, *J. Cell Biol.* **56**, 614 (1974).

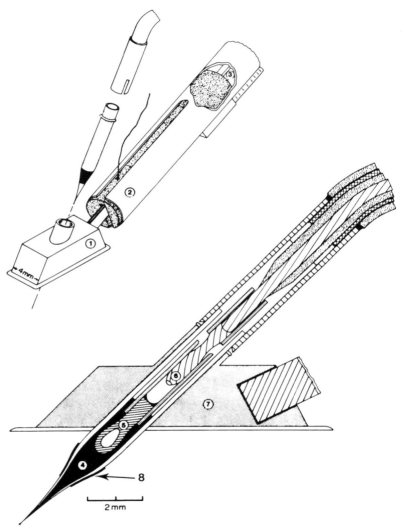

FIG. 2. Scale drawing of the one-dimensional vibrating probe with construction details: (1) Lucite boat; (2) 6-gauge stainless steel tube; (3) bender (piezoelectric) power cable; (4) Cerro-tru solder filling inside of pipette; (5) No. 9 sewing needle eye; (6) gold-plated, braided pin (Malco No. 096-0524-0000, Chicago, IL); (7) Lucite boat with cover glass meniscus setter attached to bottom; (8) reference on shank of glass electrode housing [R. Nuccitelli, *Mod. Cell Biol.* **2**, 451 (1983)].

During operation, the boat and the platinum electrode are caused to vibrate, in one plane, over a known distance (usually 20–25 μm) by varying the power supplied to a piezoelectric bender element via a voltage oscillator (sinusoidal signal). (The frequency at which the probe is vibrated is a function of the resonance frequencies of the specific system. For our recent studies on *Chara* we routinely use ~290 Hz.) This vibration converts any steady potential difference between these two extremes of the probe's excursion into a sinusoidal voltage output, with a peak-to-peak amplitude equal to the dc voltage difference. The processing of this signal is achieved using a phase-sensitive lock-in amplifier that is driven by the same reference signal used to drive the piezoelectric bender element. Thus, it is extremely important during calibration of the probe that the phase angle of the lock-in amplifier be adjusted to that of the probe.

Phase angle adjustment, calibration, and barrier artifact check can all be performed using a simple parallel plate calibration chamber. This rectangular chamber consists of two Ag/AgCl plates of fixed cross-sectional areas, separated by a solution of known conductivity (usually 60 mM NaCl). Thus, a known current can be passed between the plates (say, 100 mA m^{-2}) to permit system calibration. Barrier artifacts, or erroneous signals not due to electric fields generated by the biological specimen, must be tested for by using an inert object of approximately the same dimensions as the biological material. For *Chara,* a solid glass rod, 1 mm in diameter and 1 cm in length, serves this purpose adequately. If a new probe does not respond to positional changes with respect to this glass rod, then it can be used to study the extracellular ionic current patterns of *Chara.*[51,52]

Because of its sensitivity to spurious vibrations, the vibrating probe, as well as the compound inverted microscope (Zeiss or Nikon) used to view the *Chara* cell and the micromanipulator that adjusts the position of the probe, must be supported by a vibration-damped table (Model 1201, Kinetic Systems, Brighton, MA).

Chara internodal cells to be used for vibrating probe studies should be placed in a container of experimental solution so that cellular debris can be removed from the cut ends of the branch and adjoining internodal cells; chloroplasts, in particular, tend to get caught up in the tip of the vibrating probe. Cleaning is achieved using a Pasteur pipette to which a partial vacuum has been applied. Great care should be taken during this clean procedure so that the *Chara* cells are not perturbed. A silk thread (type Orizuru 9; Kanagawa Co., Kanagawa, Japan) is then attached to each end of the internodal cell (actually tied to the dead cell wall fragment of the adjoining internodal cells). These silk threads are then used to suspend the *Chara* internodal cell in the center of a Plexiglas chamber (9 mm deep, 30 mm wide, and 56 mm long) filled with bathing medium. Small blocks of Plexiglas (height 4.5 mm) are placed beneath the threads, near the ends of the cell, to stabilize the cell's position.

The Plexiglas chamber containing the suspended *Chara* cell is placed on the stage of the inverted microscope, and the cell is illuminated using a bifurcated fiber optic light source (20 W m^{-2}; Model MK II, Ehrenreigh Photo Optical Industries, Garden City, NY). The microscope is focused (200×) on the equatorial plane of the *Chara* cell, and then a high-resolution X-Y-Z micropositioner (Model H-2, Line-Tool, Allentown, PA) is used to lower the vibrating probe, such that the tip is in sharp focus within this equatorial plane and is positioned some 300 μm from the cell surface. Power to the piezoelectric bender is then switched on and adjusted to give the desired vibration amplitude (Δr). Range selection on the lock-in amplifier depends on each individual experiment; however, a short time constant (30 msec) should be employed to obtain the true signal amplitude. The vibrating probe is calibrated to reference, or background, at a distance of 1 cm from the cell surface before being brought into a position 50 μm from the wall.

It is best to have the microscope stage motorized (stepping motor, Model KD 3402-004; Hurst Manufacturing Corp., Princeton, IN; coupled to Model SL 10 adjustable speed controller; Minarik Electric Co., Los Angeles, CA) so that the entire length of the *Chara* cell can be traversed past the vibrating probe. An appropriate speed is about 1 cm of cell length per minute.

The theory that underlies the conversion of the measured voltage gradient into the putative current density which would exist within the center and direction of vibration may not be as straightforward as suggested previously.[60] This issue of interpretation will be addressed shortly. In any event, Ohm's law is invoked, and in this situation, if the amplitude of vibration (Δr, in μm) is small relative to the size of the cell, the current density (I, in A m^{-2}) is given by

$$I = -\frac{1}{\rho}\frac{\Delta V}{\Delta r}\hat{a}_r \qquad (18)$$

where ΔV is the voltage difference (detected between the extremes of vibration), $\hat{a}_r$ the unit vector in the radial direction, and ρ (ohm-m) the specific resistance of the bathing medium. An example of typical experimental values obtained on *Chara* cells within the center of an OH$^-$ band is as follows: ΔV, 120 μV; ρ, 34.7 ohm-m; Δr, 2.0 × 10^{-5} m. From Eq. (18),

[60] J. Ferrier and W. J. Lucas, *Biophys. J.* **49**, 803 (1986).

the current density can be calculated as

$$I = \frac{1}{34.7 \text{ (ohm-m)}} \times \frac{120 \times 10^{-6} \text{ (RMS volts)} \times 2.83}{2.0 \times 10^{-5} \text{ (m)}} \quad \text{A m}^{-2}$$
$$= 0.49 \quad \text{A m}^{-2}$$

It is important to realize that the current density falls off rapidly as a function of distance from the *Chara* cell surface.[51] In order to determine the current density crossing the plasmalemma, it is necessary to extrapolate the measured current densities to the surface on the basis of the fall-off of the field with distance from the surface. An empirical curve that fits the experimental data is generally obtained in order to determine the surface current. We should add a note of caution here: the fall-off profile for the inward current (OH$^-$ efflux) was found to be a function of the amplitude of vibration.[51]

If the experimental tissue is functioning normally (i.e., not perturbed by the manipulation involved in cleaning and mounting the *Chara* cell, nor by the vibrating probe per se), this technique can provide much information on the spatial distribution of membrane transport function as well as on temporal aspects of transport activation.[51,52]

Two- and Three-Dimensional Vibrating Probes. The 1-D vibrating probe provides current density information only for the one vector position, that being the center and direction of vibration. If a probe could be vibrated in such a way that both the normal and tangential components of the electric field could be detected at the same time, the current density could be computed in the 2-D plane of vibration. Freeman *et al.*[61] developed such a 2-D probe system by using two miniature loudspeakers mounted at right angles. However, this method had an undesirable side effect, in that the speaker coils produced an appreciable electromagnetic field, which complicated signal processing. This problem was avoided by using two piezoelectric bender elements mounted at right angles and vibrated out of phase (90°) so that the platinum black electrode moved in a circular pattern. This new 2-D approach was codeveloped by Nuccitelli[62] and Freeman *et al.*[63] A commercial system has recently become available (Model NR-2000, Vibrating Probe Co.).

The theory and operating details of this advanced system are very

[61] J. A. Freeman, P. B. Manis, G. J. Snipes, B. N. Mayes, P. C. Samson, J. P. Wikswo, Jr., and D. B. Freeman, *J. Neurosci. Res.* **13**, 257 (1985).
[62] R. Nuccitelli, *in* "Ionic Currents in Development " (R. Nuccitelli, ed.), p. 13. Liss, New York, 1986.
[63] J. A. Freeman, P. B. Manis, P. C. Samson, and J. P. Wikswo, Jr., *in* "Ionic Currents in Development" (R. Nuccitelli, ed.), p. 21. Liss, New York, 1986.

similar to those described for the 1-D probe. In the 2-D system, a stationary platinum black reference electrode is mounted on the front of the vibrator assembly, via a miniature braided wire pin. A similar pin is used to attach the measuring platinum black electrode. These electrodes differ from those used in the 1-D system in that they are unshielded wire probes, constructed from a 90% platinum – 10% iridium wire that has been coated with a layer of insulation (Parlene, 1 μm thick; Nova Tran Corp., Clear Lake, MI). The insulation is removed from the electrode tip by local heating, and platinum black is electroplated onto the tip until the desired size is obtained. These new electrodes are a big improvement over the glass microcapillaries. The wire electrodes can be bent to fit any application, and new platinum black tips can be plated, *in situ,* using the NR 2000 system manufactured by the Vibrating Probe Company .

Signal processing involves the following: (1) The outputs from the two electrodes are connected to a low-noise differential preamplifier. (2) The preamplified signal is fed into a two-phase lock-in amplifier that detects both the in-phase and quadrature components of the signal. (3) The two outputs of the lock-in samplifier are then digitized by an analog-to-digital converter (Labmaster TM40 PGH, Tekmar Co., Solon, OH) and analyzed by an IBM PCXT computer. (4) The computer calculates the averaged current density vector, and the actual values (vectors) can be stored on hard disk as well as be displayed on a video screen for real-time experimental analysis. Associated software for the IBM PCXT is commercially available. Further details of these new 2-D and 3-D systems can be found in the proceedings of a recent conference in which technical advances and special applications of the vibrating probe were reviewed.[64]

Effects of Probe Mixing. As mentioned above, a problem may exist in terms of extrapolating from the voltage gradient, as measured by the vibrating probe, to the physiological processes occurring at the *Chara* plasmalemma. In an earlier section we discussed the relative importance of ∇C_{OH} and ∇E, in terms of the flux leaving an OH$^-$ band. Extracellular electric field measurements indicate that $\nabla C_{OH} \gg \nabla E$, yet when the vibrating probe is used large extracellular "currents" can be measured.[51,57]

This problem has been recently addressed by Ferrier and Lucas,[60] who showed that when the experimental medium contains an ionic species whose diffusion coefficient is significantly greater (D_h) than all other ionic species present (D_i), a non-Ohmic condition exists. In this situation, the current density is given by

$$I = -(1 - \beta)Fz_hD_h \nabla C_h - \sigma \nabla E \tag{19}$$

[64] R. Nuccitelli, ed., "Ionic Currents in Development." Liss, New York, 1986.

where $\beta = D_i/D_h$ and $\sigma = \Sigma_i D_i C_i (z_i F)^2/RT$ is the electrical conductivity of the medium. Clearly, if $\beta \simeq 1.0$, Ohm's law holds. However, if high mobility ions such as H^+ or OH^- carry the current, there will be a significant deviation from Ohm's law, since β will be much less than 1. Under the bathing medium conditions used in our *Chara* experiments, $\beta = 0.17$ for protons and 0.31 for OH^-. Results of a computer numerical analysis study, based on Eqs. (14) and (19), indicated that the convection loops produced by the vibrating probe have a *profound* effect on the activity gradients, and hence on the electric potential gradient, in the case where β is not close to 1.0. If the vibrating probe measures the electric potential gradient within one of the convection loops, which appears to be the case judging from the observations made by Jaffe and Nuccitelli,[59] then the measured electrical gradient will be increased *relative* to the gradient *without* convection. Based on their simulation studies, Ferrier and Lucas suggested that the electrical gradient could be almost 9 times greater than the gradient present without convection.[60] Thus, the "current" density obtained with the vibrating probe depends, in a complex manner, on the convection velocity and on the specific properties of the ionic species carrying the flux ("current"). Clearly, this complexity must always be borne in mind when using this technique.

Ion Fluxes during Electrophysiological Transients

Rapid Measurement of Changes in External Ion Concentration. There has been considerable interest in identifying and quantifying the ionic fluxes which occur during the characean action potential. Early studies[65,66] simply reported measurements of isotopic efflux of K^+ and Cl^- after repetitive electrical stimulation of cells in small volumes of CPW. In most cases, conventional liquid membrane-based ion-selective electrodes simply do not respond fast enough to record the changes in ion concentration. However, Oda[67] developed an apparatus which not only enables simultaneous estimation of K^+ and Cl^- efflux during a single action potential, but also gives an indication of the time course of these fluxes.

Figure 3 is a diagrammatic representation of the apparatus used by Oda. The cell, 50–75 mm long, is preconditioned in CPW overnight, then for 30 min in CPW in which the K^+ and Cl^- salts are replaced by Na^+ and NO_3^-, before being placed in the apparatus. Each of the three compartments A, B, and C are milled out of a Plexiglas block. Solution is perfused (3.6 ml/min) into A through glass tubing (3 mm i.d.), thence through a

[65] L. J. Mullins, *Nature (London)* **196**, 986 (1962).
[66] A. B. Hope and G. P. Findlay, *Plant Cell Physiol.* **5**, 377 (1965).
[67] K. Oda, *Plant Cell Physiol.* **17**, 1085 (1976).

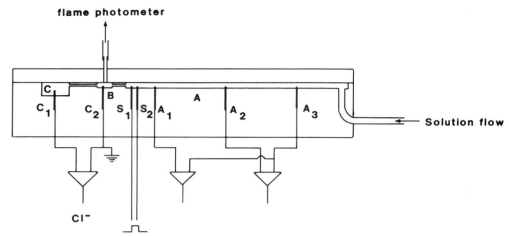

FIG. 3. Apparatus for measurement of net K^+ and Cl^- fluxes during an action potential.[67] The cell is placed in compartment A. Ag/AgCl electrodes A_1, A_2, and A_3 record the action potential, which is initiated by a current pulse between electrodes S_1 and S_2. [Cl^-] in compartment B is monitored by Ag/AgCl electrodes in compartments B and C. Dimensions of the compartments (length $\times$ depth $\times$ width, mm): A, $80 \times 2 \times 2$; B, $5 \times 1.5 \times 1.5$; C, $10 \times 5 \times 20$.

5-mm length of 1 mm i.d. tubing to B, and out to an atomizer burner of a flame photometer (Hitachi Type-EPU) via silicon rubber tubing and glass tubing (1 mm i.d.) inserted through the plate. The K^+ concentration is monitored in the effluent by the flame photometer. Five Ag/AgCl electrodes are inserted in A as shown, the distance A_1-A_2 and A_2-A_3 being 25 mm. The two S electrodes are for electrical stimulation, and the three A electrodes, for recording the action potential. The electrodes labeled C are also of Ag/AgCl and are located in compartments B and C to record the change in Cl^- concentration in compartment B. The solution in compartment C is static, and contact with B is made through glass tubing 10 mm long and 0.3 mm i.d. The internodal cell is positioned in compartment A.

On application of a suprathreshold stimulus (e.g., 30 μA for 10 msec) an action potential is propagated which can be recorded extracellularly by the A electrodes. The outputs from the three amplifiers (two signaling the action potential and one, the change in Cl^- concentration), together with the output of the flame photometer, are recorded on a four-channel oscillograph (Yokogawa Electric Works Type-2901 "Photocorder").

The Cl^- electrode can be calibrated by perfusion of standard solutions through the apparatus, and exhibits the theoretically predicted response:

$$E = 58 \log([Cl^-] + 19)/19 \tag{20}$$

where E is the electrode output in mV, $[Cl^-]$ the Cl^- concentration in μM, and 19 the constant derived from the solubility product of AgCl. The operational lag in response time of the apparatus is about 1 sec. The K^+ and Cl^- concentrations appear to rise simultaneously, and to about the same extent (17.8 and 18.1 μmol m^{-2} impulse^{-1}, respectively). More recent work,[68] using a similar method, however, has indicated that the Cl^- flux during an action potential is not obligatory.

Transient Changes in Cytosolic Free Ca^{2+}. Imbalance between the K^+ and Cl^- fluxes, as well as electrophysiological data,[69] led to the conclusion that Ca^{2+} influx might provide some of the charge compensation, for K^+ efflux, during the action potential. Cytoplasmic free Ca^{2+} can be measured in intact or perfused *Chara* with either intracellular Ca^{2+} microelectrodes (see above) or the photoprotein aequorin.[70,71] A detailed discussion of the methods for preparation of aequorin and quantification of the light signal is given elsewhere.[70,72,73] Injection into intact cells[70] is accomplished by first centrifuging the cytoplasm to one end of an internodal cell and then reducing turgor in Ca^{2+}-free CPW plus 200 or 250 mM sucrose. Desalted aequorin is dissolved in 150 mM KCl, 10 mM Na$_2$TES, pH 7.5, and about 50 nl of the solution is introduced into the endoplasm with a micropipette. The cell is then returned to CPW. The light emission takes a considerable time (up to 1 hr) to stabilize, perhaps because of Ca^{2+} entry during injection.

The resting level of cytosolic free calcium in *Chara* is 220 nM. During an action potential, a transient rise to 6.7 μM is observed in intact cells.[70] The major sources of error in the aequorin method are a lack of information concerning cytosolic free Mg^{2+}, which has a profound influence on the calibration curve, and poor mixing during cell lysis, which necessitates a large correction of the estimate for the peak emission signal.

[68] M. Kikuyama, K. Oda, T. Shimmen, T. Hayama, and M. Tazawa, *Plant Cell Physiol.* **25**, 965 (1984).

[69] M. J. Beilby, *Plant, Cell Environ.* **7**, 415 (1984).

[70] R. E. Williamson and C. C. Ashley, *Nature (London)* **296**, 647 (1982).

[71] M. Kikuyama and M. Tazawa, *Protoplasma* **117**, 62 (1983).

[72] C. C. Ashley and A. K. Campbell, "Detection and Measurement of Free Ca^{2+} in Cells." North-Holland Publ., Amsterdam, 1979.

[73] J. R. Blinks, W. G. Wier, P. Hess, and F. G. Prendergast, *Prog. Biophys. Mol. Biol.* **40**, 1 (1982).

Investigation of Factors Regulating Ion Transport

Adenine Nucleotide Levels in Chara

Numerous biophysical studies on perfused *Chara* have indicated that the plasmalemma-bound H^+-translocating ATPase is strongly influenced by the cytosolic levels of ATP and ADP.[74] Several different protocols can be used to measure the adenine nucleotide levels in *Chara*. Until recently, the most widely used method was a modification of the luciferin–luciferase assay.[75] At present, the method of choice is chromatographic analysis of ATP, ADP, and AMP by high-performance liquid chromatography (HPLC).[76]

At the conclusion of an experimental treatment, individual *Chara* cells are blotted on a paper towel, to remove excess moisture, and then quickly frozen in liquid nitrogen. Frozen cells are lyophilized (Dura Dry Freeze Dryer, FTS Systems Inc., Stone Ridge, NY), weighed, and then ground in 2.0 μl of ice-cold 1.0 M $HClO_3$ using a ground-glass homogenizer. In addition, 2 mg ml^{-1} of Polycar AT (Polysciences, Warrington, PA) is added to the extract. The homogenizer is rinsed with 1.0 ml of 50 mM potassium phosphate buffer (pH 7.5) and, after allowing 20 min for complete extraction and protein denaturation, the rinse and extract are combined. This mixture is then centrifuged at 25,000 g for 30 min at 40°, and the supernatant is adjusted to pH 7.5 with 5 N KOH. After waiting 20 min to allow for the precipitation of potassium perchlorate, the extract is again centrifuged for 30 min at 25,000 g. The supernatant can then be stored at $-70°$ until assayed.

Extracts are analyzed for AMP, ADP, and ATP by first forming fluorescent 1,N-ethenoadenosine derivatives by reacting the nucleotides with chloracetaldehyde and then separating them using an HPLC system. In order to maximize precision and reduce the breakdown of ATP, it is necessary to modify the procedure of Yoshioka and Tamura[77] and Preston[78] by using phosphate instead of acetate buffer. Chloroacetaldehyde (10 μl) is added to 500 μl of extract (pH 7.5) in Pyrex test tubes. The tubes are covered with glass marbles and placed in a boiling water bath for 5 min. Caution is needed with this step, as prolonged boiling results in breakdown of ATP and ADP. Nucleotide standards are handled in a similar manner, except that 100 μl of 0.1 M potassium phosphate buffer (pH 7.5) is also

[74] M. Tazawa, T. Shimmen, and T. Mimura, *Annu. Rev. Plant Physiol.* **38**, 95 (1987).

[75] W. J. Lucas, C. Wilson, and J. P. Wright, *Plant Physiol.* **74**, 61 (1984).

[76] C. W. Wilson, J. W. Oross, and W. J. Lucas, *Planta* **164**, 227 (1985).

[77] M. Yoshioka and Z. Tamura, *J. Chromatogr.* **123**, 220 (1976).

[78] M. R. Preston, *J. Chromatogr.* **275**, 178 (1983).

added prior to boiling. [Chloroacetaldehyde is obtained by distilling a solution of chloroacetaldehyde dimethyl acetal (100 g) (Aldrich, Milwaukee, WI) in 500 ml of 50% sulfuric acid (w/v) as described by Secrist *et al.*[79]]

Chromatography can be performed using a standard HPLC system (Beckman Series 344 Gradient Liquid Chromatograph) equipped with an Ultrasphere IP column (4.6 mm diameter, 250 mm long). The eluting peaks are detected by using a fluorescence detector (Model 157; Beckman Instruments) and analyzed on an Altex Chromatopac C-^{R}1B data processor (Beckman Instruments). The etheno derivatives of AMP and ADP can be separated using a mobile phase consisting of 10% acetonitrile and 90% of a solution composed of 50 mM tetrabutyl-ammonium hydroxide (Fisher Scientific, Pittsburgh, PA) and 50 mM, KH_2PO_4 (concentrated phosphoric acid is used to adjust the pH value to 2.5). ATP is analyzed separately using a 30% acetonitrile and 70% mixture of the above-detailed components. Fluorescence is measured using a 280 nm excitation filter and a 350–470 nm emission filter to eliminate detection of cytidine derivatives. A 20 μl injection volume and a flow rate of 1 ml min^{-1} are suitable for all analyses.

Peaks are identified using retention times of standards. Further confirmation of the peaks in the *Chara* cell extracts can be obtained by enzymatic analysis. Use of pyruvate kinase and adenylate kinase (Sigma) in combination, or pyruvate kinase alone, results in the disappearance of the peaks which have been identified as AMP and ADP, respectively, on the basis of retention times. With these treatments there is a concomitant increase in the ATP peak. The putative ATP peak can be removed by addition of hexokinase or calcium phosphate to the extract. These enzymatic conversions are performed using the procedure of Kimmich *et al.*[80]

Using the above-mentioned procedures, the ATP, ADP, and AMP levels can be determined and expressed on a pmol mg dry weight^{-1} basis. By weighing the *Chara* cells, prior to freezing in liquid nitrogen, and by assuming that the cytosol occupies 7% of the cell volume, these nucleotides can also be expressed on a millimolar basis.

Cytoplasmic pH

Cytoplasmic pH (pH$_c$) is a crucial factor which can be shown, by intracellular perfusion, to control the activity of both the primary[81] and

[79] J. A. Secrist III, J. R. Barrio, N. J. Leonard, and G. Weber, *Biochemistry* **11**, 3499 (1972).
[80] G. A. Kimmich, J. Randles, and J. S. Brand, *Anal. Biochem.* **69**, 187 (1975).
[81] M. Tazawa and T. Shimmen, *Bot. Mag.* **95**, 147 (1982).

secondary[82] transport systems of *Chara*. It is therefore important that accurate estimates of pH_c are available for intact cells, especially where mediation of pH_c is suspected in the effect of environmental conditions on ion transport. Liquid membrane-based H^+ microelectrodes for intracellular use have been discussed in an earlier section. Nevertheless, for those whose manual dexterity and patience do not extend to ion-selective electrode technology, two alternative methods are available. The principles underlying application of weak acid distribution and of ^{31}P nuclear magnetic resonance to measurement of intracellular pH have both been outlined in previous volumes of this series[83,84]: the methods are considered here only from the practical aspect of their use with *Chara*.

Weak Acid Technique with 5,5-Dimethyl-2,4-oxazolidinedione (DMO). After overnight pretreatment in CPW, batches of six to eight internodal cells are rinsed for 2 min and transferred to CPW containing [^{14}C]DMO at $30-60 \ \mu M$. After incubation for between 1.5 and 3 hr, the cells are individually removed, washed for 2 min at the pH of the DMO-containing solution, to remove [^{14}C]DMO from the cell wall, and a measured sample of vacuolar sap obtained. The sap samples and the remainder of the cell (cytoplasm, plus some sap) are assayed separately for radioactivity by liquid scintillation counting. The cytoplasmic [^{14}C]DMO is obtained by subtraction of the vacuolar from the whole cell concentrations, assuming the cytoplasm to occupy 7% of the intracellular space. Cytoplasmic pH can then be calculated, assuming concentration equilibrium of DMO free acid across the plasma membrane, by applying the Henderson–Hasselbalch equation as

$$pH_c = pK + \log[(C_c/U_o) - 1] \tag{21}$$

where pK is the pK for DMO (6.38), C_c the total DMO concentration in the cytoplasm, and U_o the external concentration of DMO in the free acid form (given also by the Henderson–Hasselbalch equation as $C_o/[1 + 10^{(pH_o - pK)}]$, with C_o being the total external DMO concentration and pH_o, the external pH).

However, the vacuolar accumulation ratio (C_v/U_o) is frequently observed to be less than 1, indicating net efflux of DMO^- across the plasma membrane, driven by the highly negative membrane potential. Thus, pH_c can also be calculated by assuming, rather, that DMO free acid equilibrates across the tonoplast, where the electrical potential is an order of magnitude lower than that at the plasma membrane. In that case, a relationship

[82] D. Sanders, *J. Membr. Biol.* **53**, 129 (1980).

[83] H. Rottenberg, this series, Vol. 55, p. 547.

[84] E. Padan and S. Shuldiner, this series, Vol. 125, p. 337.

analogous to Eq. (21) is applied:

$$pH_c = pK + \log[(C_c/U_v) - 1] \tag{22}$$

with U_v being the concentration of undissociated free acid in the vacuole. A reliable determination of U_v can be made, analogously to that of U_o, if the pH of the vacuolar sap sample is measured independently using a miniature pH electrode. However, if the vacuolar pH (pH_v) is 5.0 or less (as has been measured[85]), then C_v can be taken as equal to U_v, thereby precluding the need for individual determinations of pH_v. However, independent spot checks of pH_v are advisable before the simplification is applied. At an external pH of 6.3, a value of $pH_c = 7.75$ is obtained by application of Eq. (21), while application of Eq. (22) yields estimates about 0.1 pH unit higher.[86] pH_c falls by 0.2–0.3 units in response to darkness.

Clearly a major problem with the technique is poor time resolution: the half-time for equilibration of DMO is 15–20 min, and long incubation periods have to be used. Smith[87] has reported that the time resolution can be vastly improved because exchange of DMO across the tonoplast is considerably more rapid than across the plasma membrane. Thus, pH_c can be calculated from Eq. (22) even before the cell achieves equilibrium with the bathing medium. This enables estimates of pH_c to be obtained after as little as 5 min exposure.

^{31}P Nuclear Magnetic Resonance. Because of the highly vacuolate nature of *Chara* internodal cells, closely packed intact cells must be spun for prolonged periods (> 9 hr) in order to obtain a peak for cytoplasmic P_i.[88] In the study conducted by Mimura and Kirino,[88] a pH_c of 7.84 was obtained when cells were illuminated, with a value of 7.3 being measured in darkness. Since no provision was made for sample aeration, their cells would have been partially anaerobic in the dark, and this might be why a large drop in pH_c was observed. Future studies using this technique must incorporate aeration and utilize high-frequency spectrometers.

Reaction Kinetic Modeling of Fluxes

Cytoplasmic perfusion enables direct experimental control of solution composition on both sides of the plasma membrane, and hence the kinetics of transport can be studied in some detail with respect to variation of external *and* internal ligand concentrations. The aim of reaction kinetic

[85] J. A. Raven and F. A. Smith, *J. Exp. Bot.* **29**, 853 (1978).
[86] F. A. Smith, *J. Exp. Bot.* **35**, 43 (1984).
[87] F. A. Smith, *J. Exp. Bot.* **37**, 1733 (1986).
[88] T. Mimura and Y. Kirino, *Plant Cell Physiol.* **25**, 813 (1984).

modeling is to provide a framework within which the kinetics of transport might be interpreted, and, ultimately, to gain insight into the partial reactions of the carrier which catalyzes transport.

A detailed analysis has been undertaken for 2 H^+/Cl^- symport at the plasma membrane of *Chara*.[14] It is assumed that the reaction cycle of the carrier can be represented by a scheme of the type shown in Fig. 4, in which the ligand-binding sites are exposed alternately to each side of the membrane. For the two ligands H^+ and Cl^-, four separate ordered binding models can be specified for influx with Cl^- either binding to and dissociating from the carrier first (first on–first off: FF, as in Fig. 4), binding first and dissociating last (first on–last off: FL), binding last and dissociating last (last on–last off: LL), or binding last and dissociating first (last on–first off: LF). Each of these variants comprises two congeners: that for which the excess positive charge carried by the second H^+ is carried on the loaded form of the carrier (FF+, FL+, etc.) and that for which the charge on the second H^+ is neutralized by the negative charge on the unloaded form of the carrier (FF−, FL−, etc.).

All analyses of reaction kinetic models have been performed assuming that ligand concentration effects on the kinetics of transport are mediated solely via mass action at the appropriate binding sites for transport, and that membrane electrical potential difference influences only those reactions responsible for transmembrane charge translocation. (These assump-

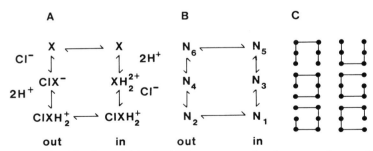

FIG. 4. Reaction kinetic model for 2 H^+/Cl^- symport at the plasma membrane of *Chara*. (A) Carrier (X) binds to Cl^- and 2 H^+ with binding sites exposed to the external surface of the membrane and undergoes a conformational change to release the ligands internally. The model (one of eight variants of ordered ligand binding) is first on–first off with respect to inward transport of Cl^-, and carries inward current on the loaded form of the carrier (FF+). (B) General representation of ordered binding carriers, defining carrier state densities (N_j) in terms of the state number ($j = 1–6$). The 12 unidirectional rate constants are defined according to the reaction states participating (e.g., k_{13} is the rate constant for the transition from state 1 to state 3). (C) The six King and Altman diagrams for the ordered binding models.

tions appear to be the simplest which can be made and are, of course, subject to modification if kinetic data fall outside the realm of description by all eight models.)

The kinetics of isotope transport through ordered binding models is described for initial rate measurements by a rate equation which can be derived by the method of King and Altman.[89] Since the internal concentration of isotope ($[^{36}Cl^-]_i$) is 0, the influx of labeled Cl^-, $^*\phi_{Cl}$, is given for the FF+ model as

$$^*\phi_{Cl} = {^*N_1}k_{13} \tag{23}$$

where $^*N_1$ is the concentration of carrier bound to $^{36}Cl^-$ at the inner surface of the membrane (Fig. 4) and k_{13} the rate constant of the partial reaction from state N_1 to state N_3. The proportion of carrier state 1 that is isotopically labeled is not immediately known. For purposes of calculation of the isotopic flux it is therefore easier to express Eq. (23) in terms of carrier states not bound to isotope, i.e., the carrier state 6 to which Cl^- binds externally. The following rate equations can then be solved for N_6, given steady state of the carrier:

$$\frac{d^*N_1}{dt} = 0 = {^*N_2}k_{21} - {^*N_1}(k_{12} + k_{13}) \tag{24}$$

$$\frac{d^*N_2}{dt} = 0 = {^*N_1}k_{12} + {^*N_4}k_{42} - {^*N_2}(k_{21} + k_{24}) \tag{25}$$

$$\frac{d^*N_4}{dt} = 0 = {^*N_2}k_{24} + {^*N_6}k_{64} - {^*N_4}(k_{42} + k_{46}) \tag{26}$$

in which the N and k symbols have analogous meanings to those in Eq. (23). Substitution of Eq. (24)–(26) into Eq. (23) leads to

$$^*\phi_{Cl} = \frac{N_6 k_{64} k_{42} k_{21} k_{13}}{k_{46} k_{24}(k_{12} + k_{13}) + k_{21} k_{13}(k_{46} + k_{42})} \tag{27}$$

It simply remains, then, to express the concentration of carrier state N_6 in terms of the rate constants of the cycle. The King and Altman method for expressing N_6 as a fraction of total carrier concentration (N_T) can be summarized as follows. The six-state scheme in Fig. 4B is redrawn in Fig. 4C as six possible permutations of five contiguous partial reactions. Each diagram can be related to any of the six states as the product of the five rate constants directed toward that state. The carrier state concentration N_6 can then be expressed as a proportion of N_T by the ratio of the six King and

[89] E. L. King and C. Altman, *J. Phys. Chem.* **60**, 1375 (1956).

Altman diagrams relating to N_6 and the 36 diagrams describing all six states, i.e.,

$$N_6 = N_T \frac{6KA_6}{36KA_T} \qquad (28)$$

in which

$$KA_6 = k_{13}k_{35}k_{56}[k_{46}(k_{21} + k_{24}) + k_{42}k_{21}]$$
$$+ k_{46}k_{24}k_{12}[k_{31}(k_{53} + k_{56}) + k_{35}k_{56}] \qquad (29)$$

and $36KA_T$ can be split into 21 k_{64}-independent terms and 15 k_{64}-dependent terms, respectively, as follows:

$$k_{46}(k_{24} + k_{21})[k_{65}k_{53}(k_{13} + k_{31}) + k_{13}k_{35}(k_{65} + k_{56})]$$
$$+ (k_{42}k_{21}k_{13} + k_{46}k_{24}k_{12})[k_{65}(k_{35} + k_{53}) + k_{35}k_{56}] + k_{53}k_{65}k_{12}k_{31}$$
$$\times (k_{46} + k_{42} + k_{24}) + k_{46}k_{24}k_{12}k_{31}(k_{53} + k_{56} + k_{65}) + k_{65}k_{42}k_{21}k_{31}k_{53}$$

and

$$k_{64}\{(k_{56} + k_{53})[k_{42}k_{21}(k_{13} + k_{31}) + k_{12}k_{31}(k_{42} + k_{24})]$$
$$+ k_{35}k_{56}[(k_{42} + k_{24})(k_{12} + k_{13}) + k_{42}k_{21}] + k_{21}k_{13}k_{35}(k_{42} + k_{56})\}$$

Note that the derivation does not include the assumption that the transmembrane reactions are rate-limiting, which results in unnecessarily restrictive conclusions concerning kinetic behavior. Substitution of Eq. (28) into Eq. (27) yields the full rate equation. The electrochemical gradients of the ligands are automatically incorporated in the full rate equation: concentration terms are subsumed in the ligand binding reactions, which vary linearly with concentration (Cl^-) or as the square of concentration (H^+), and the effect of membrane potential on the charge transit reactions is incorporated in the form of a symmetric Eyring barrier. Analogous equations can also be derived for the other three ligand binding orders.[90] The behavior of the equations can now be examined with respect to variation of components of the electrochemical gradients.

The full rate equation always describes a Michaelis–Menten relationship as a function of external Cl^- concentration ($[Cl^-]_o$). This can be seen quite simply because k_{64}, which subsumes $[Cl^-]_o$, is present as a multiplier in the numerator and in some, but not all, terms in the denominator. However, it is convenient to simplify the equation for the purpose of designing experiments to determine binding order, since in their complete form, the rate equations for all models can predict a wide variety of kinetic

[90] D. Sanders, U.-P. Hansen, D. Gradmann, and C. L. Slayman, *J. Membr. Biol.* **77**, 123 (1984).

responses. Simplification can be achieved experimentally by setting some of the reaction constants to zero, and making others very large. For $2H^+/Cl^-$ symport in *Chara*, experiments were performed at saturating external $[H^+]$ (i.e., k_{42} very large) and with membrane electrical potential rather negative (k_{21} large, k_{12} small for the FF+ model). In these conditions, the two Michaelian parameters simplify to

$$\phi_{max} = \frac{k_{13}k_{35}k_{56}}{(k_{31} + k_{13})(k_{53} + k_{56}) + k_{35}(k_{13} + k_{56})} \tag{30}$$

$$K_m = \frac{k_{53}k_{65}(k_{31} + k_{13}) + k_{13}k_{35}(k_{56} + k_{65})}{k_{64}^o[(k_{31} + k_{13})(k_{53} + k_{56}) + k_{35}(k_{13} + k_{56})]} \tag{31}$$

where k_{64}^o is the rate constant from N_6 to N_4 with $[Cl^-]_o$ extracted. If intracellular chloride ($[Cl^-]_i$) is now varied in perfused cells, Eqs. (30) and (31) predict linear noncompetitive inhibition if k_{31} is greater than k_{13} and $k_{35}(k_{13} + k_{56})$. Similarly, in the absence of internal Cl^-, variation of intracellular H^+ ($[H^+]_i$) can result in noncompetitive inhibition if k_{53} is greater than k_{35} and k_{56}. In both these conditions, noncompetitive inhibition is observed experimentally.[14] However, none of the other orders for ligand binding, or an alternative site for charge translocation, is compatible with these kinetics. Thus, on this basis, it is possible to conclude that Cl^- binds to the carrier first at the external surface, dissociates first at the inner surface, and that charge translocation is on the loaded form of the carrier. These algebraic methods therefore demonstrate the utility of the reaction kinetic modeling approach in the design of experiments on perfused *Chara*.

[29] Light-Induced Hyperpolarization in *Nitella*

By ROGER M. SPANSWICK

Introduction

An electrogenic H^+-translocating ATPase is central to our understanding of the transport of ions and small organic molecules across the plasma membrane of plant cells in terms of a chemiosmotic scheme.[1,2] The influence of this electrogenic pump on the electrophysiological properties of the

[1] J. A. Raven, this volume [25].
[2] L. Reinhold and A. Kaplan, *Annu. Rev. Plant Physiol.* **35,** 45 (1984).

plasma membrane of freshwater algal cells is relatively much greater than that of the Na^+,K^+-ATPase in animal cells, mainly because the passive ion fluxes, which act to short-circuit the pump, are much smaller in algal than in animal cells. As a result, the electrical properties of the membrane reflect predominantly the characteristics of the electrogenic pump. One interpretation of the behavior of the membrane is that the electrogenic pump operates in a region of its current–voltage relationship in which it demonstrates the property of electrical conductance.[3,4] Thus, measurements of membrane conductance as well as membrane potential are important in characterizing the pump. This is illustrated in the light-induced hyperpolarization of the membrane potential of *Nitella translucens*[3,5] which is accompanied by a marked decrease in membrane conductance.

Although the electrogenic pump has been demonstrated to be dependent on ATP as an energy source, using both inhibitor studies[6] and intracellular perfusion,[7,8] the light-induced hyperpolarization in intact cells is not accompanied by a significant change in the cellular ATP level,[6,9-11] and the mechanism by which the pump is activated remains to be elucidated.[12] While the kinetics of H^+ transport are more easily investigated in membrane vesicles isolated from higher plants than in intact cells,[13] the characean internodal cells, with their cylindrical geometry and the relative ease with which electrophysiological measurements can be made, provide a system in which the electrical properties of the pump can be investigated in much greater detail than in higher plants. Determination of the optimal conditions for the demonstration of the light-induced hyperpolarization in *Nitella* involves both empirical and theoretical considerations.

Electrophysiological Techniques

Membrane Potential. The membrane potential in plant cells is usually measured between a microelectrode placed with its tip in the vacuole and a

[3] R. M. Spanswick, *Biochim. Biophys. Acta* **288,** 73 (1972).

[4] R. M. Spanswick, *Annu. Rev. Plant Physiol.* **32,** 267 (1981).

[5] R. M. Spanswick, *Biochim. Biophys. Acta* **332,** 387 (1974).

[6] D. W. Keifer and R. M. Spanswick, *Plant Physiol.* **64,** 165 (1979).

[7] T. Shimmen and M. Tazawa, *J. Membr. Biol.* **37,** 167 (1977).

[8] M. J. Morse and R. M. Spanswick, *Biochim. Biophys. Acta* **818,** 386 (1985).

[9] B. Penth and J. Weigl, *Planta* **96,** 212 (1971).

[10] R. M. Spanswick and A. G. Miller, *in* "Transmembrane Ionic Exchanges in Plants" (M. Thellier, A. Monnier, M. Demarty, and J. Dainty, eds.), p. 239. CNRS, Paris/Rouen, 1977.

[11] Y. Takeuchi and U. Kishimoto, *Plant Cell Physiol.* **24,** 1401 (1983).

[12] R. M. Spanswick, *in* "Plant Membrane Transport: Current Conceptual Issues" (R. M. Spanswick, W. J. Lucas, and J. Dainty, eds.), p. 305. Elsevier/North-Holland Biomedical Press, Amsterdam, 1980.

[13] H. Sze, *Annu. Rev. Plant Physiol.* **36,** 175 (1985).

reference electrode placed in the bathing medium. However, it is important to be aware that this measures the sum of the potentials across the
plasmalemma and tonoplast. While the greatest part of the potential difference is across the plasmalemma, in *Nitella* there is a potential of about
18 mV across the tonoplast, the vacuole being positive relative to the
cytoplasm.[14] Although the potential difference across the tonoplast remains relatively constant under most conditions, it is advisable to use a
second microelectrode to verify that this is so for a particular experimental
treatment.

Standard techniques for plant cells[15] include the fabrication of micropipettes under relatively low heat (about 18 A on a Kopf vertical puller with
no extra pull from the solenoid, for example) so that the shank has a sharp
taper to provide mechanical strength and the electrode tip is sufficiently
fine (about 1 μm) to permit penetration of the cell wall without breakage.
The microelectrodes can be filled rapidly if they are made from glass
containing a microfiber attached to the interior wall (WPI or Haer). A
filling solution consisting of 3 M HCl adjusted to pH 2 with HCl reduces
electrode tip potentials by suppressing the negative charges on the glass
surface.[16]

The reference electrode can be made by hand-pulling pipettes from
standard 1 or 2 mm outside diameter borosilicate (Pyrex) tubing and
breaking the glass to give a tip diameter of about 10 μm. This is large
enough to prevent the establishment of tip potentials. However, to avoid
leakage of KCl into the bathing solution it is advisable to fill the tip
2–3 cm of the pipette with 2% agar dissolved in 3 M KCl and place the
reference electrode in the bathing solution on the downstream side of the
preparation. The electrodes are completed by connecting the pipettes to
identical metallic electrodes. This is most easily accomplished by using
commercial pipette holders which incorporate, for example, a silver–silver
chloride junction.

The electrodes are connected to a conventional electrophysiological dc
amplifier with a high input impedance, and the output is recorded using a
chart recorder. If the amplifier is single ended (one input grounded),
connect the reference electrode to the ground terminal. It is important to
make sure that the preparation is grounded at only one point. Therefore, if
the membrane conductance is being measured simultaneously (see below)
it is necessary to use a pulse generator with an output isolated from ground
if the potential recording amplifier is single ended. Alternatively, if the

[14] R. M. Spanswick and E. J. Williams, *J. Exp. Bot.* **15**, 422 (1964).
[15] M. J. Beilby, this volume [27].
[16] J. Riemer, C.-J. Mayer, and G. Ulbrecht, *Pfluegers Arch.* **349**, 267 (1974).

current is passed via an external electrode that is grounded, a differential amplifier must be used to measure the membrane potential.

Membrane Conductance. The membrane conductance (the reciprocal of the resistance) is determined by passing a small current across the membrane and measuring the resulting change in the membrane potential.[15] The current is usually derived from a square-pulse generator that is capable of giving pulses of about 1-sec duration since the large surface area of *Nitella* cells can result in a large RC time constant. This length of time is necessary to charge the membrane capacity and for the potential to reach a steady level. *Nitella* cells are sufficiently large for the current to be passed across the membrane by inserting a second microelectrode into the cell and placing a separate "current" electrode (a silver chloride-coated silver wire) next to the cell in the bathing solution.

The pulse generator is connected to the "current microelectrode" via a large electrical resistor (100 MΩ) which serves to limit the flow of current and, being larger than the resistance of the microelectrode, minimizes changes in the magnitude of the current due to variation in the electrode resistance. The current is measured by placing a resistor of accurately known value (in the range 10–100 kΩ) in series with the external current electrode. The current passing through the circuit may then be determined by measuring the voltage generated across this resistance during the current pulse using a good potentiometric recorder or an oscilloscope.

Since the cell is long and the sap has a finite resistance, it behaves analogously to a "leaky cable." In other words, the current crossing the membrane, and hence the resulting change in membrane potential, varies with distance from the current injection microelectrode. Fortunately, it has been shown that the conductance can be calculated from Ohm's law with reasonable accuracy if the placement of the microelectrodes is done properly.[17,18] The current microelectrode should be inserted at the midpoint of the cell, and the potential-measuring microelectrode should be inserted at a distance $0.42l$ from the current electrode, where the length of the cell is $2l$.

Since we are concerned here with relatively slow changes in potential and conductance, the pulse generator can be set to give pulses about once per minute. To avoid triggering the action potential, the pulses should be in the hyperpolarizing direction but should produce a potential change of only a few millivolts. It is important to be aware that a time-dependent conductance change termed the "hyperpolarizing response"[19] may appear under conditions in which the pump is inhibited[5] and may lead to an

---

[17] J. Hogg, E. J. Williams, and R. J. Johnston, *Biochim. Biophys. Acta* **150**, 518 (1968).
[18] J. R. Smith, *Aust. J. Plant. Physiol.* **10**, 329 (1983).
[19] U. Kishimoto, *Annu. Rep. Sci. Workshop, Fac. Sci., Osaka Univ.* **7**, 115 (1959).

underestimate of the membrane conductance if its occurrence is not recognized.

Conditions for Demonstration of Light-Induced Hyperpolarization

A basic problem in attributing the light-induced hyperpolarization to the action of an electrogenic pump arises from the fact that, in experiments using the original "artificial pond water" as a bathing medium, the membrane potential was in the range of possible diffusion potentials. Thus, in the absence of evidence to the contrary, observed changes in membrane potential could be produced by light-induced changes in passive ion permeabilities, which would affect the diffusion potential, and/or activation of an electrogenic pump. However, by minor manipulation of the bathing medium it is possible to remove this ambiguity.

Ion Concentrations. In the original artificial pond water (APW), consisting of 0.1 mM KCl, 1.0 mM NaCl, and 0.1 mM CaCl$_2$, the membrane potential across the plasma membrane of *Nitella translucens* in the light[14] (-138 mV) is more positive than the K$^+$ equilibrium potential (-178 mV) calculated from the Nernst equation, based on the measured cytoplasmic potassium concentration of 119 mM. The potential generated by the passive diffusion of ions across the membrane (E_D) may be described by the Goldman equation:

$$E_D = (RT/F) \ln[(P_K K_o + P_{Na} Na_o + P_{Cl} Cl_i)/(P_K K_i + P_{Na} Na_i + P_{Cl} Cl_o)] \quad (1)$$

where P_K, P_{Na}, and P_{Cl} are the permeability coefficients for K$^+$, Na$^+$, and Cl$^-$ respectively; K_o, K_i, etc., are the internal and external ion concentrations; and R, T, and F have their usual meanings. Examination of this equation shows that in *Nitella translucens* the negative limit of the diffusion potential is set by the Nernst potential for K$^+$, E_K, this being the most negative of the equilibrium potentials.

In some organisms, the presence of an electrogenic pump is clearly indicated by the fact that the membrane potential is more negative than the negative limit of the diffusion potential. This is true of *Neurospora crassa*[20] and *Elodea canadensis*.[21] The same situation can be achieved for *Nitella translucens* in the light by modifying the bathing medium.[3] One important factor is to increase the K$^+$ concentration in APW. There is a limit to which this can be done because above 1 mM the potential decreases sharply, and the membrane conductance and potassium fluxes

[20] C. L. Slayman, *J. Gen. Physiol.* **49**, 69 (1965).
[21] R. M. Spanswick, *Planta* **102**, 215 (1972).

increase. This indicates that there has been an increase in the passive permeability of the membrane to K^+. Under these conditions the membrane potential is very close to the value of E_K calculated from the K^+ concentrations on either side of the membrane, making it impossible to detect the presence of the electrogenic pump. However, it is possible to use this phenomenon to determine the value of E_K when the K^+ concentration is, say, 10 mM in the external solution, and to use this value to calculate the value of E_K for other values of the external K^+ concentration. This makes it possible to determine the value of E_K in a particular cell relative to the membrane potential without correcting for the tip potential of the microelectrode.[3]

It was found[3] that the K^+ concentration could be increased from 0.1 to 0.5 mM with only a few millivolts change in the membrane potential, while the effect on the value of E_K is to make it more positive by about 40 mV. This, combined with the changes described below, permits a clear hyperpolarization of the membrane potential beyond the negative limit of the diffusion potential to be observed. Note that the fact that the membrane potential does not respond to changes in the external K^+ concentration below 1 mM provides an indication that the potential is not controlled simply by passive ion diffusion.

External pH. The demonstration of a strong dependence of the membrane potential of *Nitella* on external pH[3,22,23] provided another parameter that could be varied to increase hyperpolarization of the membrane potential.[3] The pH of unbuffered APW in equilibrium with atmospheric CO_2 is about 5.5. The membrane potential hyperpolarizes as the pH is increased, but the increase tends to be transient above pH 7. A pH of 6.0 gives a good compromise with regard to hyperpolarization and stability. The pH of the APW can be controlled conveniently with 1 mM of an appropriate zwitterionic buffer, and the notation APW6 is used, for example, to indicate that the pH of the APW is 6.0.[3]

Carbon Dioxide. Even with the external pH controlled, there tends to be instability of the membrane potential that is dependent on the rate of flow of the external solution past the cell.[3] By trial and error, this was found to result from the presence of varying amounts of CO_2 in the external solution. Depolarization of the membrane potential can be produced by addition of CO_2 to the solution (Fig. 1), and a stable hyperpolarization in the light can be maintained by scrubbing the solutions with CO_2-free air before use.

Light. By using a CO_2-free solution buffered at pH 6 and containing

[22] H. Kitasato, *J. Gen. Physiol.* **52,** 60 (1968).
[23] R. M. Spanswick, *J. Membr. Biol.* **2,** 59 (1970).

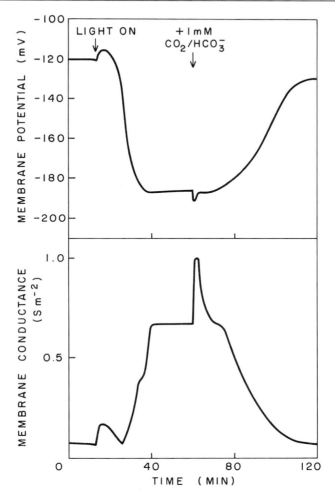

FIG. 1. Typical time course of the membrane potential and conductance in *Nitella translucens* produced by the transition from dark to light, and subsequently by the addition of 1 mM CO_2/HCO_3^- to the external solution (APW6 plus 0.4 mM KCl; Table I).

0.5 mM K^+ (APW6 plus 0.4 mM KCl; Table I), a light-induced hyperpolarization of the membrane potential can be demonstrated in *Nitella translucens* (Fig. 1). In the dark, the potential is close to the value of E_K, and the membrane conductance is low. Illumination results in a small transient depolarization of the membrane potential followed by a hyperpolarization of about 40 mV which takes the potential significantly beyond the negative limit of the diffusion potential set by E_K. The hyperpolarization is accom-

TABLE I
COMPOSITION OF APW6 PLUS 0.4 mM KCl[a]

Ion or compound	Concentration (mM)
Na$^+$	1.05[b]
K$^+$	0.5
Ca^{2+}	0.1
Mg^{2+}	0.1
Cl$^-$	1.4
MES^c	1.0

[a] Based on the solution used by Spanswick.[3]
[b] In practice, the concentration of Na$^+$ varies slightly because the solution is made up to contain 0.5 mM NaCl, bubbled with CO_2-free air, and then adjusted to pH 6.0 with freshly prepared NaOH.
[c] MES, 2-(N-Morpholino)ethanesulfonic acid.

panied by a marked increase in the membrane conductance. Since this effect could not be accounted for by a stimulation of the passive fluxes in the light, it was postulated[3] that the conductance change resulted from the activation of an electrogenic pump with conductance. It is then possible to account for the light-induced hyperpolarization in terms of an increase of the conductance of the electrogenic pump (g_P) using an equivalent circuit for the membrane in which the pump is in parallel with the passive ion channels in the membrane,[4] and with reasonable values for the diffusion potential [E_D, given by Eq. (1)], the electromotive force (EMF) of the electrogenic pump (E_P), and the passive conductance of the diffusion channels (g_D). The equation is

$$E_m = (g_P E_P + g_D E_D)/(g_P + g_D) \qquad (2)$$

It can be seen that an increase in g_P relative to g_D causes the potential to shift from E_D toward E_P and vice versa.

A convenient light source is a 150-W flood lamp separated from the preparation by water and glass heat filters. An irradiance of 10 W m^{-2}, or a photon flux density (400–700 nm) in the range 55 to 65 μmol m^{-2} sec^{-1}, was sufficient to saturate the response.

Dependence of Light-Induced Hyperpolarization on ATP Hydrolysis

Stimulation of ions fluxes by light was originally taken to indicate a dependence on energy derived from photosynthesis. However, it has been necessary to modify this conclusion as a result of direct measurements of the levels of ATP present in the cell under different conditions. The first measurements of ATP concentrations in *Nitella* cells demonstrated quite clearly that there was no significant difference in the levels in the light and dark.[9] This has since been confirmed with other species.[6,10,11] An alternative explanation would be that the electrogenic pump does not depend on ATP hydrolysis as an energy source. However, inhibitors that lower the ATP level also inhibit the electrogenic pump, and it has been shown directly using the intracellular perfusion technique that the hyperpolarization can be maintained using ATP as a sole energy source.[7,8]

Application of Luciferase Assay to Single Nitella Cells. The amount of ATP in a single internodal *Nitella* cell can be measured using a variant of the firefly luciferase assay.[6,24] The most important feature of the assay is that it can be performed using a scintillation counter in the noncoincident mode. This is achieved by using a low concentration of the luciferase in the assay and delaying the count until the initial burst of radiation from the reaction and the luminescence from the plastic vial has decayed. The radiation emitted between 30 and 60 sec after the initiation of the reaction is found to be linearly proportional to the amount of ATP added to the assay.

The ATP concentration in the cytoplasm is calculated knowing that the ATP concentration in the vacuole is below the limits of detection and assuming that the cytoplasm occupies 5% of the cell volume. This assumption is probably the greatest source of uncertainty in the assay, but no convenient method is available for estimating the cytoplasmic volume in individual cells.

Intracellular Perfusion. Intracellular perfusion may be used in an attempt to control the internal ATP concentration in characean cells. This may be done by continuous perfusion,[7] which is technically demanding, or by perfusion followed by ligation.[25] In the ligation method it is necessary to include EGTA in the perfusate to destroy the vacuolar membrane and gain access to the plasmalemma.

While it is possible to demonstrate hyperpolarization of the membrane

---

[24] E. Schram, *in* "The Current Status of Liquid Scintillation Counting" (E. D. Bronsome, Jr., ed.), p. 129. Grune & Stratton, New York, 1970

[25] M. Tazawa, M. Kikuyama, and T. Shimmen, *Cell Struct. Funct.* **1**, 165 (1976).

potential in the light in perfused cells, there are some technical problems that have arisen in using this system. One is that the magnitude of the membrane potential is often less than that in the intact system. This problem has been overcome by including lead acetate or lead nitrate in the external medium.[7] Lead acetate has been shown to reduce the permeability of the plasmalemma to K^+.[26] However, it also has secondary effects on the electrogenic pump, giving short-term activation in the dark but long-term inhibition in both the light and the dark (M. J. Morse and R. M. Spanswick, unpublished). A second problem is that the ATP level is not maintained in the perfused and ligated cells.[8] It has been claimed[27] that inclusion of an ATP-regenerating system in the perfusion medium leads to maintenance of ATP concentrations at 80–100% of the original perfusate, but supporting measurements were not provided.

There is also a conceptual problem relating to attempts to compare the K_m for the hyperpolarization, obtained by varying the ATP concentration, with the K_m for ATP hydrolysis obtained with isolated membrane preparations.[27] This would require a fixed stoichiometry between the rate of ATP hydrolysis and the rate of charge (H^+) transfer through the electrogenic pump, as well as a membrane in which the hyperpolarization was proportional to the current through the electrogenic pump. This is not the case in a system such as *Nitella* in which the membrane conductance is affected by the conductance of the electrogenic pump.[8] Because both the conductance and the EMF of the electrogenic pump vary with the ATP level, the hyperpolarization of the membrane potential, given by

$$E_a = g_P(E_P - E_D)/(g_D + g_P) \tag{3}$$

is not simply related to the K_m for ATP hydrolysis or the flux through the pump.

Conclusion

The light-induced hyperpolarization in *Nitella* can be reproduced reliably under the conditions given here. The electrogenic pump is clearly dependent on ATP hydrolysis, in agreement with information available for the H^+-transport ATPase from higher plant plasma membranes.[13] However, the mechanism by which light activates the pump or by which 1 mM CO_2/HCO_3^- at pH 6 in the light inhibits the pump[3,10] is still unclear. One effect of 1 mM CO_2/HCO_3^- is to reduce the cytoplasmic pH from 7.5 to

[26] M. J. Morse and R. M. Spanswick, *in* "Membrane Transport in Plants" (W. J. Cram, K. Janáček, R. Rybová, and K. Sigler, eds.), p. 336. Akademia-Verlag, Prague, 1984.
[27] T. Mimura, T. Shimmen, and M. Tazawa, *Planta* **157**, 97 (1983).

6.0.[10] However, this alone is not sufficient to inhibit the pump since a similar reduction in the cytoplasmic pH using 5 mM dimethyloxazolidine-dione (a permeant weak acid) does not produce a depolarization; indeed, under some conditions in which the membrane has depolarized, possibly because of alkalinization of the cytoplasm,[12] the weak acid treatment can produce repolarization of the membrane. There appear to be differences of only 0.2–0.3 units in the cytoplasmic pH between light and dark,[28] so pH is probably not the major factor involved in the hyperpolarization. The inhibitory action of low concentrations of calmodulin inhibitors on the hyperpolarization in *Chara corallina* suggest the possibility that activation of the ATPase may involve the action of kinases.[29] This is entirely speculative, however, and further investigation is needed.

Although attention has been confined here to conditions in which a stable hyperpolarization of the membrane potential can be obtained, it should be noted that transient changes in the membrane potential may also be used to provide useful information. For example, the membrane potential in *Chara corallina* is strongly hyperpolarized by brief dark treatments in the presence of bicarbonate.[30] While the magnitude of the membrane potential in the "pump" state is usually consistent with a stoichiometry of 2 H^+/1 ATP,[12,31] the potentials measured by Lucas[30] appear to be consistent only with a stoichiometry of 1 H^+/1 ATP.

Another approach, which has been pioneered by Hansen,[32,33] is the analysis of oscillations in the membrane potential, either spontaneous or entrained to sinusoidal variations in irradiance. Using network analysis it is possible to deduce the existence of feedback controls, and progress is now being made in identifying the processes underlying these phenomena.[34]

[28] F. A. Smith, *J. Exp. Bot.* **150**, 43 (1984).
[29] M. J. Beilby and E. A. C. MacRobbie, *J. Exp. Bot.* **153**, 568 (1984).
[30] W. J. Lucas, *Planta* **156**, 181 (1982).
[31] N. A. Walker and F. A. Smith, *in* "Transmembrane Ionic Exchanges in Plants" (M. Thellier, A. Monnier, M. Demarty, and J. Dainty, eds.), p. 255. CNRS, Paris/Rouen, 1977.
[32] U.-P. Hansen, *Biophysik* **7**, 223 (1971).
[33] U.-P. Hansen, *J. Membr. Biol.* **41**, 197 (1978).
[34] J. Fisahn, E. Mikschl, and U.-P. Hansen, *J. Exp. Bot.* **174**, 34 (1986).

[30] ATP-Driven Chloride Pump in Giant Alga
Acetabularia

By D. GRADMANN

Introduction

Most work on the electrogenic Cl⁻ pump in *Acetabularia* is based on *in vivo* investigations rather than on biochemical studies. Therefore, this chapter is mainly concerned with *in vivo* methods for identification and characterization of this enzyme. In particular, electrophysiological techniques are emphasized.

The genus *Acetabularia* comprises unicellular marine green algae. The most frequently used species is *Acetabularia mediterranea*. Like related species, the mature cells of *A. mediterranea* consist of an approximately cylindrical stalk (~ 0.4 mm in diameter and 50 mm in length) with a basal rhizoid (containing the nucleus in earlier developmental states) and an apical, flat cap (~ 10 mm in diameter) in which cysts with gametes are formed in a late developmental state. The giant size and simple geometry of (especially) younger cells without a cap enable microsurgery on individual cells and comfortable impalement by microelectrodes for electrophysiological studies. The experiments referred to in this chapter were done with *A. mediterranea* and *A. crenulata*,[1] the latter having a somewhat larger stalk diameter (~ 1.0 mm). Additional advantages for *in vivo* ion-transport studies (in particular of the electrogenic Cl⁻ pump) on *Acetabularia* are (1) the high ionic strength of the natural environment, allowing the series resistance due to unstirred layer to be ignored, (2) the exceptionally large Cl⁻ fluxes through the pump which dominate the apparent transport properties of the membrane under normal conditions, and (3) the possibility of using a convenient isotope (³⁶Cl) as a tracer.

Culturing and Handling

Since *Acetabularia* was introduced as a standard object in plant physiology, culturing methods have been continuously improved. Hämmerling,[2] Beth,[3] and Keck[4] have described culturing in Erdschreiber medium.

[1] D. Gradmann, *Planta* **93**, 323 (1970).

[2] J. Hämmerling, *Arch. Protistenkd.* **97**, 7 (1944).

[3] K. Beth, *Z. Naturforsch., B: Anorg. Chem., Org. Chem., Biochem., Biophys., Biol.* **8B**, 334 (1953).

[4] K. Keck, *Methods Cell Physiol.* **1**, 189 (1964).

Shepard[5] and Schweiger *et al.*[6] introduced synthetic media. For culturing we use now a very convenient, commercial growth medium with "hw-seasalt" (Wiegandt GmbH & Co., K. G. Krefeld, FRG), supplemented with 5 mM Na$_2$HPO$_4$ and 47 mM NaNO$_3$. It has been analyzed in detail and test by Lüttke.[7] Under laboratory conditions the life cycle of *A. mediterranea* is about 6 months: 2 months for growth at 21° and a 12 hr/12 hr light/dark cycle and 4 months for dormancy of the cysts in darkness. Explicit instructions for handling the cells for reproduction are given by Schweiger *et al.*[6]

For experiments involving controlled changes of the external medium, simple mixtures are made from stock solutions which are isotonic to the growth medium (1.1 os$M$). A standard mixture of artificial seawater for experiments has the following composition (in mM): NaCl (460), KCl (10), CaCl$_2$ (10), MgSO$_4$ (28), MgCl$_2$ (15), buffered with 10 mM Tris-HCl to pH 8.0.

Microsurgery

When individual cells are handled mechanically, hold the hard rhizoid carefully with a pair of soft elastic steel forceps to avoid injury to the tender body of the cells. Under a stereo microscope or even by eye, ligatures can be performed with women's hair or fibers of dental floss. Ligated segments must be cut off from the rest of the cell in order to ensure proper ligatures.

For cell preparations without vacuoles and/or with depleted cytoplasm, cells are clamped by the tough basal end to a holder, on which the cells are gently centrifuged at about 600 *g* for 30 min at room temperature.[8,9] After incubating the cells (for ~1 hr) in neutral red (about 0.2 mM), their condition, in particular the integrity of the membranes, can easily be examined by the red staining of the acid vacuoles: injured cells loose their staining. When the cytoplasm (~10% of the cell volume) with the chloroplasts is spun down to the apical end, this part is free of vacuoles for about 1 hr, after which new vacuoles begin to form.[10] The vacuole-free segments can be tied off and cut from the remainder of the cell. In the large basal part (~90% of the cell length) neutral red staining of the vacuole indicates the proper state of the preparation. Parts (of ~10 mm length) of these

[5] D. C. Shepard, *Methods Cell Physiol.* **4**, 49 (1970).
[6] H.-G. Schweiger, P. Dehm, and S. Berger, *in* "Progress in Acetabularia Research" (C.L.E. Woodcock, ed.), p. 319. Academic Press, New York, 1977.
[7] A. Lüttke and F. Grawe, *Br. Phycol. J.* **19**, 23 (1984).
[8] H. D. W. Saddler, *J. Exp. Bot.* **21**, 345 (1970).
[9] C. Freudling and D. Gradmann, *Biochim. Biophys. Acta* **552**, 358 (1979).
[10] H. Mummert, Ph.D. Thesis, University of Tübingen (1979).

segments which are depleted of cytoplasm ($\sim 15\%$ of normal cytoplasm content[11]) can also be isolated for particular purposes (see below).

Demonstration of an Electrogenic Cl⁻ Pump

The voltage range of the diffusion potential of *Acetabularia* is given by the distribution of the major ions outside (cf. external medium) and inside (about 400 mM K⁺, 70 mM Na⁺, and 500 mM Cl⁻ as determined by tracer efflux kinetics[8,10,12]). The most negative equilibrium potential for passive ion diffusion is the Nernst potential for K⁺, E_K of about -90 mV. Simple intracellular voltage recordings (Fig 1) demonstrate, however, that the resting potential is much more negative (about -170 mV), clearly indicating the operation of an electrogenic pump. Since measurements on vacuole-free preparations show the same resting potential, this voltage must be attributed to the plasmalemma only. Cable analytical studies[13] confirm that the tips of inserted microelectrodes are located in the cytoplasm rather than in the vacuole. This fact renders the interpretation of electrophysiological measurements with glass microelectrodes in *Acetabularia* rather convenient with respect to the difficulties which arise when the two membranes (plasmalemma *and* tonoplast) are measured in series.

In order to examine which ion species is the substrate of the electrogenic pump, the response of the voltage to changes in the composition of the bathing medium is recorded. Typical results are represented schematically in Fig. 1. Most importantly, variations of the external pH between pH 5 and 9 have essentially no effect on the voltage. Similarly, changes in external K⁺, Na⁺, or Ca²⁺ have very little effect. This indicates that neither a proton pump, which is widely distributed in glycophyte membranes, nor a Na⁺/K⁺ pump or a Ca²⁺ pump, which are typical for animal membranes, can account for the striking hyperpolarization beyond E_K. However, metabolic inhibition [such as temperatures below $10°$, 3×10^{-4} M 2,4-dinitrophenol (DNP) or 3×10^{-5} M carbonyl cyanide 3-chlorophenylhydrazone (CCCP)] and/or removal of external Cl⁻ (either by $\sim 90\%$ substitution of Cl⁻ by other anions, such as Br⁻, I⁻, F⁻, SO_4^{2-}, and benzene sulfonate or by simple dilution of the ionic strength with isotonic sorbitol solution), equally depolarize the membrane to about -80 mV, i.e., to a membrane voltage which is consistent with the thermodynamic conditions for diffusion potentials according to the Goldman equation.

[11] V. Goldfarb, D. Sanders, and D. Gradmann, *J. Exp. Bot.* **35**, 626 (1984).
[12] H. Mummert, U.-P. Hansen, and D. Gradmann, *J. Membr. Biol.* **62**, 139 (1981).
[13] D. Gradmann, *J. Membr. Biol.* **25**, 183 (1975).

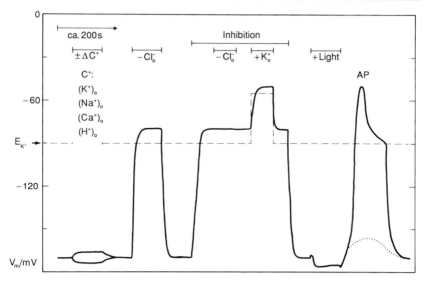

FIG. 1. Representative responses of V_m to changes in environmental conditions. The left-hand side demonstrates the operation of an electrogenic Cl^- pump; the right-hand side, the effect of pulses of white light (~ 100 W m^{-2}), showing the light-off transient depolarization small (dotted) or with all-or-none characteristics (solid, action potential, AP).

Electrical Properties

For the determination of the electrical properties of charge translocating systems in a biomembrane such as ion pumps or ion diffusion pathways, voltage-clamp experiments are very powerful. The basic voltage-clamp circuit and the measured parameters are illustrated in Fig. 2.

Voltage-Clamp Circuit. The dashed box in Fig. 2 represents a differential amplifier (I) with high input resistance (FET, $\sim 10^{12}$ Ω). It is mounted very close to the measuring chamber and allows the membrane voltage V_m to be recorded by a chart recorder (and/or oscilloscope) and be connected to the summation point Σ of the voltage-clamp amplifier (II). Here, V_m and control voltages for voltage steps (V_s) and pluses (V_p) are added. The resulting sum is inverted and amplified within the voltage range ($V_{mx} = \pm 15$ V) of the circuitry. The resulting voltage causes a current to be injected into the cell by another microelectrode with the series resistance R_s (for low R_s of about 200 KΩ, electrodes with broken tips can be used) and causes V_m there to change until the voltage at the summation point is zero, i.e., when V_m equals the chosen command voltage by that amount. The corresponding clamp current across the membrane is measured by a current–voltage converter (amplifier III, FET) against the virtual ground

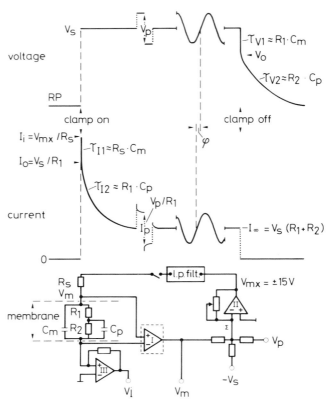

FIG. 2. Voltage clamp. Bottom: basic measuring circuit with equivalent circuit of the *Acetabularia* membrane to be investigated; top: schematic time course of voltage and current during a voltage-clamp experiment. For symbols, see text.

as V_1. Since the closed circuit may ring, an active low-pass filter can be used for damping and/or the amplification of the clamp amplifier (II) can be changed. The best compromise between damping (to avoid ringing) and amplification (to gain velocity of the feedback control) must be found empirically, because the feedback circuit is very sensitive to the experimental conditions and even to the individual preparation.

When the clamp circuit is closed (clamp on), V_m will change immediately from the resting potential (RP) by a step (V_s), and the initial current I_i will decline to I_o, recharging the membrane capacity C_m from RP to the command voltage V_s. The velocity of this recharging is essentially determined by the apparatus, i.e., by V_{mx} and R_s, but also by C_m. The time course of the current, starting at I_o, represents the electrical membrane properties. For *Acetabularia*, these can be represented by two resistances (R_1 and R_2) in series where R_2 is shunted by a (apparently huge) capacitance C_p which can be assigned to the operation of the electrogenic pump. While C_p is recharged, I_o declines to the steady-state current I_∞. When small and short voltage pulses V_p are superimposed to the step V_s, corresponding currents I_p can be recorded and yield different values of the slope $G_1 = 1/R_1 = I_p/V_p$ for different steady-state values of V_m. The time course of changes in G_1 has not been resolved yet; i.e., it is much faster (msec) than the slow (sec) recharging for C_p. If sine waves of various frequencies instead of rectangular pulses are superimposed, G_1 can be analyzed in more detail (see below).

When the clamp circuit is opened (clamp off), the current becomes zero immediately equivalent to application of a current step ("current clamp") from I_∞ to zero. Repolarization of V_m from V_s to *RP* occurs by two clearly distinct exponentials, a fast one to V_o with the time constant $R_1 C_m$, followed by a slow one with the time constant $R_2 C_p$. The representation of changes in V_m resulting from small and short current pulses is omitted in Fig. 2. It should be pointed out that the linear equivalent circuit of the membrane in Fig. 2 holds only for small voltage changes. When the investigations are carried out over a wide range of V_m, the individual elements of the equivalent circuit will change with respect to time and voltage, and the time course of voltage changes may become very complex.

Preparation. In order to circumvent cable problems, the measurements should be performed on "spherical" cells, i.e., on ligated and isolated cell segments shorter than the electrical length constant of the 2 mm cable (minimum). This will provide approximately homogeneous current densities i (in A m^{-2}) with respect to the membrane area of the preparation. On the other hand, stable electrical records are much more difficult to accomplish in such spheres compared with intact cells.

Characteristic Properties. Measurements of the "late" membrane currents as a function of the membrane voltage, $i_\infty(V_m)$, under normal conditions yield *N*-shaped curves (Fig. 3A). When the pump is inhibited, only rectification is observed with a rather sharp bending in the close vicinity of E_K. This "passive" current can essentially be assigned to K$^+$ diffusion. The current difference between the total membrane current and the passive current displays a striking maximum around -140 mV with intrinsically negative slope conductance for more positive voltages. This current is assigned to the pump. Similarly, stimulation of the pump (e.g., by photosynthetically active light[14]) increases these currents. For voltages more negative than E_K, the K$^+$ currents are small. Thus, as a first approxima-

[14] C. Schilde, *Planta* **71**, 184 (1966).

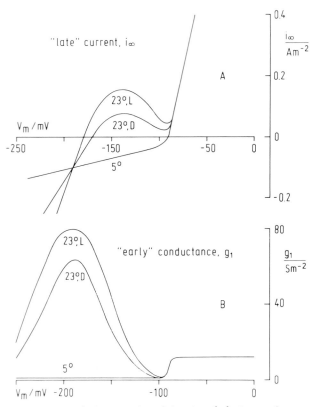

FIG. 3. Steady-state electrical properties of the *Acetabularia* membrane under various conditions: control, 23°, D (darkness or dim light); metabolic stimulation, 23°, L (bright white light); and metabolic inhibition, 5°.

tion, the total current can be taken as the pump current in this voltage range. For V_m more positive than E_K, the pump currents have not been determined, owing to the large K^+ conductance. However, inhibition of K^+ diffusion by tetraethylammonium (TEA^+) at a concentration of a few millimolar, allows the observation of the pump currents with little contamination from diffusion currents (H.-G. Klieber, unpublished results).

The behavior of G_1 with respect to V_m, pump inhibition, and stimulation is given by Fig. 3B. For V_m around -180 mV, G_1 shows an enormous peak which can also be assigned to the pump, because it is abolished under metabolic inhibition.

Steady-State I/V Curves of the Pump. Depending on the experimental conditions, three different types of current–voltage relationships of the pump $i_p(V_m)$ can be determined (Fig. 4) by subtracting current between pairs of membrane current–voltage curves, measured with the pump either stimulated or inhibited, and comparing with the control state. (1) $i_0(V_m)$ yields straight lines, but with different slopes G_1, depending on the starting voltage. The voltage at which two $i_0(V_m)$ curves (measured from different resting potentials in the light and dark) intersect is the equilibrium voltage of the pump (E_p: about -190 mV), if the changes affect the pump only. (2) $i_\infty(V_m)$ is much smaller. Two curves derived under different metabolic conditions also intersect at about -190 mV (E_p). The difference currents are approximately linear for voltages more negative than E_p; around -140 mV they display the striking maximum mentioned above. (3) Integration of the bell-shaped $G_1(V_m)$ curves (Fig. 3B) over V_m yields families of sigmoid current–voltage curves. Since the currents through G_1

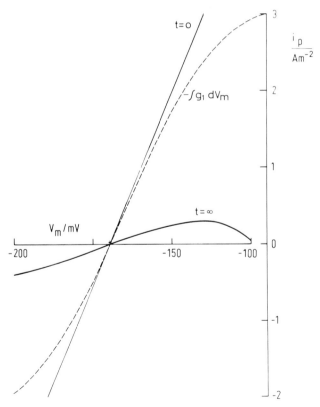

FIG. 4. Current–voltage relationships of the electrogenic pump in *Acetabularia*.

are peculiar to the pump, the integration constant for the sigmoid current–voltage curves has to be chosen to yield zero current at E_p.

For reaction kinetic interpretation (see next section), the last type of current–voltage curves of the pump is of special interest, since the linear curves contain little information and the curves with negative slope require more complicated models. In particular, the behavior of those current–voltage curves with respect to changes in external Cl^- concentration[15] and to time[16] (see next paragraph) provides essential information on pump kinetics.

Non-Steady-State Properties. When sinusoidal changes of V_m rather than rectangular pulses are superimposed on V_s (cf. Fig. 2), the amplitudes and phase shifts of the corresponding currents as determined over a wide frequency range provide more detailed information about the electrical membrane properties. Especially at frequencies higher than 1 kHz, the behavior of G_1, which is peculiar to the pump (see Fig. 3B), reveals interesting details which are represented by a refined equivalent circuit of G_1 comprising a conductance G_0 in parallel with another conductance G_s and a capacitance C_s in series with G_s (see Fig. 5). G_0 is the steady-state conductance of G_1; C_s reflects the number of charged pump molecules in the membrane and G_s their mobility in the membrane. Voltage sensitivity has been determined for G_0 and for C_s but not for G_s. Maximum values are around E_p[16]: G_0 about 80 S m^{-2}, G_s about 80 S m^{-2}, and C_s about 50 mF m^{-2}. These values can be converted to a density of about 60 nmol m^{-2} of charged pump molecules in the membrane and to a maximum turnover rate of about 500 sec^{-1}.

Reaction Kinetic Models

The reaction scheme of the electrogenic Cl^- pump as shown in Fig. 6A serves as the present working hypothesis. Current–voltage relationships of such models with one voltage-sensitive reaction (for the translocation of the charged transporter–substrate complex) can be represented by a reduced model with two states: X_i and X_o with the respective probabilities N_i and N_o which add up to the entire number $N = N_i + N_o$ of the transporter molecules under consideration. This model comprises four rate constants: two voltage-sensitive rate constants

$$k_{io} = k_{io}^0 \exp(zVF/2RT) \tag{1a}$$

$$k_{oi} = k_{oi}^0 \exp(-zVF/2RT) \tag{1b}$$

[15] D. Gradmann, J. Tittor, and V. Goldfarb, *Philos. Trans. R. Soc. London, Ser. B* **299**, 447 (1982).

[16] J. Tittor, U.-P. Hansen, and D. Gradmann, *J. Membr. Biol.* **75**, 129 (1983).

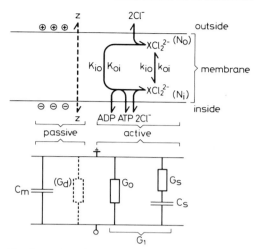

FIG. 5. Reaction kinetics and equivalent circuits for high frequencies (kHz) of the *Acetabularia* membrane with the electrogenic Cl⁻ pump; for G_1, see Fig. 3; for reaction scheme, see Fig. 6.

(with k^0_{io} and k^0_{oi} being the respective rate constants at zero voltage, F, R, and T having their usual thermodynamic meaning, and the factor $\frac{1}{2}$ representing a symmetrical Eyring barrier for the transition of charge) and two voltage-insensitive gross rate constants K_{io} and K_{oi} which summarize the voltage-insensitive part of the reaction system.

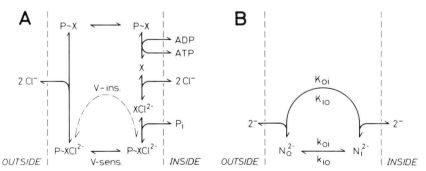

FIG. 6. Reaction schemes of the electrogenic Cl⁻ pump in *Acetabularia* after Mummert *et al.*[12] (A) Explicit scheme with six states and six distinct reversible reactions. (B) Reduced, two-state model for description of electrical properties.

Several relationships can be derived from this model[17]: the current–voltage relationship is

$$I(V) = zFN\frac{k_{io}K_{oi} - k_{oi}K_{io}}{k_{io} + k_{oi} + K_{io} + K_{oi}} \tag{2}$$

with the voltage sensitivity in k_{io} and k_{oi} [cf. Eqs. (1a) and (1b)]. The steady-state conductance G_0 is the derivative of Eq. (2)[16]:

$$G_0 = N\frac{z^2F^2}{2RT}\frac{k_{oi}(k_{io} + K_{io}) + k_{io}(k_{io} + K_{io})(K_{io} + K_{oi})}{(k_{io} + k_{oi} + K_{io} + K_{oi})^2} \tag{3}$$

The non-steady-state behavior of the system[17] is given by G_0 in parallel with the two elements G_s and C_s in series, where

$$G_s = G_0\frac{k_{io} + k_{oi}}{K_{io} + K_{oi}} \tag{4}$$

$$C_s = \frac{G_0}{(k_{io} + k_{oi} + K_{io} + K_{oi})}\frac{k_{io} + k_{oi}}{K_{io} + K_{oi}} \tag{5}$$

When the system is embedded in a membrane, the membrane capacitance C_m and a background conductance G_d (as from ion diffusion) must be considered as well for the analysis of the actual measurements. Since G_d is usually small compared to G_0 (the steady-state value of G_1, cf. Fig. 3), G_d may be ignored. The transfer function for the impedance of the entire equivalent circuit is

$$\vec{R} = \frac{1}{G_0}\frac{1 + p\tau_n}{1 + p(\tau_1 + \tau_2) + p^2\tau_1\tau_2} \tag{6}$$

where $\vec{R} = |\vec{R}|\exp(\varphi\sqrt{-1})$ is the complex impedance describing both magnitude $|\vec{R}|$ and phase angle $\varphi = \arctan(ImR/ReR)$, and $p = 2\pi f\sqrt{-1}$ with f being the frequency.[17] Using a series of small sine waves instead of rectangular pulses superimposed on the voltage steps V_s (cf. Fig. 2), the ratio between the voltage and the current amplitudes gives $|\vec{R}|$. The phase angle φ between the sine waves of the voltage and the current can also be determined from the measurements. These data can be fitted to Eq. (6) and yield at each steady-state level of V_m numerical values for the steady state conductance G_0, the "zero" $1/(k_{io} + k_{oi} + K_{io} + K_{oi})$ at $p = 1/\tau_n$, as well as the two values $(\tau_1 + \tau_2)$ and $\tau_1\tau_2$ for the two "poles" at $p = -1/\tau_1$ and $p = 1/\tau_2$. From these numerical values the four elements of the equivalent

[17] U.-P. Hansen, D. Gradmann, D. Sanders, and C. L. Slayman, *J. Membr. Biol.* **63**, 165 (1981).

circuit can be determined: G_0 as it stands, $C_m = G_0 \tau_1 \tau_2 / \tau_n$, $C_s = G_0(\tau_1 + \tau_2 - \tau_n - \tau_1\tau_2/\tau_n)$, and $G_s = C_s/\tau_n$.

Fitting these results to Eqs. (2)–(5) yields good coincidence (including voltage sensitivity) with the model for the following absolute parameter values: $N = 60$ nmol m^{-2}, $z = -2$, $k_{io}^0 = 0.125$, $k_{oi}^0 = 5 \times 10^5$, $K_{io} = 500$, and $K_{oi} = 500$ sec^{-1}. By this application of sine wave analysis to the measurements, the electrodynamic properties of the pump including their nonlinear voltage sensitivities are essentially described at the molecular level of the two-state model reaction kinetics.

Voltage-Sensitive Cl⁻ Fluxes: Prediction and Verification

With the parameters as determined by electrical measurements, the model gives the voltage-sensitive, unidirectional efflux through the pump as

$$\phi_{io}(V_m) = N \frac{k_{io}(K_{oi} + k_{oi})}{k_{io} + k_{oi} + K_{io} + K_{oi}} \tag{7}$$

provided substrate binding and debinding are fast reactions compared to the reorientation of the (charged and uncharged) binding site of the transporter.[12]

For the case of the electrogenic Cl⁻ pump in *Acetabularia* these predictions can be examined by tracer efflux experiments with ^{36}Cl⁻-equilibrated cells. The very large fluxes ($\sim 10^{-5}$ mol m^{-2} sec^{-1}) and the large size of the cells allow recordings of the Cl⁻ efflux with a temporal resolution of one sample per 30 sec on individual cells. V_m and efflux can be recorded simultaneously from single cells, for instance, during the time course of spontaneous or triggered (by a depolarizing current or by a strong light-off stimulus) "action potentials" which last several minutes[18] (cf. Fig. 1). From such experiments the Cl⁻ efflux can be plotted versus V_m. Subtraction of the Cl⁻ efflux determined under conditions when the pump is inhibited ($\sim 2 \times 10^{-6}$ mol m^{-2} sec^{-1}) from the total (voltage-sensitive) efflux results in that Cl⁻ efflux which passes through the pump. This voltage-dependent Cl⁻ efflux coincides with the above prediction (Fig. 7). This coincidence provides additional and very strong evidence for Cl⁻ being the transported substrate of the electrogenic pump.

ATP Synthesis by Reversal of the Pump *in Vivo*

Initially ATP was assumed to drive the pump. Very strong evidence for this hypothesis would be the demonstration of ATP synthesis under condi-

---

[18] D. Gradmann, *J. Membr. Biol.* **29**, 23 (1976).

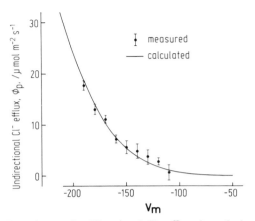

FIG. 7. Voltage dependence of unidirectional Cl⁻ efflux through the pump,[12] related to membrane surface area. The data measured are plotted with the curve predicted from Eq. (7) using parameters from electrical measurements.

tions when the pump is thermodynamically forced to operate in reverse. Since the current–voltage relationship shows, under normal conditions, appreciable negative currents at V_m more negative than the reversal potential (about -190 mV, see Fig. 4), reversibility of the pump is anticipated. In fact, when a strong outward directed Cl⁻ gradient is imposed on the membrane by depleting external Cl⁻ from about 500 mM (as inside) to zero, an increase of the level and ^{32}P labeling of ATP can be measured[19] (cf. Fig. 8). The experimental procedure is as follows.

The cytoplasm in the stalk must be reduced by centrifugation (to about 15%). Otherwise the membrane-related effects are not significant compared with the high background of overall ATP metabolism (cf. Fig. 8). Samples of $(1-10)$ individual cells are transferred to various media containing ^{32}P (specific activity about 5×10^{15} cmp (mol⁻¹) for about 10 sec, washed, and frozen in petroleum ether cooled by liquid N_2. Pieces about 10 mm in length of the cytoplasm-depleted stalk are cut out and used for analyses of the amounts and the specific radioactivities of ATP and P_i. Separation of ATP and P_i from individual neutralized HClO₄ extracts is accomplished on DEAE-cellulose columns equilibrated with 0.1 M HCl and eluated with increasing concentrations of LiCl[20]: 0.3 M for P_i and 1.5 M for ATP.

[19] V. Goldfarb, D. Sanders, and D. Gradmann, *J. Exp. Bot.* **35,** 645 (1984).
[20] R. P. Magnussen, A. R. Portis, Jr., and R. E. McCarty, *Anal. Biochem.* **72,** 653 (1976).

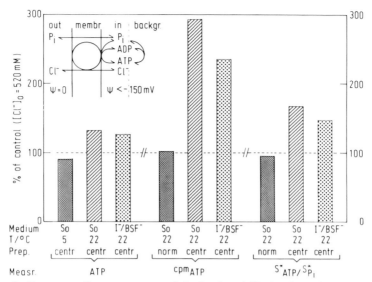

FIG. 8. Summary of measurements of the level and ^{32}P labeling of ATP in normal cytoplasm-depleted stalk segments. Percentages of results obtained under a strong outward electrochemical gradient for Cl^- (within 10 sec, zero Cl^- outside) are compared to results from normal conditions (~ 500 mM Cl^- outside), set to 100%. The inset shows a scheme of the working hypothesis.

For determination of the effects of Cl^- gradient, external Cl^- can be removed either by isotonic sorbitol (So) or by substitution of Cl^- with I^- or benzene sulfonate (BSF^-). Since under So conditions external K^+ is diluted as well, V_m can be expected to stay rather negative because of an increased amount of E_K (about -150 mV measured under these conditions). Under I^- and BSF^- conditions (summarized in Fig. 8), external K^+ remains constant and an E_K of about -90 mV is maintained, yielding a smaller electrochemical outward gradient for Cl^- than under So conditions.

Typical results are depicted in Fig. 8, with a scheme of the hypothesis shown in the inset. The ATP level is elevated in Cl^--free medium except at low temperatures, when the pump is known to be inhibited. ^{32}P labeling of ATP is dramatically increased when a strong outwardly directed Cl^- gradient is applied. This effect is revealed only by cytoplasm depleted cells but not by normal ones. Since this effect might only be due somehow to stimulated ^{32}P uptake, the increased radioactivity in ATP may only appear to be due to increased phosphorylation of ADP to ATP when just the specific radioactivity of internal P_i is increased. Therefore, the specific radioactivity of ATP must be related to the actual specific radioactivity of cytoplasmic P_i. The results of these relevant ratios as illustrated by the last three columns in Fig. 8 are, in fact, less dramatic but still highly significant. The effects on centrifuged cells appear more pronounced throughout for So conditions when the outward electrochemical gradient is anticipated to be larger compared with I^- and BSF^- conditions.

These results confirm the hypothesis that in the plasmalemma of *Acetabularia* there is an ATPase which can be forced to synthesize ATP by an outward electrochemical gradient for Cl^-. In comparison with the electrophysiological results and with the measurements of $^{36}Cl^-$ tracer fluxes, this ATPase is considered to be identical with the electrogenic Cl^- pump.

[31] Transport in Isolated Yeast Vacuoles: Characterization of Arginine Permease

By THOMAS BOLLER, MATHIAS DÜRR, and ANDRES WIEMKEN

Introduction

In most plant or fungal cells, the central vacuole is by far the largest intracellular organelle. Typically, it occupies 90% of the intracellular space and, therefore, contains 90% or more of the total solute content of a cell. The solute composition of the vacuolar sap may be entirely different from that in the rest of the cell; in addition, it may rapidly change in the course of growth and development or in response to environmental change.[1-4]

It is the vacuolar membrane, the tonoplast, which controls the accumulation and release of solutes in the vacuolar sap and thereby also greatly influences the cytoplasmic solute concentration. Knowledge of the transport processes across the tonoplast is essential not only for an appraisal of vacuolar storage but also for an understanding of metabolic regulation in the cytoplasm.

We have chosen yeast cells to study transport processes across the tonoplast. In yeast, the vacuole comprises only about 20-25% of the cell volume. Yet, it contains a large pool of amino acids and, in particular,

---

[1] P. Matile, *Annu. Rev. Plant Physiol.* **29**, 193 (1978).

[2] C. A. Ryan and M. Walker-Simmons, this series, Vol. 96, p. 580.

[3] A. M. Boudet, G. Alibert, and G. Marigo, *in* "Membranes and Compartmentation of Plant Functions" (A. M. Boudet, G. Alibert, G. Marigo, and P. Lea, eds.), p. 29. Oxford Univ. Press, London and New York, 1984.

[4] T. Boller and A. Wiemken, *Annu. Rev. Plant Physiol.* **37**, 137 (1986).

most of the free arginine in the cell.[5] Here, we describe a method to study transport in intact isolated yeast vacuoles and its application to characterize an arginine permease present in the tonoplast. The method is based on that described by Boller *et al.*[6,7] A similar technique has been used to examine transport of *S*-adenosylmethionine[8] and of purines[9] in yeast vacuoles. To our knowledge, data on transport with intact isolated yeast vacuoles have not been published since. However, an alternative method, employing tonoplast vesicles instead of intact vacuoles, has been developed.[10-12] At the end of this chapter, we compare the results of studies on intact isolated vacuoles and on tonoplast vesicles.

Preparation of Yeast Vacuoles for Transport Studies

Yeast vacuoles can be released from protoplasts either by osmotic shock[13] or by mechanical[5,13] or polybase-induced[14] lysis in isotonic media. For transport studies under near-natural conditions, vacuoles should be prepared in isotonic media since vacuoles obtained by osmotic shock do not retain micromolecules.[13] We have used vacuoles obtained by mechanical lysis or by polybase-induced lysis with equal success.[7] Polybase-induced lysis gives the best yield and is described in detail below.

Organism

We obtain the best preparations of vacuoles from *Saccharomyces cerevisiae* strain LBH 1022, a diploid strain deposited in the American Type Culture Collection under ATCC No. 32167. The yeast is inoculated from agar slants into a synthetic medium (Table I).[15] Cultures of 300 ml are grown in 2,500-ml Erlenmeyer flasks on a rotary shaker (180 rotations/min) at 28° for 30–60 hr, until they reach the early stationary phase of growth. Each culture is diluted with 1,000 ml fresh prewarmed medium

[5] A. Wiemken and M. Dürr, *Arch. Microbiol.* **101,** 45 (1974).

[6] T. Boller, M. Dürr, and A. Wiemken, *Eur. J. Biochem.* **54,** 81 (1975).

[7] T. Boller, Ph.D. Dissertation ETH Zürich, No. 5928. Buchdruckerei Wattwil, Wattwil, Switzerland (1977).

[8] J. Schwencke and H. de Robichon-Szulmajster, *Eur. J. Biochem.* **65,** 49 (1976).

[9] M. Nagy, *Biochim. Biophys. Acta* **558,** 221 (1979).

[10] Y. Ohsumi and Y. Anraku, *J. Biol. Chem.* **256,** 2079 (1981).

[11] T. Sato, Y. Ohsumi, and Y. Anraku, *J. Biol. Chem.* **259,** 11505 (1984).

[12] T. Sato, Y. Ohsumi, and Y. Anraku, *J. Biol. Chem.* **259,** 11509 (1984).

[13] A. Wiemken, *Methods Cell Biol.* **12,** 99 (1975).

[14] M. Dürr, T. Boller, and A. Wiemken, *Arch. Microbiol.* **105,** 319 (1975).

[15] H. K. von Meyenburg, *Pathol. Microbiol.* **31,** 117 (1968).

TABLE I

CULTURE MEDIUM FOR Saccharomyces cerevisiae

Main nutrients	mg/liter	mM	Trace elements[a]	mg/liter	μM	Buffer	mM	Vitamins[b]	mg/liter	μM
Glucose	5000	28	$ZnSO_4 \cdot 7H_2O$	20	70	Disodium citrate	50	(+)Biotin	0.1	0.4
NH_4Cl	1080	20	$FeCl_3 \cdot 6H_2O$	5.6	20.7	(adjusted to pH 5.1)		Calcium pantothenate	1.0	2.1
KH_2PO_4	2000	14.7	$CuSO_4 \cdot 5H_2O$	1.0	4.0					
$MgSO_4 \cdot 7H_2O$	200	0.8	HBO_3	0.2	3.3					
$CaCl_2 \cdot 2H_2O$	100	0.7	$MnSO_4 \cdot 6H_2O$	0.2	0.8					
			$Na_2MoO_4 \cdot H_2O$	0.02	0.1					
			Citric acid	20	95					

[a] Prepared as 1000-fold concentrated stock.
[b] Prepared as 100-fold concentrated stock.

containing 2.0% glucose instead of 0.5% glucose and then grown for an additional period of 4 hr.

We have prepared vacuoles from a number of different strains of *Saccharomyces cerevisiae* and from *Candida utilis*. For each new strain, even for new growth conditions, the procedure described below must be changed slightly to obtain optimal yields of vacuoles. The concentration of the osmoticum may be varied between 0.5 and 1.0 M. Because the density of vacuoles depends on growth conditions, the density of the layers in the step gradient may require adjustment, either by changing the mixing ratios of sorbitol and sucrose or by addition of the osmotically nearly inert Ficoll (Pharmacia).[5,16,17]

Reagents

Buffered sorbitol: 0.6 M sorbitol containing 5 mM Tris–piperazine-N,N'-bis(2-ethanesulfonic acid) (PIPES) buffer, pH 6.8

Pretreatment medium: 0.5 M sorbitol containing 0.14 M cysteamine–HCl and 25 mM Tris–PIPES buffer, pH 6.8

2.5% cytohelicase–sorbitol: 0.6 M sorbitol containing 25 mg/ml lyophilized cytohelicase (snail gut enzyme, obtained from Industrie Biologique Française, Clichy, France) (note that the product currently marketed under the name of "helicase" is unsuitable for the preparation of yeast protoplasts)

1% DEAE-dextran–sorbitol: 0.6 M sorbitol containing 10 mg/ml DEAE-dextran (MW 500,000, obtained from Pharmacia) and 5 mM Tris–PIPES, pH 6.8

0.1% DEAE-dextran–sorbitol: mixture of 1 part 1% DEAE-dextran–sorbitol and 9 parts buffered sorbitol

Buffered sucrose: 0.6 M sucrose containing 5 mM Tris–PIPES buffer, pH 6.8 30% and 35% sucrose: 30 and 35% household sugar (used as bottom layer and displacement solution, respectively, in the zonal rotor)

We have used Tris–PIPES or Tris–MES (2-N-morpholinoethanesulfonic acid) as membrane-impermeable buffers for transport experiments.[6,7] A buffer with 2-amino-2-methyl-1,3-propanediol (AMPD) instead of Tris is also suitable for the isolation of vacuoles.[16,17] We have not tested this buffer in transport assays.

---

[16] A. Wiemken, M. Schellenberg, and K. Urech, *Arch. Microbiol.* **123**, 23 (1979).
[17] F. Keller, M. Schellenberg, and A. Wiemken, *Arch. Microbiol.* **131**, 298 (1982).

Preparation of Protoplasts

The method for preparation of protoplasts is a slight modification of the procedure used by Wiemken and Dürr.[5] The cultures are rapidly chilled in an ice-water bath and harvested by centrifugation (5 min, 2,000 g). The sediments are washed by two cycles of resuspension and centrifugation, once in ice-cold water and once in buffered sorbitol. The wet sediment is weighed. For each gram wet weight, 4 ml pretreatment medium is added. The cells are incubated in this medium at 30° for 20 min.

The cell suspension is then diluted with an equal volume of buffered sorbitol, harvested by centrifugation, and resuspended in 2.5% cytohelicase – sorbitol (1 ml per g wet weight). The preparation is incubated in a shaking water bath at 30° for 90 min. Within this time period, over 99% of the cells are transformed in protoplasts. It should be verified, by microscopical observation, that the protoplasts are separated from each other and that they burst when immersed in water. The protoplast suspension is transferred into a 250-ml centrifuge tube. Buffered sucrose (100 ml) is carefully layered under the protoplast suspension. The step gradient thus formed is centrifuged for 20 min at 2,000 g (4°). The two layers of supernatant are aspirated off with a Pasteur pipette connected to a vacuum pump. The sediment containing the purified protoplasts is gently resuspended in ice-cold buffered sorbitol.

Preparation and Purification of Vacuoles

Step 1: Preparation of Gradient for Purification.[18] An isotonic density gradient consisting of mixtures of buffered sorbitol and sucrose is prepared in the Beckman 14 Ti zonal rotor as detailed in Fig. 1. For preparations on a smaller scale, similar gradients may be prepared in centrifuge tubes for the SW25 swinging-bucket rotor.[16,17]

Step 2: Determination of Amount of Polybase Used for Lysis.[14,16] In a preliminary assay, the amount of DEAE-dextran needed to lyse all protoplasts is determined. Samples of 10^8 protoplasts in 1 ml buffered sorbitol are pipetted into plastic vessels (glassware is unsuitable because it reduces the efficiency of the DEAE-dextran treatment in an unpredictable manner) and mixed with 10, 20, 40, 60, 100, 150, and 300 μl 0.1% DEAE-dextran – sorbitol. The samples are incubated in an ice-water bath for 1 min and then in a shaking water bath at 30° for 5 min. The suspension is gently mixed by aspiration through a Pasteur pipette to dissolve aggregates. The samples are examined by phase-contrast microscopy, and the minimal

---

[18] M. Dürr, T. Boller, and A. Wiemken, *Biochem. Biophys. Res. Commun.* **73**, 193 (1976).

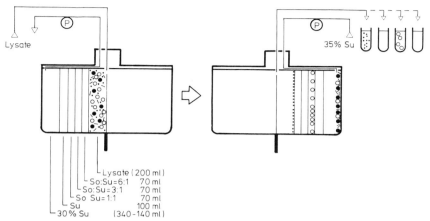

FIG. 1. Purification of vacuoles in the zonal rotor. The lysate containing intact vacuoles (open circles), some surviving intact protoplasts (filled circles), and various membrane vesicles and organelles including lipid particles (dots) and mitochondria (dashes) is layered on a step gradient consisting of mixtures of buffered sorbitol (So) and buffered sucrose (Su) as indicated. The pump (P) serves to load and unload the zonal rotor. After centrifugation (40 min, 40,000 rpm, 4°), lipid particles float on top of the gradient, vacuoles form a band on top of the layer of buffered sucrose, and surviving protoplasts, mitochondria, other membrane particles, and some dense vacuoles sediment to the bottom. The gradient is displaced with a 35% sucrose solution pumped to the outer edge of the rotor and fractionated to isolate the pure vacuoles.

amount of DEAE-dextran causing >95% lysis of the protoplasts is determined. The amount is of the order of $50-150 \, \mu g$ per 10^8 protoplasts and varies from preparation to preparation.

Step 3: Lysis of Protoplasts.[14] Up to 100 ml of the protoplast suspension containing up to 5×10^8 protoplasts/ml is mixed in an ice-water bath with 10 ml buffered sorbitol containing the calculated amount of DEAE-dextran causing complete lysis. Rapid and homogeneous mixing is important. After 1 min, the preparation is pulled a few times through a 10-ml pipette to dissolve aggregates. The preparation is cooled in an ice-water bath.

Step 4: Centrifugation.[18] The lysate is layered on top of the density gradient and centrifuged at 40,000 rpm (100,000 g) for 40 min (4°). The band at the interphase above the buffered sucrose containing pure vacuoles is collected, diluted with 5 volumes of 0.6 *M* sorbitol, and centrifuged at 2,000 g (4°) for 40 min. The supernatant is aspirated off. The sediment containing pure vacuoles is gently resuspended in a few milliliters of buffered sorbitol.

Yield

In a typical run, a preparation based on one Erlenmeyer flask containing 10 g (wet weight) cells yields approximately 3×10^{10} protoplasts. From these, about 10^{10} vacuoles are obtained. The vacuoles have a mean diameter of 3 μm. Thus, the total tonoplast surface obtained in a typical preparation is 0.28 m^2, and the total volume of all 10^{10} vacuoles is about 0.14 ml. These vacuoles contain approximately 100 μmol of amino acids (i.e., 700 mM in terms of concentration) but only about 1 mg protein, of which about one-half is associated with the membrane.

Measurement of Transport in Intact Isolated Vacuoles

The system used to measure uptake of labeled amino acids is illustrated in Fig. 2.[6,7]

Step 1: Preparation of Density Gradients. The gradients are prepared as shown in Fig. 2B. For each measurement of transport to be performed, four gradients are prepared and put into the buckets of the swing-out rotor of a refrigerated centrifuge at 4°.

Step 2: Preparation of Incubation Vessels. For each measurement of transport, four incubation vessels are prepared. Each receives 0.1 ml buf-

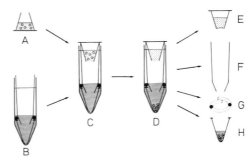

Fig. 2. Centrifugation technique for transport assays. (A) Incubation vessel (1.5-ml Eppendorf tube with the tip cut off). (B) Centrifugation tube with inserts F–H (see below), containing a 6:1 mixture of buffered sorbitol and buffered sucrose at the bottom and a 9:1 mixture of the same solutions at the top. (C) To terminate the incubation, the incubation vessel is put on top of the density gradient. The narrow tip prevents liquid flowing out from the incubation vessel when turned upside down. (D) The gradients are centrifuged for 20 min at 2,500 g (4°). (E) The incubation vessel containing most of the radioactivity not taken up is removed. (F) The connection piece (made of a trimmed 5-ml plastic tip for Gilson pipettes) is removed after the upper part of the gradient has been aspirated off. (G) Metal spring facilitating circulation of liquid into the connection piece during aspiration. (H) Reception vessel (1.5-ml Eppendorf tube) containing a sediment of pure vacuoles.

fered sorbitol containing 5,000–10,000 Bq of the radioactive amino acid and effectors as needed. The incubation vessels are mounted on a stand in a shaking water bath, usually at 25°, and preincubated for at least 5 min.

Step 3: Preincubation of Vacuoles. The vacuole suspension is adjusted to $0.5–1.0 \times 10^8$ vacuoles, divided into 1.1-ml aliquots, and kept on ice. Exactly 5, 4, 3, and 2 min before the start of the assay, one aliquot of vacuoles each time is transferred to the incubation bath.

Step 4: Incubation of Vacuoles. Exactly 5 min after the start of the preincubation, i.e., at times 0, 1, 2, and 3 min of the assay, 1.0 ml of the vacuole suspension is taken from the respective aliquots and added to one of the incubation vessels.

Step 5: Termination of Incubation. At time 3.1–3.5 min of the assay, each of the four incubation vessels is quickly turned upside down and positioned on top of one of the density gradients in the centrifuge buckets. The centrifuge is started immediately and run at 2,500 g for 20 min (4°).

Step 6: Recovery of Centrifuged Vacuoles. As shown in Fig. 2, the gradient is dismantled and the supernatant layers are aspirated off. The sediment containing the centrifuged vacuoles is resuspended in 1 ml water.

Step 7: Measurements and Calculation of Transport Rate. From each resuspended sediment, a sample of 0.2 ml is used for the measurement of radioactivity and one of 0.4 ml for the determination of the amino acid content with ninhydrin.[19] The amino acid content is determined in each sample as a measure for the amount of vacuoles recovered. (Generally, ~60–80% of the amino acids initially present in the vacuole suspension are recovered in the sediment.) The amount of radioactivity taken up is normalized according to the amino acid content of the sediment. The uptake rate is calculated from a linear regression of the amount taken up as a function of the incubation time and expressed as pmol radioactive amino acid nmol^{-1} total amino acids min^{-1}. From a number of preliminary experiments, it is known that the incubation time corresponds to the time from the start of the incubation to the start of the centrifuge *plus* 1 min, the mean time required in the centrifuge to remove the vacuoles from the incubation solution.

Characterization of Vacuolar Arginine Permease

We have used the technique described above to characterize a vacuolar arginine permease in some detail.[6,7]

---

[19] E. W. Yemm and E. C. Cocking, *Analyst* **80**, 209 (1955).

General Properties

The vacuolar arginine permease catalyzes the uptake of radioactive arginine from the incubation medium in exchange for arginine present in the vacuolar sap. This can be demonstrated with double-labeling experiments in which vacuoles are preloaded with [³H]arginine, repurified, and then incubated with [¹⁴C]arginine. For each mole [¹⁴C]arginine taken up, 1 mol [³H]arginine is lost from the vacuolar sap (see Fig. 3 below).[6,7,20]

The vacuolar arginine permease has a pH optimum in the range of pH 7.0–7.5. Its temperature dependence indicates that the apparent energy of activation is 39.1 kJ mol^{-1}. Transport of arginine is saturable with an apparent K_m of 30 μM.[6,7]

Specificity

Transport of L-arginine is competitively inhibited by D-arginine ($K_i = 60 \mu M$), by L-histidine ($K_i = 320 \mu M$), and L-canavanine ($K_i = 600 \mu M$).[6,7] A number of other substrate analogs inhibit arginine transport in isolated vacuoles. The vacuoles and the protoplasts from which they are derived behave differently with regard to the effect of various substrate analogs. For example, argininic acid inhibits arginine transport in vacuoles but not in protoplasts; arginine hydroxamate inhibits arginine transport in protoplasts but not in vacuoles.[6,7]

Isolated vacuoles transport the basic amino acids histidine, lysine, and ornithine with saturation kinetics (Table II).[7] The transport of all these amino acids is inhibited in a similar way by nonradioactive arginine or nonradioactive histidine as is transport of radioactive arginine. Thus, the same permease probably catalyzes the transport of arginine, histidine, lysine, and ornithine.

Effects of Metabolic Inhibitors and of ATP

Radioactive arginine is taken up from the incubation medium in the absence of any energy source. Neither uncouplers nor the sulfhydryl reagent N-ethylmaleimide inhibit arginine transport in isolated vacuoles (Table III).[7] These reagents strongly inhibit arginine uptake in protoplasts (Table III).

Arginine transport in isolated vacuoles is stimulated about 2-fold by 1 mM ATP, added in the absence of Mg^{2+} (Table III). At first sight, this might be taken as an indication of an active transport process. However, a

---

[20] T. Boller, *in* "The Physiological Properties of Plant Protoplasts" (P. E. Pilet, ed.), p. 76. Springer-Verlag, Berlin and New York, 1985.

TABLE II
TRANSPORT OF BASIC AMINO ACIDS IN ISOLATED
INTACT YEAST VACUOLES

| Amino acid | K_m of transport (μM) | Inhibition (%) of transport by | |
		Arginine[a]	Histidine[a]
Arginine	30	73	38
Histidine	270	72	58
Lysine	1,300	64	51
Ornithine	3,000	74	52

[a] The transport rate of 0.1 mM of the labeled amino acid is determined in the absence and presence of 1 mM unlabeled arginine or histidine.

more careful analysis of the ATP effect shows that this is not the case: Using vacuoles whose internal arginine pool is labeled with [^{3}H]arginine, it can be demonstrated that ATP stimulates the efflux of arginine to the same extent as the uptake (Fig. 3). Thus, ATP does not cause active transport but only stimulates exchange under these conditions.[7,20] The same is true for Na$^+$ and NADPH (Fig. 3).

Why is arginine retained in the isolated vacuoles? Isotonically prepared

TABLE III
EFFECT OF UNCOUPLERS AND ATP ON ARGININE TRANSPORT IN ISOLATED INTACT
VACUOLES AND PROTOPLASTS

| Effector | Rate of transport of 100 μM arginine (% of control) | |
	Vacuoles	Protoplasts
None	100	100
2,4-Dinitrophenol (DNP), 1 mM	104	14
m-Chlorocarbonyl cyanide phenylhydrazone (CCCP), 5 μM	100	14
NaN$_3$, 1 mM	134	16
N-Ethylmaleimide (NEM), 1 mM	94	22
ATP	180	n.d.[a]

[a] n.d., Not determined.

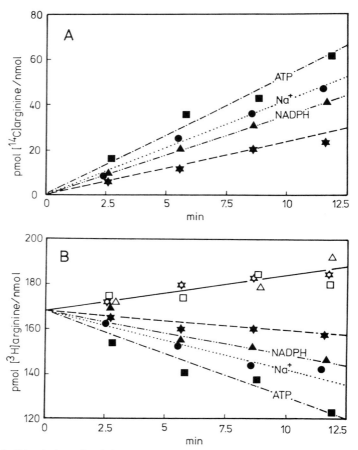

FIG. 3. Stimulation of arginine transport in isolated intact vacuoles by ATP, Na+, and NADPH. Isolated vacuoles are incubated with [³H]arginine at 0° for 18 hr to label the vacuolar arginine pool and repurified. These [³H]arginine-loaded vacuoles are then incubated in the presence (filled symbols) or in the absence (open symbols) of 150 μM [¹⁴C]arginine without further additions, with 1 mM ATP plus 4 mM Na+, with 4 mM Na+, or with 1 mM Na₄NADPH. All solutions are adjusted to pH 6.8 with Tris or PIPES. (A) Uptake of [¹⁴C]arginine; (B) efflux of [³H]arginine. Data are expressed per nmol total amino acids in the vacuoles. Note that in the absence of exogenous arginine, the [³H]arginine content of the vacuoles (open symbols) appears to increase. This is due to a decrease in total amino acid content of the vacuoles during incubation.

vacuoles contain large amounts of polyphosphate[21]; there is an approximately equimolar amount of phosphate groups in the polyanion polyphosphate and in arginine, a cation.[22] The large polyphosphate molecules cannot leave the vacuole. Arginine, the main accompanying cation, is therefore retained as well and remains ionically trapped inside the vacuole except if it is exchanged against another cation.[1,7,21]

The observation that vacuolar arginine transport is a stoichiometric exchange of arginine (and other basic amino acids[7]) across the tonoplast can therefore be explained in two ways.[7] Either the permease is intrinsically an exchange transport system, or the permease simply catalyzes a facilitated diffusion of arginine across the membrane, and the stoichiometric exchange is a consequence of the need of balanced electrical charges inside the vacuole.

Effect of Proteolytic Enzymes

The transport properties of isolated vacuoles can be influenced by treatments with proteolytic enzymes.[18] Chymotrypsin and trypsin do not affect arginine transport.[7] However, pretreatments of vacuoles with thermolysin in the presence of 0.5 mM Ca^{2+} or with a low concentration of Pronase (2 μg/ml) cause an activation of the transport systems.[18] A higher concentration of Pronase (50 μg/ml) inactivates the arginine permease in a time-dependent manner. The inactivation of the permease is prevented when its substrate, arginine, is included in the pretreatment with Pronase. Leucine (2 mM), which is not a substrate for arginine permease, does not prevent the inactivation. Lysine, histidine, and D-arginine partially protect the transport system. The extent of protection provided by these amino acids is well correlated with the extent of inhibition they exert on arginine transport.[7,18] These results indicate that the arginine permease is a membrane protein. A Pronase-sensitive region is exposed on the cytoplasmic surface of the membrane, but bound substrate protects this site from Pronase attack.

Arginine Transport in Intact Isolated Vacuoles from Different Yeast Strains and from Neurospora

The kinetic properties of the vacuolar arginine permeases of different yeast strains[7] and of *Neurospora crassa*[23] are summarized in Table IV. All

[21] K. Urech, M. Dürr, T. Boller, A. Wiemken, and J. Schwencke, *Arch. Microbiol.* **116**, 275 (1978).

[22] M. Dürr, K. Urech, T. Boller, A. Wiemken, J. Schwencke, and M. Nagy, *Arch. Microbiol.* **121**, 169 (1979).

[23] E. Martinoia, U. Heck, T. Boller, A. Wiemken, and P. Matile, *Arch. Microbiol.* **120**, 31 (1979).

TABLE IV

VACUOLAR ARGININE PERMEASE IN DIFFERENT YEAST STRAINS AND *Neurospora*

Kinetic properties of arginine transport

Organism	Vacuoles		Protoplasts	
	K_m (μM)	V_{max} (pmol nmol^{-1} min^{-1})	K_m (μM)	V_{max} (pmol nmol^{-1} min^{-1})
Saccharomyces cerevisiae				
Strain LBG H 1022[a]	30	2.5[b]	1.8	2.3
Strain Σ1278b (wt)[c]	65	2.4	1.6	4.8
Strain 2512c *(gap-1)*[c]	65	2.8	1.1	1.3
Strain MG168k3				
(agp-1,gap-1)[c]	70	3.7	75[d]	0.5[d]
Neurospora crassa[e]	80	0.4	4.3	1.3

[a] Data from Ref. 6.

[b] A value of 2.5 pmol nmol^{-1} min^{-1} corresponds to approximately 250 nmol mg^{-1} total vacuolar protein min^{-1}, to about 500 nmol mg^{-1} membrane protein min^{-1}, or 15 nmol m^{-2} membrane sec^{-1}.

[c] The vacuoles, from cells grown on proline as nitrogen source, are obtained by a slightly modified procedure. Strain 2512c and MG168k3 are derived from Σ1278b (wild type); strain 2512c lacks the general amino acid permease, strain MG168k3 lacks both the general amino acid permease and the specific arginine permease.[24]

[d] Transport is probably due to the approximately 15% of contaminating vacuoles observed in the protoplast preparation.

[e] Data from Ref. 18.

these fungi possess vacuolar arginine permeases with apparent K_m values of 30–80 μM. *Saccharomyces cerevisiae* strain Σ1278b has been studied extensively with regard to amino acid transport across the plasma membrane using genetic techniques. Mutants defective in arginine transport have been described.[24] Interestingly, mutants defective in arginine transport at the plasma membrane have the same characteristics of vacuolar arginine transport as the wild type (Table IV).[7] This indicates that the vacuolar arginine permease is independent of the arginine-transporting permeases of the plasma membrane.

[24] M. Grenson, C. Hou, and M. Crabeel, *J. Bacteriol.* **103**, 770 (1970).

Comparison of Studies on Intact Isolated Vacuoles and Isolated Vacuolar Membrane Vesicles

Ohsumi and Anraku[10] and Sato *et al.*[11,12] have extensively studied transport in isolated vacuolar membrane vesicles. These vesicles are obtained from vacuoles isolated by osmotic shock,[10,25] and they do not retain all the vacuolar ingredients like arginine and polyphosphate.[10] However, the membrane apparently reseals while vesicles are prepared. The vesicles are "right-side-out"[10] and can be considered models of vacuoles without vacuolar sap.

The transport properties of these vacuolar membrane vesicles are studied by a filtration technique.[10,11] The vesicles are incubated in a total volume of 100 μl in the presence of 3,500 Bq of a radioactive amino acid, 20–25 mM Tris–MES (pH 6.9–7.4), 4 mM MgCl$_2$, 20–25 mM KCl, and 0.5 mM ATP. The incubation is stopped by the addition of 5 ml ice-cold buffer, followed by rapid filtration through a 0.45-μm membrane filter. The transport rates are expressed in nmol per min per mg membrane protein and are of the order of 50–500 nmol per min per mg membrane protein.[10,11] The arginine transport system found by this technique depends on MgATP. Protonophores like SF6847 (1 μM) block arginine transport and cause a rapid efflux of previously accumulated arginine.[10] Thus, arginine transport in vacuolar membrane vesicles is active; the authors suggest that it is an arginine–H$^+$ antiport,[10] driven by a proton-motive force generated by the vacuolar membrane ATPase.[25,26] Sato *et al.*[11] have described and kinetically defined a total of seven different active amino acid transport systems in vacuolar membrane vesicles. In addition, a similar transport system is present for Ca^{2+}.[27]

On the contrary, the arginine permease of intact isolated vacuoles described above[6,7] is a passive transport system. It operates as an exchange transport in the absence of an exogenous energy source and is not affected by protonophores (Table III). It is most similar to the single passive transport system found in isolated tonoplast vesicles, a so-called arginine–histidine exchange system.[12] This system is independent of ATP and has a much higher affinity for arginine than the active transport system. It can be observed only in rather complicated experiments in which the vesicles are preloaded with nonradioactive histidine and then rapidly diluted into a solution of radioactive arginine. The exchange transport leads to an equi-

[25] Y. Kakinuma, Y. Ohsumi, and Y. Anraku, *J. Biol. Chem.* **256**, 10859 (1981).
[26] E. Uchida, Y. Ohsumi, and Y. Anraku, *J. Biol. Chem.* **260**, 1090 (1985).
[27] Y. Ohsumi and Y. Anraku, *J. Biol. Chem.* **258**, 5614 (1983).

librium so rapidly that it can be observed only for a few minutes at lowered temperatures.[12]

It is obvious that such a passive transport system can be discovered and analyzed more easily in intact isolated vacuoles which retain their contents. Characteristically, all solute transport systems characterized to date in intact isolated yeast vacuoles are passive transport systems. In contrast, the isolated tonoplast vesicles are more suitable for studies of active transport systems. There, the lack of internal pools is a big advantage since it reduces or eliminates the possible contribution of passive exchange transport systems to the observed transport rates.

Conclusion

Intact isolated yeast vacuoles are excellent models to study intracellular transport. Like isolated tonoplast vesicles, isolated vacuoles offer the opportunity of measuring solute fluxes across the tonoplast directly. Yet, they are isolated without osmotic shock, retain their solutes, and reflect the natural state of the vacuole more closely than vacuolar membrane vesicles. The potential of isolated vacuoles for transport studies has not yet been fully exploited. In particular, it will be interesting to explore active, ATP-driven transport in intact vacuoles and to compare it with the transport properties in vacuolar membrane vesicles.

[32] Metabolite Levels in Specific Cells and Subcellular Compartments of Plant Leaves

By MARK STITT, ROSS McC. LILLEY, RICHARD GERHARDT, and HANS W. HELDT

Photosynthesis requires close cooperation between different subcellular compartments or even between different cells, and the levels of metabolites in these metabolic compartments must be measured if the regulation of these processes is to be understood. Three different approaches are available, namely, nuclear magnetic resonance (NMR),[1] microdissection and microassay of material,[2] and destructive fractionation of bulk material. Some metabolites have been measured by NMR during respiratory metab-

[1] J. K. M. Roberts, *Annu. Rev. Plant Physiol.* **35**, 375 (1984).
[2] W. H. Outlaw, Jr., *Annu. Rev. Plant Physiol.* **31**, 299 (1980).

olism of leaves,[3,4] but NMR measurements on photosynthesizing leaves have not yet been possible. Microdissection has been successfully used to study metabolism in palisade and spongy mesophyll cells,[5] and in stomata,[6] but has not yet been applied at the subcellular level, apart from a study of vacuoles in centrifuged leaf material.[7] Several methods are now available which allow leaf material to be destructively fractionated, and in this chapter we describe four techniques[8-11] developed in our laboratory. Related techniques have also been developed by other groups.[12-16]

Destructive fractionation involves disrupting an intact biological system and then separating and quenching a series of experimental fractions, enriched in material deriving from different metabolic compartments. The degree of enrichment in each fraction is assessed by measuring the activities of marker enzymes, which are confined to a single compartment *in vivo*. The *in vivo* distribution of the metabolites can then be quantitatively estimated. This approach requires that the marker enzyme activities and metabolite levels do not alter during the fractionation and quench, and it is therefore essential that any fractionation procedure is accompanied by a demonstration that the sum of the metabolite contents or enzyme activities in the experimental fractions is the same as that in the unfractionated tissue. It also requires that metabolites and enzyme activities do not redistribute between different fractions during the fractionation procedure.

Leaf material is difficult to fractionate because a typical plant cell contains several different, mechanically fragile subcellular organelles and is surrounded by a mechanically strong plant cell wall. Further, the turnover rates of metabolites during leaf metabolism can be exceptionally rapid,[17-19] making it essential that a rapid quench is achieved. There is no method

[3] C. Foyer, D. A. Walker, C. Spencer, and B. Mann, *Biochem. J.* **202,** 429 (1982).
[4] M. Stitt, W. Wirtz, R. Gerhardt, H. W. Heldt, C. Spencer, D. A. Walker, and C. Foyer, *Planta* **166,** 354 (1985).
[5] W. H. Outlaw, Jr., C. L. Schmuck, and N. E. Tolbert, *Plant Physiol.* **58,** 186 (1976).
[6] W. H. Outlaw, Jr. and M. C. Tarczynski, *Plant Physiol.* **74,** 424 (1984).
[7] D. B. Fisher and W. H. Outlaw, Jr., *Plant Physiol.* **64,** 481 (1979).
[8] W. Wirtz, M. Stitt, and H. W. Heldt, *Plant Physiol.* **66,** 187 (1980).
[9] R. McC. Lilley, M. Stitt, G. Mader, and H. W. Heldt, *Plant Physiol.* **70,** 965 (1982).
[10] R. Gerhardt and H. W. Heldt, *Plant Physiol.* **75,** 542 (1984).
[11] M. Stitt and H. W. Heldt, *Planta* **164,** 179 (1985).
[12] S. P. Robinson and D. A. Walker, *Arch. Biochem. Biophys.* **196,** 319 (1979).
[13] R. Hampp, *Planta* **150,** 291 (1980).
[14] G. Kaiser, E. Martinoia, and A. Wiemken, *Z. Pflanzenphysiol.* **107,** 103 (1982).
[15] K.-J. Dietz and U. Heber, *Biochim. Biophys. Acta* **767,** 432 (1984).
[16] R. C. Leegood, *Planta* **164,** 163 (1985).
[17] M. Stitt, W. Wirtz, and H. W. Heldt, *Biochim. Biophys. Acta* **593,** 85 (1980).
[18] M. Stitt, R. McC. Lilley, and H. W. Heldt, *Plant Physiol.* **70,** 971 (1982).
[19] T. D. Sharkey, *Bot. Rev.* **51,** 53 (1985).

available which allows intact organelles or specialized cells to be isolated rapidly in a direct way from leaves, but two less direct approaches are available.

In one approach, the barrier posed by the cell wall is first removed by preparing protoplasts. Protoplasts are fragile and their plasmalemma is easily disrupted to release intact organelles. Two different methods will be described which allow organelles to be rapidly separated and quenched following disruption of the protoplasts. An essential precondition for work with protoplasts is that they show rates of photosynthesis which are comparable with the parent plant material. However, even rapidly photosynthesizing protoplasts are a disturbed system in which the balance between the light reactions and the dark reactions has been disturbed,[4] and many important problems pertaining to control of carbon partitioning and "sink" control of photosynthesis and sucrose synthesis cannot be answered because the normal route of sucrose export has been lost. Also, protoplasts cannot be used to study metabolite levels in photosynthetic systems where there is a close cooperation between different cell types, as the interaction is lost immediately after the cells are separated.

For this reason, alternative approaches starting with whole leaves are also necessary. Rather than attempting to isolate whole organelles or cells, leaves are frozen in liquid N_2 and then broken to small fragments which are enriched in material from a given compartment. These fragments are physically separated under conditions when metabolic activity or redistribution of metabolites is prevented, and subsequently their metabolism is quenched. One such approach is to use nonaqueous density gradient centrifugation to separate chloroplastic,[15,20,21] cytosolic, and even vacuolar material.[10] Another involves sieving homogenized maize leaf material in liquid N_2 to separate bundle sheath and mesophyll cell material.[11] The exclusion of water, or the use of extremely low temperatures, prevents metabolic activity during the fractionation procedures.

Subcellular Fractionation of Protoplasts by Silicone Oil Centrifugation

Introduction

Silicone oil centrifugation allows chloroplasts to be separated from the remainder of the protoplast and to be quenched within 2–3 sec of disrupting the protoplast.[8] The separation takes place in a microfuge tube containing, from the top downward, the suspension of protoplasts, a nylon net,

[20] U. Heber, M. G. Pon, and M. Heber, *Plant Physiol.* **66**, 355 (1963).
[21] C. R. Stocking, *Plant Physiol.* **34**, 56 (1959).

an air space, a layer of silicone oil, and perchloric acid. The protoplasts are disrupted as they are centrifuged through the nylon net, and the chloroplasts continue on through the silicone oil and are quenched in perchloric acid. A droplet of acid enters through a fine hole in the lid of the centrifuge tube to acidify the supernatant, which contains material deriving from the cytosol, mitochondria, and vacuoles. An essentially similar technique has been developed by Robinson and Walker.[12]

Plant Material

Spinach is grown for 5–6 weeks in hydroponic culture.[22] Peas are grown for 10–14 days in vermiculite, and barley in a greenhouse with supplementary lighting. Protoplasts are prepared from spinach,[8] barley,[4] and pea[4] by a procedure modified slightly from that used by Edwards *et al.*[23] Leaf segments are cut for barley leaves, while epidermis is removed for spinach and pea. For pea, the osmoticum is reduced to 0.1 M for all steps, and the incubation time reduced to 2.25 hr. Protoplasts are stored at 4° at a concentration of 200 μg chlorophyll (Chl)/ml.

Reaction Medium

Protoplasts are diluted to 50 μg Chl/ml in a medium containing (final concentration) 0.42 M sorbitol and either 50 mM (2-N-morpholinoethanesulfonic acid, pH 7.0), 3 mM NaHCO$_3$, 10 mM KCl, and 0.5 mM MgCl$_2$ *or* 50 mM HEPES (N-2-hydroxyethylpiperazine-N'-2-ethanesulfonic acid, pH 7.6), 10 mM NaHCO$_3$, and 10 mM CaCl$_2$. The metal ions and neutral pH are chosen to inhibit photosynthesis by free chloroplasts in the protoplast preparation (see below). The suspension is stirred every minute to prevent sedimentation of the protoplasts. The protoplast suspension is transferred into the microcentrifuge tubes immediately before fractionation. Incubation is carried out at 20°, and the silicone oil centrifuge tubes (see below) are preequilibrated at 20°. At higher temperatures the silicone oil layer and the protoplast suspension may invert during centrifugation, unless the density of the silicone is increased.

Fractionation of Protoplasts

Polypropylene microfuge tubes (Sarstedt, 400 μl volume) are prepared (see Fig. 1) containing successively 20 μl of 10% (v/v) HClO$_4$, 70 μl of

[22] R. McC. Lilley and D. A. Walker, *Biochim. Biophys. Acta* **368**, 269 (1974).
[23] G. E. Edwards, S. P. Robinson, N. J. C. Tyler, and D. A. Walker, *Plant Physiol.* **62**, 313 (1978).

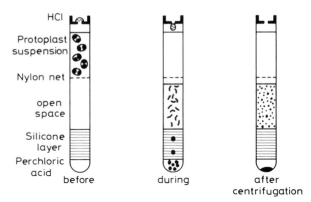

Fig. 1. Rapid separation and immediate quenching of chloroplast and extrachloroplast fractions from spinach.[8]

silicone oil (19 ml AR 200, 1 ml AR 20, Wacker Chemie), an air space of approximately 120 μl, and a nylon net of 17 μm (spinach) or 7 μm (pea, barley) (Züricher Beuteltuchfabrik AG, Rüschlikon, Switzerland) held securely in place by a 14 mm length of polypropylene tubing (3 × 4 mm, width 0.55 mm). The tube is used to force a square (9 × 9 mm) of nylon net into the microcentrifuge tube, the net periphery being tightly clamped between the tubing and the microfuge tube. A smear of silicone grease applied to the outside of the tubing improves the seal. The protoplast suspension is held in the space over the nylon net. The centrifuge tube lid, which has a fine pore, contains 10 μl 0.3 N HCl in the well.

The microfuge tubes are centrifuged 9 sec in a Beckman microfuge (Model 152) at 20°. In this centrifuge, the tubes are held horizontally and can be illuminated from above throughout the fractionation. The protoplasts disrupt as they pass through the nylon net. Intact chloroplasts continue downward through the silicone oil into 10% $HClO_4$ (F_1). The extrachloroplastic components remain above the oil, and their pH is reduced to 2.5–3.5 by the droplet of HCl which enters the tube through the pore in the lid. Immediately after the centrifuge stops, the tube is cut through under the nylon net, and a 90-μl aliquot of the extrachloroplast supernatant is removed and deproteinized by adding it to 12.5 μl 60% (v/v) $HClO_4$ (F_2). The centrifuge tube end is stored at 4° until the end of the experiment when it is cut through a second time, at the base of the silicone oil layer. The chloroplast pellet and 10% (v/v) $HClO_4$ are removed from the tip and combined with 80 μl water used to rinse the tip interior.

Separation of Protoplasts and Medium

The metabolites in the extrachloroplast supernatant (F_2) are derived from the cytosol, vacuoles, and mitochondria of the protoplasts, but also from the medium outside the protoplasts. As the medium can contain significant levels of metabolites which derive from the contents of broken protoplasts and from the activity of free intact chloroplasts which export triose phosphate to the medium during photosynthesis, it is essential to measure the level of metabolites in the medium, if the true extrachloroplast content of the protoplasts is to be estimated. For this, polypropylene tubes are prepared containing 20 μl 10% (v/v) $HClO_4$ and 70 μl silicone oil (75 ml AR 20 and 180 ml AR 200). The protoplast suspension (100 μl) is pipetted above the silicone oil and centrifuged for 6 sec in a Beckman microfuge (Model 152). Immediately after the centrifuge comes to rest, 85-μl aliquots of the supernatant are added to 12.5 μl 60% (v/v) $HClO_4$ (F_3). The protoplast sediment (F_4) is stored at 4° until the end of the experiment and then resuspended as for the chloroplast sediment.

Measurement of Marker Enzyme Distribution

The activity of marker enzymes cannot be measured in the samples whose metabolism is quenched in $HClO_4$. Instead, parallel fractionations are carried out in tubes where the 10% $HClO_4$ is replaced by 25% (w/w) sucrose and HCl omitted from the lid. The sucrose is dissolved in the same buffer as that used to suspend the protoplasts. The sediments are resuspended and the tip washed with 80 μl of the medium used to suspend the protoplasts before measuring the activity of marker enzymes. Marker enzymes are assayed directly in the supernatant.

Phosphoenolpyruvate carboxylase (PEPCX) was taken as a marker for the cytosol. Although NADP-glyceraldehydephosphate (GAP) dehydrogenase can be used as a marker for the chloroplast, it is preferable to use the metabolite ribulose 1,5-bisphosphate (RuBP). The extent of contamination from chloroplasts in the extrachloroplast fraction F_2 varies between 10 and 35%, and using RuBP as the marker allows the cross-contamination to be measured in each individual fractionation.

Calculations of Metabolite Distribution

The distribution of metabolites and enzyme activities in fractions F_1–F_4 can be used to evaluate the level of metabolites in the chloroplast (C) and extrachloroplast (E) compartment. "Compartment" refers to a subcellular region present in the intact protoplast, and "fraction" to an experimentally produced sample. The data required are the metabolite amounts

(A_1, A_2, A_3), the RuBP content (stromal marker), and PEP carboxylase activity (cytosol marker) in the fractions F_1 (chloroplast sediment), F_2 (extrachloroplast supernatant), and F_3 (medium), respectively. A typical fractionation is shown in Table I. First, the cross-contamination is estimated.

$$a = \frac{RuBP_2 - RuBP_3}{RuBP_1 + RuBP_2 - RuBP_3}$$

$$b = \frac{PEPCX_1}{PEPCX_1 + PEPCX_2 - PEPCX_3}$$

The value for a varies between 0.35 and 0.10, depending on the tissue, and can be estimated individually for every sample. The value for b is determined for each protoplast type and condition. Typically, b is very low, being between 0.02 and 0.04. The following equations can then be solved.

$$C + E = A_1 + A_2 - A_3$$
$$A_1 = (1 - a)C + b(A_2 - A_3)$$
$$C = \frac{A_1 - b(A_2 - A_3)}{1 - a}$$
$$E = A_1 + A_2 - A_3 - C$$

This calculation assumes that cross-contamination from the metabolites in the medium (F_3) is negligible in the chloroplast fraction (F_1). Experiments showed that only 0.2% of the medium is carried through the silicone oil with the chloroplasts,[9] which is 10-fold lower than the cytoplasmic contamination measured by the distribution of PEP carboxylase.

Applications and Limitations

This technique allows fractions to be prepared which are highly enriched in chloroplast and extrachloroplast material, and which are quenched within 2–3 sec of disrupting the protoplasts. Many cytosolic pools turn over fairly slowly,[17] and reactions will be further slowed by the dilution of metabolites and enzymes after disrupting the protoplasts. In contrast, chloroplasts remain intact during the fractionation, and their pools turn over in 1 sec or less during photosynthesis.[17] The method therefore relies on chloroplast metabolism having continued for 1–2 sec after protoplast disruption without large changes in metabolite pools occurring. Comparison of the overall metabolite levels in whole protoplasts with those after silicone oil fractionation showed that large changes did not occur,[8] but small shifts in chloroplast metabolite pools cannot be excluded. A second limitation of the method is that the extrachloroplast fraction contains material deriving from vacuoles and mitochondria, as well as the

TABLE I

DISTRIBUTION OF SUBCELLULAR MARKERS PEP CARBOXYLASE AND RIBULOSE 1,5-BISPHOSPHATE FOLLOWING SILICONE OIL FRACTIONATION OF SPINACH LEAF PROTOPLASTS, AND DISTRIBUTION OF DIHYDROXYACETONE PHOSPHATE BETWEEN CHLOROPLAST AND EXTRACHLOROPLAST FRACTIONS

Enzyme markers	Distribution in experimental fractions				Distribution in subcellular compartments		
	F_1 sediment	F_2 supernatant	F_3 supernatant	F_4 sediment	Intact protoplast	Chloroplast	Extra-chloroplast
PEP carboxylase (μmol/mg Chl-hr)	1.3	55	5.2	51.0	51.1	—	48.8
Ribulose 1,5-bisphosphate (nmol/mg Chl)	17.0	6.1	1.2	21.5	21.9	21.9	—
Dihydroxyacetone phosphate (nmol/mg Chl)	14.5	79.2	42.8	44.5	50.9	17.4	33.4

cytosol. This means that this technique cannot be used to provide information on the cytosolic levels of metabolites known to be present in the vacuole (e.g., sucrose, malate, P_i) or the mitochondria (e.g., adenine nucleotides).

It is essential that contamination of the extrachloroplast fraction by metabolites from the medium surrounding the protoplasts should be minimized and accurately assessed. Clearly, the contamination is decreased in protoplast preparations of high intactness. However, most of the metabolites in the medium may not derive directly from broken protoplasts, but are produced during photosynthesis by a small number of free chloroplasts. These can generate more metabolites in the medium than are present in the cytosol. The latter contains small pools which turn over rapidly, while metabolites exported from free chloroplasts accumulate in the medium. For example, if only 10% of the protoplasts have broken to release chloroplasts which photosynthesize at the very low rate of 5 μmol CO_2 mg^{-1} Chl hr^{-1}, they would produce about 11 nmol triose phosphate in 4 min. This is one-half the size of the total cytoplasmic pool. It is essential to prevent such free chloroplasts from photosynthesizing. This can be achieved by including divalent ions in the medium[8] or by using neutral or slightly acid pH.[4] The low levels of metabolites in the medium can then be assessed by separating intact protoplasts from the medium by centrifuging them through silicone oil.

A basically similar technique can be used to fractionate protoplasts from different plant sources, provided the aperture of the nylon net is decreased for species like barley or pea, where the protoplasts are smaller and can pass without breaking through a 17-μm net. The fractionation technique can also be used in varying conditions, e.g., different temperatures, provided the silicone oil mixture is adjusted so that the viscosity and density allow chloroplasts to pass through rapidly, but there is no inversion with the protoplast suspension during centrifugation.

Subcellular Fractionation of Wheat Leaf Protoplasts with Quench Times Shorter Than 100 msec

Introduction

Membrane filtration provides an alternative, and faster, way of disrupting, fractionating, and quenching protoplasts. The principle is that rupture of the protoplasts, separation of organelles, and metabolic quenching are performed sequentially in a flow system. Protoplasts are disrupted by forcing them through a nylon net, behind which membrane filters are placed which selectively remove the chloroplasts or mitochondria. The organelles retained on the filters are not analyzed, because they cannot be quenched rapidly enough. Instead, the filtrates, from which selected organelles have been almost completely removed, are immediately quenched. By maximizing flow rates, and minimizing the dead volume between each of the three stages, the time for each protoplast to complete the process can be reduced to about 100 ms.[9] This technique has been used to determine adenine nucleotide levels in the chloroplasts, mitochondria, and cytosol of wheat leaf protoplasts.[18]

Plant Material

Suitable cultivars of wheat *(Triticum aestivum)* include "Timmo" (Rothwell Plant Breeders, Lincoln, U.K.) and "Olympic" (Arthur Yates Co., Milperra, N.S.W. 2214, Australia). Plants are grown for 7 to 11 days in hydroponic culture[24] in a greenhouse (with supplementary lighting in winter), and harvested in the morning. Protoplasts are prepared by a method modified from Edwards *et al.*[23] as described in detail in Lilley *et al.*[9] Particular attention is given to washing both the cut leaf sections and the protoplasts to remove phenols which are present at high levels in wheat and can impair yield and photosynthetic activity. The washing of protoplasts is achieved by using a raffinose–sorbitol flotation gradient with a large volume. Calcium is excluded from the final stages of the protoplast preparation as it impairs the activity of some enzymes and interferes with the filtration of protoplasts extracts (see below).

Reaction Medium

Protoplasts are stored at a concentration of 100 μg Chl/ml and then diluted to 25 μg Chl/ml in the following medium: 0.32 M sorbitol, 0.1 M raffinose, 50 mM HEPES, 10 mM KCl, 0.5 mM MgCl$_2$, 1.0 mM NaHCO$_3$, 0.2% (w/v) poly(vinylpyrrolidine), and 0.06% bovine serum albumin adjusted to pH 6.8 with KOH at 20°. This medium is isopycnic with respect to the protoplasts, avoiding the need for constant stirring to keep the protoplasts in suspension.

Fractionation of Protoplasts

The protoplast suspension (2–2.5 ml) is slowly drawn into a transparent plastic syringe (2 ml disposable, maximum volume about 2.7 ml) modified by inserting an 0.85 mm diameter Polythene tube into its outlet. The syringe is attached directly to a 25 mm diameter membrane filter holder

[24] P. J. Randall and D. Bouma, *Plant Physiol.* **52,** 299 (1973).

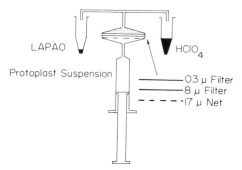

FIG. 2. Diagram of the syringe, filter holder, and collecting tube.[9]

(Sartorius GmbH, Goettingen, FRG) and the assembly is held vertically with the membrane filter holder uppermost (Fig. 2). The compartment of the filter holder on the exit side is packed with glass beads (1 mm diameter) to reduce the dead volume to 200 μl. A capillary T-piece attached to the outlet leads to two 5-ml microcentrifuge tubes. One contains 50 μl 70% (v/v) $HClO_4$ to quench metabolism. The other, for measurement of marker enzyme distribution, contains 5 μl detergent (2.4% Aminoxide WS-35; Theo Goldschmidt, AG, Munich) and is immediately snap-frozen in 50-μl aliquots at $-85°$. This procedure allows exact determination of cross-contamination in each individual fractionation.

The filter holder contains a 17-μm mesh nylon net (Scrynel, Züricher Beuteltuchfabrik) on the inlet side to rupture the protoplasts. Membrane filters (Sartorius) of the pore size and sequence specified below are placed behind this net for separation of the released organelles. In all cases, polyester membrane separators (Sartorius) are inserted between filters and between the nylon net and a filter. The following combinations are used: (1) Nylon net alone: filtrate (F_{tot}) contains the unfractionated disrupted protoplast suspension (vacuolar, cytosolic, mitochondrial, and chloroplast material); (2) nylon net and 8-μm membrane filter (cellulose nitrate): filtrate (F_1) contains mainly vacuolar, cytosolic, and mitochondrial material; (3) nylon net, 8-μm (cellulose nitrate), and 0.45-μm (cellulose acetate) membrane filters: filtrate (F_2) contains mainly vacuolar and cytosolic material.

To illustrate the principle of the procedure, the marker enzyme distributions obtained in a typical fractionation are shown in Table II. The marker enzyme activities in the F_{tot} filtrate do not differ significantly from those in intact protoplasts. The F_1 filtrate contains predominantly the cytosolic marker PEP carboxylase (80–95%) and the mitochondrial marker fumarase (fumarate hydratase) (65–80%), but only 5–15% of the

TABLE II

DISTRIBUTION OF MARKER ENZYMES IN FILTRATES COLLECTED WITH DIFFERENT
MEMBRANE FILTER COMBINATIONS[a]

| | Filter construction | | | Syringe closure time (sec) | Total activity recovered (%) in filtrate | | |
| Fraction | Nylon net | Membrane filter | | | NADP-GAP dehydrogenase | PEP carboxylase | Fumarase |
		8 μm	0.45 μm				
F_{tot}	+			0.7 – 1.0	100	100	99
F_1	+	+		0.7 – 1.0	17.3	93.2	69
F_2	+	+	+	0.7 – 1.0	13.4	93.3	6
F_3	+	+	+	90	5.4	10.1	1

[a] Protoplasts were fractionated and filtered as described. The results of a typical fractionation are expressed as percentage of total activity in unbroken protoplasts. The total enzymatic activities were as follows: NADP-GAP dehydrogenase, 1100; PEP carboxylase, 2; fumarase, 15 μmol/mg Chl-hr. Data from Ref. 9.

chloroplast marker NADP-GAP dehydrogenase, showing that most of the chloroplasts are retained on the 8-μm membrane filter. The F_2 filtrate contains comparable activities of PEP carboxylase and NADP-GAP dehydrogenase, but only 2 – 10% of fumarase, showing that the 0.45-μm membrane filter is effective in retaining the mitochondria. The somewhat lower yield of PEP carboxylase in the presence of the membrane combinations is due to a small proportion of the protoplasts being filtered off intact within the filter apparatus (see below).

For a syringe containing 2.3 ml of protoplast suspension, the optimal closure time is 0.8 sec. Satisfactory syringe closure can be made by hand, but a mechanical device (Fig. 3) permits a more accurate closure time and enables the three filtrations to be done simultaneously. This device is driven by an electronically variable electric hand drill, which provides reproducible, load-independent speeds. A rotating threaded shaft (a) drives a nut upward (b) which itself is prevented from rotating by the solenoid (c). A support (d) connects the upward moving nut to a plunger with a circular base (e). The resulting upward movement of the plunger closes three syringes (f) (of which in Fig. 3 only two are seen) at the same time. In order to attain a constant motor speed when driving the syringe pistons, the syringes are initially fitted about 3 mm above the plunger. The emptying of the syringes stops abruptly when a microswitch (g) reaches the adjustable stop (h). The power to the drill is then turned off, and the solenoid (c) simultaneously retracts, releasing the nut (b). The ball bearing (i) then enables the nut to rotate freely. This stops the flow of protoplast suspension

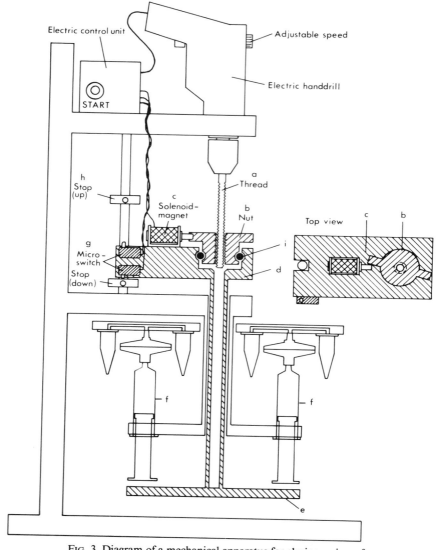

FIG. 3. Diagram of a mechanical apparatus for closing syringes.[9]

through the filter instantly by disconnecting the motor drive from the plunger movement.

During the entire procedure, the syringes are kept in a thermostatted water bath (20°). When required the syringes and the filters are illuminated obliquely from at least two directions with white light from tungsten quartz–iodine light sources of approximately 180 W/m^2.

Extracts of Unfractionated Protoplasts

Extracts are also prepared for assay of enzymes and substrates from directly frozen or quenched protoplasts, respectively. For enzymes, protoplasts are pipetted directly (from the prevailing illumination conditions) into a microcentrifuge tube in an aluminum block at −85°. For substrates, protoplasts (300 μl) are injected rapidly from a plastic syringe (outlet, 0.85 mm diameter) into an illuminated (or darkened) microcentrifuge tube containing 50 μl 70% (w/v) HClO$_4$.

Separation of Protoplasts and Medium

Intact protoplasts are filtered from the medium (F_{med}) by passage through the F_2 filter system, but using a syringe with an outlet 2.8 mm in diameter and a closure time of 90 sec. This very slow flow rate ensures that the protoplasts, held on the membrane filter, remain intact. The very small proportion of enzyme marker activities in the medium (F_{med}) indicates that protoplast breakage in the reaction mixture is low.

The separation of protoplasts from the medium by this procedure is more satisfactory than by the silicone oil centrifugation procedure.[8] It may be noted that the activity of PEP carboxylase in the medium is higher than that of NADP-GAP dehydrogenase and fumarase. Broken protoplasts in the suspension will have released many of their organelles intact, and any such free intact organelles will be retained on the filters.

Calculations of Subcellular Metabolite Levels

The levels of metabolites, e.g., ATP, actually present in the compartments of the cytosol (C), the mitochondrial matrix (M), and the chloroplast stroma (S) can be calculated from those found in the filtrates F_{tot}, F_1, and F_2. The data used in the calculations (A_T, A_1, A_2) represent the amounts of metabolite in these filtrates after subtraction of the corresponding amounts measured in the protoplast medium (F_{med}). Similarly, the enzyme activity in the medium is first subtracted from the enzyme activities in the various filtrates, before using these values to correct for cross-contamination. The proportion of the total marker enzyme activity from intact protoplasts recovered in the filtrates F_1 and F_2 is represented by the following symbols: activity in ($F_1 - F_{med}$)/($F_{tot} - F_{med}$): fumarase, u; NADP-GAP dehydrogenase, v; PEP carboxylase, w; activity in ($F_2 - F_{med}$)/($F_{tot} - F_{med}$): fumarase, x; NADP-GAP dehydrogenase, y; PEP carboxylase, z.

For the distribution of metabolites and marker enzymes among the three compartments, the following three equations can be written:

$$A_{tot} = C + S + M$$
$$A_1 = wC + vS + uM$$
$$A_2 = zC + yS + xM$$

Solved for C:

$$C = \frac{\left(\dfrac{xA_{tot} - A_2}{y - x}\right) + \left(\dfrac{A_1 - uA_{tot}}{v - u}\right)}{\left(\dfrac{w - u}{v - u}\right) + \left(\dfrac{x - z}{y - x}\right)}$$

Solved for S and M:

$$S = C\left(\frac{u - w}{v - u}\right) + \frac{A_1 - uA_{tot}}{v - u}$$
$$M = A_{tot} - S - C$$

The results of a typical calculation for ATP are shown in Table III. In addition to the final figures calculated for the subcellular ATP distribution for whole protoplasts, the amounts present in each filtrate are also shown.

TABLE III
CALCULATION OF SUBCELLULAR ATP CONTENTS FROM
DISTRIBUTION OF MARKER ENZYMES AND ATP IN FILTRATES[a]

	Amount of ATP (nmol/mg Chl)			
		Calculated subcellular source		
Filtrate	Measured in filtrate	Stroma	Cytosol	Mitochondria
F_{tot}	63.3			
F_1	42.6	2.2	37.3	3.7
F_2	38.6	1.5	34.4	0.3
F_{med}	2.4	0	0	0
Subcellular content		18.3	37.1	5.4

[a] Same experiments as in Table II; the calculation of subcellular ATP levels using the marker activities shown in Table II is described in the text. In order to show the cross-correction involved, the source of ATP is also given for F_1 and F_2. Data from Ref. 9.

Factors Affecting the Filtration

To remove an organelle by filtration, the pore size of the filter has to be small enough to retain the organelle, but the organelles rupture if the pores are too small. For example, chloroplasts are almost totally disrupted in filters with a pore size less than 1.2 μm (data not shown). Polycarbonate surface filters were found to be unsuitable, either allowing whole organelles to pass through or disrupting the organelles totally (data not shown). It seems that the pores must be large enough to allow the organelles to penetrate into the matrix of depth-type filters, where they are retained without rupture. Retention of the organelles at the surface of a filter causes rupture owing, presumably, either to higher local solution flow rates into individual pores or to an increasing pressure difference across the filter caused by the organelles blocking pores. For this reason the 8-μm membrane filter is included in the F_2 filter combination to remove the chloroplasts before removing the mitochondria with the 0.45-μm filter.

Because the separation is very dependent on the speed of filtration, the use of a power-driven device to close the syringes at a predetermined rate is advisable (Fig. 3). The optimal closure time is 0.8–0.9 sec for syringes containing 2.3 ml protoplast suspension at 25 μg Chl/ml. Faster closure leads to increased disruption of chloroplasts on the filter, while slower closure increases the retention of protoplasts in the filter apparatus, because more protoplasts are retained on the nylon net or pass through the net without being ruptured and are retained on the filters. Protoplast breakage can be aided by a narrow syringe aperture, but when the exit is smaller than 0.8 mm, turbulence is increased and more chloroplasts break during the filtration. Separation is improved when the medium density is increased by addition of raffinose until it is isopynic with the protoplasts. The chlorophyll concentration and the volume passed through the filter are also critical; in standard conditions, 1.5–1.8 ml of suspension (20–25 μg Chl/ml) can pass through the membrane filters before these saturate and organelle breakage increases. It should be emphasized that the speed of closure, syringe, design, protoplast concentration, and volume all interact in determining the efficiency of the separation.

The composition of the medium is chosen to prevent nonspecific binding of organelles, substrates, or enzymes onto the negatively charged membrane filter material. Poly(vinylpyrrolidine) prevents binding of PEP carboxylase and mitochondria. Where possible, cellulose acetate filters are used instead of cellulose nitrate because the latter bind more material. Bovine serum albumin is included not only to prevent nonspecific binding but also to protect the marker enzymes against inactivation by proteases and phenolic compounds. It might be noted that binding is much stronger on filters with a smaller pore size, presumably owing to the larger internal surface area. The concentrations of divalent cations must be carefully selected, as higher concentrations increase binding of mitochondria onto the 8-μm filter. Control experiments showed that under our conditions there is no binding of adenine nucleotides when they were present at levels resembling those in a disrupted protoplast suspension.[9] When a protoplast extract was prepared by passing protoplasts through a nylon net, identical activities of PEP carboxylase are found before and after passage through a F_2 filter combination.[9] After disrupting of the organelles with 0.12% Aminoxide detergent before passage through a F_2 filter combination, there was no retention of GAP dehydrogenase or of fumarase on the filters.[9]

Comments on the Method

This technique is designed to quench metabolism rapidly after rupturing the protoplasts. At a flow rate more than 2 ml/sec and a total dead volume in the filtration apparatus between the nylon net and the outlet of about 200 μl, the transit time for the disrupted cell material between the nylon net and the entry into the concentrated $HClO_4$ is less than 100 msec. This is more than 10 times faster than can be achieved by centrifugal silicone oil filtering techniques for separation of chloroplasts and cytosol[8,12,13] and 300–600 times faster than for separation of mitochondria.[13] The total amounts of ATP, ADP, and AMP in whole protoplasts quenched by direct injection into concentrated $HClO_4$ were comparable with the summed levels from the chloroplasts, cytosol, and mitochondria, determined from the filtration procedure for both illuminated and darkened protoplasts.[9]

Apparently, there are no substantial changes in the adenine nucleotide levels during the fractionation procedure. Small changes of AMP were found, but accurate measurement of AMP using the luciferase method is difficult if the ATP and ADP levels are much higher than that of AMP, because with the luminescence measurements employed the AMP value is obtained by subtracting the sum of ATP plus ADP from the total sum of the adenine nucleotides. Further, no large changes of added ATP, ADP, or AMP during filtration by the F_{tot}, F_1, or F_2 systems occurred, showing that there is no significant metabolism of the adenine nucleotides released to the medium during filtration.[9]

The chloroplasts in the disrupted protoplasts suspension (F_{tot}) are largely intact, while the NADP-GAP dehydrogenase in the filtrates obtained after passage through membrane filters (F_1, F_2) derives mostly from broken chloroplasts (data not shown). The cross-correction for the contribution of chloroplast-derived adenine nucleotides to filtrates F_1 and F_2 would be inaccurate if the adenine nucleotide levels contributed by intact and broken chloroplasts differed. Since stromal ATP has a half-life of about 100 msec in the light, it is particularly susceptible to alterations. However, the levels of ATP, ADP, and AMP in a normally prepared F_{tot} filtrate containing intact chloroplasts are very similar to the levels found in a protoplast filtrate in which the chloroplasts had been immediately disrupted by placing a 1-μm polycarbonate surface filter immediately behind the 17-μm nylon net.[9] This experiment provides additional evidence that the adenine nucleotide levels inside intact chloroplasts do not change greatly during the fractionation.

While these experiments make it unlikely that there are large alterations in the stromal and cytosolic adenine nucleotides during the fractionation, the evidence is not decisive for the mitochondria. As only about 10% of the total adenine nucleotides occur in the mitochondria, these could change without leading to large deviations in the total adenine nucleotide pool in the protoplasts. However, the rapid metabolic quench should limit the extent of any such changes. A typical respiration rate of 0.15 μmol CO_2 mg^{-1} Chl min^{-1} [25] would support the synthesis of about 1 nmol ATP mg^{-1} Chl in 0.1 sec. After protoplast disruption, the first effect of mitochondrial metabolism is likely to be a restriction of release of ATP in exchange for ADP via the adenine nucleotide translocator[26] owing to the very large dilution of the cytosolic ADP by the medium. The possibility, therefore, of a small rise in the mitochondrial ATP during the fractionation cannot be rigorously excluded.

These considerations emphasize the importance of rapidly quenching mitochondrial metabolism, if the measured matrix adenine nucleotides are to resemble the *in vivo* levels, and imply that slower techniques using silicone oil centrifugation[13] are basically unsuitable when the individual adenine nucleotides are to be assayed in the separated mitochondria. Even when isolated animal mitochondria are centrifuged through silicone oil directly into $HClO_4$, some changes in the adenine nucleotide pool have been demonstrated.[27] The likelihood of alterations when mitochondria are released from a cell into a diluted extract appear even greater than for isolated mitochondria, which at least remain in the same incubation medium during the centrifugation, especially as plant cells contain large amounts of malate, mostly in the vacuole,[10] which becomes available to mitochondrial respiration in the time between cell disruption and the quench.

[25] D. H. MacLennan, H. Beevers, and I. L. Harley, *Biochem. J.* **89**, 316 (1963).

[26] M. Klingenberg, *in* "Enzymes of Biological Membranes" (A. Martonosi, ed.), Vol. 3, p. 383. Plenum, New York, 1976.

[27] F. Brawand, G. Folly, and P. Walter, *Hoppe-Seyler's Z. Physiol. Chem.* **358**, 1183 (1977).

The faster separation by membrane filtration also provides advantages compared with the available silicone oil centrifugation techniques when metabolism in the chloroplast stroma and cytosol is being studied. This is especially so in the case of the cytosol, where a relatively uncontaminated fraction is obtained almost instantaneously. It should be noted that in this method, the cytosol actually designates those compartments of the cell other than the chloroplasts and mitochondria and will include, in particular, the vacuole. A measurement of cytosolic metabolite levels also requires a correction to be made for the metabolites free in the medium, which otherwise would be included in the value for the cytosol[8] (see above). The metabolites in the medium can be accurately evaluated after removing the protoplasts without significant breakage using a slow filtration.

For chloroplasts, as of course also for mitochondria, the membrane filtration method has the disadvantage that the values are obtained by subtraction from the cytosol, so that the experiments require accurate analysis and repetition. However, in all those cases where significant alterations in metabolite levels could occur during a slower separation, the method offers great advantage in sacrificing the purity of the individual fractions in order to quench these fractions faster.

Nonaqueous Fractionation of Lyophilized and Ultrasonicated Leaf Material Using Nonaqueous Gradients

Introduction

Leaf material is quenched in liquid N_2 and then homogenized, lyophilized, ultrasonicated, and subjected to density gradient centrifugation in nonpolar (hexane–carbon tetrachloride) fluids. The gradient is divided into several fractions, and the distribution of metabolites and marker enzymes is measured in each fraction.[10] The distribution of metabolites between the chloroplast stroma, cytosol, and vacuole can then be evaluated. The principle of this technique is that the metabolites and proteins in a particular region of the cell will aggregate together as the leaf material is lyophilized. Following ultrasonication to disrupt the leaf material into small particles, material deriving from different regions of the cell can be separated because the density of these particles varies, reflecting the varying protein, lipid, carbohydrate, and ionic composition of the subcellular compartments from which they derive. Until lyophilization is complete, the leaf material is kept deep frozen to prevent biochemical reactions, and subsequent alteration of metabolites or redistribution of metabolites or proteins is prevented by the nonpolar conditions.

Freeze Stop, Lyophilization, and Storage

The fractionation procedure is shown as a flow diagram in Fig. 4. Leaf metabolism is stopped by plunging leaves into liquid N_2. After removing large ribs, the material is ground to a homogeneous coarse powder in a porcelain mortar (20 cm diameter) containing about 100 ml liquid N_2. The suspension of powder in liquid N_2 is tipped into a plastic beaker, and immediately placed in a lyophilizer, which has been precooled to less than $-50°$. The material is dried at $-50°$ and 0.03 Torr for 100 hr, before ventilating the lyophilizer with dry N_2 gas. The plastic beaker is closed with an air-tight seal while still in the lyophilizer and transferred to a precooled ($-35°$) desiccator containing phosphorus pentoxide drying agent and a moisture indicator (Sicapent, Merck, Darmstadt, FRG). The dried leaf material is stored in the desiccator at $-35°$ until use.

Ultrasonication and Density Gradient Centrifugation

About 200–300 mg of dry leaf powder (one heaped teaspoon) is transferred at $-35°$ into 20 ml of a carbon tetrachloride–heptane mixture [CCl_4/C_7H_{16} 66 : 34 (v/v), density 1.28 g/cm^3] and ultrasonicated for a total of 90 sec (Branson sonifer, B15). To prevent overheating, the suspension is ultrasonicated in a polyallomer tube standing in an insulated holder con-

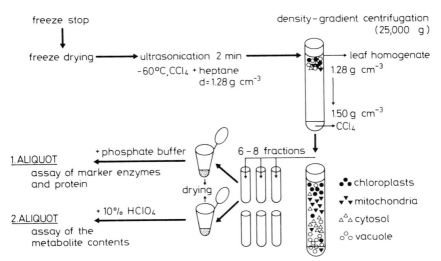

FIG. 4. Schematic representation of the nonaqueous fractionation procedure of subcellular metabolite assay in leaves.

taining a mixture of dry ice and heptane as coolant. Also to prevent overheating, the ultrasonication takes place as a pulse of 5 sec followed by a 15-sec pause, this cycle being repeated 18 times. The suspension is then poured through a layer of quartz wool contained in a filter to remove any remaining coarse material before diluting it 3-fold with heptane and centrifuging for 2 min at 3000 g. The clear supernatant is discarded, and the sediment is resuspended in 3 ml of a CCl_4/C_7H_{16} mixture, 1.28 g/cm^3, and well shaken.

Two 200-μl aliquots are withdrawn (for determination of enzyme activity and metabolites in the unfractionated material), and the remainder (2.5 ml) is layered onto a density gradient, which has been prepared in cellulose nitrate tubes (101 $\times$ 15 mm). A 1-ml cushion of CCl_4 is followed by an exponential gradient (12 ml volume) decreasing from 1.50 to 1.28 g/ cm^3. The gradient is mixed using a mixing vessel where the chamber next to the outlet (A) is sealed with a bung after receiving 5 ml of a 1.50 g/cm^3 mixture of CCl_4/C_7H_{16}. The other chamber (B) receives 10 ml of a 1.28 g/ cm^3 CCl_4/C_7H_{16} mixture. A magnetic stirrer is switched on in chamber A, and the connection between the two chambers is opened. The entire contents of chamber B pass through chamber A, whose volume remains constant, and the density of the mixture flowing into the centrifuge tube decreases exponentially. This density gradient is normally prepared immediately before the dried material is added to CCl_4/C_7H_{16} and ultrasonicated.

Throughout the fractionation procedure, it is essential to avoid contamination by condensing water vapor. For this reason, all solvents are dried and stored over molecule sieve beads (0.4 nm pore, 2 mm bead diameter, Merck) at 4° in flasks which have ground glass stops. During ultrasonication, the polyallomer tube is fitted with a bung containing a hole through which the tip of the ultrasonifier passes, to minimize condensation of water within the cold tube. The ultrasonic tip is dried thoroughly. Following ultrasonication, all steps are carried out at 4° using apparatus which has been thoroughly dried. Nonpolar solvents are transferred using pipettes which are stored at 4° to prevent condensation inside the pipettes during cycles of warming and cooling. Contamination by water is revealed by a clumping of the lyophilized material, which should appear as a very fine homogeneous powder. When a density gradient is contaminated with water, the top phase remains deep green and aggregated material collects at the top of the gradient.

The gradients are centrifuged 2.5 hr at 25,000 g (swing-out rotor SW28, ultracentrifuge TGA 65, Kontron), during which time the material distributes isopycnically in the gradient. The contents of the centrifuge tube are removed from the top in fractions of 1–2 ml. At the top of the gradient is

1 – 1.5 ml of a clear yellow-colored liquid, containing no enzyme activity or metabolites, which is discarded. A sediment which precipitates at the base of the density gradient requires intensive shaking (30 – 60 sec) to resuspend it. This material consists, at least partly, of crystallized salts from the vacuole. Each of these 6 – 8 fractions is divided into two unequal portions, consisting of one-third of the volume (for assay of marker enzymes, chlorophyll, and protein) and two-thirds (for assay of metabolites). The volumes taken at this point must be accurately determined for the calculation. The divided portions (and two aliquots taken from the material applied to the gradient) are all diluted 3-fold with heptane and centrifuged for 8 min at 18,000 g (Model 5412, Eppendorf, Hamburg, FRG). The supernatant is discarded, except for the last 200 μl which is used to resuspend the sediment by swirling it with calcined quartz (Merck). The samples are immediately placed in a desiccator containing a silica gel drying agent (Blaugel, Merck) and paraffin (solidification point 51° – 53°), dried for 18 hr at room temperature at 40 Torr, and then extracted for assay of enzymes or metabolites.

Sample Preparation for Measurement of Protein and Enzyme Activity

After adding 500 μl of 0.25 mM potassium phosphate (pH 7.5), 0.5 mM dithiothreitol (DTT), 0.5 mM EDTA, the tubes are shaken vigorously 3 times for 30 sec (followed each time by a break of 1 min in which the tubes are cooled) using a modified disintegrator (Zellhomogenisator, MSK, Braun, Melsungen, FRG), before adding another 500 μl of 100 mM potassium phosphate (pH 7.5), 0.5 mM DTT, 0.5 mM EDTA then. From the fine suspension, 700 μl is removed and centrifuged 2 min (Microfuge 152, Eppendorf), and marker enzyme activities are measured in the remaining supernatant. The remaining 300 μl of suspension is used for protein determination. Marker enzymes (NADP-GAP dehydrogenase, stroma; PEP carboxylase, cytosol; α-mannosidase, vacuole) are assayed as in the Appendix. Protein is measured as in Lowry,[28] always including a control in which extract is added after formation of the color complex to adjust for a small absorption arising from chlorophyll in the extracts.

Sample Preparation for Assay of Metabolites

After adding 550 μl ice-cold 10% (v/v) HClO$_4$, tubes are shaken (see above), left to stand 15 min at 4°, and then centrifuged 2 min (Centrifuge 5412, Eppendorf). The supernatant is neutralized to pH 7 – 7.5 by addition

[28] O. H. Lowry, N. J. Rosebrough, A. L. Farr, and R. J. Randall, *J. Biol. Chem.* **193**, 265 (1951).

of a 5 M KOH – 1 M triethanolamine solution. After standing for 15 min to allow KClO$_4$ to precipitate, the tubes are recentrifuged and the supernatant taken for substrate assays (see Appendix).

Calculations of Metabolite Contents of Subcellular Compartments

As Fig. 5 shows, the density gradient results in partial purification of the chloroplast, cytosol, and vacuolar material. The chloroplasts are the lightest material, with the cytosol occurring in the middle and the vacuolar material mainly in the heaviest fraction. Other experiments showed that material from the mitochondria distributes between stromal and cytosolic material. In the following, an evaluation procedure is given for a case where a metabolite is restricted to two subcellular compartments, e.g., the chloroplast stroma and the cytosol.

In these calculations, the enzyme activities and metabolite contents are expressed in terms of the protein content of the fraction. The measured metabolite content in a given fraction (M_{tot}) derives partly from the stroma (M_s) and partly from the cytosol (M_c):

$$M_{tot} = M_s + M_c \tag{1}$$

The ratio between the metabolite levels in the stroma (M_s) and in the cytosol (M_c) and the activities of the corresponding marker enzymes NADP-GAP dehydrogenase and PEP carboxylase are constant in each fraction:

$$a = \frac{M_s}{\text{GAPDH}} \tag{2}$$

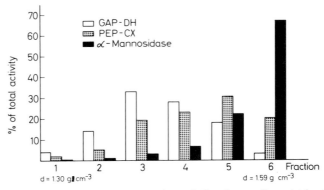

FIG. 5. Distribution of marker enzymes and metabolites in a gradient obtained from dried and homogenized spinach leaves harvested at the end of a 9-hr illumination period. Data from Ref. 10.

$$b = \frac{M_c}{\text{PEPCX}} \tag{3}$$

Substituting Eq. (1) by Eqs. (2) and (3) yields

$$M_{\text{tot}} = a\text{GAPDH} + b\text{PEPCX} \tag{4}$$

which gives

$$\frac{M_{\text{tot}}}{\text{PEPCX}} = a\frac{\text{GAPDH}}{\text{PEPCX}} + b \tag{5}$$

Equation (5) is a linear function, with two variables

$$y = \frac{M_{\text{tot}}}{\text{PEPCX}}, \qquad x = \frac{\text{GAPDH}}{\text{PEPCX}}$$

For an evaluation, these two variables are determined from the corresponding assays of the metabolite amount and enzyme activity in each fraction. The plot yields a straight line if the analysis is correct, with the intercept at the y axis equal to b and the slope equal to a. The metabolite level in the stromal and cytosolic compartment can then be evaluated by multiplying the coefficients a and b with the summed activity of GAPDH [from Eq. (2)] and PEPCX [from Eq. (3)] in the gradient, respectively.

$$M_s = a\text{GAPDH}$$
$$M_c = b\text{PEPCX}$$

For an algebraic evaluation of the metabolite distribution between two compartments, a determination of metabolites and marker enzymes in two fractions would be sufficient. Our evaluation by regression analysis of measurements in more fractions than arithmetically required, has the advantage of minimizing analytical errors. The accuracy of results can be further improved by averaging the results from several replicate gradients, using the same material. However, the fractions obtained from different gradients are not homologous, owing to the distribution of material in gradients being variable, and before averaging several gradients, it is necessary to normalize their fractions with respect to the protein content. To do this, each gradient is divided into 96 hypothetical fractions of equal protein content. Thus, the higher the protein content of an experimental fraction, the more hypothetical fractions into which it is divided. The metabolite contents and marker enzyme activity measured in each experimental fraction are equally assigned between the hypothetical fractions assigned to the experimental fraction. When this is done for each individual gradient, 96 homologous hypothetical fractions are obtained. Each 16 consecutive fractions can be summed for each individual gradient, to yield 6 fractions, which are now comparable for each gradient and can be averaged and evaluated as described above.

It is, in principle, possible to evaluate the distribution between three compartments by assaying three marker enzymes in three fractions in analogy to the calculations above. Although this evaluation mode is sensitive to analytical errors, it may be suitable for an initial check carried out to establish in which compartments a certain metabolite is distributed. In many cases it is then possible to neglect those compartments in which only traces of a certain metabolite occur and, thus, select a suitable two-compartment system.

On the Method

As described in Gerhardt and Heldt,[10] control experiments confirm the general reliability of the technique. The amount of metabolites and the activity of enzymes do not change during the separation of material in a density gradient. Multiple measurements of different markers, including not only enzymes but also metabolites known to be confined to one compartment, showed that metabolites and enzymes do not redistribute during the fractionation.

The method yields reliable information about the distribution of metabolites which are restricted to two compartments, but the analysis becomes more limited by analytical errors if a metabolite is distributed between more than two compartments. Our approach also assumes that all cells in the leaf are homologous and ignores the presence of various cell types in a leaf, as well as the presence of a cell wall. Phosphorylated metabolites should be mainly restricted to the palisade and spongy mesophyll, but sugars and ions could have a more complicated distribution, with significant amounts concentrated in the vascular tissue or the cell wall. Linear regression analysis reveals cases in which a more complicated distribution is present, as the points derived from the various fractions no longer lie on a straight line.

The method described here is time consuming, but it is the only available approach for obtaining information simultaneously about the stroma, cytosol, and vacuolar contents in leaves. Another approach to nonaqueous fractionation involves repeatedly washing chloroplasts in nonpolar gradients.[15,20,21] This method is well suited to experiments where multiple measurements of stromal metabolites in leaves are required[15] as it is less time consuming than our method. However, the measurements of metabolite contents in other cellular compartments are less reliable than in the approach described here.

Intercellular Fractionation of Maize Leaves by Partial
Homogenization and Filtration under Liquid Nitrogen

C_4 plant photosynthesis involves cooperation between two different kinds of cells.[29] The bundle sheath cells contain the Calvin cycle, while the mesophyll cells contain additional photochemical capacity and PEP carboxylase, which catalyzes the preliminary fixation of CO_2 into organic acids. The bundle sheath is more resistant to mechanical homogenization than the mesophyll, and this difference was exploited in earlier studies where differential homogenization in buffers was used to study the intercellular distribution of enzymes.[29] The mechanical properties of the cells remain sufficiently different under liquid N_2 to allow their separation by an analogous procedure at $-180°$, where alterations in metabolites can be prevented.[11] After homogenizing under liquid N_2, the leaf powder is tipped in liquid N_2 through nylon nets with different apertures, so that particles are separated on a size basis. The larger particles are enriched in bundle sheath, and the smaller particles are enriched in mesophyll.

Plant Material

Maize (*Zea mays* L. cv. Vorlo) was grown in a greenhouse (24° day, 15° night) with 14 hr per day of supplementary lighting provided by 150-W tungsten and 250-W mercury lamps. Fully grown leaves from 4- to 6-week-old plants were used.

Homogenization and Filtration

The fractionation is summarized in a flow diagram (Fig. 6). Leaves (1.5–2 mg Chl) are plunged into liquid N_2 and then thoroughly homogenized in a mortar and pestle which is prechilled to $-180°$ and contains a small amount of liquid N_2. This homogenization is best achieved at a stage where the liquid N_2 has almost evaporated, so that it forms a fine film between the frozen leaf particles which can be abraded against each other. The resulting powder is diluted to about 100 ml with liquid N_2 and tipped successively through 200-, 80- and 40-μm nylon nets. The nylon nets are held by rubber bands across the top of a 50-ml glass beaker, whose base stands in a polystyrene holder to provide insulation. The nylon nets are cut large enough so that they sag down into the glass beaker. During the filtration, the material remaining on the nylon net is stirred and washed through by repeatedly adding small amounts of liquid N_2. The residue is

[29] M. Hatch and C. B. Osmond, *Encycl. Plant Physiol., New Ser.* 3, 144 (1976).

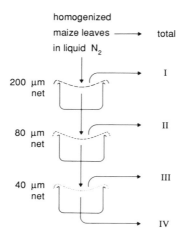

homogenized
maize leaves ⟶ total
in liquid N₂

200 μm net — I

80 μm net — II

40 μm net — III

IV

FIG. 6. Schematic representation of the separation of bundle sheath and mesophyll material by filtration of homogenized maize leaf material through nylon nets.

then washed off with liquid N_2, while the filtrate sieved in a similar manner on the next nylon net. Fractions are collected corresponding to the starting material, namely, the residues left on the 200-, 80-, and 40-μm nets, and the filtrate which passes through the 40-μm net. All fractions are stored at $-85°$. Aliquots are removed for working up using spatula prechilled to $-180°$.

Preparation of Extracts for Measurement of Marker Enzymes

Aliquots (20–100 μg Chl) are rapidly stirred into 1 ml of buffer at 4°, containing 50 mM Tris (pH 7.3), 5 mM MgCl$_2$, 10 mM DTT, 1% albumin, and 0.1% Triton X-100. After homogenization in a mortar and pestle, and in a glass homogenizer (Potter-Elvehjem, Zurich, Switzerland), 100-μl aliquots are measured immediately, or snap-frozen by pipetting them into plastic 1.5-ml centrifuge tubes held tightly in a machined aluminum block which has been precooled to $-180°$.

Preparation of Extracts for Assay of Metabolites

Powder is transferred to liquid N_2 in a pestle and mortar, and 800 μl of 10% (v/v) HClO$_4$ is added. The HClO$_4$ is prefrozen and reduced to a fine powder before addition. The mixture is continually homogenized as it thaws, allowed to stand 20 min at 4°, and centrifuged (2 min, Eppendorf 5412 centrifuge). The pellet is rinsed with 250 μl of 2% (v/v) HClO$_4$ and recentrifuged. The combined supernatants are neutralized by addition of

5 M KOH – 1 M triethanolamine solution, left at 4° for 15 min, and re-centrifuged, and the supernatant is then frozen in liquid N_2 and stored at −85° until assay.

Estimation of Intercellular Distribution

Table IV shows a typical distribution of marker enzymes for the mesophyll [PEP carboxylase, pyruvate phosphate dikinase (pyruvate, ortho-phosphate dikinase)] and bundle sheath [NADP-malic enzyme, RuBP carboxylase (ribulose-biphosphate carboxylase)]. The mesophyll is enriched in the 40-μm filtrate (fraction IV) while the bundle sheath is enriched in material retained on the 40- and 80-μm nets (fractions III and II, respectively). The material retained on the 200-μm net (fraction I) is not enriched.

The distribution of metabolites is estimated by linear regression analysis in analogy to the procedure shown above. The enzyme activity in each fraction is expressed as a percentage of the total in the summed fractions, as is the amount of each metabolite. Then the ratios of (mesophyll marker/bundle sheath marker) and (unknown metabolite/bundle sheath metabolite) in each fraction are plotted against each other. The intercept at the y axis yields the fraction of the metabolite present in the bundle sheath, because at this point the ratio of PEP carboxylase to NADP-malic enzyme is zero, representing pure bundle sheath uncontaminated by mesophyll. An example is shown in Fig. 7, where PEP carboxylase and NADP-malic enzyme are selected as marker enzymes for the mesophyll and bundle sheath, respectively. The results of three separate gradients are plotted. As enzyme activity and metabolite content are expressed as a percentage of the total, the results have already been normalized so that several replicate fractionations can be simultaneously evaluated. In this case, about 10% of the dihydroxyacetone phosphate (DHAP) is in the bundle sheath, and 90% is attributable to the mesophyll. In contrast, the majority of the phospho-glyceric acid (PGA) is attributable to the bundle sheath.

Comments on the Method

This evaluation of the subcellular distribution of a metabolite assumes that the metabolite is restricted to the bundle sheath and mesophyll cells and that the marker enzymes and metabolites are always present at the same ratio in a given cell type. These assumptions are oversimplifications as maize leaves contain a significant proportion of nonphotosynthetic cells,[30] which are fragile, and their metabolites will probably segregate with

---

[30] R. C. Leegood, *Proc. Int. Congr. Photosynth., 6th, 1983* Vol. 3, p. 441 (1984).

TABLE IV

MARKER ENZYME DISTRIBUTION IN FRACTIONS OBTAINED AFTER HOMOGENIZATION AND FILTRATION IN LIQUID N_2[a]

Enzyme	Amount in each fraction (% of total)				Initial activity recovered (%)	Activity (μmol mg^{-1} Chl hr^{-1})
	I	II	III	IV		
PEP carboxylase	38	29	14	18	96 ± 9	975 ± 14
Pyruvate phosphate dikinase	40	31	13	18	112 ± 19	151 ± 38
NADP-malic enzyme	37	35	20	9	108 ± 11	162 ± 37
Ribulose-bisphosphate carboxylase	38	35	19	9	97 ± 5	151 ± 20

[a] Maize leaves were homogenizes in liquid N_2 and filtered successively through nylon nets of aperture 200, 80, and 40 μm. Fractions I, II, III, and IV denote the material retained on the 200-, 80-, and 40-μm nets and the material passing through the 40-μm net, respectively. Aliquots were extracted and assayed as described in the text, and activities are expressed as a percentage of the total recovered. The summed activity was also expressed as a percentage of the amount in a sample of the untreated homogenate to monitor recovery through the fractionation procedure. Data from Ref. 11.

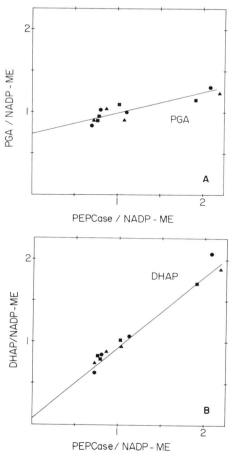

FIG. 7. Regression plots of the distribution of enzymes and metabolites in fractions obtained by homogenization and filtration of maize leaves in liquid nitrogen. Data from three experiments in Ref. 11.

those of the mesophyll cells. It is also likely that there will be concentration gradients for a metabolite within a given cell type. Nevertheless, this method can be used to demonstrate that there are substantial concentration gradients for selected metabolites between bundle sheath and mesophyll cells of maize leaves,[11,31] which decrease or disappear when photosynthesis is stopped.[31]

The results obtained by this method are also strikingly similar to those

[31] M. Stitt and H. W. Heldt, *Biochim. Biophys. Acta* **808,** 400 (1985).

obtained in a separate technique developed by Leegood.[16,30] Here, leaves are "rolled" to prepare a small droplet of sap, which is almost entirely derived from the mesophyll cells. This sap is drawn into $HClO_4$ and quenched 1–2 secs after disrupting the leaf. The rolling method quenches metabolism more slowly, and only measures metabolites in a small section of the mesophyll, but it gives a higher enrichment of mesophyll than obtained by filtration under liquid N_2.

Despite the limitations of both methods, the agreement in the results they yield provides support for the reliability of both approaches. For metabolite pools which turn over rapidly (e.g., adenine nucleotides, pyruvate, as well as PEP in the mesophyll) the method with a rapid quench in liquid N_2 and subsequent fractionation by filtration is desirable. For large pools (PGA, triose phosphate) or pools which turn over more slowly (hexose phosphate) the rolling method is equally satisfactory.

Appendix: Analytical Methods

Enzyme Assays

NADP-Glyceraldehyde-3-phosphate dehydrogenase [EC 1.2.1.13, glyceraldehyde-3-phosphate dehydrogenase ($NADP^+$) (phosphorylating)]: 20–50 μl extract in a final volume of 700 μl, containing 100 mM HEPES (pH 8), 20 mM KCl, 2 mM EDTA, 30 mM $MgCl_2$, 5 mM dithiothreitol (DTT), 5 mM ATP, 0.25 mM NADPH, 74 units (U) phosphoglycerate kinase, 5 mM 3-phosphoglycerate. Measure oxidation of NADPH at 340 nm.

Phosphoenolpyruvate carboxylase (4.1.1.31): 20–70 μl extract in a final volume of 700 μl containing 50 mM glycylglycine, 10 mM $MgCl_2$, 4 mM K_2CO_3 (final pH 7.9), 0.3 mM NADH, 6 U malate dehydrogenase, 0.75 mM phosphoenolpyruvate (PEP). Absorbance is measured at 340 nm with 400 nm as a reference in a dual-wavelength photometer. For increased sensitivity in work with protoplasts a radioactive assay can be used: 10–50 μl extract is assayed in a final volume of 250 μl containing 50 mM glycylglycine (pH 7.9), 10 mM $MgCl_2$, 2 mM [^{14}C] K_2CO_3 0.35 mM NADH, 10 U malate dehydrogenase, 2 mM PEP. The assay is started by adding extract and stopped after 15 min by adding 200 μl 0.5 N HCl. The solution is transferred to a scintillation vial, dried down at 50°–70°, redissolved in 100 μl 0.1 N HCl, and scintillation fluid added. Blanks are carried through without PEP.

Citrate synthase [EC 4.1.3.7, citrate (si)-synthase]: 100 μl sample in a final volume of 700 μl containing 100 mM triethanolamine (pH 8.5), 3 mM L-malate, 0.22 mM acetylpyridine adenine dinucleotide (APAD),

18 U malate dehydrogenase. The malate dehydrogenase-catalyzed reaction is allowed to come to equilibrium, and the citrate synthase reaction then started by addition of 0.22 mM acetyl-coenzyme A. APAD is reduced as more malate is oxidized to replace the oxaloacetate consumed in the citrate synthase reaction. This is followed at 366 nm (reference beam 440 nm) in a dual-wavelength photometer.

Fumarase (EC 4.2.1.2, fumarate hydratase): 60 μl of extract in a test volume of 600 μl containing 84 mM potassium phosphate (pH 7.5) and 46 mM L-malate. The absorbance increase is measured at 240 nm (reference beam 320 nm) on a dual-wavelength photometer. For assay of mitochondrial markers like fumarase or citrate synthase it is essential to ensure that the mitocondria have been entirely disrupted. For nonaqueous fractionation this occurs during the ultrasonication. In experiments with protoplasts it is essential either to ultrasonicate the extracts or to disrupt the mitochondria with detergents. The type and concentration of detergent should be selected to avoid inhibiting the enzyme.

α-Mannosidase (EC 3.2.1.24): 100 μl in a final volume of 900 μl containing 50 mM citrate buffer (pH 4.5), and 5 mM p-nitrophenyl-D-mannosid. Incubate 45 min at 37° and then add 500 μl 0.8 M boric acid (pH 9.8) to stop the reaction before measuring the extinction at 405 nm ($E = 18.5$ cm^2/μmol). As a control, extract is added after the boric acid.

Metabolite Assays

Many metabolites are present at low concentrations in leaves, and only a limited amount of plant material can be fractionated by the methods we describe. For this reason, sensitive metabolite assays are necessary. One approach, using enzymatic cycling,[2] has been reviewed recently. Here, we describe a series of continuous enzymatic substrate assays which allow direct measurement of metabolites in extracts prepared as described in this chapter. It is, however, a precondition that these measurements are carried out using a dual-wavelength photometer, which is able to reliably register absorbance changes of 0.005 E or lower. Such sensitive measurements are not possible with single- or double-beam instruments. Two kinds of dual-wavelength photometer are now available. The convential dual-wavelength photometer has a light source which allows choice of any wavelength between 200 and 800 nm and is also equipped to scan. However, less expensive instruments, which are at least as good for metabolite assays, are also available in which the light source is a mercury vapor lamp and the wavelength is selected by differential filters.

For optimal measurements, buffers and extracts should be free of dust and particles, if necessary by membrane filtration. Extracts can also be freed of interfering absorbing aromatic compounds by treatment with activated charcoal, except when nucleotides are measured, as these bind to the charcoal. In general, the amount of NAD(P)(H) included is adjusted to avoid a large excess as this leads to increased technical problems arising from mixing artifacts and nonspecific oxidation of NADH or reduction of NADP. However, the NADP should be in at least 3-fold excess, as rising NADPH/NADP ratios inhibit glucose-6-phosphate dehydrogenase. Assays involving NAD^+ reduction are avoided because commercial enzymes often contain traces of malate dehydrogenase which, in combination with the very high concentrations of malate in plant extracts, can lead to artifacts.

All coupling enzymes are freed from $(NH_4)_2SO_4$ by a 2-min centrifugation, the supernatant being discarded and the sediment dissolved in buffer. Enzymes are added in a final volume of $0.5-2$ μl on a disposable plastic spatula whose end has been bent at $90°$ in a cool flame, so that it just passes in a half-microcuvette. The enzyme is added by making one pass with this adapted spatula up and down the cuvette; more vigorous stirring can lead to artifacts owing to introduction of air bubbles. Lyophilized proteins are dissolved in a mixture of buffer and 50% glycerol. All enzymes can be stored at $-85°$, following snap-freezing of small aliquots (under 100 μl) in liquid N_2, but slower cooling or storage at warmer temperature can lead to loss of activity. The amount of enzyme added is adjusted so that, ideally, the reaction is complete in $2-10$ min, but less than 10% of the substrate is removed in the time required to add the enzyme to the cuvette and recommence recording. As extracts from different sources can act to inhibit enzymes during the assay, the precise amount of coupling enzyme added may require modification, depending on the material being used.

Glucose 6-P, fructose 6-P, glucose 1-P, UDPglucose: Extracts ($20-100$ μl) in 600 μl total volume containing 100 mM Tris-HCl (pH 8.1), 5 mM MgCl$_2$, 0.25 mM NADP. Add sequentially 0.7 U glucose-6-phosphate dehydrogenase, 0.7 U glucose-6-phosphate isomerase, 0.2 U phosphoglucomutase, 15 mM Na$_4$P$_2$O$_7$, 0.2 U UDPglucose pyrophosphorylase (UTP-glucose-1-phosphate uridylyltransferase). The Na$_4$P$_2$O$_7$ is added at a late stage of the assay because commercial preparations of glucose-6-phosphate isomerase can contain enough pyrophosphatase to hydrolyze pyrophosphate during the $10-15$ min required to assay glucose 6-P, fructose 6-P, and glucose 1-P.

Dihydroxyacetone P, glyceraldehyde 3-P, and fructose 1,6-bisphosphate: Extract ($20-100$ μl) in 600 μl containing 100 mM Tris-HCl (pH 8.1), 5 mM MgCl$_2$, 15 μM NADH. Add sequentially 0.2 U glyceraldehyde-3-phosphate dehydrogenase, then 0.8 U glyceraldehyde-3-P dehydrogenase and 2.3 U triose-phosphate isomerase together, then 0.1 U aldolase (fructose-bisphosphate aldolase). The glyceraldehyde-3-P dehydrogenase is added at low activity to measure dihydroxyacetone P, and the activity is subsequently raised so that the following reactions occur at an adequate rate.

3-Phosphoglycerate: Extract (20–100 μl) in 600 μl containing 100 mM Tris-HCl (pH 8.1), 5 mM MgCl$_2$, 15 μM NADH, 1.1 mM ATP. Add together, 18 U phosphoglycerate kinase and 18 U glyceraldehyde-3-P dehydrogenase.

Glucose, fructose, sucrose: Extract (2–20 μl) in 600 μl containing 100 mM imidazol-HCl (pH 6.9), 1.5 mM MgCl$_2$, 0.5 mM NADP, 1.1 mM ATP, and 2 U glucose-6-P dehydrogenase. Add sequentially, 2 U glucose-6-phosphate isomerase, 0.5 U hexokinase, and 20 U invertase (β-fructofuranosidase). With this assay it is particularly important to check solutions and enzymes for contaminating sugars, especially when small amounts of sucrose are being measured.

Malate: Extract (10–100 μl) in 600 μl containing 100 mM triethanol-amine-HCl (pH 8.5), 0.33 mM APAD, 0.33 mM acetyl-coenzyme A, 6 U malate dehydrogenase. Start by adding 0.3 U of citrate synthase. In this case absorption is measured at 366 nm (reference 440 nm) ($c = 9.2 \times 10^3$ cm^2/μmol).

Phosphate: Phosphate is measured in a two-step assay to decrease blanks, which is suitable for enzymatic assay of P$_i$ in the micromolar range.[8]

Ribulose 1,5-Bisphosphate: Assay by incorporation of $^{14}CO_2$ in the presence of ribulose-1,5-bisphosphate carboxylase as described by Wirtz et al.[8] When small amounts of ribulose 1,5-bisphosphate are measured, it is advisable to purify the NaH$^{14}CO_3$. Commercial preparations contain a small residue of acid-stable radioactive material, which leads to high blank values. The purification is achieved by acidifying the NaH$^{14}CO_3$ with weak acid in a closed container and redissolving the $^{14}CO_2$ in an excess of freshly prepared NaOH. For assay of ribulose 1,5-bisphosphate it is important that internal standards (2–5 nmol) are used to calibrate the assay, as extracts can contain significant and varying amounts of bicarbonate, which leads to errors if the theoretical specific activity of the H$^{14}CO_3^-$ is used to calculate the amount of ribulose 1,5-bisphosphate present.

Adenine nucleotides: Adenine nucleotides can be measured by luciferase or photometric methods. The latter is more sensitive but suffers from the disadvantage that ADP and AMP are not measured directly, but are only obtained by subtraction from the ATP value. This makes their precise measurement difficult when the majority of the adenine nucleotides are present as ATP, as is usually the case in the cytosol.

For measurements of samples by the luciferase technique,[9] samples are diluted with 25 mM HEPES (pH 7.75), 4 mM MgSO$_4$, 3 mM KCl, and

0.2 mM PEP to give a final adenine nucleotide concentration of 5–10 nM. The extracts (200 μl) are assayed immediately for ATP or incubated 30 min with 4 U/ml pyruvate kinase, or 45 min with 4 U/ml pyruvate kinase and 24 U/ml myokinase (adenylate kinase) before assay for ADP and AMP, respectively, by the luciferase method (for details, see Ref. 9). All enzymes are freed from $(NH_4)_2SO_4$ by centrifugation, and all assay solutions are pretreated with activated charcoal and filtration (0.45 μm, cellulose acetate; Sartorius) before use. Measurements with and without internal standard (20–40 pmol ATP) are made in quadruplicate.

For spectrophotometric assay, samples are taken undiluted. ATP is measured in a volume of 600 μl containing extract (100 μl) plus 100 mM Tris-HCl (pH 8.1), 5 mM $MgCl_2$, 0.25 mM NADP, 1 mM glucose, 0.7 U glucose 6-P dehydrogenase, 0.7 U glucose-6-phosphate isomerase. The endogenous glucose 6-P and fructose 6-P must first be removed before adding 0.6 U hexokinase. The glucose-6-phosphate isomerase is added to remove fructose 6-P, so that the assay of ATP proceeds with improved reaction kinetics, as fructose 6-P inhibits glucose 6-P dehydrogenase when present in excess over glucose 6-P. The glucose concentration in the assay should not be increased, as this leads to a higher blank activity owing to a slow reaction of glucose 6-P dehydrogenase directly with glucose. ADP and AMP are measured in a volume of 600 μl containing extract (100 μl), 100 mM Tris-HCl (pH 8.1), 2 mM $MgCl_2$, 10 mM KCl, 60 μM NADH, 1.5 mM PEP, 2.5 U lactate dehydrogenase. After the endogenous pyruvate has been removed, add, sequentially, 4 U pyruvate kinase and 15 U myokinase. Myokinase and NADH preparations can contain small amounts of AMP, which must be measured in a control assay.

Acknowledgments

The development of the methods described in this chapter has been made possible by grants from the Deutsche Forschungsgemeinschaft.

[33] Isolation of Plant Vacuoles and Measurement of Transport

By EWALD KOMOR and MARGARET THOM

Plant cells are usually highly vacuolated, the vacuoles eventually comprising more than 90% of the cell's volume. The function of plant vacuoles is integrated into the control of cell metabolism by specific storage or release of compounds, including enzymes, in response to cytoplasmic signals. It is in this context that the physiology of vacuoles and their transport properties have become increasingly interesting during the past few decades.[1,2] Progress on vacuolar physiology is, however, still very much dependent on the isolation of intact, functional vacuoles, and so far successful isolation procedures have been developed for only a relatively few plant species. Any description of the methodology of transport in vacuoles depends entirely on the methodology of vacuole isolation suitable to the plant species of interest. A detailed review of successful isolation procedures must therefore precede the description of transport measurements per se.

Vacuole Isolation Procedures

Direct Harvest of Vacuoles

Some plant families such as the Papaveraceae and Euphorbiaceae contain highly elongated laticiferous cells which stretch along the stem and leaves of the plants. Incision of the leaf or stem allows recovery of cell contents (latex) and the fractionation of the cellular organelles, including the vacuoles. This specialized technique has been used for *Hevea brasiliensis,* the coutchouc tree, for the study of vacuoles (called lutoids in such species).[3,4] The procedure is as follows:

An incision is made in the tree bark, and the latex is collected in ice-cooled flasks. The latex is diluted 4-fold in isotonic medium (0.3 M mannitol, 50 mM triethanolamine, pH 2.5, 2 mM mercaptoethanol) and centrifuged at 39,000 g for 60 min at 4°. The lutoid-containing pellet is washed twice in the same medium and pelleted as above. The purified lutoids can be stored at 4° for several hours, or they can be lyophilized.

The lyophilized material is typically suspended in ice-cold 25 mM 2-(N-morpholino)ethanesulfonic acid (MES), 25 mM 3-(N-morpholino) propanesulfonic acid (MOPS), 25 mM N-2-hydroxyethylpiperazine-N'-2-ethanesulfonic acid (HEPES), 300 mM mannitol, and 5 mM mercaptoethanol at pH 6.0 in a ratio of 1 g lyophilized material per 100 ml medium. The suspension is gently homogenized in a glass tissue grinder and centrifuged at 10,000 g for 3 min. After two washes in the same homogenization medium, the tonoplast vesicles can be used for uptake experiments. The yield of vacuoles, which comprise 10–20% of the latex, is

---

[1] P. Matile, *Annu. Rev. Plant Physiol.* **29,** 193 (1978).
[2] R. A. Leigh, *Physiol. Plant.* **57,** 390 (1983).
[3] S. Pujarniscle, *Physiol. Veg.* **6,** 27 (1968).
[4] B. Marin, M. Marin-Lanza, and E. Komor, *Biochem. J.* **198,** 365 (1981).

higher than by any other vacuole isolation procedure, and, in the case of *Hevea,* vacuoles can be obtained in close to kilogram quantities.

In the Papaveraceae, the methodology is similar to that of *Hevea.* The latex of *Papaver solemniferum*[5] or *Chelidonium majus*[6] can be harvested by decapitating the capsules or the stems and pipetting the expelled latex from the cut ends into ice-cooled centrifuge tubes. The latex is diluted with 3 times its volume of cold 0.4 M mannitol in 50 mM phosphate buffer, pH 7. The diluted latex is centrifuged at 1000 g for 10 min at 0° – 4° and the pellet is washed and resuspended in the same medium and by the same procedure.

Vacuole Isolation by Tissue Slicing

A very simple method for vacuole preparation in small-scale quantities has been developed by Grob and Matile[7] for *Armoracia* (horseradish) roots: Plasmolyzed tissue (tissue pieces are incubated for at least 5 min in 1.5 M sucrose) is gently sliced with a razor blade that is inserted at a 10° angle into a Plexiglas plate. The slices drop through a slit in the plate into a reservoir of ice-cold medium (0.55 M NaCl, 50 mM Tris-HCl, pH 7.6, and 5 mM EDTA). Slicing the plasmolyzed tissue releases the vacuoles into the medium. Usually 12 g of tissue is sliced into 28 ml of medium. The yield is highest with sharp blades, with minimal pressure exerted on the tissue, and with minimal sliding of the tissue on the Plexiglas plate. Slices of 0.2 mm thickness were optimal for horseradish roots.

The crude vacuole preparation is layered on a two-step gradient of 0.7 M NaCl and 25% urografin (N-methylglucamine) mixed at a ratio of 2 : 3 (v/v) for the bottom (separating) layer and 7 : 1 for the upper (washing) layer above. A 7-min centrifugation at approximately 500 g yields vacuoles at the interface between the two gradient layers. These are collected, diluted 1 : 1 with 25% urografin, and prepared for a second gradient centrifugation. The diluted vacuoles are overlayered first with 0.7 M NaCl and 25% urografin mixed in a ratio of 3 : 7 and next a small layer of 0.7 M NaCl. A 7-min centrifugation at 500 g floats the vacuoles to the interface between NaCl and NaCl plus urografin, where they are recovered in a purified form. The yield of vacuoles after slicing was between 5 and 10% of the amount in the tissue, of which 25% was recovered after the second centrifugation step, so that the final yield was 1 – 3%.

Large-scale preparations using a motor-driven slicing apparatus were

[5] J. W. Fairbairn and M. Djote, *Phytochemistry* **9,** 739 (1970).
[6] P. Matile, B. Jans, and R. Rickenbacher, *Biochem. Physiol. Pflanz.* **161,** 447 (1970).
[7] K. Grob and P. Matile, *Plant Sci. Lett.* **14,** 327 (1979).

developed for the isolation of *Beta* (red beet and sugar beet) vacuoles.[8-10] Fresh beet root or tissue plasmolyzed for 1 hr in 1 M sorbitol is sliced into ice-cold 1 M sorbitol, 5 mM EDTA, 25 mM mercaptoethanol, and 50 mM Tris-HCl, pH 8.0, with the tissue rotating at 530 rpm over the razor blades. The medium is filtered on a stainless steel sieve and cheesecloth, and the root residues are passed through the slicing procedure a second time. The combined filtrates are pelleted at 1300 g for 20 min. The pellet is resuspended in 20 ml of 15% metrizamide in a medium consisting of 1.2 M sorbitol, 1 mM EDTA, 25 mM mercaptoethanol, and 25 mM Tris–MES, pH 8, and is overlayered first with 5 ml of 10% metrizamide in the same medium then with 2 ml of medium. A 10-min centrifugation at 430 g yields a fraction of intact vacuoles at the interface between the medium and 10% metrizamide. Approximately 10^7 to 10^8 vacuoles were obtained from 450 g of tissue.

Vacuole Isolation from Protoplasts

Vacuoles from different plant species have been successfully prepared via isolation of protoplasts followed by lysis of the plasmalemma either by chemical or by mechanical means. These methods give relatively high yields and are applicable for many plant species and organs from which protoplasts can be obtained. The disadvantage is that viable protoplast preparations are a prerequisite for isolation of vacuoles.

Polybase-Induced Lysis of Protoplasts. This method was originally introduced by Dürr *et al.*[11] for isolation of yeast vacuoles and then modified for plant vacuoles.[9] It essentially consists of controlled lysis of the plasmalemma with polylysine or DEAE-dextran for a short, defined time, so that vacuoles are liberated. The method as described for *Bryophyllum* leaf cells[12] is as follows. A suspension of protoplasts (5×10^4 cells/ml) in 0.7 M mannitol and 10 mM HEPES–NaOH, pH 7.3, is incubated with DEAE-dextran at a final concentration of 0.5 mg ml^{-1}. An equal amount of dextran sulfate to bind the excess DEAE-dextran is added 15–30 sec after the addition of polybase. All operations are performed at room temperature. The release of intact vacuoles is complete within 15 min. The crude preparation is layered on a discontinuous gradient of 5 and 10% Ficoll in 0.7 M mannitol, and the high density of *Bryophyllum* vacuoles allows them to settle.

[8] R. A. Leigh and D. Branton, *Plant Physiol.* **58**, 656 (1976).
[9] R. A. Leigh, T. ap Rees, W. H. Fuller, and J. Banfield, *Biochem. J.* **178**, 539 (1979).
[10] S. Doll, F. Rodier, and J. Willenbrink, *Planta* **144**, 407 (1979).
[11] M. Dürr, T. Boller, and A. Wiemken, *Arch. Mikrobiol.* **105**, 319 (1975).
[12] C. Buser and P. Matile, *Z. Pflanzenphysiol.* **82**, 462 (1977).

A modification of this method that consists of lysis and separation in one step was worked out by Boudet et al.[13] for *Melilotus alba*. Protoplasts (2×10^6) suspended in 5 ml medium (25 mM MES–Tris, pH 6.5, in 0.7 M mannitol) are layered on a discontinuous gradient consisting of (from the bottom): 2 ml 20% Ficoll in 25 mM HEPES–Tris, pH 8, in 0.7 M mannitol; 2 ml 5% Ficoll and 4 mg ml^{-1} dextran sulfate in 25 mM HEPES–Tris, pH 8, in 0.7 M mannitol; and 5 ml 2% Ficoll and 4 mg ml^{-1} DEAE-dextran in 25 mM MES–Tris, pH 6.5, in 0.7 M mannitol. The gradient is centrifuged at 2000 g at 17° for 30 min, and the vacuoles above the 20% Ficoll cushion are harvested.

The polybase-induced lysis of protoplasts was also successfully used for isolation of vacuoles from *Catharanthus, Nicotiana, Papaver, Daucus, Datura, Rauwolfia*,[14] *Phaseolus*,[15] and *Beta* (sugar beet).[16] The yield of this method, based on the number of recovered protoplasts, ranged from 25 to 45%.

Lysis of Protoplasts by Osmotic Shock. Lysis of protoplasts by osmotic shock with subsequent release of vacuoles was first applied to yeast cells[17] and later adapted for *Tulipa* and *Hippeastrum* petals[18]; *Lycopersicon* (tomato),[19] *Nicotiana*,[20] *Hordeum* (barley), and *Vigna* (cowpea) leaves[21]; *Malus* (apple) fruit[22]; and leaves of crassulacean plants.[23] All these methods of preparation contain minor modifications, adapted for the individual plant material. As an example, the method for barley and cowpea will be described.[21] Protoplasts suspended in 2.5 ml of 0.5 M sucrose, 18 mM potassium phosphate, 7 mM citric acid, pH 5.8, and 0.1% bovine serum albumin (BSA) are added to 5 ml of 0.145 M sucrose, 25 mM MOPS, pH 7.5, plus 0.1% BSA. The change in osmotic concentration (and perhaps also of pH) leads to the release of vacuoles. The crude suspension is mixed with an equal volume of 10% Ficoll in 0.25 M mannitol and then overlayered with 3 ml 3.5%, 3 ml 2.5%, and 1 ml 0% Ficoll in 0.25 M mannitol, 25 mM MOPS, and 0.1% BSA. Centrifugation at 80 g for 10 min floats

---

[13] A. M. Boudet, H. Canut, and S. Alibert, *Plant Physiol.* **68**, 1354 (1981).
[14] B. Deus-Neumann and M. H. Zenk, *Planta* **162**, 250 (1984).
[15] T. Boller and N. Vogeli, *Plant Physiol.* **74**, 442 (1984).
[16] R. Schmidt and R. J. Poole, *Plant Physiol.* **66**, 25 (1980).
[17] P. Matile and A. Wiemken, *Arch. Mikrobiol.* **56**, 148 (1967).
[18] J. J. Wagner and H. W. Siegelman, *Science* **190**, 1298 (1975).
[19] M. Walker-Simmons and C. A. Ryan, *Plant Physiol.* **60**, 61 (1977).
[20] J. A. Saunders, *Plant Physiol.* **64**, 74 (1979).
[21] J. B. Ohlrogge, J. L. Garcia-Martinez, D. Adams, and L. Rappaport, *Plant Physiol.* **66** 422 (1980).
[22] S. Yamaki, *Plant Cell Physiol.* **25**, 151 (1984).
[23] R. Kringstad, W. H. Kenyon, and C. C. Black, Jr., *Plant Physiol.* **66**, 379 (1980).

vacuoles to the upper layer and to the 0–2.5% Ficoll interface. The vacuole preparation can be further washed by repeated flotation centrifugation. Based on vacuole counts, the yields were 5–15% of the number of protoplasts. For crassulacean plants, yields of up to 50% were reported.[23]

Protoplast Lysis by Shearing Force. Externally applied shearing force to break the plasmalemma and to release the vacuole provides an alternative mechanical method of vacuole isolation. The shearing force can be exerted by high-speed centrifugation, as successfully applied for *Pisum* (pea) mesophyll protoplasts,[24] *Saccharum* (sugarcane) parenchyma,[25] *Zea* coleoptiles,[26] and *Nicotiana* leaves.[27] The technique was originally introduced by Lörz *et al.*[28] As an example, the method for sugarcane storage parenchyma is described.[25] A 5-ml suspension of protoplasts (in 10 mM HEPES, pH 5.6, and 0.5 M mannitol) is layered over a 1-ml cushion of 12% Ficoll in 0.5 M mannitol and centrifuged for 30 min at 200,000 g. Vacuoles are recovered at the 0–12% Ficoll interface and washed 3 times in buffered 0.5 M mannitol with 5-min centrifugations at 50 g each time. The vacuole yield was 20–30% based on protoplast number.

A faster method of mechanical rupture of the plasmalemma was developed by Martinoia *et al.*[29] for *Hordeum* (barley) mesophyll. Protoplasts ($\sim 5 \times 10^6$) are suspended in 6 ml of cold 0.4 M sorbitol plus "supplementing compounds" [1 mM MgCl$_2$, 2 mM EDTA, 10 mM NaCl, 0.5 mM KH$_2$PO$_4$, 30 mM HEPES, pH 7.8, 30 mM KCl, 0.5% BSA, 0.1% poly(vinylpyrrolidone)] which have been dissolved in osmotically adjusted 12.5% Percoll. This suspension is gently taken up into a 10-ml syringe (without needle) and then pressed vigorously through a narrow, long cannula (10 cm × 0.7 mm), thereby breaking the plasmalemma and liberating the vacuoles. The crude vacuole preparation is transferred to a centrifuge tube and overlayered with 5 ml of the same medium as above (but without Percoll), and with 2 ml of the same medium with 0.4 M glycine–betaine instead of sorbitol. Centrifugation at 100 g for 2 min then at 1100 g for 2 min will float the vacuoles to the interface between the two uppermost layers. These vacuoles can be further washed by resuspending in 8% Percoll and overlayering with 1 ml of 0.4 M glycine-betaine, 30 mM MES, pH 5.6, and "supplementing compounds" followed by centrifugation for 2 min at 650 g. The vacuole yield was 25–40% based on protoplast number.

[24] M. Guy, L. Reinhold, and D. Michaeli, *Plant Physiol.* **69**, 61 (1979).
[25] M. Thom, A. Maretzki, and E. Komor, *Plant Physiol.* **69**, 1315 (1982).
[26] S. Mandala and L. Taiz, *Plant Physiol.* **78**, 104 (1985).
[27] E. Pahlich, R. Kerres, and H. J. Jager, *Plant Physiol.* **72**, 590 (1983).
[28] H. Lörz, J. T. Harms, and I. Potrykus, *Biochem. Physiol. Pflanz.* **169**, 617 (1976).
[29] E. Martinoia, U. Heck, and A. Wiemken, *Nature (London)* **289**, (1981).

Nishimura and Beevers[30] have isolated vacuoles from *Ricinus* (castor bean) endosperm by gently slicing the endosperm (14 halves) with a razor blade and incubating the slices in a cellulase-containing medium in 0.7 *M* mannitol. After a 5-hr incubation, the enzyme solution is removed, and 7 ml of 0.7 *M* mannitol is added. The flask is briefly swirled, and the suspension is filtered through 35-mesh nylon cloth. These procedures are repeated several times with fresh mannitol solutions until 40 ml filtrate is obtained. The filtrate is centrifuged for 2 min at 100 *g*, and the pellet is resuspended in 3 ml 0.7 *M* mannitol containing 5 m*M* EDTA, pH 7. This suspension is filtered through absorbent cotton (0–5 mm) then rinsed with 5 ml 0.7 *M* mannitol containing 5 m*M* EDTA. The filtrate is centrifuged for 2 min at 100 *g* and the vacuoles in the pellet are washed twice with the medium as above. The yield was 2×10^5 vacuoles per 3–4 g tissue.

Preparation of Tonoplast Membranes

For many applications, e.g., determination of enzymatic characteristics, study of the mechanism of bioenergetic processes, or membrane protein isolation, it is necessary to remove intravacuolar solutes from tonoplast membranes. Tonoplast membranes can be prepared either from isolated intact vacuoles or by separation of microsomal membranes on gradients.

Preparation from Vacuoles. Tonoplast vesicles can be easily obtained from whole vacuoles by osmotic lysis, sonication, or mechanical force. For instance, for *Ricinus* (castor bean) vacuoles,[30] sonication of 0.7×10^6 vacuoles in 0.5 ml medium (150 m*M* Tricine–KOH, pH 7.5, 15% sucrose, 0.1 m*M* EDTA) for 10 sec at 0° yielded tonoplast vesicles. For *Nicotiana* vacuoles,[27] sonication of 20 sec at 0° followed by subsequent suspension in 10 ml of 40% sucrose and centrifugation in a continuous sucrose gradient (5–40%) at 82,000 *g* for 15 hr is used. The tonoplast vesicles band at 29% sucrose.

Saccharum (sugarcane) tonoplast is obtained by a combination of mechanical disruption and osmotic lysis. A 1-ml suspension of vacuoles is placed into 10 ml of medium containing 10 m*M* Tris–MES, pH 7.0, and 0.25 *M* mannitol. Vacuoles are broken with approximately 30 strokes in a glass homogenizer and centrifuged over a 10% dextran (T70) cushion at 100,000 *g* for 1 hr. The tonoplast vesicles band immediately above the dextran layer.

Preparation from a Crude Microsomal Fraction. Preparation of microsomal membranes and separation of the membranes on gradient have been

[30] M. Nishimura and H. Beevers, *Plant Physiol.* **62**, 44 (1978).

reported to yield a tonoplast-enriched fraction from *Beta* (red beet),[31] *Avena* (oat),[32] and *Zea* (corn)[33] roots and from corn coleoptiles.[34] As an example, the method for beet root is described.[31] Peeled and chilled beet root (200 g) is blended for 30 sec with 200 ml medium containing 250 mM sucrose, 25 mM potassium metabisulfite, 2–3 g Chelex resin (Bio-Rad), and 2–3 g Amberlite XAD-4. The procedure is repeated with a second 200 g of beet. The homogenate is then filtered through 8 layers of cheese-cloth, and the combined filtrate is centrifuged for 20 min at 80,000 g. The pellet is resuspended with a homogenizer in 10 ml of 124 mM sucrose, 5 mM dithioerythritol (DTE), and 15 mM MES, pH 6.8. The membrane suspension is treated with 50 ml of 0.3 M KI, 5mM DTE, and 15 mM MES, pH 6.8, and is sedimented at 80,000 g for 20 min. The pellet is resuspended in 4 ml of homogenizer medium, and a 1-ml aliquot is layered on an 11-ml gradient of 15–45% sucrose containing 5 mM DTE and 15 mM MES, pH 6.8. The gradient is centrifuged for 3 hr at 80,000 g. Tonoplast vesicles banded at 25–28% sucrose. For routine preparations, 4 ml of the KI-treated membrane suspension is layered over a simpler two-step gradient consisting of 4 ml 10% sucrose and 4 ml 23% sucrose; tonoplast banded at the 10–23% sucrose interface. The tonoplast can be frozen in liquid nitrogen and stored at −70°.

General Comments on Vacuole Isolation

Isolation of vacuoles is presently an empirical art specifically adapted and designed for each individual species or organ of a plant. The described methods are those which worked best for the particular plant material, and often a suitable method for one species is a total failure for another. The same holds for composition of homogenization media, pH, ionic strength, etc. In a few cases, comments are found in the literature describing which ingredients have been tried and, in some instances, explanations as to why certain ingredients are more advantageous than others.[7,8]

Generally the osmotic strength of the medium has to be hypertonic, never hypotonic, but the extent of hypertonicity seems of less importance for isolation than for uptake measurements. All isolation methods which involve protoplast preparation, i.e., with plasmolyzed cells, may be impractical to compare control and water-stressed plants, since plasmolysis could trigger the physiological responses of water stress. The use of sucrose as osmoticum for sucrose transport measurement is not advisable. Sucrose,

[31] R. J. Poole, D. B. Briskin, Z. Kratky, and R. M. Johnstone, *Plant Physiol.* **74,** 549 (1984).

[32] K. A. Churchill and H. Sze, *Plant Physiol.* **71,** 610 (1983).

[33] F. M. Dupont, A. B. Bennett, and R. M. Spanswick, *Plant Physiol.* **70,** 1115 (1982).

[34] A. Hager and M. Helmle, *Z. Naturforsch., C: Biosci.* **36C,** 997 (1981).

sorbitol, and mannitol may contain trace amounts of hexose. Even a 0.1% contamination of a 1 M osmoticum will mean 1 mM hexose, which is well within the range of the K_m of hexose uptake systems. Grob and Matile[7] found that for plasmolysis sucrose gives a better final yield of vacuoles than does sorbitol, KCl, or NaCl. On the other hand, NaCl and KCl are found to be better as osmotica in the homogenization medium. They concluded that the type of osmoticum is more important than the osmotic strength.

Bovine serum albumin is often included in the medium to remove cytoplasmic impurities,[22] although this effect of BSA is sometimes not found.[7,8,23,35] However, BSA cannot be used in experiments where protein determination is needed. Grob and Matile[7] found that EDTA prevents aggregation, but EDTA has also been reported to lyse vacuoles.[23] Thiol reagents are reported necessary in some cases,[31] without effect in others,[7,8,23] or even to be disadvantageous[7]; Mg^{2+} or Ca^{2+} salts are without effect,[7,23] or harmful at high concentration.[23] The favorable pH is reported to be 6 to 8.[7]

Stability of vacuoles is in the range of several hours at $0°-4°$ at pH values near 7; Tris-HCl and phosphate stabilize better than Tris–maleate. High centrifugation force (above 1000 g) or vigorous stirring or shaking are generally harmful (also, rotary shaking is more gentle than reciprocating shaking).

The following enzymes are generally found exclusively associated with vacuoles: α-mannosidase,[13,15,29] Cl^--stimulated ATPase,[20] and chitinase.[15] Acid phosphatase is sometimes reported to be exclusively in vacuoles[19,20,25,29,35] and sometimes not.[7,8,15,22] Most commonly α-mannosidase or acid phosphatase is used as a vacuole marker, since both are soluble enzymes and their release into the supernatant can be used as monitors of vacuole lysis. No definitive marker, other than an ATPase that is insensitive to vanadate and oligomycin and is stimulated by chloride in some plants, has been identified for tonoplast preparations. Boller and Kende[36] reported that, for all vacuole isolation methods via protoplasts, it is important to purify the protoplasts from cell debris, since these tend to adsorb hydrolases from the fungal cell wall-degrading enzymes.

Measurement of Transport

Incubation Conditions for Uptake Measurement

For uptake experiments, vacuoles should be suspended in a "cytoplasmic-like environment" in terms of osmolarity, pH, and ionic balance. The

[35] I. J. Mettler and R. T. Leonard, *Plant Physiol.* **64**, 1114 (1979).
[36] T. Boller and H. Kende, *Plant Physiol.* **63**, 1123 (1979).

osmotica commonly used are mannitol, sorbitol, NaCl, sucrose, or betaine. The concentration of osmoticum should maintain vacuoles at or near the physiological state. In red beet vacuoles, an increase in the osmotic concentration of the incubation medium from 0.75 to 1.2 M decreased the rate of vectorial sucrose synthesis via UDPglucose by 50%.[37]

Choice of buffer type and concentration is empirical, but ions that could dissipate the electrochemical gradient, e.g., K^+, should be avoided. Additions of Ca^{2+} may be necessary in some systems, and a salt mixture which closely simulates the cytoplasmic concentration should be used until the ion dependence of the system is known. Addition of sulfhydryl reagent may be necessary, but with this, as well as with any other addition to the incubation medium, care should be taken that the chemical is not inhibitory for uptake.

Cytoplasmic pH as well as the pH in which the vacuoles are most stable should be considered in selection of the buffer pH. However, neither of these conditions necessarily represents the optimum pH of the uptake system. In sugarcane vacuoles, the apparent optimum pH for glucose uptake is 5.5, while the cytoplasmic pH is about 7.2.[38] It is not clear whether the optimum of 5.5 is actually an optimum for the uptake system or is merely an indication of the greater stability of the vacuoles at that pH.

For convenience, uptake measurements are conducted at room temperature (25°). However, it should be remembered that transport systems can show significant variation with temperature. Furthermore, vacuoles are extremely fragile organelles; therefore, uptake should be measured as soon as possible after isolation. A slow rate of agitation and measurement of uptake over a short time should be used to obtain linear uptake rates.

Uptake Measurements

Uptake measurement can be initiated by the addition of a vacuole suspension to the incubation medium containing the substrate or by addition of substrate to a vacuole suspension. Vacuole density in the incubation medium depends on the uptake rate. Approximately 5×10^5 vacuoles per sampling point have been used. The concentration of substrate depends on the kinetics of the system, and when a radioactive substrate is used, the specific activity needed will depend on the expected uptake velocity.

The amount of substrate taken up into the vacuolar space is relatively small compared to the volume of the suspending medium; therefore, the uptake rate cannot be followed by measuring a decrease in substrate con-

[37] M. Thom, R. A. Leigh, and A. Maretzki, *Planta* **167,** 410 (1986).
[38] M. Thom, E. Komor, and A. Maretzki, *Plant Physiol.* **69,** 1320 (1982).

centration in the medium. All uptake experiments require a separation of the vacuoles from the incubation medium, either by filtration or centrifugation.

Filtration can be through paper, glass, or membrane-type filters. The major disadvantages of filtration are the relatively long time required (30 sec or more) and the loss of substrate subsequent to uptake incurred through vacuole breakage even though extremely gentle techniques are used.

Centrifugation methods include centrifugal washing through a nonaqueous layer of silicone oil or other material either by sedimentation or flotation. Vacuoles can also be separated from the incubation medium by a rapid (10 sec) centrifugation in a microcentrifuge followed by a rinse in incubation medium without substrate. The disadvantage of this method is that the extravacuolar space is a large fraction of the total pellet space and, therefore, a considerable amount of substrate is included. However, if the intravacuolar substrate concentration is high relative to that of the extravacuolar space and if the substrate concentration in the pellet is measured as an increase over time, relatively accurate uptake rates can be achieved. Centrifugal washing through silicone oil requires that the densities of silicone oil, incubation medium, and vacuoles differ. This method provides rapid separation of vacuoles from incubation medium and, in addition, little or no adhering substrate is carried over into the vacuoles. Care must be taken that the density of the silicone oil, which is affected by temperature, does not change during the course of an experiment. In addition, in some cases, the density of the vacuoles in a population varies so that a portion of the sample remains at the incubation medium–silicone oil interface.

An elegant procedure for the separation of vacuoles was described by Kaiser.[39] A 100-μl aliquot of the uptake suspension is transferred to a microcentrifuge tube and mixed with 200 μl of 10% (w/v) Ficoll (made up in 0.4 M sucrose, 20 mM sodium phosphate buffer, pH 7.8, 1 mM EDTA, and 1 mg ml^{-1} BSA). The suspension is overlayered with 700 μl of silicone oil AR 200, 400 μl of 0.4 M sorbitol, and 100 μl of 0.4 M betaine solution, both made up in 20 mM buffer, EDTA, and BSA. The tube is centrifuged for 1 min in a Beckman Microfuge B. Vacuoles band at the sorbitol–betaine interface. All liquids above the silicone oil layer are removed, and one aliquot is taken for determination of radioactivity and another for quantitation of vacuoles.

Measurements of uptake usually involve the use of labeled substrate where determination of radioactivity in the sample can be easily per-

---

[39] S. Kaiser and U. Heber, *Planta* **161**, 562 (1984).

formed. In some instances, i.e., when sucrose, glucose, and fructose are used as substrate, the vacuoles can be rapidly extracted and the concentrations of sugars can be determined using extremely sensitive fluorometric, enzyme-linked methods. Caution must be advised since large errors can occur because the determination involves the subtraction of the amount of substrate taken up into a large preexisting intravacuolar pool.

The procedure for UDPglucose (yielding sucrose) and for hexose uptake by sugarcane vacuoles are given below. The procedure for amino acid transport would be identical.

Procedure for Measurement UDPGlucose-Dependent Vectorial Sucrose Synthesis in Sugarcane Vacuoles[40]

1. Place into a 25-ml Erlenmeyer flask: 4.5 ml White's basal salt medium [0.72 mM MgSO$_4$, 0.85 mM Ca(NO$_3$)$_2$, 1.4 mM Na$_2$SO$_4$, 0.79 mM KNO$_3$, 0.65 mM KCl, 0.12 mM NaH$_2$PO$_4$] containing 0.5 M mannitol; 50 μl 20 mM UDPglucose solution (final concentration 200 μM); and 10 μl UDP[^{14}C]glucose containing approximately 4×10^5 dpm.
2. Initiate uptake by adding 0.5 ml vacuole suspension containing approximately 5×10^6 vacuoles/ml.
3. Withdraw a 1-ml sample after 1-, 3-, 5-, and 7-min incubations at 25° on a rotary shaker at 75 rpm.
4. Place the vacuole suspension over a 0.5-ml layer of silicone oil AR 200 in a 1.5-ml polypropylene tube and centrifuge for 30 sec in a Beckman Microfuge B.
5. Remove the incubation medium and silicone oil by aspiration. Care should be taken at this point to prevent any liquid adhering on the side of the tube from running into the pellet.
6. Cut the tube just above the pellet and place the tube containing the pelleted vacuoles into a filmware tube (Nalgene) containing 3 ml scintillation fluid, heat seal, and shake the tube vigorously to resuspend the pelleted vacuoles.
7. Determine the radioactivity in the pellet and an aliquot of the incubation medium.
8. Calculate uptake rate as follows:

$$\text{Specific activity of substrate} = \frac{\text{cpm of substrate ml}^{-1} \text{ media}}{\text{amount of substrate ml}^{-1} \text{ media}}$$

[40] It was found recently that incubation of vacuoles with UDPglucose gives rise to $\beta 1-3$ glucans and not sucrose [J. Preisser and E. Komor, *Plant Physiol.* **88**, 259 (1988); A. Maretzki and M. Thom, *Plant Physiol.* **88**, 266 (1988)].

$$\text{Amount of substrate taken up} = \frac{\text{cpm taken up ml}^{-1} \text{ suspension}}{\text{specific activity}}$$

$$\text{Rate of uptake} = \frac{\text{amount taken up ml}^{-1} \text{ suspension}}{\text{time interval (no. vacuoles ml}^{-1} \text{ suspension)}}$$

Procedure for Measuring Glucose Uptake into Isolated Vacuoles. Glucose uptake into sugarcane vacuoles is measured by a procedure similar to that described above except that the incubation medium contains 4.5 ml White's basal salt medium with 0.5 M mannitol, 50 μl 5 mM 3-O-methylglucose (final concentration 50 μM), and 20 μl 3-O-methyl [^{14}C] glucose (approximately 1 × 10dpm).

General Comments

Transport activities have been based on a series of measurements such as protein, intravacuolar volume, vacuole counts, marker enzyme activity, and intravacuolar marker compounds (phenolics, alkaloids, and dyes). Although none of these means of measurement can be regarded as ideal, some are definitely invalid when a comparison with transport activities of other plant material is required. For instance, betaine or alkaloids are absent in most plants. Therefore, it is helpful to determine for each type of vacuole preparation the relationship among protein, intravacuolar volume, vacuole number, and intravacuolar marker compound. Determinations based on vacuole counts are easily subject to personal bias, as shown by Buser and Matile,[12] since small vacuoles are often disregarded.

Acknowledgment

Published as Paper No. 610 from the Experiment Station, Hawaiian Sugar Planters' Association.

Section III

Transport in Single-Cell Eukaryotes: Fungal Cells

[34] Kinetic Studies of Transport in Yeast

By Arnošt Kotyk

A minimum of three sets of transport systems exist in yeast, those in the plasma membrane, those in the mitochondrial inner membrane, and those in the tonoplast, the vacuolar membrane.[1-6] In addition, transport may occur across any intracellular organelle membrane, such as the Golgi body, endoplasmic reticulum, and small lysosomes. However, the majority of transport systems, and certainly those with the highest capacity, are localized in the outer, plasma, membrane, and it is the function of these systems that is predominantly studied when working with whole cells or their protoplasts.

Although among the simplest eukaryotic cells, yeasts comprise a variety of species and, among them, a broad range of transport mechanisms, from diffusion through pores to mediated carrier diffusion, to uphill, thermodynamically active transport, to group translocation, and, most likely, pinocytosis. The two categories of the list that are amenable to kinetic analysis are (1) nonspecific permeation, often called simple diffusion, and (2) specific transport, often associated with carriers, i.e., membrane proteins catalyzing transmembrane movement of various solutes. From a brief characterization of the two categories it should follow what experimental approaches should be chosen to distinguish between them and to characterize them in more detail.

Nonspecific Permeation

Irrespective of the mechanism used (such as passage through hydrophilic pores, use of "kinks" in the liquid crystalline lattice of the membrane, dissolving in membrane lipids, or flow through true statistical pores caused by membrane rupture), the kinetics of nonspecific permeation

[1] A. Kotyk and J. Horák, in "Yeast Cell Envelopes" (W. N. Arnold, ed.), Vol. 1, p. 49. CRC Press, Boca Raton, Florida, 1981.
[2] A. Tzagoloff, "Mitochondria," p. 199. Plenum, New York, 1982.
[3] P. Matile, Annu. Rev. Plant Physiol. 29, 193 (1982).
[4] G. Svihla, J. L. Dainko, and F. Schlenk, J. Bacteriol. 85, 399 (1963).
[5] L. A. Okorokov, L. P. Lichko, and I. S. Kulaev, J. Bacteriol. 144, 661 (1980).
[6] Y. Kakinuma, Y. Ohsumi, and Y. Anraku, J. Biol. Chem. 256, 10859 (1981).

METHODS IN ENZYMOLOGY, VOL. 174

follows a modified Fick's first law of diffusion. Thus, the rate of translocation of a solute is defined by

$$J_s = P(s' - s'')$$ (1)

where P is a permeability constant and s' and s'' are, respectively, the
concentrations of solute S outside and inside the cell. The initial rate of
uptake, when $s'' = 0$, is then

$$J_{s_0} = Ps'$$ (2)

and hence is linearly proportional to the analytical concentration in the
external medium.

If the uptake is allowed to proceed to equilibrium, its rate becomes zero
and $s' = s''$. If, however, the solute is charged, its equilibrium distribution
will follow the membrane potential $\Delta\psi$, thus

$$s'' = s' \exp(-nF\Delta\psi/RT)$$ (3)

where n is the number of positive charges on the solute, R is the gas
constant, F is the Faraday constant and T is the absolute temperature.

Systems operating by simple diffusion are practically insensitive to pH
within reasonable limits, for a great majority of yeasts between pH 3 and 8.
However, the effect of temperature may be substantial, the apparent activation energies attaining values of 80–120 kJ/mol, this being apparently
due to the stripping of hydration shells from molecules and ions as they
enter the membrane milieu. Only when bulk (volume) flow of the aqueous
solution is possible (a highly unlikely event in normal cells) would the
activation energy drop to 2–10 kJ/mol, corresponding to diffusion in a
continuous medium.

The nonspecific permeation mechanisms are unaffected by metabolic
inhibitors, such as iodoacetamide, antimycin, and 2-deoxy-D-glucose, by
proton conductors, such as nitrophenols, salicylic acid, and carbonyl cyanide phenylhydrazones (correctly phenylhydrazonopropanedinitriles).
They are very little affected by heavy metal ions, such as UO_2^{2+}, Ln^{3+}, and
Th^{4+}, at low concentrations (e.g., 10^{-5} M).

For a diffusion-type system the characteristic variable that can be calculated is P, the permeability constant. Its dimensions depend on the units
used to express the flow, J_s. If J_s is expressed in moles per gram dry weight
(of cells or protein) per minute, and s in moles per liter, P will come out in
liters per gram per minute. If the area of the cell membrane is known, J_s
can be expressed in mol cm^{-2} sec^{-1}, and P emerges in cm sec^{-1} which is a
dimension identical with that of a diffusion coefficient (cm^2 sec^{-1}) divided
by the thickness of the membrane.

Specific Transport

Transport systems involving membrane proteins are invested with an enzyme-type specificity (L versus D isomers, α versus β anomers, and the like) and are characterized by being saturable, an obvious consequence of the fact that a finite number of specific sites are available for binding the transported solute and for its subsequent translocation. Unless the transport protein contains two or more cooperating binding sites (a rare phenomenon in yeasts)[7] the fundamental kinetic equation of transport is of the type

$$J_s = \frac{As' - Bs''}{C + Ds' + Es'' + Fs's''} \tag{4}$$

which is formally identical with a Haldane-type equation of enzyme kinetics (s'' here is equivalent to the product of an enzyme reaction) in which the enzyme can have two forms, analogous here with the carrier facing the outside and the inside of the membrane. This is in fact the essence of the carrier-type mechanisms: The binding site of the carrier (or transporter or translocase) can be accessible from either one or the other side of the membrane but not simultaneously from both (a distinction from a specific pore or channel).

Data from Initial Rates of Transport

The initial rate of uptake by a specific transport system (again for $s'' = 0$) is then defined by

$$J_{s_0} = As'/(C + Ds') \tag{5}$$

an expression identical with the Michaelis–Menten equation in enzyme kinetics where $J_{max} = V = A/D$ and K_m (the half-saturation constant) is C/D. These two principal parameters of any saturable system are readily determined by any of the procedures known from enzyme kinetics (plots according to Lineweaver and Burk,[8] Hofstee,[9] Hanes,[10] or Eisenthal and Cornish-Bowden[11]) (Fig. 1). Similarly, the J_{max} and K_m of the reverse (outflow) process [B/E and C/E of Eq. (4)], can be determined from the initial rates of outflow at different intracellular concentrations.

[7] S. Janda, A. Kotyk, and R. Tauchová, *Arch. Microbiol.* **111,** 151 (1976).
[8] H. Lineweaver and D. Burk, *J. Am. Chem. Soc.* **56,** 658 (1934).
[9] B. H. J. Hofstee, *Nature (London)* **184,** 1296 (1959).
[10] C. S. Hanes, *Biochem. J.* **26,** 1406 (1932).
[11] R. Eisenthal and A. Cornish-Bowden, *Biochem. J.* **139,** 715 (1974).

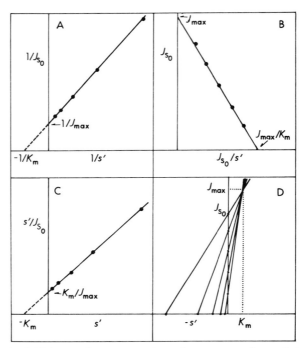

FIG. 1. Plotting initial rates of transport against solute concentration to derive K_m and J_{max} according to Lineweaver and Burk (A),[8] Hofstee (B),[9] Hanes (C),[10] and Eisenthal and Cornish-Bowden (D).[11]

It should be noted that even for the case that $A = B$, corresponding to carrier (mediated or facilitated) diffusion with no energy requirement, the kinetic parameters of the oppositely directed flows may be different, the only constraint being that $J_{\overrightarrow{max}} K_{\overleftarrow{m}} / J_{\overleftarrow{max}} K_{\overrightarrow{m}} = 1$.

Data from Net Rates of Transport

One can determine all the constants, including the interactive one which characterizes the $s's''$ term, in a single uptake experiment.[12] This is based on a modified version of Eq. (4):

$$J_s = \frac{J_{\overrightarrow{max}}(s' - s''/K_{eq})}{K_{\overrightarrow{m}}(1 + s''/K_{\overleftarrow{m}}) + s'(1 + s''/K_i)} \tag{6}$$

where $J_{\overrightarrow{max}} = A/D$, $K_{\overrightarrow{m}} = C/D$, $K_{\overleftarrow{m}} = C/E$, K_i (the interactive constant) =

---

[12] J. Cuppoletti and I. H. Segel, *J. Theor. Biol.* **53**, 125 (1975).

C/F, and K_{eq} (the equilibrium constant) $= A/B = (s''/s')_{\infty} = K_{\overleftarrow{m}}J_{\overrightarrow{max}}/K_{\overrightarrow{m}}J_{\overleftarrow{max}}$ (the Haldane expression).[13] It should be observed that in this general case of transport proceeding uphill the final equilibrium ratio s''/s' may be arbitrarily different from unity, the only constrains being of energetic nature.

Setting $\Delta s = s' - s''/K_{eq}$ and inverting Eq. (6) one obtains

$$1/J_s = (K_{\overrightarrow{m}}/J_{\overrightarrow{max}})(1 + s''/K_{\overleftarrow{m}}) + (s''/K_{eq}J_{\overrightarrow{max}})(1 + s''/K_i)(1/\Delta s) \\ + (1/J_{\overrightarrow{max}})(1 + s''/K_i) \tag{7}$$

from which all the kinetic constants can be determined. The way to proceed experimentally is to follow the uptake of several different concentrations of S, chosen in such a way that the value of K_m lies in their range. (Therefore, it is advisable to determine K_m beforehand.) Each time curve of uptake should be determined accurately for a period long enough to achieve about two-thirds of the final equilibrium value, and then the equilibrium value itself should be determined which serves to obtain K_{eq}. The uptake curves should then be analyzed at several values of s'' so chosen that they cut across as many uptake curves as possible. At each of the intersections of the uptake curve with the horizontal line of the chosen s'' both the J_s (slope of the uptake curve) and the actual s' should be determined. This must be done very carefully with a view to the fact that the actual s' may be less than the initial one, owing to the uptake of S by cells; this caution becomes important at higher cell suspension densities and with intensely uphill-transporting systems.

The results of these calculations are plotted as $1/J_s$ against $1/\Delta s$ for the different selected values of s''. The plot (Fig. 2) yields a family of intercepts with the ordinate which are defined as

$$\text{intercept}_y = s''/J_{\overrightarrow{max}}K_i + 1/J_{\overrightarrow{max}} \tag{8}$$

Now a plot of the intercepts against s'' will yield $1/J_{\overrightarrow{max}}$ as the intercept with the ordinate and $-K_i$ as the intercept with the abscissa.

Similarly, the intercepts (in Fig. 2) with the abscissa are defined as

$$\text{intercept}_x = K_{\overrightarrow{m}}(1 + s''/K_{\overleftarrow{m}})/(1 + s''/K_i) + s''/K_{eq} \tag{9}$$

From this formula (using any of the intercept$_x$ values) K_m can be calculated as all the other constants have by now been determined. The last unknown constant, $J_{\overleftarrow{max}}$, can then be obtained from the Haldane equilibrium expression.

This procedure can be simplified if we are dealing with a highly active transport where the accumulation ratio s''/s' $(= K_{eq})$ is very large so that, in

---

[13] J. B. S. Haldane, "Enzymes." Longmans, London, 1930.

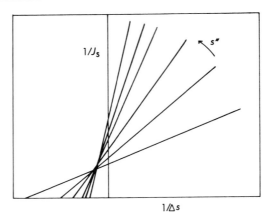

FIG. 2. Auxiliary plot of reciprocal rate versus reciprocal concentration difference to obtain intercepts with ordinate and abscissa.

Eq. (6), the expression s''/K_{eq} can be neglected with respect to s'. Then the inverted Eq. (7) is rewritten as

$$1/J_s = (K_{\overleftarrow{m}}/J_{\overleftarrow{max}})(1 + s''/K_{\overleftarrow{m}})(1/s'') + (1/J_{\overrightarrow{max}})(1 + s''/K_i) \qquad (10)$$

and the data derived from the uptake curves are plotted as $1/J_s$ against $1/s'$. Intercepts with the ordinate plotted against s'' again yield $1/J_{\overrightarrow{max}}$ and $-K_i$ while slopes of the lines plotted against s'' yield $-K_{\overleftarrow{m}}$ as the intercept with the abscissa.

Characteristics of Saturable Transport

Like any enzyme reaction, carrier-mediated transports may be pronouncedly pH dependent. The pH optima observed in yeast range from pH 4.5 for some acidic amino acids to pH 8 for some basic amino acids, with the transport of sugars, vitamins, and purines being most effective at pH 6.

Temperature affects the maximum rates of transport in a typically biphasic manner. The Arrhenius plot generally displays two linear segments, corresponding to two activation energies, presumably below and above the phase transition of membrane lipids. The actual values of activation energy above the transition temperature lie between 50 and 100 kJ/mol, those below the transition temperature may be twice as large.

A variety of inhibitors act on protein-mediated transport. Heavy metal ions, even at micromolar concentrations, will affect all kinds of saturable transport, both the energetically passive mediated diffusion and the energetically active uphill transport. In addition to these, all active uptake is inhibited by compounds interfering with energy-providing reactions, both glycolytic (2-deoxy-D-glucose) and oxidative (antimycin), as well as by substances increasing the passive permeability to driving ions, particularly protons (see next section).

Substances related structurally to the transported solute behave as competitive inhibitors, depending on the range of specificity of the given carrier. In fact, such related compounds are often used to study the transport of a "typical" substrate of a transport system, if they are not metabolized within the cell as the primary substrate of the system may be (6-deoxy-D-glucose instead of D-glucose; 2-aminoisobutyric acid instead of L-alanine).

A feature unique to carrier transport is the presence of countertransport, a kinetic phenomenon where a solute may be transported uphill even if no energy is fed into the system, merely at the expense of an oppositely oriented gradient of the same (differently labeled) solute or of a related one, sharing the same carrier system. There are two fundamental ways of demonstrating countertransport (Fig. 3).

In the first method of demonstrating countertransport, cells are preincubated with a relatively low concentration of labeled solute and, after equilibrium has been reached, a relatively high dose of nonlabeled solute is added.[14] This brings forth a transient outflow of the labeled solute from cells which, particularly in a mediated diffusion system (such as that for monosaccharides in *Saccharomyces cerevisiae* or *Saccharomyces uvarum*), represents a flow out of cells against a concentration gradient. This approach is not ideal for two reasons: (1) In an uphill transport, labeled solute will flow out of cells in this arrangement because of both countertransport and a lower accumulation ratio observed at higher concentrations (see next section). (2) If the added nonlabeled solute causes osmotic shrinking of the cell, the preloaded labeled solute will flow out because of the volume decrease and this may simulate countertransport.

In the second method, cells are preincubated with a relatively high concentration of nonlabeled solute and then transferred (after centrifugation or filtration) to a low concentration of labeled solute.[15] Labeled solute then enters cells with an overshoot. In this case no osmotic effects are involved, and even in actively accumulating systems one can distinguish between the curve after preincubation with nonlabeled solute and the one obtained in cells directly placed in a low concentration of labeled solute (no overshoot or shoulder).

[14] T. Rosenberg and W. Wilbrandt, *J. Gen. Physiol.* **41**, 289 (1957).
[15] D. M. Miller, *Biophys. J.* **5**, 417 (1965).

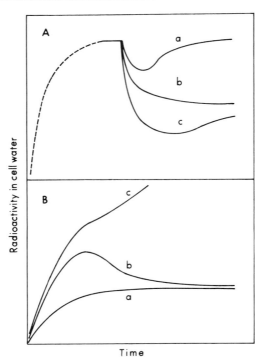

FIG. 3. Demonstration of countertransport in a carrier system. (A) Labeled solute is left to accumulate up to equilibrium, whereupon (a) the same nonlabeled solute is added in the case of mediated diffusion, (b) a metabolized solute is added which competes with the labeled one, again in mediated diffusion (e.g., glucose is added to cells equilibrated with labeled D-xylose), (c) the same nonlabeled solute is added in the case of active transport.[14] (B) Cells are preincubated (a) in buffer, (b) in nonlabeled solute in the case of mediated diffusion, (c) in nonlabeled solute in the case of active transport; cells are separated from medium and, at time zero, labeled solute is added.[15]

Active Transport Mechanisms

Using the classical distinction between primary and secondary active transport, it is clear that the yeast plasma membrane contains a single primary active mechanism in its H^+-ATPase (to be dealt with in [35] in this volume). This is a system where the free energy of a chemical reaction, splitting of ATP, is directly channeled into a translocation of H^+ out of cells. All other active transports are secondary in that they utilize the electrochemical potential gradient of H^+ built up by the ATPase. In some yeasts one may encounter group-translocation processes, and there are a number of uptake systems with unclear energy coupling; however, these are not treated here explicitly.

In a secondary, H^+-driven, active transport (H^+ symport) Eq. (4) will be valid quite generally, the various coefficients A through F being defined by rather complicated expressions[16] including concentrations of H^+ at both membrane sides. The expressions for K_m and J_{max} will differ, depending on the sequence of binding of solute S and of H^+ and on whether the system can be considered to be in local equilibrium or not. The carrier cycle patterns shown in Scheme I illustrate the point.

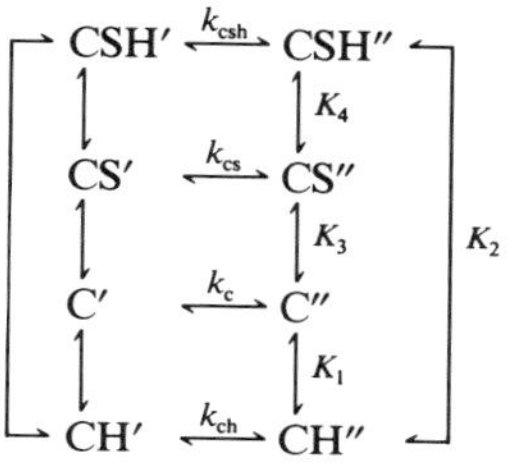

$$\begin{array}{cccc}
& \text{CSH}' \xrightarrow{k_{csh}} \text{CSH}'' & \\
& \downarrow \qquad\qquad \downarrow K_4 & \\
& \text{CS}' \xrightarrow{k_{cs}} \text{CS}'' & \\
& \downarrow \qquad\qquad \downarrow K_3 & K_2 \\
& \text{C}' \xrightarrow{k_c} \text{C}'' & \\
& \downarrow \qquad\qquad \downarrow K_1 & \\
& \text{CH}' \xrightarrow{k_{ch}} \text{CH}'' &
\end{array}$$

SCHEME I

In a local equilibrium system all the free carrier molecules (C) and their complexes (CS, CH, CSH) at each membrane side will be mutually equilibrated through $K_1 - K_4$, the translocation constants k_c, k_{cs}, k_{ch}, and k_{csh} being generally lower than the first-order constants constituting the equilibrium constants and, hence, rate-limiting. It is these constants that may be sensitive to the transmembrane potential, specifically those that represent the movement of charge, i.e., k_{ch} and k_{csh}. Following rate theory concepts of Eyring,[17] the forward constants describing charge movement are multiplied by $\exp(-F\Delta\psi/2RT)$ where F is the Faraday constant (96.5 kC mol^{-1}), $\Delta\psi$ the membrane potential (in V), R the gas constant (8.314 J mol^{-1} K^{-1}), and T the absolute temperature (in K); the backward constants are divided by the same expression. If the local equilibrium assumption is not made all the partial steps must be expressed by rate (not equilibrium) constants.

It is rather probable that in the above scheme CH and CS cannot move across the membrane or change their exposure from one membrane side to the other, the first because it would short-circuit the H^+ concentration difference across the membrane and thus waste the energy stored in it without transporting the substrate, the second because it would decrease the thermodynamic efficiency of the system by an internal slipping (solute could be transported without energy expenditure but only up to a diffusion

[16] A. Kotyk, in "Structure and Properties of Cell Membranes" (G. Benga, ed.), Vol. 2, p. 1. CRC Press, Boca Raton, Florida, 1985.
[17] P. Läuger, J. Membr. Biol. 57, 163 (1980).

equilibrium, not uphill). For that reason, one of four simplified models is assumed to operate (Scheme II).

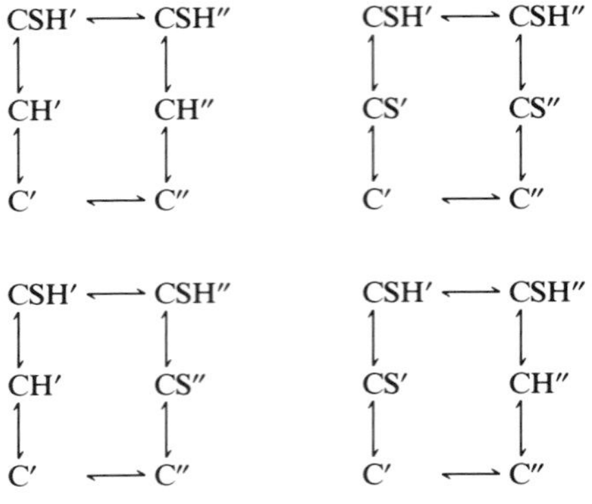

$$
\begin{array}{ccc}
\text{CSH}' \longleftrightarrow \text{CSH}'' \\
\updownarrow \qquad \updownarrow \\
\text{CH}' \qquad \text{CH}'' \\
\updownarrow \qquad \updownarrow \\
\text{C}' \longleftrightarrow \text{C}''
\end{array}
\qquad
\begin{array}{ccc}
\text{CSH}' \longleftrightarrow \text{CSH}'' \\
\updownarrow \qquad \updownarrow \\
\text{CS}' \qquad \text{CS}'' \\
\updownarrow \qquad \updownarrow \\
\text{C}' \longleftrightarrow \text{C}''
\end{array}
$$

$$
\begin{array}{ccc}
\text{CSH}' \longleftrightarrow \text{CSH}'' \\
\updownarrow \qquad \updownarrow \\
\text{CH}' \qquad \text{CS}'' \\
\updownarrow \qquad \updownarrow \\
\text{C}' \longleftrightarrow \text{C}''
\end{array}
\qquad
\begin{array}{ccc}
\text{CSH}' \longleftrightarrow \text{CSH}'' \\
\updownarrow \qquad \updownarrow \\
\text{CS}' \qquad \text{CH}'' \\
\updownarrow \qquad \updownarrow \\
\text{C}' \longleftrightarrow \text{C}''
\end{array}
$$

SCHEME II

The full expressions for the transport rate by any of the carrier models in Scheme II are extremely complicated. The kinetic constants [corresponding to those of Eq. (6)] have been computed, however, and the "forward" constants especially, K_m and J_{max}, contain such information as can be used to determine which of the models is applicable to a given system. If the system is in local equilibrium as has been shown to be true in the few cases where an analysis of binding and dissociation rates was possible, changing the driving ion concentration (here the pH) and the membrane potential (a somewhat more difficult task with yeast that is best accomplished by increasing the concentration of K^+) will elicit different responses of the parameters K_m and J_{max} (Table I). However, if a steady state can only be assumed to exist (i.e., unchanging concentrations of the individual carrier complexes during the experiment) the analysis is not straightforward.[18] The membrane potential is seen to increase the maximum rate in all cases and decrease the half-saturation constant in practically all cases.

Although in principle this approach has a high diagnostic value in practice it is hampered by the fact that pH plays a role that is catalytic in nature and has nothing to do with active transport. These pH effects on rates may then obscure the effects issuing from the active uptake mechanism. This diagnostic approach is fully applicable in cases of Na^+-driven transports, such as in animal cells and some bacteria.

In contrast to simple or mediated diffusion systems, active transports are characterized not only by K_m and J_{max} but by their K_{eq} values, equiva-

[18] D. Sanders, U.-P. Hansen, P. Gradmann, and C. L. Slayman, *J. Membr. Biol.* **77**, 123 (1984).

TABLE I
EFFECTS OF RAISING $[H^+]_{out}$, $[H^+]_{in}$, AND $\Delta\psi$ ON TRANSPORT PARAMETERS

Kinetic scheme	$[H^+]_{out}$		$[H^+]_{in}$		$\Delta\psi$	
	J_{max}	K_m	J_{max}	K_m	J_{max}	K_m
Local equilibrium						
C + H + S	—	↓	↓	↓ ↑	↑	↓
C + S + H	↑	↓	—	—	↑	↓
Both C + H + S and C + S + H	↑	↓ ↑	↓	↓ ↑	↑	↓
Steady state						
C + H + S	↑—	↓ ↑	↓	↑ ↓	↑	↓
C + S + H	↑—	↓ ↑	↓	↓ ↑	↑	↓ ↑
C + H + S, getting on and C − H − S, coming off	↑—	↓ ↑	—	↑	↑	↓ ↑

lent to the final accumulation ratio of solute, $(s''/s')_\infty$. This is usually considered as a constant that can be calculated from the individual J_{max} and K_m values and to reflect not the mechanism but the energy available to the system. In H^+-driven secondary transport the energy available is related to the electrochemical potential gradient of protons, defined as

$$\Delta\tilde{\mu}_{H^+} = -RT \ln([H^+]''/[H^+]') - F\Delta\psi \qquad (11)$$

If this potential gradient is fully utilized by a secondary transport system, i.e., if there is an obligatory coupling between the carrier and the source of energy, all of its free energy can appear in the accumulation ratio of the H^+-driven solute. Thus,

$$n\Delta\tilde{\mu}_{H^+} = \Delta G = -RT \ln(s''/s') \qquad (12)$$

where n is the number of protons transported per solute molecule. From this then

$$s''/s' = ([H^+]'/[H^+]'')^n \exp(-nF\Delta\psi/RT) \qquad (13)$$

Equation (13) is equivalent to the ratio of the product of individual rate constants going clockwise to the product of constants going counterclockwise in any of the models in Scheme II, provided that the constants contain concentrations of H^+ and the expression for half-potential, $\exp(-nF\Delta\psi/2RT)$, as appropriate. If, however, there is internal slipping, i.e., the translocation constant k_{cs} of Scheme I is nonzero, the value of s''/s' will be smaller and may tend toward unity if, at higher concentrations of solute, the formation of CSH cannot cope with the amount of CS formed and translocated.[19]

It is a common observation that the accumulation ratio of a solute decreases with increasing s' (Fig. 4). While curve b depicts a case with internal slipping (or a parallel diffusional pathway; see next section), curves c and d reflect a situation where at high values of s' it holds that $s'' < s'$ which cannot occur in an uncoupled or pump-and-leak system. The coupling here is obligatory, and the behavior must be explained in a different way.

Let us suppose that the availability of energy (e.g., tapping the H^+ gradient) is limited so that the local concentration of the energy source h is affected by the amount that has interacted with the carrier (in proportion to ch and csh). Then $h = h_{tot} - ch - csh$. On computing the relative concentrations ch/c_{tot} and csh/c_{tot} from a steady-state analysis of any of the above models, a cubic equation in s'' can be set up which makes it possible to determine the s''/s' ratio numerically, if not analytically.[20] The solutions allow one to conclude that in systems where the sequence of occupation of the carrier is C + H + S curve c of Fig. 4 will be obtained, if the sequence is C + S + H curve d will be arrived at.[21,22]

Additional Factors Affecting Transport Parameters

It should be clear by now that temperature, pH, presence of inhibitors, and, in the case of active transport, concentrations of driving ions, influence the kinetic constants that are measured in any given experiment. Several other factors, however, may come into play and alter the numerical values in a substantial manner.

Presence of Unstirred Layers

Near the boundary with a separate phase, whether solid or liquid, an aqueous solution will not be moved by convection (stirring, shaking, and the like) because of cohesive forces within the water medium. Any molecules or ions in that region will be able to move only through diffusion. In the case of a static system, say, polymer beads or dead cells in suspension, this is trivial as the concentration of a solute will have reached equilibrium throughout the system by diffusion even if it had not been in equilibrium initially. However, if an independent movement of solute particles occurs

---

[19] A. Kotyk and K. Janáček, "Cell Membrane Transport," 2nd ed., p. 149. Plenum, New York, 1975.

[20] A. Kotyk and K. Janáček, "Membrane Transport: An Interdisciplinary Approach." Plenum, New York, 1977.

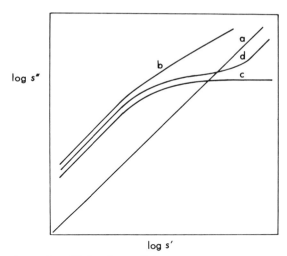

FIG. 4. Dependence of equilibrium intracellular concentration on extracellular concentration under conditions of mediated diffusion (a), active transport with internal slipping or a parallel leak (b), and tightly coupled active transport where the energy-providing ligand binds before solute (c) or after it (d).

across the phase boundary, e.g., into or out of a cell across its outer envelope, the concentration at the membrane (s_m) will be different from that in the bulk solution (s_b) by the following expressions (for inward movement):

$$s'_m = s'_b - J_s/P' \tag{14a}$$
$$s''_m = s''_b + J_s/P'' \tag{14b}$$

where J_s is net flow across the membrane and P is the permeability constant of the unstirred layer, i.e., the diffusion coefficient D of the solute particle in the unstirred layer, divided by δ, the unstirred layer thickness (Fig. 5).

Both in simple diffusion and in specific transport the presence of unstirred layers affects the transport parameters.[20] Thus, in simple diffusion,

$$P_{true} = 1/(1/P_{apparent} - 1/P' - 1/P'') \tag{15}$$

i.e., the true P is greater than the measured P if P' is relatively small.

[21] A. Kotyk and R. Stružinský, *Biochim. Biophys. Acta* **470**, 484 (1977).
[22] J. Horák and A. Kotyk, *Biochim. Biophys. Acta* **858**, (1986).

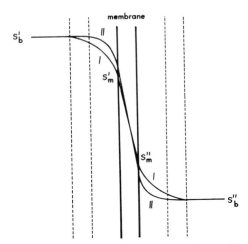

FIG. 5. Schematic representation of the solute concentration profile across a membrane where net transport of S proceeds from left to right, the unstirred layers extending approximately to the dashed vertical lines. Profile I corresponds to a relatively fast, profile II to a slow transmembrane movement.

Likewise, in specific, carrier-mediated transport it holds that[23]

$$J_s = P[0.5(K_m + s_b + J_{max}/P) \pm \sqrt{0.25(K_m - s_b + J_{max}/P)^2 + s_b K_m}] \quad (16)$$

and again, the experimentally determined K_m is greater than the true one by

$$K_{m_{app}} = K_{m_{true}} + 0.5 J_{max}/P \quad (17)$$

(J_{max} is not affected by unstirred layers).

A way to determine qualitatively an effect of unstirred layers is to observe whether increased agitation of a suspension decreases the apparent K_m or not. Caution must be exercised that increased stirring does not change K_m values by providing more oxygen for active transport energization. A quantitative determination of unstirred layer thickness is shown in Fig. 6.[24] Measuring the net flow rate J_s starting from a very high concentration inside cells against different extracellular concentrations s_b', including a zero concentration (then J_∞), and plotting as shown will yield the permeability of the external unstirred layer. Knowing the diffusion coefficient D of the examined solute in water one can calculate the unstirred layer thickness from $\delta = D/P$.

[23] D. Winne, *Biochim. Biophys. Acta* **298**, 27 (1973).
[24] W. R. Lieb and W. D. Stein, *Biochim. Biophys. Acta* **373**, 178 (1974).

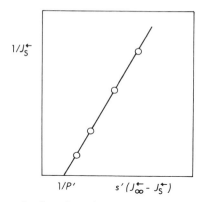

FIG. 6. Graphical determination of unstirred layer permeability. Detailed explanation is given in the text.

Effect of Membrane Surface Potential

All known membranes contain an excess of negative charges on their surface, owing to dissociation of carboxyl and phosphate groups of lipids and proteins. The total number of charges roughly averages 10 mC m^{-2}, the excess of negative charges amounting to some 10^8 per cell. This surface charge σ attracts oppositely charged ions from solution, giving rise to an adsorption layer. As there are more positive ions in the adsorption layer than correspond to the negative surface charge, a further, diffusion, layer is formed, enriched with negative charges. The whole electrical diffusion double layer (Gouy–Chapman layer) is about 1–2 nm thick, depending on the ionic strength of the solution, by $\kappa = 3 \times 10^{-10}/\sqrt{\mu}$ where μ, the ionic strength, is defined by $\frac{1}{2} \Sigma\, c_i z_i^2$, c being the concentration, z the charge on the ion i. In yeast this lies then well within the cell wall. The nonhomogeneous distribution of ions around the cell gives rise to a surface potential ψ_s, which is directly proportional to the charge density σ, by

$$\psi_s = 4\pi\sigma r\kappa/\epsilon(\kappa + r) \tag{18}$$

where r is the cell radius, κ the thickness of the Gouy–Chapman layer, and ϵ the dielectric constant of the solution (it is 0.75 nF m^{-1} for water).

In the presence of a surface potential the concentration of an ion at the membrane (s_m) is given by

$$s_m = s_b \exp(-nF\psi_s/RT) \tag{19}$$

For negatively charged surfaces this means that cations will occur at higher concentrations at the membrane than in the bulk, while the opposite holds

TABLE II
EFFECT OF DECREASING SURFACE POTENTIAL ON
APPARENT KINETIC PARAMETERS FOR ION AND
ION-DRIVEN TRANSPORT

Transport and order of binding	K_m	J_{max}
Cation alone	↑	—
Anion alone	↓	—
Neutral solute with cation		
Cation first	↑	—
Solute first	↑	↓
Cation with anion		
Anion first	↑	—
Cation first	↑	↑
Anion with cation		
Cation first	↓	—
Anion first	↓	↓

for anions.[25] This will have a pronounced effect on the kinetics of ion transport and of all ion-driven transports (Table II).[26]

To check the effect of surface potential qualitatively it is best to determine the transport parameters in the presence of different concentrations of ions, especially those with higher valencies, such as Fe^{3+} and Cr^{3+}, as these increase the ionic strength and hence diminish the thickness of the Gouy–Chapman layer and, from Eq. (18), the membrane potential.

Existence of Parallel Uptake Mechanisms

The existence of parallel uptake mechanisms is relatively frequent in yeasts. When two saturable transport systems operate in parallel one can distinguish them provided that their kinetic constants, especially the J_{max} values, differ sufficiently.[27] Using a Lineweaver–Burk plot (Fig. 1), the two linear segments can be evaluated as shown in Fig. 7. The K_m values of the two parallel systems are given by[28]

$$K_{m1,2} = 0.5[\alpha + \beta \pm \sqrt{(\alpha + \beta)^2 - 4\beta\gamma}]$$
$$J_{max1} = (K_{m1} - \beta)/I_1(K_{m1} - K_{m2})$$
$$J_{max2} = 1/I_1 - J_{max1}$$

where $\alpha = S_1/I_1$, $\beta = (S_1 - S_2)/(I_2 - I_1)$, and $\gamma = S_2/I_2$.

[25] A. P. R. Theuvenet and G. W. F. H. Borst-Pauwels, *J. Theor. Biol.* **57**, 313 (1976).
[26] G. M. Roomans and G. W. F. H. Borst-Pauwels, *J. Theor. Biol.* **73**, 453 (1978).
[27] W. Gross, P. Geck, K.-L. Burkhardt, and K. Ring, *Biophysik* **8**, 271 (1972).
[28] J. L. Neal, *J. Theor. Biol.* **35**, 113 (1972).

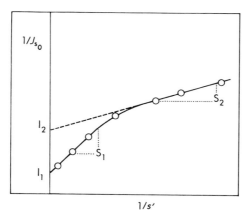

FIG. 7. Graphical determination of kinetic parameters of two saturable transport systems operating in parallel. Symbols are explained in the text. I_1 and I_2 are intercepts; S_1 and S_2 are slopes of the linear segments.

The presence of a pump-and-leak system, i.e., one saturable (and potentially uphill) transport and one diffusional pathway, is easily detected even in a direct plot of J_{s_0} against s' where, at high values of s', the curve has a constant positive slope. This slope, projected to lower values of s', can be subtracted from the total curve and the saturable transport can be defined (Fig. 8).

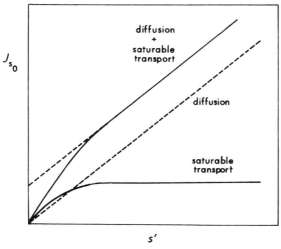

FIG. 8. Separation of diffusional and saturable pathways of transport in a pump-and-leak system. The difference between the upper (total) curve and the lower dashed line is plotted as saturable transport.

Effect of Suspension Density

A recent investigation[29] shows that a variety of processes in yeast are negatively affected by the density of cell suspension in a pronounced manner. The processes involve both mediated diffusion (of monosaccharides in baker's yeast) and active transport (of amino acids and sugars in other yeast species). The constant affected by suspension density is J_{max} not K_m, and the effect is probably caused by CO_2 produced by cells. The inhibition is generally noncompetitive in nature (Fig. 9). In addition to the kinetic effects there appears to be a more complicated effect of suspension density on the accumulation ratio. Hence, a true J_{max} must be calculated by extrapolation to zero suspension density and may prove to be several times higher than values given in the literature.[29a]

Strategy for Complete Kinetic Characterization of Transport System in Yeast

If the transport of a solute in a yeast species is to be investigated and fully defined by kinetic means, the following points should be addressed.

1. The solute taken up must not be metabolized by cells if a complete characterization of the system is desirable. Otherwise only the initial rate data would be relevant. This should be checked (a) by measuring evolution of labeled CO_2 from a labeled substrate taken up, (b) by observing incorporation into proteins if an amino acid is examined (to prevent incorporation, work must be performed with 0.4–0.6 mM cycloheximide; even then, however, metabolism of the amino acid tested must be ruled out), or (c) by conducting suitable chromatographic analyses of a cell extract after uptake of a labeled substrate.

2. The solute must be present in solution, not adsorbed, within cells. The easiest way to check this is to observe solute accumulation in yeast cells shrunk to varying degrees by exposure to different concentrations of NaCl.[30] If the amount of solute accumulated is proportional to the amount of cell water, intracellular adsorption is unlikely.

3. To distinguish between simple diffusion and carrier-mediated transport, it is best to check the saturability of the system at high solute concentrations even if this method is not fully reliable in cases where the K_m value of the studied system lies at or above 0.5 M. The membrane may become altered unspecifically by high osmotic values of the medium. An auxiliary procedure is to test competition by high concentrations of related substances that may possess a lower K_m with respect to the carrier. Another

[29] S. Janda and A. Kotyk, *Folia Microbiol. (Prague)* **30**, 465 (1985).
[29a] A. Kotyk, *Yeast,* **3**, 263 (1987).

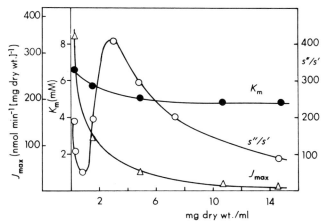

Fig. 9. Dependence of transport parameters of 6-deoxy-D-glucose on the suspension density of *Rhodotorula glutinis*. The accumulation ratio s''/s' was obtained initially with 0.2 mM substrate.

check would be the effect of heavy metals or metabolic inhibitors, which should be absent with simple diffusion.

4. If the system is one of simple diffusion, its P value should be determined, taking into account the possible effects of unstirred layers, temperature dependence, and, if an uptake of ions is examined, surface potential.

5. If the mechanism is one of carrier transport one must make sure that work is done with a single system. If it so happens that a diffusional pathway operates in parallel, its capacity must be accurately determined (see Fig. 8) and subtracted from all values measured. If two saturable transports are involved several procedures are possible: (a) use of a mutant, if available, that lacks one of the transport systems; (b) inhibition of one of the systems if a specific inhibitor exists; (c) working at a concentration range well removed from the K_m of the other system (this, however, prevents one from working over the entire concentration range, and hence a full characterization of the system is not possible).

6. Once it has been established that a single saturable system is involved, the kinetic parameters obtainable from initial rate measurements should be determined. Such parameters include K_m and J_{max}, preferably determined using different analytical concentrations of solute with the same amount of label, rather than diluting a concentrated labeled solution; specificity of the system (a) by using related substances as potential com-

[30] A. Kotyk and A. Kleinzeller, *Folia Microbiol. (Prague)* **8**, 156 (1963).

petitive inhibitors; sensitivity to various inhibitors of transport and metabolism; temperature dependence, which is best done with J_{max} as the number of first-order rate constants involved there is less than when testing either K_m or J_s at an intermediate concentration; pH dependence, preferably of J_{max} as well as of K_m, in order to assess the intrinsic dissociation constants of the carrier–solute–H^+ complexes.[31] It is now that one should test the effect of surface potential, unstirred layers, and suspension density. If quantitative determination of any of these effects proves impossible they should be described qualitatively and the exact conditions of measurement stated, namely, ionic composition of the medium, rate of agitation and size and shape of the incubation flask, and dry weight per milliliter of suspension.

7. If the transport is active, as should follow from its proceeding uphill, its being blocked by metabolic inhibitors, and (for a secondary active transport) its being sensitive to surface potential, the remaining kinetic parameters of Eq. (6) should be determined as described in the corresponding text. The detailed sequence of carrier cycle steps could be determined from the influence of pH_{in} and pH_{out} on K_m and on J_{max} (Table I). An additional test is provided by the dependence of the accumulation ratio s''/s' on s' (cf. Fig. 4). The effect of suspension density on the accumulation ratio is easily determined in this connection.

8. The source of energy for the transport system can be estimated using several simple tests. (a) A transient rise in pH_{out} on adding the tested solute, (b) a depolarization of the membrane under the same conditions (as described in [39] in this volume), and (c) a dependence of rate (not accumulation ratio) on surface potential would indicate a dependence on the electrochemical potential gradient of protons. Absence of the first two indicators is not yet full evidence against the involvement of the electrochemical potential as a stimulation of H^+-ATPase activity by the addition of an H^+-symported solute could set off the changes in pH and $\Delta\psi$ attendant on proton symport of a solute.

9. If a dependence of transport on $\Delta\tilde{\mu}_{H^+}$ can be definitely excluded, other sources of energy should be sought, such as polyphosphate or ATP splitting, either to phosphorylate the carrier or to phosphorylate the transported solute itself. If this latter mechanism should operate (e.g., with monosaccharide phosphorylation of some yeasts[32]) the simplest way to demonstrate it is by following the specific labeling of intracellular free sugar and sugar phosphate immediately after labeled sugar has been added

[31] A. Kotyk, in "Structure and Properties of Cell Membranes" (G. Benga, ed.), Vol. 2, p. 11. CRC Press, Boca Raton, Florida, 1985.
[32] H. T. A. Jaspers and J. van Steveninck, Biochim. Biophys. Acta 469, 292 (1977).

to the medium. If higher specific activity in the phosphate precedes that in the free sugar, a phosphorylative transfer is likely.

10. All the parameters obtained heretofore may now be tested in a mutant strain, a related species, or in the same species under different conditions of aerobiosis,[33] preincubation with energy sources,[34] starvation, and the like. Use of protoplasts can provide interesting new insights into the mechanism examined.

Techniques Used in Yeast Transport Studies

Specific Discontinuous Measurement

During the 1970s and 1980s the availability of radioactively labeled compounds has increased to such extent that representatives of all groups of compounds may be easily obtained from commercial sources.

The common procedure in measuring transport is to incubate a yeast suspension with a labeled compound in a flask (placed in a constant-temperature bath) that is agitated to keep the cells from sedimenting—the usual frequency of such reciprocating shakers is 1–2 Hz—and remove samples from the flask at suitable intervals. These may be as short as 10 sec if the uptake is very rapid or as leisurely as 1 min if it is slow, the main concern being to obtain at least three, but preferably more, points lying on a straight line.

To keep the gaseous phase at an oxygen concentration different from that in air one can supplement the flasks with the appropriate gas: oxygen for highly aerobic or argon for anaerobic conditions. Nitrogen can be used in place of argon, although it may exert side effects on yeast metabolism and, moreover, is not usually available at sufficiently high purity. To remove the last traces of oxygen, either the gas can be passed through solutions of alkaline pyrogallol (20% in 20% KOH) or 2 M CrCl$_2$ in acetic acid (after reduction of CrCl$_3$ with zinc amalgam), or, alternatively, a stick of white phosphorus may be placed in a separate compartment of the incubation vessel. This latter procedure, however, is not always safe as phosphorus is volatile and its vapors may be toxic for some yeast species.

Some authors prefer to flush continuously the incubation vessel with the appropriate gas so as to ensure a constant atmosphere and sufficient stirring of the suspension. This is indeed a fine procedure for short experiments, such as measuring the initial rates of transport over a few minutes, but it becomes unacceptable over longer periods of time because of loss of

[33] R. Hauer and M. Höfer, *J. Membr. Biol.* **43**, 335 (1978).
[34] A. Kotyk, J. Horák, and A. Knotková, *Biochim. Biophys. Acta* **698**, 243 (1982).

water in the suspension owing to increased evaporation and because some yeast suspensions (e.g., *Rhodotorula*) come up in froth, the bubbles containing many cells so that an unpredictable fraction is lost during incubation.

It goes without saying that to obtain meaningful results the suspension must be homogeneous and the samples removed from it identical. With most yeasts this is no problem; however, some species, and especially protoplasts, tend to creep along the incubation vessel walls and form there a kind of sediment where the moving liquid cannot reach them and wash them back. If a yeast suspension flocculates it should be discarded.

Stopping the incubation at suitable times and sufficiently rapidly is the major step in all experiments. Cells can be prevented from further transport in three basic ways: (1) filtration, (2) centrifugation, and (3) dilution in a suitable solution.

Filtration is the method used in most laboratories. A sample, preferably 0.1–0.5 ml not to waste labeled material, is transferred from the suspension onto a membrane filter (of the type supplied by Millipore, Sartorius, and many others, in wide ranges of pore diameters) with a pore diameter of 0.6 μm for small cells, such as *Rhodotorula,* up to 1 μm for average and large yeast species. The filter, usually 25 mm in diameter, is held in position over a sintered glass disk or a stainless steel mesh with a cylindrical funnel clamped to a suction flask connected to a vacuum pump. Suction is applied to the filter immediately after the suspension sample has been transferred to the filter. If the filter is under suction before the sample is placed on it the suspension will not spread evenly, and the procedure is difficult to standardize. Either hand- or foot-activated stopcocks for the vacuum system are now available.

Some cells and particularly protoplasts are difficult to filter as they clog the pores. Glass-fiber or other fibrous filters may be obtained from most firms which will perform the filtration without difficulty. For a filter to be considered efficient, a 0.2-ml sample of baker's yeast suspension at about 5 mg dry weight/ml, should pass through the filter in less than 1 sec.

Once the suspension sample has been filtered the cells on the filter must be washed to remove the remaining external medium. This is achieved best with one 1.5-ml portion (two 1.5-ml portions for thicker suspensions) of ice-cold water or buffer, a procedure that, using the above suspension sample, will take no more than 5 sec (12–15 sec with two washings) after some practice has been gained. It is an ill-advised procedure to wash the cells with larger volumes of water or buffer or to repeat the washing several times, and it is a completely wrong approach to wash with a solution of the nonlabeled species of the solute whose transport (in the labeled form) is examined. Some kinds of transport are extremely rapid and even the first seconds of exposure to a label-free solution may cause losses of previously accumulated solute. The presence of nonlabeled solute in the washing solution will further accelerate the outflow of the label (cf. the section on countertransport above).

On the other hand, one should run a filtration control without cells to see how much radioactivity is retained by the filter. If this is more than 20% of the lowest count another procedure should be used.

The filter with cells is then transferred to a scintillation vial for counting. If work is done under rigidly standardized conditions it is not necessary to dry or otherwise treat the filters before placing them in the scintillation cocktail.

If samples must be taken at intervals shorter than 15–20 sec (the minimum time required for filtering, washing, and removing the filter and placing another in position) a manifold with several filtration positions which can be actuated separately should be used (as made by Millipore and others).

Centrifugation of the suspension is the method resorted to if the cells cannot be filtered at all (I know of no such case among yeasts) or if too much radioactivity is adsorbed on the filtering device, as the case is, e.g., with the lipophilic cations used to estimate the membrane potential. Centrifugation has two serious drawbacks. (1) It takes considerable time to spin down the cells *and* wash them at least once: perhaps 40 sec in a high-speed centrifuge with an efficient brake. (2) The washing here is less efficient as much more of the original medium stays in and on the cell sediment and even on the walls of the centrifuge tube.

One way to speed up the procedure is to centrifuge through a layer of silicone oil of suitable density (1.04–1.08 g/ml) which is either placed directly on the bottom of the centrifuge tube or layered over a heavier solution, such as perchloric acid at 2–4 M. This approach requires a good deal of skill to prevent leaks of the suspension sample around the silicone oil layer, and, even with some practice, it is messy, particularly at the stage of removing the cells from the centrifuge tube.

However, centrifugation is the method of choice in cases when speed is of no concern and when one counts radioactivity in the supernatant rather than in the sediment. This gives clean results without complications arising from washing and adsorption, but it requires that enough of the solute is internalized prior to sampling, i.e., an active transport is a prerequisite. In calculating the amount of solute taken up one must be careful to take into account the volume of cells when comparing the sample radioactivity with that of a control without cells.

Diluting the sample on removal from the solution is also occasionally used to halt transport. Transport inhibitors are added to the solution, e.g., uranyl nitrate, and the volume is quite large in comparison with that of the sample. The dilution method has only one advantage over the other two approaches: one can take samples in a very rapid sequence and analyze them subsequently. However, it suffers from the danger of losing some of the accumulated solute as none of the inhibitors used acts immediately and none stops transport completely.

Specific Continuous Measurement

The ideal of all transport workers is to be able to follow the uptake or outflow of a substance from cells continuously. Several devices have been designed for this purpose, their principle being to keep the yeast suspension (as thick as possible) or sediment mixed with beads of an inert support separated by a membrane (made of suitable porous artificial polymer or hardened cellulose) from a flow of water or buffer that circulates or simply permeates through the cells kept within a column or pocket. It is particularly at the otflow from this system that one can follow the efflux of an absorbing or fluorescent substance or of a penetrating radioactive emitter (such as ^{24}Na or ^{42}K). In a circulatory arrangement, the technique can be particularly useful for distinguishing between uptake of different optical isomers and uptake of a natural mixture when a refractometer is placed in the circuit.[35]

Nonspecific Techniques

If a continuous record of uptake is essential without any danger of mistaking the substance taken up for another one can use osmotically sensitive cells, in particular protoplasts, and follow their volume changes as a solute is added.[36] The procedure may be a simple densitometric determination, measuring the light passing through a suspension which is generally diminished by the light scattered from the cells. Since, in the size range of yeast cells and with the given intrinsic absorbance of the cells, more light is scattered by smaller cells than by large ones, the overall absorbance of a suspension increases as cells shrink (Fig. 10).

Cell or protoplast size can be monitored by more sophisticated techniques, such as cytometric measurements in a computer-assisted counter, electron spin resonance measurements using a suitable paramagnetic probe, or measurements based on laser light dispersion. However, these methods have never been used to follow the transport of solutes in yeast

[35] R. Ehwald, P. Sammler, and H. Göring, *Folia Microbiol. (Prague)* **18**, 102 (1973).
[36] A. Kotyk and D. Michaljaničová, *Folia Microbiol. (Prague)* **14**, 62 (1969).

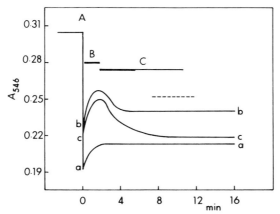

FIG. 10. Absorbance record of a suspension of *Saccharomyces cerevisiae* protoplasts in a 1.28 os$M$ Na_2HPO_4–KCl–sodium citrate solution, pH 6.4, following addition of 0.2 M lactose (a; no uptake), 0.2 M L-arabinose (b; uptake into part of cell volume), or 0.2 M D-arabinose (c; uptake into entire cell water volume). The dashed line corresponds to absorbance following addition of buffer alone. A, Dilution of suspension; B, shrinking of cells due to loss of water; C, increase in cell volume as sugar enters protoplasts.

and, moreover, owing to their costliness, can hardly be recommended for routine and serial measurement of transport.

Final Note

Perhaps more than in work with animal cells and well-defined bacterial strains, it is essential when working with yeast to maintain standard conditions throughout the entire procedure, including such details as the size of inoculum, provenience of yeast extract and other composite sources of vitamins and trace elements, time required for washing the harvested cells, length and intensity of aeration during starving the yeast, size and shape of incubation flasks, and washing or centrifugation procedure. Failure to maintain constant pH throughout the transport experiment or to ensure that the cell suspension density is the same has often led to discrepancies among results of different authors.

[35] Proton Extrusion in Yeast

By ARNOŠT KOTYK

General Mechanisms

One of the characteristic features of viable yeast cells is their ability to acidify the external medium. Teleonomically speaking, this process can serve two purposes: (1) to maintain a large pH difference between the cell interior and the outside medium, which can serve potentially as a source of energy in H^+-driven transport; and (2) to keep the external acidity high to ward off potentially pathogenic or competing bacteria.

The acidification process takes place either spontaneously, without addition of metabolic substrate, or following such an addition. Spontaneous acidification has been little studied although it represents a major energetic phenomenon in resting yeast of all types. As a crumb of pressed baker's yeast or the sediment after harvesting laboratory-grown yeast is transferred to water or a dilute buffer, the pH of the aqueous suspension drops very rapidly to as low as 4.5, depending on the type of yeast and the volume into which the yeast is suspended (Fig. 1). The yeast loses this ability to acidify the external medium as the transfer is repeated several times or as the pH is returned to the original pH by titration with an alkaline solution; likewise, the acidification decreases as cells are starved or aged, this serving as part of a test for yeast "viability" in industrial applications.[1]

Attempts to associate the instantaneous acidification with a particular enzyme process have failed so far, but it appears that in baker's yeast acidification is mainly due to the release of CO_2, which is hydrated in water to $H^+ + HCO_3^-$, or to direct release of HCO_3^-, which is in equilibrium with dissolved CO_2, as the acidity drop can be abolished by flushing the suspension with an inert gas, such as argon.[2] In *Rhodotorula*, CO_2 may not be involved.[3] Lack of involvement of the plasma membrane H^+-ATPase (see below) is also indicated by the fact that addition of diethylstilbestrol, an ATPase inhibitor, in fact increases the extent of acidification.

---

[1] M. Opekarová and K. Sigler, *Folia Microbiol. (Prague)* **27**, 395 (1982).
[2] M. Opekarová and K. Sigler, *Cell. Mol. Biol.* **31**, 195 (1985).
[3] R. Hauer, G. Uhlemann, J. Neumann, and M. Höfer, *Biochim. Biophys. Acta* **649**, 680 (1981).

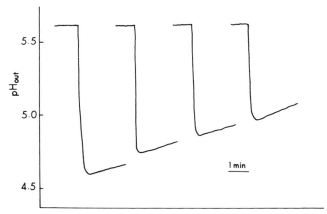

Fig. 1. Repeated acidification of the external medium on transferring the same batch of filtered *Saccharomyces cerevisiae* cells to distilled water.

The decrease of pH is extremely rapid; if it is entirely due to CO_2 it would represent a production of as much as 10 μl CO_2 per mg dry weight per min (in rich baker's yeast), i.e., 50–100 times as much as in prolonged endogenous respiration.

One system that is certainly functional even in the absence of substrate is the K^+–H^+ exchange which can transport various alkaline ions in exchange for H^+ coming out of cells and vice versa.[3] However, this cannot account for the initial acidification as at that time both H^+ and K^+ are released from cells.[2]

A more pronounced acidification and certainly one that has been recognized for a long time[4] is that following the addition of a metabolizable substrate (particularly sugar) to a yeast suspension. The source of the H^+ is no doubt the metabolism of the sugar, giving rise to various acids, either phosphorylated, such as phosphoglyceric acid in the main pathway of glycolysis, or free, such as succinic acid in the Krebs cycle or glyceric acid in the anaplerotic catabolic pathway. This may be demonstrated by the use of various mutants and by growing yeast on different substrates (Fig. 2).[5,6] The number of apparent protons appearing in the medium per glucose molecule consumed ranges from 2 to 5, depending on yeast strain and sugar concentration. The final pH attained in thick suspensions is as low as 3.0.

[4] E. J. Conway and E. O'Malley, *Biochem. J.* **40,** 59 (1946).
[5] K. Sigler, A. Knotková, and A. Kotyk, *Biochim. Biophys. Acta* **643,** 572 (1981).
[6] K. Sigler, C. Pascual, and S. Romay, *Folia Microbiol. (Prague)* **28,** 363 (1983).

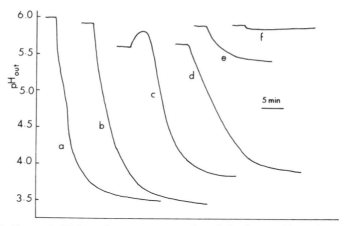

FIG. 2. External pH in various yeast suspensions following addition of metabolizable substrate. a, *Saccharomyces cerevisiae* CCY 21-4-60 after addition of D-glucose aerobically; b, the same, after addition of D-glucose anaerobically; c, the same, after addition of maltose, following a 2-hr induction with that sugar, aerobically; d, *S. cerevisiae* CCY 21-4-19, a respiration-deficient mutant, after addition of glucose; e, *S. cerevisiae* CCY 21-4-60, after addition of ethanol aerobically; f, *S. cerevisiae* 7A-32, a mutant lacking phosphomannose isomerase (mannose-6-phosphate isomerase), after addition of glucose, aerobically.

The net production of acidity can be shown in a cell-free extract to which a metabolic sugar has been added. Although the pH of the extract is buffered powerfully enough to remain constant (just as the intracellular pH is not appreciably changed), the titratable acidity increases significantly.[7]

There are at least four mechanisms whereby the external acidity can be increased after adding suitable substrate to a yeast suspension:

1. The first abrupt drop of pH after adding exogenous substrate is probably due to release of CO_2 from endogenous sources and may be analogous to the acidification observed without substrate. During this period the sugar substrate is removed from the medium but not fully catabolized so that the entire CO_2 production stems from endogenous sources, in analogy with the stimulation observed in many other yeasts. For example, D-xylose added to *Rhodotorula gracilis* is not metabolized for 20–40 min but gives rise to CO_2 production exactly equalling five-sixths of that generated by glucose, but entirely from endogenous sources.[8]

[7] K. Sigler, A. Kotyk, A. Knotková, and M. Opekarová, *Biochim. Biophys. Acta* **643**, 583 (1981).
[8] M. Höfer, A. Betz, and J.-V. Becker, *Arch. Mikrobiol.* **71**, 99 (1970).

Whatever the intensity of the CO_2 release may be the actual pH attained by this process can never reach the observed values of pH 3, as follows from considering the solubility of CO_2 in water at 30° (0.665 ml CO_2/ml H_2O) and its hydration and dissociation to the first degree of H_2CO_3 (pK_a 6.35). Even in a 100% CO_2 atmosphere the lowest possible pH would be 3.94; in an atmosphere of 1% CO_2 it would be 4.94. Hence, other processes must participate in the acidification.

2. The monovalent cation transporter (exchanger) system probably operates during some periods of the acidification process. It may exchange external K^+ for intracellular H^+, or both ions can move in parallel.

3. At later periods after addition of substrate (perhaps 5 – 10 min at 30° in baker's yeast) organic acids begin to appear in the medium, mainly succinic, malic, lactic, and acetic acid.[9] The production is enhanced anaerobically and is greater in the absence of K^+. The highest levels were found under these conditions, with succinate at about 60 μmol per g dry weight of yeast. The transport of these acids appears to be passive, most likely in the undissociated form, moving down their concentration gradient. During the maximum production of organic acids (anaerobically without K^+) the external pH might in fact approach the value of 3 so that no other H^+-extruding system should be required. However, the low pH values are attained even faster aerobically, in the presence of K^+, and certainly before the peak of organic acid concentration has been reached. Hence, another source still must be sought.

4. The major system, extruding H^+ from yeast in a net fashion, is the plasma membrane H^+-ATPase (EC 3.6.1.35). The enzyme, isolated from *Schizosaccharomyces pombe*[10] and from *Saccharomyces cerevisiae*,[11] is an E_1E_2-type ATPase with a single polypeptide chain of 100,000 molecular weight; its pH optimum lies at 6.0 – 6.5. The enzyme shows absolute specificity for ATP, the K_m for MgATP being 1 mM. It is essentially activated by Mg^{2+}, Mn^{2+}, or Co^{2+} and further slightly stimulated by K^+ or NH_4^+. It is inhibited by miconazole (although much less than the mitochondrial H^+-ATPase), vanadate, dicyclohexylcarbodiimide, *p*-hydroxymercuribenzoate, triphenyltin, diethylstilbestrol, and the antibiotic Dio-9.

The process of H^+ transport is an oriented vectorial reaction, starting with ATP hydrolysis on the internal membrane face and terminating with H^+ dissociation on the outside face, the stoichiometry (H^+/ATP) being most probably 1 : 1. Prior to ATP splitting, the β-carboxy group of an

[9] K. Sigler, A. Knotková, J. Páca, and M. Wurst, *Folia Microbiol. (Prague)* **25**, 311 (1980).
[10] J. P. Dufour and A. Goffeau, *J. Biol. Chem.* **253**, 7026 (1978).
[11] F. Malpartida and R. Serrano, *J. Biol. Chem.* **256**, 4175 (1981).

aspartic acid is phosphorylated in the peptide chain, the phosphorylation rate by ATP being 200 times greater than that by inorganic phosphate.[12]

A peculiar property of the ATPase is that it can be activated severalfold *in situ* by incubating yeast cells with glucose, mannose, fructose, and, after adaptation, with galactose or maltose.[13,14] This activation apparently entails a conformational change of the enzyme protein as it persists even after the lengthy isolation procedure. Some of its properties are also altered: it is much more sensitive to vanadate, it becomes partly sensitive to oligomycin, and its K_m(ATP) shifts to 0.3 mM, its pH optimum to 6.7.

It is not quite clear whether the ATPase can function as an exchange transport protein for H^+ and K^+ as some investigators suggest.[15] Similarly, the degree of electrogenicity of the ATPase is still debated. Although the H^+-ATPase is generally accepted as the principal ΔpH and $\Delta\psi$ generator of the yeast cell, in analogy with its established role in the fungus *Neurospora crassa*,[16] the fact remains that the membrane potential difference $\Delta\psi$ of several yeasts tested, such as *Saccharomyces cerevisiae*, *Lodderomyces elongisporus*, and *Rhodotorula glutinis*, is the same or higher in the absence of glucose than in its presence although in its absence the ATPase may not be working at all or at a much lower rate than in its presence.

Techniques Used

pH Determination

The external pH of a yeast suspension is generally estimated with a glass electrode connected to a suitable recorder, the yeast suspension being placed in a jacketed, constant-temperature chamber on a magnetic stirrer. For work in gas phases different from air the suspension should be flushed well before the beginning of measurement and then a gentle stream of the desired gas should flow over the surface of the suspension.

A useful technique for measuring the fluxes of H^+ is one where the pH is "clamped" in a titrigraph arrangement so that the amount of hydroxide that must be added to keep the pH constant is monitored.[5] A rather different approach must be used to determine the intracellular pH, which is often of importance in assessing the regulatory functions of pH_{out} as well as

[12] A. Goffeau, A. Amory, A. Villalobo, and J.-P. Dufour, *Ann. N.Y. Acad. Sci.* **402**, 91 (1982).
[13] R. Serrano, *FEBS Lett.* **156**, 11 (1983).
[14] H. Sychrová and A. Kotyk, *FEBS Lett.* **183**, 21 (1985).
[15] A. Villalobo, *J. Biol. Chem.* **257**, 1824 (1982).
[16] G. A. Scarborough, *Proc. Natl. Acad. Sci. U.S.A.* **73**, 1485 (1976).

pH_{in} on various shifts of ions. The techniques commonly used are the following.

1. Distribution of a weak acid, such as bromphenol blue[17] or ^{14}C-labeled acetic or propionic acid,[18] according to the pH outside and inside, using the formula

$$pH_{in} = pH_{out} + \log[(c_{in}/c_{out})(1 + 10^{pK_a - pH_{out}}) - 10^{pK_a - pH_{out}}]$$

The procedure is based on measuring the concentration of the probe externally and internally, which is best accomplished by computation from the difference between $c_{out(initial)}$ and $c_{out(final)}$ and recording pH_{out} in the usual way. This is a relatively simple technique, but adsorption or active uptake of the probe may be a problem. Likewise, some of the dyes used may not readily penetrate into cells of some yeast species. A detailed account of a useful continuous method for measuring intracellular pH of unicellular organisms or vesicles is given in Ref. 19.

2. Measurement of the pH-dependent chemical shift of ^{31}P in NMR spectroscopy. This technique has been described in detail in Ref. 20.

3. Measurement of electron spin resonance spectra of pH-sensitive extrinsic spin probes.[21]

4. Use of pH-dependent fluorescence of some dyes, such as quinine, chromotropic acid, or fluorescein.[22] This method has proved superior to the others in terms of economy and reliability, combined with the bonus of making it possible to prepare maps of intracellular pH.[23-24a]

Procedure. Fluorescein diacetate is dissolved in acetone to 10 mM (solution A). A cell suspension at about 1–10 mg dry weight/ml is prepared and incubated under the desired experimental conditions. This may be done conveniently in the quartz cuvette used for fluorescence measurement. Solution A is added to the suspension to achieve a concentration of 20–50 μM. In 10 min the dye will have penetrated into cells and reached equilibrium between the external and intracellular solutions after being hydrolyzed by a cell esterase. Fluorescein is a species that exhibits pH-de-

[17] A. Kotyk, *Folia Microbiol. (Prague)* **8**, 27 (1963).
[18] H. Rottenberg, this series, Vol. 55, p. 547.
[19] S. Ramos, S. Schuldiner and H. R. Kaback, this series, Vol. 55, p. 680.
[20] G. Navon, R. G. Schulman, T. Yamane, T. R. Eccleshall, K. B. Lam, J. J. Baronofsky, and J. Marmur, *Biochemistry* **18**, 4487 (1979).
[21] R. J. Mehlhorn and I. Probst, this series, Vol. 88, p. 334.
[22] J. Slavík, *FEBS Lett.* **140**, 22 (1982).
[23] J. Slavík, *FEBS Lett.* **156**, 222 (1983).
[24] J. Slavík and A. Kotyk, *Biochim. Biophys. Acta* **766**, 679 (1984).
[24a] A. Kotyk and J. Slavík, "Intracellular pH and Its Measurement." CRC Press, Boca Raton, Florida, 1989.

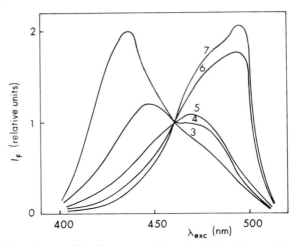

FIG. 3. pH dependence of the fluorescence intensity (I_F) of fluorescein after excitation with light of different wavelengths, measured at 520 nm and normalized at 460 nm. Numbers at curves denote pH.

pendent fluorescence at substantially greater intensity than fluorescein diacetate (Fig. 3). Moreover, fluorescein crosses yeast cell membranes very slowly so that most of it remains in cells for minutes or tens of minutes; consequently, fluorescence can be read without haste.

The spectrofluorometer cuvette is placed in a special adapter so that the excitation beam does not strike it normal to the front face but at an angle of about 30°. The fluorescence is recorded at a right angle to the incident beam and, in this arrangement, is not disturbed by reflected light, inner filter effects, or light scattering. This arrangement can be used even with opaque suspensions as the fluorescence is recorded from the surface suspension layer only. The cuvette is then illuminated with light at 435 nm while the fluorescence reading is taken at 512 nm and, within 10 sec, with light of 490 nm, while another fluorescence reading is taken at 512 nm. The logarithm of the ratio of intensities is compared with the calibration curve shown in Fig. 4, and the intracellular pH is read directly from it.

If for some reason fluorescein diacetate penetrates into cells to a small extent it is advisable to centrifuge or filter the cells after a 20-min preincubation with the dye and resuspend them in the same volume of the original medium. The essential advantage of the technique lies in the fact that the amount of fluorescein in cells is of no importance as long as it gives a reliable fluorescence signal. It is the ratio of fluorescence intensities, not the intensity itself, that is measured and used for pH estimation.

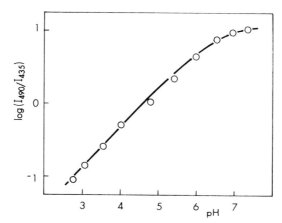

FIG. 4. Calibration curve for intracellular pH.

To obtain information on the pH values near the plasma membrane one can analyze the detailed topography of intracellular pH down to compartments about 200 nm on a side, as imposed by the laws of optics. The following procedure was found to be useful.

A yeast cell suspension incubated in the presence of fluorescein diacetate that contains sufficient fluorescein intracellularly is placed on a microscope slide. A Fluoval fluorescence microscope with a high-pressure 200-W mercury lamp, thermal filter, UV filter, and high-quality interference filters with maximum transmission at 425 and 466 nm and a yellow–green cutoff filter OG 4 (C. Zeiss, Jena, GDR) is used in our laboratory. A suitable yeast cell is then photographed in sequence after illumination through the two interference filters, and the photographs are developed to equal blackness. Because of its superior densitometric proportionality the Agfapan-Vario XL film (Agfa, FRG) is used. The two photographs are then compared digitally dot by dot using the computer-assisted Texture Analysis System of Leitz (FRG), and the difference in optical density for each dot is evaluated against a calibration curve prepared as follows. Fluorescein solutions in buffers of different pH placed in a shallow well on a microscope slide are illuminated through the same excitation filters with maximum transmission at 425 and 466 nm, and the fluorescence intensities of the whole microscope field are compared. The calibration curve obtained is shown in Fig. 5.

Taking all extraneous influences into account, such as buffer composition, ionic strength, viscosity of the medium, and uncertainty of the densitometric evaluation, the pH values obtained lie within ±0.2 pH units of

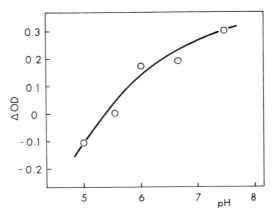

FIG. 5. Difference in the optical density (ΔOD) between two negatives exposed with 425 and 466 nm excitation filters, dependent on pH.

the true value. Their relative positions within the range of pH 5–7 are accurate to 0.03 pH unit.

Spectrophotometry of Proton Release

A method for determining the acidification rate, of limited use but highly sensitive particularly at low pH values, is that based on the color change of a suitable acid–base indicator, such as methyl red.[25] For reactions where proton liberation occurs at a pH not very distant from the pK of the indicator (5.0 for methyl red) the change in absorbance of the dye can be used as a quantitative measure of proton concentration.

Procedure. The indicator solution is made by dissolving 50 mg methyl red in 10 ml of 20 mM NaOH and diluting to 250 ml with deionized water. If an excess of NaOH occurs, this must be neutralized with HCl by carefully adding 20-μl portions of 0.1 M HCl to a known volume of the indicator solution in a spectrophotometric cuvette until a change of absorbance at 540 nm occurs. The amount of HCl added up to the detection of the absorbance change is used for calculating the amount of HCl required to neutralize the NaOH in the 250 ml stock solution. A calibration curve is then prepared to determine the value of absorbance which corresponds to a given quantity of H$^+$ ions. A change in A_{540} of 0.055 is roughly equivalent to 10 nmol H$^+$ released by the yeast (Fig. 6).

[25] C. Pascual and A. Kotyk, *Anal. Biochem.* **123,** 201 (1982).

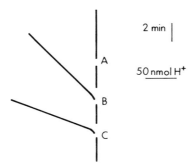

FIG. 6. Changes in indicator absorbance caused by proton liberation from *S. cerevisiae* 196-2 after addition of different sugars (4 μmol sugar to 2.5 ml indicator solution containing 95 mg fresh weight of yeast cells). A, Lactose; B, D-mannose; C, D-glucose.

Conductometry

Conductometry is a highly nonspecific method when applied to measuring the conductivity of yeast suspensions, but it may provide auxiliary support to more specific techniques. Its principle lies in measuring the current in an aqueous solution flowing between two inert (platinum) electrodes that are supplied with an alternating current at 300 Hz through a Wheatstone bridge. The conductivity is registered as the difference potential between experimental and reference conductance. At the frequency used the cells are completely unresponsive to the field and do not contribute to the current passing between the electrodes owing to ions present in the solution.[26]

Commercial conductometers designed for measuring the purity of water may be conveniently used. A 5-ml glass chamber is used in our laboratory. It is placed on a magnetic stirrer which allows simultaneous recording of conductivity, pH, and K^+ concentration, using a selective glass electrode.

In addition to its fast response the conductometric device has the advantage that it records changes of conductivity linearly over a broad range of ion concentrations. Thus, it is possible to observe a conspicuous increase of conductivity arising from H^+ ions extruded from cells following glucose addition, even at pH 3.5 when the pH changes in terms of pH milliunits and the pH meter shows no difference beyond background noise.

[26] A. Kotyk and R. Metlička, *Stud. Biophys.* **110,** 205 (1986).

Estimation of Organic Acids

There are a number of specific ways of determining the presence of acids, such as succinic, lactic, and malic acid, in the filtrates of yeast suspensions, based on manometry,[27] enzyme-catalyzed color reactions,[28] etc., but apparently the method of choice is gas chromatography (GC).[29]

Procedure. Samples of yeast suspension filtrate or supernatant (0.5 ml) are freeze-dried. The lyophilized material (about 2 mg) is dissolved in a mixture of 0.4 ml bis(trimethylsilyl)-trifluoroacetamide and 0.1 ml trimethylchlorosilane, containing 0.5 μg D-arabinitol. If the sample contains much glucose or a similar sugar which, after freeze-drying, dissolves only with difficulty in the silylation agents, the mixture is combined with 0.4 ml absolute pyridine. After 2 hr at room temperature, the samples can be used directly for GC analysis.

The apparatus used in our laboratory is a Perkin-Elmer F21 gas chromatograph with a flame ionization detector and a Siemens Compensograph electronic recorder. The column is 3×2000 mm in size and is packed with Chromosorb G-AW-DMCS (particle size 0.15–0.17 mm) as the support and 2.5% SE-52 silicone rubber as the stationary phase. The mixture, either experimental or of identically prepared standards, is separated in a temperature interval of 80–220° at a rate of 3°/min, using 240° for the sample injector and nitrogen as carrier gas at a rate of 20 ml/min. Samples 1–4 μl in volume are generally suitable.

Continuous Estimation of Carbon Dioxide

The amount of dissolved carbon dioxide is most conveniently estimated with a specific CO_2 electrode, the function of which is based on measuring pH by a glass electrode surrounded by a membrane which specifically permits the passage of CO_2 molecules to the interior. The acidity caused by the dissolved CO_2 arises through reaction with water and subsequent dissociation of carbonic acid,

$$CO_2 + H_2O \rightleftharpoons H_2CO_3 \rightleftharpoons H^+ + HCO_3^-$$

As the dissociation is extremely rapid, one may consider the overall equilibrium to be expressed by

$$K_{eq} = [CO_2]/[H^+][HCO_3^-] = 2.12 \times 10^6 \qquad M^{-1}$$

[27] A. Kleinzeller, "Manometrische Methoden." Fischer, Jena, 1965.
[28] H.-U. Bergmeyer, "Methods in Enzymatic Analysis." Verlag Chemie, Weinheim, 1983.
[29] M. Wurst, K. Sigler, and A. Knotková, *Folia Microbiol. (Prague)* **25,** 306 (1980).

which, in distilled water, yields

$$pH = 3.19 - 0.5 \log[CO_2]$$

In spite of the applicability of this theoretical relationship, commercial CO_2 electrodes must be calibrated for the particular conditions of measurement as indicated by the supplier.

The measurement of CO_2 concentration using such electrodes can be done continuously, but there may be some lag involved due to the diffusion of CO_2 to the electrode, and care should be exercised to take suitable corrections for the effects of higher concentrations of ions. The lower limit of sensitivity of the electrodes is about $10^{-5} M CO_2$, approximately the value expected in distilled water left to equilibrate with the atmosphere.

[36] Ion Transport in Yeast Including Lipophilic Ions

By George W. F. H. Borst-Pauwels

Introduction

Studies dealing with cation uptake in yeast cells focus mainly on two major problems: the mechanism of ion transport and the energization of this transport. When dealing with problems of ion transport, mechanism elucidation can be afforded by a thorough study of the kinetics of ion transport.[1] In order to study energy transduction, however, it is essential first to gain knowledge of the membrane potential by determining the distribution of lipophilic ions between cells and medium. In this chapter we discuss the problems in methodology that one may encounter in obtaining data suitable for a proper kinetic analysis and in obtaining reliable data about the membrane potential.

Determination of Ion Uptake in Yeast

The uptake of an ion by a yeast cell can be determined in several ways. First, the uptake may be calculated from the decrease in the concentration of the ion in the medium. A correction should be made for ions binding to the cell wall or cell membrane and for cation accumulation in the cell wall space owing to the negative Donnan potential of this space. As the ion uptake into the cell proceeds, the ion concentration in the medium decreases. As a result, extracellular binding and accumulation in the Donnan

---

[1] G. W. F. H. Borst-Pauwels, *Biochim. Biophys. Acta* **650**, 88 (1981).

METHODS IN ENZYMOLOGY, VOL. 174

space are reduced. This decrease in ion concentration of the medium will be counteracted, and, consequently, the initial uptake rate of the ion when calculated simply from the time course of the decrease of its concentration in the medium will always be underestimated.

Second, for a more accurate determination of initial rates of uptake the cells should be filtrated through a suitable filter disk, and both the adhering medium and the externally adsorbed ions removed. This can be done effectively by mixing the yeast sample with the washing medium just before filtration.[2-4] A suitable procedure[2] for determining uptake of radioactive ions is as follows: A glass cylinder (2.5 cm external diameter and 10 cm height) is placed on top of a sintered glass funnel containing a filter disk of 2.7 cm diameter. The cylinder is kept in place by means of a heavy metal ring. Just before filtering the yeast sample, 20 ml of an ice-cold solution of 50 mM MgCl$_2$ (for monovalent cation uptake[2]) or 50 mM EDTA, pH 8.5 (for polyvalent cation uptake[4]) is added to the cylinder; immediately thereafter 2 ml of the yeast sample is added. Suction is started just before addition of the yeast suspension. After filtration to apparent dryness, 1 ml of ice-cold distilled water is added. After reaching apparent dryness the yeast cells may be thoroughly dried by addition of 2 ml of acetone. With nonlipophilic ions no detectable loss of radioactivity is caused by washing with acetone. Uptake of the ions involved is calculated according to Eq. (1).

$$Q = \frac{5AR(S_t)}{4AT(DW)} \qquad \text{mmol kg}^{-1} \qquad (1)$$

DW is the dry weight of the residue (in grams) obtained on centrifuging 5 ml of the yeast suspension and washing the cells once with distilled water. AR and AT are the radioactivities of the yeast on the filter papers and of 0.5 ml of the suspension, respectively. S_t is the total concentration of the substrate ion in the suspension (in millimolar).

The concentration of substrate ion in the medium (S_m) reached after equilibration of the ion between the cell wall space and medium before appreciable uptake into the cell may be somewhat lower than S_t.[5] The value of S_m can be calculated from Eq. (2).

$$S_m = \frac{(AS + AR/4)S_t}{AT} \qquad (2)$$

[2] G. W. F. H. Borst-Pauwels, P. Schnetkamp, and P. van Well, *Biochim. Biophys. Acta* **291**, 274 (1973).
[3] M. Boutry, F. Foury, and A. Goffeau, *Biochim. Biophys. Acta* **464**, 602 (1977).
[4] G. M. Roomans, A. P. R. Theuvenet, T. P. R. van den Berg, and G. W. F. H. Borst-Pauwels, *Biochim. Biophys. Acta* **551**, 187 (1979).
[5] M. Borbolla and A. Peña, *J. Membr. Biol.* **54**, 149 (1980).

AS is the radioactivity of 0.5 ml of yeast supernatant obtained on centrifuging an aliquot of the yeast suspension almost immediately after addition of the radioactive substrate ion. Correction for ion uptake into the cells is made by determining *AR* at the same time. If *AR* is relatively high, S_m will be overestimated, as the accumulation of the ion in the cell wall will decrease during ion uptake (see above). In that case S_m can be better determined with nonmetabolizing cells.

Third, for ion uptake in protoplasts, cellulose filters are not well suited, and Millipore filters are preferable. Another suitable procedure[6] consists of centrifuging 0.5 ml of yeast suspension through a 0.2-ml layer of silicone oil covered by 0.5 ml of an ice-cold solution of either 50 mM MgCl$_2$ or 50 mM EDTA in Eppendorf vials for 15 sec in a microfuge at 10,000 g. After centrifugation the buffer is removed by aspiration. One milliliter ice-cold stop buffer is added, and this is again removed by aspiration. The oil layer is also removed, and the precipitate is resuspended in 0.6 ml distilled water; 0.5 ml of this suspension is dissolved in 4.5 ml scintillation fluid for determination of the radioactivity. Silicone oils suitable for this procedure are Alcatel 214 from VG Instruments (Nieuw Vennep, The Netherlands)[6] and MS 702/AR 50 3/1 from Wacker Chemie (Munich).[7] If the buffer anion forms complexes with the substrate cation, which may occur, e.g., on applying succinate-, citrate-, or phosphate-containing buffers,[8] appropriate corrections should be applied for the decrease in free cation concentration.

Finally, for studies of ion uptake under approximately constant conditions Rothstein and Bruce[9] developed the so-called column technique. This technique can also be applied for studying cation release from yeast as it prevents reentry of the released cation.[10] A yeast suspension containing 300–1600 mg (wet weight) of yeast is placed in a 30-ml Büchner-type funnel provided with a sintered glass of medium porosity. The medium is withdrawn by gentle suction. The surface of the yeast pad is covered by a filter paper 3 cm in diameter and a wad of glass wool to prevent channeling within the packed yeast as the medium passes through the yeast under pressure (either air or nitrogen) at a rate of 2–7 ml/min. It is not necessary to use a buffer to keep the pH constant between pH 2 to 5 even in the presence of glucose. The pH can be adjusted with HCl.

[6] R. A. Gage, W. van Wijngaarden, A. P. R. Theuvenet, G. W. F. H. Borst-Pauwels, and A. J. Verkleij, *Biochim. Biophys. Acta* **812**, 1 (1985).
[7] M. E. Bianchi, M. L. Carbone, and G. Lucchini, *Plant Sci. Lett.* **22**, 345 (1981).
[8] K. B. Yatsimirskii and V. P. Vasil'ev, "Instability Constants of Complex Compounds." Van Nostrand, Princeton, New Jersey, 1966.
[9] A. Rothstein and M. Bruce, *J. Cell Comp. Physiol.* **51**, 145 (1958).
[10] A. Rothstein, *J. Gen. Physiol.* **64**, 608 (1974).

Kinetics of Uptake, Derivation of Initial Rates of Uptake

For obtaining information about the mechanism of transport, initial rates of ion uptake are generally needed. Because rate equations normally deal with influx rates rather than net uptake rates, influx rates are determined at varying concentrations of substrate ion or competing ion. They should be referenced to the concentration of free substrate ion in the medium near zero time but after the ion is equilibrated between cell wall and medium [S_m in Eq. (2)]. This also applies to concentrations of competing ions. The uptake should be determined at relatively short time intervals, in many cases at 10-sec intervals, otherwise deviations from linearity may occur in the time course of uptake, thus complicating the calculation of the initial rates of uptake.

Deviations from linearity may be due to the following:

1. Progressive decrease in the concentration of free substrate ion in the medium owing to uptake of the ion into the cell. This complication can be avoided for the most part by reducing the yeast concentration.

2. Saturation of the cell with the substrate ion. On loading the cell with the substrate ion, efflux of the ion from the cell progressively contributes to the net uptake rate. Initial rates of uptake (v_0) may be obtained by a first-order fit [see Eqs. (3) and (4)].

$$\ln(A_e - A_t) = \ln(A_e - A_0) - kt \tag{3}$$

$$v_0 = k(A_e - A_0) \tag{4}$$

A_0, A_t, and A_e are the ion concentrations in the cell at zero time, at time t, and at equilibrium distribution, respectively. k is a rate constant.

3. Biphasic uptake isotherms. Sr^{2+} and Ca^{2+} uptake isotherms are complicated because uptake of these two ions into the cytosol rapidly reaches a steady state while at the same time the net influx rate also decreases rapidly. On the other hand, uptake of the ions from the cytosol into vacuoles persists over a relatively long period at a rather slow rate.[11,12] In these cases the rapid initial uptake may be easily overlooked, and initial rates of uptake may therefore be underestimated.

4. Changes in cell pH. Rb^+ and K^+ influx into the yeast causes alkalinization of the cells.[13] Since the rates of Rb^+ and K^+ uptake decrease with

---

[11] B. J. W. M. Nieuwenhuis, C. A. G. M. Weyers, and G. W. F. H. Borst-Pauwels, *Biochim. Biophys. Acta* **649**, 83 (1981).

[12] Y. Eilam, H. Lavi, and N. Grossowicz, *J. Gen. Microbiol.* **131**, 623 (1985).

[13] J. P. Ryan and H. Ryan, *Biochem. J.* **128**, 134 (1972).

increasing cell pH,[13,14] the influx rate of the cations will decrease during accumulation of the ions into the yeast cell. On the other hand, phosphate uptake leads to acidification of the cells.[15] The rate of phosphate uptake shows an optimum at pH 6.8 in the cell.[16] A decrease in the pH during phosphate uptake from a value exceeding pH 6.8 will lead to a transient increase in phosphate uptake. If, however, the initial cell pH is lower than 6.8, the influx rate will show a decrease only as the cells become more acidified. If the pH of the medium is high, deviations from linearity in the uptake isotherm occur after a 30-sec incubation of the cells with phosphate.

5. Feedback regulation of monovalent cation uptake by the cellular cation content. Besides the increase in the pH of the cell, the increase in monovalent cation content of the cells occurring during Rb^+ or K^+ uptake leads to a decrease in the influx rate of ^{86}Rb on loading the cells with either K^+ or Rb^+.[17]

Changing Cation Composition of Cells

Normally yeast cells have a high K^+ content and a low Na^+ content. It is possible to prepare cells in which most of the K^+ is replaced by another cation. Small amounts of cellular Na^+ can be replaced by K^+. Cells in which the cation content has been changed are important in studies of the efflux mechanism of the ions involved.

The procedures are as follows: Yeast [5% (w/v)] is washed with distilled water and shaken at room temperature for successive 2-hr periods in 200 mM NH_4Cl, KCl, $RbCl$, or $CsCl$ for making NH_4^+, K^+, Rb^+, or Cs^+ yeast or in 200 mM sodium citrate or lithium citrate for preparing Na^+ or Li^+ yeast, in the presence of 5% (w/v) glucose.[18-21] The yeast is centrifuged and washed every 2 hr. The pH is adjusted to 7.4 with 200 mM Tris-Cl. Ca^{2+} yeast is prepared from Na^+ yeast by incubating the cells in 0.2 M calcium acetate and 5% (w/v) glucose.[19]

[14] A. P. R. Theuvenet, G. M. Roomans, and G. W. F. H. Borst-Pauwels, *Biochim. Biophys. Acta* **469,** 272 (1977).

[15] M. Cockburn, P. Earnshaw, and A. A. Eddy, *Biochem. J.* **146,** 705 (1975).

[16] G. W. F. H. Borst-Pauwels and P. H. J. Peters, *Biochim. Biophys. Acta* **466,** 488 (1977).

[17] A. W. Boxman, A. P. R. Theuvenet, P. H. J. Peters, J. Dobbelmann, and G. W. F. H. Borst-Pauwels, *Biochim. Biophys. Acta* **814,** 50 (1985).

[18] E. J. Conway, H. Ryan, and E. Carton, *Biochem. J.* **58,** 158 (1954).

[19] E. J. Conway and H. M. Gaffney, *Biochem. J.* **101,** 385 (1966).

[20] E. J. Conway and P. T. Moore, *Biochem. J.* **57,** 523 (1954).

[21] E. J. Conway and J. Breen, *Biochem. J.* **39,** 368 (1941).

Determination of Distribution of Ions Inside Cell

The polybase DEAE-dextran specifically disrupts the plasmalemma and leaves the other internal membranes intact.[22] In the presence of 0.7 M sorbitol, cell organelles like mitochondria and vacuoles will still remain intact, and only cytosolic solutes are extracted from the cell. The procedure[22] is as follows: 2×10^8 cells are collected on a membrane filter and washed 5 times with 10 mM 2-(N-morphilino)ethanesulfonic acid (MES)–Tris buffer, pH 6, at 0° (or 3 times with 20 mM $MgCl_2$ for studying Ca^{2+} compartmentalization[12]) and 3 times with buffer to which 0.7 M sorbitol is added. The filters with cells are resuspended in 3 ml of buffered sorbitol in plastic tubes. Plastic tubes are used because DEAE-dextran-treated cells stick to glass walls. One-half milliliter of 0.2% (w/v) DEAE-dextran is added, and the suspension is mixed immediately. After 30 sec at 0° the cell suspension is filtered and washed 3 times with 1 ml buffered sorbitol. The total filtrate (6 ml) contains the ions extracted from the cytosol. For extracting the vacuoles the remaining cells are washed 3 times with 1 ml of 60% v/v methanol and 3 times with 1 ml distilled water. The total filtrate (6 ml) contains the ions which are present in the vacuoles and which are not firmly bound to cellular constituents. The bound ions can be released by incubating the cells obtained after extraction with methanol with 3 ml of 0.5 M $HClO_4$ at 60° for 10 min and washing the cells 3 times with 1 ml distilled water.[12]

Suitable Conditions for Studying Repression or Derepression of High-Affinity Phosphate–Proton Cotransport System

The rate of phosphate uptake via the derepressible high-affinity phosphate–proton cotransport system greatly depends on the cell pH.[16] Both K_m and V_{max} show an optimum at pH 6.8. The quotient V_{max}/K_m, however, is independent of the cell pH. The quotient of the rate of uptake and the phosphate concentration at infinitely low phosphate, which can be approximated by applying tracer amounts of ortho[^{32}P]phosphate, equals V_{max}/K_m. Under these conditions the contribution of the low-affinity carrier is also negligibly small. Therefore, the relative rate of phosphate uptake is a measure of the concentration of the high-affinity carrier, and no corrections for changes in cell pH should be applied.

Corrections for Effect of Changes in Cell pH on Cation Uptake

Uptake of both monovalent and divalent cations depends on the pH of the cell.[4,13,14] In kinetic studies of cation uptake, corrections should be

---

[22] V. Huber-Wälchli and A. Wiemken, *Arch. Microbiol.* **120**, 141 (1979).

applied for pH effects. For this purpose a standard curve of uptake rate versus cell pH can be made by varying the pH at a fixed substrate concentration. The exact concentration is not important, since the relative effect of cell pH is independent of concentration. The pH can be varied[13,14] by altering the time of preincubation of the cells with glucose, by changing the gas phase (air or nitrogen), by changing the substrate (glucose, ethanol, or propanol), or by adding a weak acid (propionic or butyric acid) up to concentrations which do not affect glycolysis.

Accounting for Effects of Changes in Surface Potential on Ion Uptake

Ion uptake greatly depends on the surface potential (ψ) of the cells.[1] Owing to the normally negative surface potential of the cells, cations are concentrated at the cell membrane interphase and anions are expelled from this interphase. This leads to an increase in the apparent affinity of cations for the carrier and to a decrease in the apparent affinity of anions. One can account for the effect of ψ on ion uptake by replacing the bulk aqueous phase concentrations of both substrate ion and competing ions (s) by the interphase concentrations s_0 given by Eq. (5).

$$s_0 = s \exp(-Fz\psi/RT) \tag{5}$$

z is the valency of the ion, and R, F and T have their usual meaning.

In *Saccharomyces cerevisiae* the surface potential near transport sites is approximately 10 times higher than the zeta potential determined by means of microelectrophoresis.[23] This is mainly due to the fact that the charges on the cell membrane are not uniformly distributed. A reasonable estimate of ψ can be obtained from studies of the binding of the cationic dye 9-aminoacridine (9AA) to yeast cells.[24] Care should be taken that uptake of 9AA into the cells is minimalized. At low pH 9AA can be translocated into the yeast cells by means of the derepressible thiamin carrier. This uptake can be reduced by using nonmetabolizing cells. At high pH uptake of the nonprotonated form of 9AA occurs. Incubation of the cells with 9AA should therefore be as short as possible. Equilibration between free 9AA and bound 9AA is normally established within about 20 sec. The relative amount of 9AA bound to the cell membrane is given by Eq. (6).

$$(F_0 - F_m)/F_m = B/F_m = a \exp(-F\psi/RT) \tag{6}$$

a is a temperature-dependent constant. F, R, and T have their usual

[23] A. P. R. Theuvenet and G. W. F. H. Borst-Pauwels, *Biochim. Biophys. Acta* **734**, 62 (1983).
[24] A. P. R. Theuvenet, W. M. H. van de Wijngaard, J. W. van de Rijke, and G. W. F. H. Borst-Pauwels, *Biochim. Biophys. Acta* **775**, 161 (1984).

meaning. F_m and F_0 are the fluorescence intensities of 9AA determined at the excitation/emission pair 400/454 nm in the supernatant obtained after mixing equal volumes of 10% (w/v) yeast in distilled water and 2 μM 9AA in the desired buffer or in 150 mM CaCl$_2$ in 0.15 N HCl, respectively. In the latter mixture binding of 9AA to the cells is almost completely prevented. The final pH should be maintained at approximately 1.5. Provided that the zeta potential is zero when ψ is zero, the value of a can be obtained by plotting B/F against the zeta potential and extrapolating to zero potential.[24] Then ψ can be calculated from Eq. (6).

Extracellular and Intracellular Binding of Monovalent Cation

Cations can be trapped inside the cell owing to binding to intracellular constituents. A measure for this intracellular binding can be obtained by determining the distribution of a cation between permeabilized cells and medium after correction for binding to extracellular constituents and for the contribution of the Donnan potentials of the extracellular and intracellular space to the ion distribution. The latter contribution can be determined by means of ^{86}Rb, which is not supposed to bind appreciably to yeast cell constituents.

Extracellular Binding of a Cation. Ten milliliters of 10% (w/v) nonmetabolizing yeast in distilled water is mixed with 10 ml of 500 mM KCl in 100 mM HEPES–Tris of appropriate pH containing 30 mM deoxyglucose, 30 μM antimycin, and carrier-free radioactive cation L or [86]Rb. Immediately after establishing distribution equilibrium, the cells are centrifuged and the radioactivity of 1 ml of the supernatant (AS) is determined. The remainder of the supernatant is removed and the radioactivity of the residue determined (AR). The relative amount of cation bound to the cell walls and outer cell membrane (index e) is given by Eq. (7).

$$B_{L,e} = (AR/AS)_{L,e} - (AR/AS)_{Rb,e} \tag{7}$$

By varying the time of preincubation of the cells one can control whether the equilibrium distribution is to be established and whether appreciable uptake of L into the cells is to occur.

Intracellular Binding of the Ions. The same procedure as described above is applied to a 10% (w/v) yeast suspension which is boiled for 1 min before use. The relative amount of L trapped inside the boiled cells (index i) is found by subtracting $B_{L,e}$ from the relative amount of L bound to the boiled cells, $(AR/AS)_{L,b}$, and making appropriate corrections for accumulation of free cation L in both the cell wall space and the intracellular space by subtracting the relative amount of Rb$^+$ accumulated into the boiled cell residue [Eq. (8)].

$$B_{L,i} = (AR/AS)_{L,b} - (AR/AS)_{Rb,b} - B_{L,e} \qquad (8)$$

Boiling the cells may unmask potential binding groups inside the cell which would lead to an overestimation of intracellular binding.[25]

Determination of Membrane Potential by Microelectrodes

Vacata et al.[26] have shown that on inserting a microelectrode into the large yeast *Endomyces magnusii* a rapid transient increase in negative membrane potential (measured from inside to outside) occurs, which is followed by a small, rather stable, but negative potential. This relatively stable potential was considered to be the true membrane potential of the cell, whereas the potential at the peak was considered to be an artifact. Bakker et al.,[27] however, have shown that the stable potential is independent of the pH of the medium and is not affected by the addition of protonophores at low pH of the medium, whereas the peak potential depends greatly on the medium pH and collapses on adding a protonophore. Therefore, the peak potential reflects the real membrane potential, and the subsequent low potential is an artifact.

Determining the membrane potential consists essentially of puncturing a single cell by means of a suction pipette with a glass electrode filled with either 3 M KCl or 0.5 M K$_2$SO$_4$ (40 or 100 MΩ resistance, respectively). Even on applying a compensation for the capacity of the electrode, the membrane potential may be underestimated as the potential after reaching the peak value may decline too rapidly ($t_{0.5}$ 1 msec) for accurate measurement. For a further discussion of this methodology, see [39] in this volume.

Determination of Membrane Potential with Quaternary Phosphonium Compounds

In yeast cells smaller than *E. magnusii* only indirect methods for the determination of membrane potential can be applied. The most common method consists of determining the equilibrium distribution of the lipophilic cation tetraphenylphosphonium (TPP) between cell and medium

[25] A. W. Boxman, J. Dobbelmann, and G. W. F. H. Borst-Pauwels, *Biochim. Biophys. Acta* **772**, 51 (1984).

[26] O. Vacata, A. Kotyk, and K. Sigler, *Biochim. Biophys. Acta* **643**, 265 (1981).

[27] R. Bakker, J. Dobbelmann, and G. W. F. H. Borst-Pauwels, *Biochim. Biophys. Acta* **861**, 205 (1986).

and calculating the membrane potential on applying the Nernst equation. [Eq. (9)].

$$E = (RT/F) \ln(C_{TPP,cell}/C_{TPP,medium}) \qquad (9)$$

Two different methods are applied for the study of TPP distribution. In the first, the concentration of TPP in the medium is determined either directly in the suspension by means of a TPP-sensitive electrode[28] or in the supernatant obtained by centrifuging the yeast suspension[29] by means of liquid scintillation of [^{14}C] or [^{3}H]TPP. The uptake of TPP is calculated from the difference in the concentration of TPP added initially and the TPP concentration at equilibrium distribution. In this method corrections should be made for the amount of externally bound TPP. In the second procedure, filtering the cells and washing them as described above for monovalent cation uptake remove the greater part of externally bound radioactive TPP, and the uptake of TPP in the cells can be determined directly.

Complications in the interpretation of data regarding the steady-state distribution of phosphonium compounds may arise for several reasons:

1. Intracellular binding of phosphonium compounds. In *Rhodotorula glutinis* the equilibrium distribution of TPP between cells and medium greatly exceeds that of triphenylmethylphosphonium (TPMP).[29] This may be traced to differences in the extent of binding of these two cations, indicating that negative membrane potentials calculated from the equilibrium distribution of TPP between cells and medium are overestimated. At this stage no reliable method for the determination of the binding correction exists. The magnitude of the binding correction greatly depends on the method applied[25] (see Table I).

2. Internal compartmentation of phosphonium compounds. The uptake of TPMP in *S. cerevisiae* under aerobic conditions is much greater than under anaerobic conditions owing to accumulation of TPMP into the mitochondria.[30] This leads to an overestimation of the membrane potential across the plasmalemma.[31] It is quite probable that this also applies to TPP uptake. A second complication may arise because of compartmentation if the electrical potential across the tonoplast measured from vacuole

[28] T. Shinho, N. Kamo, K. Kurihara, and Y. Kobatake, *Arch. Biochem. Biophys.* **187**, 414 (1978).
[29] R. Hauer and M. Höfer, *J. Membr. Biol.* **43**, 335 (1978).
[30] M. Midgley and C. L. Thompson, *FEMS Microbiol. Lett.* **26**, 311 (1985).
[31] R. J. Ritchie, *Prog. Biophys. Mol. Biol.* **43**, 1 (1984).

TABLE I
VALUES OF BINDING CORRECTIONS FOR
CALCULATION OF MEMBRANE POTENTIAL OF
S. cerevisiae STRAIN DELFT II

Method applied	Correction terms (mV)
TPP binding to boiled cells	75 ± 1
TPP binding to frozen and thawed cells	49 ± 7
TPP uptake in depolarized cells	19 ± 9

to cytosol is positive as suggested.[32,33] This will give rise to an underestimation of the electrical potential across the plasmalemma.

3. The existence of pathways other than simple diffusion for the uptake of quaternary ammonium or phosphonium compounds. Dimethyldibenzylammonium (DDA) and TPMP have affinity for the derepressible thiamin carrier of some strains of yeast.[34] On incubating the cells with glucose the carrier is derepressed, and a great increase in the rate of influx of the two cations is found. The equilibrium distribution of DDA between cells and medium shows a pH dependence quite different from that of TPP (which is not translocated into the yeast cells by means of the thiamin carrier). Therefore, apparent membrane potentials calculated from DDA or TPMP distribution are not reliable in this strain of yeast. In *S. cerevisiae* strain D27 TPMP uptake does not proceed by means of the thiamin carrier.[30] It is hypothesized that in eukaryotic cells an efflux system for TPMP is also operative, giving rise to an underestimation of the membrane potential.[35]

4. Effect of the growth phase of the yeast cells on TPP uptake. The phase of growth of the yeast cells may have a tremendous effect on TPP uptake. In the exponential phase, TPP uptake is very small until the glucose added to the medium is consumed, whereupon TPP uptake increases greatly.[36]

[32] L. P. Lichko and L. A. Okorokov, *FEBS Lett.* **174**, 233 (1984).

[33] Y. Kakinuma, Y. Ohsumi, and Y. Anraku, *J. Biol. Chem.* **256**, 10859 (1981).

[34] P. W. J. A. Barts, H. A. Hoeberichts, A. Klaassen, and G. W. F. H. Borst-Pauwels, *Biochim. Biophys. Acta* **597**, 125 (1980).

[35] R. J. Ritchie, *J. Membr. Biol.* **69**, 57 (1982).

[36] P. Eraso, M. J. Mazón, and J. M. Gancedo, *Biochim. Biophys. Acta* **778**, 516 (1984).

5. Occurrence of all-or-none effects. Most inhibitors of membrane ATPases act by way of an all-or-none process in yeast.[37] Some cells abruptly lose their K^+ almost completely and concomitantly shrink to a great extent, whereas the remainder gradually lose K^+ and also shrink in a gradual way. If TPP distributes to different extents among the medium and the two "cell size" populations, a mean membrane potential only is obtained, and no information about the membrane potentials of each single cell population is available.

Indications of an all-or-none K^+ loss can be obtained by determining the cell-size distribution of an originally single cell-size population after incubating the cells with a membrane ATPase inhibitor.[37] Yeast (40 g wet weight) is suspended in 1000 ml distilled water in a 1000-ml cylinder and kept overnight at 4°. Then the upper 100 ml is carefully collected without disturbing the suspension. It should be verified that this upper layer contains only small-sized cells and consists of a single cell-size population. For cell-size determinations, 0.1 ml yeast suspension containing 10^6 cells is added to 10 ml 0.9% (w/v) NaCl. Cell sizes are determined in this suspension by means of a Coulter counter (Model ZF) equipped with a size-distribution analyzer (Model P64). Maximal shrinkage of cells is found in yeast samples to which 125 μM cetyltrimethylammonium bromide is added before dilution with 0.9% NaCl. An all-or-none effect will lead either to a second peak in the cell-size distribution pattern or to an increase in the relative width (RW) of the cell-size distribution curve. RW equals the width of the curve at half-maximal height divided by the mean cell size. When approximately 50% of the cells are shrunk according to an all-or-none process RW has its optimal value. When all cells are shrunk to the maximal extent, RW becomes of the same order of magnitude as the RW of control cells.

A second method for detecting all-or-none K^+ loss consists of determining the distribution of the K^+ content of single cells by means of electron-probe X-ray microanalysis.[38] The procedure is as follows: Samples of yeast are washed 2 times with distilled water and resuspended in distilled water. The cell suspension is sprayed through a glass capillary tube of 0.5 cm internal diameter onto polished pure carbon disks of 12 cm diameter and 0.8 mm thickness over 5 sec at a distance of 20 cm. The plates are air-dried and stored under vacuum. They are coated with a carbon layer to improve conductivity and kept under vacuum. X-Ray analysis is performed with a scanning electron microscope provided with an energy-dis-

[37] G. W. F. H. Borst-Pauwels, A. P. R. Theuvenet, and A. L. H. Stols, *Biochim. Biophys. Acta* **732**, 186 (1983).
[38] G. A. J. Kuypers and G. M. Roomans, *J. Gen. Microbiol.* **115**, 13 (1979).

persive spectrometer. In addition to the K^+ content of the cell, the P and S contents are determined. Relative K^+ contents are expressed as K/P or K/S. These relative K^+ contents show less scattering than the absolute K^+ contents. Though the P contents can be analyzed more accurately than the S contents, determination of the S content is sometimes preferred since some ATPase inhibitors cause a significant loss of P but not of S. All-or-none K^+ loss results in the occurrence of two populations of K/P or K/S contents instead of one as is found in controls.

Application of Carbocyanines for Determination of Membrane Potential

Various fluorescent cationic carbocyanines which are distributed between cells and medium according to the Nernst potential are applied for detecting qualitative changes in the magnitude of the membrane potential.

3,3′-Dipropylthiodicarbocyanine [diS-C$_3$-(5)]. Accumulation of the dye 3,3′-dipropylthiodicarbocyanine into the cytosol or mitochondria is accompanied by quenching.[39,40] To a 0.5% (w/v) yeast suspension in an appropriate buffer is added 0.4 μM dye dissolved in methanol.[40] Energization of the mitochondria is suppressed by means of 8 μg/ml antimycin. Eventually, energy transduction from ATP formed glycolytically to mitochondrial ATP is impaired by 300 μM bongkrekic acid. Hyperpolarization is accompanied by a decrease in fluorescence intensity, and depolarization has the reverse effect. Complications may arise from the fact that the dye can adhere to the wall of the cuvettes which also leads to quenching. Fluorescence is measured at 622 and 670 nm for excitation and emission, respectively. In *Saccharomyces fragilis* there are indications that the probe depolarizes the cells at concentrations exceeding 0.25 μM in a 0.1% (w/v) yeast suspension.[41]

3,3′-Dipropylthiacarbocyanine [diS-C$_3$-(3)]. Accumulation of 3,3′-dipropylthiacarbocyanine into mitochondria also causes quenching.[42] Accumulation into the cytosol, however, leads to an increase in fluorescence intensity. The dye (0.25 μM) is added to a 1.25% (w/v) buffered yeast suspension. The fluorescence intensity is measured at the wavelength pair 540–590 nm. Uptake of dye by the cells is determined by measuring the fluorescence intensity of the supernatant obtained on centrifuging the yeast

[39] A. A. Eddy, R. Philo, P. Earnshaw, and R. Brocklehurst, *FEBS Symp.* **42**, 250 (1977).
[40] L. Kováč and L. Varečka, *Biochim. Biophys. Acta* **637**, 209 (1981).
[41] P. J. A. van der Broek, K. Christianse, and J. van Steveninck, *Biochim. Biophys. Acta* **692**, 231 (1982).
[42] A. Peña, S. Uribe, J. P. Pardo, and M. Borbolla, *Arch. Biochem. Biophys.* **231**, 217 (1984).

suspension. This intensity is increased by the addition of 0.1% sodium dodecyl sulfate (SDS) or Triton X-100. Inside the cells the dye is strongly adsorbed to intracellular constituents.

Application of Lipophilic Anions for Determination of Membrane Potential

Anions may also be distributed between the cell and medium according to the Nernst potential, allowing for the calculation of this potential. Appreciable accumulation allowing a quantitative determination of the membrane potential is expected only when membrane potentials are positive.

Application of Rhodanide. As pointed out by Ritchey[31] one should be very cautious when applying rhodanide as a probe for negative membrane potentials. HCNS has a pK of 2.17. Therefore, at low external pH entrance of the undissociated acid into the cells may occur. If the pH of the cell is higher than the pH of the medium, the anion will accumulate in the cells. Consequently, accumulation of rhodanide at low pH[43] is not necessarily due to the existence of a positive membrane potential between cells and medium. On the other hand, at relatively high external pH uptake of rhodanide into, for example, vacuolar membrane vesicles[33] or inside-out plasma membrane vesicles[44] may give valuable information about the magnitude of the positive potential.

Application of Oxanol. The fluorescent anion oxanol V shows quenching of fluorescence when it is accumulated into vacuoles[45] or inside-out plasmamembrane[46] vesicles in response to a positive potential. The fluorescence is measured at the wavelength pair 580–640 nm.

[43] M. Höfer and A. Künemund, *Biochem. J.* **225**, 815 (1984).
[44] G. A. Scarborough, *Biochemistry* **19**, 2925 (1980).
[45] L. A. Okorokov, T. V. Kulakovskaya, L. P. Lichko, and E. V. Polorotova, *FEBS Lett.* **192**, 303 (1985).
[46] A. Peña, S. M. Clement, and M. Calahorra, *in* "Proceedings of the Third Small Meeting on Yeast Transport and Bioenergetics at Mook" (G. W. F. H. Borst-Pauwels and A. P. R. Theuvenet, eds.), p. 9. Ars drukkerij B. V., Roermond, The Netherlands, 1985.

[37] Sugar Transport in Normal and Mutant Yeast Cells

By VINCENT P. CIRILLO

Introduction

Most studies of sugar transport in yeasts have been carried out with baker's yeast, *Saccharomyces cerevisiae,* which exhibits a variety of sugar transport processes, namely, facilitated diffusion, active transport, and, possibly, group translocation by transport-associated phosphorylation.[1,2] Monosaccharides which can be phosphorylated are transported by a high- and low-affinity process (Fig. 1).[3-5] The high-affinity process depends on the activity of specific, cognate kinases (i.e., hexokinase or glucokinase for D-glucose, hexokinase for D-fructose, and galactokinase for D-galactose), whereas the low-affinity process is independent of kinase activity. Thus, D-glucose is transported by the high-affinity processes in cells containing either hexokinase or glucokinase, whereas D-fructose or D-galactose are transported by high-affinity processes only in cells containing hexokinase or galactokinase, respectively. Nonphosphorylated analogs, except for 6-deoxy-D-glucose, are transported by a single, low-affinity process. In the absence of their cognate kinases, the normally phosphorylated sugars are transported like their nonphosphorylated analogs, namely, by a low-affinity process. 6-Deoxy-D-glucose, unlike other nonphosphorylated analogs, is also transported by a kinase-dependent high-affinity process and a kinase-independent, low-affinity process.[6] Thus, the role of the kinases in high-affinity transport is complex and may be involved in a process other than merely phosphorylation of its substrate.[4]

The high- and low-affinity processes for monosaccharide uptake in *S. cerevisiae* may differ with respect to mechanism as well as kinetics. The low-affinity process is clearly due to a constitutive, facilitated diffusion mechanism; however, the high-affinity process may be mediated by a group translocation mechanism, namely, a transport-associated phos-

---

[1] A. Kotyk, *in* "Membranes and Transport" (A. N. Martonosi, ed.), Vol. 2, p. 55. Plenum, New York, 1982.

[2] V. P. Cirillo, *in* "Current Developments in Yeast Research" (G. G. Steward and I. Russell, eds.), p. 229. Pergamon, New York, 1981.

[3] R. Serrano and G. DelaFuente, *Mol. Cell. Biochem.* **5,** 161 (1974).

[4] L. F. Bisson and D. G. Fraenkel, *Proc. Natl. Acad. Sci. U.S.A.* **80,** 1730 (1982).

[5] J. M. Lang and V. P. Cirillo, *J. Bacteriol.* **169,** 2926 (1987).

[6] L. B. Bisson and D. G. Fraenkel, *J. Bacteriol.* **155,** 995 (1983).

phorylation.[7-9] While the latter mechanism remains controversial for *S. cerevisiae,* there is very strong evidence for its role in the uptake of 2-deoxy-D-glucose by *S. fragilis.*[10]

The uptake of maltose in *S. cerevisiae,* on the other hand, is a proton-symport-mediated, active transport process[11] dependent on the activity of the yeast plasma membrane proton-transporting ATPase. Active transport of sugars by proton symport has also been demonstrated for other yeasts, notably the uptake of D-xylose, L-rhamnose, and D-arabinose by *Rhodotorula gracilis.*[12]

Cell Growth and Preparation

Many studies on sugar transport can been carried out with commercial baker's yeast purchased as a pressed yeast cake from a bakery. The cells in a fresh, pressed yeast cake are virtually 100% viable and can be maintained in a refrigerator at 4° for over 1 week. Cell suspensions of baker's yeast are washed by centrifugation in distilled water at 3000 g in a refrigerated centrifuge at 4°. The washes are repeated until the supernatant is clear. The packed yeasts may be stored at 4° for 2–3 days. Stored yeast should be rewashed before use. Yeast suspensions for sugar transport are usually prepared in distilled water at cell densities based on wet weight. Heavy cell suspensions, approximately 25% by volume, are prepared in distilled water from packed cell suspensions and transferred to tared plastic centrifuge tubes, and the wet weight of the cells is determined from the weight of the tube after centrifugation at 3000 g for 15 min. The cells are then made up to 4–40% suspensions (wet weight/volume) for use in transport experiments.

When commercial baker's yeast is not appropriate for the study at hand, yeasts are easily grown in the laboratory on any scale from test tube to fermenter. All wild-type and mutant yeasts can be grown on rich, undefined media containing yeast extract, peptone, and appropriate sugars or other carbon sources. The most commonly used medium contains 1% yeast extract, 2% peptone, and 2% sugar and is usually designated as YPx media; for example, YPD medium refers to dextrose (= D-glucose)-containing YP medium. Yeasts are usually grown on YP media in conical flasks on rotary shakers at 30°. Supplemented with 2% agar, YP media can

[7] J. Van Steveninck, *Biochim. Biophys. Acta* **274,** 575 (1972).
[8] S. A. Meredith and A. H. Romano, *Biochim. Biophys. Acta* **497,** 745 (1977).
[9] A. J. Franzusoff and V. P. Cirillo, *Biochim. Biophys. Acta* **688,** 295 (1982).
[10] H. T. A. Jaspers and J. Van Steveninck, *Biochim. Biophys. Acta* **506,** 370 (1975).
[11] R. Serrano, *Eur. J. Biochem.* **80,** 97 (1977).
[12] P. C. Misra and M. Hofer, *FEBS Lett.* **52,** 95 (1975).

be used to maintain yeast stock cultures and as starter cultures for inoculation of liquid media. Concentrated chemically defined Wickerham media[13] are commercially available as yeast nitrogen base and yeast carbon base from Difco. These media are prepared by filter sterilization and supplemented with a carbon or nitrogen source, respectively. They can also be solidified with 2% agar.

The cells are grown in shake flasks at 30° overnight, harvested, washed by centrifugation as described above, and suspended at 4–40% wet weight per volume. The relative activity of the high-affinity D-glucose transport system is a function of the extracellular glucose concentration and the sugar concentration of the medium at the time of harvest since the activity is repressed at high D-glucose concentrations and derepressed at low concentrations.[14]

Sample Protocol

Transport is customarily studied in washed cells suspended in distilled water or buffered salt solutions. Radiolabeled sugars are added to the cell suspensions, and samples are filtered and washed on glass microfiber or membrane filters. Glass microfiber filters are less expensive and allow much higher filtration rates than membrane filters. The following is a sample protocol for a single time point at an external sugar concentration of 1 mM. With an apropriate increase in volume, multiple samples can be taken for a time course study. For initial velocities, incubation times of 5 and 15 sec are used for metabolizable and nonmetabolizable sugars, respectively.

Materials

40% (wet weight/volume) cell suspension in distilled water in an ice bath

2 mM sugar solution (1 μCi ^{14}C/μmol)

Distilled water in an ice bath set up with an automatic syringe to dispense 10-ml samples of water

Test tubes (12 × 75 mm) with minimagnetic stirring bars

Magnetic stirrer (for incubations at temperatures other than ambient, the submersible TRI-R model is very convenient)

Vortex mixer

Glass microfiber filters (e.g., Reeve-Angel or Whatman)

---

[13] L. T. Wickerham, *J. Bacteriol.* **52,** 95 (1945).
[14] L. B. Bisson and D. G. Fraenkel, *J. Bacteriol.* **159,** 1013 (1984).

25-cm filter holder attached to a vacuum flask (the Millipore holder with a stainless steel screen gives very rapid filtration rates)
Scintillation vials and scintillation fluid
Automatic pipettes to dispense 10- and 100-μl samples

Procedure

1. Add 60 μl vortexed 40% cell suspensions to test tubes in a water bath at room temperature or 30°.
2. Maintain cells in suspension with magnetic stirrer.
3. Temperature equilibrate cell suspensions and sugar solutions for 5 min.
4. At zero time add 60 μl of 2 mM labeled sugar solution, resulting in a reaction mix of 20% cell suspension in a 1 mM sugar solution in a final volume of 120 μl.
5. Vortex and incubate for the desired time.
6. Add 10 ml ice-cold distilled water over filters on filtration apparatus with vacuum turned off.
7. Transfer 100-μl samples to distilled water over the filters.
8. Filter cells by turning on vacuum.
9. Wash cells with 5 ml ice-cold distilled water.
10. Transfer filters with cells to scintillation vials.
11. Transfer 10 μl of the reaction mix to a separate scintillation vial labeled "Specific Activity." (This volume of external medium is equal to the volume of intracellular water in the cells contained in 100 μl of a 20% suspension.)
12. Add scintillation fluid and count.

Determining the C_i/C_o Ratio

Uptake can be expressed as amount of sugar per unit of cell wet weight, dry weight, protein, cell number, or cell water; however, uptake expressed as amount of substrate per unit of cell water immediately provides the substrate concentration gradient between the cell and the extracellular medium (see below). The following relationships between cell water and the other measures of cell concentration for *S. cerevisiae* have been determined by the author but will vary with yeast strain and growth conditions:

Wet weight	1 g
Cell water	500 μl
Dry weight	200 μg
Protein	100 μg
Cell number	1×10^{10}

In the experimental protocol described above (Step 7), the 100-μl samples contain 20 mg wet weight of cells which represents 10 μl of cell water. The ratio of radioactivity (counts/min) in the filtered cells and in the 10 μl of suspension medium (designated as "Specific Activity" in Step 11 above) gives the sugar concentration ratio, C_i/C_o, between cells and the medium. The intracellular concentration can then be obtained by multiplying the extracellular concentration by the C_i/C_o ratio.

A C_i/C_o ratio significantly greater than one for a nonmetabolized, nonderivatized sugar represents preliminary evidence for active transport. Confirmation of active transport requires (1) evidence that the sugar is in fact *not* derivatized and (2) evidence that the observed accumulation against a gradient is energy dependent. The former is established by autoradiography or autofluorography of cell extracts,[9] the latter by inhibition by uncouplers (e.g., 2,4-dinitrophenol, 60–250 μM, or carbonyl cyanide *m*-chlorophenylhydrazone, 50 μM).[11]

Transport Kinetics

Serrano and DelaFuente[3] were the first to recognize that the uptake of metabolizable monosaccharides exhibits complex kinetics which could be resolved into two components representing a high- and low-affinity component. A thorough genetic analysis of the high-affinity component has been carried out by Bisson and Fraenkel,[4,14] who showed that the high-affinity component is dependent on the presence of cognate kinases. Mutants lacking the cognate kinases exhibit only the low-affinity component. Figure 1 illustrates these effects in an Eadie–Hofstee plot of glucose uptake by a wild-type strain of *S. cerevisiae* (Fig. 1A) and a kinaseless mutant (Fig. 1B). Such studies emphasize the necessity of using a wide range of substrate concentrations and appropriate analysis of the data to discover the effects of inhibitors and mutations on the complex kinetics of sugar uptake in yeast.

Sugar Transport Mutants

Transport mutants have been reported for galactose,[15-17] glucose,[18] and maltose[19] for *S. cerevisiae*. The genes for each of these transporters have

[15] H. C. Douglas and Condie, *J. Bacteriol.* **68,** 662 (1953).
[16] S.-C. Kuo and V. P. Cirillo, *J. Bacteriol.* **103,** 679 (1970).
[17] J. F. Tschopp, S. D. Emr, C. Field, and R. Schekman, *J. Bacteriol.* **166,** 313 (1986).
[18] L. F. Bisson, L. Neigenborn, M. Carlson, and D. G. Fraenkel, *J. Bacteriol.* **169,** 1656 (1987).
[19] M. G. Goldenthal, J. D. Cohen, and J. Marmur, *Curr. Genet.* **7,** 195 (1983).

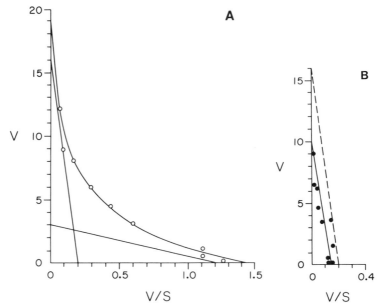

Fig. 1. Eadie–Hofstee plot of initial rate of D-[^{14}C]glucose uptake from external D-glucose concentrations ranging from 0.6 to 200 mM. (A) Uptake by wild-type yeast cells resolved into components of high (shallow slope) and low (steep slope) affinity. (B) Uptake by kinaseless yeast mutants. The dashed line is a reproduction of the low-affinity component of wild-type uptake in (A). (From Ref. 5, with permission.)

been cloned and sequenced: *GAL2* for galactose,[20] *SNF3* for glucose,[21] and *MAL61* for maltose.[22] The *LAC12* gene for lactose transport in *Kluyveromyces lactis*[23] has also been cloned and sequenced. All of these transporters belong to a common transporter superfamily which includes the *Escherichia coli* arabinose and xylose permeases, the human Hep-G2/erythrose and fetal muscle transporters, and the rat brain, liver, and adipocyte transporters (see Ref. 22 for further references).

[20] K. Szkutnicka, J. F. Tschopp, L. Andrews, and V. P. Cirillo, *J. Bacteriol.* **171**, 4486 (1989).
[21] J. L. Celenza, L. Marshall-Carlson, and M. Carlson, *Proc. Natl. Acad. Sci. U.S.A.* **85**, 2130 (1988).
[22] Q. Cheng and C. A. Michels, *Genetics* (in press).
[23] Y.-D. Chang and R. C. Dickson, *J. Biol. Chem.* **263**, 16696 (1988).

[38] Transport of Amino Acids and Selected Anions in Yeast

By A. ALAN EDDY and PAMELA HOPKINS

Introduction

The absorption of diverse amino acids by *Saccharomyces* species or *Candida* species has been widely studied in terms of mechanism and genetics. The respective mechanisms of phosphate and nitrate uptake have likewise been explored, especially, as with the amino acids, in relation to the notion that transport is driven by the proton gradient across the yeast plasma membrane.[1] Several techniques have been used for this purpose, notably (1) assay of the progress of absorption of the solute labeled with ^{14}C; (2) comparison of the uptake of solute with the simultaneous net flow of protons into the yeast and concomitant outflow of K^+; (3) study of the fluorescence changes occurring during solute uptake in the presence of voltage-sensitive carbocyanine dyes[2]; (4) selection of mutant yeast strains with altered transport characteristics.

The possibility of studying solute transport in vesicles derived from yeast plasma membrane has not been realized owing to the difficulty of preparing stable vesicles from this organism. Nevertheless, the proton pump of the plasma membrane has been reconstituted into liposomes, and, furthermore, intact vesicles derived from the vacuolar membrane have been shown to transport amino acids.[3]

Assaying Progress of Amino Acid Absorption

There is an extensive literature showing that the rate of amino acid uptake by yeast depends not only on the genetic constitution of the organism but also on its physiological state, reflecting, among other factors, whether the cells are in exponential growth and whether deamination of the amino acid occurs.[4] For instance, activity of the general acid permease tends to be repressed in the presence of NH_4^+, a preferred nitrogen source for yeast, and expressed when proline or, with certain yeasts, glutamate is the source of nitrogen.[5] Similarly, growth of yeast in a typical nutrient

[1] A. A. Eddy, *Adv. Microb. Physiol.* **23**, 1 (1982).
[2] A. A. Eddy and P. G. Hopkins, *Biochem. J.* **231**, 291 (1985).
[3] Y. Ohsumi and Y. Anraku, *J. Biol. Chem.* **256**, 2079 (1981).
[4] J. R. Woodward and V. P. Cirillo, *J. Bacteriol.* **130**, 714 (1977).
[5] W. E. Courchesne and B. Magasanik, *Mol. Cell. Biol.* **3**, 672 (1983).

broth leads to partial repression of several amino acid permeases. In the following procedures the storage, cultivation, and assay of the yeast cells are described.

Stock Cultures. Stock cultures are maintained at 2° on nutrient agar slopes (MYGP agar): 1% (w/v) glucose, 0.3% (w/v) malt extract, 0.5% (w/v) mycological peptone, 0.3% (w/v) yeast extract, and 2% (w/v) agar. Cultures (10 ml) for use are prepared in a similar medium (MYGP) lacking agar at 25°, with a portion of the native culture being transferred at intervals of 2–3 days to fresh growth medium.

Preparation of Washed Yeast for Assay. A portion of the yeast grown in the liquid MYGP medium at 30° is transferred to a defined mineral salts solution comprising the following (which are sterilized before being mixed): solution A, 8 ml; glucose, 2 g in 10 ml of water; $(NH_4)_2SO_4$, 0.4 g in 10 ml of water, or an alternative nitrogen source; water to 200 ml final volume. This sterile solution is mixed with 2 ml of solution B. Alternative sources of nitrogen are 3 mM proline or 3 mM sodium glutamate. Solution A is kept frozen and contains, per liter: $MgSO_4 \cdot 7H_2O$, 12.5 g; KH_2PO_4, 40 g; trace element solution C, 25 ml. Solution B is sterilized separately and contains, per liter: *meso*-inositol, 320 mg; calcium D-pantothenate, 320 mg; thiamin-HCl, 64 mg; pyridoxin-HCl, 64 mg; nicotinic acid, 64 mg; *p*-aminobenzoic acid, 32 mg; D-biotin, 0.2 mg; folic acid, 0.1 mg. Solution C contains, per liter: boric acid, 0.1 g; $ZnSO_4 \cdot 7H_2O$, 40 mg; ammonium molybdate, 20 mg; $MnSO_4 \cdot 7H_2O$, 40 mg; $CuSO_4 \cdot 5H_2O$, 45 mg; $FeSO_4 \cdot 7H_2O$, 25 mg.

The yeast cultures are agitated, with access to air, for 17–40 hr in a reciprocal shaker (180 cycles/min) at 30° and harvested in the mid-log phase (A_{610} about 0.4). The yeast is collected at 4° by centrifugation and washed with water (3×20 ml per 300 mg of yeast dry weight). Such preparations are stable for several hours at 4°

Assay of Glycine Uptake in Suspension of Washed Yeast

The washed yeast cells (5–50 mg dry weight) are suspended in 5 ml of 5 mM Tris base brought to pH 5 with citric acid and containing 1% (w/v) glucose, 2 mM KCl. After incubation for 5 min at 30°, glycine uptake is initiated by addition of the amino acid (0.05–5 μmol; labeled with ^{14}C, 0.1 μCi). Portions of the cell suspension (0.1–1.0 ml) are withdrawn at intervals, and yeast cells are separated in one of two ways. (1) A portion containing less than 1 mg of yeast is filtered (Whatman GF/A glass fiber filters) and washed with water (3×2 ml). (2) Alternatively, the sample (100 μl) is mixed with water (1.3 ml) at 0°, the yeast being separated by centrifugation in an Eppendorf microfuge at 10,000 g for 1 min. The yeast is then washed with water (1.3 ml) at 4°.

The yeast collected as a pellet (in centrifugation procedures) or on the glass filter (after filtration) is then assayed for ^{14}C in a liquid scintillation counter. Study of the initial rate of glycine uptake as a function of the extracellular glycine concentration is done under conditions where a negligible fraction of the glycine available is absorbed into the yeast.

Assay of Proton Uptake and Potassium Ion Outflow in Presence of Glycine[6]

Stoichiometrical amounts of protons are absorbed along with glycine or other amino aids into washed yeast cells, their detection and assay being contingent on the following four conditions.[7] (1) Ejection of absorbed protons through the proton pump is prevented. This is achieved by depletion of cellular ATP in the presence of antimycin and 2-deoxyglucose. (2) A compensating outflow of K^+ from the depleted yeast can occur. One way of demonstrating this is to add a proton conductor such as 0.1 mM 2,4-dinitrophenol. (3) The rate of amino acid uptake and of the accompanying protons by the depleted yeast is sufficiently large to be detected in the presence of the basal inflow of protons. As the latter is typically about 1 – 2 nEq of protons/min/mg of cellular dry weight at pH 4.8, at least a similar and preferably a faster amino acid uptake is required. (4) The amount of yeast present in the assay system is sufficiently large to absorb the amount of amino acid which causes a detectable influx of protons. In the recommended procedure utilizing 50 mg dry weight yeast, about 0.05 – 0.2 μmol of amino acid is absorbed in 1 – 2 min. Amino acid absorption by the yeast cells depleted of ATP apears to be limited by internal acidification, due in part to the operation of the amino acid proton symport itself. After about 10 min, the depleted yeast cells kept at pH 4.8 thus lose the ability to absorb amino acid rapidly, owing to the spontaneous uptake of protons in exchange for cellular K^+.

Procedure. The yeast suspension (5 – 8 ml) is put in a water-jacketed vessel at 30° and stirred magnetically. The vessel also contains a glass pH electrode, a K^+-selective glass electrode, and a calomel electrode, which serves as reference electrode for both ions. If a combined pH and reference electrode is used (e.g., Type GK 2321C, Radiometer A/S, Copenhagen, Denmark), the reference can also serve for the monitoring of K^+. Problems arising from the leakage of K^+ from the reference electrode into the system

---

[6] R. Brocklehurst, D. Gardner, and A. A. Eddy, *Biochem. J.* **162,** 591 (1977).
[7] A. Seaston, G. Carr, and A. A. Eddy, *Biochem. J.* **154,** 669 (1976).

can be overcome by using NaCl instead of KCl as a bridging solution in the electrode. The electrical output from the electrodes is fed to two high-impedance voltmeters and then to a suitable chart recorder with two pens displaying the two signals as a function of time. The width of the chart recorder paper typically corresponds to 0.3 pH unit, the changes in pH associated with the absorption of an amino acid representing about 0.03 pH unit. In order to minimize electrical artifacts arising from stirring of the yeast suspension, the electrodes are cautiously moved into optimum positions over the stirrer, which are found by trial. [An alternative approach is to use larger volumes (15 ml) of yeast suspension, where the positioning of the electrodes is less critical.]

1. Proton uptake and K^+ outflow are assayed by suspending the yeast (50 mg dry weight) in 5 mM Tris (5–8 ml) adjusted to pH 4.8 with citric acid and containing 0.2 mM KCl. One minute later energy metabolism is inhibited by the addition of 5 μg of antimycin and 5 mM 2-deoxyglucose. Provided energy metabolism is inhibited, the spontaneous uptake of protons into the yeast then settles within 1–2 min to a steady rate, the basal rate. The outflow of K^+ behaves similarly. (It is of some interest that certain yeast preparations that have been exposed to glucose for 1 hr or more, in the absence of a source of nitrogen for growth, exhibit a high rate of endogenous energy metabolism which is inhibited by deoxyglucose and antimycin only after a lapse of 10 min or more.)

2. A small volume (20 μl) of a solution of glycine (0.05–0.2 μmol labeled with ^{14}C) is added next, and the electrode traces are observed. Typically the basal rate of proton uptake is accelerated 2- to 5-fold for 0.5–2 min, and then proton uptake reverts to near basal level. Meanwhile the outflow of K^+ accelerates in parallel. Artifacts associated with the addition of the reagents are detected in controls lacking yeast.

3. Portions (0.1 ml) of the cell suspension are withdrawn, and the cells are separated and washed by filtration on a glass-fiber filter (see above) before being assayed for ^{14}C content.

4. The buffering power of the main yeast suspension in the jacketed vessel is meanwhile calibrated by the addition of 0.5 μmol of HCl and 0.5 μmol of KCl in turn. Each of these causes an immediate displacement of the respective chart trace, which is carefully measured. The amounts of acid and K^+, respectively entering and leaving the yeast, in a given interval of time, can thus be related directly to the corresponding displacements of the traces on the chart recordings. A useful precaution is to calibrate the pH and K^+ electrode responses both before and after adding amino acid. The average value of each of the two calibrations is used in computing the results.

5. In a typical assay the basal rate of proton uptake increases from A units to $(A + B)$ units in the presence of glycine which is itself absorbed at a rate of C units. After $0.5 - 1$ min all the amino acid is absorbed, and the rate of proton uptake falls to A units again. On the assumption that the basal flow of protons continues unchanged in the presence of the amino acid, the proton stoichiometry is the ratio B/C. This value is close to 2 in *Saccharomyces cerevisiae* when glycine is absorbed through the general amino acid permease.[7] A number of other genetically distinct amino acid symport mechanisms exhibit a stoichiometry of about 1 proton per amino acid equivalent absorbed.[1] Recent results call into question this simple interpretation of the latter observations, which may conceivably reflect the relatively low rate of absorption of the amino acids.[8] Because all these yeasts can absorb more than 95% of two or three successive small portions (0.1 μmol) of amino acid, added under the recommended conditions, the total proton absorption that each of these portions causes is a convenient measure of the stoichiometry of the system.

6. During glycine uptake the stoichiometry ascertained using a K^+ electrode is approximately the same as the proton stoichiometry. This is in contrast to the uptake of either the lysine cation or that of arginine, which results in an additional equivalent of K^+ leaving the yeast cells in excess of the proton inflow. Similarly, proton absorption with glutamate is larger than the displacement of K^+ by an amount equal to the equivalent of glutamate absorbed.

7. It has been suggested that membrane depolarization during glycine absorption might lead to serious underestimation of the true magnitude of the proton stoichiometry of the symport.[9,10] The argument is that during electrogenic solute uptake, the basal flow of cosubstrate ions (assumed to be electrogenic) is smaller than its flow before the solute is added, so that the actual flow through the symport is underestimated by the procedure in Step 5 above. However, in a system where the inflow of H^+ is neutralized by an outflow of K^+ the relevant correction is unlikely to lower the apparent stoichiometry to 1 from a true value of 2 or 3 as has been suggested. Thus, the same workers observed with a certain yeast that a basal proton flow of about 1 unit increased to about 3.3 units in the presence of glycine, the latter being absorbed at a rate of about 2.3 units.[10] The apparent stoichiometry is therefore about 1 proton per glycine mole-

[8] A. A. Eddy and P. Hopkins, *Biochem J.* **251**, 115 (1988).

[9] A. A. Eddy, *in* "Intestinal Permeation" (M. Kramer and F. Lauterbach, eds.), p. 332. Excerpta Medica, Amsterdam, 1977.

[10] A. Ballarin-Denti, J. A. den Hollander, D. Sanders, C. W. Slayman, and C. L. Slayman, *Biochim. Biophys. Acta* **778**, 1 (1984).

cule absorbed. If the true stoichiometry were 2 protons per glycine the basal flow would need to be -1.3 units in the presence of glycine, an inversion of the direction of the basal proton flow, which implies that the proton gradient itself is inverted. Such a notion is quite inconsistent with the role of H^+ in simultaneously driving the observed uptake and concentration of glycine into the very same yeast preparation. A simple extension of this argument shows that the proton stoichiometry in this experiment was certainly not larger than $3.3/2.3 = 1.4$. A detailed analysis of these factors has been reported.[8]

Use of Chemostat to Obtain Substantial Amounts of Yeast Exhibiting High Levels of Permease Activity

Demonstration of the coupling between proton uptake and inflow of phosphate, sulfate, or nitrate depends on the preparation of yeast cultures exhibiting activities of the respective permeases that are much larger than those found when the respective nutrients are available in excess of the requirements for growth.[2] When the yeast (*S. cerevisiae* or *Candida utilis*) is grown in a chemostat under conditions where provision of phosphate, for instance, limits the growth rate, the subsequent rate of uptake of phosphate by the washed yeast cells can approach 10 nmol/min/mg dry weight and is associated with a proton uptake of up to 15 times the basal rate. Analogous observations of accelerated proton uptake in the presence of nitrate have been made with preparations of *Candida utilis* grown with nitrate as the limiting nutrient in a chemostat.[2] Phosphate-limited chemostat cultures[11] are fed at about 60 ml/hr with a modified mineral salts nutrient medium containing 0.1 mM KH_2PO_4 and 10 mM KCl instead of the standard amount of KH_2PO_4 (see above). The yeast culture is maintained at pH 4.7 by the addition of 0.1 M KOH. It is stirred at 250 rpm and given 0.8 liter/min sterile air. The outflowing yeast is collected at $0°$ over 15 hr, washed with water, and proton absorption in the presence of phosphate assayed exactly as described above for glycine.

[11] M. Cockburn, P. Earnshaw, and A. A. Eddy, *Biochem. J.* **146,** 705 (1975).

[39] Accumulation of Electroneutral and Charged Carbohydrates by Proton Cotransport in *Rhodotorula*

By MILAN HÖFER

Introduction

The uptake of most nutrients, either organic molecules or inorganic ions, in prokaryotic and eukaryotic microorganisms as well as in cells of higher plants depends on the free energy of a proton electrochemical potential difference, $\Delta\tilde{\mu}_{H^+}$. Transport systems driven by $\Delta\tilde{\mu}_{H^+}$ are called secondary active and have been reviewed recently.[1-4] It is well known that each of the two components of $\Delta\tilde{\mu}_{H^+}$, the membrane potential ($\Delta\psi$) and the concentration gradient of H$^+$(ΔpH) across the plasma membrane, is alone capable of driving secondary active transport.

The obligatory aerobic yeast *Rhodotorula glutinis* (*Rhodosporidium toruloides,* mating type a, ATCC 26194 = CBS 6681) accumulates electroneutral monosaccharides and acyclic polyols as well as positively charged sugar amines and negatively charged glucuronate intracellularly by means of a common plasma membrane-bound carrier driven by $\Delta\tilde{\mu}_{H^+}$.[5-7] Thus, this H$^+$-cotransport system (H$^+$ symport) offers the possibility of testing how the carrier itself and with it the energetics of the translocation process respond to the different natural charge of the substrate. Recently, using the lipophilic cation tetraphenylphosphonium, it became possible to measure the membrane potential $\Delta\psi$[8] and, using the ^{31}P NMR spectra of the intracellular inorganic phosphate, to determine the H$^+$ concentration gradient ΔpH.[9] By varying the physiological conditions (see section on energetics of substrate accumulation below), it is possible to manipulate each of the two components of $\Delta\tilde{\mu}_{H^+}$. Thus, the impact of either $\Delta\psi$ or ΔpH on the translocation process can be ascertained.

The important advantage of *Rhodotorula* being an obligate aerobe is the possibility of suspending its energy metabolism completely by intro-

[1] R. J. Poole, *Annu. Rev. Plant Physiol.* **29**, 437 (1978).
[2] I. C. West, *Biochim. Biophys. Acta* **604**, 91 (1980).
[3] A. A. Eddy, *Adv. Microb. Physiol.* **23**, 2 (1982).
[4] A. Kotyk, *J. Bioenerg.* **15**, 307 (1983).
[5] R. Klöppel and M. Höfer, *Arch. Microbiol.* **107**, 329 (1976).
[6] C. Niemietz, R. Hauer, and M. Höfer, *Biochem. J.* **194**, 433 (1981).
[7] C. Niemietz and M. Höfer, *J. Membr. Biol.* **80**, 235 (1984).
[8] M. Höfer and A. Künemund, *Biochem. J.* **225**, 815 (1984).
[9] M. Höfer, K. Nicolay, and G. T. Robillard, *J. Bioenerg.* **17**, 175 (1985).

ducing anaerobic conditions. In this respect, no poisoning agents having uncertain side effect(s) are necessary. Moreover, a mutant resistant to the polyene antibiotic nystatin has been isolated and characterized.[10-12] It contains no ergosterol and exhibits a distinctly lower plasma membrane permeability for K^+ and H^+.

The limiting disadvantage of *Rhodotorula* is the severe difficulties in preparing protoplasts,[13,14] the starting material for membrane fractionation. In Appendix 1 a method for preparation of yeast plasma membranes from another yeast, *Metschnikowia reukaufii*,[15] is described. Finally, an electrophysiological method for direct measurement of plasma membrane potential in yeast cells using glass microelectrodes and glass microfunnels to immobilize a single cell is described in Appendix 2. This method has recently been applied in experiments with another obligatory aerobic yeast, *Pichia humboldtii*.[16]

Cultivation of Cells

Rhodotorula glutinis is kept on 2% agar slants containing malt extract, peptone, and glucose (1, 0.5, and 2% w/v, respectively) in a refrigerator and inoculated every 4–6 weeks. Yeast is grown aerobically at 30° on a gyratory shaker in the following medium: 0.066% NH_4NO_3, 0.1% K_2HPO_4, 0.05% NaCl, 0.1% $MgSO_4 \cdot 7H_2O$, 0.025% $CaCl_2 \cdot 2H_2O$, 0.005% $FeCl_3 \cdot 6H_2O$, 0.03% yeast extract (Difco), and 2.5% D-glucose. The pH is adjusted to 5.4 with HCl. A 1- to 4-day-old subculture is used to inoculate 160 ml fresh medium in a culture flask of 1000-ml capacity. The cells are harvested by centrifugation after 24 hr, washed 3 times with distilled water, and aerated for 4–20 hr in distilled water as a 5% suspension (fresh weight/volume). The generation time amounts to 100 min, the average yield to 3–5 g fresh weight/flask (inoculum ~ 10^6 cells/ml culture medium). This procedure has been routinely used in our laboratory to cultivate other yeast species, e.g., *M. reukaufii* and *P. humboldtii*, except that the pH of the growth medium is adjusted to 3.0 and 4.9, respectively.

---

[10] M. Höfer, O. W. Thiele, H. Huh, D. H. Hunneman, and M. Mracek, *Arch. Microbiol.* **132**, 313 (1982).

[11] A. Künemund and M. Höfer, *Biochim. Biophys. Acta* **735**, 203 (1983).

[12] M. Höfer, H. Huh, and A. Künemund, *Biochim. Biophys. Acta* **735**, 211 (1983).

[13] M. von Hedenström and M. Höfer, *Arch. Microbiol.* **98**, 51 (1974).

[14] M. von Hedenström and M. Höfer, *Arch. Microbiol.* **98**, 59 (1974).

[15] B. Aldermann and M. Höfer, *J. Gen. Microbiol.* **130**, 711 (1984).

[16] M. Höfer and A. Novacky, *Biochim. Biophys. Acta* **862**, 372 (1986).

Measurement of Transport Kinetics and Substrate Accumulation

Generally, transport experiments are carried out with a 2.5% cell suspension (fresh weight/volume) in 0.12 M potassium phosphate or Tris–citrate buffer having various pH values (4–8) at 28°–30°. The initial rates of transport are determined either from the increase of the intracellular substrate concentration (see section on filtration method below) or from the decrease of the extracellular substrate concentration (see section on centrifugation method below). The cell dry weight proves to be a dependable reference value for quantifying experimental results.

Filtration Method.[17] Samples of a cell suspension (1 ml) are withdrawn at intervals and filtered through membrane filters (pore size 0.8 μm). The cells on membrane filters are washed twice with ice-cold water and then extracted with 2 ml distilled water in a boiling water bath for 20 min. After cooling, the extracts are deproteinized by subsequent treatment with equal volumes (1 ml) of 5% $ZnSO_4$ and 5% $Ba(OH)_2$, filled up to volume, centrifuged, and aliquots of the supernatant used for substrate determination. When ¹⁴C- or ³H-labeled substrates are used, either aliquots of the supernatants or whole membrane filters with the cell pellet are directly dissolved in scintillation cocktail and measured for radioactivity in a liquid scintillation counter. Both methods lead to comparable results.

Centrifugation Method.[17] Samples of cell suspension (1 ml or less) are centrifuged in a small-volume high-speed centrifuge at 13,000 g for 10–15 sec at room temperature to separate cells quickly from incubation medium. Aliquots of the supernatants are diluted as suitable for substrate determination or used for counting of radioactivity without dilution (cf. filtration method).

Determination of Intracellular Water Volume. For calculations of the intracellular substrate concentrations it is necessary to determine the ratio of cellular dry weight to intracellular water volume. The latter is measured by means of the difference between ³H₂O and hydroxyl[¹⁴C]methylinulin contents of the cell pellet after incubation of yeast cells with the two labeled compounds and separation of cells by membrane filtration (cf. filtration method). Hydroxylmethylinulin serves as an impermeable substance for estimation of the extracellular space (a modified method of Harold *et al.*[18]). The values calculated from the difference in ³H and ¹⁴C radioactivity absorbed by the cells, using a 10% (fresh weight/volume) cell suspension, amount to an average of 2.0 μl of intracellular water per mg dry weight of yeast.

---

[17] M. Höfer and P. Dahle, *Eur. J. Biochem.* **29**, 326 (1972).
[18] F. M. Harold, E. Pavlasova, and J. R. Baarda, *Biochim. Biophys. Acta* **196**, 235 (1970).

Determination of Electrochemical Proton Gradient across Plasma Membrane

The energy stored in the form of an electrochemical gradient of H^+ across the plasma membrane of R. *glutinis* is given by

$$\Delta\tilde{\mu}_{H+} = F(\Delta\psi + 0.060\Delta pH)$$

at $28° - 30°$, where F is the Faraday constant, $\Delta\psi$ the membrane potential (in V), and $\Delta pH = pH_o - pH_i$. $\Delta\tilde{\mu}_{H+}$ is maintained fairly constant over the range of physiological pH_o values between 3.5 and 6, amounting from -17 to -18 kJ mol^{-1}.[9] Above a pH_o of 6 it gradually decreases to -11 kJ mol^{-1} at pH_o 8. The energy stored in $\Delta\tilde{\mu}_{H+}$ is sufficient to energize a 1000-fold accumulation of a transport substrate ($\Delta G° = +17$ kJ mol^{-1}) as observed for D-xylose.[19]

pH Gradient across the Plasma Membrane. Unbuffered aqueous suspensions of R. *glutinis* cells maintain the pH_o of the suspension (1 – 5% fresh weight/volume) constant between 3.8 and 4.5 depending on the particular cell batch used. Several subsequent washings with distilled water do not significantly change the steady-state pH_o.[20] The acidification is completed within a few minutes. To enhance the ionic strength of the aqueous cell suspension during pH_o measurements 0.1 mM $CaCl_2$ is added. When measuring the pH_o, however, care should be taken to use a pH electrode which does not leak K^+ out of the reference electrode since R. *glutinis* cells readily take up K^+ from the suspension in exchange for intracellular H^+.[21] The equilibrium $[K^+]_o$ amounts to $3 - 10$ μM at an average $[K^+]_i$ of 160 mM in fully energized cells. Increasing $[K^+]_o$ above this level induces a K^+ uptake and a stoichiometric extrusion of H^+, eventually leading to a decrease of pH_o.[20,21] The generation of ΔpH is an energy-requiring process. Introducing anaerobic conditions effects an alkalinization of unbuffered cell suspensions and a stoichiometric efflux of intracellular K^+. The passive ion translocation across the plasma membrane is enhanced by adding the uncoupler carbonyl cyanide *m*-chlorophenylhydrazone (proportionally in the concentration range between 1 and 80 μM), indicating that the basic plasma membrane permeability to K^+ is considerably higher than that to H^+.[11]

Determination of Intracellular (Cytoplasmic) pH_i. Whereas extracellular pH_o can be easily measured directly by a glass electrode, for measurements of pH_i only indirect methods were available until recently. As long

[19] A. Kotyk and M. Höfer, *Biochim. Biophys. Acta* **102**, 410 (1965).
[20] P. C. Misra and M. Höfer, *FEBS Lett.* **52**, 95 (1975).
[21] R. Hauer, G. Uhlemann, J. Neumann, and M. Höfer, *Biochim. Biophys. Acta* **649**, 680 (1981).

as pH_i is to be determined in an unbuffered cell suspension, the method using the polyene antibiotic nystatin delivers dependable values.[21] Nystatin interacts with the ergosterol of the plasma membrane to produce nonspecific pores which permit the equilibration of protons (and other ions) across the membrane. Because the inner mitochondrial membrane contains very little sterol it is not affected by nystatin.[22] Hence, this method gives values which should be closer to the true cytoplasmic pH_i than any of those obtained by other indirect methods (see also [34] in this volume).

To an unbuffered 2.5% (fresh weight/volume) cell suspension add 10 μM nystatin (dissolved in 1,3-propanediol). The pH of the suspension is followed until a constant value is reached (3–5 min). One should bear in mind, however, that values obtained by this method are also shifted toward acidic pH owing to the massive influx of protons into the cytoplasm.

The most reliable values of cytoplasmic pH_i can be obtained by using ^{31}P NMR spectra of the cytoplasmic inorganic phosphate.[9] An important advantage of this method is its applicability also in strongly buffered cell suspensions at all pH_o values desired.

In the procedure, a 3.3-ml portion of cell suspension (0.5 g fresh weight/ml) is put into a 10-mm NMR tube. D_2O is added to a final concentration of 10% and, in addition, 20 μl of antifoam (33151, BDH). Aerobic samples are oxygenated continuously during the data-collecting process. The cytoplasmic pH_i is determined from the chemical shift position of the cytoplasmic inorganic phosphate resonance, as described by den Hollander *et al.*[23] The chemical shift position of added glycerophosphoryl-choline is used as a reference. For measurements in the anaerobic state, *R. glutinis* cells are not oxygenated; instead, nitrogen is bubbled through the suspension and the NMR tube is sealed. When cell suspensions are buffered, 125 mM Tris–citrate is used. The pH_o is adjusted to the desired value by small additions of either solid citric acid or Tris base.

Measurement of Membrane Potential $\Delta\psi$. The most dependable method to determine an electrical potential difference across biological membrane is the electrophysiological technique using glass microelectrodes. However, its application to yeast has failed so far because of the small size of yeast cells and the difficulties of immobilizing a single cell. Recent electrophysiological experiments with *P. humboldtii*[16,24] are described in Appendix 2 (p. 650, see also [36] in this volume).

---

[22] M. von Hedenström and M. Höfer, *Biochim. Biophys. Acta* **555**, 169 (1979).

[23] J. A. den Hollander, K. Ugurbil, T. R. Brown, and R. G. Shulman, *Biochemistry* **20**, 5871 (1981).

[24] H. C. Lichtenberg, H. Giebeler, and M. Höfer, *J. Membr. Biol.* **103**, 255 (1988).

For $\Delta\psi$ measurements in yeast suspensions either fluorescent probes[25] or lipophilic cations are used.[26] Neither is without pitfalls. The difficulties with fluorescent probes lie in calibrating the response of fluorescence to the actual membrane potential and, more severe, calibrating the sensitivity of the fluorescence response to effects independent of $\Delta\psi$. The latter may include effects of pH and of ion concentration in the cell suspension and interaction of the probe with various agents added to influence cell metabolism and even with the glass wall of the cuvette. Very little is known about the translocation of the probes across the plasma membrane, the effect of the cell surface charge, and whether the probe molecules are translocated at all. Recent experiments in this laboratory[27] have proved that there is no translocation of the cyanine dye 3,3'-dipropylthiodicarbocyanine [dis-C₃-(5)] across a black-film membrane when an external voltage is applied. On the other hand, liposomes prepared from pure phosphatidylcholine absorbed cyanine dye molecules only because of the hydrophobic interaction between the dye and the phospholipid bilayer. Hence, a change in fluorescence of a fluorescent probe does not necessarily indicate an actual membrane potential or its change.

In *R. glutinis* we have obtained reproducible and dependable results using the lipophilic cation tetraphenylphosphonium (TPP⁺) for negative $\Delta\psi$ and the anion thiocyanate (SCN⁻) for positive $\Delta\psi$.

In the procedure to measure negative $\Delta\psi$ with TPP⁺, because TPP⁺ is nonspecifically adsorbed to the surface of membrane filters the centrifugation method is used to measure the accumulation of the lipophilic cation. Cell suspensions (2.5% fresh weight/volume) are incubated in 20 mM Tris–citrate buffer of a given pH with 5 μM [³H]tetraphenylphosphonium chloride (730 GBq/mol). Samples of 0.4 ml are withdrawn at 1-min intervals and the radioactivity of the supernatants after centrifugation determined by a liquid scintillation counter. All counts per minute (cpm) values are corrected for quench by means of an external standard and the channel ratio method.[28] At the end of each experiment (after 15–20 min), triplicate samples of the cell suspension (100 μl) are taken and counted for total radioactivity added. The decrease of radioactivity in the supernatants is the measure of TPP⁺ taken up by the cells. $\Delta\psi$ is calculated by means of the Nernst equation for diffusion potentials (in mV at 28°–30°):

$$\Delta\psi = 60 \log([TPP^+]_o/[TPP^+]_i)$$

[25] A. S. Waggoner, this series, Vol. 55, p. 689.
[26] E. A. Liberman and V. P. Topali, *Biofizika* **14**, 452 (1969).
[27] U. Wolk and M. Höfer, *Biochem. Int.* **14**, 501 (1987).
[28] J. M. Brewer, A. I. Pesce, and R. B. Ashworth, "Experimental Techniques in Biochemistry," p. 229. Prentice-Hall, Englewood Cliffs, New Jersey, 1974.

where $60 = 2.3RT/F$ and the subscripts o and i correspond to extra- and intracellular spaces, respectively.

Recently, $\Delta\psi$ values calculated from the steady-state distribution of TPP⁺ were compared with direct microelectrode measurements in *Pichia humboldtii*.[24] The values obtained by the two methods agreed within 10 mV under all physiological conditions tested. TPP⁺ uptake in the yeast cells was determined by the filtration method using Schleicher & Schüll filters (Type 602h) which do not adsorb TPP⁺.

In the procedure to measure positive $\Delta\psi$ with SCN⁻, the accumulation of thiocyanate is measured by the membrane filtration technique (see section on filtration method). One micromolar potassium thio[¹⁴C]-cyanate (2.1 TBq/mol) is added to a cell suspension (2.5% fresh weight/volume) in 20 mM Tris–citrate buffer of a given pH. Samples of 1.0 ml are taken at 2-min intervals and filtered; the filters with the cell pellet are dissolved in 10 ml of a scintillation cocktail and counted for radioactivity. At the end of each experiment (after 15–20 min) a portion of cell suspension is centrifuged and the extracellular radioactivity determined in the supernatant. The values of the extracellular and calculated intracellular SCN⁻ concentrations are substituted into the Nernst equation

$$\Delta\psi = -60 \log([SCN^-]_o/[SCN^-]_i)$$

to determine $\Delta\psi$ (in mV at 28°–30°)

The pH profile of $\Delta\psi$ in *R. glutinis* is shown in Fig. 1. The smooth

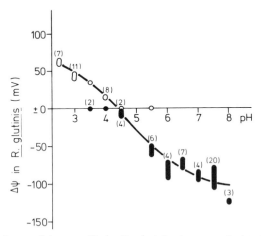

FIG. 1. Dependence of $\Delta\psi$ on pH_o in *R. glutinis*. $\Delta\psi$ was calculated by means of the Nernst equation from the steady-state accumulation of TPP⁺ (for negative $\Delta\psi$ at $pH_o > 4.5$) or SCN⁻ (for positive $\Delta\psi$ at $pH_o < 4.5$). The height of the symbols represents the statistics (±S.E.) of a given number of measurements (in parentheses).

transition of $\Delta\psi$ from negative values at pH > 4.5, as indicated by TPP$^+$ accumulation, to positive values at pH < 4.5, as indicated by SCN$^-$ accumulation (at pH 4.5, $\Delta\psi = 0$, no accumulation of either TPP$^+$ or SCN$^-$), demonstrates the suitability of both TPP$^+$ and SCN$^-$ as true indicators of the actual electrical potential difference across the plasma membrane of *R. glutinis*.[8]

Evidence for a H$^+$–Substrate Cotransport (Symport)

Transport systems catalyzing the accumulation of a substrate in cotransport with protons and driven by the electrochemical H$^+$ gradient have to fulfill the following three criteria: (1) The onset of substrate uptake induces a (transient) alkalinization of slightly buffered (or unbuffered) cell suspension.[29] (2) Since the H$^+$ cotransport is basically electrogenic, the onset of substrate uptake dissipates (transiently) the membrane potential.[30] (3) There is a fixed stoichiometry between the amounts of substrate taken up and H$^+$ cotransported independent of physiological conditions (e.g., temperature).[29]

It should be pointed out, however, that these criteria become detectable only when there is a delay in the response of the onset of H$^+$ cotransport by the plasma membrane-bound ATPase, generating the electrochemical H$^+$ gradient. In *R. glutinis* this delay amounts to 30–60 sec before the H$^+$ pump starts to compensate for the H$^+$ influx (the cause of the transient nature of the effects). It should be mentioned that during this delay time, the measurable uptake of H$^+$ is electrically compensated for, e.g., by a stoichiometric efflux of intracellular K$^+$ (1 : 1) as it is the case in *R. glutinis*. The transient efflux of K$^+$ is a further characteristic feature of a H$^+$–substrate cotransport.[31]

Measurement of (Transient) Alkalinization of Cell Suspensions. Best results, i.e., the highest velocities of H$^+$ cotransport, are obtained with cell cultures grown for 15–17 hr and aerated for 8 hr, or grown for 20 hr and aerated for 24 hr. The H$^+$ cotransport is more pronounced when the aerated cell suspension is washed once by centrifugation immediately before the experiment (to decrease the buffering capacity).

In the procedure, a 2.5% (fresh weight/volume) aqueous cell suspension containing 0.1 mM CaCl$_2$ (to increase the ionic strength of the medium) is stirred on a magnetic stirrer in a water-jacketed plastic vessel maintained at constant temperature. The pH of the cell suspension is monitored by a pH

[29] M. Höfer and P. C. Misra, *Biochem. J.* **172**, 15 (1978).
[30] R. Hauer and M. Höfer, *J. Membr. Biol.* **43**, 335 (1978).
[31] R. Hauer and M. Höfer, *Biochem. J.* **208**, 459 (1982).

electrode (GK201) connected to a pH meter (PHM62) and a Servograph recorder (REC61; all three from Radiometer Deutschland, Krefeld, FRG). The H$^+$ cotransport is started by the addition of chosen concentration of substrate.

In order to calculate the velocity and amount of H$^+$ taken up by cells, the pH change is calibrated by repeated additions of known aliquots of H$^+$ (10–50 μl of 10 mM HCl) at the end of each run. The buffering capacity of aqueous cell suspension may be varied considerably by changing its physiological conditions (e.g., pH). For experiments at pH values higher than 5, the cell suspension is titrated with 10 mM Ca(OH)$_2$ to the desired value. The initial velocity of H$^+$ cotransport is determined by extrapolating the initial linear part of the pH electrode response over an interval and calculating the amount of H$^+$ taken up per mg dry weight of cells per minute using the ΔpH calibration data.

Measurement of K$^+$ Efflux Compensating for H$^+$ Cotransport. As already mentioned, the measurable uptake of H$^+$ induced by the onset of sugar transport is compensated for by a stoichiometric efflux of K$^+$ from the cells. To measure the K$^+$ efflux, samples (1 ml) are withdrawn from the above cell suspension at 20-sec intervals before and after adding sugar and sedimented for 12 sec at 13,000 g in an Ecco-Quick centrifuge (Gollatz, Berlin, FRG). The supernatants are assayed for K$^+$ by atomic absorption spectrophotometry (Perkin-Elmer Bodenseewerk, Überlingen, FRG).[31]

Determination of H$^+$–Substrate Stoichiometry. In order to calculate the stoichiometry of substrate uptake and H$^+$ cotransport, the initial velocity of substrate uptake is measured as follows: From the aqueous cell suspensions used to measure H$^+$ cotransport (see above) induced by the addition of a substrate, samples (1.0 ml) are taken at intervals (30 sec) during the pH recording and treated as described above (see section on filtration method).

Measurement of $\Delta\psi$ Dissipation. Since the H$^+$–substrate cotransport is an electrogenic process, the onset of substrate transport should (transiently) depolarize the plasma membrane. This has been tested in *R. glutinis* as follows: A 4% (fresh weight/volume) cell suspension in 60 mM Tris–citrate buffer, pH 7.5, is incubated with 5 μM [^{3}H]TPP$^+$ as described above (see section on measurement of membrane potential). Samples are taken at intervals over a period of 4 min, then portions of the incubation suspension are separated and added to different concentrations of the tested substrate (either in solid state or in highly concentrated solutions to prevent any considerable dilution of radioactivity). Again, samples are taken at intervals. All samples and at least three aliquots of the incubation suspension are counted for radioactivity. The reduced velocity of TPP$^+$ uptake after addition of the particular substrate concentration is extrapolated from the uptake curves for each substrate concentration. Relating the reduced TPP$^+$ uptake rates to that of a control (TPP$^+$ without substrate) in arbitrary units allows one to calculate the half-maximal depolarizing concentration of the substrate. It should correspond to the half-saturating constant of transport of the particular substrate under the same experimental conditions.[30,31]

Number of Carriers Catalyzing Carbohydrate Transport

Numerous electroneutral and charged carbohydrates, e.g., monosaccharides, both hexoses[19,32] and pentoses,[19,33] disaccharides,[34] acyclic polyols,[5,35] cationic amino sugars,[6] as well as anionic glucuronate,[7] are intracellularly accumulated by *R. glutinis*. There is some controversy in the literature as to whether all substrates share a common plasma membrane-bound carrier.[32,36,37] So far available evidence has been based on kinetic data, which will never be able to solve this problem conclusively. Search for mutants defective in one of the transport systems (if there are more) can provide the only answer. However, our isolates of a glucose transport-defective mutant, selected according to 2-deoxy-D-glucose resistance, all turned out to be defective in D-glucose phosphorylation and not in its transport.[38]

Very recently, we succeeded in isolating a mutant, designated R_{33}, which is defective in D-glucose transport. This mutant is resistant to 2-deoxy-D-glucose and grows in media containing D-glucose as carbon source considerably slower than wild-type cells. An important characteristic of the mutant is the loss of its ability to accumulate D-xylose (which wild-type cells accumulate up to 1000-fold[19]). Kinetic studies revealed that the carrier affinity and not the maximal transport velocity is affected by the mutation. In contrast, the transport of D-fructose (the growth substrate for R_{33}) is equal in the mutant and in the wild strain. These experiments[39] provide evidence for the existence of at least two different carriers for monosaccharides (the glucose and the fructose carriers) in *R. glutinis* plasma membrane.

There is no doubt that some transport systems are inducible, like the

[32] S. Janda, A. Kotyk, and R. Tauchova, *Arch. Microbiol.* **111**, 151 (1976).
[33] M. Höfer, *Arch. Mikrobiol.* **80**, 50 (1971).
[34] S. Janda and M. von Hedenström, *Arch. Microbiol.* **101**, 273 (1974).
[35] R. Klöppel and M. Höfer, *Arch. Microbiol.* **107**, 335 (1976).
[36] M. Höfer and A. Kotyk, *Folia Microbiol. (Prague)* **13**, 197 (1968).
[37] M. Höfer and P. Dahle, *Eur. J. Biochem.* **29**, 326 (1972).
[38] D. Mahlberg, M. Höfer, and A. Täuber, *J. Gen. Microbiol.* **131**, 479 (1985).
[39] B. Milbradt and M. Höfer, *J. Gen. Microbiol.* (submitted).

one for acyclic polyols[35] or the low-affinity transport system for D-xylose.[40] The latter was induced by prolonged starvation of cells. It appeared to be electroneutral and independent of $\Delta\tilde{\mu}_{H^+}$.[31]

Besides the problem described above, evidence has been provided that in *R. glutinis* some monosaccharides, namely, D-glucose, D-galactose, and D-xylose, share a constitutive accumulating transport system with charged glucose derivatives: the cationic glucosamine[6] and the anionic glucuronate.[7] All the substrates mentioned are accumulated in *R. glutinis* cells by a carrier mechanism driven by $\Delta\tilde{\mu}_{H^+}$. The criteria for the carbohydrates sharing a common transport system are: (1) Mutual competitive inhibition of glucosamine and glucuronate transport by monosaccharides and vice versa. In addition, glucuronate and glucosamine exhibited a reciprocal competitive inhibition as well. The inhibition constant for the particular inhibitor correlated with the half-saturation constant for its transport, thus demonstrating a true competition of substrate and inhibitor for a common carrier and not for, e.g., a common energy source. (2) The phenomenon of exchange transport. Monosaccharides brought about a stoichiometric outflow of glucuronate from preloaded cells, as did glucosamine from cells preloaded with D-xylose.[6] The aldopentose has proved to be advantageous for such experiments since it is taken up rapidly (with a half-saturation constant of about 1 mM) and metabolized only after induction of a catabolic enzyme (30 min).[37,41]

Energetics (Driving Forces) of Substrate Accumulation

Rhodotorula glutinis accumulates monosaccharides (up to 1000-fold, D-xylose at pH 4.5[19]), positively charged glucosamine, and negatively charged glucuronate (both up to 100-fold, at pH 6.5[6] and 4.0,[7] respectively) by a common carrier mechanism driven by $\Delta\tilde{\mu}_{H^+}$. Thus, this transport system offers the possibility of testing how the carrier itself and with it the energetics of the translocation process respond to the different natural charges of substrates. The H$^+$ cotransport with differently charged carbohydrates in *R. glutinis* differs distinctly from similar work with amino acids,[3,42] since the uptake of amino acids is mediated by more than one transport system.

The basic principle of a H$^+$ cotransport consists of reversible binding of one or more H$^+$ ions to the membrane-bound carrier protein. This makes the carrier–substrate–H$^+$-translocating system responsive to the electrical

[40] M. E. Alcorn and C. C. Griffin, *Biochim. Biophys. Acta* **510**, 361 (1978).
[41] M. Höfer, A. Betz, and A. Kotyk, *Biochim. Biophys. Acta* **252**, 1 (1971).
[42] D. F. Niven and W. A. Hamilton, *Eur. J. Biochem.* **44**, 517 (1974).

field of the membrane potential and/or to the force of the H^+ concentration gradient. It is less important whether the carrier molecule itself is electroneutral or negatively charged since in both cases one step in the carrier-translocating cycle (either the flow of CSH^+ or that of C^-) is electrogenic. Since charged carbohydrates contribute to the net charge of the translocated molecular species of the carrier system, the possible events summarized in Fig. 2 may be operating during H^+ symport in *R. glutinis*.

Estimation of Apparent Carrier Dissociation Constant. The apparent half-saturation constant of the membrane-bound carrier with protons, pK_a, can be calculated from the pH_o dependence of transport velocity at a

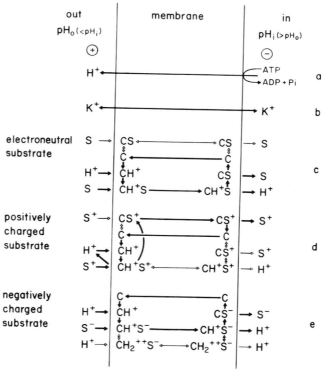

Fig. 2. Scheme of actual and possible translocation events in the plasma membrane of *R. glutinis*. (a) ATP-driven electrogenic H^+ pump generating a membrane potential (inside negative) and a pH gradient (inside alkaline); (b) charge translocation compensating K^+ flow (probably through specific, $\Delta\psi$-controlled channels); (c–e) possible events during uptake of electroneutral substrate (S), cationic S^+, and anionic S^-, respectively. Thick arrows mark translocation steps suggested on the basis of available data; C represents the common carrier for all three groups of transported substrates.

given substrate concentration, provided the protonated carrier is the only actual catalytic species of the transport system at the particular substrate concentration (preferably at the level of its K_T value).[29] Under such conditions, the decreasing initial rate of monosaccharide uptake with increasing pH$_o$ can be taken as a measure of that portion of carrier molecules which become deprotonated on the outside of plasma membrane. At the half-maximal velocity, 50% of the carrier molecules occur in the deprotonated form, and consequently this pH$_o$ corresponds to the apparent pK_a of the carrier. Hence, plotting the reciprocal values of the initial uptake rate against the reciprocal values of H$^+$ concentration in the cell suspension (in a rather narrow range of pH$_o$ where the uptake rate displays the largest change) gives a straight line intercepting the abscissa at the negative reciprocal value of pK_a (Lineweaver–Burk plot). In *R. glutinis* using 2 mM D-xylose as transport substrate, the pK_a of the carrier was estimated to be 6.7.[29]

Recently, passive influx of monosaccharides measured in deenergized cells, either by applying uncouplers or by introducing anaerobic conditions, was also shown to be pH$_o$ dependent, thus confirming the interaction of substrate molecules only with the protonated carrier molecules. Moreover, the half-maximal passive influx was measured at pH$_o$ 6.5, thus verifying the above value of carrier pK_a (M. Höfer, unpublished results). These results support the earlier conclusion that the unprotonated carrier does not participate in transport catalysis in *R. glutinis*.[7] The features of H$^+$–carbohydrate transport, described below, are also in line with this conclusion. The influx of any substrate tested was diminished at pH$_o$ above 7 irrespective of the actual size of the driving force. Obviously, the protonated carrier is only capable of transport catalysis.

Electroneutral Monosaccharides. The dependence of the maximal initial rate of monosaccharide uptake on changes in driving forces, $\Delta\psi$ and ΔpH, as well as the entire $\Delta\tilde{\mu}_{H^+}$, is shown for D-xylose in Fig. 3. The component driving forces were changed by varying the pH$_o$ of the cell suspension.[9] The monosaccharide transport follows neither the $\Delta\psi$ nor ΔpH curves, but that of $\Delta\tilde{\mu}_{H^+}$, thus indicating that both components of $\Delta\tilde{\mu}_{H^+}$ energize the transport of electroneutral monosaccharides. The transport has been shown to be electrogenic, the stoichiometry being 1 H$^+$ per transported monosaccharide molecule.[29,30] A stoichiometric efflux of K$^+$ compensated for the mass influx of protons during the initial period of the electrogenic symport.[31] The steeper decline of the transport rate as compared with the reduction of $\Delta\tilde{\mu}_{H^+}$ at pH$_o$ above 6.5 reflects the decreasing population of the protonated carrier molecules, the actual catalytic moiety of the H$^+$ cotransport.[7]

These results demonstrate the equivalency of the two component driv-

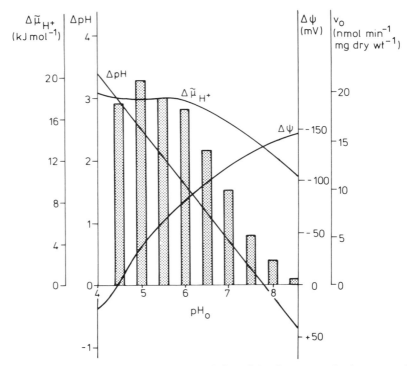

FIG. 3. Dependence of the maximal initial influx of the electroneutral substrate D-xylose (bar graph) on variations of $\Delta\tilde{\mu}_{H^+}$ and the individual components $\Delta\psi$ and ΔpH (continuous lines) promoted by changing the pH_o of *R. glutinis* suspensions as indicated. The values of $\Delta\psi$, ΔpH, and $\Delta\tilde{\mu}_{H^+}$ were taken from Ref. 9.

ing forces, $\Delta\psi$ and ΔpH, for the accumulation of monosaccharides. The recently observed effect of high tetraphenylphosphonium concentrations (5 mM), which dissipated $\Delta\psi$ without interfering with ΔpH or the plasma membrane-bound H^+-ATPase, is in line with the above conclusion: the uptake of monosaccharides was almost unaffected at pH_o 4.5, where $\Delta\tilde{\mu}_{H^+}$ consists predominantly of ΔpH, though it was considerably inhibited at pH_o 7, where $\Delta\psi$ is the prevailing component of $\Delta\tilde{\mu}_{H^+}$.

In addition, comparison of the kinetic parameters in fully energized and depolarized cells provided evidence that the glucose carrier in *R. glutinis* bears a negative charge. A computer fitting of kinetic data to elementary functions derived from a model constructed on the basis of some simplifying premises for ordered (either $C + H^+ + S$ or $C + S + H^+$) and random reaction mechanisms led to the conclusion that the binding se-

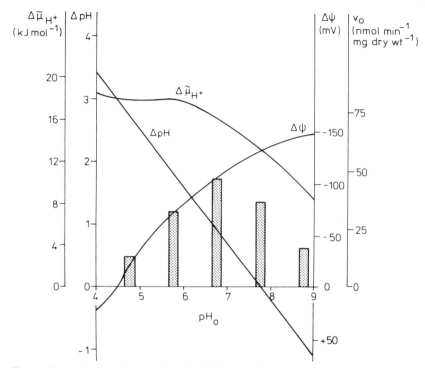

Fig. 4. Dependence of the maximal initial influx of the cationic substrate glucosamine (bar graph) on variations of $\Delta\psi$, ΔpH, and $\Delta\tilde{\mu}_{H^+}$ as in Fig. 3.

quence of formation of the ternary carrier – H⁺ – substrate complex follows a random mechanism.[43]

Cationic Sugar Amines. Among various sugar amines tested, glucosamine and mannosamine are accumulated intracellularly while galactosamine is taken up according to its diffusion equilibrium. On the other hand, *N*-acetylglucosamine does not interact with the transport system at all (no uptake, no competitive inhibition of glucosamine uptake).[6] Contrary to the case for monosaccharides, the cationic sugar amines display a distinctly different pattern of maximal initial influx dependence on $\Delta\psi$ and ΔpH, as shown for glucosamine in Fig. 4. The influx of glucosamine increased with increasing $\Delta\psi$ up to pH 6.7, then declined as the population of protonated carrier and, in addition, that of protonated glucosamine (pK_a 7.7) decreased. Since the transport of glucosamine was electrogenic (the onset of

[43] J. Severin, P. Langel and M. Höfer, *J. Bioenerg.* **21**, No. 3 (1989).

glucosamine uptake induced a strong dissipation of $\Delta\psi$), it was driven solely by the electrical component of $\Delta\tilde{\mu}_{H^+}$, i.e., by the membrane potential. Accordingly, no H^+ was cotransported; instead, a stoichiometric efflux of both H^+ and K^+ compensated for the mass influx of positive charges linked to glucosamine uptake. The lipophilic cation tetraphenylphosphonium, at 5 mM concentration, inhibited the accumulation of glucosamine, thus confirming that $\Delta\psi$ is the sole driving force for glucosamine uptake.[6]

Anionic Glucuronate. A still different pattern of dependency of maximal initial influx on variations of $\Delta\psi$ and ΔpH was obtained for the negatively charged glucuronate (pK_a 2.9). As depicted in Fig. 5, the maximal initial influx closely followed the ΔpH curve, demonstrating that the proton concentration gradient was the sole driving force of glucuronate uptake. There was virtually no measurable uptake of glucuronate above pH$_o$ 6 in spite of further increasing $\Delta\psi$. The uptake of glucuronate was accompanied by a stoichiometric cotransport of 1 H^+ per glucuronate

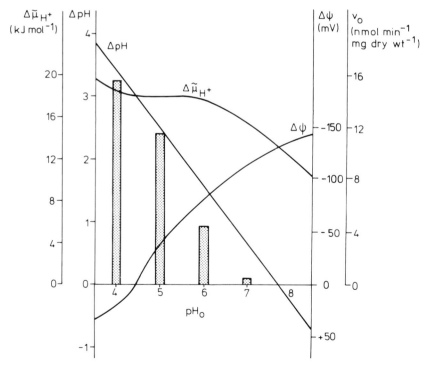

Fig. 5. Dependence of the maximal initial influx of the anionic substrate glucuronate (bar graph) on variations of $\Delta\psi$, ΔpH, and $\Delta\tilde{\mu}_{H^+}$ as in Fig. 3.

molecule.[7] However, the H$^+$–glucuronate cotransport was electroneutral owing to charge compensation between H$^+$ and glucuronate$^-$. Consequently, the influx was independent of $\Delta\psi$ and driven solely by ΔpH.

Interaction of Driving Force with Carrier Function. Organic acids of low molecular weight are assumed to penetrate freely through biological membranes in their protonated, i.e., uncharged form, and to disperse between cell interior and suspension medium according to the actual ΔpH. It is quite unlikely that acetate or propionate, which have completely different chemical structures compared to monosaccharides or glucuronate, will interfere with the carbohydrate transport on the level of substrate binding to the carrier. In fact, pyruvate and malate were shown to be translocated in *R. glutinis* by a transport system different from the carbohydrate carrier described here.[44] On the other hand, the uptake of the anionic glucuronate and of organic acids is energized by the same ΔpH. On this energetic level an interference is to be expected. When added in a mixture, organic acids (0–5 mM) display a clearly noncompetitive inhibition of glucuronate transport. Thus, it is the velocity of translocation and not the effective affinity of the carrier for its substrates (in this case glucuronate) that is affected by depleted driving force (ΔpH).[7]

These results were obtained in cell suspensions buffered with 0.12 M Tris–citrate at pH$_o$ 4 (filtration method) where $\Delta\psi$ was virtually zero. Thus, any effect of $\Delta\psi$ on transport parameters can be excluded even though the carrier was completely protonated at this pH$_o$ value. However, this does not imply that $\Delta\psi$, in addition to its energetic impact, is generally without influence on the half-saturation constant of transport, K_T. Under conditions where $\Delta\psi$ exists, e.g., in experiments on the pH dependence of transport, $\Delta\psi$ may and very likely does influence K_T as well, as demonstrated for another H$^+$ cotransport system.[45]

Alkali Metal Ion Transport

Energy-Dependent K$^+$ Uptake in Exchange for H$^+$. Unbuffered aqueous suspensions of *R. glutinis* cells (2.5% fresh weight/volume) spontaneously acidify, attaining a pH$_o$ value as low as 4. When the pH$_o$ of a cell suspension is increased by, e.g., KOH (an increase of about 0.5 pH), a rapid resumption of the original pH$_o$ follows.[20] Under these conditions, the extrusion of protons is compensated for by a stoichiometric uptake of K$^+$. The H$^+$–K$^+$ exchange can also be induced by adding salts, e.g., KCl. Other

[44] M. Höfer and J.-U. Becker, *Zentralbl. Bakteriol., Parasitenkd., Infektionskr. Hyg., Abt. 1: Orig., Reihe A* **220,** 374 (1972).
[45] W. G. M. Schwab and E. Komor, *FEBS Lett.* **81,** 157 (1978).

alkali metal ions and NH_4^+ but not earth alkali ions, e.g., Ca^{2+}, can serve as counterions for H^+ extrusion. The stoichiometry is always $1:1$. The H^+–cation exchange has been demonstrated mostly in cells aerated for about 24 hr and washed by centrifugation before the experiment.[21]

The H^+–K^+ exchange exhibited a half-saturation constant for K^+ of 20 μM at pH_o 4, at which it also proceeded with maximal velocity. The H^+–K^+ exchange was dependent on supply of metabolic energy since application of uncouplers [e.g., 10 μM carbonyl cyanide m-chlorophenyl-hydrazone (CCCP)] or introduction of anaerobic conditions reversed the flow of both H^+ and K^+. The stoichiometry of the reversed fluxes was also maintained at $1:1$ (see section on plasma membrane permeability for H^+ and K^+ below). Because of the strict stoichiometry of the H^+–K^+ exchange observed in a large number of experiments, conducted under different physiological conditions, the pH dependence of this process can be measured by determining the influx of K^+ alone. Consequently, the measurement can be conveniently carried out in buffered suspensions. The pH dependence of the exchange process displayed a sharp optimum at pH_o 4, where $\Delta\psi = 0$. On the other hand, there was no measurable uptake of K^+ at pH_o 8, where the highest $\Delta\psi$ values have been measured. Hence, the H^+–K^+ exchange proceeds independently of $\Delta\psi$ and, considering its strict $1:1$ stoichiometry, is electroneutral. Consequently, its molecular mechanism is different from the electrogenic H^+-translocating ATPase.[21]

This electroneutral H^+–K^+ exchange should not be mistaken for the electrogenic K^+ efflux compensating for the electrogenic H^+ cotransport during the initial stage of H^+ substrate symport (cf. section on measurement of K^+ efflux compensating for H^+ cotransport).

Plasma Membrane Permeability for H^+ and K^+. Energized cells of *R. glutinis* maintain considerable H^+ and K^+ gradients across the plasma membrane. At an outside pH_o of 4.5 the intracellular pH_i is 7.5, corresponding to a H^+ concentration gradient of 10^3. There is, however, an even higher, oppositely directed gradient of K^+. The intracellular K^+ concentration in *R. glutinis* is on average 160 mM, which is counterbalanced by 8 μM K^+ in suspension media, representing a K^+ concentration gradient of 2×10^4.

When nitrogen is passed into a closed cuvette containing yeast suspension, the cells cannot maintain the ion gradients, and consequently K^+ flows out of and H^+ into the cells (passive H^+–K^+ exchange). In order to find out which of the two ion permeabilities determines the velocity of the exchange process, the uncoupler CCCP as a specific protonophore was used to enhance the passive plasma membrane permeability for H^+. Under these conditions, the H^+ influx was stimulated proportionally to the rise of CCCP concentration (in the range 1–80 μM) with a concomitant stoichiometric increase of K$^+$ efflux; the ratio of H$^+$ and K$^+$ fluxes was under all conditions 1.01 ± 0.04 (±S.E., $n = 18$).[11] Thus, the rate of passive ion exchange was determined by the lower plasma membrane permeability for H$^+$, and the K$^+$ efflux was always sufficiently high to compensate for the CCCP-induced H$^+$ influx. Consequently, by measuring the passive H$^+$-K$^+$ exchange, the plasma membrane permeability for H$^+$ can be determined.

In the procedure, a 5% (fresh weight/volume) unbuffered aqueous cell suspension in a closed acrylic glass cuvette (30 ml of volume) is kept at constant temperature under a flow of nitrogen (99.9% pure). The pH$_o$ of the cell suspension is measured continuously by a glass electrode, and the record is calibrated at the end of the experiment by adding known amounts of HCl to a given volume of cell suspension. The K$^+$ concentration is determined in the supernatants of the cell suspension (centrifugation method) using atomic absorption spectrophotometry. The ion fluxes are calculated in nmol/mg dry weight/min, intracellular concentrations using the ratio of 2 μl intracellular water volume per mg dry weight (see section on determination of intracellular water volume). The permeability coefficient, P_{H^+}, is calculated from

$$P_{H^+} = H^+ \text{ influx}/([H^+]_o - [H^+]_i)$$

The P_{H^+} value was 3.48 ± 0.33 × 10^{-3} min^{-1} and was increased to 1.24 ± 0.24 × 10^{-2} min^{-1} (±S.E., $n = 13$) in the presence of 25 μM CCCP.[11]

Electrically Gated K$^+$ Channel. An important problem is whether the passive H$^+$-K$^+$ exchange occurs through a single electroneutral system in the plasma membrane or through two independent systems coupled electrically. The observed stimulation of the exchange by the protonophore CCCP strongly indicates an electrical coupling of the two ion fluxes. However, there is a difficulty with this conclusion. The existing concentration gradient of K$^+$ does not lead to a generation of a diffusion potential across the plasma membrane, e.g., under anaerobic conditions. Actually, $\Delta\psi$ is the first gradient dissipated by introducing anaerobic conditions.[30] The plasma membrane is obviously impermeable for electrogenic K$^+$ leak under deenergized conditions. Only electrically compensated H$^+$-K$^+$ exchange can occur. An intracellular binding or compartmentation, e.g., in mitochondria, is very unlikely since application of the polyene antibiotic nystatin (10 μM), which interacts with plasma membrane sterols only,[22] induced a rapid and extensive (nonstoichiometric) outflow of K$^+$ from *R. glutinis* cells. The most simple and hence most reasonable explanation of this apparent discrepancy is the postulate of an electrically gated, self-regulated K$^+$ channel, closing whenever electrogenic K$^+$ flux would generate an electrical potential difference.[11]

Effect of DCCD on Plasma Membrane. N,N'-Dicyclohexylcarbodiimide (DCCD) is very likely an inhibitor of *R. glutinis* plasma membrane H^+-ATPase since it deenergized the plasma membrane, when applied for 30 min or longer, without affecting either cell respiration or intracellular ATP content.[21] However, it displays an obviously multiple action, as its immediate effect on *R. glutinis* is a hyperpolarization of $\Delta\psi$. DCCD (1 mM) also caused a reduction of the passive H^+–K^+ exchange. Thus, DCCD interacts (1) with the H^+-ATPase on long-term incubation and (2) with some protein(s) of the plasma membrane, other than the H^+-ATPase, which mediates the passive ion fluxes. The latter effect immediately follows the application of the agent.[11]

Note: Ionophores other than uncouplers (valinomycin, gramicidin, nonactin, up to 100 μM each) were without effect on H^+ and K^+ fluxes or gradients in *R. glutinis*.[10,11,46]

Final Remarks

The obligatory aerobic yeast *Rhodotorula glutinis* proved to be exceptionally advantageous for kinetic and energetic characterization of the proton cotransport system. Using various charged carbohydrates, the accumulation of which is mediated by one common carrier system, the equivalency of the two driving forces, $\Delta\psi$ and ΔpH, of the electrochemical H^+ gradient has been unequivocally demonstrated. The H^+–amino sugar cotransport further provided evidence that the transient alkalinization and the dissipation of $\Delta\psi$ are the primary effects of the H^+ cotransport and not secondary phenomena caused by other membrane-bound processes (cf. Ref. 4).

Rhodotorula glutinis, being an obligatory aerobe, is especially suited for investigation of energy-independent transport processes. Thus, passive plasma membrane permeability for inorganic ions could be extensively studied under anaerobic conditions. Moreover, *R. glutinis* is among the first yeasts where TPP^+ was successfully used as $\Delta\psi$ probe under variety of physiological conditions.

Appendix 1: Isolation of Plasma Membrane Fractions Using Cationic Silica Microbeads

This method has been routinely used in our laboratory to fractionate membranes from the facultative anaerobic yeast *Metschnikowia reukaufii* (CBS 5834).[15,47] It was originally developed for use in *Saccharomyces*

---

[46] A. Künemund, Ph.D. Thesis, University of Bonn (1984).
[47] H.-U. Gläser and M. Höfer, *J. Gen. Microbiol.* **132**, 2615 (1986).

cerevisiae.[48] The principle of the method is to enhance the plasma membrane density relative to other intracellular membranes by attachment of silica microbeads. The plasma membranes with attached silica microbeads were suitable for reconstitution in phospholipid vesicles (proteoliposomes). The reconstituted plasma membrane vesicles generated both a proton gradient (acidic inside) and an electrical potential difference (positive inside) on addition of MgATP to vesicle suspensions.[49]

The starting material is a purified protoplast suspension. When the presence of plasma membranes (PM) is to be traced during membrane fractionation, PM can be radioactively labeled by dansylation.

In the procedure, a 25% (v/v) suspension of purified protoplasts in 20 ml of 1 M sorbitol containing 0.3 M potassium phosphate buffer (pH 9.2) is incubated with 0.75 MBq [^{3}H]dansyl chloride for 30 min at 20° and washed 6 times in the buffer by centrifugation (Sorvall RC5-B, SS34 rotor, 3000 g, 10 min, 4°). For attachment of microbeads, a 25% (v/v) suspension of protoplasts is incubated in 1 M sorbitol containing 25 mM sodium acetate buffer (pH 4.5) and 0.3% (w/v) silica microbeads for 15 min at 4°. Protoplasts are washed from unbound microbeads by centrifugation (Sorvall RC5-B, SS34 rotor, 1000 g, 10 min, 4°) in acetate buffer.

For lysis of protoplasts, protoplasts with attached microbeads are lysed in a buffer containing 5 mM Tris-Cl, 1 mM EGTA, 0.2 mM MgCl$_2$, 0.33 mg/ml dextran sulfate, 0.3 M sorbitol (pH 7.0–7.2) and centrifuged at 1800 g for 5 min. In order to ensure complete lysis, the pellet is resuspended in 5 mM Tris-Cl buffer containing 2 mM EGTA and 0.4 mM MgCl$_2$ (pH 8.0) and centrifuged 3 times to wash out residual mitochondrial membranes. The final pellet consisting of the purified plasma membranes (10% contamination with mitochondrial protein as estimated by cytochrome-c oxidase activity) is resuspended in 5 mM Tris-Cl containing 0.3 mM MgCl$_2$ (pH 7.2) and stored at $-35°$ for up to 3 days without a significant loss of ATPase activity.

Note: The cationic silica microbeads (Na$^+$ stabilized) were kindly supplied by Dr. B. Jacobson, University of Massachusetts, Amherst.

For isolation of the mitochondrial fraction the supernatant after the 1800 g centrifugation (the first centrifugation step following protoplast lysis) is used. The supernatant is first centrifuged at 4300 g for 10 min (Sorvall RC5-B, SS34 rotor, 4°); then the pellet discarded and the supernatant centrifuged at 30,000 g for 20 min. The pellet containing almost pure mitochondrial fraction (<3% contamination by plasma membranes ac-

[48] R. Schmidt, B. Ackermann, Z. Kratky, B. Wassermann, and B. Jacobson, *Biochim. Biophys. Acta* **132,** 421 (1983).
[49] H.-U. Gläser and M. Höfer, *Biochim. Biophys. Acta* **905,** 287 (1987).

cording to the specific radioactivity of PM) is suspended in the same buffer as the PM fraction.

Appendix 2: Electrophysiological Measurements on Yeast Cells

Membrane potentials are most dependably measured by electrophysiological techniques using glass microelectrodes. When the technique is applied to yeast cells, however, severe difficulties occur owing to the small size and the mobility of cells in suspension.

In the following, a successful procedure is described for measuring reproducible membrane potentials in the yeast *Pichia humboldtii* under various physiological conditions.[16,24] Cells of this yeast have proved to be large enough to allow insertion of glass microelectrodes without destroying cell integrity. Individual cells have been immobilized by capture in the neck of a fine glass microfunnel.

In the procedure, *P. humboldtii* is grown (see above) and harvested after 2–4 days. The yeast grows in chains of two to four cells, the biggest of which reach a size of 12–15 μm. This has proved to be of special advantage since individual chains can be firmly anchored by the smaller cells inside the neck of a microfunnel, thus exposing the biggest one to the approach of the microelectrode (Fig. 6).

To immobilize the cells, special fine microfunnels are prepared from glass micropipettes pulled from 1B100-4 Kwik-Fil glass capillaries (WP Instruments, New Haven, CT) by means of a conventional micropipette puller. First, a small balloon is blown at the tip of a micropipette and then broken off by stretching the balloon (breakage occurs through its cooling when the heat source is switched off). The neck of a usable microfunnel must be less than 10 μm in diameter. Individual cell chains are drawn into the microfunnel from cells sedimented on the slant of an acrylic glass wedge located in the measuring chamber by means of a plastic syringe (5 ml volume) connected to the microfunnel by plastic tubing. The microfunnel is operated by means of a Leitz micromanipulator, and the movement is controlled under a microscope fitted with a long working-distance objective (magnification: $\times$25).

The measuring chamber is constructed from acrylic glass in such a way as to make a horizontal approach for the microfunnel and the measuring microelectrode possible from each of the two sides of the chamber (Fig. 7). The chamber is covered by a cover glass so that the bathing solution (see below) is kept in the chamber by adhesion despite both sides being open. Yeast cells, in a drop of suspension, are pipetted onto the slant of the wedge and allowed to sediment (within 30 sec). Then, bathing solution from a container located above is slowly circulated through the chamber, eventu-

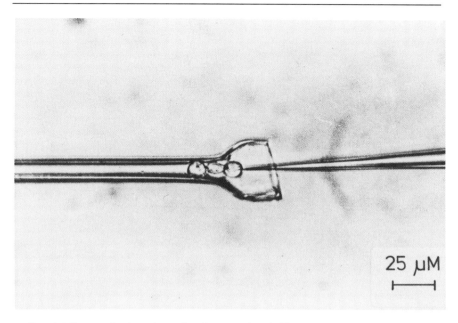

Fɪɢ. 6. Microscopic view of a chain of three *P. humboldtii* cells anchored in the neck of a microfunnel. The cell exposed to the outside has been punctured with a microelectrode.

ally dropping into a collecting flask in front of the microscope. Continuous flow is facilitated by connecting the first reservoir, the measuring space, and the collecting flask with filter paper strips in a flat groove. Individual cell chains are withdrawn from the wedge by the microfunnel and transferred to the middle of the measuring space without the cells leaving the bathing solution. The measuring microelectrode is administered from the opposite side of the chamber (at 180° angle).

For puncturing the cells, measuring microelectrodes are pulled from 1B100F Kwik-Fil glass capillaries and filled with 3 M KCl[16] or 0.1 M KCl.[24] Their electrical resistance is 10–20 MΩ with tip potentials of less than −10 mV. The microelectrode is impelled forward by means of a Narishige hydraulic microdrive mounted to a Leitz micromanipulator. It is important not to touch the microfunnel with the tip when inserting the microelectrode into the microfunnel with anchored yeast cells, otherwise the very sharp tip, less than 0.5 μm, gets broken (without being perceptible to the observer). Yeast cells are small and very flexible owing to cell turgor so that reliable punctures are obtained only by trial. Frequent substitution of used microelectrodes after a few unsuccessful punctures is made.

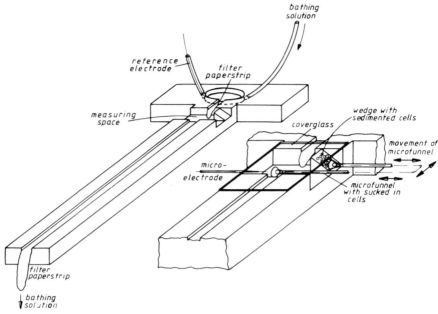

FIG. 7. Scheme of the measuring chamber (shown actual size). At right, a twice-enlarged segment of the measuring chamber shows the measuring space with the wedge on which slant cells have been sedimented. A chain of cells has been drawn into the microfunnel, and the microfunnel transported to the middle of the measuring space. The microelectrode has been impelled from the left.

Successful insertion of the microelectrode into the yeast cell is indicated by a sudden increase of negative potential, which in many cases is stable over several minutes (Fig. 8). Frequently, however, only peaklike responses are recorded, indicating deadly injuries to cells by the penetrating micro-electrode. Such injuries lead to a sudden appearance of a distinct netlike structure throughout the cytoplasm. In addition, injured cells immediately lose their turgor. As a matter of fact, many simply vanish into the micro-funnel under continuously applied suction. In some cases, a process of plasma membrane resealing is observed during which the original $\Delta\psi$ of the peak is reached again.

The reference electrode consists of a polyethylene tube (2 mm diameter) filled with 2% agar containing $3\ M$ KCl, in which a silver wire coated with silver chloride is submerged. It is inserted in a fitting boring in the side wall of the measuring chamber and connected electrically with the measuring electrode by the bathing solution [10 mM Tris–phosphate or Tris–

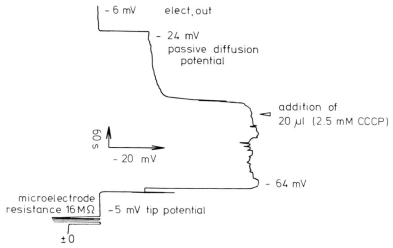

FIG. 8. Reproduction of a recording showing a stable $\Delta\psi$ measurement in a *P. humboldtii* cell. The recording pen traced from down upwards. At the arrow, the uncoupler CCCP was added, leading to a rapid dissipation of $\Delta\psi$ down to the value of the passive diffusion potential.[16]

citrate, 1 mM Ca(NO$_3$)$_2$, 1 mM MgSO$_4 \cdot$7H$_2$O, and 0.1 mM KCl, pH as required]. All substances modifying the physiological state of yeast cells are added additionally. The measuring and reference electrodes are connected to an electrometer amplifier (either WP Instruments, Model 701; or Keithly 604, Munich; both of input impedance higher than 10^{12} Ω) and the measured potential difference recorded on a pen chart recorder.

Recently, $\Delta\psi$ values from microelectrode measurements were compared with those calculated from the steady-state distribution of the lipophilic cation tetraphenylphosphonium (cf. above). The agreement was within 10 mV under all physiological conditions tested, except at pH above 7.[24]

Concluding Remarks. The electrophysiological approach to investigation of electrogenic plasma membrane-bound processes in yeast cells as representatives of the most simple eukaryotic organisms is, at present, limited to relatively large cells such as those of *P. humboldtii.* However, the results obtained with this yeast demonstrated that $\Delta\psi$ values determined by indirect methods based on the steady-state distribution of lipophilic ions also delivered principally dependable values, provided that the plasma membrane permeability for the indicator ion is high enough to allow for its rapid equilibration and that there is no extensive binding of the ion to cell structure.

[40] Proton–Potassium Symport in Walled Eukaryotes: *Neurospora*

By CLIFFORD L. SLAYMAN and GERALD N. ZUCKIER

Introduction

It is remarkable that the basic physiological mechanisms responsible for active transport of potassium in plants and plantlike microorganisms are not yet clearly defined. This is so despite the fact that, since the 1950s, uptake of K^+ by animal cells has been convincingly shown to occur via three distinct mechanisms, which operate under different (but overlapping) conditions. Resting potassium uptake in animal cells occurs by chemically coupled exchange for sodium through the so-called sodium pump, which, given the normal K^+ leaks, is able to concentrate that ion 30- to 50-fold in the cytoplasm. But under the special condition of membrane hyperpolarization (e.g., when the sodium pump is strongly activated), net potassium influx can also occur via several different kinds of K^+-specific channels. In addition, the circumstance of acute volume loss activates a multiple-ion transporter, which simultaneously transmits K^+, Na^+, and Cl^- (in the ratio 1:1:2) down their common chemical gradient; the net salt movement, compensated by isosmotic water, reexpands cell volume.

For plant cells a few facts about potassium transport are well established even though no general model exists. In metabolizing cells, for example, K^+ uptake can be compensated by Na^+ loss, by H^+ loss, or by uptake of several different anions, depending on the history of the cells and the particular ions available.[1] While potassium does stimulate[2] the plasma membrane ATPase in plants (usually responsible for H^+ pumping, not for Na^+ pumping), it does so only weakly and without the synergistic enhancement which characterizes the sodium pump. And finally, under conditions of rapid net K^+ uptake, there is often stoichiometrically equivalent net extrusion of protons.[3] Two continuing problems for interpretation of such ion-flux measurements, however, have been (1) an absence of ion-specific inhibitors, which would permit separation of distinct, but coordinately functioning, transport systems, and (2) lack of quantitatively comparable electrical data, which could identify the actual charge flux associated with net K^+ entry.

[1] A. Rodriguez-Navarro, M. R. Blatt, and C. L. Slayman, *J. Gen. Physiol.* **87**, 649 (1986).
[2] R. Leonard and C. Hotchkiss, *Plant Physiol.* **58**, 331 (1976).
[3] W. Lin and J. B. Hanson, *Plant Physiol.* **58**, 276 (1976).

Increasing use of dissociated cultured plant cells and/or protoplasts, over the next few years, will certainly relieve the second problem and facilitate precise descriptions of charge and coupled-ion movements. A prototype for possible future experiments on cultured plant cells has been developed in the large-celled mycelial fungus, *Neurospora,* which can be forced to grow in a spherocyte form by treatment with ethylene glycol.[4] Study of high-affinity potassium uptake by these cells, including quantitative comparisons of net and unidirectional K^+ flux, net H^+ flux, K^+-linked current, and membrane potential, has led to the idea that "active" potassium transport in *Neurospora* occurs via a K^+–H^+ symport driven by the membrane potential and by the combined chemical potential difference for the two ion species.[1,5]

Handling of *Neurospora* Spherocytes

Growth of Cells. Standard methods for preparing shaking-culture cells of *Neurospora* have been described earlier in this series.[6] Such cells are slender filaments ($2-3$ μm in diameter) not suitable for electrophysiological work, but useful for background flux experiments. Larger spherocytes (> 12 μm diameter), suitable for both types of experiments, are produced by a recipe derived from Bates and Wilson.[4] Coniadiating mycelial cultures are grown $5-7$ days under moderate illumination, at $25°$ and on 2% agar–Vogel's minimal medium[7] (see below) plus 2% sucrose. Conidia are harvested by shaking the cultures with sterile distilled water (SDW), pouring through sterile gauze or cotton, centrifuging the conidia (500 g, 5 min), rinsing the pellet 2 times with SDW, and resuspending in SDW to a final density of $\sim 10^9$ conidia/ml. Conidia are then inoculated into sterile liquid cultures at densities of $3-10 \times 10^6$/ml, either in Vogel's medium (VM) containing 1% glucose and 18% (v/v) ethylene glycol (EG), or in one of two low-K^+ media (Table I) plus glucose and EG. [The EG concentration for growing swollen conidia (spherocytes) without germ tubes is critical and seems to vary somewhat from one batch of cells to the next. Therefore, it is helpful to prepare and inoculate media containing a range of EG concentrations, from 15 to 20% (v/v); this can be done conveniently by making up the highest concentration and diluting aliquots with SDW.] For small quantities of spherocytes, as typically needed for electrophysiological experiments, cultures are set up in small petri plates (e.g., 3 mm diameter containing $2-3$ ml of suspension) or slanted test tubes. For the larger

[4] W. Bates and J. Wilson, *J. Bacteriol.* **117,** 560 (1974).

[5] M. R. Blatt, A. Rodriguez-Navarro, and C. L. Slayman, *J. Membr. Biol.* **98,** 169 (1987).

[6] C. L. Slayman and C. W. Slayman, this series, Vol. 55, p. 656.

TABLE I
COMPOSITIONS OF GROWTH MEDIA AND FLUX BUFFERS[a]

Component	Growth media			Flux buffers		
	AmVM	NaVM	VM	CaMES	NaDMG	KDMG
Na^+	0^b	61.8	25	0	25	0
K^+	0.3	0.3	36.8	0–0.2 (as needed)	0–0.2 (as needed)	25
NH_4^+	30	25	25	0	0	0
Ca^{2+}	0.1	0.1	0.1	1.5	1	1
Mg^{2+}	0.8	0.8	0.8	0	0	0
Cl^-	0.3	0.5	0.5	0^c	2	2
NO_3^-	0	25	25	0	0	0
SO_4^{2-}	0.8	0.8	0.8	0–0.1 (added with K^+)	0–0.1 (added with K^+)	0
Phosphate	15	36.8	36.8	0	0	0
Citrate	8.4	8.4	8.4	0	0	0
MES[d]	0	0	0	10	0	0
DMG[d]	0	0	0	0	20	20
Glucose (%)	1	1	1	0^e	0	0
EG (%)	~18	~18	~18	0^f	0	0

[a] All concentrations except those designated (%) are given in mM; glucose is in % (w/v), EG in % (v/v). Solutions are buffered at pH 5.80 ± 0.5, by phosphate–citrate for the growth media, and by titration of MES or DMG acids with KOH or NaOH for the flux buffers. All chemicals are reagent grade, usually obtained from J. T. Baker (Phillipsburg, NJ) or from Aldrich (Milwaukee, WI).

[b] Components designated by "0" are simply omitted.

[c] KCl rather than K_2SO_4, can be added to raise $[K^+]_o$.

[d] MES, 2-N-Morpholinoethanesulfonic acid, No. M-0895 from Sigma (St. Louis, MO); DMG, 3,3'-dimethylglutaric acid, Sigma No. D-4379.

[e] Flux buffers should *not* contain glucose or any other potential carbon source, which would allow rapid germination of the spherocytes on removal of EG.

[f] Ethylene glycol is added to all growth media for preparing spherocytes, but to the flux buffers only for rinsing the cells free of glucose.

quantities needed in flux experiments, cultures are set up in Erlenmeyer flasks (e.g., 50- to 500-ml flasks, with an 8- to 10-mm pool in the bottom).

Cultures can be left stationary and swirled by hand once every 2–3 hr, or placed on a slow rotary shaker or tube rotator (10–20 cycles/min). Cells are grown at 25° for 2–3 days, and they should be sampled and assessed optically in the later stages. Between 60 and 84 hr, spherocytes of wild-type *Neurospora* should be 12–15 μm in diameter, with nearly homogeneous cytoplasm as viewed by bright-field at 400×. There should be no large vacuoles in the cytoplasm and few crinkled cells in the suspension; most suspensions, however, will contain unexpanded conidia (~20%) and modest small-particle debris, which can be ignored.

Removal of Ethylene Glycol. The cells must be washed of ethylene glycol before they can be used for electrical or flux measurements, and in order to accomplish this without either bursting cells or allowing germination, the following procedure is used. Growing cells are harvested on Millipore filters (Type RA, Millipore, Bedford, MA), rinsed thoroughly, and resuspended in flux buffer (see Table I) containing 18% EG but lacking glucose or other potential carbon sources. The suspension is then diluted 10-fold over a period of 3 hr by gradual addition of EG-free and glucose-free flux buffer. At the end of that time cells for flux experiments are again filtered and resuspended in EG-free, glucose-free flux buffer.

Impalement. Cells for electrical experiments are injected into the recording chamber containing appropriate buffer solution and are allowed to settle for 5–20 min. Without additional treatment, a fraction of spherocytes will adhere lightly to the chamber bottom, and the majority of suspended and unattached cells can be washed away by gentle perfusion of the chamber. More plentiful attachment can be obtained by smearing a thin coat of polylysine (1% in distilled water; $M_r > 190,000$: No. P-1399 or P-1524, from Sigma, St. Lous, MO) on the chamber bottom before addition of buffer or cells; this maneuver has no evident deleterious effect on the spherocytes. Manipulations of the cells and micropipettes from this point onward are controlled visually, using a good quality compound microscope with either dark-field (preferred) or bright-field optics, at 400× or 600× (40 × 10, or 40 × 15).

After a large and healthy looking cell (see comments above) is identified and positioned with the stage movements, its diameter is measured, and it is then drawn onto the tip of a similar-sized pipette and lifted off the chamber bottom. Suitable holding pipettes are made by pulling borosilicate capillary tubing (Kimax 54300, from Kimble Glass, Vineland, NJ) into conventional micropipettes with tapered shafts ~1.5 cm long, fusing the shaft to a hot platinum loop, cooling the loop, and then breaking off the pipette. The broken tip is then fire polished and silanated as described below for ion microelectrodes. The finished pipette is connected to a three-way stopcock and then to a vacuum line and a tank of dry nitrogen (regulated at 14 psi), and it is mounted on a micromanipulator at a slight angle (2°–5° downward).

For puncture, the recording microelectrode, mounted on a second micromanipulator, is pushed into the cell by motion directly in line with the holding pipette. After the membrane potential has stabilized, the holding pipette can often be released, so the cell is entirely held by the impaling electrode, which tends to prolong impalement lifetime by reducing incidental vibrations. Rapid and smooth changes of bath composition around the cell are accomplished by positioning a large pipette (~100 μm tip diameter, connected to a reservoir of test solution) upstream from the impaled cell and offline (above, below, or sideways) from it. Prior to a solution change, flow of test solution is started and adjusted to match the bath flow. The actual solution change is made simply by moving the pipette into line with the impaled cell.

Strains and Media

Standard spherocytes for these experiments are prepared from the wild-type strain RL21a of *Neurospora crassa* (Culture 2219 from the Fungal Genetics Stock Center, Department of Microbiology, University of Kansas Medical Center, Kansas City, KS) or from the complementary (mating-type) strain R3-8A (FGSC 2218). Other wild types of *N. crassa* (such as 74-OR23-1A, FGSC 981) are expected to yield equivalent spherocytes, but K^+ transport has not been characterized in them. The starting conidia for RL21a and R3-8A are 5–7 μm in diameter. Naturally larger conidia (9–12 μm diameter) are produced by another strain of *N. crassa*, FGSC 4474(a), and these expand into spheroplasts of 15–25 μm diameter, which are consequently somewhat easier to penetrate with microelectrodes than cells from the wild types. [A giant-conidial strain of *N. intermedia*, P27A (FGSC 5641) has also been described. It normally produces conidia 20–30 μm in diameter, which enlarge slightly on the EG regimen to form spherocytes 30–40 μm in diameter. Such spherocytes have proved less reliable in electrophysiological experiments than those produced by *N. crassa* RL21a, R3-8A, or 4474(a) and may have a complicated morphology with only pseudospherical shape.]

Three variations of growth medium and three flux buffers are useful for different aspects of these experiments. All growth media are modified from the "N" minimal medium of Vogel[7] (designated below as VM), and contain the same heavy metal trace elements and biotin. Flux buffers do not contain the trace elements or biotin. The major components of all six solutions are shown in Table I. VM and KDMG,[8] respectively, are used for growing and testing normal-K^+ spherocytes. AmVM or NaVM can be used to grow low-K^+ spherocytes, and CaMES or NaDMG can be used to test them. Growth in AmVM and testing in CaMES result in faster K^+ uptake at higher affinity than any other combination we have tried.

[7] H. J. Vogel, *Microb. Genet. Bull.* **13**, 42 (1956).
[8] C. L. Slayman and C. W. Slayman, *J. Gen. Physiol.* **52**, 424 (1968).

Electrophysiological Techniques

A prior article[6] in this series should be consulted for general apparatus: the Faraday cage, vibration isolation table, micromanipulators, general-purpose amplifiers, reference half-cells, electrode mounts, chart recorders, microscope, and recording chambers. Newer and/or more specialized components are described below. Some aspects of these techniques have recently been improved and extended by Blatt,[9,10] for use on intact plant tissues such as the leaf guard cell complex.

Microelectrodes.[11] Microelectrodes for these experiments are of the penetrating type, with multibarreled tips approximately 0.2 μm in diameter (scanning electron microscope measurements). Depending on filling solutions, external solutions, and the number of barrels, electrode tip resistances range between 50 and ~500 MΩ, which compares with 5–50 GΩ for the total cell resistance. In this circumstance, accurate voltage-clamp experiments (±0.1%) are best obtained with separate electrode barrels to pass current and record the membrane potential; two barrels can be augmented to three or four barrels for simultaneous ion injection or cytoplasmic ion activity measurements.

Micropipettes are made from 1-mm (o.d.) borosilicate capillary glass with an internal fiber (Glasswerk Hilgenberg, Malsfeld, FRG; Frederick Haer Co., Brunswick, ME; A-M Systems, Seattle, WA). A horizontal pipette puller (Model Ml, Industrial Science Associates, Ridgewood, NY) is used, on which the fixed-position clamp is replaced by a pin vice mounted on a small dc motor (~10 rpm; Edmund Scientific, Barrington, NH). Pipettes are fabricated in three stages from clusters of capillary tubing: (1) with the puller solenoid circuit opened, the glass is heated, twisted 360°, and allowed to cool; (2) the solenoid is reconnected, and a normal pull cycle is executed; and (3) behind its tapered shank, the pipette is softened with a small gas flame so the individual barrels can be bent apart for attachment to separate amplifiers. Provided that multiple separate capillaries, rather than single, divided tubing, are used, the technique yields pipettes with negligible electrical interaction between barrels. Interbarrel dc-coupling resistance ranges from 10^{13} to 10^{15} ohms, for the usual bath and filling solutions.

The small volume (0.9–4.2 pl) of individual *Neurospora* spherocytes demands careful selection of pipette-filling solutions in order to prevent cell damage by ions, particularly anions, leaking from the pipettes. (Chloride and nitrate ions above ~50 mM can cause irreversible conductance

[9] M. R. Blatt, *J. Exp. Bot.* **36**, 240 (1985).
[10] M. R. Blatt, *J. Membr. Biol.* **98**, 257 (1987).
[11] M. R. Blatt and C. L. Slayman, *J. Membr. Biol.* **72**, 223 (1983).

increases, as can butyrate and propionate above ~5 mM; formate, sulfate, and several other divalent anions can initiate action potentials.) Acetate, HEPES, or MES buffers, titrated to pH 7.0–7.2 with KOH, make satisfactory filling solutions; they can also be supplemented with 1–5 mM KCl, to stabilize the usual Ag/AgCl reference half-cell. For recordings of more than a few minutes duration, cytoplasmic cation activity (e.g., $[K^+]_i$) drifts to that of the pipette-filling solution,[1] which should therefore be set, by adjusting the total buffer concentration, to the cytoplasmic level expected for the particular growth conditions: i.e., 50–75 mM for cells grown in AmVM and incubated in CaMES, and 180–200 mM for cells grown in VM and incubated in KDMG. (Starvation per se, not just cytoplasmic dilution, is required to produce the high-affinity K^+ transport system.) Pipettes containing glass fibers can be filled with the appropriate buffer or saline solution simply by infusion from a 30-gauge syringe needle (Popper & Sons, New Hyde Park, NY) inserted through the butt of each barrel.

Ion-Sensitive Microelectrodes. Ion-sensitive microelectrodes are made via several additional steps following fabrication of the multibarrelled pipettes[12,13]: (1) drying at 160° for 15 min in a small oven (and good fume hood!); (2) silanation of the selected barrels by exposure to the vapor of tri-*n*-butylchlorosilane [(TCS; No. T18510 from Pfaltz and Bauer, Waterbury, CT); a liquid droplet, 30 μl/liter of oven volume, is placed in the oven, and allowed to react at 160° for 15 min; barrels not to be silanated are connected, via Teflon pressure tubing, to a tank of dry nitrogen and flushed at 2 psi throughout the oven treatment]; (3) flushing of the entire oven with dry air, and continued baking for 5–10 min; (4) placement of a droplet of liquid ion exchanger (LIX) into the tapered shank of the silanated barrel(s), via a slender glass pipette (the following LIX's have proved satisfactory: for K^+, No. 477317 from Dow-Corning, Midland, MI; for Cl^-, Corning No. 477913; and for H^+, No. 82500 from Fluka, Ronkonkoma, NY); (5) back-filling of the electrode shaft with 0.1 M KCl buffered (pH 7) by 10 mM MES–KOH.

The additional step of inserting a thin plug of polyvinyl chloride (PVC) ahead of the LIX is sometimes used[13] to enhance turgor resistance of the pipettes. For this purpose, cold tips are dipped into 2% PVC (w/v) dissolved in tetrahydrofuran (PVC: No. 18,956-1 from Aldrich, Milwaukee, WI; THF: Aldrich No. 24,288-8), and light suction is applied. Saline barrels must be protected during this step, either by pressurized nitrogen (see above) or by prior filling with the electrolyte solution. Before calibration and use, filled electrodes are allowed to equilibrate at room tempera-

---

[12] J. Coles and M. Tsacopoulos, *J. Physiol. (London)* **270**, 12 (1977).
[13] H. Felle and A. Bertl, *J. Exp. Bot.* **37**, 1416 (1986).

ture for at least 30 min with their tips resting in the same buffered KCl solution. The resultant resin-filled barrels are calibrated in decade solutions of KCl (10^{-4} to 10^0 M), each compared with an adjacent buffer- or saline-filled barrel; only those with slopes greater than 50 mV/decade are kept for experiments. The success rate in preparation of such electrodes is about 50%.

Amplifiers and Recorders. Each voltage and current electrode (barrel) is connected, via a salt bridge (1 M KCl) and Ag/AgCl reference junction, to a Model WPI-M701 electrometer amplifier (10^{12}–10^{13} Ω input resistance; World Precision Instruments, New Haven, CT). For voltage-clamp experiments, the membrane potential measured via one such amplifier is fed to a high-gain differential amplifier, whose output drives the current-injection circuit of a second WPI-M701, in turn connected to a separate barrel of the pipette.

For ion-injection experiments, two barrels of the pipette are connected to WPI-M701 amplifiers whose current circuits are driven by carefully matched, but inverted amplifiers (approximately unity gain), so that a single pulse to the latter generates equal currents (±0.1%) of opposite sign through the two barrels. This procedure permits large ionophoretic currents (up to 1 nA) with negligible direct effect on the cell membrane potential. Filling solutions for the two electrode barrels are chosen to give unmatched ion flows. Thus, to deposit protons into the cytoplasm, the anodal barrel is filled with (10 mM) H_2SO_4 and the cathodal barrel with (20 mM) neutralized KOAc; to inject vanadate ions, both barrels are filled with (50 mM) neutralized KOAc, but the cathodal barrel also contains sodium orthovanadate (5 mM).

For measurement of intracellular ion activities, one ion-sensitive barrel and one saline-filled barrel are connected to separate channels of an ultra-high-resistance electrometer (WPI-FD223, input resistances of 10^{15} Ω), switched to report the signal difference between the two barrels.

Signal Control and Processing. Amplifier signals for the membrane potential, the voltage clamp current, the ion-injection current (anodal), and the specific ion activities are all led to a multichannel storage oscilloscope (TEK-5113, Tekronix, Beaverton, OR) fitted with a Polaroid camera, as well as to a simple chart recorder. Amplitude and timing of the ion-injection currents are controlled manually.

The voltage-clamp regimen, however, is controlled by a microcomputer, which also records membrane potential and clamp current, and prints out the measured membrane current–voltage curves. In typical use, the clamp program produces staircases of alternating (±) 100-msec square pulses, progressively increasing in amplitude and spanning membrane potentials between −400 and +100 mV, with an adjustable total of 24 to

40 pulses. Each pair of pulses is separated by a 100-msec "baseline" clamp (at the initial resting potential), and the clamp current for each pulse is calculated as the difference between the measured pulse current and the (local) average baseline current. A full set of pulse data, with currents and matching clamp voltages, is referred to as a current–voltage scan (I–V scan) and normally requires ~10 sec to execute. (Useful scans as brief as ~1 sec can be run by decreasing the total number of pulses and shortening individual pulses to 50 msec, but interpretation of still briefer scans is restricted both by graininess and by capacitance evident in shorter pulses.) Fortran programs for running these operations have been written over a period of several years by U.-P. Hansen, J. Warncke, and J. Rose for use with Z80-based hardware, and are available on request. Some machine-language components (analog-to-digital and digital-to-analog conversions, and graphics) would need to be rewritten for each particular microcomputer system.

To analyze the electrical behavior of the potassium transporter, distinct from background properties of the spherocyte membrane, potassium is presented to the starved cells only in short pulses (~30 sec), and I–V scans are run immediately before, during, and immediately after each K^+ pulse. Smooth curves are drawn by polynomial fitting[14] or spline fitting (ASYST Scientific Program, MacMillan Software, New York, NY) through each set of current and voltage data, and the "before" and "after" curves are averaged. Potassium-stimulated current is then calculated as the difference current, at 10- or 20-mV intervals, between the "K^+-on" curve and the average "K^+-off" curve. As pointed out by Blatt,[15] this kind of procedure is valid only over a restricted range of voltages; for K^+ transport by *Neurospora* spherocytes, it is satisfactory at membrane potentials negative to −100 mV. A particularly useful value of K^+-stimulated current is given by the difference current at the resting membrane potential assumed in the presence of added K^+: it is the total current drawn by the K^+ transporter through all the other membrane current sources when the membrane potential is not clamped. In other words, it is the characteristic current of the "free-running" transport system, at the chosen test concentration (see Table II).

Ion Flux Techniques

Potassium Influx. After removal of glucose and dilution of EG, the spherocytes are filtered, resuspended at ~5×10^5 cells/ml in flux buffer, and preincubated for 30–90 min, with aeration via a fine-tipped plastic

[14] D. Marquardt, *J. Soc. Ind. Appl. Math.* **11**, 431 (1963).
[15] M. R. Blatt, *J. Membr. Biol.* **92**, 91 (1986).

pipette. The cells are then reharvested and suspended at $1-5 \times 10^6$ cells/ml in air-saturated buffer, which is partitioned into 20-ml lots (usually in 120-ml Erlenmeyer flasks) for the assays. Maximal K^+ uptake rates require continued aeration of the cell suspensions. Potassium (chloride or sulfate), at the appropriate concentration, is added to a pair of flasks to start the uptake, and 2- to 4-ml aliquots from each flask are removed at intervals (typically ~ 15, 30, 60, and 90 sec), syringe-filtered through a Millipore membrane, and analyzed for chemical potassium by means of a flame absorption spectrophotometer (Model 560, Perkin-Elmer, Norwalk, CT) or for isotopic (^{42}K) potassium by means of a Nuclear Chicago autogamma counter (G.B. Searle, Chicago, IL).

Potassium – Proton Exchange. Simultaneous net uptake of K^+ and release of H^+ are followed by means of extracellular ion-sensitive electrodes. K^+ electrodes are fabricated from valinomycin-impregnated PVC, by modification of the recipe of Kamo *et al.*[16,17] PVC at 6% (w/v) is dissolved in cyclohexanone (Sigma No. C-8390) by stirring at room temperature for several hours. [All operations with PVC solvents are conducted in a fume hood, most conveniently in this case with a shallow layer of solution (~ 10 ml total) in a small *glass* Petri dish.] Dipentyl phthalate (Pfaltz and Bauer No. 256-DO6590), $\sim 14\%$ (v/v) is added as a plasticizer, and after this is thoroughly mixed, 4 mg% valinomycin (Sigma No. V-0627) is added, and stirring is continued for $1-2$ hr. Borosilicate capillary tubing (Kimax 35402), in 1.5- to 2-inch lengths, is fire polished to form melted tips of ~ 50 μm i.d. Fifteen to twenty of these pipettes are mounted on a small aluminum block, then dried, silanated, and baked as described above for the ion *micro*electrodes. The block and pipettes are transferred to a desiccator (over anhydrous calcium chloride) to cool. Cooled pipettes are dipped into the PVC mixture, inverted, and allowed to dry overnight. (The membrane film adheres tightly in the pipette orifice, but not elsewhere, so spurious drips and streaks are easily cleaned off.)

A syringe and 27- to 30-gauge needle are used to backfill the pipettes with 0.1 M KCl, but they must be manipulated carefully to exclude bubbles and avoid repeated fillings, which will dislodge the membrane. Thereafter, the butt of each electrode is connected (via an agar bridge; 2% agar plus 0.1 M KCl) to a broken micropipette, and both are dipped into 0.1 M NaCl for storage. With this short-circuiting procedure, the electrodes are stable for several months at $10°-25°$. As with the microelectrodes, these are mounted on (or bridged to) Ag/AgCl half-cells in 0.1 M KCl, and calibrated against decade KCl solutions; electrodes with slopes less than ~ 55 mV/decade, at least to $[K^+]_o = 0.01$ mM, are discarded.

[16] N. Kamo, M. Muratsugu, R. Hongoh, and Y. Kobatake, *J. Membr. Biol.* **49,** 105 (1979).
[17] C. Shen, C. C. Boens, and S. Ogawa, *Biochem. Biophys. Res. Commun.* **93,** 243 (1980).

Commercial pH electrodes are generally satisfactory for the H^+ measurements, but are most convenient when constructed with a concentric or adjacent reference half-cell and salt bridge (e.g., Model 4098-M30, A. H. Thomas, Swedesboro, NJ), whose saline filling must be replaced with saturated NaCl in order to avoid K^+ leakage into the cell suspensions. Both sensing electrodes and the reference electrode can be connected either to commercial pH meters or to ultrahigh-resistance electrometer amplifiers (e.g., WPI-FD223), with outputs led to a simple strip-chart recorder.

Cell Sizing. Direct comparisons of chemical or isotopic flux data with membrane current data almost always require standardization for unit membrane area. EG spherocytes of *Neurospora* have quite variable diameters *and* water contents,[1] so simple mass determinations (filtering, drying, and weighing) cannot be converted reliably to membrane areas. It is therefore necessary to measure individual cell sizes, in both the flux experiments and the impalement experiments, and to compute total cell surface area per unit suspension volume for the former, and individual cell surface areas for the latter. On a macroscopic scale, spherocyte suspensions are homogeneous, and this plus the regular geometry of such cells makes electronic sizing methods feasible for flux suspensions, using parallel samples taken either before or after the flux samples. It is satisfactory, and much simpler, however, to measure individual cell diameters via a microscope eyepiece grating (400×, bright-field illumination) on calibrated volumes of flux suspension.

Summary of Properties of K^+–H^+ Symport

Simple Kinetic Properties. Simple kinetic properties of high-affinity K^+ transport in *Neurospora* depend strongly on the conditions used for K^+ starvation.[1,18] Normal 12-hr shaking-culture cells of wild-type *Neurospora* grown on limiting K^+ show a Michaelis constant ($K_{1/2}$), for net and unidirectional influx, of 2 μM or even lower. After 3 hr of carbon starvation, these same cells display a 3- to 4-fold elevation of $K_{1/2}$, and a 30% reduction of the maximal flux (J_{max}). EG spherocytes (which are partially carbon-starved) display slightly complicated K^+ uptake kinetics. If interpreted as single-saturation curves, the published data yield $K_{1/2}$ values near $[K^+]_o = 15$ μM; but if interpreted as the sum of one saturating and one nonsaturating component, $K_{1/2}$ for the former drops to 3–4 μM. Actual experimental values of kinetic parameters obtained under different conditions are listed in Table II.

[18] J. Ramos and A. Rodriguez-Navarro, *Biochim. Biophys. Acta* **815**, 97 (1985).

TABLE II

KINETIC PARAMETERS FOR HIGH-AFFINITY K$^+$ UPTAKE IN *Neurospora*[a]

Cell preparation	Growth medium	Flux medium	Michaelis constant (μM)	Maximal velocity (pmol/cm^2 sec)	Ref.
12-hr shake cultures	AmVM + 0.3 mM KCl + 1% glucose	CaMES[b] + test K$_2$SO$_4$ + 1% glucose	2.0 ± 0.5[c]	17.3 ± 1.2	1
12-hr shake cultures	AmVM + 0.3 mM KCl + 1% glucose	CaMES + test K$_2$SO$_4$ 3 hr 0 glucose	6.7 ± 1.2	12.1 ± 0.5	1
12-hr shake cultures	AmVM + 0.25 mM KCl + 1% glucose	CaMES + test RbCl + 1% glucose	6.0 (Rb)	10.4 (Rb)	18
3-Day EG spherocytes	AmVM + 18% EG + 0.3 mM KCl + 1% glucose	CaMES + test K$_2$SO$_4$ 0 glucose, 0 EG	14.9 ± 2.6	15.3 ± 1.0 (chemical flux) 30.1 ± 1.5 (membrane current)	1
OR:					
With the same data fitted as the sum of a Michaelis function and a linear "leak"[d]			3.6 ± 1.0	8.6 ± 0.7 (chemical flux) 18.6 ± 1.2 (membrane current)	

[a] All experiments were carried out with wild-type *N. crassa* strain RL21a.

[b] Starting values of [K$^+$]$_i$ were ~75 mM for all flux measurements from Ref. 1, ~120 mM for flux measurements from Ref. 18, and 50 mM for the electric current measurements.

[c] ±1 S.E. for nonlinear regression, which corresponds to ±1 S.D. for a simple mean.

[d] The apparent linear rates for the data in Ref. 1 are 0.057[K$^+$]$_o$ for current and 0.038[K$^+$]$_o$ for chemical flux (units of pmols/cm^2 sec, with [K$^+$]$_o$ in μM).

Charge Uptake, Ion Exchange, and Implied Coupling. The most important result from the kinetic measurements, however, comes by comparing K$^+$-stimulated charge influx (i.e., inward membrane current) with chemical influx of K$^+$: the two occur with the same Michaelis constant, but charge flow equals twice the K$^+$ influx. This is true for all K$^+$ concentrations, and does not depend on the formal model used to describe the fluxes. However, when simultaneous measurements of potassium-stimulated H$^+$ and K$^+$ movements are made (see above), the results from K$^+$-starved *Neurospora* (shaking-culture cells or EG spherocytes) are consistent with the many potassium-transport experiments on plants: net uptake of K$^+$ is quantitatively matched by net extrusion of H^+.[1,3] The K^+/H^+ ratio is 0.95 for initial fluxes (initial slopes of concentration versus time) and is 0.84–0.90 for total ion transfer.

These results, together with knowledge that the major ion pump in *Neurospora* drives proton efflux, defines the high-affinity K^+ transporter as a $K^+–H^+$ symport.[1] (For each K^+ taken up, two charges are taken up, and one H^+ is released; the two entering charges are matched by two H^+ pumped out, of which only one is measured; therefore one H^+ must enter with each K^+.) Finally, $K^+–H^+$ cotransport is thermodynamically sufficient for the observed cytoplasmic K^+ concentrations, whereas K^+ uniports (carriers or channels) are not. It is easy, particularly at low external pH, to create conditions of net potassium uptake in which the K^+ diffusion potential exceeds the membrane potential by $(-)$ 60–80 mV, representing a cytoplasmic concentration excess of 10- to 30-fold.

Role of the Symport in pH Regulation. The primary function of high-affinity potassium transport almost certainly is to develop and maintain normal $[K^+]_i$ against very low external concentrations. But, by allowing net $K^+–H^+$ exchange, it also can critically assist in restoring cytoplasmic pH, after an acid load.[19] Although cytoplasmic acid directly stimulates the proton pump, that alone cannot regulate pH_i, because protons are also the major reentering ionic current in *Neurospora*. So the $K^+–H^+$ symport, fed potassium, acts to supply non-H^+ inward current. It also plays a more subtle role in pH regulation: by allowing the membrane to depolarize, it further stimulates pumped proton efflux (owing to the voltage dependence of the pump). Operation of the cotransport system has been tested following ionophoretic injections of protons (from H_2SO_4) down to $pH_i = 6.2$, and normal pH_i (7.0–7.2) is restored within 6–8 min.

Electrical Kinetics of the Symport. The $K^+–H^+$ symport in *Neurospora* differs from most other ion contransport systems in being very voltage sensitive over the physiological range of membrane potentials (i.e., -100 mV to -350 mV).[5] With saturating $[K^+]_o$, for example, the symport admits 6- to 7-fold more current at a membrane potential of -320 mV than at -160 mV. Voltage and both substrates (extracellular) also show strong kinetic interaction. Increasing $(-)$ membrane potential is a "linear mixed-type" activator[20] when $[H^+]_o$ is varied and is a "hyperbolic mixed-type" activator when $[K^+]_o$ is varied. Such interactions are a ready basis for reaction-sequence modeling of the cotransport process, and lead to three interesting and testable predictions: (1) that binding/dissociation of H^+, not K^+, should lie adjacent to the step(s) of transmembrane charge transfer;

[19] M. R. Blatt and C. L. Slayman, *Proc. Natl. Acad. Sci. U.S.A.* **84,** 2737 (1987).
[20] I. H. Segal, "Enzyme Kinetics." Wiley (Interscience), New York, 1975.

(2) that the fast steps of the overall reaction should be extracellular dissociation of K^+ or H^+ (normally interpreted as reversed operation of the symport); and (3) that the slow steps of the overall reaction should be either the charge-transfer steps or intracellular dissociation of the transported ions.

[41] Isolation of Everted Plasma Membrane Vesicles from *Neurospora crassa* and Measurement of Transport Function

By Gene A. Scarborough

Introduction

The molecular mechanism of membrane transport is a subject that has occupied the attention of many investigators for many years. Of a variety of experimental systems available for the study of membrane transport, isolated membrane vesicles have proved to be among the most useful. Over the last several years, techniques for the isolation of everted plasma membrane vesicles from *Neurospora crassa* have been developed in this laboratory,[1-4] and two transport systems, an electrogenic proton-translocating ATPase[5-14] (H^+-ATPase) and a Ca^{2+}–H^+ antiporter,[15,16] have been defined and characterized in these vesicles. In this chapter, our current procedures for isolating *Neurospora* plasma membrane vesicles and assaying the transport function of these two transport systems are described.

[1] G. A. Scarborough, *J. Biol. Chem.* **250**, 1106 (1975).
[2] G. A. Scarborough, *Methods Cell Biol.* **20**, 117 (1978).
[3] P. Stroobant and G. A. Scarborough, *Anal. Biochem.* **95**, 554 (1979).
[4] K. M. Brooks, R. Addison, and G. A. Scarborough, *J. Biol. Chem.* **258**, 13909 (1983).
[5] G. A. Scarborough, *Proc. Natl. Acad. Sci. U.S.A.* **73**, 1485 (1976).
[6] G. A. Scarborough, *Arch. Biochem. Biophys.* **180**, 384 (1977).
[7] G. A. Scarborough, *Biochemistry* **19**, 2925 (1980).
[8] J. B. Dame and G. A. Scarborough, *Biochemistry* **19**, 2931 (1980).
[9] J. B. Dame and G. A. Scarborough, *J. Biol. Chem.* **256**, 10724 (1981).
[10] R. Addison and G. A. Scarborough, *J. Biol. Chem.* **256**, 13165 (1981).
[11] R. Addison and G. A. Scarborough, *J. Biol. Chem.* **257**, 10421 (1982).
[12] R. Smith and G. A. Scarborough, *Anal. Biochem.* **138**, 156 (1984).
[13] G. A. Scarborough and R. Addison, *J. Biol. Chem.* **259**, 9109 (1984).
[14] G. A. Scarborough this series, Vol. 157, p. 574.
[15] P. Stroobant and G. A. Scarborough, *Proc. Natl. Acad. Sci. U.S.A.* **76**, 3102 (1979).
[16] P. Stroobant, J. B. Dame, and G. A. Scarborough, *Fed. Proc., Fed. Am. Soc. Exp. Biol.* **39**, 2437 (1980).

METHODS IN ENZYMOLOGY, VOL. 174

Principles

Plasma Membrane Isolation. The problematic resistance of *Neurospora* to cell breakage owing to its rigid cell wall is overcome either by the use of a cell wall-less mutant or by the use of a complex mixture of commercial enzymes to degrade the cell wall of strains with wild-type morphology. In either case, cell wall-less forms are produced in an osmotically stabilized medium. The cell wall-less forms are then treated with the plant lectin, concanavalin A (Con A) and lysed by hypotonic shock. The Con A treatment stabilizes the plasma membrane against fragmentation and vesiculation, and large plasma membrane sheets are thus produced when the cells are lysed. The plasma membrane sheets are readily sedimentable, whereas most of the other constituents of the cell lysate are not, allowing extensive purification of the sheets in two low-speed centrifugation steps. The purified plasma membrane sheets are then treated with the Con A ligand α-methylmannoside, which dissociates the bulk of the Con A from the plasma membranes, whereupon they spontaneously fragment and vesiculate with the majority in an everted orientation. The preparation is then layered over a sucrose solution and centrifuged again, whereupon the plasma membrane vesicles float and most of the nonvesicular material sediments. The vesicle suspension is then diluted and the vesicles harvested by centrifugation.

Assays of Proton Translocation by the H^+-ATPase. In the absence of a permeant anion, proton translocation by the H^+-ATPase in the isolated vesicles leads to the generation of a transmembrane protonic potential difference ($\Delta\bar{\mu}_{H^+}$), which is predominated by the electrical component, $\Delta\varphi$ (inside positive).[5,7] When present in tracer amounts, the permeant anion [^{14}C]SCN$^-$ will accumulate inside the vesicles as a function of the magnitude of such a $\Delta\varphi$ as it comes to electrochemical equilibrium.[17] [^{14}C]SCN$^-$ accumulation by the vesicles can be readily measured by Millipore filtration and scintillation counting of the radioactivity retained on the filter.[5,7] The generation of $\Delta\varphi$ can also be monitored as enhancement of the fluorescence of 1-anilinonaphthalene 8-sulfonate (ANS) in a spectrophotofluorometer.[5]

In the presence of excess amonts of a permeant anion, the $\Delta\bar{\mu}_{H^+}$ generated by the H^+-ATPase is predominantly in the form of an inside-acid ΔpH.[7] The weak base [^{14}C]imidazole will accumulate inside the vesicles as a function of the magnitude of this ΔpH, for reasons that have been explained by Rottenberg *et al.*,[18] and accordingly [^{14}C]imidazole accumulation can be used to measure the ΔpH generated by the H^+-ATPase. This is done by Millipore filtration and scintillation counting as for [^{14}C]SCN$^-$

---

[17] P. Mitchell, *Theor. Exp. Biophys.* **2**, 159 (1969).
[18] H. Rottenberg, T. Grunwald, and M. Avron, *Eur. J. Biochem.* **25**, 54 (1972).

accumulation.[7] Alternatively, if the impermeant fluorescent pH indicator fluorescein isothiocyanate (FITC)-labeled dextran (FITC-dextran) is included in the medium when the Con A-stabilized plasma membrane sheets are converted to closed vesicles, some of the FITC-dextran is trapped inside of the vesicles and can be used to directly monitor the intravesicular pH in a spectrophotofluorometer.[7] Because transmembrane probe movements are not involved in this technique, proton translocation rates can be estimated.

Assay of the $Ca^{2+}-H^+$ *Antiporter.* The $\Delta\bar{\mu}_{H+}$ across the vesicle membrane generated by the H^+-ATPase can drive the formation of an equivalent $\Delta\bar{\mu}_{Ca^{2+}}$ in the presence of the $Ca^{2+}-H^+$ antiporter, for reasons that have been fully elaborated and explained by Mitchell.[17] This amounts to ATP hydrolysis-driven intravesicular Ca^{2+} accumulation, which can be measured by Millipore filtration and scintillation counting as for [^{14}C]SCN$^-$ and [^{14}C]imidazole uptakes, using $^{45}Ca^{2+}$ as the tracer.

Procedures

Isolation of Plasma Membrane Vesicles from the Cell Wall-less Strain of Neurospora. Con A-stabilized plasma membrane sheets are isolated on a large scale and stored at −70° in a concentrated glycerol solution. Plasma membrane vesicles are then prepared as needed from the stored plasma membrane sheets.[3] The growth medium used is Vogel's medium N[19] (without chloroform as a preservative) supplemented with 2% (w/v) mannitol, 0.75% (w/v) yeast extract (Difco), and 0.75% (w/v) nutrient broth (Difco), sterilized by autoclaving. Cells of the *fz;sg;os*-1 strain of *Neurospora crassa* (Fungal Genetics Stock Center Stock Number 1118) are routinely maintained by transfer of 1 ml of a 1–3 day 50-ml culture contained in a 250-ml Erlenmeyer flask into 50 ml of fresh growth medium and rotary shaking (150 rpm) at 30° for 1–3 days.

At noon before the day the plasma membrane isolation is to be carried out, three overnight 50-ml cultures are transferred into three 5-liter portions of growth medium contained in 12-liter flat-bottomed boiling flasks, followed by rotary shaking (100 rpm) at 30° for 21 hr. The optical density (650 nm, 1 cm) of such cultures is about 0.8–1.3 as calculated from the optical density of an appropriate dilution.[20] The cells are harvested by

---

[19] H. J. Vogel. *Am. Nat.* **98**, 435 (1964).

[20] The yield of cells and plasma membranes is increased significantly when the cells are grown with a stream of sterile air (approximately 1.5 liters/min) passing sequentially over the surface of each of the cultures. Thus far, no detrimental effect of this modification on the characteristics of the isolated membranes has been detected, but only a limited number of such preparations have been made.

centrifugation (650 g, 10 min) in 1-liter plastic bottles, resuspended in a total of 1 liter of ice-cold Buffer A [50 mM Tris containing 10 mM MgSO$_4$ and 0.25 M mannitol (pH 7.5 with HCl) filtered through an ethanol-washed AAWP Millipore filter], pelleted again by centrifugation (200 g, 15 min) in two 500-ml polycarbonate centrifuge bottles, and washed 4 more times by alternate resuspension in 400 ml of ice-cold Buffer A and centrifugation (200 g, 10 min).[21]

After the final wash, the cells are resuspended and combined in a total of 100 ml of room temperature Buffer A and the resulting cell suspension is mixed with 200 ml of room temperature Buffer A containing 1.5 mg/ml of Con A (Calbiochem, not corrected for salt content) in a single 500-ml polycarbonate centrifuge bottle, followed by incubation without agitation for 10 min at room temperature, during which time the cells are agglutinated by the Con A. The centrifuge bottle containing the Con A-agglutinated cells is then placed on ice for 10 min after which the cells are pelleted by centrifugation (200 g, 1 min), resuspended gently and not necessarily completely in 400 ml of ice-cold Buffer A, and once again pelleted by centrifugation (200 g, 6 min).

The cells are lysed by resuspension in 800 ml of ice-cold 10 mM Tris (pH 7.5 with HCl) containing 5 mM MgSO$_4$ and 30 mg of DNase (Worthington, code D) and homogenization (50 passes over a period of 10 min) in a Plexiglas homogenizer[22] embedded in ice in a cold room. Approximately 200 ml aliquots of the resulting cell lysate are then layered[23] over 500 ml of Buffer B [0.1 M Tris (pH 7.5 with HCl) containing 0.5 M mannitol, 1 μg/ml chymostatin[24] (Sigma), and 5 mM EDTA[24] (pH 7.5 with Tris)] in

---

[21] Unless indicated otherwise, all centrifugation steps described in this chapter are carried out in a Damon IEC CRU 5000 centrifuge with swinging-bucket heads at 4°, all supernatant fluids generated in the various steps are removed by aspiration, and all pellets are resuspended by first dispersing thoroughly in an approximately equal volume of the resuspension solution and then adding the remaining volume. Before the cell lysis step, resuspension of the cells is accomplished by a firm, but not violent, swirling motion.

[22] The homogenizer is constructed from commercially available Plexiglas rod and tubing with the use of a lathe, and chloroform as glue. The vessel consists of a tube (inside diameter 4.9 cm, outside diameter 5.7 cm, length 67 cm) plugged at one end with a disk 1.5 cm in thickness. The pestle consists of a rod (diameter 1.3 cm, length 84 cm) attached concentrically to a cylinder (diameter 4.83 cm, length 7.2 cm). With the vessel filled with H$_2$O, the drop time of the pestle is about 2.5 min at room temperature.

[23] Layering is simplified by the use of 120-ml polypropylene funnels attached via silicone rubber tubing to a Pasteur pipette with a 60° bend at the tip, through which the lysate can be directed down the wall of the centrifuge bottle with a minimum of mixing.

[24] The inclusion of the chymostatin and EDTA in this and most of the subsequent steps are relatively recent modifications of the procedure that appear to prevent proteolytic nicking and improve the transport activity of the vesicles. Thus far, no deleterious effects of including the chymostatin and EDTA have been noted, but not all of the various membrane properties have been rigorously checked.

each of four 1-liter plastic centrifuge bottles. After the lysate has been layered, the resulting two-phase systems are centrifuged at 200 g for 45 min to sediment the Con A-stabilized plasma membrane sheets, and the resulting pellets are resuspended in 400 ml of ice-cold Buffer C [10 mM Tris (pH 7.5 with HCl) containing 1 μg/ml chymostatin and 5 mM EDTA (pH 7.5 with Tris)] and homogenized again in the Plexiglas homogenizer (20 passes, 4°). Portions of 100 ml of the membrane suspension are then layered over 300 ml of Buffer B in each of four 500-ml polycarbonate centrifuge bottles, and the resulting two-phase systems are centrifuged again (290 g, 45 min). The plasma membrane sheets in the resulting pellets are then resuspended in ice-cold 10 mM Tris (pH 7.5 with HCl) containing 1 μg/ml chymostatin to a total volume of 80 ml, the resulting suspension is mixed thoroughly with 120 ml of ice-cold glycerol, and 6.7-ml aliquots of this suspension are then stored at −70° in 20-ml plastic liquid scintillation counting vials.

A small-scale version of this procedure is also used occasionally. It is similar to the above procedure with all materials reduced by a factor of 1/30; the mechanics of this small-scale procedure are similar (but not identical) tho those described below for the isolation of plasma membrane vesicles from cell wall-bearing strains of *Neurospora,* and were described in the original paper on the Con A procedure.[1]

Plasma membrane vesicles are obtained from the plasma membrane sheets as follows. A 6.7-ml aliquot of the plasma membrane sheet suspension is diluted to 70 ml with ice-cold Buffer C and the resulting mixture is centrifuged (1100 g, 15 min) to sediment the membranes. The membranes are then resuspended and combined in 4 ml of a room temperature solution of 0.5 M α-methylmannoside (recrystallized from 85° H_2O) in Buffer D [10 mM Tris (pH 7.5 with HCl) containing 1 μg/ml chymostatin and 1 mM EDTA (pH 7.5 with Tris)], and the resulting suspension is incubated with shaking at 30° for 5 min in a 45-ml polycarbonate centrifuge tube. In the small-scale version of the procedure, the pellets obtained after the second centrifugation through Buffer B are combined into 4 ml of the α-methylmannoside solution as above and are then treated similarly in this and subsequent steps.

Following the 30° incubation, the membrane suspension is diluted with 16 ml of ice-cold Buffer D in the same centrifuge tube, and 20 ml of 20% (w/v) sucrose in Buffer D is then layered under the membrane suspension using a 20-ml volumetric pipette, and the resulting two-phase system is centrifuged (255 g, 30 min) to sediment unwanted material. The turbid upper 30 ml of solution, which contains the plasma membrane vesicles, is then removed with the aid of a Pasteur pipette, diluted with 15 ml of ice-cold 10 mM 2-(N-morpholino)ethanesulfonic acid (MES) (pH 6.8 with Tris) containing 1 μg/ml chymostatin, and the vesicles pelleted by centrifugation (12,000 g, 30 min) in a high-speed centrifuge. Unless indicated otherwise, the vesicles are then resuspended in 0.75–1.5 ml of ice-cold 10 mM MES (pH 6.8 with Tris) containing 1 μg/ml chymostatin using a Pasteur pipette, and used immediately for the transport assays.

The yield of vesicle protein is in the range of 0.5–2 mg by the Lowry *et al.* assay procedure[25] using bovine serum albumin as a standard, and the specific H$^+$-ATPase activity is around 1–2 μmol of ATP hydrolyzed/mg protein/min when measured by our previously reported assay procedure.[11] These assay conditions do not yield the maximum obtainable ATPase specific activity as we have pointed out,[4] but, provided that the NaN$_3$ is omitted, the assay is relatively free of substances that interfere with the various transport assays and may therefore be preferable when comparisons between the ATP hydrolytic activity and a particular transport process are to be made.

Isolation of Plasma Membrane Vesicles from Cell Wall-Bearing Strains of Neurospora. The solid growth medium used consists of 1.5% (w/v) Bacto-agar (Difco) in Beadle and Tatum's minimal medium[26] with 1.5% (w/v) fructose replacing the sucrose. The liquid growth medium used consists of Beadle and Tatum's minimal medium adjusted to pH 6.5 with KOH and with 1% (w/v) fructose replacing the sucrose. The media are sterilized by autoclaving; in the case of the liquid medium, the fructose is autoclaved separately as a 50% (w/v) solution.

Conidia are generated by growth of conidiating strains on 50 ml of solid medium contained in loosely plugged 250-ml Erlenmeyer flasks for 7–10 days at room temperature. Such cultures can be stored in a cold room for several weeks before use, with no noticeable effect on the plasma membrane isolation procedure. Conidial suspensions are obtained by sequentially shaking each of four conidial cultures in 50 ml of sterile liquid medium minus fructose and filtering the suspension obtained through sterile glass wool to remove mycelial fragments. This process is then repeated twice, pooling all of the washes. The optical density (650 nm, 1 cm) of the conidial suspension obtained is usually around 2.5–3 as calculated from the optical density of an appropriate dilution. One hundred milliliters of the conidial suspension is then inoculated into 400 ml of liquid medium, and the conidia are germinated by rotary shaking (150 rpm) at 30° for 10 hr, during which time the optical density of the culture increases to about 1.

The germinated conidia are harvested by filtration on a Millipore filter

[25] O. H. Lowry, N. J. Rosebrough, A. L. Farr, and R. J. Randall, *J. Biol. Chem.* **193**, 265 (1951).
[26] G. W. Beadle and E. L. Tatum, *Am. J. Bot.* **32**, 678 (1945).

(AAWP 047), and the resulting cell pad is resuspended in 10 ml of Buffer I [0.75 M sorbitol in 0.1 M MES (pH 5.8 with Tris)] containing 50 mg of β-glucuronidase from *Helix pomatia* (Sigma, Type H-1) 40 mg of lytic enzyme L1 (Gallard-Schlesinger), 20 mg chitinase (Sigma), 200 mg Cellulysin (Calbiochem), 10 mg zymolyase 20 T (Miles), 50 mg α-amylase[27] (Sigma, Type II A), 650 μg penicillin, and 500 μg streptomycin. The suspension is then incubated with rotary shaking (150 rpm) at 30° for 10–13 hr, during which time the majority of the cell wall is degraded with the production of spheroplasts. The spheroplasts are collected and washed 3 times by alternate centrifugation (400 g, 6 min) and resuspension in 80 ml of Buffer II [0.75 M sorbitol in 0.1 M MES (pH 7.5 with Tris)] in four 45-ml polycarbonate centrifuge tubes, pelleted again (400 g, 6 min), and resuspended in 8 ml of room temperature Buffer II.

The resulting washed spheroplast suspension is mixed with 8 ml of 2 mg/ml Con A in room temperature Buffer II and allowed to stand for 10 min at room temperature, during which time the spheroplasts are agglutinated by the Con A. The mixture is then diluted 10-fold with ice-cold Buffer II and centrifuged at 200 g for 1 min. The pellet containing the Con A-treated spheroplasts is then resuspended in 25 ml of ice-cold 10 mM Tris (pH 7.5 with HCl) containing 5 mM MgSO$_4$, 1 mg DNase (Worthington, code D), and 1 μg/ml chymostatin, and homogenized (50 passes over 10 min) on ice in a glass–Teflon homogenizer (clearance ~0.02 cm) to lyse the spheroplasts.

The lystate is layered over two 30-ml portions of ice-cold Buffer B (see above) in 45-ml polycarbonate centrifuge tubes, and the resulting two-phase systems are centrifuged at 255 g for 30 min. The pellets are then resuspended in a total of 20 ml of ice-cold Buffer C (see above), homogenized again in the glass–Teflon homogenizer (20 passes), and equal aliquots of the resulting suspension are layered again over two 30-ml portions of ice-cold Buffer B and centrifuged at 255 g for 30 min. The pellets obtained contain the Con A-stabilized plasma membrane sheets, which are converted to vesicles as follows.

The pellets are resuspended and combined in 4 ml of 0.5 M recrystallized α-methylmannoside in 10 mM Tris (pH 7.5 with HCl) containing 1 μg/ml chymostatin in a 45-ml polycarbonate centrifuge tube, and the resulting plasma membrane suspension is incubated with shaking for 5

---

[27] The α-amylase is a recent addition to the procedure suggested by Quigley *et al.*[28] Although we have not used the α-amylase extensively, it appears to improve the specific activity of the isolated wild-type vesicles in the [¹⁴C]SCN⁻ uptake assay and to reduce the problem of proteolytic breakdown of the wild-type membrane proteins during disaggregation for sodium dodecyl sulfate–polyacrylamide gel electrophoresis.[4]

[28] D. R. Quigley, E. Jabri, and C. P. Selitrennikoff, *Exp. Mycol.* **9**, 254 (1985).

min at 30°, after which it is diluted with 16 ml of ice-cold 10 mM Tris (pH 7.5 with HCl) containing 1 μg/ml chymostatin in the same centrifuge tube. Twenty milliliters of ice-cold 20% (w/v) sucrose in 10 mM Tris (pH 7.5 with HCl) containing 1 μg/ml chymostatin is then layered below the membrane suspension as described above, and the resulting two-phase system is centrifuged at 255 g for 20 min. After centrifugation, the turbid upper 30 ml, which contains the plasma membrane vesicles, is collected and diluted with 15 ml of ice-cold 10 mM MES (pH 6.8 with Tris) containing 1 μg/ml chymostatin, and the vesicles are then pelleted by centrifugation at 12,000 g for 20 min in a high-speed centrifuge. The vesicles are then resuspended in 0.75–1.5 ml of ice-cold 10 mM MES (pH 6.8 with Tris) containing 1 μg/ml chymostatin and used immediately. The yield of vesicle protein and ATPase specific activity are comparable to those mentioned above for the cell wall-less strain of *Neurospora*.

Transport Assays

[^{14}C]SCN$^-$ Uptake. The assay mixtures contain 25 μl of the plasma membrane vesicle suspension (approximately 25 μg of vesicle protein), 5 μl of 0.4 mM [^{14}C]KSCN (New England Nuclear, specific activity 60 Ci/mol), 5 μl of 5 mM ethylene glycol bis(β-aminoethyl ether)-N,N'-tetraacetic acid (EGTA), 5 μl of 0.1 M Tris-ATP–MgSO$_4$ (pH 6.8 with Tris), and any other additions in a total volume of 50 μl. The reactions are started by the addition of the membranes to the tubes containing the other assay constituents at room temperature, allowed to proceed for the desired time period, and then terminated by the addition of 2 ml of ice-cold 10 mM MES (pH 6.8 with Tris), after which the vesicles are filtered on a prewet Millipore filter (AAWP 025) and washed with an additional 2 ml of ice-cold 10 mM MES (pH 6.8 with Tris). Filtration and washing are carried out using a standard sintered glass Millipore filter base with a glass chimney attached by a metal clamp. The entire termination and filtration process is carried out in about 2 sec. Zero-time controls are carried out by adding the termination solution to the assay tubes before the membranes, and filtering and washing as above.

The filters are dried under a heat lamp for about 15 min, placed in scintillation counting vials, and the associated radioactivity is estimated by scintillation counting in a toluene-based counting fluid containing 2,5-diphenyloxazole (5 g) and 1,4-bis([2-(4-methyl-5-phenyloxazolyl)]benzene (0.3 g) per liter. Five microliters of the [^{14}C]KSCN solution is counted similarly, and the results obtained are used to calculate the nanomoles of [^{14}C]SCN$^-$ accumulated by the vesicles. The specific activities obtained for plasma membrane vesicles from cell wall-less and cell wall-bearing strains are in the range of 3–15 nmol of SCN^- accumulated/mg protein at the steady state, which is reached after about 1 min. If desired, the concentration of the accumulated $[^{14}C]SCN^-$ can be estimated using a value of 3 μl of intravesicular space/mg protein.[7]

[^{14}C]Imidazole Uptake. The assay mixtures contain 25 μl of the plasma membrane vesicle suspension (~25 μg vesicle protein), 5 μl of 0.5 mM $[^{14}C]$imidazole (California Bionuclear Corp., specific activity 3.5 Ci/mol), 5 μl of 0.1 M NaSCN, 5 μl of 5 mM EGTA (pH 6.8 with Tris), 5 μl of 0.1 M Tris-ATP–MgSO$_4$ (pH 6.8 with Tris), and any other additions in a total volume of 50 μl. The uptake assays are carried out as described above for $[^{14}C]SCN^-$ uptake. The specific activities obtained for plasma membrane vesicles from both cell wall-less and cell wall-bearing strains are in the range of 3–12 nmol of imidazole accumulated/mg protein at the steady state, which is reached after about 2 min.

^{45}Ca^{2+} Uptake. Plasma membrane vesicles are resuspended in 10 mM MES (pH 7.3 with Tris) instead of the pH 6.8 solution described above, to a concentration of approximately 0.5 mg protein/ml, and 50-μl aliquots of the resulting suspension are preincubated at room temperature for 30 sec, after which Tris-ATP, MgSO$_4$, and NaSCN are added to give final concentrations of 10 mM in a total volume of 53 μl. Any other additions are also made at this point. The pH of all of the stock solutions used is preadjusted to 7.3 with MES or Tris. Immediately after all of the desired additions are made, 2 μl of 2.5 mM ^{45}CaCl$_2$ (New England Nuclear, specific activity 5–10 Ci/mol) is added to initiate the uptake reaction. ^{45}Ca^{2+} uptake is then measured as described above for $[^{14}C]SCN^-$ and $[^{14}C]$imidazole uptakes, except that the termination and wash solution is 10 mM MES (pH 7.3 with Tris). The specific activities obtained for vesicles from both wild-type and the cell wall-less strain are in the range of 10–20 nmol of Ca^{2+} accumulated/mg protein at the steady state, which is reached after about 10 min.

FITC-Dextran Fluorescence Quenching. Plasma membrane vesicles are isolated as described above except that FITC-dextran (5 mg/ml) is added to the α-methylmannoside solution, and the resulting solution is centrifuged (12,000 g, 30 min) and filtered through a Millipore filter (AAWP 025) before use. The FITC-dextran-containing plasma membrane vesicles (0.5–2 mg protein) are resuspended in 1.71 ml of 10 mM MES containing 0.53 mM EGTA and 10.53 mM MgSO$_4$ (pH 6.8 with Tris), brought to room temperature, and transferred to a cuvette in a spectrophotofluorometer as quickly as possible. Fluorescence is measured at 90° in the direct mode with a Farrand interference filter (peak wavelength, 515 nm; half-bandwidth, 14 nm; 40% transmission) in the path of the emission beam. Stirring is accomplished magnetically. The excitation wavelength is 475 nm and the emission wavelength is 519 nm. After the membrane suspension has been placed in the cuvette, the light emission is adjusted to 100% by varying the emission slit width. When the trace has stabilized, 90 μl of 0.2 M Tris-ATP (pH 7.5) is added through a light-tight port positioned over the cuvette with the aid of a Hamilton syringe with a long needle. Other additions are made similarly. In the presence of the permeant anion SCN^- at 10 mM, fluorescence quenching in the range of 25–75% is commonly observed, with 0% fluorescence defined as the light emission that is not quenched at pH 4.8.

ANS Fluorescence Enhancement. Plasma membrane vesicles are isolated as described above except that the MES–Tris buffer used in the final two steps of the procedure is replaced with 10 mM Tris (pH 6.8 with H_3PO_4). Incubations contain plasma membrane vesicles (~80 μg vesicle protein), ANS (20 nmol), and 10 mM Tris (pH 6.8 with H_3PO_4) in a total volume of 0.9 ml at room temperature. The reactions are initiated by the addition of 0.1 ml of 0.1 M Tris-ATP–MgSO$_4$ (pH 6.8 with Tris). Fluorescence is measured at 90° with an excitation wavelength of 366 nm and an emission wavelength of 470 nm.

Acknowledgments

The work was supported by U.S. Public Health Service National Institutes of Health Grants AM 14479, GM19971, and GM24784 and National Science Foundation Grant GB38801. The expert technical assistance of Ms. Beth Seigler is also acknowledged.

Author Index

Numbers in parentheses are footnote reference numbers and indicate that an author's work is referred to although the name is not cited in the text.

Subject Index

A

O